FLORA OF AUSTRALIA

Grevillea cravenii Makinson. Painting by Patricia Dundas.

A publication of the
AUSTRALIAN BIOLOGICAL RESOURCES STUDY, CANBERRA

FLORA OF AUSTRALIA

Volume 17A
Proteaceae 2
Grevillea

Environment Australia
Department of the Environment and Heritage

CSIRO PUBLISHING

EXECUTIVE EDITORS

Anthony E. Orchard (1993–1998)
Helen S. Thompson (1998 Ag)
Patrick M. McCarthy (1998–1999 Ag)
Annette J.G. Wilson (1999–2000 Ag)

VOLUME EDITOR

Annette J.G. Wilson

EDITORIAL ASSISTANCE

Jane Mowatt Helen Thompson

This work may be cited as:

Flora of Australia Volume 17A, Proteaceae 2, *Grevillea*. Melbourne : ABRS/CSIRO Australia (2000).

Individual contributions may be cited as:

R.O.Makinson, *Grevillea*, *Flora of Australia*, 17A: 1–460 (2000).

This book is available from:

CSIRO PUBLISHING
PO Box 1139 (150 Oxford Street)
Collingwood VIC 3066
Australia

Tel: (03) 9662 7666 Int: +(61 3) 9662 7666
Fax: (03) 9662 7555 Int: +(61 3) 9662 7555

Email: sales@publish.csiro.au

National Library of Australia
Cataloguing-in-Publication entry

Flora of Australia. Volume 17A, Proteaceae 2, *Grevillea*.

Bibliography.

Includes index.

ISBN 0 643 05969 5 (hard cover).

ISBN 0 643 05970 9 (soft cover).

ISBN 0 643 05702 1 (hard cover set).

ISBN 0 643 05695 5 (soft cover set).

1. Botany – Australia – Classification. 2. Plants – Identification. I. Australian Biological Resources Study. II. CSIRO.

Published by ABRS, Canberra/CSIRO, Melbourne
Printed in Australia by Brown Prior Anderson

This volume is dedicated to Dr Barbara Gillian Briggs and the late Dr Lawrence Alexander Sidney Johnson.

The names of Barbara Briggs and Lawrie Johnson are closely associated with the family Proteaceae in Australia. They produced two seminal studies on the evolutionary relationships and classification of the family, the first in 1963 (Johnson & Briggs, Evolution in the Proteaceae, *Austral. J. Bot.* 11: 21–61), the second in 1975 (Johnson & Briggs, On the Proteaceae - the evolution and classification of a southern family, *J. Linn. Soc. Bot.* 70: 83–182). For the second paper, they introduced a phylogenetic approach to the analysis of evolutionary relationships, and took into account the biogeographic possibilities raised by plate tectonics, the exciting geological theory that had recently been accepted. The classification in that second paper has been used as the basis for the *Flora of Australia* treatment.

The inception and development of both the *Flora of Australia* and the Australian Biological Resources Study were strongly supported by Barbara and Lawrie. As early as 1960, Lawrie was part of an ANZAAS delegation putting a case to the Prime Minister's Department for a national Flora project, and he was later appointed to the committee that eventually succeeded in this aim, the Interim Council of the Australian Biological Resources Study. Barbara was a member of the ABRS Advisory Committee 1979–1983, and was also a member of the first Flora Editorial Committee in 1980. She remained a member of the Editorial Committee until 1992 and chaired it between 1982 and 1987.

Barbara Briggs retired on 19 September 1997 but remains active in research. Unfortunately, Lawrie Johnson died on 1 August 1997, without seeing Proteaceae completed in these volumes.

CONTENTS

CONTRIBUTORS TO VOLUME 17A

Mr M.D.Barrett, Kings Park & Botanic Gardens, West Perth, Western Australia 6005.

Mr G.J.Keighery, Department of Conservation & Land Management, Woodvale Research Centre, PO Box 51, Wanneroo, Western Australia 6055.

Mr R.O.Makinson, Australian National Herbarium, Centre for Plant Biodiversity Research, CSIRO, GPO Box 1600, Canberra, Australian Capital Territory 2601.

Mr W.Molyneux, PO Box 386, Yarra Glen, Victoria 3375.

Mr V.Stajsic, PO Box 37R, Redan, Victoria 3350.

ILLUSTRATORS

Mrs D.Boyer, Bobbara Creek, Binalong, New South Wales 2584.

Mrs B.Chandler, 63 Penang Avenue, Colonel Light Gardens, South Australia 5041.

Mrs P.Dundas, 21 Wooloomooloo Road, Greenmount, Western Australia 6056.

Mr D.C.Fortescue, c/- Australian Biological Resources Study, GPO Box 787, Canberra, Australian Capital Territory 2601.

Mr D.Mackay, PO Box 729, Armidale, New South Wales 2350.

Ms C.Payne, 'Ilonka', Sawyer's Ridge Road, Reidsdale, New South Wales 2622.

Ms L.Spindler, 78 Louth Road, Greystones, Sheffield, South Yorkshire S117AW, United Kingdom.

Ms C.Wardrop, c/- Royal Botanic Gardens, Mrs Macquaries Road, Sydney, New South Wales 2000.

Miss A.Wilson, Australian Biological Resources Study, GPO Box 787, Canberra, Australian Capital Territory 2601.

PHOTOGRAPHERS

Mr P.Abell, PO Box 116, Blaxland, New South Wales 2774.

Mr M.D.Barrett, Kings Park & Botanic Gardens, West Perth, Western Australia 6005.

Dr R.Boden, c/- Australian National Botanic Gardens, GPO Box 1777, Canberra, Australian Capital Territory 2601.

Mr J.Briggs, c/- Australian National Botanic Gardens, GPO Box 1777, Canberra, Australian Capital Territory 2601.

PHOTOGRAPHERS

Mr J.Brock, PO Box 38543, Winnellie, Northern Territory 0821.

Dr M.D.Crisp, c/- Australian National Botanic Gardens, GPO Box 1777, Canberra, Australian Capital Territory 2601.

Mr M.Fagg, Australian National Botanic Gardens, GPO Box 1777, Canberra, Australian Capital Territory 2601.

Dr D.B.Foreman, 50 Benjamin Drive, Lara, Victoria 3213.

Mr A.S.George, 'Four Gables', 18 Barclay Road, Kardinya, Western Australia 6163.

Mr B.Gray, 48 Loder Street, Atherton, Queensland 4883.

Ms D.Greig, c/- Australian National Botanic Gardens, GPO Box 1777, Canberra, Australian Capital Territory 2601.

Mr D.L.Jones, Australian National Herbarium, Centre for Plant Biodiversity Research, GPO Box 1600, Canberra, Australian Capital Territory 2601.

Mr T.Low, 6 Henry Street, Chapel Hill, Queensland 4069.

Mr A.Lyne, c/- Australian National Botanic Gardens, GPO Box 1777, Canberra, Australian Capital Territory 2601.

Mr R.O.Makinson, Australian National Herbarium, Centre for Plant Biodiversity Research, CSIRO, GPO Box 1600, Canberra, Australian Capital Territory 2601.

Mr D.McGillivray, 2 Amulla Close, Point Clare, New South Wales 2250.

Mr P.Ollerenshaw, c/- Australian National Botanic Gardens, GPO Box 1777, Canberra, Australian Capital Territory 2601.

Mr J.Plaza, National Herbarium of New South Wales, Royal Botanic Gardens, Mrs Macquaries Road, Sydney, New South Wales 2000.

Mr I.R.H.Telford, c/- Australian National Botanic Gardens, GPO Box 1777, Canberra, Australian Capital Territory 2601.

INTRODUCTION

Volume 17A of the *Flora of Australia* is the second of the books describing the Proteaceae. This book contains the genus *Grevillea* (Tribe Grevilleeae). Introductory chapters describing important features of the Proteaceae in Australia, including relationships, morphology, palaeobotany, pollination biology and uses, are contained in Volume 16.

Grevillea is the third-largest genus of Australian flowering plants, with all but 7 species endemic to Australia. It has been the subject of two recent treatments (McGillivray & Makinson, 1993, and Olde & Marriott, vol. 1, 1994; vols 2 & 3, 1995). This treatment recognises 357 species (452 taxa), divided into 33 informal groups. In view of the size of the genus, keys to groups for each State except W.A. and Tasmania are provided, as well as the main Key to Groups.

Scope and Presentation of the *Flora*

The geographical area covered by the *Flora* includes the six Australian States, the Northern Territory, the Australian Capital Territory and immediate offshore islands. Other Australian and State-administered territories such as Christmas Is. and Lord Howe Is. are excluded, but the occurrence in those territories of species included in the *Flora* is added to the notes on distribution. Complete Floras of the oceanic islands are in Volumes 49 and 50.

Descriptions and discussion in the *Flora* are concise and supplemented by important references, synonymy, and information on type collections, chromosome numbers, distribution, habitat, and published illustrations. Descriptions are based on Australian material except for some taxa not confined to Australia for which the collections in Australian herbaria are inadequate. Synonymy is restricted to names based on Australian types or used in Australian literature. Misapplied names are given in square brackets together with an example of the misapplication. Alien taxa established in one or more localities, other than under cultivation, are considered naturalised and are included and asterisked (*).

Families are arranged in the system of A.Cronquist, *An Integrated System of Classification of Flowering Plants* (Columbia University Press, New York, 1981). Within families, genera and species are arranged to show natural relationships as interpreted by contributors. Although relationships cannot be shown adequately in a linear sequence, such an arrangement in a *Flora* assists comparison of related taxa. Infraspecific taxa are keyed out under relevant species. Up to seven collections are cited for each species and infraspecific taxon.

Maps showing distribution in Australia are arranged in the same sequence as the descriptions and are grouped together at the end of the main text (pp. 461–492). The term 'Malesia' is sometimes used in the notes on geographical distribution for species with occur widely in the region covered by *Flora Malesiana*, i.e. Malaysia, Singapore, Indonesia, the Philippines, New Guinea and adjacent islands.

Type citations under taxa in the main body of the text reflect the authors' belief in their current status (holotype, isotype, syntype, etc) and where they are held. In cases where the type specimen has not been examined, this is indicated by *n.v.* These type statements are not to be interpreted as lectotypifications. Where lectotypifications have been made previously, these are cited with *fide*, followed by a reference to the author and place of publication (or, sometimes, to a secondary reference). Any formal lectotypifications required for this volume, as in previous parts of the *Flora*, are confined to the Appendix.

New taxa and lectotypifications are included in an Appendix where they are formally published in accordance with the *International Code of Botanical Nomenclature* (Koeltz Scientific Books, Königstein, 1994). Abbreviations, contractions and notes on format are listed after the Appendix.

A key to families of flowering plants and a glossary of technical terms are provided in Volume 1 of the *Flora*. Supplementary glossaries are included in each volume as necessary.

Acknowledgments

There are 31 contributors, illustrators and photographers to Volume 17A. Their co-operation is gratefully acknowledged.

The Australian National Botanic Gardens slide collection provided a number of the colour photographs used in this volume.

The Librarians at the Australian National Botanic Gardens were ever cheerful in assisting to locate references.

The co-operation of referees, usually working to tight deadlines, is also acknowledged.

Many of the line illustrations used herein were co-operatively produced with the Royal Botanic Gardens, Sydney, and used also in the monographic book *Grevillea* by D.J.McGillivray & R.O.Makinson (1993).

The production of this volume would not have been possible without the substantial assistance of the Australian Commonwealth, State and University Herbaria. Their willingness to provide staff time and resources for this project of national importance is an outstanding example of co-operation between the States and the Commonwealth. Overseas institutions have also assisted preparation of the Volume with loans of specimens and by making facilities available to contributors and illustrators.

The Director, ABRS Publication Section (Tony Orchard) acknowledges with great pleasure the input by staff of the Australian Biological Resources Study. Their work, invisible in the final product, is onerous, often tedious, but essential in maintaining the high standards that this series has achieved. This work has been prepared for publication during a period of considerable instability within the ABRS organisation, as shown by the long list of editorial contributors cited above. Principal editorial credit for this volume goes to Annette Wilson who has been volume editor throughout. Checking and bromiding of illustrations was undertaken by Jane Mowatt. Other ABRS staff were involved at various stages in editing, indexing and collating text.

The editorial staff acknowledge with thanks the cheerful co-operation of the author of this mammoth work, Bob Makinson.

The co-operation of CSIRO Publishing in bringing this book to press is gratefully acknowledged.

Plate 1. *Grevillea montis-cole* subsp. *montis-cole*
Photograph — M.Crisp (ANBG).

Plate 2. *Grevillea hookeriana* subsp. *apiciloba*
Photograph — D.L.Jones.

Plate 3. *Grevillea concinna* subsp. *lemanniana*
Photograph — P.Ollerenshaw (ANBG).

Plate 4. *Grevillea renwickiana*
Photograph — J.Briggs (ANBG).

Plate 5. *Grevillea pteridifolia*
Photograph — B.Gray.

Plate 6. *Grevillea cravenii*
Photograph — M.Barrett.

Plate 7. *Grevillea nana* subsp. *nana*
Photograph — D.McGillivray.

Plate 8. *Grevillea excelsior*
Photograph — A.S.George.

Plate 9. *Grevillea hodgei*
Photograph — D.L.Jones.

Plate 10. *Grevillea goodii*
Photograph — I.R.Telford (ANBG).

Plate 11. *Grevillea asparagoides*
Photograph — c/- ANBG.

Plate 12. *Grevillea juncifolia* subsp. *juncifolia*
Photograph — D.L.Jones.

Plate 13. *Grevillea pluricaulis*
Photograph — J.Brock.

Plate 14. *Grevillea decora* subsp. *decora*
Photograph — B.Gray.

Plate 15. *Grevillea heliosperma*
Photograph — J.Brock.

Plate 16. *Grevillea decurrens*
Photograph — D.L.Jones.

Plate 17. *Grevillea pungens*
Photograph — J.Brock.

Plate 18. *Grevillea aurea*
Photograph — D.L.Jones.

Plate 19. *Grevillea singuliflora*
Photograph — J.Plaza.

Plate 20. *Grevillea agrifolia* subsp. *agrifolia*
Photograph — D.L.Jones.

Plate 21. *Grevillea angulata*
Photograph — D.L.Jones.

Plate 22. *Grevillea wickhamii* subsp. *aprica*
Photograph — D.L.Jones.

Plate 23. *Grevillea longicuspis*
Photograph — J.Brock.

Plate 24. *Grevillea microstyla*
Photograph — M.Barrett.

Plate 25. *Grevillea huegelii*
Photograph — M.Fagg (ANBG).

Plate 26. *Grevillea ripicola*
Photograph — D.Foreman.

Plate 27. *Grevillea pinaster*
Photograph — D.Foreman.

Plate 28. *Grevillea patentiloba* subsp. *patentiloba*
Photograph — D.Foreman.

Plate 29. *Grevillea tripartita* subsp. *macrostylis*
Photograph — P.Ollerenshaw (ANBG).

Plate 30. *Grevillea hakeoides* subsp. *hakeoides*
Photograph — D.Foreman.

Plate 31. *Grevillea diversifolia* subsp. *diversifolia*
Photograph — D.Foreman.

Plate 32. *Grevillea scapigera*
Photograph — A.S.George.

INTRODUCTION TO GREVILLEA

R.O.Makinson[1]

HISTORY

The first *Grevillea* species known to have been collected was the taxon later named *G. mucronulata* R.Br., gathered at Botany Bay by Joseph Banks and Daniel Solander on Cook's voyage in 1770. Solander advanced a manuscript name of '*Leucandendroides*' for the genus, showing an immediate recognition of affinities with South African Proteaceae. Other surviving collections from later in the Cook voyage are of *G. pteridifolia*, *G. glauca*, and *G. parallela*.

Following establishment of the Sydney settlement in 1788, some collections were forwarded to Britain by early colonists. George Caley arrived in Sydney in 1800, and soon amassed significant collections of the local flora, including several grevilleas. Caley's work was soon eclipsed by that of Robert Brown, who accompanied Flinders' *Investigator* voyage of 1801–1805, and collected a vast array of Australian plants including many grevilleas. Caley was collecting for Banks and did not publish, but J.E.Smith (1794) described two Port Jackson species in *Embothrium*, (*E. buxifolium* and *E. sericeum*), a genus familiar to him from South America (and now regarded as endemic to that continent). Placement of grevilleas in *Embothrium* was followed by Cavanilles (three species, 1798) and Willdenow (one variety under *E. sericeum*, 1798).

The genus name was erected by Joseph Knight in 1809 (as '*Grevillia*'), along with several names under the new genera *Lysanthe* and *Stylurus*, and honours Charles Francis Greville (1749–1809), who was a Fellow of the Royal Society, an office-bearer of the [later Royal] Horticultural Society, and a patron of various branches of science. Knight's text was mostly written by Richard Salisbury, and some at least of it was taken from an oral presentation by Brown to the Linnaean Society (perhaps including the name, although there is evidence that this was previously coined in manuscript by Joseph Dryander on sheets of Brown collections in the Banksian Herbarium).

Brown (1810a, 1810b, 1830, 1849) named 79 taxa in *Grevillea* (synonymising *Stylurus* and *Lysanthe*) and initially *Anadenia* (which distinguished certain taxa with an erect stigma and apparently no nectary). Brown began to grapple with questions of generic and subgeneric circumscription, and his developing views of the genus were greatly assisted by many further collections made by Allan Cunningham in many parts of the continent.

Brown's spelling of '*Grevillea*', used by all subsequent workers, was proposed for conservation under the International Code of Botanical Nomenclature against Knight's '*Grevillia*' by Johnson & Garden (1949, 1950), and the act of nomenclatural conservation was published in 1959. As what appeared at the time to be a secondary matter, Johnson & Garden also proposed lectotypification of the genus on *G. aspleniifolia*, apparently on the arbitrary grounds that it was first of the four named in Knight's publication. McGillivray (1985) proposed continued conservation of Brown's orthography but a change of lectotype to *G. pteridifolia*, correctly pointing out that the Type of *G. aspleniifolia* (and indeed the taxon) 'is in serious conflict with the generic protologue', which corresponds much better to the character states exhibited by the type material of the name *Grevillea pteridifolia* Knight. The IAPT Committee for Spermatophyta (Brummitt, 1988: 444–445) rejected the proposal on the specious grounds that 'since both these species currently fall into the same section of the genus, the matter seems academic only'.

In addition to Cunningham's important collecting and unpublished documentation, much collecting and/or descriptive work was done between 1820 and the late 1840s by (among

[1] Australian National Herbarium, Centre for Plant Biodiversity Research, CSIRO, GPO Box 1600, Canberra, Australian Capital Territory 2601.

others) F.Sieber, C.Fraser, K. von Huegel, the prodigious J.Drummond, L.Preiss, W.J.Hooker and S.Endlicher (who in 1839 erected the genus *Manglesia* for the very distinctive Western Australian group of taxa subsequently reduced by Bentham (1870) to *Grevillea* sect. *Manglesia*).

Works by C.F.Meisner (1845, 1848, 1852, 1854, 1855) culminated (Meisner, 1856) in an effective synoptic revision of the whole genus (synonymising both Brown's *Anadenia* and Endlicher's *Manglesia*). One hundred and seventy six species were recognised. Meisner made particular progress in the naming and description of many insect-pollinated taxa, especially from the south-west, and in his subgeneric scheme correctly emphasised seed morphology as a source of critical classificatory characters.

From the late 1840s, Ferdinand Mueller described some 40 taxa in *Grevillea*, and from 1861 worked in uneasy collaboration with George Bentham on the *Flora Australiensis*, the fifth volume of which (Bentham, 1870) recognised 156 species of *Grevillea*. Bentham's was the first work on the genus from an avowedly evolutionary standpoint, and his subgeneric classification remains the most recent formal one.

For the next hundred years, surprisingly little work was done on the genus, with accretions of new taxa here and there but no synthetic or large-scale floristic work. A partial exception was the attempt by Charles Gardner to produce a Flora of Western Australia, never published, but the manuscript for which (including the *Grevillea* treatment) has been of use to subsequent workers. Expertise in the Western Australian Proteaceae after Gardner's death was maintained by Alex George, who published nine new grevilleas and, alone and together with Don McGillivray, greatly increased the specimen base from Western Australia.

Interest in the grevilleoid Proteaceae on a broader scale was reawakened by formative work by Lawrie Johnson and Barbara Briggs (1963, 1975) on the higher-level systematics of the family. They encouraged the main 20th century work on the genus, the revision by Don McGillivray, completed and edited by the present author (McGillivray & Makinson, 1993) and precursor papers (McGillivray, 1975, 1986).

McGillivray's accumulation of an enormous amount of highly reliable data remains a primary resource for descriptive work, and his painstaking nomenclatural work clarified most uncertainties as to names and typification. He adopted an unusually broad species concept, partly as a philosophical matter but also reflecting the available specimen base and the restricted amount of fieldwork he was able to perform, with many anomalies not being amenable to investigation at that time. He reduced many previously named species and varieties to lower (often informal) ranks, and recognised many new 'races' and 'forms', albeit with careful circumscription that has allowed their reinstatement or promotion in status by subsequent workers who incline to a narrower species concept. Even during the course of McGillivray's revision, his work coincided with, and further stimulated, greatly increased taxonomic collecting, to the point where our specimen base from poorly known areas and taxa is now very much larger than at his retirement in 1985.

Partly as a response to the broad ranking scheme by McGillivray, work originally intended as a popular and horticultural guide to the genus by Peter Olde and Neil Marriott became a popularised reworking of the revision, drawing very heavily on McGillivray's data but re-ranking many taxa and recognising many new ones (precursor papers include Olde & Marriott, 1993a, 1993b, 1994a, 1994b, 1994c, 1994d; Olde & Molyneux, 1994). Their substantive synopsis, the three volumes of *The Grevillea Book* (1994e, 1995a, 1995b), was also informed by recent work on particular Western Australian species complexes by Greg Keighery (1990, 1992) and on various taxa by other authors (Kenneally, 1988; Makinson & Albrecht, 1989; Makinson & Olde, 1991).

A summary of the Bentham (1870), McGillivray and Makinson (1993), and Olde and Marriott (1994e, 1995a, 1995b) infrageneric classifications, compared with the present scheme, is presented at Table 1 (pp. 14–16).

The present work is based on a complete review of the circumscriptions of all taxa. It accepts most of the additional (post 1993) taxa named or reinstated by Keighery, Olde & Marriott, and Molyneux, but with some modifications to circumscriptions, somewhat more attention to intergrades and a few re-rankings. A total of 452 taxa (357 species) are here recognised, including 42 new taxa. In a few cases, taxa recently named as species seem scarcely distinct at that rank from near relatives; these cases are noted in the text, but it seems both premature and unnecessarily confusing to the user (this being the third major work on the genus in 10 years) for new combinations at lower rank to be made except in really obtrusive cases. Conversely, some complexes currently treated at subspecific level, notably under *G. sarissa*, should probably have the constituent taxa raised to species status, but some elements are poorly collected and more field work is required for adequate understanding.

There remain several other unsatisfactorily resolved species complexes, requiring considerably more work. Morphological study on the complex of taxa around *G. victoriae* by V.Stajsic and W.Molyneux at MEL is continuing; they publish some new names in this treatment as precursors to a synoptic review. *Grevillea australis* R.Br. remains a single species here, but undoubtedly comprises a number of cryptic taxa, at least in Tasmania; *G. paniculata* is a similar case in Western Australia. Many of the problems in the *G. linearifolia* alliance have eased, but delimitation of some taxa is still unsatisfactory. The widespread *G. stenobotrya* contains clines and distinctive populations requiring further investigation, as do the more geographically restricted species *G. ilicifolia*, *G. alpina*, *G. rosmarinifolia* and *G. lanigera*. Some of these are amenable to isolated studies; others require attention in the context of a phylogenetic study (Makinson, in prep.).

TOWARDS A PHYLOGENY OF GREVILLEA

There have to date been no phylogenetic analyses of *Grevillea*. This section outlines possible relationships based on comparative morphology, and some phylogenetic considerations. The infrageneric classification schemes used in the past by Bentham (1870), McGillivray and Makinson (1993), and Olde and Marriott (1994e, 1995a, 1995b), compared with the present scheme, are presented at Table 1 (pp. 14–16).

Tribe Grevilleeae has three member genera: *Grevillea*, the Malesian *Finschia* Warb. (4 spp.), and the Australian endemic *Hakea* Schrad. & J.C.Wendl. (149 spp.). Of the other six tribes in subfamily Grevilleoideae, closest affinities appear to be to Macadamieae (P.Weston, pers. comm.).

Finschia is distinguished from *Grevillea* mainly by its indehiscent follicles with one thick-bodied seed. In all *Grevillea* (except *G. candicans*), the fruit is a dehiscent follicle, and often, and probably ancestrally, 2-seeded. There are *Grevillea* species with single-seeded fruit and/or thick seed-bodies, but these are secondarily derived conditions in lineages with consistently dehiscent fruits.

Hakea is distinguished by the presence of a continuous cambial layer producing secondary woody thickening in the fruit valves (Johnson & Briggs, 1975; W.R.Barker *et al.*, 1999). Typically the inner face of each valve shows clearly demarcated zones of light and dark wood. Secondary distinguishing features (with respect to *Grevillea*) include the always-glabrous ovary of *Hakea*, and the much more frequent occurrence in *Hakea* of truly terete (ungrooved) leaves, of similifacial leaves (in which ad- and abaxial leaf surfaces are very similar), of cauline, non-pedunculate conflorescences, of simpler patterns of conflorescence organisation and floral orientation, and a usual seed morphology of a flattened seed-body with an asymmetrical papery wing (although other patterns exist in a few species (W.R.Barker *et al.,* 1999). In *Grevillea* the seeds are either flattened and peripterous with the wing more or less symmetrical, or obloid-ellipsoidal and thick-bodied with the wing reduced to an apical elaiosome and a symmetrical or asymmetrical waxy border).

Even making allowance for the smaller number of taxa relative to *Grevillea*, and notwithstanding the elaboration of some organs into suites of characters with several complex states (e.g. leaf anatomy; see Catling & Gates, 1995), *Hakea* shows an altogether more limited range of morphological character states in leaves, conflorescences, flowers, seeds, and indumentum.

Towards a phylogeny of Grevillea

Limited phylogenetic analysis of *Hakea* has been performed (W.R.Barker *et al.*, 1999), but it seems to be clearly monophyletic, a conclusion based primarily on the fruit character but supported to some degree by the generalised presence of involucral bracts enclosing young vegetative and floral shoots (present in a few groups of *Grevillea* only), and the asymmetrically winged seed.

There is considerable homoplasy of character states within both *Hakea* and *Grevillea*, and between the genera. Flowers of the two genera, even in clearly derived syndromes, are sometimes almost unassignable without the employment of non-floral characters.

From a phylogenetic point of view, the purported monophyly of *Grevillea* with respect to the other two genera is increasingly under question, although as yet no large or rigorous analytical study has been completed.

Grevillea robusta has sometimes been postulated as a likely representative of the ancestral *Grevillea* stock, as opposed to other more modified lineages within the genus. This seems to have been based largely on speculation as to the origins of the genus, and of other significant lineages in the family, in the rainforest environments that were widespread in Australia during the Tertiary period. The preference of *G. robusta* for mesic, rainforest-margin environments on rich basaltic soils (features unusual in the genus) would have contributed to this line of thought, as perhaps would the lack of hybrid compatibility of *G. robusta* with other species, and its relatively unique suite of character-states and lack of close relatives except in New Caledonia (although these latter factors did not become clear until McGillivray's revision). *Grevillea robusta* is certainly isolated within the genus on these features, but it does not seem likely to be basal to the genus as a whole. It is strongly ornithophilous, and some characters associated with this (its large, orange, nectar-rich flowers and strongly secund conflorescences) are almost certainly derived.

While assessments of the age of bird lineages, including modern pollination-vector groups like the parrots and honeyeaters, are being steadily pushed back beyond -50 Ma, it nevertheless seems likely that the common ancestral stock of the Grevilleeae was insect-pollinated. This factor, together with some morphological traits of the conflorescence and flower common in *Hakea* and/or in the other tribes of the subfamily suggest a possible evolutionary root for *Grevillea* close to the group of taxa here treated as the *Hilliana* group (although not necessarily in a rainforest context). That group, and certain other groups here postulated to be derived from it (especially the *Petrophiloides* group and probably also the *Integrifolia* group), all have some features that are suggestive of a link to *Hakea*.

One such feature is the presence in these groups of most of the fundamental variations in leaf morphology (and anatomy?) to be found in the genus: entire and divided leaves, the dipleural pattern, the unifacial pattern (in both subterete and vertically flattened forms), truly terete (ungrooved) leaves, and the dorsiventral pattern that is dominant in other groups. This variability within what is here postulated to be a single subgeneric lineage parallels the complexity of fundamental leaf organisation in *Hakea*.

Another suggestive feature is the presence in both the *Hilliana* and *Petrophiloides* groups of numerous taxa with unusual activity of one form or another in the pericarp. This takes various forms: in *G. stenobotrya* the exocarp becomes tesselated and eventually exfoliates; in *G. pyramidalis* and several close relatives the fruit surface lacks obvious glands but becomes covered with a thick sticky caustic exudate; in *G. petrophiloides* and some relatives the fruit develops viscid crater-like pustules. Many taxa in the *Hilliana* group have unusually thick-walled fruits (though none have secondary thickening). These features may conceivably relate, in origin, to the presence in *Hakea* of cambial activity.

Finally, conflorescence forms in these two groups seem to be conservative and primitive. Most of the taxa have regular-cylindrical many-flowered conflorescences; paniculate branching patterns are common. They lack certain patterns of conflorescence form common elsewhere in the genus (mainly in highly ornithophilous groups), such as the secund form and the few-flowered cluster. Transverse orientation of the flowers on the rachis, rare in the genus as a whole, is not uncommon in these groups. These conflorescence features are relatively plastic and subject to reversal (instances of apparent reversion to entomophily from ornithophily, and of changes within lineages of floral orientation, are known elsewhere in the genus). Nevertheless, their dominance in this group is significant.

Given the degree of apparent homoplasy, the morphological characters identified so far are likely to provide only partial resolution of phylogeny within the tribe and genus, and complementary molecular data will probably also be required to cast light on basal relationships.

The 'groups' identified in this treatment comprise related species and are likely to prove to be phylogenetically robust, but the hierarchy indicated is not phylogenetic. An initial approach to phylogeny within the genus is underway (Makinson, in prep.).

BIOGEOGRAPHY, VEGETATION TYPES, AND EDAPHICS

The main geographical areas of species diversity in *Grevillea* are, in descending order, the south-west of Western Australia, the eastern seaboard between about East Gippsland (Vic.) and Maryborough (Qld) and inland to the Great Dividing Range, the Kimberley region of Western Australia, and the 'Top End' of the Northern Territory.

The south-west of Western Australia is defined roughly as the area SW of the 200 mm rainfall isohyet, including the South-west Botanical Province and the Southwestern (Coolgardie) Interzone, both *sensu* Beard (1980). Some 190 species occur in this region. Correlations between *Grevillea* species and soil types in this region are poorly documented as yet, but there are suites of species occurring on most of the extensive substrates: acidic to alkaline sandy soils (often over clay or laterite), calcareous sands and limestones, and laterite.

In the eastern seaboard zone some 75 species occur, nearly all endemic, and the great majority on siliceous sandstone substrates (e.g. Sydney Basin, Nowra sandstones) or in sandy riparian situations or coastal heath, and to a lesser extent on granitic substrates.

In the Kimberley region, broadly defined, some 22 species have been recorded to date, although further new taxa are to be expected. Most of these occur on acidic sandstones or quartzites or in siliceous colluvial sands, with lesser numbers on laterites, limestones, mixed volcanic substrates, and a few in basaltic soils.

The Top End, with 25 species, exhibits a similar pattern of edaphic distribution to the Kimberley.

Ultrabasic rock types are rare in Australia, but where they do occur over reasonably extensive areas there are a few associated (and more or less endemic) *Grevillea* species, in both eastern and south-western Australia. In this regard, note the occurrence of all three currently recognised New Caledonian species (several taxa, in need of revision) on serpentinite-derived substrates (Virot, 1968). Occurrence on ultrabasics is likely to reflect a generalised preference in the genus for well-drained soils with low available nutrients, rather than affinity for any specific minerals characteristic of these soils.

Few species occur in high fertility soils. In part this no doubt reflects the relative rarity of such soils in Australia. The extensive eastern Australian basaltic flows of the late Tertiary and early Quaternary periods are arguably too recent for an evolution of tolerance for these soils (and the associated vegetation types) to have developed among grevilleas of adjacent regions, but even in areas of much older basaltic or basalt-derived substrates there are few species to be found—*G. robusta*, *G. helmsiae*, and a cluster of three or four taxa in the *G. mimosoides* alliance of the monsoon tropics being exceptions.

The great majority of *Grevillea* species occur primarily as shrubs in shrubland, woodland, or open shrubby forest associations, or as relict species in areas which were formerly of these vegetation types. Relatively few species occur under dense canopy, and only a very few (*G. robusta*, *G. hilliana*, *G. baileyana*, *G. helmsiae*, all arguably representing relatively ancient lineages within the genus) occur in rainforest or 'dry rainforest scrub' associations, sometimes on basalts. The genus is entirely absent from the most extensive areas of high-altitude rainforest in far northern Queensland, and from the cool-temperate rainforests of Tasmania and Victoria, except for transgressive species from other habitats in limited open-canopy situations.

Grevillea species show little tolerance for salt-affected environments, with few occurring in foreshore coastal vegetation exposed to salt spray, and few in inland saline soils. A few taxa in the south-east and south-west show a limited tolerance for seasonally waterlogged soils, but most taxa occur in relatively well-drained situations (although this tendency must be viewed in relation also to differential site clearing and grazing, and fire effects).

Pending a comprehensive biogeographic analysis, only the most general of trends in vicariance relationships can be noted here. There is a strong trend for south-east/south-west sister taxa relationships in a number of groups, notably in the *Pteridifolia* group, (the 'toothbrush inflorescence' group, series *Hebegynae* of Bentham) and in the closely inter-related *Floribunda*, *Buxifolia* and *Longistyla* groups.

The recent discovery of *G. cravenii* and *G. maherae* in the Kimberley, the only members of the *Aspleniifolia/Hookeriana* subgroup recorded north of the Tropic, is noteworthy and argues strongly for a formerly pan-continental distribution for the subgroup.

There is also an apparent trend for vicariant sister-taxa relationships between the monsoon tropics on the one hand and the eremaean zone (*sensu* Burbidge, 1960) and south-west on the other. Thus *G. hilliana* and *G. baileyana* from the north-eastern rainforests are closely allied to *G. striata* and other taxa in the eremaean zone; *G. parallela* from the monsoon tropics has a sister relationship with *G. nematophylla* and *G. berryana* from the eremaean, and probably also with *G. polybotrya* and *G. makinsonii* in the south-west; and the monsoon-tropic members of the *Pteridifolia* subgroup of the *Pteridifolia* group (*G. pteridifolia, G. formosa*) have their closest relationships with sister taxa in the south-west and eremaean.

Occasional cases of apparently recent and probably very rapid dispersal over substantial distances do occur. The occurrence of *G. anethifolia*, a member of the highly distinctive and otherwise entirely south-western *Triloba* group (Section *Manglesia sensu* Bentham), on low-amplitude sand dunes in the mallee communities of south-western Australia, the Eyre Peninsula of South Australia, and the Griffith region of south-western New South Wales, is very likely attributable to transcontinental eastward movement of the species, with aeolian sand movements, during recent (Quaternary) periods of extreme aridity. There is no reason to suppose that Section *Manglesia* or close relatives have ever occurred in eastern Australia in any other context.

For the most part, however, *Grevillea* species are likely to be slow dispersers. Most taxa have relatively large, heavy, unwinged seeds, suitable for only local dispersal (by gravity, ants, and fresh water). There is, however, a strong trend in the tropics and (?consequently) to some extent in the south-west, for seeds to be flatter, lighter, and winged, although the wing is rarely robust and is unlikely to facilitate wind-dispersal over very long distances.

Affinities of the non-Australian taxa are of interest. The three New Caledonian species (*G. exul*, *G. meisneri* and *G. gillivrayi*) all seem to have closest affinity to the Australian *G. robusta*, which otherwise has no close relatives and no recorded incidence of hybridisation with other species (despite its extensive use in horticulture, and the known propensity to hybridisation of many species). The New Guinean endemic *G. papuana* and the Sulawesi endemic *G. elbertii* both have their closest affinities probably with Queensland's *G. helmsiae*; all three are monsoon savanna or dry-rainforest scrub species, and the alliance of which they are all members (the *Hilliana* group) may prove to be basal in the genus.

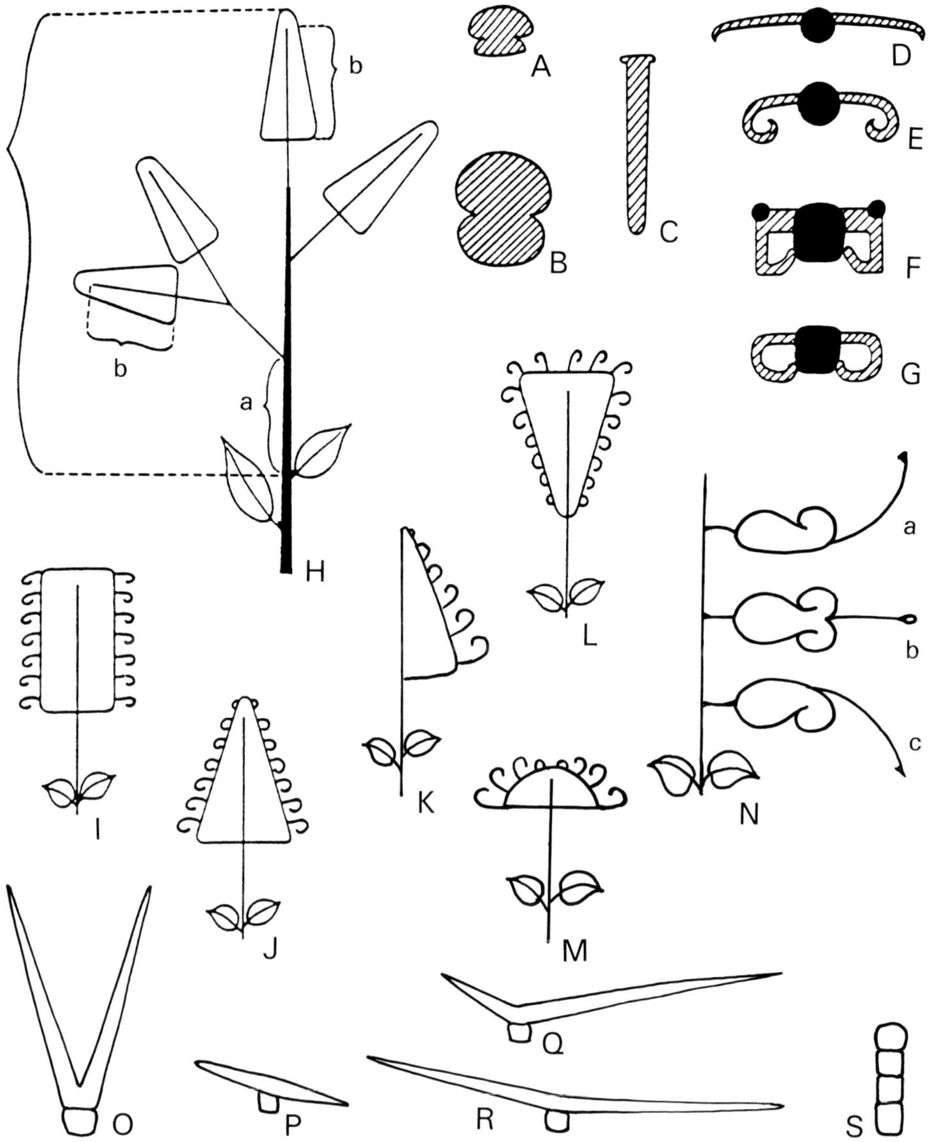

Figure 1. *Grevillea*. Morphological terms. **A–G**, leaves (transverse sections). **A**, **B** dorsiventral dipleural leaves. **C**, unifacial leaf. **D–G**, dorsiventral leaves: **D**, margin slightly recurved, undersurface on either side of the midvein wholly exposed; **E**, margin smoothly recurved, undersurface on either side of midvein partly exposed; **F**, margin angularly revolute (= refracted) to midvein, lower surface enclosed; **G**, margin smoothly revolute to midvein, enclosing lower surface. **H–M**, inflorescences. **H**, branched inflorescence: **a**, primary peduncle; **b**, unit conflorescence (= ultimate flowering branch including flowers and buds. Its central axis is the ultimate floral rachis). **I–M**, flower opening order and inflorescence shape: **I**, synchronous, cylindrical; **J**, acropetal, conical; **K**, acropetal, secund; **L**, basipetal, obconical; **M**, acropetal, umbelloid. **N**, flower position on floral rachis: **a**, acroscopic (ventral suture facing towards apex); **b**, transverse; **c**, basiscopic (ventral suture facing towards base). **O–S**, hairs. **O–R**, biramous (2-armed) hairs - single short 'stalk cell' and transverse (usually medifixed) elongate terminal cell, sub-epidermal basal cell not shown; **S**, simple erect hair, terminal cell often glandular. Drawn by: **A–J**, **L**, ?D.Fortescue; **K**, **M**, **N–S**, A.Wilson.

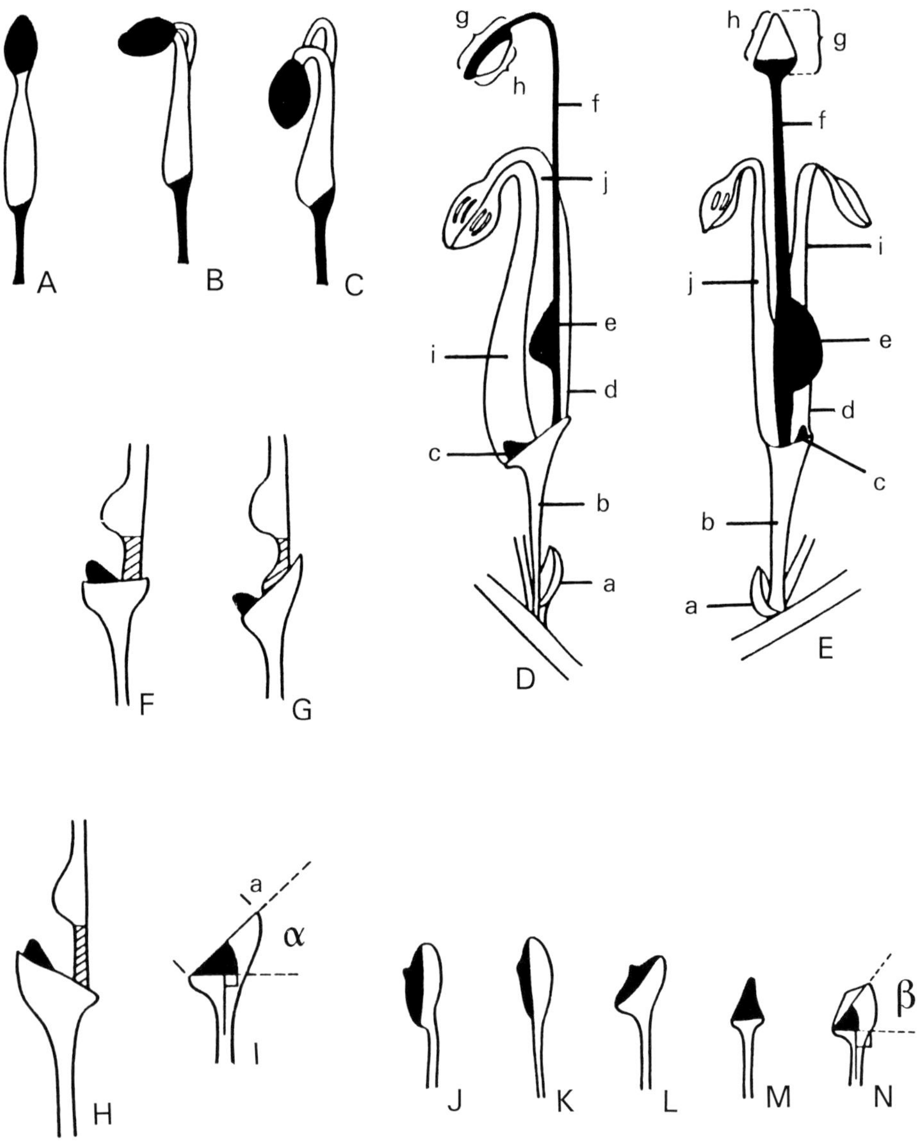

Figure 2. *Grevillea.* Morphological terms. **A–C**, buds with limb orientation (limb shown in black). **A**, limb erect; **B**, limb nodding, style exserted at dorsal side near anthesis; **C**, limb decurved or declined, style exserted near anthesis. **D–E**, flowers (half perianth removed, showing the two most common flower patterns). **D**, style and limb decurved (or declined), perianth strongly zygomorphic, dorsal and ventral tepals obviously differentiated; **E**, style and style-end erect, limb decurved; perianth more or less symmetrical, dorsal and ventral tepals ±isomorphic; **a**, floral bract (subtending flower pair); **b**, pedicel; **c**, nectary (black); **d**, stipe (black); **e**, ovary (black); **f**, style (black); **g**, style-end; **h**, pollen presenter (white); **i**, ventral tepal; **j**, dorsal tepal. **F–I**, torus, stipe (hatched) and nectary (black). **F**, torus transverse, stipe not adnate to torus. **G**, torus oblique, stipe partially adnate to torus; **H**, torus reversely oblique, stipe not adnate; **I**, determination of angle of obliquity (α) of torus; **a**, dorsal edge of torus rim. **J–N**, pollen presenters (black). **J**, lateral, convex, base not concurrent with style; **K**, lateral, base concurrent with style; **L**, oblique; **M**, erect, conical; **N**, determination of angle of obliquity (β) of pollen presenter. Drawn by ?D.Fortescue.

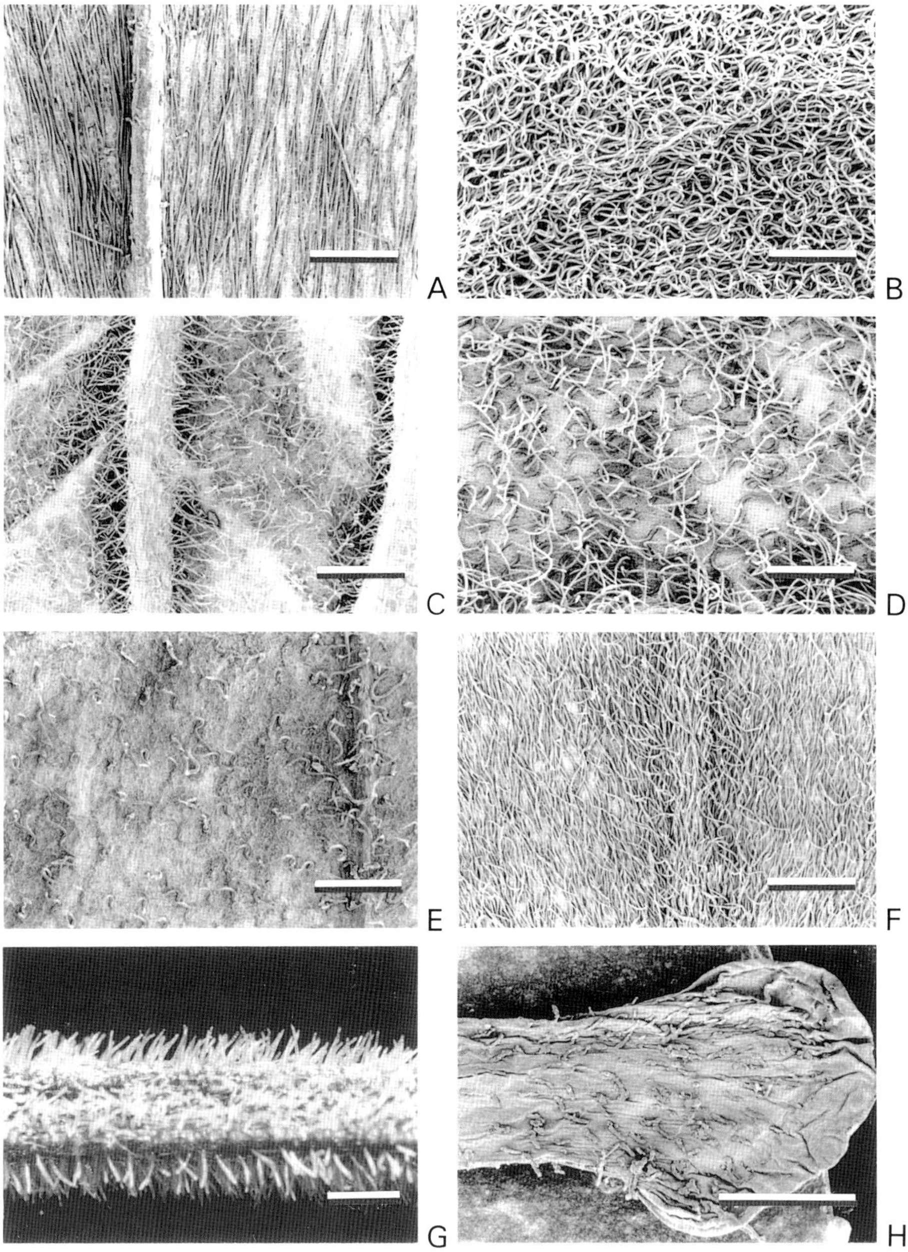

Figure 3. *Grevillea* indumental state and hair types. **A–F**, leaf lower surfaces. **A**, sericeous: biramous hairs, apressed, mutually aligned (*G. sericea* subsp. *sericea*, anon. ANBG 3002, CANB). **B**, tomentose: short curled biramous hairs (*G. oxyantha* subsp. *oxyantha*, M.Doherty *s.n.*, CBG9809209, CANB). **C**, open tomentose: biramous hairs ascending to erect (*G. vestita* subsp. *vestita*, E.M.Canning CBG026720, CANB). **D**, loosely villous: biramous hairs erect, long (*G. mollis*, R.O.Makinson 1448, CANB). **E**, open tomentose: mixed ascending biramous hairs and twisted simple erect glandular hairs (*G. rhyolitica* subsp. *rhyolitica*, P.Gilmore 4320, CANB). **F**, subsericeous: biramous hairs appressed (*G. parvula*, D.E.Albrecht 2679, CANB). **G**, branchlet: pilose, erect simple hairs (non-glandular) (*G. infundibularis*, A.S.George 10589, PERTH). **H**, style end: minute erect hairs (*G. sericea* subsp. *sericea*, R.O.Makinson *s.n.*, 26 Dec. 1981, CANB). Scale bars: **A–F, H** = 0.5 mm; **G** = 1 mm. Photographs: **A–F**, **H**, E.Hines & Celia Miller; **G**, Cathy Miller.

definition is at variance with usage by R.M.Barker *et al.* (1999) for *Hakea*, in which for some taxa 'rachis' apparently includes the portion below the lowest flower scar, i.e. the peduncle.

Floral orientation is diagnostically important, but can be hard to judge from pressed material or from inflorescences which are lax or few-flowered or in which the flowers have long pedicels. The terms used here (*acroscopic, basiscopic, transverse*) relate to the orientation (with respect to the rachis) of the flower's single (dorsiventral) plane of symmetry, and (through the qualifying terms *adaxially* and *abaxially*) to whether the ventral suture of the perianth (and ovary) is inclined towards the rachis or away from it (see the Glossary and Figure 2). An abaxially acroscopic flower thus has the plane of floral symmetry aligned along the rachis, with the ventral suture of the perianth (and the ventral bulge of the ovary and fruit) facing in the same general direction as the distal end of the rachis, but with the flower somewhat reclinate and the ovary bulge thus directed outwards from the rachis axis itself. Conversely, an adaxially basiscopic flower has the plane of symmetry aligned along the rachis, with the ventral suture facing towards the basal end of the rachis and the flower declined so the ovary bulge is directed inwards towards the rachis.

Exsertion of the style from the dorsal suture of the late flower bud before anthesis and the release of the style end from the limb of the bud, occurs in most species. Among taxa with zygomorphic perianths, certain groups (e.g. *Wickhamii* group, *Adenotricha* group) have flowers with a style that is very short, scarcely exceeding the perianth after flower opening; in these taxa the late bud has the dorsal suture of the perianth gaping to expose part of the style, but the style is not or scarcely exserted. The species with regular perianths (*Rudis* group, *Triloba* group, *Petrophiloides* group) usually have the styles neither exposed nor exserted (rarely slightly so).

The torus of the flower is the region at the summit of the pedicel upon and around which are positioned the tepals and pistil. It is more or less equivalent to 'thalamus' or 'receptacle' as used by some authors in other families. The toral rim is the annular or elliptic edge to which the tepal bases are attached and surrounding the cavity in the apex of the pedicel that houses the nectary and the base of the pistil. The angle of the plane formed by the toral rim, relative to the axis of the pedicel, determines toral obliquity; where the plane is perpendicular to the pedicel, the torus is said to be transverse (= 'straight' of Olde & Marriott, 1994, 1995a, 1995b). See Figure 2 for a diagrammatic illustration of the straight, oblique, and reversely oblique states.

The pollen-presenter shows considerable variation in form, and this affects the shape and posture of the style-end as a whole (the pollen-presenter being technically only the surface, usually shiny, on which the stigma is located and to and from which pollen transfer is effected). Shape and posture relative to the subtending style are diagnostically important. See Figure 2 for the method of determining these traits. Bird-pollinated taxa typically have a subdiscoid or hoof-like style-end with the pollen-presenter more or less round and flat to shallowly concave or shallowly convex; insect-pollinated taxa typically have a weakly to very strongly conical, or even fusiform, style-end, although there are exceptions in both pollination syndromes.

POISONOUS AND IRRITANT PROPERTIES

Some *Grevillea* species and hybrids are known to cause contact dermatitis in susceptible individuals, the severity of the reaction varying widely. Apart from mechanical irritation by hairs, a minor problem, the main cause appears to be the presence in at least some grevilleas of resorcinols and related phenolic compounds, which have irritant or allergenic properties. The same class of compounds occurs in *Toxicodendron* species. Dermatological or allergic reactions are most commonly reported for *G. banksii* and *G. pteridifolia*, or species related to them or hybrids based on them, although this may be partly an artifact of the popularity of these species and hybrids in cultivation; Cirigottis *et al.* (1974) document the occurrence of phenolics in the two above-named species. The artificial hybrid *G.* 'Robyn Gordon' (*G. bipinnatifida* × *banksii*) is a common offender (Menz *et al.*, 1986). *Grevillea robusta* has been adversely reported in the timber industry (Cannon *et al.*, 1970, and see entries for 'poisonous and irritant properties' in Harwood, 1989), as has *G. hilliana* (Cirigottis *et al.*, 1974).

A few species are known to be cyanogenic. Smith & White (1920) record strong positive HCN reactions from flowers and fruits of *G. banksii* and *G. robusta*.

ACKNOWLEDGMENTS

This *Flora* treatment would not have been possible without the prior taxonomic work, and enormous compilation of highly reliable data, performed by Mr Donald J. McGillivray (retired, formerly of the National Herbarium of New South Wales). For those data, and for much guidance and friendship over the years, my deep appreciation.

Many others have contributed in various ways. I am indebted in particular to Dr Barbara Briggs, the late Dr Lawrie Johnson, Bill Molyneux, Alex George, Peter Olde and Dr Peter Weston for various forms of encouragement and collaboration, and to Bill Molyneux, Sue Forrester, and Ian Telford for much help in the field and in grappling with difficult groups.

In addition I am grateful to the following people: for assistance with fieldwork and/or provision of collections in northern Australia: Jenny Chappill and Carol Wilkins (University of W.A.); Clyde Dunlop (DNA); Judy West and Jo Palmer (CANB); Matt Barrett (Kings Park and Botanic Garden, Perth); Lindy Cayzer (CANB); Murray Fagg (ANBG) and Rosemary Purdie (Murray-Darling Commission); Kevin Kenneally (W.A. Dept of CALM); Rick, Anne and Mandy Jane of Bushtrack Safaris, Derby, W.A.; Peter Lacy (Mount Elizabeth Station, W.A.); Hugh Nicholson (Terania Creek, N.S.W.); and Robyn Maher (Derby, W.A.). For collaboration, advice, and/or field assistance in southern Australia: Bob Coveny, Peter Hind, Ken Hill, Karen Wilson, Peter Weston and Jaime Plaza (all NSW); Bill and Robyn Barker (AD); Faye Davies, Andrew Lyne and Laurie Adams (CANB); Andrew Ford (QRS); Greg Keighery and Sue Patrick (W.A. Dept of CALM); David Albrecht (now of DNA); Siegy Krauss (now of University of W.A.); Peter Richards, Judy Scott and Tony Auld (all N.S.W. Natl Parks & Wildlife Service); John Nightingale (ANBG); Jeremy Bruhl (University of New England); Rob Whelan (University of Wollongong); Kevin Thiele (Vic.); Geoff Butler (Canberra); Phil Gilmour (Thora, N.S.W.); Ray Brown (Illawarra Grevillea Park, N.S.W.); Val Stajsic (Vic.); Tony Rodd (Sydney, N.S.W.); Anna-Louise Quirico (Sydney, N.S.W.); Emma Jefferson (Cooranbong, N.S.W.); Bro. Howard (Little Brothers of St Francis, Tabulam, N.S.W.); Don and Betty Wood (Canberra, A.C.T.); Wayne, Matt & Corey (a beach somewhere); and Bronwyn Harrison (Sydney, N.S.W.). For technical assistance at CANB: Andrew Slee, Lindy Cayzer and Jim Mant; and at NSW: Mark Oakwood and Sue Gould. For microscopy: Cathy Miller (CPBR); Eric Hines (CSIRO); and Celia Miller (CSIRO).

The directors of the following herbaria kindly provided loans of specimens and/or direct access to their collections, for which I am grateful: AD, BM, BRI, DNA, E, G, HO, K, LY, MEL, NE, NSW, P, PERTH, PR. Catherine Jordan of the Australian National Botanic Gardens Library (A.C.T.) provided much-needed assistance in sourcing protologues and other literature, as did successive ABLOs at Kew, U.K. (Barry Conn, Ken Hill, Don Foreman). Thanks also go to flora permit sections of conservation agencies in N.S.W., Vic., W.A., N.T. and Qld, and of forestry agencies in N.S.W., Qld and Vic., and to managers and traditional owners of Kakadu National Park, N.T. for access to some sensitive areas.

I am grateful to Tony Orchard (ABRS), Lawrie Johnson (NSW), Lyn Craven (CANB) and Alan Radcliffe-Smith (K) for assistance with Latin.

Special thanks to the Australian Biological Resources Study and CANB for financial support of my posting in 1995–1996 as Australian Botanical Liaison Officer at the Royal Botanic Gardens, Kew, where much of the write-up was completed, and to the Director and staff at K.

Special thanks also to Annette Wilson (ABRS) whose editorial tolerance was stretched but didn't break.

Table 1. Comparison of informal groups recognised in this treatment with groups used by McGillivray & Makinson (1993) and Olde & Marriott (1994), and with the formal infrageneric classification used by Bentham (1870). Species numbering used in the text is included, in parentheses, after the Makinson group name.

Makinson 2000	**Olde & Marriott** 1994e, 1995a, 1995b	**McGillivray & Makinson** 1993	**Bentham** 1870 **Section**	**Series**
Pteridifolia Group				
Aspleniifolia/Hookeriana Subgroup (1–47)	Group 35 *p.p.*	Group 1.2 (with further subdivision) Group 21 (misc.) *p.p.* (*G. nana*)	1. *Eugrevillea*	2. *Hebegynae* (except *G. cirsiifolia*)
Ramosissima Subgroup (48–50)	Group 34	Group 1.2 *p.p.*	9. *Conogyne p.p.*	
Pteridifolia Subgroup (51–55)	Group 35 *p.p.*	Group 1.1 *p.p.*	1. *Eugrevillea*	2. *Hebegynae p.p.*
Banksii Subgroup (56–60)	Group 35 *p.p.*	Group 1.1 *p.p.*	1. *Eugrevillea*	2. *Hebegynae p.p.*
Bipinnatifida Subgroup (61–65)	Group 35 *p.p.*	Group 6	1. *Eugrevillea*	2. *Hebegynae p.p.*
Polyacida Subgroup (66)	Group 32	Group 21 (misc.) *p.p.*		
Rubicunda Group (67–68)	Group 31	Excluded as '*genus nov.*'		
Goodii Group (69–73)	Group 19	Group 15	3. *Plagiopoda p.p.*	
Heliosperma Group (74–80)	Group 5	Group 11 *p.p.* Group 21 (misc.) *p.p.*	4. *Calothyrsus p.p.*	
Shiressii Group (81–82)	Group 18	Group 16	3. *Plagiopoda p.p.*	
Agrifolia Group (83–88)	Group 11	Group 11 *p.p.*	4. *Calothyrsus p.p.*	
Wickhamii Group (89–92)	Group 9	Group 11 *p.p.*	4. *Calothyrsus p.p.*	
Adenotricha Group (93–97)	Group 10	Group 11 *p.p.*	4. *Calothyrsus p.p.*	
Huegelii Group (98–99)	Group 6	Group 7 *p.p.*	4. *Calothyrsus p.p.*	
Robusta Group (100)	Group 20	Group 7 *p.p.*	4. *Calothyrsus p.p.*	

Makinson 2000	Olde & Marriott 1994, 1995a, 1995b	McGillivray & Makinson 1993	Bentham 1870	
			Section	**Series**
Thelemanniana Group (101–116)	Group 14	Group 14	1. *Eugrevillea*	1. *Leiogyne p.p.*
		Group 21 (misc.) *p.p.* (*G. olivacea* only)		
Oncogyne Group (117–124)	Group 12	Group 13	1. *Eugrevillea*	1. *Leiogyne p.p.*
Hakeoides Group (125–136)	Group 16	Group 19	8. *Lissostylis*	3. *Occidentales p.p.*
	Group 36	Group 21 (misc.) *p.p.*		
	Group 37			
Acacioides Group (137–139)	Group 17	Group 3	8. *Lissostylis*	3. *Occidentales p.p.*
Acuaria Group (140–146)	Group 15	Group 2.2.2 *p.p.*	3. *Plagiopoda p.p.*	
Linearifolia Group				
Speciosa Subgroup (147–151)	Group 21 *p.p.*	Group 2.2.1.2 *p.p.*	8. *Lissostylis*	1. Puniceae *p.p.*
Diffusa Subgroup (152–156)	Group 21 *p.p.*	Group 2.2.1.2 *p.p.*	8. *Lissostylis*	2. Sericeae *p.p.*
Linearifolia Subgroup (157–174)	Group 21 *p.p.*	Group 2.2.1.2 *p.p.*	8. *Lissostylis*	2. Sericeae *p.p.*
Costata Subgroup (175–177)	Group 21 *p.p.*	Group 2.2.1.2 *p.p.*		
Victoriae Subgroup (178–191)	Group 21 *p.p.*	Group 2.2.1.1 *p.p.*	8. *Lissostylis*	1. Puniceae *p.p.*
Aspera Group (192–194)	Group 13	Group 2.2.1.1 *p.p.*	3. *Plagiopoda p.p.*	
		Group 2.2.2 *p.p.*		
Floribunda Group				
Floribunda Subgroup (195–204)	Group 25 *p.p.*	Group 2.1.1.1.1 *p.p.*	2. *Ptychocarpa p.p.*	
Arenaria Subgroup (205–210)	Group 25 *p.p.*	Group 2.1.1.1.1 *p.p.*	2. *Ptychocarpa p.p.*	
Rosmarinifolia Subgroup (211–216)	Group 25 *p.p.*	Group 2.1.1.1.2	2. *Ptychocarpa p.p.*	
Lavandulacea Subgroup (217–229)	Group 25 *p.p.*	Group 2.1.1.3	3. *Plagiopoda p.p.*	
Fasciculata Subgroup (230–241)	Group 23	Group 2.1.1.2	3. *Plagiopoda p.p.*	
Buxifolia Group (242–252)	Group 22	Group 2.1.2	7. *Eriostylis*	

Makinson 2000	**Olde & Marriott** 1994, 1995a, 1995b	**McGillivray & Makinson** 1993	**Bentham** 1870	
			Section	**Series**
Longistyla Group (253–263)	Group 26	Group 12 Group 21 (misc.) *p.p.*	3. *Plagiopoda*	
Marriottii Group (264–266)	Group 27	Group 20		
Cirsiifolia Group (267)	Group 33	Group 1.2.5	1. *Eugrevillea*	2. *Hebegynae p.p.*
Pythara Group (268)	Group 28			
Pterosperma Group (269–272)	Group 30	Group 18	1. *Eugrevillea*	2. *Hebegynae p.p.*
Petrophiloides Group (273–276)	Group 7	Group 4	10. *Anadenia p.p.*	
Hilliana Group (277–301)	Group 3 Group 4	Group 7 *p.p.* Group 8 *p.p.*	4. *Calothyrsus p.p.* 5. *Cycladenia* 6. *Cycloptera* 9. *Conogyne p.p.*	
Trifida Group (302–320)	Group 8	Group 9 *p.p.*, Group 21 (misc.) *p.p.*	8. *Lissostylis* 9. *Conogyne p.p.*	3. *Occidentales p.p.*
Integrifolia Group (321–328)	Group 2	Group 8 *p.p.*	10. *Anadenia p.p.*	
Triloba Group (329–349)	Group 1	Group 9 *p.p.*	11. *Manglesia*	
Rudis Group (350–354)	Group 41	Group 5	10. *Anadenia p.p.*	
Eryngioides Group (355)	Group 38	Group 21 (misc.) *p.p.*	8. *Lissostylis p.p.*	3. *Occidentales p.p.*
Prostrata Group (356)	Group 40	Group 21 (misc.) *p.p.*		
Quercifolia Group (357)	Group 39	Group 21 (misc.) *p.p.*	4. *Calothyrsus p.p.*	

REFERENCES

Barker, W.R., Barker, R.M. & Haegi, L. (1999), Introduction to *Hakea*, *Flora of Australia* 17B: 1–30. ABRS/CSIRO, Melbourne.

Barker, R.M., Haegi, L. & Barker, W.R. (1999), *Hakea*, *Flora of Australia* 17B: 31–170. ABRS/CSIRO, Melbourne.

Beard, J.S. (1980), A new phytogeographic map of Western Australia, *Western Australian Herb. Res. Notes* 3: 37–58.

Bentham, G. (1870), *Grevillea*, *Flora Australiensis*. Vol. 5, pp. 417–489. L.Reeve & Co., London.

Brown, R. (1810a), On the Proteae of Jussieu, *Trans. Linn. Soc., London* 10: 15–226.

Brown, R. (1810b), *Prodromus Florae Novae Hollandiae et Insulae van-Dieman...* Vol. 1. J.Johnson & Co., London.

Brown, R. (1830), *Supplementum Primum Prodromi Florae Novae Hollandiae: Exhibens Proteaceas Novas quas in Australasia...* J.Johnson & Co., London.

Brown, R. (1849), *in* C.Sturt, Narrative of an Expedition into Central Australia. T. & W.Boone, London.

Brummitt, R.K. (1988), Report of the Committee for Spermatophyta: 35, *Taxon* 3: 444–450.

Burbidge, N.T. (1960), The Phytogeography of the Australian Region, *Austral. J. Bot.* 8: 75–209.

Cannon, J.R., Chow, P.W., Metcalfe, B.W. & Power, A.J. (1970), The structure of robustol, a novel phenol from *Grevillea robusta* A.Cunn., *Tetrahedron Letters* 4: 325–328.

Catling, D.M. & Gates, P.J. (1995), Leaf anatomy in *Hakea* Schrader (Proteaceae), *Bot. Jahrb. Syst.* 117: 153–172.

Cavanilles, A. (1798), *Icones et Descriptiones Plantarum ...* volume 4. Typographia Regia, Madrid.

Cirigottis, K.A., Cleaver, L., Corrie, J.E.T., Grassby, R.G., Green, G.H., Mock, J., Nimgirawath, S., Read, R.W., Ritchie, E., Taylor, W.C., Vadasz, A. & Webb, W.R.G. (1974), Chemical studies of the Proteaceae. VII. An examination of the woods of 17 species for resorcinol derivatives, *Austral. J. Chemistry* 27: 345–355.

Garden, J. & Johnson, L.A.S. (1949), Proposals for conservation of the names of three Australian genera, *Austral. Herb. News* 5: 3–5.

Garden, J. & Johnson, L.A.S. (1950), Valid publication and *nomina alternativa*, *Austral. Herb. News* 7: 1–4.

Harwood, C.E. (ed.) (1989), *Grevillea robusta*: an annotated bibliography. International Council for Research in Agroforestry, Nairobi.

Hewson, H.J. (1988), *Plant indumentum, a handbook of terminology*. AGPS, Canberra.

Johnson, L.A.S. & Briggs, B.G. (1963), Evolution in the Proteaceae, *Austral. Syst. Bot.* 11: 21–61.

Johnson, L.A.S. & Briggs, B.G. (1975), On the Proteaceae – the evolution and classification of a southern family, *Bot. J. Linn. Soc.* 70: 83–182.

Keighery, G. (1990), Taxonomy of the *Grevillea brachystylis* species complex (Proteaceae), *Nuytsia* 7: 125–131.

Keighery, G. (1992), Taxonomic review of the *Grevillea drummondii* Meissn. species group (Proteaceae), *Nuytsia* 8: 225–230.

Kenneally, K.F. (1988), *Grevillea donaldiana* (Proteaceae), a new plant species from the Kimberley division of Western Australia, *Western Australian Nat.* 17: 111–117.

Knight, J. (1809), *On the cultivation of the plants belonging to the Natural Order Proteae*. W.Savage, London.

Trib. 6. GREVILLEEAE

Proteaceae trib. *Grevilleeae* Endl., *Gen. Pl.* 340 (1837), as *Grevilleae*

Type: *Grevillea* R.Br. ex Knight

Peduncles usually present. Floral bracts present or absent. Flowers zygomorphic, with diagonal orientation. Pollen-presenter well developed. Hypogynous glands present or ±absent. Ovules 2, laterally inserted, hemitropous. Fruit folliculate or indehiscent. Seeds 1 to many. $n = 10$.

A tribe of three genera, one (*Hakea*) confined to Australia, another (*Grevillea*) substantially so, but extending to Sulawesi, New Guinea and New Caledonia, and the third (*Finschia*) confined to the western Pacific but extra-Australian.

41. GREVILLEA

R.O.Makinson[1]

Grevillea R.Br. ex Knight, *Cult. Prot.* xvii, 120 (1809), as *Grevillia, nom. cons.*; named for Charles Francis Greville (1749–1809), a Fellow of the Royal Society and of the Linnaean Society, and a Vice-President and Treasurer of the (later Royal) Horticultural Society.

T: not designated. Conserved Type: *G. aspleniifolia* Knight

Stylurus Salisb. ex Knight, *Cult. Prot.* xvii, 115 (1809). T: *Embothrium buxifolium* Sm. (*Stylurus buxifolius* (Sm.) Knight); lecto, *fide* D.J.McGillivray & R.O.Makinson, *Grevillea* 399 (1993).

Lysanthe Salisb. ex Knight, *Cult. Prot.* xvii, 116 (1809). T: *Embothrium sericeum* Sm. (*Lysanthe sericea* (Sm.) Knight); lecto, *fide* D.J.McGillivray & R.O.Makinson, *Grevillea* 399 (1993).

Anadenia R.Br., *Trans. Linn. Soc. London* 10: *clavis inter* 48, 49, 166 (1810). T: *Anadenia pulchella* R.Br.; lecto, *fide* D.J.McGillivray & R.O.Makinson, *Grevillea* 400 (1993).

Manglesia Endl., *in* S.F.L.Endlicher & E.Fenzl, *Nov. Stirp. Dec.* 25 (1839). T: *Manglesia vestita* Endl.; lecto, *fide* D.J.McGillivray & R.O.Makinson, *Grevillea* 400 (1993).

Lissostylis (R.Br.) Spach, *Hist. Nat. Vég.* 10: 402 (1841), *nom. illeg.* (*Lysanthe* in synonymy). T: *Embothrium sericeum* Sm.

Ptychocarpa (R.Br.) Spach, *Hist. Nat. Vég.* 10: 402 (1841). T: *Grevillea mucronulata* R.Br.

Eriostylis (R.Br.) Spach, *Hist. Nat. Vég.* 10: 402 (1841), *nom. illeg.* (*Stylurus* in synonymy). T: *Embothrium buxifolium* Sm.

Plagiopoda (R.Br.) Spach, *Hist. Nat. Vég.* 10: 402 (1841). T: *Grevillea goodii* R.Br.

Conogyne (R.Br.) Spach, *Hist. Nat. Vég.* 10: 402 (1841). T: *Grevillea anethifolia* R.Br.

Cycloptera (R.Br.) Spach, *Hist. Nat. Vég.* 10: 402 (1841). T: *Grevillea heliosperma* R.Br.

Erect to prostrate shrubs or occasionally trees, sometimes lignotuberous or root-suckering. Leaves sessile or petiolate, entire or toothed, or divided with up to three orders of division; margins flat to revolute, sometimes obscure or absent; venation pinnate-reticulate to parallel. Flowers hermaphrodite, pedicellate or sessile, usually paired, with a common bract, aggregated into conflorescences. Perianth usually zygomorphic with limb decurved, sometimes regular with limb erect. Tepals 4 or rarely (*G. donaldiana*) 3, separating after anthesis to release style-end. Outer and inner surfaces of perianth glabrous or hairy. Anthers sessile in limb-cups of each tepal. Hypogynous gland (nectary) integral or sometimes annular, rarely absent or 4-merous. Gynoecium (pistil) glabrous or hairy; ovary superior, sessile or stipitate; ovules 2; pollen-presenter erect or oblique to lateral on style, discoid or conical. Fruit a follicle or rarely (*G. candicans*) an achene, 2- or 1-seeded, caducous or persistent, glabrous or hairy; pericarp crustaceous to bony-textured. Seeds either peripterous

---

[1] Australian National Herbarium, Centre for Plant Biodiversity Research, CSIRO, GPO Box 1600, Canberra, Australian Capital Territory 2601

with a usually flat-ellipsoidal seed-body, or unwinged ellipsoidal with or without a terminal subtriangular waxy elaiosome, or rarely hemispherical.

Chromosome number: n = 10 (H.P.Ramsay, *Austral. J. Bot.* 11: 1–20, 1963, for 26 taxa, unvouchered); n = c. 22 (G.D.Carr & G.McPherson, *Ann. Missouri Bot. Gard.* 73: 486–489, 1986, for *G. meisneri* (New Caledonia) only); n = 10 (J.M.Hart & M.J.Henwood, *Telopea* 7(1): 65–76, 1996, for *G. sphacelata*, *G. phylicoides* and the 2 subspecies of *G. buxifolia* recognised in this treatment).

A predominantly Australian genus of 362 species. Outside Australia, three species in New Guinea (1 endemic), three endemic species in New Caledonia, and one endemic species in Sulawesi. In Australia 357 species, with 355 endemic. Occurs in most habitats, but only a few species alpine, few in swamps, and few salt-tolerant. Many species prefer siliceous acid substrates. Several species are used by Aboriginal cultures for food (nectar and rarely seeds or gums), medicines (gums and bark extracts), glues (gum exudates), and timber for tools and weapons. European uses include ornamental horticulture (many species and hybrids), and limited exploitation of timber and various agroforestry uses (especially *G. robusta*, now widely planted overseas in the tropics). A few species, including *G. banksii* and derived hybrids, and rarely *G. robusta*, may cause contact dermatitis following contact with the foliage or sawdust.

J.Knight, *Cult. Prot.* 115–121 (1809); R.Brown, On the Proteaceae of Jussieu, *Trans. Linn. Soc. London* 19: 15–226 (1810); R.Brown, *Prodr.* 374–380 (1810); R.Brown, *Suppl. Prodr. Fl. Nov. Holl.* 16–25 (1830); C.D.F.Meisner, Ordo CLXIV. Proteaceae (1), *in* A.L.P.P. de Candolle, *Prodr.* 14: 209–482 (1856); G.Bentham, *Fl. Austral.* 5: 417–489 (1870); D.J.McGillivray & R.O.Makinson, *Grevillea* (Proteaceae) - a taxonomic revision (1993); P.M.Olde & N.R.Marriott, *Grevillea Book*, vol. 1 (1994), vols 2, 3 (1995).

KEY TO GROUPS

Keys in this volume will usually require the use of magnification up to 20×, or (for some diagnostic hair features) up to 40×.

This key to groups uses, where possible, characters that are fairly evident and likely to be available on flowering plants or specimens, rather than characters which may be more rapidly diagnostic but more obscure (e.g. floral orientation in short-rachis taxa or nectary features), or less likely to be present (e.g. fruit and seed characters). Homoplasy is also rife in the genus, so few groups key out in toto on any one or few character states.

Accordingly, this key to Groups is not fully articulated with keys to species within the Groups; some repeat scoring of features may be necessary when moving to species-level keys.

Additional Keys to Groups are provided (see pp. 451–460) for the following States and Territories: Northern Territory, South Australia, Queensland, New South Wales and Victoria. Tasmania has only one currently recognised species (*G. australis* complex). Western Australia has 29 of the 33 Groups represented, making significant abbreviation of the Key to Groups impractical for that State.

1 Perianth actinomorphic with limb (apical knob) of bud erect just before anthesis

2 Pistil > 10 mm long; unit conflorescences ±cylindrical **Petrophiloides Group** (p. 355)

2: Pistil < 10 mm long; unit conflorescences cylindrical or otherwise

3 Ovary with hairs **Rudis Group** (p. 440)

3: Ovary glabrous

4 Style ±dilated into a flask-shaped swelling; pedicels as long as, to much longer than, the pistil **Triloba Group** (p. 414)

4: Style slender, not dilated except at style-end; pedicels shorter than pistil — **Integrifolia Group** (p. 406)

1: Perianth zygomorphic, with limb (apical knob) of bud deflexed to revolute just before anthesis

5 Pollen-presenter very narrowly conical to subulate and erect on the erect style, *and* pistil ⩾ 14 mm long, *and* unit conflorescences regular and cylindrical *and* leaf margins flat to incurved or obscure; leaves or leaf lobes often similar on both surfaces (or terete)

6 Unit conflorescences decurved to pendulous; leaves entire, ±elliptic, 15–65 mm wide; fruits massive, 25–40 mm long, glabrous, lacking pustular craters or viscid secretions — **Hilliana Group** (*G. glauca*) (p. 378)

6: Unit conflorescences erect or rarely decurved; leaves toothed to divided, or if entire then either linear to terete and < 5 mm wide, or spathulate and 10–35 mm wide; fruits 7–20 mm long, with glandular hairs and/or pustular viscid craters — **Petrophiloides Group** (p. 355)

5: Pollen-presenter discoid to clavate and oblique to lateral on the decurved to reflexed style, or if erect then broadly conical to cylindrical, very rarely narrowly conico-cylindrical; pistil 3–60 mm long; unit conflorescences cylindrical or not; style decurved or reflexed or erect; leaf margins flat or incurved or recurved or revolute; upper and lower surfaces of leaves or leaf lobes similar or dissimilar (or leaves terete)

7 Ovary with hairs (sometimes few and minute), sessile or stipitate

8 Stipe of ovary 7–13 mm long (including adnate to the very oblique to lateral torus), *and* leaves entire

9 Ovary densely villous, with biramous hairs only; inner surface of perianth with a dense profuse beard; leaves ⩽ 16 mm long, < 5 mm wide (south-western W.A.) — **Pythara Group** (p. 352)

9: Ovary sparsely pubescent, with minute erect simple hairs only; inner surface of perianth pubescent or papillose, lacking a profuse beard; leaves 10–210 mm long, 5–30 mm wide (eastern Australia) — **Shiressii Group** (p. 124)

8: Stipe of ovary either < 5 mm long, or if longer then leaves divided or toothed; torus transverse to very oblique

10 Inner surface of perianth glabrous below limb segment (ignore hairs behind anthers)

11 Flowers ±sessile (pedicels 0–0.5 mm long), basiscopic; nectary divided into 4 separate erect lobes, spaced evenly around the torus (leaves with linear lobes; pistils 5–10 mm long; ovary sessile; flowers on rachises 20–70 mm long) (N.T.) — **Rubicunda Group** (p. 108)

11: Flowers pedicellate (pedicels > 1 mm long), usually acroscopic, sometimes transverse on rachis or basiscopic but then flowers with pedicels ⩾ 2.5 mm long; nectary integral, arcuate to linguiform or annular (never in 4 separate lobes); ovary sessile or stipitate

12 Ovary with a dense indumentum obscuring the ground tissue

13 Pistil ⩽ 8 mm long, *and* pollen-presenter ±lateral on the style, *and* habit prostrate; nectary absent; tepals remaining apically joined in a small coherent cup after release of style-end, with the lower portions separating and widely everting (south-western W.A.) — **Cirsiifolia Group** (p. 351)

13: Pistil either > 8 mm long with pollen-presenter variously oriented or pistil 3–8 mm long and then pollen-presenter erect-conical; nectary evident or rarely obscure or absent; habit prostrate to erect; tepals either remaining loosely coherent and everting or not, or independently lax

14 Unit conflorescences secund, sometimes shortly or broadly so — **Pteridifolia Group** (p. 39)

14: Unit conflorescences regular and ±cylindrical, or conico-cylindrical

15 Pistil 3–8 mm long; pollen-presenter erect, conical; indumentum of ovary and fruit pale with dark red-brown to purplish markings — **Pteridifolia Group** (*Ramosissima* subgroup) (p. 91)

15: Pistil ⩾ 9 mm long; pollen-presenter usually oblique, rarely erect and conical; indumentum of ovary and fruit without or rarely with dark red-brown to purplish markings

16 Upper surface of leaves and lobes with several longitudinal ridges, lacking a distinct midvein; flowers oriented transversely to rachis (ventral sutures of perianth at right angles to rachis); flower white to cream — **Pterosperma Group** (p. 352)

16: Upper surface of leaves and lobes with a distinct midvein (sometimes also an intramarginal ridged vein on either side); flowers oriented basiscopically (ventral suture of perianth facing towards base of rachis) or rarely transversely or acroscopically; flowers white to cream, red, or orange — **Pteridifolia Group** (p. 39)

12: Ovary indumentum open to sparse (ground tissue visible), or confined to the dorsal side

17 Unit conflorescences regular-umbelloid

18 Ovary papillose or pubescent with simple-erect glandular hairs; prostate shrubs — **Prostrata Group** (p. 446)

18: Ovary mostly glabrous, with scattered biramous hairs on dorsal side of stipe apex and style base only; decumbent to erect shrubs — **Hakeoides Group** (p. 174)

17: Unit conflorescences secund, or regular and ±cylindrical (sometimes shortly so)

19 Ovary with simple-erect glandular hairs only; unit conflorescences secund (tropical northern Australia) — **Heliosperma Group** (p. 116)

19: Ovary with non-glandular hairs only; unit conflorescences ±cylindrical (W.A., S of the tropics) — **Hilliana Group** (p. 361)

10: Inner surface of perianth with hairs or papillae below the limb segments (elsewhere than behind anthers)

20 Style cobra-like (dilated in apical 2/3); fruit muricate-echinate, the projections with glandular tips — **Quercifolia Group** (p. 447)

20: Style even, slender, not dilated except at or just below the style-end; fruit smooth or variously ornamented, if muricate-echinate then tips of projections not glandular

21 Unit conflorescences cylindrical

22 Flowers acroscopic (ventral suture of perianth and concave curve of style facing towards apex of rachis)

23 Ovary ±sessile; leaf surfaces dissimilar, the lower surface conspicuously hairy; seed not winged (temperate south-eastern Australia to central Qld) — **Floribunda Group** (p. 268)

23: Ovary clearly stipitate; leaf surfaces often ±similar, the hairs of the lower surface inconspicuously minute, or absent; seed winged (tropical N.T.)

24 Ovary stipe 0.5–1.0 mm long; pistil 6.5–9 mm long, scarcely projecting beyond perianth after release of style-end; leaves broadly obovate-rhombate, > 30 mm wide — **Wickhamii Group** (*G. velutinella*) (p. 138)

24: Ovary stipe 1.5–5 mm long; pistil 8–24 mm long (if < 12 mm then leaves narrowly ±elliptic, ≤ 15 mm wide); pistil markedly projecting beyond perianth after release of style-end **Agrifolia Group** (p. 126)

22: Flowers either basiscopic (ventral suture of perianth and concave curve of style facing towards base of rachis) or transversely oriented (ventral suture and style curve facing laterally outwards from rachis)

25 Flowers oriented transversely on rachis, unit conflorescences dense; pistils 10–22 mm long

26 Ovary densely villous; floral rachis villous to lanate; pedicels 2–6 mm long; ovary stipe densely hairy (at least on dorsal side), sometimes stipe obscure **Pterosperma Group** (p. 352)

26: Ovary sparsely to loosely hairy; floral rachis ±subsericeous; pedicels 1–2 mm long; ovary stipe glabrous, short but evident **Hilliana Group** (*G. berryana*) (p. 368)

25: Flowers basiscopic (ventral suture of perianth and concave curve of style facing towards base of rachis); unit conflorescences ±loose; pistils (20–) 25–55 mm long

27 Style with appressed hairs in lower half only, or glabrous throughout; simple leaves and lobes of divided leaves < 3 mm wide, with revolute margins **Longistyla Group** (p. 340)

27: Style with ±spreading hairs over most of its length; simple leaves or lobes of divided leaves 6–30 mm wide, with margins flat or shortly recurved **Goodii Group** (*G. venusta*) (p. 115)

21: Unit conflorescences either secund, or ±regular and then umbelloid, subglobose, or 1–few-flowered cluster (never cylindrical)

28 Ovary indumentum open to sparse, ground tissue visible between hairs, or ovary with a short basal or apical tuft of hairs only

29 Floral rachises < 10 mm long; unit conflorescences erect, regular and umbelloid or a loose short cluster; leaves entire and linear to subterete or narrowly elliptic, or deeply divided with linear to subterete lobes

30 Stipe of ovary stout, almost as thick as ovary (south-eastern Australia) **Floribunda Group** (*Rosmarinifolia* subgroup) (p. 299)

30: Stipe of ovary slender, much narrower than ovary (south-western Australia)

31 Style with minute hairs or papillae, sometimes in apical 1–3 mm only **Marriottii Group** (p. 349)

31: Style completely glabrous

32 Leaves subpectinate, with 7–30 lobes; prostrate shrub **Prostrata Group** (p. 446)

32: Leaves with 2–5 lobes or entire; ±erect shrub **Hakeoides Group** (p. 174)

29: Floral rachises 10–160 mm long; unit conflorescences erect to decurved, not umbelloid; leaves deeply divided with linear lobes, or shallowly toothed or lobed, or entire

33 Unit conflorescences erect and long-secund; floral rachises 100–160 mm long; pistils 40–52 mm long; leaves deeply divided with many linear lobes **Heliosperma Group** (p. 116)

33: Unit conflorescences decurved, secund or conico-cylindrical or ovoid; floral rachises 10–100 mm long; pistils 5–20 mm long; leaves shallowly toothed or lobed, or rarely entire

34 Pistils 5–10 mm long, scarcely projecting beyond perianth after release of style-end — **Wickhamii Group** (p. 133)

34: Pistils 13–20 mm long, projecting markedly beyond perianth after release of style-end — **Agrifolia Group** (p. 126)

28: Ovary densely hairy, ground tissue not visible between hairs

35 Stipe of ovary (including portion adnate to inner wall of torus) 5–11 mm long; torus lateral on pedicel or almost so — **Goodii Group** (p. 111)

35: Stipe of ovary < 5 mm long, free from toral wall, or ovary sessile; torus transverse on pedicel or oblique at not more than 45°

36 Some or all leaves toothed or divided

37 Leaves deeply divided, with lobes strongly spreading to divaricate, the rachis decurved or deflexed at nodes, *and* pistils 18–38 mm long

38 Ovary shortly but clearly stipitate; outer surface of perianth glabrous or with hairs — **Longistyla Group** (p. 340)

38: Ovary sessile; outer surface of perianth with hairs — **Pteridifolia Group** (p. 39)

37: Leaves shallowly toothed, or more deeply divided but then with lobes mutually aligned on either side of the straight to upcurved leaf rachis (not divaricate), and/or pistils ⩽ 15 mm long

39 Leaves with shallow ±triangular teeth, or more deeply divided but then with the ultimate lobes (linear or not) < 15 mm long

40 Leaves usually all toothed or divided, rarely a few entire; teeth or lobes 3–19 per leaf; unit conflorescences usually loose irregular or semi-secund clusters, 2–many-flowered, or flower solitary; flowers red or deep to pale pink — **Longistyla Group** (p. 340)

40: Leaves usually mixed entire and toothed or divided; teeth or lobes 2 or 3 per leaf; unit conflorescences regular, umbelloid (style arching in towards rachis apex), 10–26-flowered; flowers white — **Marriottii Group** (p. 349)

39: Leaves deeply divided with linear lobes; most or all ultimate lobes > 20 mm long

41 Pistils 25–52 mm long; perianth and style predominantly red, red-orange or pink; conflorescence a loose cluster, sometimes shortly subcylindrical (eastern & south-western Australia) — **Longistyla Group** (p. 340)

41: Pistils 13–22 mm long; perianth whitish, style light green; conflorescence dense, subglobose (eastern & south-western Australia) — **Pterosperma Group** (p. 352)

36: All leaves entire

42 Either outer surface of perianth glabrous except for a white villous beard on limb, or pistils > 30 mm long and the conflorescences terminal — **Longistyla Group** (p. 340)

42: Outer surface of perianth either sparsely to densely hairy throughout, or entirely glabrous; pistils 5–38 mm long, if > 30 mm long then conflorescences axillary or cauline

43 Style-end with an incurled to erect or retrorse dorsal appendage; appendage 1–4 mm long

44 Style-end scarcely projecting beyond the perianth after release of style-end; flowers predominantly bright red **Floribunda Group** (p. 268)

44: Style-end projecting well beyond the perianth after release of style-end; flowers predominantly white, grey, or fawn-pink **Buxifolia Group** (p. 327)

43: Style-end lacking a dorsal appendage

45 Unit conflorescences dense, regular, umbelloid (style curved in towards apex of rachis); pistils 6–13 mm long

46 Style densely hairy throughout, at least on dorsal side, the hairs ±obscuring the ground tissue **Buxifolia Group** (p. 327)

46: Style glabrous or papillose or very sparsely to loosely hairy, with ground tissue clearly visible between hairs **Marriottii Group** (p. 349)

45: Unit conflorescences dense to loose, or occasionally 1- or 2-flowered, regular or not, not umbelloid; pistils 4–38 mm long

47 Tepals remaining coherent (except along dorsal suture) after release of style-end and everting over most of their length to form a small platform ventral to the style, the white hairs of the inner surface of the tepals conspicuously displayed; styles densely hairy; pistils 6–13 mm long **Buxifolia Group** (p. 327)

47: Tepals after release of style-end either separating and independently recoiled, or remaining loosely connate and ventral to the style but not everting (except sometimes at apex), and not displaying the hairs of the inner surface; style densely to sparsely hairy or rarely glabrous; pistils 4–38 mm long **Floribunda Group** (p. 268)

7: Ovary glabrous

48 Stipe of ovary ⩾ 7 mm long, including portion adnate to inner wall of toral cavity; torus sometimes very oblique, to almost lateral (co-linear) on the pedicel

49 Leaves entire *and* > 5 mm wide

50 Conflorescences usually paniculately branched; unit conflorescences cylindrical, many-flowered; floral rachises > 20 mm long **Hilliana Group** (p. 361)

50: Conflorescences usually simple, or basally 2-branched; unit conflorescences short loose clusters or single-flowered; floral rachises < 10 mm long **Shiressii Group** (p. 124)

49: Leaves divided, or entire and then < 5 mm wide

51 Floral rachis < 10 mm long; conflorescences short clusters, often axillary or cauline (shrubs of southern semi-arid Australia) **Huegelii Group** (p. 144)

51: Floral rachis > 40 mm long; conflorescences elongate, conico-cylindrical or secund, terminal (shrubs or small trees of tropical monsoon Australia) **Heliosperma Group** (p. 116)

48: Stipe of ovary ⩽ 6 mm long, usually free from toral wall (occasionally adnate but then usually < 4 mm long); torus transverse to moderately oblique (⩾ 45°) on pedicel, or rarely reversely oblique

52 Pistil strongly reflexed at ovary (style reclinate at 60–90° to line of stipe), *and* pistil > 15 mm long

53 Floral rachis < 10 mm long; conflorescences simple or basally branched short clusters, often axillary or cauline (shrubs of southern semi-arid Australia) **Huegelii Group** (p. 144)

53: Floral rachis > 30 mm long; conflorescences paniculately several-branched, terminal (shrubs or small trees of tropical monsoon Australia) — **Hilliana Group** (p. 361)

52: Pistil not strongly reflexed at ovary (lower part of style ±co-linear with stipe), or if reflexed then pistils < 8 mm long

54 Pistil 9–12 mm long *and* style cobra-like (dilated in apical ²/₃) and sharply inflexed above ovary; pollen-presenter lateral on style; flowers basiscopic; leaves with triangular teeth or lobes (prostrate pink-flowered shrub, south-western W.A.) — **Quercifolia Group** (p. 447)

54: Pistil 5–40 mm long; style slender except at very tip, not inflexed immediately above ovary (but sometimes reflexed); pollen-presenter erect or oblique or lateral on style; flowers acroscopic or basiscopic or irregularly oriented; leaves entire or variously divided (prostrate to erect shrubs or trees)

55 Conflorescences simple or several branched, erect, and with the ultimate floral rachises > 30 mm long

56 Inner surface of perianth glabrous

57 Unit conflorescences distinctly secund *and* pistils 15–30 mm long (northern and eastern Australia)

58 Leaves with the teeth or lobes rigid, pungent; pistil 15–18 mm long; shrub to 2 m tall — **Heliosperma Group** (p. 116)

58: Leaves with the lobes pliable, not pungent; pistil 20–30 mm long; tree 8–40 m tall — **Robusta Group** (p. 148)

57: Unit conflorescences either regular (cylindrical or obovoid-cylindrical) or, if secund then pistils < 10 mm long

59 Leaves deeply and divaricately divided, with ±rigid, linear or subterete lobes < 2 mm wide — **Trifida Group** (p. 386)

59: Leaves entire, or shallowly to deeply divided, the teeth or lobes either not linear to subterete or not divaricate, usually pliable

60 Leaves *either* entire *or* deeply divided and then with the ultimate lobes linear or subterete and > 20 mm long and < 3 mm wide

61 All leaves entire

62 Leaves 3–75 mm wide, strap-like, elliptical, falcate, obovate, spathulate, or subrotund — **Hilliana Group** (p. 361)

62: Leaves < 3 mm wide, linear or subterete

63 Leaves either lacking continuous longitudinal grooves, or with 7–17 such grooves — **Hilliana Group** (p. 361)

63: Leaves with 2 continuous longitudinal grooves only, on lower surface

64 Outer surface of perianth glabrous; most conflorescences simple, occasionally basally few-branched, the branches strongly ascending; unit conflorescences strongly acropetal, usually appearing conical — **Trifida Group** (*G. obliquistigma*) (p. 397)

64: Outer surface of perianth usually with few to many hairs, rarely glabrous; most or all conflorescences paniculately 5–20-branched, the branches ±spreading; unit conflorescences subsynchronous or weakly acropetal, usually appearing cylindrical — **Hilliana Group** (p. 361)

61: Some or all leaves divided

65 Leaf margin not evident, leaves unifacial, leaves or segments subterete or laterally flattened, the upper and lower surfaces indistinguishable, lacking an obvious lower surface and lacking continuous longitudinal grooves (semi-arid zone and monsoon tropics) — **Hilliana Group** (p. 361)

65: Leaf margin evident, sometimes thickened, flat to recurved or revolute; leaves dorsiventral, upper and lower surfaces distinguishable, lower surface often differing in venation and/or indumentum, sometimes lower surface concealed by the revolute margins except for the midvein and then 2-grooved

66 Leaf lower surface with the lamina on either side of the midvein at least partly exposed

67 Pistils > 10 mm long — **Hilliana Group** (*G. parallela*) (p. 367)

67: Pistils < 8 mm long — **Trifida Group** (p. 386)

66: Leaf lower surface 2-grooved, the lamina on either side of the midvein enclosed by the revolute margin

68 Outer surface of perianth glabrous; most conflorescences simple, occasionally basally few-branched, the branches strongly ascending; unit conflorescences strongly acropetal, usually appearing conical — **Trifida Group** (p. 386)

68: Outer surface of perianth usually with few to many hairs, rarely glabrous; most or all conflorescences paniculately 5–20-branched, the branches ±spreading; unit conflorescences subsynchronous or weakly acropetal, usually appearing cylindrical — **Hilliana Group** (p. 361)

60: Leaves toothed or shallowly to deeply divided, if deeply divided then either the ultimate lobes not linear or subterete and > 3 mm wide, or the lobes linear but then < 20 mm long

69 Conflorescences borne on long erect leafless scapes to 1.5 m long; foliage all at ground level; leaves undulate with 5–15 broad obtuse nonpungent lobes; pollen-presenter lateral on a deflexed style-end; unit conflorescences markedly basipetal (south-western W.A.) — **Eryngioides Group** (p. 444)

69: Conflorescences borne on short peduncles, close to or within the foliage; foliage not confined to ground level; leaves flat to divaricately lobed but not undulate, with few to many pungent or nonpungent lobes; pollen-presenter erect to oblique; style-end not deflexed; unit conflorescences acropetal, subsynchronous or basipetal

70 Either pistils > 10 mm long, or ultimate leaf lobes ⩾ 50 mm long (tall erect shrubs or small tree of tropical Australia) — **Hilliana Group** (p. 361)

70: Pistils < 10 mm long; ultimate leaf lobes ⩽ 30 mm long (shrubs of south-western W.A.) — **Trifida Group** (p. 386)

56: Inner surface of perianth with few to many hairs

71 Pistils ⩾ 25 mm long

72 Unit conflorescences strongly secund; lower surface of leaf exposed on either side of midvein, glabrous — **Thelemanniana Group** (*G. ripicola*) (p. 160)

72: Unit conflorescences either regular and ±cylindrical, or ±secund but then with the lower surface of leaf fully enclosed by the revolute margins on either side of midvein, 2-grooved

73 Leaves entire, or deeply divided with the leaf rachis ±straight and the lobes mutually aligned, not divaricate

74 Pistils 25–34 mm long — **Hilliana Group** (p. 361)

74: Pistils ≥ 35 mm long

75 Leaves entire, 8–30 cm long; leaf margins recurved or loosely revolute, with lower surface of lamina exposed on either side of midvein; ovary and fruit smooth (small tree, northern Qld) — **Hilliana Group** (*G. coriacea*) (p. 367)

75: Leaves entire or divided, 2–15 cm long; leaf margins tightly revolute to midvein of (2-grooved) lower surface of leaf; ovary and fruit with lumps or ridges (shrubs, south-western W.A.) — **Oncogyne Group** (p. 163)

73: Leaves divided, with rachis angularly deflexed at nodes, and lobes spreading and divaricate

76 Unit conflorescences ±secund; conflorescences simple or 2- or 3-branched — **Trifida Group** (*G. dielsiana*) (p. 404)

76: Unit conflorescences cylindrical; conflorescences usually 3–9-branched — **Hilliana Group** (*G. annulifera*) (p. 382)

71: Pistils $<$ 25 mm long

77 Pistils ≤ 12 mm long

78 Leaves divaricately lobed, with leaf rachis either flexuose or deflexed at one or more nodes — **Trifida Group** (p. 386)

78: Leaves entire, or toothed or lobed but the lobes not divaricate; leaf rachis neither flexuose nor deflexed at nodes

79 Unit conflorescences regular, umbelloid or ovate; longest ultimate floral rachises $<$ 10 mm long — **Trifida Group** (p. 386)

79: Unit conflorescences regular and cylindrical to conico-cylindrical, or secund; longest ultimate floral rachises ≥ 10 mm long

80 Unit conflorescences secund; flowers pale to deep pink; pistils 4–6 mm long — **Trifida Group** (*G. leptobotrys*) (p. 404)

80: Unit conflorescences regular; flowers white to cream or yellow or very rarely pale pink; pistils 4–12 mm long

81 Longest leaves ≤ 30 mm long — **Trifida Group** (*G. crithmifolia*) (p. 405)

81: Longest leaves $>$ 30 mm long

82 Most or all conflorescences paniculately 5–20-branched; unit conflorescences regular-cylindrical; longest ultimate floral rachises 60–250 mm long — **Hilliana Group** (p. 361)

82: Conflorescences simple or basally 1–3 (–5)-branched; unit conflorescences regular, cylindrical to conico-cylindrical; longest ultimate floral rachises usually $<$ 60 mm long, rarely to 150 mm long

83 Leaves (and lobes if present) subterete or linear, ≤ 3 mm wide

144 Stylar hairs confined to area immediately above ovary; leaves linear, < 2 mm wide

145 Unit conflorescences umbelloid to ovoid; floral rachises < 10 mm long — **Hakeoides Group** (p. 174)

145: Unit conflorescences cylindrical; floral rachises 20–60 mm long — **Hilliana Group** (*G. berryana*) (p. 368)

144: Stylar hairs (sometimes minute) over most of style or near the apex only (sometimes confined to the back of the pollen-presenter); leaves linear to ovate or obovate or elliptic

146 Stipe of ovary swollen, almost as thick as ovary, with a tuft of hairs at its base on ventral side — **Floribunda Group** (*Rosmarinifolia* subgroup) (p. 299)

146: Stipe of ovary slender, glabrous

147 Styles exposed but not or scarcely exserted from dorsal side of late bud, and scarcely projecting beyond the perianth after release of style-end; conflorescences decurved; pistils 7–11 mm long — **Aspera Group** (p. 264)

147: Styles exserted from dorsal side of late bud, and projecting well beyond the perianth after release of style-end; conflorescences erect to decurved; pistils 4–35 mm long — **Linearifolia Group** (p. 196)

139: Styles completely glabrous

148 Some or all leaves toothed or divided

149 Pistils < 10 mm long

150 Unit conflorescences cylindrical — **Trifida Group** (p. 386)

150: Unit conflorescences regular-umbelloid or ±secund

151 Tepals separating and independently recoiled, not forming a ventral platform — **Hakeoides Group** (p. 174)

151: Tepals flaring open whilst remaining connate at the tips, to display the inner surfaces, and forming a circular platform held ventral to the style

152 Leaves secund, 3–7 cm long, with 9–31 very regular, mutually aligned, simple ±oblong lobes; floral rachis 5–13 mm long — **Prostrata Group** (p. 446)

152: Leaves secund or not, 3–29 cm long, with often irregular toothing or lobing (secondary and tertiary division sometimes present); lobes often divaricate; ultimate lobes often triangular; floral rachis 30–50 mm long — **Trifida Group** (*G. leptobotrys*) (p. 404)

149: Pistils ⩾ 10 mm long

153 Unit conflorescences very regular-umbelloid, with styles arching inwards towards the rachis apex — **Hakeoides Group** (p. 174)

153: Unit conflorescences various, not umbelloid

154 Ovary (and fruit) with prominent ridges or peaks

155 Pistils < 40 mm long, *or* peduncles flattened and flexuose, *or* torus reversely oblique (the ventral rim higher in side view than the dorsal rim), *or* peduncles trailing on ground beyond foliage, *or* leaves rotund and stem-clasping, forming a shallow funnel — **Oncogyne Group** (p. 163)

155: Not conforming in respect of pistil length or peduncles or torus or leaves	
156 Outer surface of perianth with biramous non-glandular only (usually appressed)	**Thelemanniana Group** (p. 150)
156: Outer surface of perianth with both biramous and simple erect glandular hairs	**Oncogyne Group** (p. 163)
154: Ovary smooth; fruit smooth or faintly colliculose	
157 Unit conflorescences cylindrical; ultimate floral rachises 90–220 mm long; rainforest tree of eastern Australia	**Hilliana Group** (p. 361)
157: Unit conflorescences secund, or short irregular clusters; ultimate floral rachises 10–50 (–110) mm long; shrubs or small trees of monsoon tropics or south-western Australia	
158 Pistils 6–11 mm long; seed not winged (south-western Australia)	**Hakeoides Group** (*G. manglesioides*) (p. 182)
158: Pistils 15–25 mm long; seed winged all around (monsoon tropics)	**Heliosperma Group** (p. 116)
148: All leaves entire	
159 Unit conflorescences cylindrical, > 100-flowered; floral rachis 90–220 mm long	**Hilliana Group** (*G. hilliana*) (p. 365)
159: Unit conflorescences variously shaped, if cylindrical then < 50-flowered and floral rachis ⩽ 50 mm long	
160 Stipe of ovary swollen, almost as wide as ovary, with a tuft of hairs at its base	**Floribunda Group** (*Rosmarinifolia* subgroup) (p. 299)
160: Stipe of ovary slender, glabrous	
161 Unit conflorescences regular, umbelloid, with styles all arched in towards the rachis apex	
162 Pollen-presenter erect, conical; pistil 5–8 mm long	**Trifida Group** (p. 386)
162: Pollen-presenter oblique on style, ±discoid, sometimes convex but never conical; pistil 4–15 mm long	
163 Conflorescences borne on long scapes above foliage	**Acacioides Group** (*G. endlicheriana*) (p. 189)
163: Conflorescences borne within or close to foliage	**Hakeoides Group** (p. 174)
161: Unit conflorescences regular and then ±cylindrical, or secund, or an irregular cluster or 1–4-flowered, or rarely pseudo-umbelloid but then with styles arching outward	
164 Conflorescences borne on long scapes above foliage	**Acacioides Group** (p. 188)
164: Conflorescences borne within or close to foliage	
165 Pistils ⩽ 15 mm long	
166 Pollen-presenter erect, conical; pistil 5–8 mm long	**Trifida Group** (p. 386)
166: Pollen-presenter oblique on style, ±discoid, sometimes convex but never conical; pistil 4–20 mm long	
167 Unit conflorescences 1–6-flowered	

168 Styles exposed but not or scarcely exserted from dorsal side of late bud, and scarcely projecting beyond the perianth after release of style-end ..... **Aspera Group** (*G. pauciflora*) (p. 265)

168: Styles exserted from dorsal side of late bud, and projecting well beyond the perianth after release of style-end ..... **Acuaria Group** (*G. punctata*) (p. 191)

167: Unit conflorescences > 10-flowered

169 Pistils 4–6 mm long; conflorescences lax; tepals flaring to create a circular platform ventral to the style ..... **Trifida Group** (*G. leptobotrys*) (p. 404)

169: Pistils 6–15 mm long; conflorescences dense; tepals not flaring widely

170 Flowers basiscopic; unit conflorescences erect (south-eastern Qld) ..... **Hilliana Group** (*G. helmsiae*) (p. 364)

170: Flowers acroscopic; unit conflorescences erect or decurved (S.A. and south-western W.A.)

171 Styles exposed but not or scarcely exserted from dorsal side of late bud, and scarcely projecting beyond the perianth after release of style-end ..... **Aspera Group** (*G. pauciflora*) (p. 265)

171: Styles exserted from dorsal side of late bud, and projecting well beyond the perianth after release of style-end ..... **Hakeoides Group** (p. 174)

165: Pistils > 15 mm long

172 Pistils > 40 mm long ..... **Oncogyne Group** (*G. tripartita*) (p. 165)

172: Pistils 16–40 mm long

173 Unit conflorescences 1–8-flowered

174 Outer surface of perianth with both ±appressed nonglandular biramous hairs and simple erect glandular hairs; ovary and fruits with conspicuous ridges or peaks ..... **Oncogyne Group** (p. 163)

174: Outer surface of perianth with both ±appressed nonglandular biramous hairs only, or glabrous; ovary and fruits smooth or with subdued warts ..... **Acuaria Group** (p. 190)

173: Unit conflorescences > 8-flowered

175 Flowers basiscopic; unit conflorescences erect; leaves 7–40 mm wide (south-eastern Qld) ..... **Hilliana Group** (*G. helmsiae*) (p. 364)

175: Flowers acroscopic; unit conflorescences usually decurved, occasionally erect; leaves 1–10 mm wide (south-western W.A.) ..... **Thelemanniana Group** (p. 150)

Pteridifolia Group

Shrubs, rarely small trees. Leaves entire or variously divided, dorsiventral; surfaces dissimilar; margins flat to revolute. Conflorescence terminal or rarely axillary or cauline, usually simple, erect to deflexed, usually secund or sometimes subcylindrical or cylindrical or rarely hemispherical, acropetal or occasionally subsynchronous. Flowers acroscopic or (*Banksii* subgroup) basiscopic or transversely oriented. Torus transverse to oblique. Perianth zygomorphic (limb of bud decurved), usually hairy outside, glabrous inside or (rarely in *Bipinnatifida* subgroup) hairy; tepals remaining loosely coherent ventral to style after release of style-end. Pistil 3.5–60 mm long; ovary sessile to stipitate, densely hairy; style glabrous or hairy, exserted from late bud; pollen-presenter oblique or rarely transverse, flat to conical. Follicle hairy with biramous and sometimes also simple glandular hairs, the indumentum usually marked with areas of red to purple hairs; pericarp crustaceous or rarely bony. Seeds ellipsoidal and unwinged (with or without a narrow waxy border and a terminal elaiosome), or flat-ellipsoidal and peripterous, or rarely hemispherical and unbordered.

Sixty six species, pan-continental except Tas., with main concentrations of species in south-western W.A. and south-eastern temperate areas. Bird-pollinated unless otherwise noted, with mammals possibly also implicated in some tropical and south-western species. This group corresponds closely to Series *Hebegynae* of Bentham (1870).

1 Pistil < 8 mm long

2 Style hairy with hairs for at least $\frac{1}{2}$ (and usually $\frac{2}{3}$) of its length; unit conflorescences usually > 2.5 cm long **48. G. ramosissima**

2: Style glabrous over most of its length; unit conflorescences usually < 2 cm long

3 Limb of bud with a white indumentum; unit conflorescences weakly acropetal, ±cylindrical; flowers not crowded; pistil ≤ 5 mm long; pedicels < 2 mm long **50. G. triternata**

3: Limb of bud with a brown indumentum; unit conflorescences strongly acropetal, ovoid; flowers crowded; pistil 6–7.5 mm long; pedicels > 3 mm long **49. G. raybrownii**

1: Pistil > 9 mm long

4 Ovary sessile or subsessile (stipe < c. 0.5 mm long)

5 Unit conflorescences regular, cylindrical or conical

6 Leaf lobes rigid, ±divaricate, pungent, longest ultimate lobes < 15 mm long; all leaves divided; conflorescence usually decurved to pendulous, occasionally weakly erect **12. G. pachylostyla**

6: Leaf lobes pliable, not divaricate, not pungent, longest ultimate lobes ≥ 50 mm long; usually all leaves divided, sometimes some or all entire; conflorescence erect

7 Pistil < 30 mm long

8 Unit conflorescences cylindrical with flowers opening ±synchronously or sometimes those on one side first; flowers white to cream (Qld)

9 Pedicels 0.1–0.5 mm long (flowers nearly sessile); leaf lobes broadly linear to narrowly obovate, 4–12 mm wide **59. G. sessilis**

9: Pedicels 3–6 mm long (flowers clearly pedicellate); leaf lobes linear, 1.5–3 mm wide **58. G. hodgei**

8: Unit conflorescences with flower opening strongly acropetal, conical, with young buds at apex when first mature buds open at base; flowers orange or bright yellow (shrubs widespread in semi-arid zone)

10 Pedicels 1.5–4 mm long; pistils 14–22 mm long; flowers acroscopic **54. G. eriostachya**

10: Pedicels 8–20 mm long; pistils 18–27 mm long; flowers basiscopic **60. G. juncifolia**

7: Pistil ⩾ 30 mm long

11 Pedicels 0.1–0.5 mm long (flowers nearly sessile) **59. G. sessilis**

11: Pedicels 3–10 mm long (flowers clearly pedicellate)

12 Widest leaf lobes 5–15 mm wide; perianth and style usually red, or white to cream (rarely the colours mixed) **56. G. banksii**

12: Widest leaf lobes < 3 mm wide; perianth and style white to cream

13 Floral rachis (60–) 80–200 mm long; pistil 43–46 mm long **57. G. whiteana**

13: Floral rachis 20–80 mm long; pistil 26–35 mm long **58. G. hodgei**

5: Unit conflorescences secund, loosely hemispherical or a loose cluster (not cylindrical or conico-cylindrical)

14 Leaves ovate, 18–70 mm wide, with 7–25 shallow teeth; torus reversely oblique; pistil 9–11 mm long **66. G. polyacida**

14: Leaves variously shaped, entire or divided but either not ovate or not with teeth (or if doubtfully so then pistil 29–40 mm long); torus transverse to oblique; pistil 11–60 mm long

15 Pistil > 50 mm long; prostrate shrub; styles bright yellow to yellow or orange (N.T.) **52. G. formosa**

15: Pistil ⩽ 40 mm long; erect to prostrate shrubs; styles red, pink, blackish, white, or yellow to orange

16 Leaves oblong to elliptic or rhomboid in general outline; widest leaves 20–35 mm wide, with (3–) 9–17 shallow triangular teeth or lobes; leaf margins very shortly recurved; pistil 29–30 mm long; outer surface of perianth with both gingery biramous hairs and simple erect purplish glandular hairs **44. G. maherae**

16: Leaves entire, or if divided then either the division deep and the ultimate lobes linear to oblong, or leaves with secondary division, or leaves ⩽ 10 mm wide; leaf margins shortly recurved to strongly revolute (lower leaf surface mostly concealed); pistil 11–40 mm long; outer surface of perianth either with biramous hairs only or if simple erect glandular hairs present then leaf lobes ±linear

17 Pistil ⩾ 30 mm long

18 Leaves with deep primary division, the 6–18 primary lobes then with shallow ±triangular teeth 3–30 mm long and 5–10 mm wide **61. G. bipinnatifida**

18: Leaves entire or with primary division only, or if secondary division present then this also deep and ultimate lobes ±linear or subterete

19 Leaves 10–45 cm long, primary lobes on either side of rachis ascending, ±mutually aligned, not divaricate; ultimate lobes 15–25 cm long, 1–10 mm wide; leaf rachis straight or gently upcurved; floral rachis 80–220 mm long **51. G. pteridifolia**

19: Leaves 1–9.5 cm long, primary lobes on either side of rachis either ascending and mutually aligned or widely spreading and the lobing divaricate (secondary division often present); all leaf segments ⩽ 1.5 mm wide; leaf rachis straight or angularly deflexed at some or all nodes; floral rachis 15–80 mm long

20 Leaf lobes not divaricate or only weakly so, pliable, not or scarcely pungent; styles deep maroon-black to reddish

21 Floral rachis 50–70 mm long; leaf lobes 1.0–1.5 mm wide; mature follicles with indumentum dominated by short erect glandular hairs (biramous hairs mostly caducous) **39. G. calliantha**

21: Floral rachis 20–50 mm long; leaf lobes c. 0.8 mm wide; mature follicles with indumentum dominated by biramous nonglandular hairs (simple erect glandular hairs present but scattered) **38. G. crowleyae**

20: Leaf lobes divaricate, ±rigid, pungent; styles red

22 Distance between leaf attachment to stem and lowest leaf lobes ⩽ 5 mm; pedicels 7–15 mm long **64. G. batrachioides**

22: Distance between leaf attachment to stem and lowest leaf lobes > 5 mm; pedicels 3–7 (–10) mm long

23 Leaf upper surface with at least the midveins evident; pedicels 5.5–10 mm long; perianth ventrally markedly saccate, c. 3.5–5 mm wide; outer perianth surface with both biramous and simple-glandular hairs **63. G. asparagoides**

23: Leaf upper surface with all venation obscure; pedicels 3–4.5 mm long; perianth ventrally convex but not saccate, c. 2.5 mm wide; outer perianth surface with biramous hairs only **65. G. secunda**

17: Pistil < 30 mm long

24 Either all leaves entire, or some or all leaves divided and then with longest ultimate lobes ⩾ 60 mm long

25 Styles bright orange or bright yellow; base of floral rachises (at lowermost flower scar) stout, usually > 2 mm thick beneath the indumentum

26 Most or all leaves with > 10 primary lobes **51. G. pteridifolia**

26: Leaves with < 10 primary lobes

27 Floral bracts 3–5 mm long, subspathulate with a narrowly ovate head 0.5–0.9 mm wide, lacking a median keel and resin-filled duct; leaves drying green or grey-green; conflorescences usually emergent from foliage on canes **54. G. eriostachya**

27: Floral bracts 6–9 mm long, subspathulate with a broadly ovate head 1.7–4.0 mm wide, with a median keel containing a resin-filled duct; leaves drying yellow-green; conflorescences usually borne close to foliage **55. G. excelsior**

25: Styles red or deep maroon to black, rarely dull orange or pale yellow; base of floral rachises slender, < 2 mm thick beneath the indumentum

28 Leaf margins smoothly revolute

29 Erect to spreading shrub 1.5–2.5 m tall; unit conflorescences erect, borne above foliage; styles usually black or very deep maroon, rarely red or yellow; pistil 18–23 mm long **37. G. hookeriana**

29: Prostrate to low mounded shrub 0.2–0.5 m tall; unit conflorescences erect to decurved, often lax on or near ground; styles pink to red or pale orange; pistil 24–28 mm long **43. G. nana**

28: Leaf margins angularly revolute

30 All leaves entire; floral rachis 25–65 mm long **33. G. coccinea**

30: Some or all leaves divided, rarely all entire; floral rachis 45–115 mm long **29. G. tetragonoloba**

24: Some or all leaves divided; longest ultimate leaf lobes < 60 mm long

31 Leaf rachis straight to gently upcurved; leaf lobes either not divaricate, or divaricate in or above the plane of the leaf rachis only

32 Lower surface of leaf on either side of midveins exposed, tomentose with curled hairs; style white to cream, sometimes pale pink with age (south-eastern Australia)

33 Unit conflorescences erect or slightly decurved, dense, strongly secund; styles remaining white to cream after anthesis; floral rachis and branchlets usually weakly pubescent **11. G. willisii**

33: Unit conflorescences usually decurved to pendulous, occasionally weakly erect, subcylindrical to weakly secund; styles reddening rapidly after anthesis; floral rachis and branchlets usually appressed-sericeous **12. G. pachylostyla**

32: Lower surface of leaf 2-grooved, with lamina on either side of midveins enclosed or very narrowly exposed; styles red or deep maroon-black or rarely yellow or pink (south-western Australia)

34 Leaf margins angularly revolute (angularity best seen in cross-section)

35 Some or all leaves with 3–7 short (≤ 20 mm long) rigid linear lobes clustered near leaf apex **31. G. fastigiata**

35: Some or all leaves with 3–13 pliable lobes (10–130 mm long) not clustered near leaf apex

36 Styles deep maroon to black or occasionally pale red; floral rachis 20–50 mm long; low shrub 0.5–1.5 m tall **38. G. crowleyae**

36: Styles bright red; floral rachis 45–115 mm long; robust shrub 2–2.5 m tall **29. G. tetragonoloba**

34: Leaf margins smoothly revolute

37 Upper leaf surface with prominent longitudinal ridges, granulose **32. G. wittweri**

37: Upper leaf surface ±smooth, not ridged

38 Most or all leaves with secondary division; pedicels 3–4.5 mm long; styles red or reddish pink **65. G. secunda**

38: Leaves with primary division only, rarely a few secondary lobes; pedicels 1–2.5 mm long; styles usually deep maroon or black, rarely red or dull yellow

39 Leaves with ±triangular teeth or lobes crowded at apex, or deeply divided with linear lobes; pistils 18–23 mm long **37. G. hookeriana**

39: Leaves deeply divided, lobes not crowded at apex; pistils ≥ 23 mm long

40 Floral rachis 50–70 mm long; leaf lobes 1.0–1.5 mm wide; mature follicles with indumentum dominated by short erect glandular hairs (biramous hairs mostly caducous) **39. G. calliantha**

40: Floral rachis 20–50 mm long; leaf lobes c. 0.8 mm wide; mature follicles with indumentum dominated by biramous nonglandular hairs (simple erect glandular hairs present but scattered) **38. G. crowleyae**

Erratum

Flora of Australia
Volume 17A
Proteaceae 2 – *Grevillea*

Page 363, key couplet 27/27: is misplaced under key lead 26.
The couplet should appear immediately after lead 26:, as couplet 26A/26A:, with lead 26A still yielding *G. nematophylla*.

31: Leaf rachis angularly deflexed at some or all nodes; leaf lobes strongly divaricate

41 Leaves with primary division only

42 Leaf lobes crowded at apex of leaf

43 Floral rachis rusty-villous; fruit at maturity with predominantly simple glandular hairs **31. G. fastigiata**

43: Floral rachis white-villous; fruit at maturity with mainly nonglandular biramous hairs **30. G. rigida**

42: Leaf lobes not crowded at apex of leaf

44 Low mounded shrub 0.2–0.5 m tall; conflorescence erect to decurved, often lax on or near ground; pistils 24–28 mm long **43. G. nana**

44: Robust erect to spreading shrub 1.5–3 m tall; conflorescence erect, not borne near ground; pistils 17–25 mm long

45 Style orange; pistil 17–22 mm long; seed peripterous (arid zone of northern central W.A.) **53. G. spinosa**

45: Style red; pistil 22–25 mm long; seed not peripterous (near S coast of W.A.) **30. G. rigida**

41: Some or all leaves with secondary division

46 Lower leaf surface exposed on either side of midvein, tomentose with curled hairs; conflorescence decurved to pendulous; styles white to cream, reddening only after anthesis (south-eastern Australia) **12. G. pachylostyla**

46: Lower leaf surface enclosed by revolute margins; conflorescence erect or decurved; styles red or black to blackish maroon (south-western Australia)

47 Style black or black-maroon; inner face of seed with a projecting elliptic waxy honeycombed 'ruff' **36. G. armigera**

47: Styles red or deep pink; inner face of seed with a narrow waxy border, not elaborated into a 'ruff'

48 Conflorescence erect; pedicels 1–2 mm long; perianth greenish to fawn-pink **32. G. wittwerii**

48: Conflorescence decurved to deflexed; pedicels 3–5.5 mm long; perianth red

49 Edge-veins evident on upper leaf surface; outer perianth surface with both biramous and simple-erect glandular hairs **62. G. maxwellii**

49: Edge-veins not evident on upper leaf surface; outer perianth surface with appressed biramous hairs only **65. G. secunda**

4: Ovary clearly stipitate (sometimes shortly so), with stipe > 0.5 mm long

50 Outer surface of perianth glabrous

51 Low spreading shrub ≤ 1 m tall; perianth and style orange (W.A.) **42. G. tenuiloba**

51: Spreading to erect shrub 2–2.5 m tall; perianth translucent-cream to purplish mauve to pink; style cream at base, becoming pale pinkish or mauve above, with a green tip (N.S.W.) **10. G. rivularis**

50: Outer surface of perianth sparsely to densely hairy, at least on limb

52 Style with an indumentum of long fine spreading to ascending hairs (biramous but sometimes appearing simple and basifixed) over at least basal half (sometimes to apex)

53 Pollen-presenter erect, very narrowly conical or cylindrical with a bulbous base; style becoming glabrous in apical 1/2–1/3 **46. G. dryandroides**

53: Pollen-presenter oblique on style, flat to convex; style with spreading indumentum persistent to apex **47. G. thyrsoides**

52: Style glabrous over most of its length, or with only appressed biramous hairs in basal third, or minute erect simple hairs (lacking long spreading hairs)

54 Most or all leaves deeply divided, *and* with lower surface on either side midveins of leaf and lobes enclosed by the revolute leaf margins

55 Divided leaves with primary division only

56 Leaf lobes strongly divaricate, with foliage usually entangled; prostrate to low mounded shrubs ⩽ 1.5 m tall

57 Pistil 27–32 mm long; leaf lobes subterete with a marginal groove along each side **41. G. aneura**

57: Pistil 19–28 mm long; leaf lobes smoothly revolute but not subterete, with the 2 marginal grooves clearly situated on the topologically lower surface

58 Unit conflorescences 6–17 cm long, usually held flat at ground level around periphery of shrub; pistil 24–28 mm long; fruit 8–12 mm thick (from side to side); seed 11–15 mm long, hemispherical with a spongy testa (SW of W.A.) **43. G. nana**

58: Unit conflorescences 2–5 cm long, usually held above ground level; pistil 19–25 mm long; fruit 5–6 mm thick; seed c. 10 mm long, flattened-ellipsoidal; testa not spongy; (south-eastern Australia) **22. G. ilicifolia**

56: Leaf lobes usually ±mutually parallel, with foliage not usually entangled; prostrate or low mounded to erect shrubs to 4 (–6) m tall

59 Limb of flower bud glabrous, or occasionally with a very few appressed hairs; torus oblique at > 45°; perianth and style orange; low spreading to prostrate shrub (SW of W.A.) **42. G. tenuiloba**

59: Limb of flower bud with a ±dense indumentum; torus oblique at < 45°, or transverse; perianth and style various colours, occasionally orange; prostrate to erect shrubs

60 Prostrate to low mounded or weakly erect shrub ⩽ 1 m tall; style apex bright green, contrasting with main style colour (south-eastern Australia) **22. G. ilicifolia**

60: Erect shrubs 1–4 (–6) m tall; style apex similar in colour to rest of style or a little paler (SW of W.A.)

61 Style ±glabrous; pistil < 22 mm long; stipe of ovary (1.5–) 2–5 mm long **27. G. cagiana**

61: Style with appressed biramous hairs for c. 4–8 mm above ovary, and with minute simple hairs on the ventral side towards apex; pistil ⩾ 22 mm long; stipe of ovary < 2 mm long **28. G. baxteri**

55: Some or all divided leaves with secondary (and sometimes tertiary) division

62 Ultimate leaf lobes usually wavy or curved, pliable not pungent; ovary ±sericeous with all hairs closely appressed **22. G. ilicifolia**

62: Ultimate leaf lobes ±straight, ±rigid, pungent; ovary villous with many hairs ascending to spreading

63 Longest ultimate leaf lobes < 2 cm long; upper surface of lobes with midvein ±evident; leaf margins angularly refracted; pollen-presenter moderately oblique on style; floral rachis usually straight **40. G. treueriana**

63: Longest ultimate leaf lobes > 2 cm long; upper surface of lobes with venation obscure; leaf margins smoothly revolute; pollen-presenter very oblique to lateral on style; floral rachis often gently incurved **41. G. aneura**

54: Most leaves either simple and entire or toothed, or if more deeply divided then the lower surface on either side of the midvein(s) exposed

64 Most or all leaves simple and entire, (sometimes occasional leaves bifid or trifid); widest leaves < 4.5 mm wide

65 Anthers vestigial or absent; style-end scarcely wider than style (Vic.) **20. G. williamsonii**

65: Anthers present, plump; style-end distinctly wider than style

66 Some or all conflorescences decurved to deflexed

67 Stipe of ovary ⩽ 2.5 mm long; unit conflorescences 1–3 cm long **34. G. concinna**

67: Stipe of ovary 2.5–5 mm long; unit conflorescences 1.8–4.5 cm long **35. G. beardiana**

66: All conflorescences erect

68 Pistil 28–33 mm long **35. G. beardiana**

68: Pistil < 24 mm long

69 Floral bracts < 2 mm long; pollen-presenter oblique on style; stipe of ovary 1.4–5 mm long **27. G. cagiana**

69: Floral bracts > 2 mm long; pollen-presenter erect on style, or almost so; stipe of ovary 0.4–1.6 mm long **33. G. coccinea**

64: Most leaves divided or toothed, or entire and then ⩾ 5 mm wide

70 Pistil < 19 mm long *and* some or all leaves with at least some secondary division

71 Leaf lower surface obscured by a dense thick indumentum

72 Floral bracts > 5 mm long; leaves deeply divided; leaf lobes pliable **2. G. beadleana**

72: Floral bracts < 3 mm long; leaves shallowly to deeply divided; leaf lobes stiff to rigid

73 Stipe of ovary > 1.5 mm long; leaves toothed to shallowly lobed, occasionally only the lowermost teeth or lobes 2–3-fid **21. G. infecunda**

73: Stipe of ovary < 1.5 mm long; leaves shallowly to deeply lobed, most or all leaves with two orders of division

74 Unit conflorescences dense, mostly erect, strongly secund; floral rachis tomentose to woolly; style-end just below pollen-presenter < 0.4 mm wide **11. G. willisii**

74: Unit conflorescences somewhat loose, most or all decurved to pendulous, weakly secund; floral rachis silky; style-end just below pollen-presenter > 0.5 mm wide **12. G. pachylostyla**

71: Leaf lower surface glabrous, or clearly visible through an thin, open to sparse indumentum

75 Peduncle glabrous, thin, wiry; unit conflorescences usually pendent, occasionally decurved **18. G. floripendula**

75: Peduncle hairy over at least part of its length, stout; unit conflorescences pendent to decurved or erect

76 Upper surface of leaf with a close, evenly distributed indumentum of curled to wavy hairs **19. G. dryophylla**

76: Upper surface of leaf glabrous, or with hairs scattered, irregularly patchy or restricted to the midvein area only

77 Floral bracts ⩽ c. 1 mm long and shorter than the pedicels; ultimate leaf lobes mostly 1–2 mm wide **17. G. microstegia**

77: Floral bracts ⩾ 1.5 mm long, as long as or longer than the pedicels; ultimate leaf lobes mostly ⩾ 3 mm wide

78 Floral bracts ⩾ 5 mm long; peduncle villous **15. G. obtecta**

78: Floral bracts < 5 mm long; peduncle shortly tomentose or subsericeous, at least towards apex

79 Leaves shallowly divided; primary sinuses extending less than halfway to midrib; plant rhizomatous **21. G. infecunda**

79: Leaves deeply divided, primary sinuses extending more than halfway to midrib; plants not rhizomatous

80 Widest ultimate leaf lobes ⩽ 7 mm wide; lower surface of leaf glabrous except for scattered hairs mainly along veins **16. G. montis-cole**

80: Widest ultimate leaf lobes > 8 mm wide; lower surface of leaf with a sparse but even indumentum of appressed wavy hairs **24. G. steiglitziana**

70: Pistil either ⩾ 19 mm long, or < 19 mm long and then the leaves simple and entire or with primary division only

81 Lower surface of leaf glabrous, or with a sparse to dense indumentum of closely appressed hairs, the hairs straight to wavy (not tightly curled)

82 Lower surface of leaf glabrous or nearly so, if hairs present then concentrated on and alongside the main veins; leaves with deep primary division; leaf lobes rigid and pungent

83 Pistil ⩽ 15 mm long; floral bracts 5–10.5 mm long **15. G. obtecta**

83: Pistil > 15 mm long; floral bracts < 5 mm long

84 Limb of perianth (late bud) villous with a beard of straight silky hairs; peduncles robust and straight, usually > 1 mm thick; stipe of ovary 1.5–2.5 mm long; fruit borne on sharply incurved stipe **9. G. acanthifolia**

84: Limb of perianth tomentose to subvillous with curled to wavy hairs; peduncles flexible and usually curved to sinuate, < 1 mm thick; stipe of ovary 2–4 mm long; fruit ±erect on stipe **16. G. montis-cole**

82: Lower surface of leaf with a more or less evenly distributed, sparse to dense indumentum; leaves entire or with shallow to deep primary division; leaf lobes (if present) rigid to pliable, pungent or not pungent

85 Outer surface of perianth tomentose to villous or loosely so, the hairs ascending to spreading, straight to curled, not mutually aligned along the perianth

86 Pistil < 18 mm long; floral bracts > 5 mm long, conspicuous and persistent to anthesis **15. G. obtecta**

86: Pistil > 18 mm long; floral bracts < 5 mm long, usually inconspicuous, falling early or persistent to anthesis

87 Anthers vestigial or absent; style-end scarcely wider than style (Vic.) — **20. G. williamsonii**

87: Anthers present, plump; style-end distinctly wider than style

88 Hairs on ovary spreading, multi-directional, lacking regular alignment; floral bracts (0.9–) 1.5–4 mm long, usually persistent at or near anthesis; style-end abruptly contracted into style with dorsal side of style-end concave — **13. G. aquifolium**

88: Hairs on ovary appressed (occasionally ascending), regularly aligned; floral bracts ⩽ 1 mm long, falling in bud stage; style-end tapering smoothly into style with dorsal side of style-end convex — **22. G. ilicifolia**

85: Outer surface of perianth subsericeous, the hairs appressed, straight, mutually aligned along the perianth

89 All leaves entire

90 Most leaves ⩾ 25 mm wide; plant prostrate — **8. G. laurifolia**

90: Most leaves < 20 mm wide; plant an erect to spreading shrub — **4. G. longifolia**

89: Some or all leaves toothed or lobed

91 Pistil ⩽ 17 mm long

92 Leaves bipinnatifid, the primary division more than half-way to the midrib; primary lobes (3–) 5–7, always with some secondary lobing — **24. G. steiglitziana**

92: Leaves regularly dentate with 5–25 small marginal teeth (rarely shallowly pinnatifid), usually lacking secondary division (occasionally a secondary tooth)

93 Erect to spreading shrub; pistil 14–15 mm long; stipe of ovary 0.6–1.0 mm long; style lilac-pink with a pale yellow apex — **5. G. wilkinsonii**

93: Prostrate, often mat-forming shrub; pistil 16–19 mm long; stipe of ovary 1.8–2.7 mm long; style deep burgundy red or rarely dull pink or orange to yellow, with a green tip — **26. G. repens**

91: Pistil > 17 mm long

94 Most or all leaves > 12 cm long, < 20 mm wide — **4. G. longifolia**

94: Most or all leaves < 12 cm long, and/or > 20 mm wide (width including lobes)

95 Leaves regularly and shallowly dentate, with 5–19 (–25) small teeth of < 3 mm amplitude, spaced ±evenly around margins; floral rachis sericeous or sparsely so; prostrate often mat-forming shrub — **26. G. repens**

95: Leaves dentate or pinnatifid to pinnatipartite, with 3–18 teeth or lobes of > 4 mm amplitude, either evenly spaced or mainly in apical half; floral rachis subsericeous (appressed hairs) or tomentose to subvillous (ascending to spreading hairs); prostrate to spreading or weakly erect shrubs

96 Lower surface of leaf with the ground-tissue on either side of the midvein visible through an open indumentum

97 Prostrate rhizomatous shrub; pedicels 3–8 mm long; leaves usually 6–14 cm long — **25. G. renwickiana**

97: Decumbent to weakly upright shrubs (sometimes rhizomatous); pedicels 1–3 mm long; leaves 2.5–7 cm long

98 Hairs of leaf lower surface straight; leaves shallowly and coarsely toothed or lobed (basal lobes sometimes bifid); anthers poorly developed with sparse pollen **21. G. infecunda**

98: Hairs of leaf lower surface usually curved to wavy; leaves with deep primary division, most primary lobes again 2–5-fid **24. G. steiglitziana**

96: Lower surface of leaf with the ground-tissue on either side of the midvein concealed by a dense indumentum (10× magnification)

99 Floral bracts 2–3 mm long, usually persistent at anthesis; anthers poorly developed, with sparse pollen **21. G. infecunda**

99: Floral bracts ⩽ 2.5 mm long, usually falling before anthesis; anthers well-developed with abundant pollen

100 Floral bracts ⩽ 1 mm long; fruit borne erect on a straight stipe; leaves shallowly to very deeply divided; pollen-presenter oblique at > 45° **22. G. ilicifolia**

100: Floral bracts 1–2.5 mm long; fruit borne on an incurved stipe; leaves shallowly divided; pollen-presenter oblique at ⩽ 45° **23. G. scortechinii**

81: Lower surface of leaf with a sparse to dense indumentum of ascending to spreading hairs, the hairs straight to wavy or tightly curled

101 Most leaves ⩽ 15 mm wide **1. G. aspleniifolia**

101: Most leaves ⩾ 25 mm wide

102 Lower surface of leaves with only scattered ascending to spreading hairs, concentrated on and beside the veins, rest of surface glabrous or nearly so **16. G. montis-cole**

102: Lower surface of leaves with a more or less evenly distributed, dense to sparse indumentum

103 Leaves divided with the ultimate leaf lobes ±narrowly oblong to linear or narrowly obovate

104 Leaves with (11–) 19–36 spreading soft simple straight primary lobes; lower surface appearing densely villous, with a two-layered indumentum, the lower layer of dense-packed appressed pale hairs, underlying a looser upper layer of spreading long straight to wavy hairs **3. G. caleyi**

104: Leaves with (2–) 5–13 (–21) spreading to ascending, simple to 5-partite primary lobes; lower surface densely to openly tomentose with wavy to curled hairs, indumentum not two-layered

105 Floral bracts 5–8 (–10.5) mm long, usually persistent to anthesis; hairs of leaf lower surface straight to wavy or loosely curled **15. G. obtecta**

105: Floral bracts < 1 mm long, falling in bud stage; hairs of leaf lower surface tightly twisted or curled **22. G. ilicifolia**

103: Leaves entire, or divided with the ultimate leaf lobes or teeth broadly to narrowly triangular or -ovate

106 Pistil < 17 mm long

107 Floral bracts 5–8 (–10.5) mm long; leaves usually 5–12 cm long; perianth villous outside **15. G. obtecta**

107: Floral bracts < c. 3 mm long; leaves usually 3–6 cm long; perianth subsericeous to shortly tomentose outside **14. G. bedggoodiana**

106: Pistil ⩾ 18 mm long

108 All leaves entire

109 Most or all leaves < 5 cm long; lower surface of leaf with a usually open indumentum, the ground-tissue visible between the hairs (at 10× magnification); prostrate to sprawling shrub to 1 m tall **13. G. aquifolium**

109: Most or all leaves < 7 cm long; lower surface of leaf with a dense indumentum, ground-tissue not visible between the hairs; spreading to erect shrubs or small trees 1–8 m tall

110 Erect shrub or tree to 8 m tall; floral bracts oblong-ovate to angularly ovate, 2.5–3.0 mm long, falling when buds are 4–5 mm long; stipe and ovary subsericeous **6. G. barklyana**

110: Spreading to erect shrub 1–4 m tall; floral bracts depressed-ovate to subtriangular, 0.8–2.3 mm long, persistent at least to anthesis; stipe and ovary tomentose to subvillous **7. G. macleayana**

108: Some or all leaves toothed or divided

111 Some or all leaves > 15 cm long

112 Erect shrub or tree to 8 m tall; floral bracts oblong-ovate to angularly ovate, 2.5–3.0 mm long, falling when buds are 4–5 mm long; stipe and ovary subsericeous **6. G. barklyana**

112: Spreading to erect shrub 1–4 m tall; floral bracts depressed-ovate to subtriangular, 0.8–2.3 mm long, persistent at least to anthesis; stipe and ovary tomentose to subvillous **7. G. macleayana**

111: All leaves < 12 cm long

113 Outer surface of perianth with mixed pale biramous hairs and conspicuous purple simple erect glandular hairs; pistils 29–40 mm long; upper and lower leaf surfaces slightly discolorous but with similar short indumentum of curled hairs (far north-western Australia)

114 Pistils 29–30 mm long **44. G. maherae**

114: Pistils 33–40 mm long **45. G. cravenii**

113: Outer surface of perianth with pale biramous hairs only, lacking purple glandular hairs; pistils 18–26 mm long; upper and lower leaf surfaces strongly dissimilar (upper surface occasionally with short curled hairs, but clearly differing from lower surface) (south-eastern Australia)

115 Outer surface of perianth tomentose to villous with ascending to spreading, straight to wavy hairs

116 Hairs on ovary multi-directional, lacking regular alignment; floral bracts 1.5–4 mm long, usually persistent at or near anthesis; style-end abruptly contracted into style with dorsal side of style-end concave **13. G. aquifolium**

116: Hairs on ovary appressed (rarely ascending), regularly aligned; floral bracts < 1 mm long, falling in bud stage; style-end tapering smoothly into style with dorsal side of style-end convex **22. G. ilicifolia**

115: Outer surface of perianth with appressed straight hairs

117 Anthers poorly developed with sparse pollen; leaves shallowly divided; lower surface of leaves with more or less straight hairs, forming an open to moderately dense indumentum **21. G. infecunda**

117: Anthers plump, well-developed, with abundant pollen; leaves shallowly to deeply divided; lower surface of leaves with tightly curled hairs, forming a dense indumentum **22. G. ilicifolia**

Asplenifolia/Hookeriana Subgroup

Erect to prostrate shrubs. Leaves entire or with up to 3 orders of division. Conflorescence usually simple, erect to deflexed, rarely incurved, secund or rarely semi-cylindrical, acropetal. Perianth hairy or glabrous outside, glabrous inside. Pistil 10–30 (–40) mm long; ovary sessile to stipitate; style usually glabrous, or with biramous and/or minute erect simple hairs; pollen-presenter oblique to occasionally erect, flat to broadly conical, rarely (*G. dryandroides*) narrowly conico-cylindrical. Follicle not (or only weakly) laterally compressed; surface with biramous hairs, sometimes also with erect simple glandular hairs, the indumentum with red-brown to purple markings. Seeds unwinged, usually ellipsoidal, usually with a terminal elaiosome, sometimes waxy-bordered or ornamented along one or both margins, rarely (*G. nana*) hemispherical with a thick spongy testa.

A group of 47 species, in temperate south-eastern and south-western Australia, with one taxon in the eremaean zone, and two in the Kimberley, W.A.

1. Grevillea aspleniifolia Knight, *Cult. Prot.* 120 (1809)

T: 'Many plants of this species were raised 3 years ago by Mr Colville, who received the seeds from Port Jackson ...' [protologue]; neo: New South Wales: Port Jackson [verso], *s.d.*, *coll. unknown*; neo: BM, *fide* D.J.McGillivray, *Telopea* 1: 25 (1975).

G. aspleniifolia R.Br., *Trans. Linn. Soc. London* 10: 175 (1810), *nom. illeg. non* Knight (1809). T: Port Jackson [verso], New South Wales, *s.d.*, *coll. unknown*; neo: BM, *fide* D.J.McGillivray, *Telopea* 1: 25 (1975).

G. aspleniifolia var. *typica* Domin, *Biblioth. Bot.* 89: 34 (1921), *nom. illeg.* (type var.).

G. aspleniifolia var. *shepherdiana* F.Muell., *Victorian Naturalist* 10: 151 (1894); *G. shepherdii*, Maiden & Betche, *Census N.S.W. Pl.* 59 (footnote) (1916), as *G. Shepherdi*. T: Cole R., near Jervis Bay, N.S.W., 1893, [*J.?*]*Shepherd*; holo: MEL.

[*G. longifolia auct. non* R.Br.: A.M.Blombery & B.Maloney, *Prot. Sydney Reg.* t. 39 (1981); M.Baker *et al.*, *Nat. Pl. Sydney Reg.* 79 (1986)]

Illustrations: A.M.Blombery & B.Maloney, *Prot. Sydney Reg.* t. 39 (1981); M.Baker *et al.*, *Nat. Pl. Sydney Reg.* 79 (1986), as *G. longifolia*; P.M.Olde & N.R.Marriott, *Grevillea Book* 2: 42 (top centre & 29A, B) (1995).

Shrub 1–5 m high, to 4 m across. Leaves linear to narrowly ovate, 15–25 cm long, 3–15 (–25) mm wide, usually coarsely and irregularly toothed or serrate, sometimes almost entire; teeth subtriangular, not pungent; margins recurved to revolute; lower surface usually partly exposed, densely tomentose with curled hairs. Unit conflorescence erect, secund; floral rachises 30–50 (–100) mm long. Flower colour: perianth purplish with a grey to white indumentum; style pink to purplish pink with a green tip. Perianth villous to subsericeous outside. Pistil (10.5–) 15–25 mm long; ovary stipitate; style glabrous. Follicle 11–12 mm long, sericeous.

Occurs in eastern N.S.W., in the southern Blue Mtns and Woronora Plateau, and S to Bungonia; mainly along the Nattai, Wollondilly, Kowmung, Cox's and Woronora Rivers. Grows in eucalypt woodland, usually on slopes or ridges in open rocky situations, in skeletal sandy or loam soils on sandstone or shale. Regenerates from seed. Flowers mainly July–Nov. Map 1.

N.S.W.: The Peak, Yerranderie, *J.L.Boorman NSW92328* (BRI, NSW, SYD); Burragorang Valley, 3 km WSW of Donohoes [Donohues?] Flat, *B.G.Briggs 1121 & L.A.S.Johnson* (BM, NSW); Nattai R. via Hilltop, *E.Cheel NSW20616* (NSW); Wanganderry Ck, 19 km S of Yerranderie, *E.F.Constable 5670* (NSW); 5 km S of Bungonia, 0.8 km SE of Tolwong Mines Rd and Dog Den Ck, *T. & J.Whaite 3476* (NSW).

Sometimes confused with *G. longifolia*, which has subsericeous branchlets angular in cross-section, and a sericeous leaf under-surface. *Grevillea asplenifolia* has tomentose branches rounded or very slightly angular in cross-section, and the lower surface of the leaf tomentose with curled hairs.

2. Grevillea beadleana McGill., *New Names Grevillea* 2 (1986)

T: 4 km SSW of Chaelundi Falls, N.S.W., 13 June 1982, *J.B.Williams NSW151305*; holo: NSW; iso: K, US.

Illustrations: P.M.Olde, *Austral. Pl.* 13(108): 364 (1986); J.W.Wrigley & M.Fagg, *Banksias, Waratahs & Grevilleas* 223 (1989); D.J.McGillivray & R.O.Makinson, *Grevillea* 52, col. pl. & fig. 9 (1993); P.M.Olde & N.R.Marriott, *Grevillea Book* 2: 56 (bottom right), 57 (41A, B) (1995).

Dense to open shrub, 0.8–2.5 m high, to 2.5 m across. Leaves pinnatipartite and partly bipinnatifid (rarely an odd small leaf entire and elliptic), 8–16.5 cm long, 50–105 mm wide; primary lobes 5–16, simple to 6-fid; ultimate lobes subtriangular to narrowly oblong, 0.5–4 cm long, not pungent; margins shortly recurved; lower surface densely tomentose with curled hairs. Unit conflorescence erect, secund; floral rachis 30–50 mm long. Flower colour: perianth grey to purplish with pale hairs; style deep burgundy to scarlet or pale pinkish mauve. Perianth loosely subvillous outside. Pistil 15–18.5 mm long; ovary stipitate; style glabrous. Follicle 9–10.5 mm long, tomentose.

Occurs in north-eastern N.S.W. in at least three disjunct populations: on gorges of the upper Guy Fawkes R. N of Ebor; on the Mole R. catchment c. 70 km W of Tenterfield; and in low hills c. 25 km SW of Grafton. A collection from the Apsley R. gorge near Walcha in 1887 has not been replicated. Grows in shallow sandy-loam soils over granite leucoadamellite or (Grafton population) sandstone. Regenerates from seed. Flowers spring. Map 2.

N.S.W.: Oakey Ck, tributary of Mole R., *R.Coveny 14679 et al.* (A, AD, B, BISH, BRI, CANB, COLO, K, L, LE, MEL, MO, NGB, NSW, PE, PERTH, PRE); Walcha, 1887, *Capt. Crawford 31* (MEL); 5 km S of Old Glen Innes to Grafton road, *N.Taws 432* (CANB, MEL).

Grevillea beadleana has rather prominent floral bracts which are 5.7–6.2 mm long and persistent to anthesis. There is some minor variation in leaf division and flower colour. Plants of the Guy Fawkes population tend to have numerous secondary leaf lobes and burgundy styles; those of the Mole R. population also have extensive secondary lobing and more scarlet styles. The Grafton population is known from only fragmentary fresh specimens, but has more sparing secondary lobing (as does the Walcha collection) and pale mauve-pink styles; this sub-coastal form probably deserves subspecific status when adequate type material becomes available. Variation in habit between sites is probably due to ecotype and fire history.

This species is recognised as 'Endangered' in J.D.Briggs & J.H.Leigh, *Rare or Threatened Australian Plants* (1995).

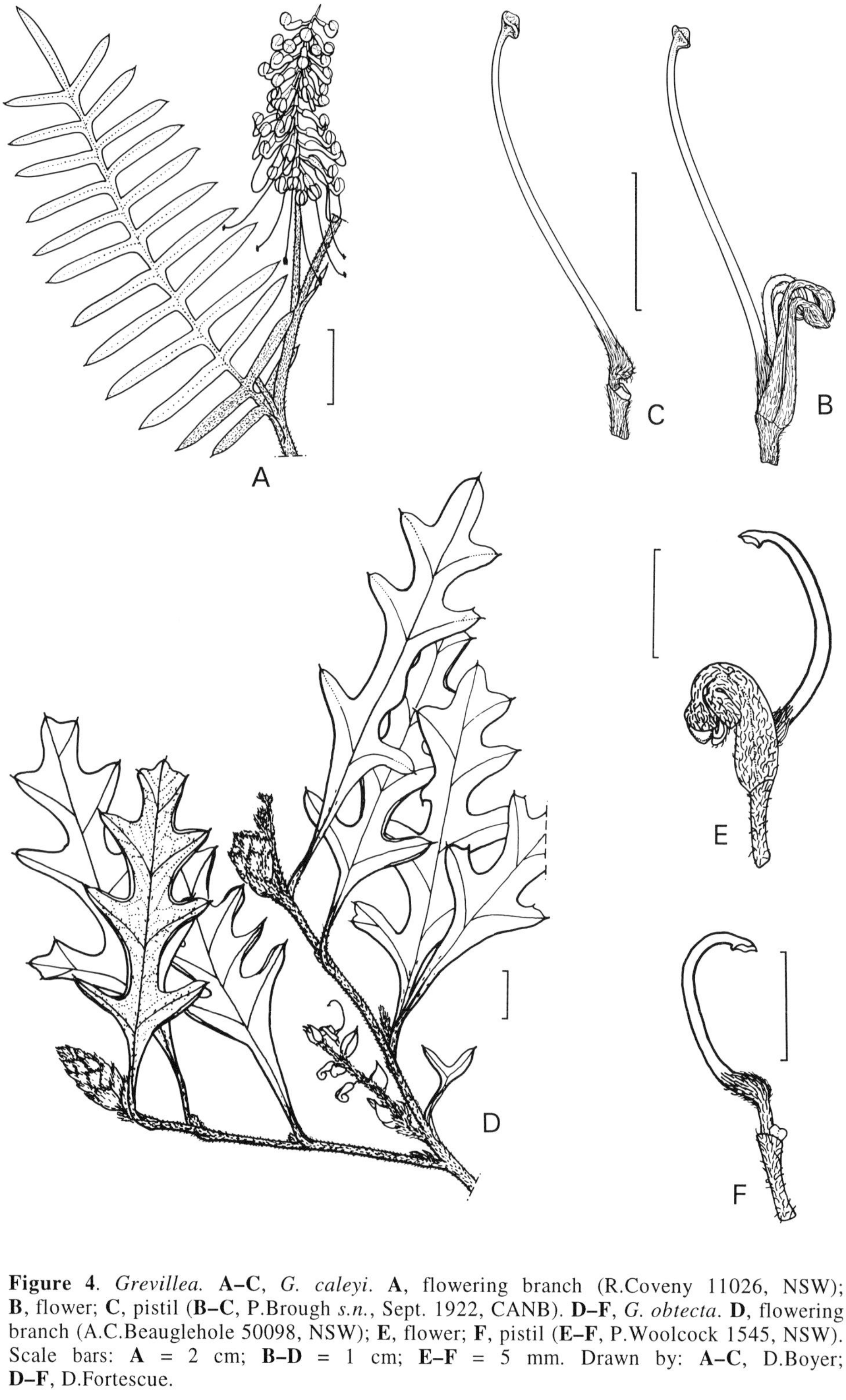

Figure 4. *Grevillea.* **A–C**, *G. caleyi.* **A**, flowering branch (R.Coveny 11026, NSW); **B**, flower; **C**, pistil (**B–C**, P.Brough *s.n.*, Sept. 1922, CANB). **D–F**, *G. obtecta.* **D**, flowering branch (A.C.Beauglehole 50098, NSW); **E**, flower; **F**, pistil (**E–F**, P.Woolcock 1545, NSW). Scale bars: **A** = 2 cm; **B–D** = 1 cm; **E–F** = 5 mm. Drawn by: **A–C**, D.Boyer; **D–F**, D.Fortescue.

3. **Grevillea caleyi** R.Br., *Suppl. Prodr. Fl. Nov. Holl.* 22 (1830)

Stylurus caleyi (R.Br.) Pedro, *Bol. Soc. Estud. Moçam.* xxv. 91: 9 (1955) *n.v.* T: near Sea Sight Hill, near Port Jackson, N.S.W., Feb. 1805, *G.Caley*; lecto: BM, *fide* D.J.McGillivray & R.O.Makinson, *Grevillea* 410 (1993).

Illustrations: A.M.Blombery & B.Maloney, *Prot. Sydney Reg.* t. 38 (1981); M.Baker *et al.*, *Nat. Pl. Sydney Reg.* 15 (1986); D.J.McGillivray & R.O.Makinson, *Grevillea* 55 (1993); P.M.Olde & N.R.Marriott, *Grevillea Book* 2: 79 (bottom right), 80 (62A, B) (1995).

Open spreading shrub, 1–4 m high. Leaves pinnatisect or pinnatipartite, elliptic in overall outline, 5–18 cm long, 30–75 mm wide, with (11–) 19–36 spreading linear to narrowly obovate lobes; lobes 1.5–3.5 cm long, usually simple or occasionally the basal ones bifid, not pungent; margins recurved; lower surface villous, with hairs 2-layered. Unit conflorescence erect, secund; floral rachis 40–80 mm long. Flower colour: perianth fawn; style maroon to red with a green tip. Perianth villous outside. Pistil 25–28 mm long; ovary stipitate; style glabrous. Follicle 17–21 mm long, tomentose. *n* =10, H.Ramsay, *Austral. J. Bot.* 11: 5 (1963). *Caley's Grevillea.* Fig. 4A–C.

Occurs in a small area of the north-eastern suburbs of Sydney, N.S.W. A record of a single plant from North Brother (on N.S.W. north coast near Laurieton) at CANB is highly unlikely to represent a natural population. Restricted to woodland-heath associations on ridgetops at 200 m alt., in laterised gravelly soils. Regenerates from seed. Flowers mainly Aug.–Dec. Map 3.

N.S.W.: Middle Harbour, Nov. 1897, *J.L.Boorman NSW92240* (NSW); 'Pines', Mona Vale, *H.K.C.Mair 150* (CANB); Kuringai Chase, *R.D.Melville 501* (AD, BRI, K, NSW).

Grevillea caleyi has a characteristic 'herringbone' leaf division; it can be distinguished from similar-leaved forms of *G. ilicifolia* and *G. obtecta* by the 2-layered indumentum on its lower leaf surface, the lower layer of dense-packed appressed pale hairs underlying a looser upper layer of spreading long straight to wavy hairs. *Grevillea obtecta* has an open indumentum of ascending (rarely appressed) curved to curled hairs. *Grevillea ilicifolia* has the lower surface subsericeous with straight hairs or densely pubescent-tomentose with curled hairs.

This species is recognised as 'Endangered' in J.D.Briggs & J.H.Leigh, *Rare or Threatened Australian Plants* (1995).

4. **Grevillea longifolia** R.Br., *Suppl. Prodr. Fl. Nov. Holl.* 22 (1830)

G. aspleniifolia var. *longifolia* (R.Br.) Domin, *Biblioth. Bot.* 89: 34 (1921). T: brush close to the river a little above the Cataract, near Port Jackson, N.S.W., July 1807, *G.Caley*; holo: BM; iso: BRI, NSW.

[*G. aspleniifolia auct. non* Knight: G.Bentham, *Fl. Austral.* 5: 435 (1870), *p.p.*]

Illustrations: A.M.Blombery & B.Maloney, *Prot. Sydney Reg.* 63 (1986); D.J.McGillivray & R.O.Makinson, *Grevillea* 57 (1993); P.M.Olde & N.R.Marriott, *Grevillea Book* 2: 244 (centre right), 245 (203A–C) (1995).

Erect to spreading shrub, 1.5–6 m high. Leaves narrowly ovate to narrowly elliptic or almost linear in general outline, 7.5–22 cm long, 5–25 mm wide, entire or serrate to pinnatifid; teeth (when present) obliquely triangular, 3–6 mm long, sometimes almost pungent; margins shortly recurved; lower surface mostly exposed, sericeous with straight hairs. Unit conflorescence erect, secund; floral rachis 45–75 mm long. Flower colour: perianth pinkish fawn; style coral-pink to red or rarely pale orange, with a green tip. Perianth subsericeous outside. Pistil 21–24 mm long; ovary stipitate; style glabrous. Follicle 13–16 mm long, sericeous.

Occurs in N.S.W., in the southern half of the Sydney Basin and on the Woronora Plateau, between Springwood and the Woronora R. Grows in moist areas of sclerophyll forest, often beside creeks in sandy soils on Hawkesbury sandstone. Flowers July–Jan. Regenerates from seed. Map 4.

N.S.W.: Kings Fall, *s.d.*, *A.Cunningham* (BM, *NSW13782*); Woolwash, E of Campbelltown, 7 Aug. 1950, *N.C.Ford* (NSW); Woronora R., 5 Sept. 1897, *W.Forsyth* (AD).

Sometimes confused with *G. aspleniifolia*: see under that species for differences. This species is recognised as 'Rare' in J.D.Briggs & J.H.Leigh, *Rare or Threatened Australian Plants* (1995).

Occurs in the Blue Mtns region of N.S.W., from Newnes S to about Richlands. Grows in sandy or clay soils, near swamps or on slopes or ridges, on sandstone or sandy-shale substrates, usually in open eucalypt woodland or dry sclerophyll forest. Regenerates from seed. Flowers mainly Sept.–Jan. Map 8.

N.S.W.: Megalong Ck, Megalong Valley, *L.A.S.Johnson 206* (NSW); Richlands to Wombeyan Caves, *C.W.E.Moore 2636* (CANB, NSW); Mt Cameron Track, 1.5 km W of Dinner Ck, 19 km N of Bell, *J.Pickard 730 & Black* (NSW); off Chapman Pde, 8 km from Faulconbridge, *J.Pulley 858* (CANB).

Natural hybrids occur with *G. acanthifolia* subsp. *acanthifolia*, and were described as *G. gaudichaudii* R.Br. ex Gaudich. See also J.McLuckie, *Proc. Linn. Soc. New South Wales* 55: 386–412 (1930).

A smaller-flowered form occurs in the E of the range at lower elevations, in the Wentworth Falls to Valley Heights area (e.g. *J.Pulley 858* (CANB)); it has flowers with pistils 13–15 mm long, more compact secund to semicylindrical conflorescences, and deep maroon flowers.

9. Grevillea acanthifolia A.Cunn., *in* B.Field (ed.), *Geogr. Mem. New South Wales* 328, plate [*s.n.*] (1825)

T: 'Peaty bogs on the Blue Mtns and banks of Cox's River' [protologue], N.S.W., 1817, *A.Cunningham*; lecto: Blue Mountains, N.S.W., 1817, *A.Cunningham*; lecto: K, *fide* D.J.McGillivray & R.O.Makinson, *Grevillea* 402 (1993); ?isolecto: G-DC.

Illustrations: A.M.Blombery & B.Maloney, *Prot. Sydney Reg.* t. 42 (1981); D.J.McGillivray & R.O.Makinson, *Grevillea* 59 (1993); P.M.Olde & N.R.Marriott, *Grevillea Book* 2: 8 (2A), 9 (2B) (1995).

Spreading to erect shrub, 0.5–3 m high. Leaves deeply bipinnatifid to bipinnatisect, 4–9 cm long, 30–70 mm wide, with 2–7 entire to 5-fid primary lobes; ultimate lobes rigid, triangular or oblong to linear, 0.5–2 cm long, usually divaricate, pungent; margins recurved to revolute; lower surface exposed to ±enclosed, glabrous or sometimes with appressed hairs along veins. Unit conflorescence erect or decurved, secund; floral rachis usually 20–100 mm long. Perianth ±villous outside (especially about the limb, giving the buds a very woolly appearance). Pistil 20–28 mm long; stipe 1.5–2.5 mm long; style glabrous. Follicle 9–11 mm long, villous to velutinous.

Occurs in mountains of eastern N.S.W. There are three geographically disjunct subspecies. *Grevillea acanthifolia* is most closely related to *G. rivularis*, from which it may be distinguished by the hairs on the perianth outer surface (perianth glabrous outside in *G. rivularis*). *Grevillea acanthifolia* also has the follicles borne on sharply incurved stipes.

1 Pollen-presenter convex to broadly conical, clearly shorter than wide, oblique to transverse on style; leaf lower surface glabrous or with scattered appressed usually dark hairs on the veins only; leaf lobes triangular or subulate to linear; irregular shrub to 1.5 m tall

2 Unit conflorescence usually > 5 cm long, strongly acropetal and appearing conico-secund; leaves mostly bipinnatifid to bipinnatipartite with triangular lobes **9a.** subsp. **acanthifolia**

2: Unit conflorescence usually < 3 cm long, weakly acropetal and appearing oblong-secund; leaves mostly bipinnatisect with oblong-subulate to linear lobes **9b.** subsp. **stenomera**

1: Pollen-presenter conical, about as long as wide, transverse on style; leaf lower surface usually with pale appressed hairs beside main veins; leaf lobes linear; robust erect shrub to 3 m tall **9c.** subsp. **paludosa**

9a. Grevillea acanthifolia A.Cunn. subsp. **acanthifolia**

G. acanthifolia Sieber ex Spreng., *Syst. Veg.* 4(2), *Cur. Post.* 46 (1827), *nom. illeg. non* A.Cunn. (1825); *G. acanthifolia* Sieber ex Schult. & Schult.f., *Mant.* 3: 281 (1827), *nom. illeg.* T: Australia [N.S.W.], *Sieber Herb. No. 32*; syn: A *n.v.*, B *n.v.*, G, K, L *n.v.*, LE *n.v.*, MEL, NY *n.v.*

Illustrations: N.C.W.Beadle *et al.*, *Fl. Sydney Reg.* t. 10 (1972), as *G. acanthifolia*; E.R.Rotherham *et al.*, *Fl. & Pl. New South Wales & S Queensland* 34, t. 63 (1975); J.W.Wrigley & M.Fagg, *Banksias, Waratahs & Grevilleas* 156 (1989); D.J.McGillivray & R.O.Makinson, *Grevillea* 59 (1993).

Shrub 1–1.5 m tall. Leaves mostly bipinnatifid to bipinnatipartite, 4–9 cm long, ultimate lobes usually triangular, rarely linear; lower surface mostly exposed and glabrous or with scattered hairs on veins only. Unit conflorescence erect, strongly acropetal, appearing conico-secund; floral rachis 40–90 mm long. Flower colour: perianth silver to reddish grey; style mauve-pink to red. Pistil 22–28 mm long. Pollen-presenter transverse to oblique, convex.

Occurs in the upper Blue Mtns, N.S.W., between about Bell and Woodford. Grows in swampy areas, on wet rock shelves or on stream banks, in acid peaty or sandy soils. Some tendency to resprout from the base. Peak flowering Sept.–Dec. Map 9.

N.S.W.: Woodford, 3 Apr. 1948, *I.Bowden NSW4983* (NSW); Zig Zag Railway cutting 1.5 km NE of Lithgow, *E.Constable 7187* (NSW); Cox's R. 8 km WSW of Lithgow, *J.H.Maiden & R.H.Cambage NSW92381* (NSW); Mt Cameron Track, 1.5 km W of Dinner Ck, 19 km N of Bell, *J.Pickard 729 & Black* (NSW).

This taxon hybridises naturally with *G. laurifolia*. The hybrid, known as *G.* × *gaudichaudii* R.Br. ex Gaudich., is very common in horticulture. See typification and notes under *G. laurifolia*.

9b. Grevillea acanthifolia subsp. **stenomera** (F.Muell. ex Benth.) McGill., *Telopea* 1: 23 (1975)

G. acanthifolia var. *stenomera* F.Muell. ex Benth., *Fl. Austral.* 5: 439 (1870). T: head of Macleay R., N.S.W., *C.Moore s.n.*; holo: K; iso: (*C.Moore 85*) MEL, NSW.

Illustrations: P.M.Olde & N.R.Marriott, *Grevillea Book* 2: 10 (top right & 4A, B) (1995).

Spreading, straggling or prostrate shrub, to 1.5 m tall and 2 m across. Leaves mostly very deeply pinnatipartite to bipinnatisect, 3–6 cm long; ultimate lobes linear to oblong-subulate; lower surface often mostly enclosed, glabrous or with scattered hairs on veins only. Unit conflorescence erect to decurved, weakly acropetal, appearing oblong-secund; floral rachis usually 20–30 (–40) mm long. Flower colour: perianth grey to fawn; style bright pink to red with a green tip. Pistil 20–24 mm long. Pollen-presenter oblique, convex to broadly conical.

Occurs in north-eastern N.S.W. along the eastern fall of the tablelands, from E of Tenterfield to Werrikimbe Natl Park, E of Walcha. Grows in heath or woodland, often beside swamps or streams. Regeneration probably from seed only. Flowers Oct.–Nov. Map 10.

N.S.W.: Guy Fawkes, Nov. 1913, *J.L.Boorman NSW92459* (NSW); Dingo Ck, Werrikimbe Natl Park, *P.Hitchcock 10* (NSW); Oakey Ck, between Wollomombi and Ebor, *D.J.McGillivray 3172 & R.Coveny* (NSW); Bullock Ck, Point Lookout Rd, *R.J.Millington 43* (NSW); 22.5 km along Poverty Point/Timbarra road S of Bruxner Hwy, *K.L.Wilson 3215* (NSW).

This subspecies is recognised as 'Rare' in J.D.Briggs & J.H.Leigh, *Rare or Threatened Australian Plants* (1995).

9c. Grevillea acanthifolia subsp. **paludosa** Makinson & Albr., *Muelleria* 7: 89 (1989)

T: c. 2.5 km WNW of Mt Wog Wog trig, Nalbaugh Natl Park, N.S.W., 22 Feb. 1987, *D.E.Albrecht & P.Gilmour*; holo: MEL.

Illustrations: R.O.Makinson & D.E.Albrecht, *Muelleria* 7: 90 (1989); P.M.Olde & N.R.Marriott, *Grevillea Book* 2: 9 (3A, B) (1995).

Robust shrub to 3 m tall and 5 m wide. Leaves deeply bipinnatipartite to bipinnatisect, 4–6 cm long; ultimate lobes linear; lower surface mostly enclosed, usually with silky appressed hairs on lamina beside veins. Unit conflorescence erect, weakly acropetal, weakly conico-secund; floral rachis 30–50 mm long. Flower colour: perianth grey to fawn; style pale to bright mauve-pink or occasionally reddish, with a green tip. Pistil 21–24 mm long. Pollen-presenter transverse, conical.

Occurs in far south-eastern N.S.W., where known only from small populations on the Nalbaugh Plateau, W of Eden. Grows on hummocks in sphagnum swamp in peaty acid soils over quartzitic gravel, in dense wet heath associations, and beside streams on margins of wet sclerophyll forest. Flowers Sept.–Dec. Map 11.

N.S.W.: plateau between Mt Wog Wog and White Rock Mtn, 'Long Swamp', catchment of White Rock R., *R.O.Makinson 322* (BRI, CANB, MEL, NE, NSW).

This subspecies is recognised as 'Endangered' in J.D.Briggs & J.H.Leigh, *Rare or Threatened Australian Plants* (1995).

10. Grevillea rivularis L.A.S.Johnson & McGill., *Telopea* 1: 23 (1975)

T: Carrington Falls, N.S.W., 1960, *E.F.Constable NSW92463*; holo: NSW.

Illustrations: J.W.Wrigley & M.Fagg, *Banksias, Waratahs & Grevilleas* 313 (1989); D.J.McGillivray & R.O.Makinson, *Grevillea* 61, col. pl., 62, fig. 11 (1993); P.M.Olde & N.R.Marriott, *Grevillea Book* 3: 141 (top left & 112A–C) (1995).

Dense spreading shrub, 2–2.5 m high. Leaves bipinnatipartite, 3–6 cm long, 30–60 mm wide, with 3–9 primary lobes each 3–5-partite; ultimate lobes linear to very narrowly triangular, 1–3 cm long, 1–2.5 mm wide, rigid, pungent; margins revolute; lower surface mostly or completely enclosed except for midvein, glabrous or with scattered appressed hairs. Unit conflorescence usually decurved, secund; floral rachis 50–60 mm long. Flower colour: perianth translucent cream to purplish mauve or pink; style cream at base, becoming pale pinkish or mauve above, with a green tip. Perianth glabrous outside. Pistil 27–32 mm long; ovary stipitate; style glabrous. Follicle 8–10 mm long, loosely villous. *Carrington Falls Grevillea.*

Occurs in N.S.W., restricted to the Carrington Falls area SE of Moss Vale. Grows in moist shrubby creekside sites on sandstone. Probably regenerates from seed only. Flowers Sept.–Apr. Map 12.

N.S.W.: Carrington Falls, *R.Coveny 12179 et al.* (B, K, NSW, PERTH, RSA); Carrington Falls, *R.Pullen 2524* (CANB).

This species is recognised as 'Vulnerable' in J.D.Briggs & J.H.Leigh, *Rare or Threatened Australian Plants* (1995).

11. Grevillea willisii R.V.Sm. & McGill., *Muelleria* 3: 102 (1975)

G. willisii R.V.Sm. & McGill. subsp. *willisii sensu* D.J.McGillivray & R.O.Makinson, *Grevillea* 63, 64 (1993). T: Bundara R. bridge on Omeo Hwy, 16 km NW of Omeo (direct), 32 km by road, Vic., 2 Dec. 1966, *R.V.Smith 66/647*; holo: MEL; iso: A *n.v.*, B *n.v.*, CANB, K, MEL, NSW.

[*G. ramosissima auct. non* Meisn.: J.Stirling, *Trans. & Proc. Roy. Soc. S. Australia* 6: 40 (1883)]

Illustrations: R.V.Smith, *Muelleria* 3: 105, pl. 7, 107, pl. 8 (1975); D.J.McGillivray & R.O.Makinson, *Grevillea* 63 (1993); P.M.Olde & N.R.Marriott, *Grevillea Book* 3: 232 (bottom right), 233 (190A–C) (1995).

Spreading to erect shrub 2.0–4.5 m high, to 3 m across. Leaves pinnatipartite, 2.5–14 cm long, 20–65 mm wide, with 5–19 primary lobes each simple to 5-fid; ultimate lobes narrowly to broadly triangular to oblong-elliptic, 0.5–2.5 cm long, 3–17 mm wide, sometimes pungent; margins recurved to shortly revolute; lower surface mostly exposed, densely tomentose with curled hairs. Unit conflorescence erect or slightly decurved, dense, secund; floral rachis 30–85 mm long. Flower colour: perianth greenish white to pale fawn-yellow on curve and limb, inside blackish; style white to cream (not reddening with age). Perianth tomentose outside. Pistil 11–18 mm long; stipe 0.3–1.1 mm long. Follicle 8–11 mm long, softly tomentose. *Omeo Grevillea, Rock Grevillea.*

Occurs in the eastern highlands of Vic., on the upper Murray R. system, in the catchments of the Mitta Mitta R., Nariel Ck, and Wheelers Ck. Grows in open shrubby associations on rocky bluffs and near streams in granitic sandy loam soils on granite. Regenerates from seed. Flowers Sept.–Jan. Map 13.

Vic.: bank of Nariel Ck, c. 0.4 km upstream from Staceys Bridge, 23 Oct. 1955, *C.K.Butler* (MEL); Wheelers Ck area at Corryong, 27 Mar. 1973, *G.F.Luck* (MEL); Bundarra R. bridge on Omeo Hwy, *R.V.Smith 65/20* (K, MEL).

Two forms may be distinguished. The 'shorter-leaved (Type) form' is a spreading shrub with leaves 2.5–6.5 cm long with 5–11 primary lobes, and floral rachises 3–5.5 cm long. It occurs to the E and N of Omeo. The 'longer-leaved form' is known from only two collections S of Corryong. It is an erect shrub to 4.5 m tall, with leaves to 14 cm long with 9–19 primary lobes, and floral rachises 8–8.5 cm long. This form probably deserves subspecific status. The flowers are scented, and pollinator is unknown.

Grevillea willisii is similar to and a sister-species of *G. pachylostyla*, which has usually smaller leaves (2–4 cm long), a subsericeous indumentum on the branchlets, floral rachis and perianth; a looser, conico-cylindrical and often pendulous unit conflorescence; styles more strongly incurved and turning pink to red soon after anthesis; and the style-end just below the pollen-presenter more than 0.5 mm wide in dry material (0.8–0.9 mm wide in fresh). *Grevillea willisii* has larger leaves (usually 3–6, rarely to 14 cm long); a tomentose to woolly indumentum on the branchlets, floral rachis and perianth; an erect, dense, and strongly secund unit conflorescence; styles remaining cream after anthesis; and the style-end just below the pollen-presenter less than 0.4 mm wide in dry material (0.4–0.5 mm wide in fresh). Both species have stiff to rigid leaf lobes with evident venation above, and short pedicels and floral bracts.

Collections possibly intermediate between *G. willisii* and *G. pachylostyla* are known from Mt Stradbroke and Boundary Ck near Gelantipy.

This species is recognised as 'Rare' in J.D.Briggs & J.H.Leigh, *Rare or Threatened Australian Plants* (1995).

12. Grevillea pachylostyla (McGill.) Olde & Marriott, *Grevillea Book* 1: 186 (1994)

G. willisii subsp. *pachylostyla* McGill., *New Names Grevillea* 16 (1986). T: Reedy R. Gorge, Nunniong Plateau area, Vic., 13 Nov. 1964, *J.H.Willis*; holo: MEL.

Illustrations: P.M.Olde & N.R.Marriott, *Grevillea Book* 3: 71 (top left & 51A, B) (1995).

Sprawling or mounded to spreading shrub to 1.5 m tall. Leaves bipinnatifid, 2–4 cm long, 20–30 mm wide, with 5–11 usually 2- or 3-fid primary lobes; ultimate lobes triangular or narrowly so, 5–12 mm long, 2–3 mm wide, pungent; margins refracted; lower surface woolly-pubescent with tightly curled hairs. Unit conflorescence decurved to pendulous or sometimes erect, conico-cylindrical to weakly secund; floral rachis 30–50 mm long. Flower colour: perianth cream on curve and limb, black below and inside; style cream rapidly becoming pink to red after anthesis. Perianth sericeous outside. Pistil 16–18.5 mm long; ovary sessile or subsessile; style glabrous. Follicle c. 12 mm long.

Occurs in eastern Vic., on the upper Buchan R. and its tributaries, possibly once extending to Buchan. Grows in rocky situations near streams on granite. Probably regenerates from seed only. Flowers Oct.–Feb. Map 14.

Vic.: upper Buchan R., NW of West Tree, via Buchan, 11 Feb. 1950, *L.Hodge* (MEL); in old bed of a river at Buchan, 15 Dec. 1919, *M.McRae* (MEL); Reedy R. valley below Brumby Point, Nunniong Plateau area, 13 Nov. 1964, *J.H.Willis* (MEL 501463).

Grevillea pachylostyla is similar to and a sister-species of *G. willisii*. For differences see notes under the latter species. Collections possibly intermediate between the two species are known from Mt Stradbroke and Boundary Ck near Gelantipy.

13. Grevillea aquifolium Lindl., *in* T.L.Mitchell, *Three Exped. Australia* 2: 178 (1838)

G. aquifolium var. *truncata* Meisn., *in* A.L.P.P. de Candolle, *Prodr.* 14: 378 (1856), as *α truncata*, *nom. illeg.* (type var.). T: interior of Australia, 1836, *Major Mitchell's Exped. 232*; lecto: CGE *n.v.*, *fide* D.J.McGillivray & R.O.Makinson, *Grevillea* 404 (1993).

G. variabilis Lindl., *in* T.L.Mitchell, *Three Exped. Australia* 2: 178 (1838). T: interior of New Holland. Mt William [Victoria] 1836, *Major Mitchell's Exped. 238*; holo: CGE *n.v.*; probable iso: K.

G. aquifolium var. *attenuata* Meisn. *in* A.L.P.P. de Candolle, *Prodr.* 14: 378 (1856), as β *attenuata*. T: Mt William, interior of N.S.W., [Vic.], Mitchell Exped. 1835, *[R.?]Cunningham 244*; holo: NY *n.v.*; iso: K, MEL.

Illustrations: G.R.Cochrane *et al.*, *Fls & Pls Victoria* 46, t. 102 (1968); J.W.Wrigley & M.Fagg, *Banksias, Waratahs & Grevilleas* 206 (1989); P.M.Olde & N.R.Marriott, *Grevillea Book* 2: 32 (bottom right), 33 (22A, B), 34 (22C–G) (1995).

Erect-spreading to decumbent or prostrate shrub to 1.5 (–4) m tall. Leaves very variable, (2–) 3–7 (–10.5) cm long, 10–45 mm wide, usually ovate to oblong in gross outline and pinnatifid or toothed with 2–19 spreading lobes or teeth, sometimes cuneate or rhomboid with 2–5 teeth near apex, rarely entire and elliptic to rhomboid; primary lobes sometimes with a single secondary tooth on distal edge; ultimate lobes triangular to narrowly so or ovate, 5–17 mm long, 2–8 mm wide, usually pungent; margins flat to shortly recurved; lower surface densely to openly tomentose with straight to curled semi-appressed to spreading hairs. Unit conflorescence decurved or sometimes erect, secund; floral rachis 15–50 mm long. Flower colour: perianth grey, green, cream, or dull orange-pink; style usually red, or sometimes pink, orange, or dull yellow. Perianth loosely to densely tomentose outside, often villous on limb. Pistil 21–26 mm long; ovary stipitate; style glabrous. Follicle 9.5–13 mm long, tomentose. *Holly Grevillea.*

Occurs in lower south-eastern S.A., and in western Vic. from The Grampians area NW to Little Desert, also in the Portland area and in Otway Ra. E of Forrest. Grows in a wide range of habitats usually in woodland or dry to wet heath or mallee associations, in siliceous or sometimes calcareous sandy soils. Some forms regenerate from root-suckers. Flowers Sept.–Dec., rarely as late as Apr. Map 15.

S.A.: 40 km direct W of Mt Gambier near Buck Lake Natl Park, S of S end of L. Bonney, *A.C.Beauglehole 33419* (MEL, NSW); W Dairy Ra., 2.5 km ESE of 'Sugarloaf Hill', 22 km E of Robe, *M.D.Crisp 3818* (CANB, NSW). Vic.: c. 32 km S of Nhill ... c. 5–8 km NNW of Nurcoung, *H.I.Aston 979* (A, BRI, CANB, MEL); 2.5 km NE of Silverband Falls, Grampians, *R.Melville 1783* (K, MEL, NSW); near Golton Gorge, Grampians, on track to Dadswell Bridge, *M.E.Phillips CBG017657* (CANB, MEL, NSW).

This species is very variable especially in habit and leaf form. Numerous local forms have been described (see, for example, Olde & Marriott, *Grevillea Book* 2: 32, 33 (1995)), but most intergrade geographically or overlap morphologically, and formal subspecific taxa are not yet adequately defined.

Commonly confused with *G. ilicifolia* and other closely related species. *Grevillea aquifolium* has floral bracts 0.9–4.1 mm long, 1.4–3.0 mm wide, tomentose on outer surface, and falling at or near anthesis; perianth tomentose outside; ovary subvillous with spreading hairs; style-end abruptly contracted into the style; fructual stipe markedly incurved. *Grevillea ilicifolia* has floral bracts 0.3–0.9 mm long, 0.4–0.8 mm wide, with appressed hairs on the outer surface, and falling when buds are less than 3 mm long; perianth usually appressed-silky outside; ovary subsericeous; style-end more gradually tapering into the style; and stipe of fruit ±straight. Other potential misidentifications are dealt with under relevant species, below. See *G. bedggoodiana*, *G. montis-cole*, *G. microstegia* and *G. floripendula*.

14. Grevillea bedggoodiana J.H.Willis ex McGill., *New Names Grevillea* 3 (1986)

T: Little Hard Hills, Enfield Forest Park, 23.5 km SSW of Ballarat, Vic., 23 Oct. 1978, *A.C.Beauglehole 60988A*; holo: NSW.

Illustrations: P.M.Olde & N.R.Marriott, *Grevillea Book* 2: 59 (centre top & 43A, B) (1995).

Prostrate to decumbent shrub to 0.5 m high. Leaves ovate, oblong or angular-obovate in general outline, 1–7 cm long, (10–) 25–35 mm wide, toothed to pinnatifid with 5–9 simple subtriangular teeth or lobes to c. 1 cm long, pungent; margins almost flat to shortly recurved; lower surface openly or occasionally densely tomentose with curled hairs. Unit conflorescence erect to decurved, secund to subsecund; floral rachis 20–65 mm long. Flower colour: perianth green becoming pale pink; style pale greenish pink becoming darker. Perianth subsericeous to tomentose outside with limb often subvillous. Pistil 12–16.5 mm long; ovary stipitate; style glabrous. Follicle 8.5–10 mm long, tomentose. *Enfield Grevillea.*

Occurs in Vic. in the Enfield to Smythesdale area near Ballarat. Grows in eucalypt (Messmate, Brown Stringybark) woodland in sandy clay. Mode of regeneration unknown. Flowers Oct.–Nov. Map 16.

Vic.: Doctors Rd c. 5 km SE of Smythesdale, *A.C.Beauglehole 61714* (MEL, NSW); Enfield, *D.J.McGillivray 3216 & C.Bartlett* (K, MEL, NSW); Scarsdale Forest, 1.2 km S of Enfield, 17 Oct. 1962, *B.Strange* (MEL); Enfield, W of Ballarat Rd, 26 June 1966, *J.H.Willis* (MEL).

Sometimes confused with *G. aquifolium*, which has longer pistils (21–26 mm long). J.H.Willis (*Handb. Pl. Victoria* 2: 40 (1973)) refers to *G. bedggoodiana* as a 'population in Enfield district' in discussion of the 'Elphinstone Grevillea' (*G. obtecta*). See the latter species for differences.

This species is recognised as 'Rare' in J.D.Briggs & J.H.Leigh, *Rare or Threatened Australian Plants* (1995).

15. **Grevillea obtecta** Molyneux, *Muelleria* 6: 147 (1985)

T: 6 km SW of Taradale, 1 km SW of aqueduct, Fryers Ra. State Forest, central Vic., 2 Nov. 1977, *W.M.Molyneux & S.G.Forrester*; holo: MEL; iso: CANB, K, MEL, NSW.

G. sp. 'Elphinstone Grevillea'; J.H.Willis, *Handb. Pl. Victoria* 2: 40 (1973).

Illustrations: W.R.Elliot & D.L.Jones, *Encycl. Austral. Pl.* 5: 85 (1990); D.J.McGillivray & R.O.Makinson, *Grevillea* 87, fig. 14, 88 (1993); P.M.Olde & N.R.Marriott, *Grevillea Book* 3: 58 (bottom left & 41A–C) (1995).

Prostrate, straggling or clumping shrub, to 0.5 m tall, to c. 1.2 m across. Leaves usually ovate to oblong in gross outline, 2–18 cm long, 20–60 mm wide, pinnatifid to pinnatipartite or toothed or rarely entire; primary lobes 2–21, entire to 5-fid; ultimate lobes triangular to narrowly ovate, 0.5–3 cm long, 0.2–1.5 cm wide, sometimes pungent; margins flat to shortly recurved; lower surface with an open indumentum of ascending, rarely closely appressed curved to curled hairs. Unit conflorescence erect to decurved, secund; floral rachis 35–60 mm long. Flower colour: perianth light green to yellow outside, purplish to black inside; style dull yellow to pale or dark pink with a red or green tip. Perianth loosely tomentose to villous outside. Pistil 12–16 mm long; ovary stipitate; style glabrous. Follicle 10–12 mm long, subsericeous to subvillous. *Fryerstown Grevillea, Elphinstone Grevillea, Taradale Grevillea.* Fig. 4D–F.

Occurs in Vic., in the Fryerstown to North Daylesford area. Grows in well-drained situations in dry eucalypt forest, often on south-facing gravelly shale ridges and slopes. Regenerates from suckers or seed. Flowers Aug.–Nov. Map 17.

Vic.: c. 15 km S of Castlemaine P.O., *A.C.Beauglehole 50094* (MEL, NSW); between Malmesbury and Glenluce, 20 Sept. 1961, *M.E.Phillips CBG005188* (CANB); Holcombe Parish c. 3.2 km N of Glenlyon on upper Loddon R., 18 July 1937, *J.H.Willis* (MEL).

There is much variation in leaf form. The 'Fryers Range form' is prostrate with mostly entire or lightly toothed leaves (only occasionally deeply divided) 6–12 cm long. The 'Upper Loddon form' is prostrate with very deeply divided leaves 8–18 cm long. The 'Drummond North form' forms rhizomatous clumps up to 50 cm tall, with leaves small (4–7 cm long), ovate to rhomboid, and sparingly toothed.

Distinguished by its larger floral bracts (5–8 (–10.5) mm long, usually persistent to anthesis, 3–5 mm wide) from most similar species, especially *G. bedggoodiana* (2–3 mm long, 3–4 mm wide) and *G. repens* (0.7–2 mm long, 0.2–2.2 mm wide).

This species is recognised as 'Rare' in J.D.Briggs & J.H.Leigh, *Rare or Threatened Australian Plants* (1995).

burgundy styles. The 'Musical Gully form', from NW of Beaufort, is decumbent to weakly erect in habit, with larger, often deeply divided leaves often with secondary division, and styles ranging from dull purplish or pink-red to yellowish.

Grevillea floripendula can be distinguished from *G. aquifolium* which has a denser indumentum on the leaf lower surface and a longer pistil (> 20 mm long); from *G. microstegia* which has a stouter, usually shorter, and hairy peduncle and narrower more acute leaf lobes; from *G. dryophylla* which has a hairy and stouter peduncle (> 1 mm thick); from *G. montis-cole* which has a thicker usually hairy peduncle and erect fruits (fruits on incurved stipe in *G. floripendula*) and (subsp. *montis-cole*) pistils 26–27.5 mm long, and from *G. steiglitziana* which has a usually longer pistil (15–20 mm long) and a sericeous leaf lower surface.

This species is recognised as 'Rare' in J.D.Briggs & J.H.Leigh, *Rare or Threatened Australian Plants* (1995).

19. **Grevillea dryophylla** N.A.Wakef., *Victorian Naturalist* 73: 74 (1956)

T: Kangaroo Flat, Bendigo, Vic., Nov. 1934, *A.J.Tadgell s.n.*; holo: MEL; iso: MEL [isotypes indicated at K and NSW in the protologue apparently not sent].

Illustrations: L.F.Costermans, *Native Trees & Shrubs SE Australia* 164 (1981); J.W.Wrigley & M.Fagg, *Banksias, Waratahs & Grevilleas* 226 (1989); P.M.Olde & N.R.Marriott, *Grevillea Book* 2: 140 (bottom right), 141 (114A, B) (1995).

Spreading to erect shrub, 0.3–1.5 m high. Leaves ovate in gross outline, 2–8.5 cm long, 20–70 mm wide, mostly pinnatipartite or occasionally pinnatifid or rarely simple and marginally toothed; primary lobes up to 7, each entire or 2–5-dentate; ultimate lobes triangular or rounded-oblong, 6–30 mm long, 3–15 (–20) mm wide, usually pungent; margins shortly recurved; lower surface with an open to moderately dense indumentum of evenly spaced wavy to curly hairs. Unit conflorescence erect to decurved, ±secund; floral rachis 10–40 mm long. Flower colour: perianth green tending to light brown-maroon or dull yellowish; style red to pink, dull yellow, or buff. Perianth subsericeous to tomentose outside. Pistil 13.5–15.5 mm long; ovary stipitate; style glabrous. Follicle 9–12.5 mm long, subsericeous. *Goldfields Grevillea.*

Occurs in the Western Goldfields area of Vic. bounded by Maryborough, St Arnaud, Bendigo and Castlemaine. Grows in dry sclerophyll forest (Box/Ironbark/Stringybark associations) in poor stony or gravelly soil. Regenerates from seed. Flowers Aug.–Nov. Map 22.

Vic.: c. 4 km NNW of Castlemaine, *A.C.Beauglehole 50106* (NSW); c. 8.5 km WSW of Inglewood, *A.C.Beauglehole 50183* (MEL, NSW); 13 km S of St Arnaud, *A.C.Beauglehole 65623* (MEL, NSW); 2 km S of Maryborough, *D.J.McGillivray 3222 & C.Bartlett* (MEL, NSW).

Variable in details of leaf division, and in style colour.

Grevillea dryophylla is distinguished from *G. floripendula* which has thin wiry glabrous peduncles; *G. ilicifolia* which has a longer pistil (19–25 mm long); *G. microstegia* which has narrower more acute leaf lobes (< 3 mm wide) and villous branchlets (*G. dryophylla* has subsericeous to tomentose branchlets); and *G. steiglitziana* which has the upper leaf surface glabrous or nearly so (with short curly to wavy hairs in *G. dryophylla*) and the lower leaf surface with straight to slightly wavy appressed hairs.

20. **Grevillea williamsonii** F.Muell., *Victorian Naturalist* 10: 129 (1893)

T: between Mt Abrupt and Mt Sturgeon, The Grampians, Vic., *s.d.* [1893?], *H.B.Williamson*; holo: MEL.

Illustrations: P.M.Olde & N.R.Marriott, *Grevillea Book* 3: 231 (bottom centre), 232 (189A, B) (1995).

Spreading shrub 0.6–1 m tall. Leaves elliptic to narrowly so, 1.5–4.0 cm long, 3–8 mm wide, entire or (occasionally on adults, commonly on juveniles) with 2–4 spreading triangular pungent teeth; margins shortly and angularly refracted; lower surface subsericeous to subvillous with straight hairs. Unit conflorescence erect to deflexed, secund; floral rachis 15–26 mm long. Flower colour: perianth lime-green becoming yellowish; style pale orange yellow becoming pale pink at about anthesis then darker reddish pink. Perianth sparsely tomentose outside. Pistil 18.5–21.5 mm long; ovary stipitate; style glabrous. Follicle c. 11 mm long, sparsely subsericeous.

Occurs in the Grampians Ra. in western Vic., where known only from a small area near Cassidy Gap. The species has not been relocated at the Type locality (further S at Picaninny Hill). Grows in open eucalypt woodland with a dense low shrub layer, in poor grey sand. Possibly lignotuberous. Flowers c. Aug.–Oct. Map 23.

Vic.: Grampians Natl Park, 0.74 km E of Serra Ra., *ANBG2775 per N.Marriott* (CANB); 4.36 km from intersection of Halls Gap/Dunkeld road, 15 Apr. 1993, *W.Molyneux s.n.* (MEL).

Originally known from specimens from one plant, and thought to be either extinct or a chance hybrid involving *G. aquifolium* (e.g McGillivray & R.O.Makinson, *Grevillea* 448 (1993)), this species was recently rediscovered. It grows with *G. aquifolium* and is clearly distinct from it. The flowers have withered anthers with no pollen or very shrunken grains, but there is a low level of seed set. The style-end is scarcely wider than the style.

This species is recognised as 'Endangered' in J.D.Briggs & J.H.Leigh, *Rare or Threatened Australian Plants* (1995).

21. Grevillea infecunda McGill., *New Names Grevillea* 7 (1986)

T: Gum Flat Rd between Harvester Co.'s concession and Wensleydale, Anglesea district, Vic., 25 Oct. 1969, *J.H.Willis s.n.*; holo: MEL.

[*G. ilicifolia auct. non* (R.Br.) R.Br.: C.F.W.Meissner, *Linnaea* 26: 355 (1854), *p.p.*]

[*G. aquifolium auct. non* Lindl.: J.H.Willis, *Handb. Pl. Victoria* 2: 42 (1973), *p.p.*]

Illustrations: W.R.Elliot & D.L.Jones, *Encycl. Austral. Pl.* 5: 65 (1990); P.M.Olde & N.R.Marriott, *Grevillea Book* 2: 210 (top right, 173A, B) (1995).

Decumbent to weakly upright root-suckering shrub, 0.3–1.2 m tall. Leaves ovate to rhomboid or oblong in gross outline, 3–7 cm long, 13–40 mm wide, trifid or pinnatifid with 3–16 usually entire triangular to rounded lobes or teeth, the lower ones sometimes again bifid; ultimate lobes stiff to rigid, 6–15 mm long, 7–10 mm wide, pungent; margins slightly recurved; lower surface subsericeous to tomentose. Unit conflorescence ±erect, secund; floral rachis 20–40 mm long. Flower colour: perianth greenish yellow, ageing to orange or reddish; style yellow-green or dull pink, becoming orange or reddish at anthesis. Perianth subsericeous outside, sometimes tomentose on limb. Pistil 18–26 mm long; stipe 1.8–3.7 mm long; style glabrous. Mature fruit and seed not known to be produced.

Occurs in southern coastal Vic., the only known extant population being at Anglesea, SW of Melbourne. There is a 19th century collection from red sand scrubs at Brighton (Melbourne), where it is now presumed extinct. Grows in dry sclerophyll woodland in sandy, gravelly or clay-loam soils. Regeneration is apparently exclusively by rhizomes. Flowers Oct.–Dec. Map 24.

Vic.: Anglesea ... about [4.8 km] W of township, 1 Dec. 1960, *M.Allender* (MEL); Anglesea to Wensleydale road, 12 Nov. 1961, *M.Allender* (MEL); 13 km NW of Anglesea, *A.C.Beauglehole 49561* (MEL, NSW).

Grevillea infecunda has floral bracts 1.5–3 mm long, pedicels < 1.5 mm long, and hairy peduncles and floral rachises. The anthers are poorly developed and pollen-grains of this species are normally shrunken. It has never been known to set mature fruit; occasional young fruit in cultivation abort, and the species is presumed sterile.

The name *G. integrifolia* γ *quercina* Meisn. (type not seen, 'La Trobe, c. Port-Phillip') may be applicable to this species.

This species is recognised as 'Vulnerable' in J.D.Briggs & J.H.Leigh, *Rare or Threatened Australian Plants* (1995).

22. Grevillea ilicifolia (R.Br.) R.Br., *Suppl. Prodr. Fl. Nov. Holl.* 21 (1830)

Anadenia ilicifolia R.Br., *Trans. Linn. Soc. London* 10: 167 (1810); *G. ilicifolia* var. *attenuata* R.Br., *Suppl. Prodr. Fl. Nov. Holl.* 21 (1830), as α *attenuata, nom. illeg.* (type var.). T: coast of southern Australia, Bay X [Port Lincoln, S.A.], 1802, *R.Brown Iter. Austral. 3314*; lecto: BM, *fide* R.O.Makinson, *Fl. Australia* 17A: 493 (2000); isolecto: E, G-DC, K, MEL, NSW.

G. ilicifolia var. *dilatata* R.Br., *Suppl. Prodr. Fl. Nov. Holl.* 21 (1830), as β *dilatata*. T: Kangaroo Is., [S.A.], 1823, *No. 30, W.Baxter*; holo: BM; iso: K.

G. behrii Schltdl., *Linnaea* 20: 585 (1847). T: pine forest prope Gawlertown pr. Adelaide, [S.A.], *H.H.Behr s.n.*; syn: HAL 40155 *n.v.*; possible syn: MEL 1516686.

G. lobata F.Muell., *Trans. Philos. Soc. Victoria* 1: 22 '1855' (pre 8 Nov 1854); *G. ilicifolia* var. *lobata* (F.Muell.) Benth., *Fl. Austral.* 5: 438 (1870). T: in mallee scrub along the Murray R. W of Swan Hill, Vic., Jan. [18]53, *F.Mueller*; lecto: MEL, *fide* D.J.McGillivray & R.O.Makinson, *Grevillea* 421 (1993).

G. dumetorum Meisn., *in* A.L.P.P. de Candolle, *Prodr.* 14: 378 (1856). T: Australia Felici, [N.S.W.], Mitchell Exped. 1835, *R.Cunningham 210*; holo: NY *n.v.*

G. ilicifolia var. *integrifolia* F.Muell. ex Meisn., *in* A.L.P.P. de Candolle, *Prodr.* 14: 378 (1856). T: Golway, [Vic.?] summer 1851, *C.Wilhelmi*; holo: MEL.

G. lobata var. *sturtii* Meisn., *in* A.L.P.P. de Candolle, *Prodr.* 14: 379 (1856). T: in dumetis planitei ad fl. Murray [lat. 35°, near Mannum, S.A.?], Mar. 1836, *C.Sturt*; holo: NY *n.v.*; iso: K.

G. ilicifolia var. *angustiloba* F.Muell., *Fragm.* 6: 212 (1868). T: in monte Ariples [Mount Arapiles, Vic.], 22 Sept. 1860, *coll. unknown, N61*; holo: MEL.

G. approximata Gand., *Bull. Soc. Bot. France* 66: 231 (1919). T: Victoria, 1902, *C.Walter*; holo: LY.

Illustrations: L.F.Costermans, *Nat. Trees & Shrubs S.E. Australia* 164–165 (1981); J.P.Jessop & H.R.Toelken (eds), *Fl. S. Australia* 4th edn, 1: 129, fig. 69D (1986); P.M.Olde & N.R.Marriott, *Grevillea Book* 2: 206 (centre top & 168A–C), 207 (169A, B) (1995).

Prostrate to erect shrub 0.3–2.0 m high, to 3 m across. Leaves very variable, 2–11 cm long, 25–60 (–105) mm wide, ovate to obovate in outline, often markedly cuneate in basal half, either toothed subapically with several triangular teeth (some secondary), or shallowly to very deeply lobed to first or second order, rarely entire and elliptic; primary lobes or teeth 2–13; ultimate lobes triangular to oblong, ovate, or occasionally linear and then usually wavy, 0.7–5 cm long, 0.8–8 mm wide, usually pungent; margins shortly recurved or sometimes revolute; lower surface subsericeous with straight hairs or densely pubescent-tomentose with twisted or curled hairs, mostly enclosed on very narrow lobes (c. 1 mm wide). Unit conflorescence erect or decurved, secund; floral rachis 20–50 mm long. Flower colour: perianth basally green to cream, usually grey to mauve above; style pink to red, occasionally orange or pale yellow, with a green tip. Perianth openly subsericeous outside, occasionally densely so or tomentose or glabrous. Pistil 19.5–25 mm long; ovary stipitate; style glabrous. Follicle 10.5–16.5 mm long, subsericeous or tomentose. *Holly Grevillea, Holly Bush.* Fig. 5A–N.

Occurs mainly in south-eastern S.A. (including Eyre Peninsula and Kangaroo Is.) and western inland Vic., extending patchily into N.S.W. (Griffith area, and NW of Lake Cargelligo). Grows in mallee, heath or shrubland associations in sandy, acid to calcareous soils. Regenerates from seed and in at least some populations from rhizomes. Flowers Sept.–Nov. Map 25.

S.A.: between Kingscote & Vivonne Bay, 16 Nov. 1924, *J.B.Cleland* (AD); 5 km S of Reeves Plains, *N.N.Donner 826* (AD). N.S.W.: Nericon near Griffith, *G.R.Sainty 351* (NSW). Vic.: Kulkyne Natl Park, *A.C.Beauglehole 8020* (MEL); Big Desert, c. 18 km SE of Cowangie, *M.G.Corrick 6721* (MEL).

This species is very variable in habit, leaf morphology and some floral characters (pedicel and stipe length, indumentum). Some ten leaf-shape types, corresponding to geographically definable populations, may be identified (e.g. McGillivray & R.O.Makinson, *loc. cit.*) but these are only partially correlated with other variable characters, and more research is needed before meaningful delimitation of formal infraspecific taxa is possible. Plants with very fine leaf division (lobes < 1.5 mm wide) occur in the Vic. and S.A. border region, usually around salt pans, and are sometimes known as *G. ilicifolia* var. *angustiloba*, although the type of that name does not belong to these populations.

For differences between *G. ilicifolia* and *G. aquifolium* see notes under the latter species.

Figure 5. *Grevillea*. **A–N**, *G. ilicifolia*. **A–K**, leaf variation (**A**, M.G.Corrick 6364, MEL; **B**, D.E.Symon 8937, NSW; **C**, M.D.Tindale 528, NSW; **D**, J.H.Maiden *s.n.*, Jan. 1907, NSW70276; **E**, M.E.Phillips, 1 Oct. 1965, NSW; **F**, R.F.Thorne & Hj.Eichler, 23 Sept. 1959, NSW; **G**, A.C.Beauglehole 66914, NSW; **H**, D.J.McGillivray 3237, NSW; **I**, A.C.Beauglehole 7472, NSW; **J**, D.Woolcock, 9 Dec. 1978, NSW; **K**, B.Copley 3359, AD); **L**, flowering branch; **M**, flower; **N**, pistil and half perianth (**L–N**, J.J.Ackland 67, CANB). **O–Q**, *G. cagiana*. **O**, flowering branch; **P**, flower; **Q**, pistil (**O–Q**, T.W.Stone 823, CANB). Scale bars: **A–L**, **O** = 2 cm; **M–N** = 1 cm; **P–Q** = 5 mm. Drawn by: **A–K**, ?D.Fortescue; **L–Q**, D.Boyer.

23. **Grevillea scortechinii** (F.Muell. ex Scort.) F.Muell., *Second Syst. Census Austral. Pl., Pt. 1 Vasculares* 117 (1889)

G. ilicifolia var. *scortechinii* F.Muell. ex Scort., *Proc. Linn. Soc. New South Wales* 8: 174 (17 July 1883). T: Stanthorpe, Qld, 1883, *B.Scortechini s.n.*; lecto: MEL, *fide* D.J.McGillivray & R.O.Makinson, *Grevillea* 440 (1993); ?isolecto: BRI, NSW.

G. ilicifolia var. *scortechinii* F.Muell. ex F.M.Bailey, *Syn. Queensland Fl.* 435 (before 6 Aug. 1883), *non* Scort. (1883). T: Stanthorpe, Qld, [1883?], *B.Scortechini s.n.*; holo: BRI.

Prostrate to sprawling shrub to 0.8 m tall and 1.2 m across. Leaves angular-ovate to oblong, 3–11 cm long, 20–60 mm wide, serrato-dentate to pinnatifid or rarely entire, usually with 3–14 lobes or marginal teeth, these sometimes again 2- or 3-dentate; ultimate lobes rounded, broadly subtriangular, to 2 cm long, often pungent; margins slightly recurved; lower surface sericeous. Unit conflorescence erect to decurved, secund; floral rachis 20–45 mm long. Perianth subsericeous outside. Pistil 19–30 mm long; stipe 1–2.6 mm long; style glabrous. Follicle 8.5–12 mm long, subsericeous. *Black Grevillea.*

Occurs in southern Qld and north-eastern N.S.W. There are two subspecies.

Closely related to *G. ilicifolia*, which has fruit borne on an erect straight stipe (curved in *G. scortechinii*). *Grevillea scortechinii* also has floral bracts 1–2.5 mm long, and a hairy conflorescence rachis. There is some doubt about the date of publication of this name by Bailey (*loc. cit.*); see McGillivray & Makinson (*Grevillea* 440 (1993)) for details.

Pistil c. 28–30 mm long; primary leaf lobes usually simple **23a.** subsp. **scortechinii**

Pistil c. 19–22 mm long; primary leaf lobes often 2- or 3-toothed **23b.** subsp. **sarmentosa**

23a. **Grevillea scortechinii** (F.Muell. ex Scort.) F.Muell. subsp. **scortechinii**

Illustrations: Society for Growing Australian Plants, *Hort. Guide Austral. Pl.* Set 4: t. 19 (1978); K.A.W.Williams, *Native Pl. Queensland* 3rd edn, 1: 139 (1984); P.M.Olde & N.R.Marriott, *Grevillea Book* 3: 164 (lower right & 132A, B) (1995).

Leaves with lobes usually simple; venation of upper (adaxial) surface prominent. Flower colour: perianth brownish; style dark purplish black; pollen-presenter green. Pistil 28–30 mm long; stipe 2.0–2.6 mm long.

Occurs in south-eastern Qld, known only from the Stanthorpe to Cottonvale area. Grows in sclerophyll woodland or remnant roadside associations in granitic, sandy-loamy soils. Regenerates from seed. Flowers Oct.–Nov. Map 26.

Qld: along road between Pozieres and Cottonvale, June 1971, *J.Birbeck* (BRI); 0.6 km along Pozieres Rd from New England Hwy at Cottonvale, *R.O.Makinson 1363 et al.* (CANB, K); Jollys Falls, c. 8 km N of Stanthorpe, 17 Oct. 1971, *K.Williams* (BRI).

This subspecies is recognised as 'Endangered' in J.D.Briggs & J.H.Leigh, *Rare or Threatened Australian Plants* (1995).

23b. **Grevillea scortechinii** subsp. **sarmentosa** (Blakely & McKie) McGill., *New Names Grevillea* 14 (1986)

G. sarmentosa Blakely & McKie, *Proc. Linn. Soc. New South Wales* ser. 2, 55: 587, t. XXIX-1 (1930). T: Pheasant Mtn, 3.2 km NE of Backwater, Guyra district, N.S.W., 30 Oct. 1929, *W.F.Blakely et al.*; lecto: NSW, *fide* D.J.McGillivray & R.O.Makinson, *Grevillea* 440 (1993); isolecto: MEL, PERTH, US *n.v.*

Illustrations: W.F.Blakely & E.N.McKie, *loc. cit.*; D.J.McGillivray & R.O.Makinson, *Grevillea* 95 (1993); P.M.Olde & N.R.Marriott, *Grevillea Book* 3: 164 (lower right), 165 (133A, B) (1995).

Leaves with primary lobes often bi- or tridentate; venation not conspicuous on upper surface. Flower colour: perianth brownish; style dark purplish black; pollen-presenter green. Pistil 19–22 mm long; stipe 1.0–1.6 mm long.

Occurs in montane north-eastern N.S.W., mainly in the Backwater region ENE of Guyra. Grows in sclerophyll woodland on granitic slabs and slopes in sandy-loamy soils. Regenerates probably from seed only. Flowers Oct.–Mar. Map 27.

N.S.W.: 4.8 km E of Backwater, *D.J.McGillivray 3605A & R.Coveny* (NSW); c. 30 km (direct) NE of Guyra, Pheasant Mtn, *R.O.Makinson 1453* (CANB, K, MEL, NE, NSW); Pheasant Mtn, c. 3 km NE of Backwater, *I.R.Telford 8995 & G.Butler* (CANB).

This subspecies is recognised as 'Vulnerable' in J.D.Briggs & J.H.Leigh, *Rare or Threatened Australian Plants* (1995).

24. Grevillea steiglitziana N.A.Wakef., *Victorian Naturalist* 73: 74 (1956)

T: Brisbane Ranges near Geelong Reservoir, Vic., 16 Sept. 1911, *P.R.H.St.John*; holo: MEL.

G. longistyla Gand., *Bull. Soc. Bot. France* 66: 231 (1919), *nom. illeg. non* Hook. (1848). T: Brisbane Ranges, Vic., 1 Oct. 1909, *P.R.H.St.John*; holo: LY.

Illustrations: J.W.Wrigley & M.Fagg, *Banksias, Waratahs & Grevilleas* 331 (1989); P.M.Olde & N.R.Marriott, *Grevillea Book* 3: 182 (top right & 148A, B) (1995).

Spreading shrub 0.7–2.0 m high. Leaves ovate in gross outline, 2.5–7 cm long, 25–55 mm wide, pinnatifid to pinnatipartite, with (3–) 5–7 primary lobes each usually 2–5-fid or -toothed; ultimate lobes subtriangular, 3–22 mm long, to 13 mm wide, pungent; margins flat to shortly recurved; lower surface with an open indumentum of straight to wavy appressed hairs. Unit conflorescence erect to decurved, secund; floral rachis 25–50 mm long. Flower colour: perianth greenish brown; style red. Perianth subsericeous outside. Pistil (15–) 22–27 mm long; ovary stipitate; style glabrous. Follicle 12.5–13.5 mm long, subsericeous. *Brisbane Range Grevillea.*

Occurs in Vic., where restricted to the Brisbane Ra. area between Anakie and Bacchus Marsh. Grows in dry open sclerophyll forest in rocky situations with sandy soil. Regenerates from seed. Flowers Sept.–Jan. Map 28.

Vic.: N38 ... Mt Wallace to Bacchus Marsh road, 0.5 km W of junction of Aeroplane Rd, *A.C.Beauglehole 56753* (MEL, NSW); c. 6 km E of Mt Wallace on road between Durdidwarrah and Bacchus Marsh, *D.J.McGillivray 3209 & C.Bartlett* (MEL, NSW); Brisbane Ra., N end, about 1.5 km S of Spring Ck, *T.B.Muir 818* (MEL).

Grevillea steiglitziana has pedicels 1–3 mm long and hairy rachises. It is rather similar to *G. dryophylla*; for differences see under that species.

25. Grevillea renwickiana F.Muell., *Proc. Linn. Soc. New South Wales* ser. 2, 1: 1105 (1887)

T: on heath-ground near the Little R., Braidwood district, N.S.W., [Nov. 1886], *W.Bauerlen s.n.?*; lecto: MEL, *fide* D.J.McGillivray & R.O.Makinson, *Grevillea* 438 (1993).

Illustrations: D.J.McGillivray & R.O.Makinson, *Grevillea* 97 (1993); P.M.Olde & N.R.Marriott, *Grevillea Book* 1: 82, fig. 49 (1994); P.M.Olde & N.Marriott, *op. cit.* 3: 134 (bottom centre), 135 (106A–C) (1995).

Prostrate, root-suckering shrub to 20 cm high; ramets to 1 m across; clonal units to ?30 m across. Leaves usually narrowly ovate to narrowly oblong in gross outline, (2.5–) 6–14 cm long, 15–75 mm wide, pinnatifid to pinnatipartite or sometimes serrato-dentate, with 5–18 teeth or lobes, these usually entire or the lowermost 2- or 3-toothed; ultimate teeth or lobes triangular to oblong, 0.5–3.5 cm long, 5–15 mm wide, pungent; margins recurved to shortly revolute; lower surface loosely subsericeous or with a sparse appressed indumentum. Unit conflorescence erect to slightly curved, secund; floral rachis 20–25 mm long. Flower colour: perianth greenish in late bud, becoming cream to pale pink with a purplish limb; style cream becoming pink to reddish with a green tip. Perianth loosely subsericeous outside. Pistil 26–30.5 mm long (flowers often digynous to polygynous); ovary stipitate; style glabrous. Fruit and seed apparently not produced. Plate 4.

Occurs in south-eastern N.S.W., known only from a few populations in the Braidwood to Nerriga area of the southern tablelands. Grows in open dry sclerophyll forest in sandy, loamy, or clay soils. Not known to set fruit, appearing to reproduce entirely from rhizomes. Flowers Nov.–Dec. Map 29.

(CANB, K, NSW, PERTH, US *n.v.*); 11 km by road N of Gibson, *D.J.McGillivray 3631 & A.S.George* (K, NSW, PERTH); near Boyatup Hill c. 117 km E of Esperance, *K.Newbey 6807* (PERTH).

Grevillea baxteri has pedicels 1.5–2 mm long, a shortly stipitate ovary (stipe 0.9–1.3 mm long), the style with appressed biramous hairs persisting for 4–8 mm above the ovary and with erect glandular hairs in the apical 7–10 mm on the ventral side, and an arcuate nectary barely projecting above the toral rim. It is distinguished from *G. hookeriana* which has a sessile ovary, a prominent linguiform nectary, and glabrous purplish black style; from *G. concinna* which has the conflorescence deflexed; from *G. coccinea* which has simple leaves; from *G. tetragonoloba, G. rigida* and *G. fastigiata* which have the ovary sessile or nearly so with the stipe obscure; from *G. cagiana* which has the style ventrally papillose around middle; and from *G. wittweri* which has pistil 18–21 mm long.

This species is recognised as 'Rare' in J.D.Briggs & J.H.Leigh, *Rare or Threatened Australian Plants* (1995).

29. **Grevillea tetragonoloba** Meisn., *in* A.L.P.P. de Candolle, *Prodr.* 14: 374 (1856)

T: Swan R. Colony, W.A., 1848, *J.Drummond (IV) 282*; lecto: NY (photo seen), *fide* D.J.McGillivray & R.O.Makinson, *Grevillea* 444 (1993); isolecto: CGE *n.v.*, FI *n.v.*, G, K, LE *n.v.*, MEL, NY (photo seen), P *n.v.*, TCD *n.v.*

[*G. hookeriana auct. non* Meisn.: G.Bentham, *Fl. Austral.* 5: 432 (1870), *p.p.*]

G. tetragonoloba races 'a' and 'b' only, of D.J.McGillivray & R.O.Makinson, *Grevillea* 71, 72 (1993).

Illustrations: A.S.George, *Introd. Prot. W. Australia* 52, t. 71 (1984); J.W.Wrigley & M.Fagg, *Banksias, Waratahs & Grevilleas* 332 (1989); D.J.McGillivray & R.O.Makinson, *Grevillea* 72 (1993); P.M.Olde & N.R.Marriott, *Grevillea Book* 3: 199 (top right & 162A, B) (1995).

Erect to spreading shrub, 2–2.5 m high. Leaves subsecund, often gently upcurved, 6–13 cm long, rarely entire, usually pinnatipartite to almost pinnatisect with 3–13 ascending lobes, sometimes lowermost primary lobes bisect; simple leaves and ultimate lobes linear, 3–13 cm long, 0.8–1.5 mm wide, rectangular in cross-section, often gently incurved, pungent or almost so; margins angularly twice refracted about longitudinal veins; lower surface enclosed except for midvein, pubescent in grooves. Unit conflorescence erect, secund; floral rachis 45–115 mm long. Flower colour: perianth yellowish brown to fawn; style scarlet to orange-red. Perianth subsericeous to tomentose outside. Pistil 22–25 mm long; ovary sessile or subsessile; style glabrous except for a few hairs close to ovary. Follicle 10–15 mm long, subsericeous to subtomentose.

Occurs in south-western W.A., in the areas Cape Riche to Needilup, and around Bremer Bay. Grows in mallee heath, open eucalypt woodland and shrubland, in granitic or lateritic sandy or loamy soils. Regenerates from seed. Flowers throughout the year, peaking Oct.–Mar. Map 33.

W.A.: Doubtful Is. Penin., [*Anon.*] *246, MEL64565* (MEL); near Cape Riche, *C.A.Gardner 6508* (PERTH); Bremer Bay, *C.A.Gardner 6578* (PERTH); 1 km W of Needilup, *D.J.McGillivray 3510 & A.S.George* (NSW, PERTH); Needilup Hill, *K.Newbey 731* (PERTH).

Grevillea tetragonoloba has leaf venation evident above, floral bracts ovate, elliptic or obovate, $\leqslant$ 1 mm long, floral rachis and perianth limb subsericeous to tomentose, ovary sessile to subsessile, style glabrous, and rachis, pedicels and occasionally the outer perianth surface with inconspicuous erect glandular hairs producing minute persistent waxy globules c. 0.01 mm diam. (similar to sand or pollen grains; these rarely absent). McGillivray's (*loc. cit.*) broad concept of *G. tetragonoloba* included the taxa since segregated (Olde & Marriott, *op. cit.* 1: 186, 1994) and here accepted as *G. fastigiata* and *G. rigida*, which have floral bracts $\geqslant$ 2 mm long, a villous rachis and perianth limb, style usually hairy over part of its length, and lack glandular hairs and waxy globules on rachis and pedicels. *Grevillea hookeriana* differs in its smoothly rounded leaf margins and maroon-black style; *G. wittweri* in its leaf lobes with margins rounded and usually with secondary division; and *G. baxteri, G. cagiana* and *G. beardiana* in having a clearly stipitate ovary.

Despite the recent erection of *G. rigida* and *G. fastigiata*, the species remains somewhat variable, and two forms can be recognised. The 'narrow-lobed form' corresponds to 'race a' of McGillivray & Makinson (*loc. cit.*), and to the 'fine-leaf typical form' of Olde and Marriott (*op. cit.* 3: 200 (1995)). It has leaf lobes 3–7.5 cm long, 0.8–1.4 mm wide, with ±acute somewhat pungent apices, and perianth indumentum subsericeous and rather rust-coloured at least on the limb. The 'blunt-lobed form' corresponds to McGillivray & Makinson's 'race b' and Olde & Marriott's (*loc. cit.*) 'blunt-leaf form'; it has rather coarser lobes usually 1.5–4 (–6) cm long and 1.1–1.4 mm wide, with obtuse non-pungent apices, and perianth tomentose and pale brown.

30. **Grevillea rigida** Olde & Marriott, *Grevillea Book* 1: 186 (1994)

T: 33.5 km E of Ravensthorpe on Highway 1, W.A., 11 Nov. 1986, *P.M.Olde 86/1148*; holo: NSW.

Compact spreading shrub 1.5–3 m tall. Leaves 3–6 cm long, pinnatipartite with 3–9 lobes; lobes ascending, simple, linear to broadly linear, rigid, 5–35 mm long and 1.3–2.5 mm wide, rectangular in cross-section, pungent or almost so; margins angularly twice refracted enclosing most or all of leaf lower surface except for midvein; lower surface tomentose with curled hairs. Unit conflorescence erect, secund; floral rachis 45–95 mm long. Perianth tomentose-villous outside, especially villous on limb. Pistil 22–25 mm long; ovary sessile or subsessile; style usually with few to many minute curled or erect simple hairs especially about middle on ventral side, sometimes throughout. Follicle 10–15 mm long, subsericeous with mainly biramous hairs and scattered minute erect simple ?glandular hairs.

Grevillea rigida is very closely related to *G. fastigiata*, which has the fruit at maturity with predominantly glandular hairs, and the floral rachis rusty-villous (white-villous in *G. rigida*). It is also close to *G. tetragonoloba* (see under that species for differences); and to *G. wittweri*, which has smoothly revolute leaf margins and usually secondary leaf division. In addition, *G. rigida* often has short lateral veins evident on the upper surface of the leaf lobes, a useful diagnostic feature not shared by related species. It also has pedicels 1.5–2 mm long; and floral bracts ovate, elliptic or obovate and 3–4.5 mm long.

Occurs in south-western W.A. in the Jerramungup to Ravensthorpe area in granitic loam in mallee-heath and tall shrubland associations. Regenerates by seed only. There are two subspecies.

Longest leaf lobes usually < 2 cm long; lower surface of leaf lamina beside midveins completely enclosed or sometimes slightly exposed on broader lobes (especially near sinuses); leaf lobes mostly 3–5, crowded at apex of leaf rachis, with terminal lobe recurved or reflexed **30a.** subsp. **rigida**

Longest leaf lobes > 2 cm long; lower surface of leaf lamina (except midveins) completely enclosed; leaf lobes mostly ≥ 7, usually spaced over apical half of leaf rachis; terminal lobes erect, incurved, or recurved **30b.** subsp. **distans**

30a. **Grevillea rigida** Olde & Marriott subsp. **rigida**

G. tetragonoloba 'race d', 'digitate race', of D.J.McGillivray & R.O.Makinson, *Grevillea* 71 (1993).

Illustrations: P.M.Olde & N.R.Marriott, *Grevillea Book* 3: 137 (bottom right), 138 (109A, B) (1995).

Leaves 2–4.5 cm long; lobes 2–5 (–7), 5–20 mm long, 1.3–2.5 mm wide, clustered on apical third or quarter of leaf rachis; terminal lobe recurved or reflexed; lower surface (except midveins) either completely enclosed by margins and the 2 grooves pubescent with curly hairs, or slightly exposed at sinuses on broader lobes and tomentose with curly hairs. Flower colour: perianth creamy-brown to off-white; style scarlet or orange-red throughout.

Restricted mainly to the area E and NE of Ravensthorpe, W.A., with outliers towards Lake King. Flowers mainly Aug.–Nov., sporadic in other months. Map 34.

W.A.: Woodenup Ck, c. 13 km NE of Ravensthorpe, *A.S.George 9461* (PERTH); 16 km E of Jerdacuttup R. on road to Esperance, *D.J.McGillivray 3594 & A.S.George* (CANB, E, LE *n.v.*, NSW, NY *n.v.*, PERTH).

30b. Grevillea rigida subsp. **distans** Olde & Marriott, *Grevillea Book* 1: 187 (1994)

T: West R., W of Ravensthorpe on Highway 1, W.A., 11 Oct. 1992, *P.M.Olde 92/266*; holo: NSW.

G. tetragonoloba 'race c' or 'pectinate race', of D.J.McGillivray & R.O.Makinson (*loc. cit.*).

Illustrations: P.M.Olde & N.R.Marriott, *Grevillea Book* 3: 138 (bottom right), 139 (110A, B) (1995).

Leaves 3–6 cm long; lobes (5–) 7–9, 15–45 mm long, 1.3–1.9 mm wide, spread over apical half or third of leaf rachis; terminal lobe erect or incurved or occasionally recurved, not reflexed; lower surface (except midveins) completely enclosed by margins, with the 2 grooves pubescent with curly hairs. Flower colour: perianth creamy-brown to off-white; style scarlet or orange-red throughout.

Occurs from Fitzgerald River Natl Park N almost to Ravensthorpe and W to Jerramungup, W.A. Flowers mainly Aug.–Mar. Map 35.

W.A.: West R., c. 32 km SW of Ravensthorpe, *N.N.Donner 1537* (AD); granite rise E of Phillips R. and c. 13 km by road from Ravensthorpe, *D.J.McGillivray 3581 & A.S.George* (K, NSW, PERTH, US *n.v.*); between Ravensthorpe and Ongerup, 100 km W of Ravensthorpe, 3 Sept. 1947, *J.H.Willis* (MEL).

Plants closely approaching the type subspecies on some of the usually diagnostic features do occur (e.g. in area between Ravensthorpe and Phillips R.), as do plants approaching *G. fastigiata*.

31. Grevillea fastigiata Olde & Marriott, *Grevillea Book* 1: 186 (1994)

T: 20.8 km E of Ravensthorpe, W.A., 12 Oct. 1991, *P.M.Olde 91/305*; holo: NSW.

G. tetragonoloba 'race e', 'short-lobed race', of D.J.McGillivray & R.O.Makinson, *Grevillea* 71 (1993).

Illustrations: P.M.Olde & N.R.Marriott, *Grevillea Book* 2: 162 (top left & 132) (1995).

Erect shrub 2–2.5 m tall, with fastigiate (clustered) branches. Leaves 1.8–4.5 cm long, sometimes linear, otherwise 3–7-partite with rigid linear lobes clustered at apex of leaf rachis; lobes 5–20 mm long, 0.8–1.2 mm wide, rectangular in cross-section, pungent; margins angularly twice refracted, enclosing lower surface except for midveins. Unit conflorescence erect, secund; floral rachis 45–80 mm long. Flower colour: perianth pale brown to rusty brown; style orange-red to scarlet throughout. Perianth tomentose-villous outside, especially villous on limb. Pistil 19–22 mm long; ovary sessile or subsessile; style with a sparse indumentum of minute curled or erect hairs on ventral side. Follicle c. 10 mm long, initially subsericeous with biramous hairs but near maturity these falling and being replaced with a pubescent indumentum of erect simple glandular hairs.

Occurs in south-western W.A., where it is apparently restricted to the upper Jerdacuttup R. catchment E of Ravensthorpe. Grows in mallee heath or tall shrubland in granitic loam soil. Regenerates from seed. Flowers Sept.–Mar. Map 36.

W.A.: Jerdacuttup R., *C.A.Gardner 13760* (PERTH); Phillips R., *A.J.Cough 26* (PERTH).

Very closely related to *G. rigida*, which has the floral rachis white-villous (rusty-villous in *G. fastigiata*), and the fructual indumentum mainly of biramous hairs; also close to *G. tetragonoloba* (see under that species for differences), and *G. wittweri* which has smoothly revolute leaf margins and usually secondary leaf division. *Grevillea fastigiata* also has leaf venation evident above; pedicels c. 1.5 mm long; and floral bracts ovate to broadly obovate, 2–2.5 mm long.

32. Grevillea wittweri McGill., *New Names Grevillea* 16 (1986)

T: 1 km NW of Lake Cairlocup, c. 40 km NW of Ongerup, W.A., 28 Nov. 1975, *K.Newbey 4927*; holo: PERTH; iso: CANB, NSW.

[*G. armigera auct. non* Meisn.: J.S.Beard, *Descr. Cat. W. Austral. Pl.* 2nd edn, 38 (1970), *p.p.*]

Illustrations: D.J.McGillivray & R.O.Makinson, *Grevillea* 73 (1993); P.M.Olde & N.R.Marriott, *Grevillea Book* 3: 235 (top right & 192A–C) (1995).

Shrub 1–2.2 m high, to 3 m across. Leaves (2.5–) 4–8 cm long, pinnatipartite to biternate with 3–8 primary lobes, these usually again 2–4-partite; ultimate lobes rather divaricate, linear, subterete to subelliptic in cross-section, 5–35 mm long, 0.8–1.2 mm wide, pungent; margins ±smoothly to angularly revolute, enclosing lower surface except for midveins. Unit conflorescence erect, secund; floral rachis (30–) 50–75 mm long. Flower colour: perianth greenish to fawn-pink; style crimson to light burgundy with a green tip. Perianth loosely tomentose to subvillous outside (biramous hairs), sometimes also with short erect simple glandular hairs. Pistil 18–21 mm long; ovary ±sessile; style with scattered minute erect simple hairs about middle on ventral side. Follicle 12–14 mm long, tomentose with both biramous and erect simple glandular hairs.

Occurs in south-western W.A., in the area bounded by Gnowangerup, Newdegate, Lake Johnston and Ravensthorpe. Grows in mallee shrubland and shrubland on sand plains over clay, often near salt lakes. Regenerates from seed. Flowers Sept.–Apr. Map 37.

W.A.: 27 km from Ravensthorpe towards Lake King, *E.M.Canning CBG033931* (CANB); 42 km N of Ravensthorpe, *A.S.George 6091* (PERTH); E of Hyden, 9 Oct. 1965, *F.W.Humphreys* (PERTH); 15 km W of crossroads E of Lake King, *D.J.McGillivray 3564 & A.S.George* (NSW, PERTH); 8 km WSW of L. Cronin, *K.Newbey 6219* (PERTH).

This species has the venation of the upper leaf surface obvious (5–7 distinct longitudinal veins), and the stipe obscure (0.1–0.5 mm long); these features, and the secondary division of most leaves, together or separately serve to distinguish it from *G. beardiana*, *G. cagiana*, *G. baxteri*, *G. coccinea* and *G. concinna*. It is most closely related to *G. tetragonoloba*, *G. rigida* and *G. fastigiata*, all of which have the leaf margins angularly twice-refracted with the leaf lobes rectangular in cross-section. *Grevillea hookeriana* has the venation of the leaf upper surface obscure and the styles usually maroon-black. *Grevillea wittweri* also has pedicels 1–2 mm long; and floral bracts ovate, 0.8–1.5 mm long.

This species is recognised as 'Poorly Known' in J.D.Briggs & J.H.Leigh, *Rare or Threatened Australian Plants* (1995).

33. Grevillea coccinea Meisn., *Hooker's J. Bot. Kew Gard. Misc.* 7: 76 (1855)

T: '... 7 mill. a monte Manypeak l. Tjilberup, d. 22 Novbr. 1840.' W.A., *L.Preiss. No 711*; lecto: NY *n.v.*, *fide* D.J.McGillivray & R.O.Makinson, *Grevillea* 411 (1993); isolecto: LD *n.v.*

G. coccinea Meisn., *in* J.G.C.Lehmann, *Pl. Preiss.* 1: 545 (1845), *nom. nud. sub G. concinna.*

G. hewardiana Meisn., *in* A.L.P.P. de Candolle, *Prodr.* 14: 366 (1856). T: South West Australia, W.A., *J.Drummond V: 404*; lecto: K, *fide* D.J.McGillivray & R.O.Makinson, *Grevillea* 411 (1993); isolecto: BM, CGE *n.v.*, FI *n.v.*, G *n.v.*, LE *n.v.*, MEL, P *n.v.*, TCD *n.v.*

G. concinna var. *racemosa* Benth., *Fl. Austral.* 5: 432 (1870). T: South West Australia, W.A., *J.Drummond V: 404*; lecto: K, *fide* D.J.McGillivray & R.O.Makinson, *Grevillea* 411 (1993); isolecto: BM, CGE *n.v.*, FI *n.v.*, G, LE *n.v.*, MEL, P *n.v.*, TCD *n.v.*

[*G. concinna auct. non* R.Br.: W.E.Blackall & B.J.Grieve, *How to Know W. Austral. Wildfl.* 1: 108 (1954), *p.p.*]

Decumbent or sprawling to erect shrub, 0.7–3 m high. Leaves often gently upcurved, entire, very narrowly cuneate to linear, 2.5–12.5 cm long, 1.0–4.5 mm wide; apex usually pungent; margins angularly refracted; lower surface usually enclosed except for midvein, tomentose when visible. Unit conflorescence erect, secund; floral rachis 25–65 mm long. Perianth either sericeous to shortly tomentose, or subvillous-sublanate outside. Pistil 19–23.5 mm long; stipe 0.4–1.6 mm long; style glabrous. Follicle 10.5–16 mm long, subsericeous.

Occurs in south-western W.A., from near Mt Manypeaks to the Hopetoun area. Grows in shrub or heath associations in sandy soils. Regeneration mode unknown. Two subspecies are recognised.

Grevillea coccinea has an erect pollen-presenter, a feature unusual in the subgroup.

Perianth 1–2 mm wide, sericeous to shortly tomentose outside; torus 1.0–1.5 mm across (dorsal to ventral) **33a.** subsp. **coccinea**

Perianth 3–3.5 mm wide, subvillous to sublanate outside; torus c. 1.6 mm across **33b.** subsp. **lanata**

33a. **Grevillea coccinea** Meisn. subsp. **coccinea**

Illustrations: J.W.Wrigley & M.Fagg, *Banksias, Waratahs & Grevilleas* 224 (1989); D.J.McGillivray & R.O.Makinson, *Grevillea* 74 (1993); P.M.Olde & N.R.Marriott, *Grevillea Book* 2: 93 (top right & 74A, B), 94 (74C) (1995).

Shrub 1–3 m tall. Leaves 2.5–12.5 cm long, 1–4.5 mm wide. Floral bracts 2.4–4.8 mm long. Torus 1.0–1.5 mm across. Flower colour: perianth greenish cream to pale brown becoming pinkish; style red or rarely yellow. Perianth 1–2 mm wide, sericeous to shortly tomentose outside. Pistil 19–23.5 mm long. Stipe 0.5–1.6 mm long.

Widespread S from the Ravensthorpe Ra. to the coast and W towards Mt Manypeaks, W.A. Flowers July–Dec., sporadic in other months. Map 38.

W.A.: c. 30 km due S of Reynolds Hill on road S from Jerramungup, *D.J.McGillivray 3507 & A.S.George* (CANB, K, NSW, PERTH, US *n.v.*); N side of Mt Short, *D.J.McGillivray 3576 & A.S.George* (K, NSW, NY *n.v.*, PERTH); NW corner of Fitzgerald River Natl Park, *R.A.Saffrey 1480* (PERTH); Hopetoun, *E.Wittwer 1201* (PERTH).

A 'broad-leaf form' occurs in the western part of the range, in the area from Mt Manypeaks to Bremer Bay; it has leaves usually 2–5 mm wide and the lower surface partly exposed on either side of the midvein. A 'narrow-leaved form' occurs in the eastern part of the range, from Mt Short to East Mt Barren; it has leaves usually < 1.5 mm wide and the lower surface (except for the midvein) enclosed by the margins.

33b. **Grevillea coccinea** subsp. **lanata** Olde & Marriott, *Nuytsia* 9: 277 (1993)

T: Middle Mt Barren Reserve 24048, W.A., 16 July 1970, *A.S.George 10104*; holo: PERTH.

Illustration: P.M.Olde & N.R.Marriott, *Grevillea Book* 2: 94 (74D) (1995).

Shrub 1–2 (?) m tall. Leaves 8–12 cm long, 1.8–2.5 mm wide. Floral bracts c. 6 mm long. Flower colour: perianth greenish cream to pale brown becoming pinkish; style red or rarely yellow. Perianth 3–3.5 mm wide, subvillous to sublanate outside. Torus c. 1.6 mm across. Pistil c. 20 mm long. Stipe 0.4–0.5 mm long.

Occurring in south-western W.A. from Thumb Peak to Middle Mt Barren in Fitzgerald River Natl Park. Flowers recorded July and Oct. Map 39.

W.A.: Thumb Peak, *A.S.George 7126* (PERTH).

Known from very few collections; characters may be more variable than indicated. This subspecies is recognised as 'Poorly Known' in J.D.Briggs & J.H.Leigh, *Rare or Threatened Australian Plants* (1995).

34. **Grevillea concinna** R.Br., *Trans. Linn. Soc. London* 10: 172 (1810)

T: 'In Novae Hollandiae ora australis; Lewins Land ...' [protologue]; lecto: Cape Leeuwin, Bay 1 [Lucky Bay, W.A.], 11 Jan. 1802, *R.Brown Iter. Austral. 3342*; lecto: BM, *fide* D.J.McGillivray & R.O.Makinson, *Grevillea* 411 (1993); isolecto: BM, E, G-DC, K, LE *n.v.*, MEL, NSW, P *n.v.*

Spreading to erect shrub 0.3–1.6 m high. Leaves entire and linear to narrowly cuneate or narrowly obovate, or occasionally 2- or 3 (–5)-partite with lobes similar, 2–7 cm long, 0.9–4.5 mm wide, sometimes pungent; margins usually angularly revolute and enclosing lower surface, occasionally loosely revolute with lower surface partly exposed; lower surface subsericeous to tomentose. Unit conflorescence deflexed, secund; floral rachis 10–30 mm long. Perianth subsericeous or loosely so outside. Pistil (20–) 23–26 (–30) mm long; stipe 0.9–2.5 mm long; style glabrous or sometimes minutely subpapillose about middle on ventral side. Follicle 10–14.5 mm long, loosely and shortly villous. *Elegant Grevillea.*

Occurs in south-western W.A. There are two subspecies.

Entire leaves either linear and subterete with margins smoothly revolute and enclosing lower surface (sometimes including midvein), or narrowly cuneate to narrowly obovate with margins smoothly recurved except for a sharp submarginal refraction; lower surface subsericeous with straight hairs ..... **34a.** subsp. **concinna**

Entire leaves linear or almost so and angular in cross-section, with angularly revolute margins enclosing lower surface (except midvein); lower surface with flexuose or curled hairs ..... **34b.** subsp. **lemanniana**

34a. **Grevillea concinna** R.Br. subsp. **concinna**

Illustrations: R.Sweet, *Fl. Australas.* t. 7 (1827), as *G. concinna*; D.J.McGillivray & R.O.Makinson, *Grevillea* 75, col. pl., 76, fig. 12a–d (1993); P.M.Olde & N.R.Marriott, *Grevillea Book* 1: 25 (1994); 2: 96 (top right & 76A, B), 97 (76C) (1995).

Entire leaves either linear and subterete with margins smoothly revolute, enclosing lower surface (sometimes including midvein), or narrowly cuneate to narrowly obovate, with margins smoothly recurved except submarginally and midvein exposed; lower surface subsericeous with straight hairs. Flower colour: perianth silvery to cream or yellowish green; style usually bright red to mid-pink, rarely pale pink or yellow.

Occurs in coastal south-western W.A. in a small area between Cape Le Grand and Lucky Bay, E of Esperance. Grows in exposed coastal shrubland in granitic sandy soils. Regeneration mode unknown. Flowers mainly Sept.–Dec. Map 40.

W.A.: shoulder of Mt Le Grand, *A.S.George 2224* (PERTH); Lucky Bay, *E.M.Scrymgeour 896B* (PERTH); Le Grand, 9 Oct. 1952, *N.H.Speck* (CANB); Cape Le Grand, *J.W.Wrigley CBG036895* (CANB).

34b. **Grevillea concinna** subsp. **lemanniana** (Meisn.) McGill., *New Names Grevillea* 4 (1986)

G. lemanniana Meisn., *in* A.L.P.P. de Candolle, *Prodr.* 14: 366 (1856). T: Swan R., W.A., 11 Oct. 1853, *J.Drummond V: 405*; lecto: NY *n.v.*, *fide* D.J.McGillivray & R.O.Makinson, *Grevillea* 411 (1993); isolecto: K, MEL, P.

Illustrations: W.R.Elliot & D.L.Jones, *Encycl. Austral. Pl.* 5: 42 (1990); D.J.McGillivray & R.O.Makinson, *Grevillea* 41, 76, fig. 12e–h (1993); P.M.Olde & N.R.Marriott, *Grevillea Book* 2: 97 (bottom centre & 77) (1995).

Entire leaves linear or almost so, angular in cross-section, with angular-revolute (often box-like) margins enclosing lower surface (except for midvein); lower surface with flexuose or curled hairs. Flower colour: perianth lemonish green becoming orange-brown especially towards limb; style bright red. Plate 3.

Occurs in south-western W.A. in the southern sub-coastal region from near Boxwood Hill (W of Bremer Bay) inland as far as Needilup, and E to Ravensthorpe and Mt Ragged. Grows in heath, open scrub or tall shrubland in sandy (siliceous) soils, sometimes sandy loam, sand over laterite, or granitic soil. Regenerates from seed. Flowers mostly July–Jan. Map 41.

W.A.: SW slopes of Mt Ragged, *B.Barnsley 300* (CANB, NSW, PERTH); 10 km ESE of Howick Hill, *N.N.Donner 2647* (AD, PERTH); 3 km W of Needilup, *D.J.McGillivray 3514 & A.S.George* (CANB, NSW, PERTH, RSA *n.v.*); 2 km SW of Mt Ragged, *D.J.McGillivray 3614 & A.S.George* (K, MEL, NSW, NY *n.v.*, PERTH); Point Gordon, Bremer Bay, 7 Dec. 1862, *G.Maxwell* (MEL).

Subsp. *lemanniana* varies moderately on a populational basis in habit, leaf proportions (plants with broader sometimes divided leaves occur at the extreme E of the range), and leaf surface (glaucous leaved plants are known from N of Ravensthorpe).

35. **Grevillea beardiana** McGill., *New Names Grevillea* 2 (1986)

T: 28 miles [45 km] E of Newdegate, W.A., 30 Oct. 1962, *J.S.Beard 2179*; holo: PERTH; iso: KPBG.

[*G. concinna auct. non* R.Br.: W.E.Blackall & B.J.Grieve, *How to Know W. Australian Wildfl.* 1: 108 (1954), *p.p.*; J.S.Beard (ed.), *Descr. Cat. W. Austral. Pl.* 21 (1965), *p.p.*, 2nd edn, 38 (1970), *p.p.*]

Illustrations: R.Erickson *et al.*, *Fls & Pls W. Australia* 80, pl. 220 (1973, 1979), as *G. concinna*; J.W.Wrigley & M.Fagg, *Banksias, Waratahs & Grevilleas* 223 (1989); P.M.Olde & N.R.Marriott, *Grevillea Book* 2: 57 (bottom right) 58 (42A–C) (1995).

Spreading shrub 0.3–1.5 m tall. Leaves entire, linear to very narrowly cuneate, 1–5.5 cm long, 0.7–2.0 mm wide, sometimes also a few leaves 2- or 3-partite with similar ascending lobes, sometimes pungent; margins smoothly revolute; lower surface enclosed except for midvein, packed with appressed hairs in grooves. Unit conflorescence erect to deflexed, secund; floral rachis 18–45 mm long. Flower colour: perianth bright red or orange; style bright red or orange with a green to pale tip. Perianth subsericeous outside.

Pistil 28–31.5 mm long; stipe 2.5–5.1 mm long; style glabrous. Follicle 7.5–11.5 mm long, tomentose with biramous hairs only.

Occurs in south-western W.A., from Newdegate area E to about 50 km W of Kumarl, and N to Lake Johnston. Grows in heathy associations in sandy soils over laterite, and on lateritic rises; occasionally in mallee heaths and in granitic soils. Probably regenerates from seed. Flowers mainly Sept.–Dec. Map 42.

W.A.: between Mt Madden & Baanga Hill, *E.M.Canning CBG036624* (CANB); Kumarl to Lake King road, 52 km W of Kumarl, *R.Filson 9334* (MEL, NSW); 8 km E of Newdegate, *C.A.Gardner 1759* (PERTH); 4 km SW of McDermid Rock, *T.F.Houston 210-10* (PERTH); 16–19 km from Lake King along road to Lake Grace, *A.Strid 21073* (C *n.v.*, NSW).

This species has a distinctively long ovarian stipe (2.5–5.1 mm long) which distinguishes it from several similar species. It is sometimes confused with *G. concinna* which has a stipe < 2.5 mm long and (subsp. *lemanniana* only) angularly refracted leaf margins, and with *G. cagiana* which has the leaves often divided and the pistil < 25 mm long.

Populations at the eastern end of the range have consistently narrower leaves with a reduced lamina and pronounced midvein on the lower surface; the leaves are approaching the dipleural state. In the western part of the range many populations have leaves at the broad end of the variance.

36. Grevillea armigera Meisn., *in* A.L.P.P. de Candolle, *Prodr.* 14: 373 (1856)

T: 'In colonia Swan River (Drumm. coll. 4, n. 284!) ... v.s. in herb. Shuttl.' [protologue]; lecto: S.W. Australia, [coll. 1844?], [J.] *Drummond IV: 284*; lecto: K, *fide* D.J.McGillivray & R.O.Makinson, *Grevillea* 405 (1993); isolecto: BM, CGE *n.v.*, FI *n.v.*, G, K, LE *n.v.*, MEL, NY *n.v.*, P *n.v.*, TCD *n.v.*

Illustrations: W.R.Elliot & D.L.Jones, *Encycl. Austral. Pl.* 5: 27 (1990); P.M.Olde & N.R.Marriott, *Grevillea Book* 2: 38 (bottom left & 26A), 39 (26B, C) (1995).

Shrub 0.5–3.5 m high. Leaves 2.5–5 cm long, pinnatipartite and mostly partly bipinnatipartite, with rachis angularly deflexed at each node; primary lobes 5–11, divaricate, at least the basal lobes usually again 3–5-partite; ultimate lobes linear, 0.2–2.5 cm long, 0.5–1.5 mm wide, divaricate, pungent; margins tightly and smoothly revolute; lower surface enclosed except for midvein, packed with appressed flexuose hairs in grooves. Unit conflorescence erect, secund; floral rachis 50–90 mm long. Flower colour: perianth grey, green, or pale yellow, sometimes flushed pink; style black to blackish maroon. Perianth subsericeous outside. Pistil 18–23 mm long; ovary ±sessile; style glabrous except for appressed hairs just above ovary. Follicle 11.5–14 mm long, usually tomentose to pubescent with biramous hairs, sometimes also with small simple erect hairs. *Thorny Grevillea, Prickly Toothbrushes.*

Occurs in south-western W.A., in the northern wheatbelt region from Buntine to near Dowerin. Grows in heath or shrubland usually in deep sandy soil, often yellow (occasionally gravelly) sand. Regenerates from seed. Flowers mainly June–Feb. Map 43.

W.A.: 22.5 km W of Ballidu on Ballidu to Bindi Bindi road, *K.F.Kenneally 1102* (PERTH); 2 km N of Cadoux, *D.J.McGillivray 3423 & A.S.George* (K, NSW, PERTH, PRE *n.v.*, US *n.v.*); Mortlock Creek Flora Reserve, Wongan Hills, *B.G.Muir 300* (PERTH); 16 km W of Dowerin, *K.Newbey 1965* (PERTH); 6.5 km E of Ballidu, *R.D.Royce 2202* (PERTH).

Not easily confused with other species. *Grevillea spinosa* (known in the past as *G. sp. aff. armigera*) has most leaves simple or with primary division only, the nectary erect and appressed to the ovary, and the style yellow to orange (*G. armigera* has the nectary spreading and tongue-like, with the margins entire). *Grevillea hookeriana*, *G. crowleyae* and *G. calliantha* all have leaves with primary division only and non-divaricate lobes.

37. Grevillea hookeriana Meisn., *in* J.G.C.Lehmann, *Pl. Preiss.* 1: 546 (1845)

T: Swan R., W.A., [?1844] *J.Drummond IV: 633*; lecto: NY *n.v.*, *fide* D.J.McGillivray & R.O.Makinson, *Grevillea* 420 (1993); isolecto: CGE *n.v.*, G *n.v.*, K, LE *n.v.*, MEL, P *n.v.*

Shrub 0.5–2.5 m high, to 4 m across. Leaves 1–13.5 cm long, sometimes entire and linear and 0.6–2.6 mm across, and/or variously divided, usually either deeply 2–9-partite or -sect with ascending linear lobes 0.8–1.9 mm across and basal lobes rarely again 2- or 3-partite, or narrowly cuneate and apically digitate, to c. 10 mm wide (at base of lobes) with 3–10 (–23) short ascending apical lobes or teeth; apices ±pungent; margins smoothly revolute; lower surface usually enclosed except for midvein, densely tomentose with flexuose and/or curled hairs in grooves especially near lobe sinuses. Unit conflorescence erect, secund; floral rachis 25–80 mm long. Perianth subsericeous to villous. Pistil 18–21.5 (–23) mm long; ovary ±sessile; style usually glabrous, rarely a few papillae or minute simple erect hairs near middle of ventral side. Follicle 12–18 mm long, tomentose with biramous hairs and sometimes also simple erect glandular hairs. *Red Toothbrushes, Hooker's Grevillea.*

Occurs in south-western W.A., where widespread throughout much of the wheatbelt in the area bounded by Three Springs, Mt Churchman, Coolgardie and Katanning. Grows in heath or shrubland in sandy soils.

Grevillea hookeriana is a superspecies with several inadequately resolved elements, differing mainly in leaf morphology and to some extent in habit and fruit indumentum; these generally have coherent geographical distributions. Some of these elements have been regarded as separate species in the past but there are several intergrades and the morphological diagnoses usually advanced are not sufficiently consistent to maintain species status at this time. The complex would repay closer research.

The species as a whole has leaves with venation of the upper surface obscure, smoothly rounded margins, and laminas usually entire or with primary division only (rarely the basal lobes again 2- or 3-partite); divided leaves flat to subsecund (lobes not divaricate, and leaf rachis not deflexing at each node); unit conflorescence erect; floral bracts usually acute to apiculate, ovate to narrowly rhomboid, 1.5–3.5 mm long; floral rachis tomentose to shortly villous; nectary prominent, linguiform; pistil 18–21.5 mm long; ovary ±sessile; and styles (usually) maroon to almost black. These character states serve to distinguish it from the most closely related species *G. crowleyae* and *G. calliantha* (both of which have pistils ⩾ 23 mm long), and *G. armigera* (which has leaves with secondary division, leaf rachis deflexing at each node, and floral bracts broadly ovate and ⩽ 1.5 mm long), as well as more distant relatives such as *G. baxteri*, *G. cagiana*, *G. coccinea*, *G. concinna*, *G. tetragonoloba*, *G. rigida*, *G. fastigiata*, *G. beardiana* and *G. aneura* (all of which are different in some or all of these states).

A *Grevillea* with pinnatipartite very regular 'herringbone' leaves and red styles, common in horticulture and sold for many years as '*G. hookerana*' [sic] and more recently as G. 'Red Hooks', is a sterile hybrid with a parentage probably involving *G. tetragonoloba*; it is illustrated in J.W.Wrigley & M.Fagg, *Banksias, Waratahs & Grevilleas* 243 (1989), as '*G. hookeriana* hybrid'.

Three subspecies are recognised; these intergrade considerably, with morphologically intermediate or ambiguous plants and populations being especially common in the area bounded by Wongan Hills, Merredin and Cunderdin.

1 Either most leaves entire and linear with only a few lobed, or most leaves deeply pinnatipartite with the rachis linear and 3–9 linear lobes (not crowded near apex) with only a few entire; margins tightly and smoothly revolute, usually completely enclosing the lower surface except for midveins, occasionally a narrow area of lamina exposed near sinuses …… **37a.** subsp. **hookeriana**

1: Adult leaves cuneate to narrowly (rarely broadly) so, apically lobed or digitate with 3–10 (–23) linear to oblong or subtriangular teeth or lobes all crowded on the apical few millimetres of the leaf rachis; margins tightly and smoothly revolute, enclosing the lower surface except for midveins of leaves, usually with narrow areas of lamina exposed near sinuses

2 Most leaves > 3.5 cm long; conflorescences usually partly enclosed within foliage …… **37b.** subsp. **apiciloba**

2: Leaves 1–2 cm long; conflorescences usually clearly exceeding foliage …… **37c.** subsp. **digitata**

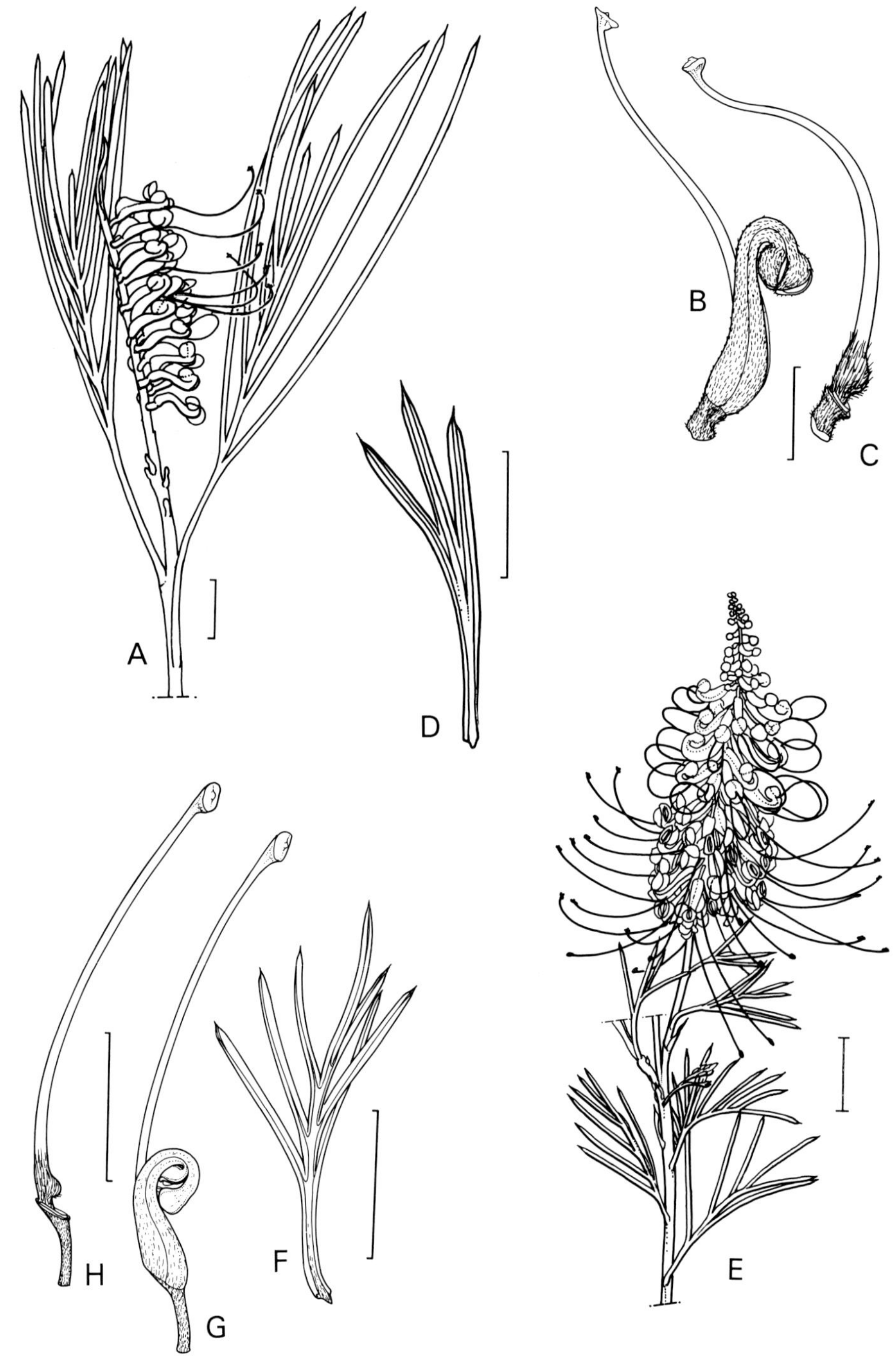

Figure 6. *Grevillea*. **A–C**, *G. hookeriana* subsp. *hookeriana*. **A**, flowering branch; **B**, flower; **C**, pistil (**A–C**, J.W.Green 5123, PERTH). **D**, *G. hookeriana* inter subsp. *apiciloba* and subsp. *digitata*, leaf (H.Staufer & R.Saffery 5325, CANB). **E–H**, *G. tenuiloba*. **E**, flowering branch; **F**, leaf; **G**, flower; **H**, pistil (M.E.Ballingal 1937, CANB). Scale bars: **A–B**, **E–H** = 1 cm; **C–D** = 5 mm. Drawn by D.Boyer.

37a. Grevillea hookeriana Meisn. subsp. **hookeriana**

G. pritzelii Diels, *Bot. Jahrb. Syst.* 35: 150 (1904). T: Tammin, W.A., Oct. 1901, *E.Pritzel 753*; lecto: B *n.v.*, *fide* D.J.McGillivray & R.O.Makinson, *Grevillea* 420 (1993); isolecto: A *n.v.*, AD, B *n.v.*, BM, E, G, K, L *n.v.*, NSW.

G. hookeriana races 'a', 'b', 'c', of D.J.McGillivray & R.O.Makinson, *Grevillea* 81 (1993).

Illustrations: W.R.Elliot & D.L.Jones, *Encycl. Austral. Pl.* 5: 63 (1990); P.M.Olde & N.R.Marriott, *Grevillea Book* 2: 199 (bottom left & 164A, B), 200 (164C, D) (1995).

Spreading shrubs 1.5–2.5 m tall. Juvenile leaves pinnatipartite. Adult leaves either entire and linear, 3.5–13.5 cm long, 1–2 mm wide, with occasional 2- or 3-partite leaves, or 3–9-partite (sometimes with occasional entire, linear leaves) with a linear base and linear lobes; primary lobes (when present) simple or rarely the lowermost 2- or 3-partite, usually the apical 3–5 lobes subdivaricate; ultimate lobes 3–7 cm long, 0.8–1.0 mm wide; margins tightly and smoothly revolute; lower surface completely enclosed except for midveins (or lamina rarely narrowly exposed near sinuses); revolute margins not or scarcely decurrent below sinuses. Unit conflorescence often exceeding foliage; floral rachis 25–80 mm long, Flower colour: perianth yellowish green to grey-fawn, rarely dull pink or dull reddish; style black, purplish black, or deep maroon, rarely red or yellow; style tip green. Follicle tomentose, usually with biramous hairs only, or occasionally (narrow-lobed form) also with a few small erect simple glandular hairs. Fig. 6A–C.

Occurs in south-western W.A. in the area bounded by Coorow, Katanning, Newdegate and Merredin. Grows in sand-heath or mixed shrubland or occasionally eucalypt woodland, usually in yellow-sand soils, occasionally on laterite. Regenerates from seed. Flowers mainly May–Nov. Map 44.

W.A.: 5 km E of Katanning, *B.Barnsley 654* (CANB, NSW, PERTH); near Kulin, *C.A.Gardner 12209* (PERTH); 28 km W of Wongan Hills on Wongan Hills to Piawaning road, *T.A.Halliday 124* (AK *n.v.*, CANB, PERTH); 18 km WNW of Coorow, *J.Taylor 975* (CANB, NSW, PERTH); c. 16 km E of Lake Grace, *D.J.E.Whibley 5350* (AD, PERTH).

Subsp. *hookeriana* is variable, and four forms may be distinguished, mainly on foliar characters; intermediates are fairly frequent.

(a) The 'long-lobed form' has leaves deeply pinnatipartite, usually with 5–7 primary lobes (occasionally some 2- or 3-partite), occasionally the lowest lobes again bipartite; longest leaves 6.5–11 cm long; ultimate lobes 1.0 mm wide, maximum length 4–6.5 cm; perianth villous outside. There is occasionally a narrow area of the lower leaf lamina visible near the sinuses, a feature rare in this subspecies but common in subsp. *apiciloba* and subsp. *digitata*. This form includes the collection *Drummond 633*, the type collection of *G. hookeriana* Meisn. It occurs in the NW of the species range, in the area between Winchester and Wongan Hills, with morphological intergrades to the 'simple-leaved form' between Dowerin and Piawaning. This form corresponds to 'race a' of McGillivray & Makinson (*loc. cit.*).

(b) The 'narrow-lobed form' has leaves deeply pinnatipartite with 3–9 primary lobes, occasionally some leaves simple, occasionally the lowermost lobes again bipartite; longest leaves 3.5–6.5 cm long; ultimate lobes 0.8–1.0 mm wide, maximum length 3–4 cm; perianth usually subsericeous to tomentose. The lower surface of the leaf on either side of the main veins is completely enclosed by the revolute margins. This form occurs in the S of the species range, in the area bounded by Pingelly–Katanning–Newdegate, with intergradation to the 'simple-leaved form' in the Corrigin area. This form corresponds to 'race b' of McGillivray & Makinson (*loc. cit.*).

(c) The 'simple-leaved form' has predominantly entire linear to very narrowly cuneate leaves, sometimes with a few leaves 2- or 3-partite; leaves 5.5–12 cm long; leaves (or lobes) 1.1–1.9 mm wide; lower leaf surface usually fully enclosed. The perianth is usually subsericeous to villous. This form occurs in the centre of the species range, in the area between Kellerberrin, Corrigin, Cunderdin and Dowerin, intergrading with the 'long-lobed form' to the north and the 'narrow-lobed form' to the south. This form includes the type of *G. pritzelii* Diels, and corresponds to 'race c' of McGillivray & Makinson (*loc. cit.*).

surface. Follicle 13–18 mm long, tomentose with biramous hairs of ovary mostly caducous and replaced with short simple erect glandular hairs (often matted with exudate). *Foote's Grevillea, Cataby Grevillea, Black Magic Grevillea.*

Occurs in south-western W.A., where it is known only from small populations near Cataby, S of Badgingarra. Grows in remnant heathy associations in sandy soils over lateritic gravel. Probably regenerates mainly from seed but has some ability to resprout from stem after low-intensity fires. Flowers mainly Aug.–Nov. Map 48.

W.A.: near Cataby, *B.J.Conn 3278 & J.A.Scott* (AD, BRI, CANB, HO, MEL, NSW, PERTH); Cataby [area], *S.Hopper 6350* (AD, BRI, CANB, DNA, HO, NSW, PERTH).

Grevillea calliantha has leaves with venation distinct or obscure above; floral bracts ovate; short pedicels 1–2 mm long; and a spreading, tongue-like nectary, projecting ventrally beyond the toral rim. It is closely related to both *G. hookeriana*, which has pistils ⩽ 23 mm long and fruits with biramous hairs remaining dominant; and to *G. crowleyae* which has a shorter conflorescence (2–5 cm long), leaf lobes < 1 mm wide, and fruits with biramous hairs remaining dominant.

40. Grevillea treueriana F.Muell., *Fragm.* 9: 123 (1875)

T: between Ooldea ('Youldeh') and the Elizabeth R. ('Beda R.'), S.A., [probably on Mt Finke, 20 June 1875], [*J.*] *Young s.n.*; holo: MEL; iso: AD.

Illustrations: J.P.Jessop & H.R.Toelken (eds), *Fl. S. Australia* 4th edn, 1: 129, fig. 69E (1986); P.M.Olde & N.R.Marriott, *Grevillea Book* 3: 206 (top left & 168A–C) (1995).

Shrub to 1.5 m high, to 2 m across. Leaves 2.5–3.5 cm long, mostly bipinnatisect, with 3–7 divaricate primary lobes, and lateral ones usually again divaricately 2- or 3-partite; rachis angularly deflexed at each node; ultimate lobes linear to subulate, 0.5–1.7 cm long, 0.7–1.0 mm wide, pungent; margins angularly revolute; lower surface enclosed except for prominent midvein, with grooves packed with flexuous to curly hairs. Unit conflorescence erect to incurved, secund; floral rachis 35–65 mm long. Flower colour: perianth bright scarlet, usually with a yellow limb; style red. Perianth subsericeous outside. Pistil 23–27 mm long; stipe c. 3 mm long; style glabrous. Follicle 10–11 mm long, pubescent. *Mt Finke Grevillea.*

Occurs in central inland S.A., known only from Mt Finke. Grows in rocky crevices in skeletal quartzitic sandy soil. Probably regenerates from seed only. Flowers mainly June–Oct. Map 49.

S.A.: S slope of Mt Finke, *B.Lay 91* (AD).

Grevillea treueriana has the leaf midvein prominent above, the floral rachis usually straight, the pollen-presenter moderately oblique and the ovary villous. It is very closely related to *G. aneura* from W.A., which has longer (to 5 cm) subterete leaf lobes and a longer (27–28 mm) pistil.

This species is recognised as 'Vulnerable' in J.D.Briggs & J.H.Leigh, *Rare or Threatened Australian Plants* (1995).

41. Grevillea aneura McGill., *New Names Grevillea* 1 (1986)

T: 16 miles [25.6 km] E of Red Lake, W.A., 2 Nov. 1967, *J.S.Beard 5422*; holo: PERTH.

Illustrations: W.R.Elliot & D.L.Jones, *Encycl. Austral. Pl.* 5: 24 (1990); D.J.McGillivray & R.O.Makinson, *Grevillea* 83, fig. 13 (1993); P.M.Olde & N.R.Marriott, *Grevillea Book* 2: 26 (bottom right), 27 (17A–C) (1995).

Dense prickly shrub 0.6–2.8 m high. Leaves 4–7.5 cm long, divaricately tripartite to biternate; primary lobes entire to 3-partite; ultimate lobes linear-subterete, 0.5–5 cm long; 0.8–1.1 mm wide, pungent; margins smoothly revolute; lower surface enclosed except for very prominent midvein; grooves ±lateral (lobes and rachis almost dipleural), packed with slender straight hairs. Unit conflorescence erect on a sometimes pendulous branchlet, sometimes gently incurved, secund to subsecund; floral rachis 35–70 mm long. Flower colour: perianth light yellow to light red or red-orange; style light orange to bright red, with a greenish tip. Perianth subsericeous outside. Pistil 27–28 mm long; stipe 3.3–4.3 mm long; style usually glabrous or sometimes with appressed hairs extending for up to 5 mm above ovary. Follicle 10–13 mm long, subsericeous. *Red Lake Grevillea.* Fig. 7A–C.

Occurs in south-western W.A., from just E of Lake King to E of Red Lake, S of Salmon Gums. Grows in heath, mallee scrub or open shrub mallee in sandy soils over laterite or deep sand, often on rises. Regenerates from seed. Flowers mostly Aug.–Jan. Map 50.

W.A.: 11.5 km WSW of Dog Rock, *M.D.Crisp 6020 et al.* (CANB, NSW); 18 km W of Norseman to Esperance road on Ravenswood Rd, *G.J.Keighery 3685* (PERTH); 21 km E of the crossroads just E of Lake King, *D.J.McGillivray 3573 & A.S.George* (NSW, PERTH); 21 km NW of Roberts Swamp, c. 51 km WNW of Grass Patch, *K.Newbey 8149* (PERTH).

This species has rigid leaf lobes with venation obscure above, the floral rachis usually curved, the pollen-presenter very oblique, and the ovary on a distinct villous stipe 3.3–4.3 mm long. It is occasionally confused with *G. armigera* and *G. asparagoides*, both of which have the ovary sessile or subsessile on a very short obscure stipe. *Grevillea aneura* is closely related to *G. treueriana*; see under that species for differences.

Grevillea aneura is recognised as 'Poorly Known' in J.D.Briggs & J.H.Leigh, *Rare or Threatened Australian Plants* (1995).

42. Grevillea tenuiloba C.A.Gardner, *J. Roy. Soc. W. Australia* 19: 81 (1934)

T: Dandaragan [as 'Dandarragan'; loc. possibly erroneous], W.A., Sept. 1932, *W.E.Blackall s.n.*; holo: PERTH; iso: K.

Illustrations: P.M.Olde & N.R.Marriott, *Grevillea Book* 3: 196 (top right & 160A), 197 (160B–D) (1995).

Low spreading shrub 0.4–1.0 m tall. Leaves 1.5–5 cm long, rather pectinate, with rachis strongly decurved, pinnatipartite (almost pinnatisect) and with (2–) 5–11 ascending to spreading lobes; lobes entire, linear-subulate to subterete, 2–25 mm long, 0.6–1.3 mm wide, pungent; margins smoothly revolute, enclosing lower surface (on lobes often including the midvein except in lower half of lobe). Unit conflorescence erect or decurved, often borne laxly on or close to ground, secund; floral rachis 55–135 mm long. Flower colour: perianth and style pale to rich orange. Perianth glabrous outside or with scattered appressed biramous hairs below curve only and denser towards base. Pistil 23–34 mm long; ovary stipitate; style glabrous except for hairs in basal 4 mm. Follicle 11–14 mm long, shortly tomentose with mixed biramous and simple erect glandular hairs. *Amber Grevillea.* Fig. 6E–H.

Occurs in south-western W.A., where currently known from the area between Wongan Hills and Jibberding. A 1932 collection (*W.E.Blackall 2809*; K, PERTH) may represent the correct type locality; the species has not been confirmed near Dandaragan. Grows in mixed *Melaleuca* shrubland, in shallow rocky granitic loam soils. Regenerates from seed. Flowers Aug.–Oct. Map 51.

W.A.: between Mullewa & Morowa, Sept. 1932, *W.E.Blackall 2809* (CANB, K, PERTH); near Morowa, *C.A.Gardner 12016* (PERTH); Jibberding, *C.A.Gardner 12057* (CANB, PERTH); 1.6 km S of hotel on Pindar to Tardun road, *J.W.Glass* (PERTH); Morowa to Mullewa c. 459 km, 17 Sept. 1964, *E.B.J.Smith* (PERTH).

Closely related to *G. nana*, and can be mistaken for the short-lobed subsp. *nana*, which has a hairy perianth limb and large fat-ellipsoidal to hemispherical pithy seed (seed flat-ellipsoidal in *G. tenuiloba*).

This species is recognised as 'Poorly Known' in J.D.Briggs & J.H.Leigh, *Rare or Threatened Australian Plants* (1995).

43. Grevillea nana C.A.Gardner, *J. Roy. Soc. W. Australia* 27: 168 (1942)

T: between Koorda and Bencubbin, W.A., *s.d.*, *W.E.Blackall 3380*; holo: PERTH.

Prostrate to low mounded dense shrub, 0.2–0.5 m tall, 1–2 m across. Leaves 10–30 cm long, pinnatipartite with 5–19 linear entire lobes; lobes either 2–13 cm long and 1.1–1.4 mm wide, or 0.3–2 cm long and 0.7–0.9 mm wide, pungent; margins tightly and smoothly revolute;

lower surface enclosed except for midvein, tomentose in grooves. Unit conflorescence erect or decurved, often lax on or near ground at base of foliage, secund; floral rachis 60–170 mm long. Perianth subsericeous to lanate outside. Pistil 24–28 mm long; stipe 0.4–1.2 mm long; style subsericeous at base with biramous hairs extending up to 8 mm from ovary, otherwise glabrous. Follicle 15–19 mm long, open-tomentose with mixed biramous and simple erect glandular hairs.

Occurs in south-western W.A. Two subspecies are recognised.

Grevillea nana is unique in the group in having almost hemispherical pithy seeds. It also has leaves obscurely veined above, and a V- or U-shaped nectary, not spreading outwards beyond toral rim. Closely related and similar to *G. tenuiloba*; see under that species for differences.

Most leaves 10–30 cm long; leaf lobes 2–13 cm long, 1.1–1.4 mm wide **43a.** subsp. **nana**

Most leaves 3–7 cm long; leaf lobes 0.3–2 cm long, 0.7–0.9 mm wide **43b.** subsp. **abbreviata**

43a. Grevillea nana C.A.Gardner subsp. nana

Illustrations: A.S.George, *Introd. Proteaceae W. Australia* 52, t. 72 (1984), as *G. nana*; W.R.Elliot & D.L.Jones, *Encycl. Austral. Pl.* 5, 84 (1990); D.J.McGillivray & R.O.Makinson, *Grevillea* 367, figs 89, 90, & col. pl. (1993); P.M.Olde & N.R.Marriott, *Grevillea Book* 3: 46 (top centre & 32A, B), 47 (32C, D) (1995).

Prostrate or low mounded shrub. Leaves mostly 10–30 cm long, the long pseudo-petiolate base (below first lobes) 4.5–13 cm long; lobes 5–13, pliable, 2–13 cm long, 1.1–1.4 mm wide. Unit conflorescence 6–17 cm long; peduncle and rachis subsericeous to villous; floral bracts 2.5–4 mm long; pedicels 3.5–6.5 mm long. Flower colour: perianth variably cream-green to pale or greenish pink to scarlet; style pink, purplish pink, red, or pale orange. Perianth subsericeous to tomentose or woolly outside. Pistil 24–28 mm long; ovary stipe 0.4–1.2 mm long. Plate 7; Fig. 7D–F.

Occurs in south-western W.A. in scattered populations in the area from Bullfinch NW to Mt Gibson and W to Manmanning. Grows in open heath or shrubland in shallow granitic soils over outcropping granite. Regenerates from seed. Flowers June–Oct. Map 52.

W.A.: 5.5 km S of Warralakin, *D.J.McGillivray 3693 & A.S.George* (K, NSW, PERTH, US *n.v.*); Manmanning, 23 Sept. 1967, *B.H.Smith per M.Smith* (PERTH); 5 km N of Koorda, Sept. 1970, *B.H.Smith* (PERTH).

There is considerable variation in conflorescence indumentum, toral obliquity, and style colour.

43b. Grevillea nana subsp. abbreviata McGill., *New Names Grevillea* 10 (1986)

T: R.P.F. [Rabbit Proof Fence] Wubin, W.A., 8 Oct. 1963, *F.Lullfitz 2473 & E.Wittwer*; holo: KPBG; iso: NA *n.v.*

Illustrations: D.J.McGillivray & R.O.Makinson, *Grevillea* 368 (1993); P.M.Olde & N.R.Marriott, *Grevillea Book* 3: 47 (bottom right & 32C, D) (1995).

Prostrate or low mounded shrub. Leaves 3–7 cm long, the pseudo-petiolate base (below first lobes) 1.5–2.5 cm long; lobes usually 7–19, rigid, 0.3–2 cm long, 0.7–0.9 mm wide. Unit conflorescence 4–7 cm long; peduncle and rachis sparsely subsericeous to sparsely tomentose; floral bracts 1.5–2.5 mm long; pedicels 3–3.5 mm long. Flower colour: perianth green or greenish fawn to grey (hairs); style pink to purplish pink. Perianth subsericeous to tomentose outside. Pistil 25–26 mm long; ovary stipe 0.4–0.5 mm long.

Occurs in south-western W.A. to the W and NW of the type subspecies, in the Wubin to Jibberding to Kirwan area. Grows in shrubby associations in shallow soils around outcropping granite. Regenerates from seed. Flowers Aug.–Oct. Map 53.

W.A.: Rabbit Proof Fence, near Bencubbin, *W.E.Blackall 3503* (PERTH); 3.2 km from Jibberding, *C.A.Gardner 12091* (PERTH); 4 km NNW of Kirwan along Rabbit Proof Fence, *D.J.McGillivray 3427 & A.S.George* (NSW).

This subspecies is recognised as 'Poorly Known' in J.D.Briggs & J.H.Leigh, *Rare or Threatened Australian Plants* (1995).

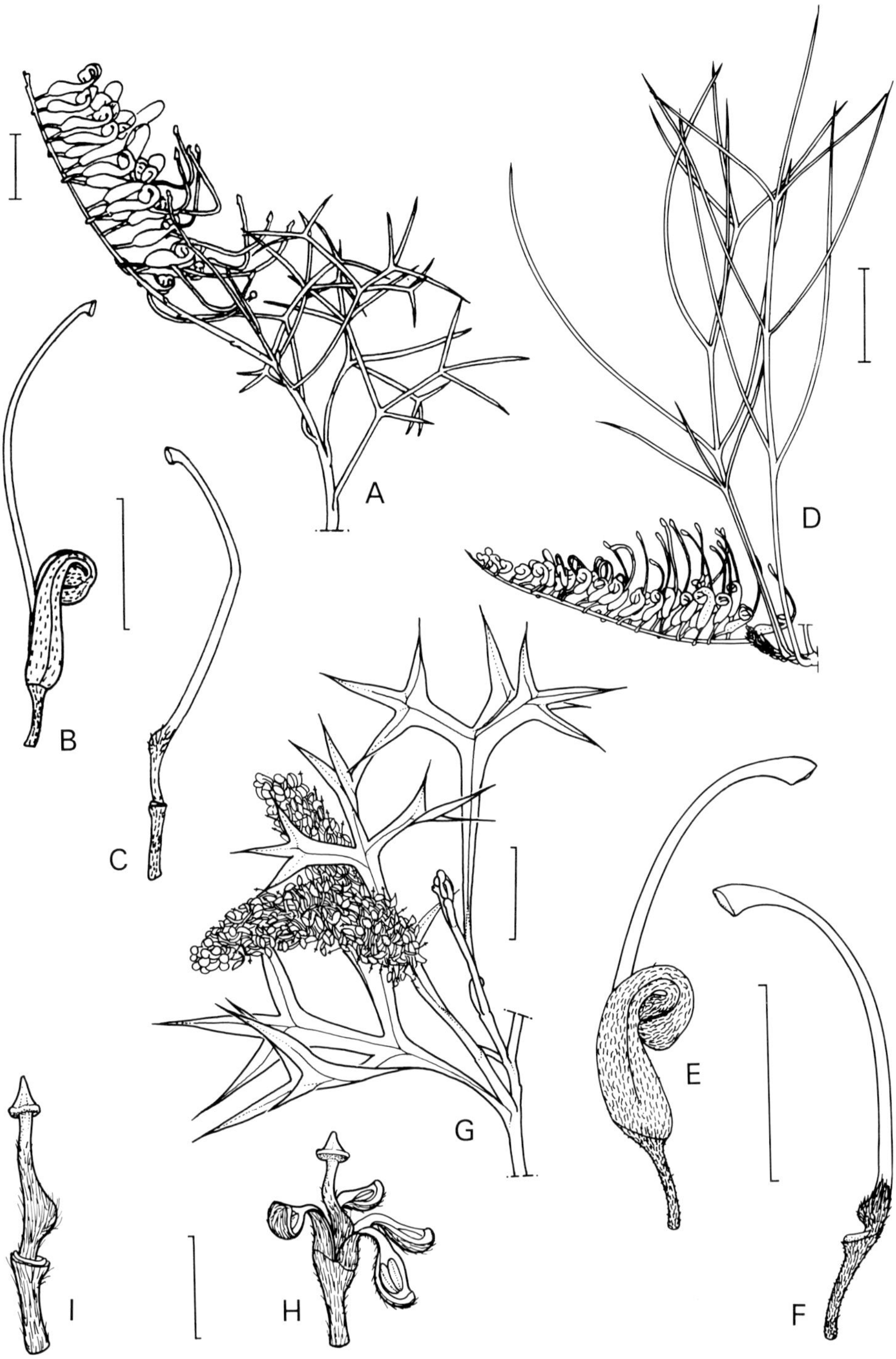

Figure 7. *Grevillea*. **A–C**, *G. aneura*. **A**, flowering branch; **B**, flower; **C**, pistil (**A–C**, M.D.Crisp 6020, CANB). **D–F**, *G. nana* subsp. *nana*. **D**, flowering branch; **E**, flower; **F**, pistil (**D–F**, D.J.McGillivray 3693 & A.S.George, NSW). **G–I**, *G. ramosissima* subsp. *ramosissima*. **G**, flowering branch; **H**, flower; **I**, pistil (**G–I**, P.J.Darbyshire 1233, CANB). Scale bars: **A–C**, **E–G** = 1 cm; **D** = 2 cm; **H–I** = 2 mm. Drawn by: **A–F**, D.Fortescue; **G–I**, D.Boyer.

44. Grevillea maherae Makinson & M.D.Barrett, *Fl. Australia* 17A: 493 (2000)

T: c. 10 km west of new Mount Elizabeth Homestead, W.A., 13 Mar. 1998, *M.D.Barrett 340*; holo: PERTH05060729 (sheet annotated 'sheet 1 of 3'). [Other sheets of this collection number are from separate plants and are excluded from Type status.]

Low multistemmed spreading to weakly erect shrub 15–50 cm tall. Leaves 40–65 mm long, (12–) 20–34 mm wide, rhomboid in gross outline, basally long-cuneate, pinnatifid with (3–) 9–13 (–15) simple subtriangular teeth or lobes spaced evenly in apical $^1/_2$–$^2/_3$ of leaf; teeth or lobes 2–6 mm long, 3–5 mm wide, pungent; margins very shortly recurved; lower surface tomentose with wavy biramous hairs. Unit conflorescence erect to gently decurved, strongly secund; floral rachis 10–20 (–35?) mm long. Flower colour: perianth pinkish red to maroon with a gingery brown limb (hairs); style red. Perianth outer surface villous on limb segments, densely tomentose below with both biramous and simple glandular hairs. Pistil 29–30 mm long; stipe 0.8–1.0 mm long; style sometimes with a few long biramous hairs mostly in basal 5–10 mm and with minute purplish glandular hairs scattered over basal 15–20 mm, glabrous above. Follicle 11–13 mm long, tomentose.

Occurs in the Kimberley region of north-western Australia, where known only from the type locality on Mount Elizabeth Stn. Grows in grassy *Eucalyptus miniata* open woodland on levels below sandstone ridges in deeper sandy-loam colluvial soils. Regenerates from seed, lignotuber, and rhizomes. Flowers Dec.–Mar. Map 54.

W.A.: Mt Elizabeth Stn, *R.Maher s.n.*, Dec. 1997 (PERTH); *loc. id.*, *R.Maher s.n.*, Jan. 1998 (PERTH); c. 17 km (direct) WSW of new Mount Elizabeth HS, *R.O.Makinson 1663 et al.* (CANB, NSW, PERTH).

Grevillea maherae is very closely related to *G. cravenii*; for differences and a discussion of wider affinities and biogeography see under that species.

45. Grevillea cravenii Makinson, *Fl. Australia* 17A: 494 (2000)

T: c. 10 km E of Purulba massif, Prince Regent River Reserve, W.A., 2 Feb. 1999, *M.D.Barrett 706*; holo: PERTH; iso: CANB, DNA, K, KPBG, MEL.

Low multistemmed spreading to weakly erect shrub, 30–50 cm tall. Leaves (30–) 50–90 (–100) mm long, (8–) 20–25 (–40) mm wide, narrowly oblong-elliptic in gross outline or occasionally obovate-cuneate or rarely broadly elliptic, dentate to pinnatifid with (3–) 7–17 marginal teeth or shallow subtriangular lobes spaced ±evenly about margins in apical $^1/_3$–$^2/_3$, often also a few scattered small leaves simple and entire; teeth and lobes 1–5 mm long, 3–8 mm wide, weakly pungent; margins flat to very shortly recurved; lower surface tomentose with ascending wavy biramous hairs. Unit conflorescence erect to gently decurved, secund to irregular; floral rachis 9–25 mm long. Flower colour: perianth purplish red to deep maroon; style (including style-end) crimson. Perianth outer surface with subvillous gingery limb (biramous hairs only), below limb with intermixed pale biramous and purplish glandular hairs. Pistil 33–40 mm long; stipe 1.0–1.5 mm long; style with scattered pale long biramous hairs in basal 2–5 mm and with scattered minute purplish glandular hairs over basal half. Follicle villous (immature only seen). *Cover*; *Frontispiece*; Plate 6.

Occurs in the Kimberley region of north-western Australia, where as yet known only from the type locality in Prince Regent Nature Reserve. Grows in grassy *Eucalyptus miniata* woodland, in sandy loam soil on levels near sandstone ridges. Regenerates from seed, lignotuber and rhizomes. Flowers probably Dec.–Mar. Map 55.

W.A.: foothills of Princess May Ra., N of Prince Regent R., *P.A.Fryxell 4722 et al.* (CANB); c. 20 km E of Prince Regent R. mouth, *L.A.Craven 9221 et al.* (CANB); Prince Regent Nature Reserve, southern foothills of Princess May Ra., *R.O.Makinson 1699 et al.* (CANB, NSW, PERTH); *loc. id.*, *R.O.Makinson 1705 et al.* (CANB, PERTH).

Grevillea cravenii is narrowly distinct from the very closely related *G. maherae*, although both species are known only from very limited fertile material; future collections may necessitate adjustment of the diagnostic features given here. *Grevillea maherae* has a strongly secund conflorescence, a tendency to somewhat deeper leaf division, a significantly shorter pistil (29–30 mm long), and a shallowly tridentate nectary (nectary very strongly tridentate in *G. cravenii*). Fruits of *G. maherae* have a tomentose indumentum of evenly intermixed pale and gingery hairs, the darker hairs not aggregated to form blotches or stripes.

The fruits of *G. cravenii* are more villous, and the biramous hairs are strongly segregated into gingery brown-ferruginous and pale areas, forming darker stripes and blotches about the dorsal side and the ventral suture.

Both species seem closely related to other taxa in the *Asplenifolia/Hookeriana* subgroup, all other members of which are restricted to southern temperate latitudes. There is a particular resemblance to the group of south-eastern species around *G. aquifolium*, which, however differ in having more clearly dissimilar indumenta on the upper and lower leaf surfaces, glabrous styles, and (in most taxa) lack glandular hairs. The plump seeds of *G. maherae* are somewhat reminiscent of those of the south-western *G. nana*. Alternatively, there may be an affinity to the *Bipinnatifida* subgroup, several species of which have mixed simple glandular and biramous non-glandular hairs on the perianth. *Grevillea polyacida* from the N.T. has leaves somewhat similar to *G. cravenii* and *G. maherae*, but differs in a number of floral characters; some relationship is possible.

46. Grevillea dryandroides C.A.Gardner, *J. Roy. Soc. W. Australia* 19: 81 (1934)

T: Ballidu, W.A., 22 Sept. 1931, *C.A.Gardner 2711*; lecto: PERTH, *fide* D.J.McGillivray & R.O.Makinson, *Grevillea* 415 (1993); isolecto: K (erroneously dated '22 Sept. 1932'), PERTH.

Clumping suckering shrub 0.1–0.5 m tall, with leafless inflorescence peduncles sometimes up to 1 m long and trailing on ground. Leaves usually 7–14 cm long, secund-pectinate with rachis gently recurved, pinnatipartite with 10–35 pairs of spreading lobes; lobes mutually aligned on either side of rachis, entire, linear to narrowly obovate, 0.6–1.6 cm long, 1.2–2.0 mm wide; apices deflexed, not pungent; margins refracted; lower surface partly or completely enclosed except midveins, densely villous with tightly curled hairs. Unit conflorescence erect on long trailing peduncle, often lax on ground, secund; floral rachis 30–100 mm long. Perianth subsericeous to subvillous outside. Pistil 17–23 mm long; stipe 0.5–1.5 mm long; style loosely villous in basal half with long ascending hairs, becoming glabrous in upper half or third. Follicle 14–16.5 mm long, tomentose with biramous hairs dominant and scattered simple erect hairs. *Phalanx Grevillea.*

Occurs in south-western W.A., in the Ballidu and Cadoux areas. Two subspecies are recognised.

This species is distinctive in the group in having an erect and very narrowly conical or cylindrical pollen-presenter with a bulbous base. It is sometimes confused with *G. thyrsoides*, which has the pollen-presenter oblique on the style and flat to slightly convex, and the style conspicuously loose-villous right to the apex with long spreading hairs.

Most leaf lobes < 10 mm long, becoming glabrous or nearly so on upper surface at maturity; pistil 17–18 mm long; stipe of ovary < 1 mm long ..... **46a.** subsp. **dryandroides**

Most leaf lobes > 12 mm long, persistently hairy on upper surface at maturity; pistil 17–23 mm long; stipe of ovary 1–1.5 mm long ..... **46b.** subsp. **hirsuta**

46a. Grevillea dryandroides C.A.Gardner subsp. dryandroides

Illustrations: P.M.Olde & N.R.Marriott, *Grevillea Book* 2: 139 (top left & 112A, B) (1995); A.Brown *et al.* (eds), *W. Australia's Threatened Flora* 96 (1998).

Sparingly root-suckering shrub to 50 cm tall, usually either solitary or forming colonies of up to 4 ramets. Leaves dull yellow-green; leaf rachis glabrous; lobes 5–10 (–15) mm long, glabrescent. Floral rachis 30–40 mm long. Pedicels 1–1.5 mm long. Flower colour: perianth pink to orange-pink with a grey-green limb; style red or pink with a green tip. Pistil 17–18 mm long; ovary stipe 0.5–0.7 mm long.

Occurs in south-western W.A., where known only from the Ballidu area, N of Wongan Hills. Grows in open heath and *Banksia* woodland, usually in sandy loam soils over clay or laterite. Regenerates from either seed or rhizomes. Flowers mainly Aug.–Dec. Map 56.

W.A.: 43 km from Wubin towards Wongan Hills, *E.M.Canning WA/68 2898* (CANB); Ballidu, reserve adjacent to Pioneer Cemetery, *M.G.Corrick 9800* (CANB, NSW).

This subspecies is now largely confined to road verges and is highly endangered. It is recognised as 'Vulnerable' in J.D.Briggs & J.H.Leigh, *Rare or Threatened Australian Plants* (1995).

48. Grevillea ramosissima Meisn., *Hooker's J. Bot. Kew Gard. Misc.* 7: 74 (1855)

Anadenia caleyi R.Br., *Suppl. Prodr. Fl. Nov. Holl.* 16 (1830), *non G. caleyi* R.Br. (1810). T: N.S.W., 'Loc. Ora orient., mont. Port Jackson, 1804. b. Caley. 1817. D. Cunningham.' [protologue]; lecto: 'Anadenia montana R.B.Barralier's Journey Mr George Caley; Embothrium ilicifolium August 11-1806'; lecto: BM, *fide* D.J.McGillivray & R.O.Makinson, *Grevillea* 437 (1993); isolecto: A *n.v.*, NSW.

Shrub, often root-suckering, 0.3–1.5 (–3.0) m high. Leaves 3–8 (–10) cm long, pinnatifid to pinnatipartite with 3–11 primary lobes, the lower ones usually again 3–5 (–7)-partite, often some secondary lobes again 2- or 3-fid; ultimate lobes subtriangular to narrowly so or ovate, 3–20 (–28) mm long, 2–7 (–10) mm wide, pungent; margins shortly and angularly recurved; lower surface mostly or completely exposed, tomentose with twisted or curly hairs, or subsericeous with straight hairs. Unit conflorescence narrowly conical to cylindrical, acropetal; floral rachis 25–85 (–100) mm long. Perianth tomentose outside. Pistil 3.5–5.3 mm long; style tomentose almost to apex with biramous hairs and occasional simple erect hairs, sparser towards apex. Follicle 7.5–10 mm long, tomentose. *Fan Grevillea.*

Occurs in eastern N.S.W. along the western fall of the Great Dividing Ra. and adjacent slopes from the Pilliga Scrub (Coonabarabran) area S to Goulburn, Canberra, Tumut, and the Walwa area of north-eastern Vic. Grows in dry sclerophyll woodland, usually in skeletal soils on sandstone, granite or porphyry. There are two subspecies.

Distinguished from the closely related *G. triternata* and *G. raybrownii* which both have the style glabrous, or with hairs only in the millimetre or so above the subsericeous ovary; they also generally have much narrower ultimate leaf lobes (usually 1–3 mm wide), and usually shorter unit conflorescences (1–3 cm long), which in *G. raybrownii* are also dense and ovoid. Possible intergrades between *G. ramosissima* subsp. *ramosissima* and *G. triternata* are known (see under the latter species for details), but the two species generally maintain their integrity in other parapatric or even fully sympatric parts of their ranges.

Lower surface of leaf tomentose with twisted or curly hairs; unit conflorescence 2.5–4 (–5) cm long **48a.** subsp. **ramosissima**

Lower surface of leaf sericeous to subsericeous with straight hairs; unit conflorescence 4.5–5.5 (–10) cm long **48b.** subsp. **hypargyrea**

48a. Grevillea ramosissima Meisn. subsp. ramosissima

Illustrations: A.M.Blombery & B.Maloney, *Prot. Sydney Reg.* 72, 73 (1992); L.F.Costermans, *Native Trees & Shrubs SE Australia* 164 (1981); P.M.Olde & N.R.Marriott, *Grevillea Book* 3: 129 (top centre & 100A, B) (1995).

Prickly shrub to 1.5 (–3) m tall, sometimes root-suckering. Leaves stiff and leathery; lower surface tomentose with twisted or curly hairs. Floral rachis (20–) 25–40 (–50) mm long. Flower colour: perianth white to cream, sometimes tinged light green or pale yellow; style cream. Fig. 7G–I.

Occurs in eastern N.S.W. on the tablelands and western slopes of the Great Dividing Ra., from the Pilliga Scrub and Coonabarabran area S to Humula near Albury. There is also a single collection from the Parlour Mtns near Guyra. Grows in shrubby understorey of eucalypt woodland in sandy or loamy soils, usually over sandstone or granitic substrates. Regenerates from seed and (at least in the S of the range—Canberra, Tumut, Humula) from rhizomes. Flowers Aug.–Nov. Map 60.

N.S.W.: Timor Rock c. 13 km W of Coonabarabran, Sept. 1908, *J.L.Boorman NSW129211* (NSW); 'The Drips' on Goulburn R., 9.9 km NNE of Ulan, *R.Coveny 9574* (B *n.v.*, K, NBG *n.v.*, NSW, PERTH, RSA *n.v.*); lower slopes of Mt Jerrabomberra, Queanbeyan, *P.J.Darbyshire 1233* (A *n.v.*, B *n.v.*, CANB, G, K, L *n.v.*, MEL, NSW, US *n.v.*); Weddin Ra. near NW corner of Weddin State Forest, *D.J.McGillivray 3145 & R.Coveny* (NSW, PERTH); on a ridge near Mt Solitary, Blue Mtns, Oct. 1950, *A.Willows NSW129196* (NSW).

Occasional apparent intergrades with *G. triternata* occur (see under that species); in other areas the species are fully sympatric without sign of intergrading (e.g. SSW of Mendooran).

48b. Grevillea ramosissima subsp. **hypargyrea** (F.Muell.) Olde & Marriott, *Telopea* 5: 778 (1994)

G. ramosissima var. *hypargyrea* F.Muell., *Fragm.* 8: 150 (1874). T: 'In montibus granitico-rupestribus ad flumen Humei' [protologue]; lecto: Humes [Murray] R., [Vic.], Jan. 1874, *F.Mueller*; lecto: MEL, *fide* P.M.Olde & N.R.Marriott, *loc. cit.*; isolecto: MEL.

Illustrations: P.M.Olde & N.R.Marriott, *Grevillea Book* 3: 129 (bottom right), 130 (101) (1995).

Prickly shrub to 2 m tall, sometimes root-suckering. Leaves stiff (relatively thin compared to the type subspecies); lower surface sericeous to subsericeous with appressed straight hairs. Floral rachis 45–55 (–100) mm long. Flower colour: perianth white to cream, sometimes tinged light green or pale yellow; style cream.

Occurs on the upper Murray R. system in far northern Vic., where known only from the Pine Mtn and Cudgewa Bluff area, and from Mt Mittamatite, with one doubtful record from the upper Ovens R. Grows in dry eucalypt woodland in granitic loamy soils among granite outcrops. Regenerates from seed and rhizomes. Flowers mainly Sept.–Dec. Map 61.

Vic.: beside Murray Valley Hwy at foot of Pine Mtn, *M.G.Corrick 5982* (MEL, NSW); Mt Mittamatite, alt. 750 m, 7 Oct. 1968, *McKinnan* (MELU).

Poorly collected, but plants from Mt Mittamatite seem to have consistently smaller leaves.

49. Grevillea raybrownii Olde & Marriott, *Telopea* 5: 774 (1994)

T: Waterboard Reserve near carpark, Welby, N.S.W., 26 Sept. 1993, *P.M.Olde 93/52 & M.Olde*; holo: NSW; iso: CANB, K *n.v.*, MEL *n.v.*, US *n.v.*

Illustrations: P.M.Olde & N.Marriott, *Telopea* 5: 775, fig. 1a–d (1994); P.M.Olde & N.R.Marriott, *Grevillea Book* 3: 131 (top right & 103A, B) (1995).

Bushy shrub c. 1.5 m high. Leaves 2.5–5 cm long, bipinnatisect; primary lobes 3–5, 2–5-partite; ultimate lobes narrowly linear to subulate, often divaricate, 0.5–2.4 cm long, 0.6–1.2 mm wide, pungent; margins tightly and angularly refracted; lower surface enclosed except for midveins, subsericeous in grooves. Unit conflorescence dense, ovoid, strongly acropetal; floral rachis 15–17 mm long. Flower colour: perianth white (hairs) below curve, with limb-hairs rusty brown; style white in basal half, pale lilac above. Perianth sericeous outside. Pistil 6–7.5 mm long; style glabrous. Follicle c. 12 mm long, sericeous.

Occurs in central-eastern N.S.W., where restricted to an area bounded by Dapto, Robertson and Berrima, possibly also Bungonia. Grows in dry sclerophyll forest in sandy, gravelly loam derived from sandstone. Regenerates from seed (only?). Flowers Aug.–Nov. Map 62.

N.S.W.: West Dapto, *R.H.Cambage 369* (CANB, NSW, SYD); 19.2 km NE of Robertson on No. 1 fire trail to the catchment area of the Avon Dam, *R.Coveny 848* (NSW); divide between the Nattai R. and Allum R., *I.Olsen 2183* (NSW); Mundamar Ck track, old road through to Joadja, 5 Apr. 1975, *T.Stead* (NSW); Bullio Tunnel, Mittagong to Wombeyan Caves road, *M.Taylor 366 et al.* (B *n.v.*, CANB, K, NSW, PERTH, RSA *n.v.*).

Very closely related to *G. triternata*, from which it is barely distinct at species level. The taxon was referred to in McGillivray & Makinson (*Grevillea* 66 (1993)) but given no informal or formal status. *Grevillea triternata* has a white indumentum over the whole perianth (including the limb), a ±loosely cylindrical to narrowly conical unit conflorescence (less strongly acropetal than in *G. raybrownii*), and pedicels 0.8–1.8 mm long (3.5–4.2 mm long in *G. raybrownii*). A recent collection (B.Wood, CANB) from Bungonia State Recreation Area corresponds well to *G. raybrownii*, but has short pedicels like *G. triternata*.

This species is recognised as 'Poorly Known' in J.D.Briggs & J.H.Leigh, *Rare or Threatened Australian Plants* (1995).

56. Grevillea banksii R.Br., *Trans. Linn. Soc. London* 10: 176 (1810)

Stylurus banksii (R.Br.) O.Deg., *Fl. Hawaiiensis*, family No. 98 (1932). T: 'In Novae Hollandiae ora orientali; Keppel Bay, Pine Port &c' [protologue]; lecto: Port I [between Facing and Curtis Islands, near Gladstone, Qld], c. Aug. 1802, *R.Brown [Iter. Austral.] 3344*; lecto: BM, *fide* D.J.McGillivray & R.O.Makinson, *Grevillea* 407 (1993); isolecto: B *n.v.*, BM, BRI, G-DC, K, MEL, NSW, P *n.v.*, P-JU *n.v.*

G. forsteri Anon [T.Moore], *Gard. Chron.* n. ser., 1: 60 (1874). T: not designated.

G. banksii var. *forsteri* Guilf., *Cat. Fibres, Papers, etc.* 41 (1888), *nom. nud.*; *G. banksii* var. *forsteri* Guilf., *Austral. Pl.* 193 (1909). T: not designated.

G. forsteri Hulle, *Rev. Hort. Belge Etrangère* 24: 2, 3 (1898), as *G. Forsterii* Hort.; G. *robusta* var. *forsteri* (Hulle) L.H.Bailey, *Stand. Cycl. Hort.* 3: 1412 (1915). T: not designated.

Stylurus banksii f. *albiflora* O.Deg., *Fl. Hawaiiensis*, family No. 98 (1932); *G. banksii* f. *albiflora* (O.Deg.) O.Deg. & I.Deg., *Fl. Hawaiiensis* family No. 98 (1959). T: Manoa valley, Oahu, Hawaii, Nov. 1931, *O.Degener & Park 4081*; holo: BISH *n.v.*

Illustrations: K.A.W.Williams, *Nat. Pl. Queensland* 3rd edn, 1: 135 (1984); P.M.Olde & N.R.Marriott, *Grevillea Book* 1: 77, fig. 42 (1994); 2: 48 (34A, B), 49 (top left & 34C–F) (1995).

Bushy to spindly erect shrub or slender tree, 2–10 m high, or rarely a prostrate to sprawling shrub. Leaves 8–30 cm long, pinnatipartite (rarely the odd leaf entire); lobes 4–12, narrowly elliptic to linear, 5–18 cm long, 5–15 mm wide, not pungent; margins shortly recurved or revolute; lower surface usually mostly exposed, subsericeous to subvillous. Conflorescence simple or few-branched; unit conflorescence erect, ±cylindrical, irregular to subsynchronous; floral rachis 50–120 (–200?) mm long. Flower colour: perianth creamy white or bright scarlet to crimson, rarely pink to apricot with a yellow limb; style usually matching perianth in colour, rarely much paler. Perianth open-tomentose to shortly pilose outside with both biramous and simple erect hairs. Pistil 32–50 mm long; style sometimes with biramous hairs for up to 4 mm above ovary, otherwise glabrous, lacking a dorsal hump immediately below style-end. Follicle 15–25 mm long, tomentose with biramous hairs mostly replaced by erect simple glandular hairs as fruit matures. *Banks' Grevillea, Byfield Waratah, Red Flowered Silky Oak, Dwarf Silky Oak.* Fig. 8D–G.

Occurs patchily in Qld from Yeppoon S to about Ipswich, usually near the coast but extending inland near Eidsvold and the upper Brisbane R. Grows in various habitats, usually in woodland or open forest, usually on flatter sites in sandy soils but sometimes in heath or on rocky slopes. Probably regenerates from seed only. Flowers Aug.–Oct. Map 69.

Qld: upper parts of the Brisbane R., *A.Cunningham 32* (K, NSW); 2 km W of Goodwood, near Childers, 13 Oct. 1973, *K.H.L.Key* (CANB); a few miles NW of Bundaberg, *M.E.Phillips CBG029964* (CANB, BRI); Ingham Rd N of Townsville, *G.G.Stewart 5* (BRI); Byfield, near Keppel Bay, *C.T.White 8154* (BRI).

Grevillea banksii has pedicels 3–10 mm long, and floral bracts < 2 mm long. There is considerable variation in habit, leaf size and lobing, indumentum, and flower colour. Red and white-flowered plants may occur together or in pure stands. This species is sometimes an aggressive coloniser on disturbed ground, and is a significant woody weed in Malagasy.

Forms based on habit and distribution are distinguishable, although these do not accommodate all populations. A 'tree form' growing to 6–10 m tall is common in the area between Maryborough and Bundaberg in flat sandy forest country. The 'Townsville form', occurring between Ingham and Townsville, is generally more shrubby and has pistil length (< 35 mm) and inflorescence length (usually < 6 cm) at the lower end of the ranges. The 'prostrate coastal form' is sporadic along the coast over much of the range, usually on exposed headlands; its habit varies from truly prostrate to decumbent, with the flower heads usually raised. The widely cultivated variant known as *G. banksii* var. *forsteri* is of unknown but probably wild origin.

Aberrant floral development is not uncommon, e.g. digyny, and a condition in which the tepals remain coherent until the whole perianth is lifted by the straightening style away from the toral rim. The species is a parent of several horticultural hybrids. There are recorded cases of contact dermatitis caused by the foliage.

57. **Grevillea whiteana** McGill., *New Names Grevillea* 16 (1986)

T: Glenwood Stn, 48.3 km SW of Mundubbera, Qld, 24 July 1974, *N.Chopping s.n.*; holo: BRI.

Illustrations: P.M.Olde & N.R.Marriott, *Grevillea Book* 3: 226 (top left & 185A, B), 227 (185C) (1995).

Erect shrub or small tree, 2–9 m high. Leaves 12–25 cm long, rather secund, pinnatisect with 10–18 ascending parallel linear lobes 5–13 cm long, 1–2.5 mm wide, not pungent; margins angularly revolute; lower surface mostly or completely enclosed except for midveins, subsericeous. Conflorescence simple or branched; unit conflorescence erect, cylindrical, subsynchronous; floral rachis (60–) 80–120 (–200?) mm long. Flower colour: perianth cream, scattered with caducous rusty hairs especially on limb; style cream. Perianth villous outside. Pistil 43–46 mm long; style glabrous, with a slight dorsal hump immediately below style-end. Follicle 14–18 mm long, tomentose with biramous hairs. *Mundubbera Grevillea.*

Occurs in south-eastern Qld, from near Mundubbera S to Boondooma; also occurs on the Mt Walsh massif near Biggenden. Grows in pale sandy soils in open eucalypt forest and on rocky slopes over rhyolite. Regenerates from seed or sometimes from a lignotuber. Flowers Mar.–Oct. Map 70.

Qld: road between Durong and Gayndah c. 40 km N of Durong, 11 June 1961, *F.Hockings* (BRI); Mt Walsh Natl Park, 15 km SW of Biggenden, 3 Sept. 1973, *J.W.Randall* (BRI); 'Narayen', 50 km W of Mundubbera, *J.C.Tothill N521* (BRI); Biggenden Bluff, *C.T.White 7268* (BRI).

Populations from the Biggenden to Mt Walsh area have large floral bracts up to 10 mm long, and plants are at least sometimes lignotuberous. Populations from the Mundubbera area have smaller bracts and apparently lack lignotubers, regenerating from seed only. The species is most closely related to *G. hodgei*, which is barely distinct at species level and differs mainly in its shorter conflorescences (2–8 cm long), its shorter pistil (26–35 mm long), and in having rusty-brown hairs more extensively distributed and more persistent on the outer surface of the perianth. Both species have a ±straight torus, and the margin of the nectary erose or toothed.

58. **Grevillea hodgei** Olde & Marriott, *Grevillea Book* 1: 185 (1994)

T: East Peak, Coochin Hills, Qld, 17 July 1992, *N.Marriott NM 92/06*; holo: NSW; iso: BRI.

[*G. whiteana auct. non* McGill.: D.J.McGillivray & R.O.Makinson, *Grevillea* 47 (1993)]

Illustrations: D.J.McGillivray & R.O.Makinson, *Grevillea* 47 (1993), as *G. whiteana*; P.M.Olde & N.R.Marriott, *Grevillea Book* 2: 198 (top centre & 163A *p.p.*, B) (1995).

Erect shrub 1–4 m high. Leaves 6–19 cm long, pinnatisect with 6–14 ascending parallel linear lobes; lobes 5–12 cm long, 1.5–2.8 mm wide, acute, not pungent; margins revolute, enclosing most or all of the sericeous lower surface on either side of midvein. Conflorescence simple, erect, cylindrical, subsynchronous or opening irregularly; floral rachis 20–80 mm long. Flower colour: perianth cream with a ±dense indumentum of persistent brown hairs especially on limb and at base; style rich cream to pale yellow. Perianth tomentose to villous outside. Pistil 26–35 mm long; style glabrous, with a slight dorsal hump immediately below style-end. Follicle 13–14 mm long, tomentose. *Coochin Hills Grevillea.* Plate 9.

Occurs in south-eastern Qld, known only from several small populations in the Beerwah area. Grows in skeletal sandy soils around exposed rocky platforms. Regenerates from seed, other modes not recorded. Flowers all months, peaking Mar.–Oct. Map 71.

Qld: Rupari Hill, 1.8 km SW of Beerwah, *R.Dowling 14* (BRI, K); Coochin Hills, Beerwah, top of NW slope, 3 June 1967, *F.D.Hockings* (BRI); East Peak, Coochin Hills, *L.S.Smith 14034* (AD, BRI, NSW, US *n.v.*).

No significant variation within the species. Barely distinct at species level from *G. whiteana*; see under that species for differences.

59. Grevillea sessilis C.T.White & W.D.Francis, *Proc. Roy. Soc. Queensland* 37: 165, Pl. X (1926)

T: Torrens Ck, Qld, *s.d.* [probably 1870s], *J.E.Young s.n.*; holo: MEL.

Illustrations: C.T.White & W.D.Francis, *loc. cit.*; Society for Growing Australian Plants, *Hort. Guide Austral. Pl.*, Set 1, 3rd edn: t. 13 (1980); K.A.W.Williams, *Nat. Pl. Queensland* 3rd edn, 1: 138 (1984); P.M.Olde & N.R.Marriott, *Grevillea Book* 3: 169 (top centre & 137A, B), 170 (137C) (1995).

Erect shrub or small tree, 2–6 m high. Leaves 5–16 cm long, pinnatipartite with 5–18 narrowly ovate or oblong or sublinear lobes, with basal lobes sometimes again 2- or 3-partite; ultimate lobes 2.5–8 cm long, 4–12 mm wide; apices blunt; margins shortly recurved; lower surface exposed, sericeous. Conflorescence simple, erect, cylindrical, subsynchronous to acropetal; floral rachis 50–180 mm long. Flower colour: perianth white; style cream-white or occasionally greenish yellow; tip green. Perianth villous outside with biramous hairs only. Pistil 17.5–31 mm long; style glabrous, with a dorsal hump on rear of style-end. Follicle 12.5–15 mm long, tomentose to villous with ascending biramous hairs, and an underlying indumentum of shorter simple erect glandular hairs.

Occurs in low mountain ranges of inland eastern Qld, from Cairns area S to Springsure and Theodore. Grows in open woodland or shrubland in shallow sandy soils over siliceous sandstone. Regenerates from seed. Flowers Apr.–Dec. Map 72.

Qld: 23 km SW of Pentland on Flinders Hwy near Dead Horse Ck, *P.Ollerenshaw 1107* (BRI, CANB, NSW); Torrens Ck, *C.T.White 8683* (A *n.v.*, BRI); Planet Ck, c. 50 km NE of Rolleston township, *R.Story & Yapp 287* (BRI, CANB, MEL, NSW, PERTH).

Grevillea sessilis has floral bracts 3.5–6.5 mm long. It is easily distinguished from the related *G. whiteana* and *G. hodgei*, which both have narrower leaf lobes (1–4 mm wide) with the margins revolute and enclosing the lower surface of the lamina except the midvein; and from *G. banksii* which has a pistil 32–50 mm long and pedicels > 1 mm long (⩽ 0.5 mm in *G. sessilis*).

60. Grevillea juncifolia Hook., *in* T.L.Mitchell, *J. Exped. Trop. Australia* 341 (1848)

T: sub-tropical Australia [Qld?], [probably 6 Oct. 1846], *T.L.Mitchell 477*; holo: K; iso: A *n.v.*, ?MEL, NY *n.v.*

Erect to spreading shrub or small tree, 1–7 m high. Leaves 8–30 cm long, ascending, entire or pinnatipartite (almost pinnatisect); lobes 2–6, ascending, ±parallel, 5–22 cm long, oblong in cross-section; entire leaves and lobes linear, 0.8–2.1 mm wide, ±pungent; margins revolute; lower surface enclosed except for midveins, hairy in grooves. Conflorescence usually branched; unit conflorescence erect, cylindrical to conical, acropetal; ultimate floral rachis 50–160 mm long. Perianth sericeous to pubescent outside with biramous hairs, or sparsely pubescent to tomentose with both biramous and erect simple glandular hairs (the latter sometimes predominating). Pistil 18–27 mm long; style sometimes villous in basal third, otherwise papillose or glabrous, lacking dorsal hump below style-end. Follicle 15–29 mm long, pubescent to subvillous or lanate with biramous hairs, usually with some erect simple glandular hairs also. *Honeysuckle Spider-flower.*

Very widespread in the Eremaean zone, in all mainland States except Vic. Grows on sandhills or sandy flats in *Eucalyptus*, *Acacia*, *Casuarina* or *Triodia* dominated situations. Two subspecies are recognised, although the division is imperfect.

Grevillea juncifolia has an oblique torus and the nectary margin entire to sinuate; the conflorescences are occasionally axillary or cauline. It may be confused with *G. eriostachya* and *G. excelsior*, both of which have distinctly secund, terminal unit conflorescences.

Leaves usually divided, sometimes some or all entire; outer surface of perianth, pedicels, and rachis pubescent to tomentose, usually with erect simple glandular hairs predominating and biramous hairs often few or absent; unit conflorescence strongly acropetal and usually conico-cylindrical **60a.** subsp. **juncifolia**

Leaves entire; outer surface of perianth, pedicels, and rachis subsericeous with ±appressed biramous hairs only; conflorescence weakly acropetal and usually cylindrical **60b.** subsp. **temulenta**

60a. Grevillea juncifolia Hook. subsp. juncifolia

G. sturtii R.Br., *Bot. Sturt's Exped. Australia* 2: App. 86 (1849). T: on sand hills in Lat. 27°S, S.A., [?Aug. or Oct. 1845], *C.Sturt s.n.*; holo: BM.

G. sturtii var. *pinnatisecta* F.Muell., *(Vic. Parl. Pap. Votes) Proc. Legisl. Ass.* 4(61): 14 (1863), also published as *Enum. Pl. Coll. Stuart's Exped.* 14 (1863). T: scrub near Forsters Range, [N.T.], [*J.McD.Stuart*]; holo: MEL.

G. mitchellii Meisn., *Hooker's J. Bot. Kew Gard. Misc.* 7: 77–78 (1855), *nom. illeg. non* Hook. (1848). T: sub-tropical Australia [?Qld], *T.L.Mitchell*; holo: NY *n.v.*, *fide* D.J.McGillivray & R.O.Makinson, *Grevillea* 423 (1993); iso: CGE *n.v.*

G. juncifolia 'south-western form', of D.J.McGillivray & R.O.Makinson, *Grevillea* 50 (1993).

Illustrations: G.M.Cunningham *et al.*, *Pl. W New South Wales* 215 (1981); J.W.Wrigley & M.Fagg, *Banksias Waratahs & Grevilleas* 244 (1989), both as *G. juncifolia*; P.M.Olde & N.R.Marriott, *Grevillea Book* 2: 219 (top right & 182A, B) (1995).

Plant usually with most or all leaves divided into linear lobes, sometimes some or all leaves entire. Unit conflorescence strongly acropetal, usually conico-cylindrical. Rachis of unit conflorescence, pedicels, and outer surface of perianth with a pubescent to tomentose indumentum largely comprised of erect simple glandular hairs. Flower colour: buds green; perianth and style bright orange or occasionally tending yellow. Plate 12.

Occurs across the Eremaean zone in all mainland States except Vic. Grows on sandplain, sandhills, stony rises and open plains, in a range of eucalypt, sclerophyll shrub and *Triodia* associations. Regenerates from seed. Flowering often opportunistic, but usually June–Nov. Map 73.

W.A.: 8 km N of the E end of the Schwerin Mural Crescent, *D.E.Symon 2420* (AD, PERTH). N.T.: 5 km NE Elkedra HS, *G.Chippendale 1547* (AD, DNA, PERTH). S.A.: edges of Simpson Desert, 13 km W of Purni Well, *D.E.Symon 3278* (AD, CANB). Qld: Boatman Stn, *S.L.Everist 2827* (BRI, CANB). N.S.W.: roadside scrub 5 km N of Yantabulla, *K.Paijmans 3290* (CANB, NSW).

Two forms are distinguished. The 'typical form', occurring over most of the subspecies range except the Eyre Peninsula, S.A., usually has divided leaves. Plants from the Simpson Desert and adjoining areas have larger flowers and fruit than other populations of this form. The 'Eyre Peninsula form' resembles the normal 'typical form' except in having all leaves entire.

There is a tendency for some plants in northern N.S.W., within the range of the 'typical form', to have the outer surface of the perianth with biramous hairs predominating, and rarely erect simple hairs altogether absent (otherwise features of subsp. *temulenta*); in leaf division however these plants correspond to subsp. *juncifolia*. They may represent a third form or a separate subspecies.

60b. Grevillea juncifolia subsp. temulenta Olde & Marriott, *Grevillea Book* 1: 185 (1994)

T: 104 km N of Kalgoorlie on road to Menzies, W.A., 16 Sept. 1989, *B.J.Conn 3149 & J.Scott*; holo: NSW; iso: AD, MEL, MO *n.v.*, PERTH. [mis-cited as '*Conn 3439*' in protologue, and in P.M.Olde & N.R.Marriott, *op. cit.* 2: 220 (1995)]

Illustrations: P.M.Olde & N.R.Marriott, *Grevillea Book* 2: 220 (top centre & 183A, B) (1995).

Leaves entire. Unit conflorescence weakly acropetal, usually cylindrical. Rachis of conflorescence, pedicels, and outer surface of perianth subsericeous with ±appressed biramous hairs only, lacking erect simple glandular hairs. Flower colour: buds green; perianth and style bright orange or occasionally tending yellow.

Occurs in W.A., widespread in the area from Pindar and Perenjori to Laverton and Queen Victoria Spring. Grows in yellow sand, sometimes with laterite, in open shrubland. Regeneration apparently from seed only. Flowers mainly c. Aug.–Dec., but probably also opportunistic. Map 74.

W.A.: road to Mt Gibson Stn, *H.Demarz 2561* (KPBG, PERTH); 35 km S of Gullewa, between Wurarga and Morowa, 28 Sept. 1950, *A.Humphries s.n.* (CANB); 8 km N of Zanthus, *R.D.Royce 5278* (PERTH).

62. **Grevillea maxwellii** McGill., *New Names Grevillea* 9 (1986)

T: Salt R. [Pallinup R.], W.A., *s.d.*, *G.Maxwell*; holo: MEL.

[*G. asparagoides auct. non* Meisn.: G.Bentham, *Fl. Austral.* 5: 439 (1870), *p.p.*]

Illustrations: P.M.Olde & N.R.Marriott, *Grevillea Book* 3: 21 (top right & 12A, B), 22 (12C) (1995); A.Brown *et al.* (eds), *W. Australia's Threatened Flora* 196 (1998).

Decumbent to spreading shrub 0.3–1.2 m high, to 1.5 m across. Leaves 2–7.5 cm long, usually secund, divaricately bipinnatipartite (sometimes incompletely so), with 2–9 primary lobes, the lower ones mostly again tripartite; ultimate lobes linear, 6–30 mm long, 0.8–1.4 mm wide, pungent; margins ±angularly revolute; lower surface enclosed except for midveins, with dense slender flexuose hairs in grooves. Conflorescence simple to 3-branched; unit conflorescence erect to deflexed, secund; floral rachis 25–50 mm long. Flower colour: perianth pinkish orange to pink-red, red shades strengthening with age, limb sometimes greenish; style deep pinkish red. Perianth basally dilated but not or scarcely saccate on ventral side, loosely tomentose outside with biramous and simple erect glandular hairs, glabrous inside. Pistil 20–24 mm long; style minutely pubescent with simple erect glandular hairs, sparser towards base and apex. Follicle c. 10 mm long, tomentose with both biramous and simple erect glandular hairs.

Occurs in south-western W.A., in the Pallinup R. catchment NW from about Chillitup. Grows in low heath, often on ridges or in rocky situations, in shallow granitic soils. Regeneration mode unknown, probably from seed only. Flowers c. Sept.–Nov. (also recorded in May). Map 76.

W.A.: Phillips Ra., *s.d., Anon., MEL63637* (MEL); on the upland plains, Salt [Pallinup] R., *s.d., Anon., MEL63638* (MEL); 6.4 km E of Ellen Peak, *K.Newbey 2435* (PERTH); Pallinup R., *G.J.Keighery 9760* (CANB, PERTH).

Digynous flowers sometimes occur. Can be confused with *G. asparagoides* (which has a broader perianth 3.5–5 mm across and longer pistils 30–37 mm long, and more rigid, more divaricate, and shorter leaf lobes) and with *G. secunda* (which lacks glandular hairs on the outer surface of the perianth, and has the leaf margins smoothly revolute with no edge-veins apparent). *Grevillea maxwellii* has the perianth 2–3 mm across, and the leaf margins angularly revolute with the edge-veins usually visible. It can be distinguished from *G. batrachioides* by its broadly ovate floral bracts, its pedicels 3–5.5 mm long, and the distance from leaf attachment to the first lobe being 18–28 mm.

This species is recognised as 'Endangered' in J.D.Briggs & J.H.Leigh, *Rare or Threatened Australian Plants* (1995).

63. **Grevillea asparagoides** Meisn., *in* A.L.P.P. de Candolle, *Prodr.* 14: 373 (1856)

T: Swan R., [W.A.] [1848], *J.Drummond 4th coll. 283*; lecto: NY *n.v.*, *fide* D.J.McGillivray & R.O.Makinson, *Grevillea* 406 (1993); isolecto: BM, CGE *n.v.*, G, K, LE *n.v.*, MEL, P.

G. asparagoides Meisn., *Hooker's J. Bot. Kew Gard. Misc.* 4: 186 (1852), *nom. nud.*

Illustrations: J.W.Wrigley & M.Fagg, *Banksias, Waratahs & Grevilleas* 206 (1989); P.M.Olde & N.R.Marriott, *Grevillea Book* 2: 39 (centre), 40 (27A–C) (1995).

Dense prickly shrub 0.5–2 m high. Leaves 1.8–3.6 mm long, deeply and divaricately tripartite to pinnatipartite or biternate; primary lobes 3–5, entire to tripartite; ultimate lobes linear to subulate, rigid, 5–20 mm long, 0.8–1.4 mm wide, pungent; margins angularly revolute; lower surface enclosed except for midveins, subsericeous in grooves. Conflorescence simple to few-branched; unit conflorescence usually decurved to deflexed or occasionally erect, loosely subsecund or shortly subcylindrical; floral rachis 25–60 mm long. Flower colour: perianth pink to red-pink, with limb paler and sometimes cream; style red. Perianth markedly saccate at base, open-tomentose outside (often subvillous on limb) with both biramous and erect glandular hairs, usually glabrous inside or with appressed hairs near curve. Pistil 30–37 mm long; style minutely pubescent with erect simple glandular hairs. Follicle 13–17 mm long, tomentose. Plate 11.

Occurs in south-western W.A., at scattered localities from near Perenjori to Wongan Hills and Bindi Bindi. Grows in sandheath or shrubland on gravelly loam or yellow or white sand. Regeneration mode unknown, probably from seed only. Flowers July to Dec. Map 77.

W.A.: 8.5 km from Wubin towards Wongan Hills, *E.M.Canning CBG029439* (CANB); Wubin, *R.Coveny 7861 & B.R.Maslin* (K, NSW, PERTH); Perenjori, *C.A.Gardner 2683* (PERTH); 31 km from Wongan Hills towards Piawaning, *M.E.Phillips CBG024125* (AD, CANB); 13 km N of Wubin, *E.Wittwer 1237* (PERTH).

Grevillea batrachioides differs in having the basal internode of the leaf ⩽ 5 mm long: in *G. asparagoides* the basal internode is usually 12–16 mm long, rarely as short as 6 mm in the Bindi Bindi area.

This species is recognised as 'Poorly Known' in J.D.Briggs & J.H.Leigh, *Rare or Threatened Australian Plants* (1995).

64. **Grevillea batrachioides** F.Muell. ex McGill., *New Names Grevillea* 2 (1986)

T: W.A., *s.d.*, *J.Dr.* [*J.Drummond*] *s.n.*; holo: MEL 63639; iso: PERTH.

Illustrations: P.M.Olde & N.R.Marriott, *Grevillea Book* 2: 52 (bottom right), 53 (37A, B) (1995); A.Brown *et al.* (eds), *W. Australia's Threatened Flora* 48 (1998).

Shrub to 2 m tall. Leaves 1–4 cm long, with rachis angularly reflexed at each node, deeply and divaricately pinnatipartite to partly bi- or tri-pinnatipartite; primary lobes 3–7; ultimate lobes linear to subulate, 0.3–2.0 cm long, 0.9–1.2 mm wide, rigid, pungent; margins revolute; lower surface enclosed except for midveins, subsericeous in grooves. Conflorescence simple, erect to decurved, loose semi-secund clusters; floral rachis 20–50 mm long. Flower colour: perianth pale creamy pink to mid-pink; style deep red. Perianth grossly saccate, loosely villous to tomentose outside with mainly biramous hairs and some simple erect glandular hairs, glabrous inside except for a few appressed hairs near curve. Pistil 30–38 mm long; style sparsely glandular-pubescent. Follicle 17–23 mm long, tomentose.

Occurs in south-western W.A., where known only from near Mt Lesueur, NNE of Cervantes. Grows in shrub associations on rock pavements in skeletal sandy soil in rock crevices. Regeneration mode unknown. Flowers Oct.–Dec.? Map 78.

W.A.: c. 10 km NE of Mt Lesueur, *P.M.Olde 91/259 & J.Cullen* (CANB, K, NSW, PERTH).

Known only from the Type specimen until recollected first in 1982 and then confirmed as a very small population in 1991. The species is recognised as 'Endangered' in J.D.Briggs & J.H.Leigh, *Rare or Threatened Australian Plants* (1995).

Grevillea batrachioides has leaves with the distance from leaf attachment to the lowest lobe being ⩽ 5 mm, which distinguishes it from *G. maxwellii* and *G. asparagoides* (in both of which the distance is usually ⩾ 10 mm). *Grevillea batrachioides* also has pedicels 7–15 mm long (3–5.5 mm long in *G. maxwellii*).

65. **Grevillea secunda** McGill., *New Names Grevillea* 14 (1986)

T: 16 km S of Queen Victoria Spring, W.A., 2 Oct. 1956, *R.D.Royce 5536*; holo: PERTH.

[*G. treueriana auct. non* F.Muell.: J.S.Beard, *Descr. Cat. W. Austral. Pl.* 2nd edn, 40 (1970), *p.p.*]

Illustrations: D.J.McGillivray & R.O.Makinson, *Grevillea* 106, fig. 16 (1993); P.M.Olde & N.R.Marriott, *Grevillea Book* 3: 165 (bottom right), 166 (134A–C) (1995).

Spreading shrub 0.3–0.7 m high. Leaves (2–) 4–9.5 cm long, usually bipinnatipartite to pinnatipartite with 5–9 primary lobes, or rarely an odd leaf linear or 2- or 3-partite; primary lobes weakly divaricate, usually again 2–5-partite; ultimate lobes rigid, subterete, 0.5–4 cm long, 0.9–1.3 mm wide, pungent; margins smoothly revolute; lower surface enclosed except for midveins, subsericeous in grooves. Conflorescence usually sparingly branched; unit conflorescence erect to decurved, secund; floral rachis 15–80 mm long. Flower colour: perianth pinkish red; style red. Perianth loosely subsericeous outside (biramous hairs only), glabrous inside. Pistil 24.5–30 mm long; style subsericeous for 4–6 mm above ovary, otherwise glabrous or with minute simple hairs. Follicle 10–13 mm long, subsericeous. Fig. 9A–C.

Occurs in southern W.A., from Comet Vale to Queen Victoria Spring, and at c. 70 km W of Menzies. Grows in open woodland or shrubland in shrub-*Triodia* associations, in yellow or red sand soils. Probably regenerates from seed only. Flowers Sept.–Dec.(?). Map 79.

W.A.: 29 km N of Cundeelee Mission, *A.S.George 5880* (PERTH); Victoria Desert Camp 57, Elder Exploring Expedition, 20 Sept. 1891, *R.Helms 11* (AD, K, MEL, NSW); Comet Vale, *J.T.Jutson 185* (NSW); near Queen Victoria Spring, 25 Aug. 1960, *A.R.Main* (PERTH).

This species has occasionally been misidentified as *G. treueriana*, which has longitudinal veins evident on the upper leaf surface and stipe of ovary c. 3 mm long. *Grevillea secunda* has obscure venation except for the midvein on the lower surface, and a sessile ovary.

This species is recognised as 'Poorly Known' in J.D.Briggs & J.H.Leigh, *Rare or Threatened Australian Plants* (1995).

Polyacida Subgroup

Erect shrubs. Leaves dentate. Conflorescence simple, erect, broadly secund, acropetal. Perianth hairy outside, glabrous inside. Pistil 9–11 mm long; ovary sessile; style with biramous hairs; pollen-presenter lateral, ±flat. Follicle weakly laterally compressed; surface indumentum with coloured markings, of biramous hairs only. Seed flat-ellipsoidal, peripterous.

One species only, endemic to the 'Top End' of N.T. Probably insect-pollinated. The torus is reversely oblique at 45° to the pedicel (ventral edge highest), a condition unique among the *Pteridifolia* Group species where the torus is normally transverse or moderately oblique.

66. Grevillea polyacida McGill., *New Names Grevillea* 12 (1986)

T: Eva Valley Station, N.T., 6 Oct. 1973, *C.S.Robinson EV 60*; holo: BRI; iso: CANB, DNA, NSW.

Illustrations: K.Brennan, *Wildfl. Kakadu* 14 (pl. 10) (1986), as *Grevillea sp.*; P.M.Olde & N.R.Marriott, *Grevillea Book* 3: 104 (top centre, fig. 79A–C) (1995).

Open, erect shrub 1–2.5 m high. Leaves ovate, (4–) 6–9 (–13) cm long, (18–) 30–60 (–70) mm wide, dentate with 15–31 pungent teeth; margins slightly sinuate, flat to slightly recurved; lower surface curled-tomentose. Unit conflorescence erect, terminal, pedunculate, broadly conico-secund; floral rachis 30–50 mm long. Flowers acroscopic. Flower colour: perianth greenish with tawny hairs; style green to brown. Perianth tomentose-villous or sublanate outside (biramous hairs only, the arms twisted and curled); tepals remaining coherent except along dorsal suture after release of style-end. Pistil 9–11 mm long; ovary tomentose; style tomentose-villous dorsally, mostly glabrous ventrally, exserted in late bud. Follicles perpendicular to or retrorse on pedicels, obloid-ellipsoidal, 13–19 mm long, tomentose, rugulose. Fig. 10A–C.

Occurs in N.T. in eastern Arnhem Land, including Kakadu Natl Park, in eucalypt woodland on sandstone. Regenerates from seed. Flowers (Mar.–) July–Oct. Map 80.

W.A.: Site 70, 10 km N of Twin Falls, Kakadu Natl Park, *L.A.Craven 6061* (CANB, NSW); Twin Falls, Kakadu Natl Park, *C.R.Dunlop 5537* (AD, BRI, CANB, DNA, MEL, NSW, PERTH); *loc. id.*, *C.R.Dunlop 5544* (DNA, NSW); *loc. id.*, *C.R.Dunlop 6732 & G.M.Wightman* (DNA, NSW); gorge between Twin Falls & Jim Jim Falls, *G.M.Wightman 1336 & L.A.Craven* (BRI, CANB, DNA, MEL, NSW).

Rubicunda Group

Shrubs. Leaves divided, dorsiventral; surfaces dissimilar; margins recurved to revolute. Conflorescences terminal, ±erect, simple or basally 2-branched; unit conflorescences conico-cylindrical to subsecund, acropetal. Flowers basiscopic. Torus transverse. Perianth zygomorphic, hairy outside (biramous hairs), glabrous inside; tepals separating and spreading widely after anthesis. Pistil 5–10 mm long; ovary sessile, densely hairy; style hairy over at least part of length, scarcely exserted from late bud; pollen presenter very oblique to lateral, discoid. Follicle 2-seeded, villous with biramous hairs (sometimes also glandular hairs in *G. dunlopii*), a short dense layer overlain by longer emergent hairs, with pale gingery indumental markings (sometimes obscure); pericarp thin crustaceous. Seed plumply ellipsoidal, subperipterous.

A small group of two species, endemic to Kakadu escarpment in N.T. Pollinator unknown. The nectary consists of four separate erect lobes, a feature not seen elsewhere in *Grevillea* except as an occasional aberration in individual flowers. The species in this group are also highly distinctive in their sessile, basiscopic flowers and buds crowded at the apex of very long floral rachises which at least in some cases seem to elongate ofter bud formation, occasionally with two interrupted acropetal flowering zones. *Grevillea rubicunda s. lat.* (including *G. dunlopii*) was regarded as a distinct unnamed genus by McGillivray (*Grevillea* 452, 465 (1993)), who excluded it from his revision. It is unclear whether the distinctive characters of this group are ancient or derived, but the latter seems more likely. It is here retained in *Grevillea* pending analysis of the whole genus. The hairy ovary (with aggregations of coloured hairs), and the glabrous inner surface of the tepals, suggest a relationship with the *Pteridifolia* Group. *Grevillea polyacida* has some points of similarity.

Leaf upper surface with curly biramous hairs only; style densely villous throughout; leaf veins decurrent and forming 3 conspicous stem ridges below leaf **67. G. rubicunda**

Leaf upper surface with both straight biramous hairs and simple erect glandular hairs; style sparsely to loosely villous; leaf veins not decurrent to stem, not forming ridges below leaf base **68. G. dunlopii**

67. Grevillea rubicunda S.Moore, *J. Linn. Soc., Bot.* 45: 210 (1920)

T: westwards of the Gulf, Table-land of the South Alligator, [N.T.], *s.d.*, *[L.]Leichhart*; holo: BM.

Illustrations: P.M.Olde & N.R.Marriott, *Grevillea Book* 3: 151 (top centre & 119A) (1995).

Low spreading shrub 0.6–2 m tall. Leaves 12–20 cm long, pinnatipartite to almost pinnatisect with 15–25 spreading to ascending simple lobes; lobes 4–10 cm long, 1.2–2 (–4) mm wide, often gently curved; upper surface loosely subsericeous to tomentose with curly biramous hairs only, soon glabrous; lower surface usually narrowly exposed on either side of midvein, usually subsericeous to subtomentose, with biramous ±straight or occasionally curly hairs. Conflorescences erect, simple or rarely 2-branched, terminal; rachis 4–20 cm long (active flowering zone 2–5 cm long), lanate with ferruginous hairs; unit conflorescence conico-cylindrical, acropetal. Flowers sessile, basiscopic. Flower colour: perianth greenish white in bud, becoming white with pinkish brown hairs, inner surface partially displayed and white; style white. Perianth loosely tomentose to villous outside, glabrous inside. Pistil 8–11 mm long; ovary villous; style densely long-villous throughout. Follicles 8–10 mm long, appressed to rachis, villous. Fig. 10D–F.

Occurs in the N.T., on the Kakadu Escarpment and in Western Arnhem Land, growing beside creeks on sandstone plateaus, in sandy soil often over rock sheet. Regenerates primarily from seed, possibly also from epicormic buds. Flowers Dec.–May. Map 81.

N.T.: top of Jim Jim Falls, *C.R.Dunlop 5649* (AD, BRI, CANB, DNA, MEL, NSW, PERTH); Jim Jim Creek, upstream of falls, *R.O.Makinson 1165* (BRI, CANB, DNA, K, NSW); near Mt Gilruth, *L.A.Craven & G.M.Wightman 8304* (CANB, DNA); c. 49 km E of Mudginberry HS, *M.Lazarides 7771* (CANB, DNA); 44 km SE [of] Oenpelli, *C.R.Dunlop 4918* (AD, BRI, CANB, DNA, K, L, NSW).

Grevillea rubicunda together with *G. dunlopii* are highly distinctive within the genus, for differences between the two see under *G. dunlopii*.

Most populations of *G. rubicunda* have branchlets and leaf lower surfaces with straight appressed hairs; some populations, for example Jim Jim Falls and the Type of the species, have a tomentose indumentum of curled hairs on these parts.

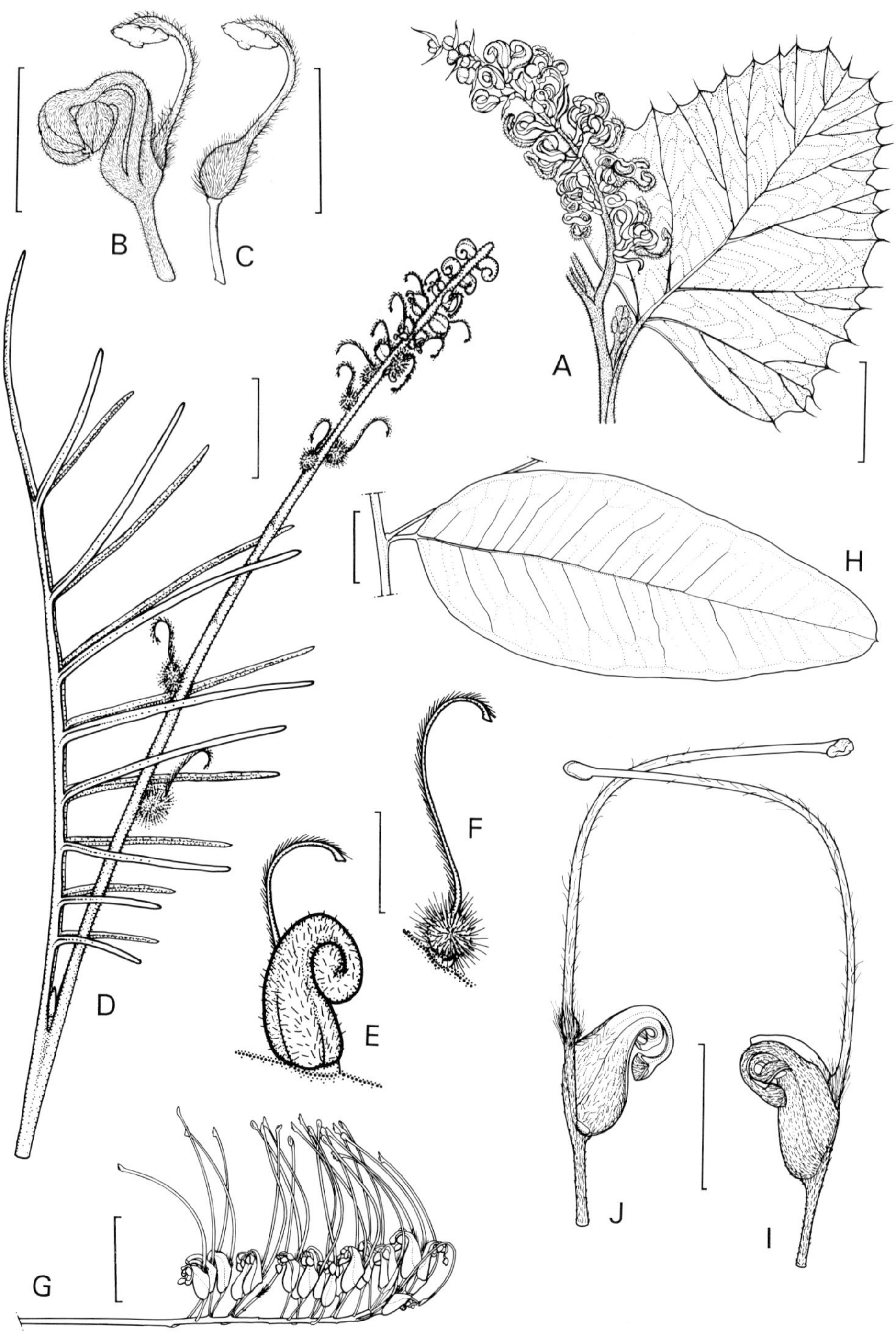

Figure 10. *Grevillea*. **A–C**, *G. polyacida*. **A**, flowering branch; **B**, flower; **C**, pistil (**A–C**, G.Wightman 1336, NSW). **D–F**, *G. rubicunda*. **D**, flowering branch; **E**, flower; **F**, pistil (**D–F**, R.O.Makinson 1165, CANB). **G–J**, *G. goodii*. **G**, inflorescence; **H**, leaf; **I**, flower; **J**, pistil and half perianth (**G**, **I–J**, N.T.Burbidge 5259, CANB; **H**, P.Martensz & R.Schodde AE410, CANB). Scale bars: **A**, **D**, **G** = 2 cm; **B–C**, **E–F** = 5 mm; **H–J** = 1 cm. Drawn by: **A–C**, C.Payne; **D–F**, L.Spindler; **G–J**, D.Boyer.

68. Grevillea dunlopii Makinson, *Fl. Australia* 17A: 494 (2000)

T: Mt Gilruth area, 13°03'S, 133°01'E, N.T., 5 June 1978, *C.R.Dunlop 4901*; holo: DNA; iso: BRI, CANB, K, NSW.

G. sp. aff. rubicunda, P.M.Olde & N.R.Marriott, *Grevillea Book* 3: 151 (1995).

[*G. rubicunda auct. non* S.Moore: J.Brock, *Top End Native Plants* 208 (1988)]

Illustrations: K.Brennan, *Wildfl. Kakadu* 23 (pl. 28) (1986), as *G. rubicunda*; J.Brock, *loc. cit.*, as *G. rubicunda*; P.M.Olde & N.R.Marriott, *Grevillea Book* 3: 151 (120A), 152 (top left & 120B) (1995), as *G. sp. aff. rubicunda*.

Spreading shrub 0.5–2 m tall. Leaves 4–6 cm long, pinnatisect with 9–17 broadly linear to narrowly elliptic simple spreading lobes; lobes 2–5 cm long, 1.8–4 mm wide; upper surface tomentose with biramous and scattered erect simple hairs; lower surface usually narrowly exposed, open-tomentose to subvillous with ±straight biramous hairs, and with simple erect glandular hairs concentrated along midvein. Conflorescences ±erect, terminal, simple; rachis 8–15 cm long (active flowering zone 3–?8 cm long), villous; unit conflorescence narrowly conico-cylindrical, acropetal; flowers sessile, basiscopic. Flower colour: perianth and style pale cream to white (pale hairs only). Perianth loosely villous outside, glabrous inside. Pistil 6–8 mm long; ovary villous; style loosely to sparsely villous, ±glabrous on ventral side. Follicles 8–10 mm long, oblique-ovoid, appressed to rachis, villous.

Occurs in the N.T., in Kakadu Natl Park and western Arnhem Land. Often beside streams or in shallow sandy pockets on sandstone escarpment. Regenerates probably from seed only. Flowers Dec.–May. Map 82.

N.T.: 8 km ESE of Jabiluka, 8 Apr. 1979, *D.Millar & L.D.Pryor* (DNA); 23.5 km SSE of Jabiluka East, *L.A.Craven 6282* (CANB, DNA); 17.5 km NNE of Jabiluka East, *L.A.Craven 6045* (CANB, DNA); East Alligator region, *J.Brock 38* (DNA); 4 km S of Cahills Crossing, *J.Russell-Smith 571* (DNA).

Grevillea dunlopii has the leaf midvein and marginal veins scarcely decurrent to the stem, with no or very slight ridges below the leaf; the floral bracts are 4–7 mm long, narrowly lanceolate, subvillous outside with mixed biramous and glandular hairs; the leaves usually drying to olive green above and khaki below; the outer surface of the perianth with only pale hairs; and the floral rachis and follicle often with simple glandular hairs as well as the biramous non-glandular type. In *G. rubicunda* the leaf veins are decurrent to the stem and form 3 conspicuous ridges below the leaf; the floral bracts are 3–4 mm long, ovate-acuminate and tomentose outside with biramous hairs only; the leaves dry to dark green above and silvery below; the outer surface of the perianth has mixed pale and brown-pink hairs; and the floral rachis and follicle lack glandular hairs.

The specimen *S.King 227* from 'Northern Kakadu' (DNA, NSW) is consistent with *G. dunlopii* but lacks glandular hairs.

Goodii Group

Shrubs or rarely small trees. Leaves entire or (*G. venusta*) sometimes 3–7-partite, dorsiventral; surfaces ±dissimilar; margins shortly recurved to flat, sometimes also undulate. Conflorescence erect, axillary or terminal, simple or sometimes basally 2- or 3-branched; unit conflorescence secund or subcylindrical or a loose short cluster, strongly basipetal to weakly acropetal. Flowers basiscopic; torus lateral. Perianth zygomorphic (limb of bud decurved), hairy on both surfaces (sometimes sparsely so outside); tepals either loosely coherent at apex in 2 lateral pairs after release of style-end, or with apical portions separating and inner surfaces everted. Pistil 25–55 mm long; ovary stipitate, densely clad with biramous hairs; style exserted from late bud; pollen-presenter lateral, convex-umbonate. Follicle sparsely to densely hairy; indumentum sometimes with reddish blotches; pericarp moderately thick, bony. Seed flat-ellipsoidal, peripterous.

A group of five species in the monsoon tropics of N.T. and northern Qld. Affinities are probably with the *Pteridifolia* Group (hairy ovary and fruit, fruit indumentum with coloured aggregations of hairs) and the *Heliosperma* Group (tendency to basiscopically oriented flowers, an oblique to lateral torus and thick-walled follicles). Bird-pollinated.

1 Unit conflorescences secund, usually > 20-flowered

2 Unit conflorescences weakly basipetal to subsynchronous; conflorescences simple (very rarely branched); prostrate or low multistemmed shrubs

3 Ovary villous (hairs ascending); style usually glabrous in apical third; conflorescences on long peduncles usually on or near ground, and usually exceeding foliage; prostrate shrub with trailing stems **69. G. goodii**

3: Ovary subsericeous (hairs ±appressed); style with at least a few hairs persisting in apical third; conflorescences usually borne clear of ground, usually not exceeding foliage; low shrub with several erect stems,usually < 1.5 m tall (rarely to 3 m) **70. G. pluricaulis**

2: Unit conflorescences strongly basipetal; conflorescences usually paniculately few-branched, sometimes simple; robust single-stemmed shrubs or small trees 1–6 m tall **71. G. decora**

1: Unit conflorescences cylindrical to shortly so, or a loose cluster; usually < 20-flowered

4 Longest ultimate floral rachises usually < 1.5 cm long; style hairs rusty brown; unit conflorescences usually 6–10-flowered; nectary linguiform, conspicuous **72. G. glossadenia**

4: Longest ultimate floral rachises 5–9 cm long; style hairs whitish; unit conflorescences often > 10-flowered; nectary arcuate, inconspicuous **73. G. venusta**

69. Grevillea goodii R.Br., *Trans. Linn. Soc. London* 10: 174 (1810)

T: 'In Novae Hollandiae ora septentrionali; Carpentaria, et Arnhem's Land' [protologue]; lecto: north coast [Australia, ?N.T.], 1803, *R.Brown Iter Austral. 3348*; lecto: K, *fide* D.J.McGillivray & R.O.Makinson, *Grevillea* 418 (1993); isolecto: B *n.v.*, BM, BRI, E, G *n.v.*, K, MEL, NSW, NY *n.v.*, P *n.v.*

G. goodii R.Br. subsp. *goodii sensu* D.J.McGillivray & R.O.Makinson, *Grevillea* 220 (1993).

Illustrations: K.Brennan, *Wildfl. Kakadu* 60, col. pl. (1986); J.Brock, *Top End Native Pl.* 202 (1988), as *G. goodii* subsp. *goodii*; P.M.Olde & N.R.Marriott, *Grevillea Book* 2: 181 (bottom left & 148A, B) (1995).

Prostrate lignotuberous shrub with trailing stems. Leaves ovate to narrowly elliptic, 5–25 cm long, 15–50 mm wide, entire, obtuse to ±truncate at base or rarely cuneate to broadly so; margins undulate, not recurved; lower surface with open to very sparse layer of appressed biramous hairs, often also with few to many erect minute papillae 0.1–0.5 mm long. Conflorescence erect, terminal or rarely axillary, simple, usually exceeding foliage, secund, weakly basipetal to subsynchronous; floral rachis 45–100 mm long, with appressed pale hairs. Flower colour: perianth light green; style pink to red. Perianth openly to sparsely sericeous outside with mixed rusty-brown and pale biramous hairs, pilose inside. Pistil 36–42 mm long; stipe 5–6 mm long; ovary villous; style loosely villous in lower third to half, nearly always glabrous in apical half to third, rarely a few hairs persisting. Follicle obloid-ovoid to subglobose, c. 15 mm long, tomentose with red-brown blotching. *Good's Grevillea.* Plate 10; Fig. 10G–J.

Occurs in the 'Top End' of N.T., from Darwin E to Oenpelli, and inland almost to Pine Creek. Grows in open eucalypt forest, savanna woodland, and low heath, in siliceous sands or loamy soil. Regenerates from seed and/or lignotuber; new stems may arise annually from the lignotuber. Flowers mainly in the Wet season, Nov.–Apr. Map 83.

N.T.: 56 km S of Darwin, 28 Nov. 1955, *W.Bateman* (MEL); c. 64 km ENE of Pine Creek township, *M.Lazarides 163 & L.Adams* (CANB, DNA, E, K, L *n.v.*, NSW, US *n.v.*); c. 22.5 km NE of Oenpelli Mission, *M.Lazarides 7752* (CANB, K); Katherine Gorge Natl Park, 7 Mar. 1977, *A.Wood* (AD, BRI, CANB, DNA, MEL, NSW, PERTH).

Very closely related to *G. pluricaulis*, which differs in always lacking minute papillae on the leaf lower surface, and in having a ±erect multistemmed lignotuberous habit 1–3 m tall, the leaf base always cuneate, exclusively axillary conflorescences not exceeding the foliage, pistils 40–55 mm long, and style with at least a few wispy hairs persisting in the upper half.

McGillivray (*in* McGillivray & Makinson, *Grevillea* 219–221 (1993)) included *G. pluricaulis* and *G. decora* within a broadly circumscribed *G. goodii*.

Over most of the range the leaves have minute papilloid trichomes on the lower surface of the lamina. Specimens from the N and W of the range (Katherine Gorge, Darwin area, and some only from the Oenpelli to Kakadu area), lack these processes or have only a very few. The type collection also lacks this feature.

70. Grevillea pluricaulis (McGill.) Olde & Marriott, *Grevillea Book* 1: 183 (1994)

G. goodii subsp. *pluricaulis* McGill., *New Names Grevillea* 6 (1986). T: NE of Finnis River Crossing, N.T., 16 Jan. 1973, *C.R.Dunlop 3121*; holo: DNA; iso: NSW.

Illustrations: J.Brock, *Top End Native Pl.* 203 (1988), as *G. goodii* subsp. *pluricaulis*; P.M.Olde & N.R.Marriott, *Grevillea Book* 3: 101 (top right & 77A), 102 (77B, C) (1995).

Shrub with several erect stems from lignotuber, 0.7–1.5 (–3) m tall. Leaves narrowly to broadly elliptic or rarely tending ovate, 15–20 cm long, 45–85 mm wide, entire, cuneate to broadly so at base; margins not recurved, undulate; lower surface with sparse (rarely dense) appressed biramous hairs, lacking underlying minute papillae. Conflorescence erect, mostly axillary, usually not or scarcely exceeding foliage, usually simple, secund, weakly basipetal; floral rachis 50–100 mm long, with appressed mostly pale hairs (some rusty brown). Flower colour: perianth light green to apricot or creamy brown; style pale orange-apricot to pale pink. Perianth openly silky outside with mostly pale hairs, pilose inside. Pistil 40–55 mm long; stipe 7–11 mm long; ovary subsericeous; style loosely villous in basal half and some wispy hairs persisting almost to apex. Follicle obloid-ellipsoidal to subglobose, c. 15 mm long, tomentose with reddish blotching. Plate 13.

Occurs in western coastal N.T. on Melville Is., and on the mainland from Delissaville SW of Darwin S to Finniss R. area, Port Keats area, and Newcastle Ra. Grows in open mixed eucalypt forest or shrubland in sandy or lateritic soils. Regenerates from seed and lignotuber. Flowers mainly in Dry season, approx. (Jan.–) Apr.–Oct. Map 84.

N.T.: NE Finnis R. Crossing, *C.R.Dunlop 3121* (BRI, DNA, NSW); Seventeen-mile Plain, Melville Is., *D.J.McGillivray 3919* (NSW); Melville Is., Pickataramoor, Feb. 1979, *G.Fitt s.n.* (BRI, CANB, DNA, NSW); Tabletop Ra., Litchfield Natl Park, 1.2 km from Tolmer Falls turnoff along road to Florence Falls, *R.W.Purdie 3416* (CANB, DNA); Port Keats, *G.S.Robinson* (DNA, NSW).

Closely related and similar to *G. goodii*, which has a trailing prostrate habit, leaf base usually obtuse to ±truncate (occasionally cuneate), conflorescences terminal and usually exceeding the foliage, pistils 36–42 mm long, style glabrous in the upper half, and often the lower leaf surface with a minute indumentum of papillae beneath the biramous hairs.

71. Grevillea decora Domin, *Biblioth. Bot.* 89: 589 (1929)

G. goodii subsp. *decora* (Domin) McGill., *New Names Grevillea* 6 (1986), *p.p.* T: 'am Fusse des Mt. Remarkable bei Pentland (Domin II. 1910)' [protologue]; syn: in colibus ad pedem montis Mt. Remarkable prope opp. Pentland, [Qld], Feb. 1910, *K.Domin 2758, 2871*; syn: PR.

Erect shrub or small tree, 1–6 m tall, single-trunked. Leaves oblong-elliptic to narrowly (rarely broadly) so or tending slightly ovate or obovate, 3–18 cm long, 15–40 (–50) mm wide, entire, cuneate at base; margins not recurved; lower surface densely sericeous. Conflorescence subterminal-axillary or terminal, erect, simple or paniculately few-branched, scarcely exceeding foliage; unit conflorescence secund, strongly basipetal; ultimate floral rachis 15–66 mm long. Perianth densely sericeous outside, pilose inside. Pistil 35–55 mm long; stipe 7–9 mm long; ovary villous; style loosely villous in basal half, glabrous above or a few hairs persisting to near apex. Follicle obliquely ellipsoidal to subglobose, 8–15 mm long, densely pale sericeous with some reddish blotching and emergent pale villous hairs.

Occurs in the ranges of inland eastern Qld, from near Miles to N of Laura. Two subspecies are recognised.

Pistils (40–) 50–55 mm long; ultimate floral rachises 3–6 cm long; branchlets and floral rachises with a rusty brown indumentum; leaf base cuneate; follicles 12–15 mm long **71a.** subsp. **decora**

Pistils 35–40 mm long; ultimate floral rachises 1.5–2.5 (–4) cm long; branchlets and floral rachises with a mostly pale indumentum (a few rusty hairs); leaf base obtuse-rounded; follicles 8–10 mm long **71b.** subsp. **telfordii**

71a. Grevillea decora Domin subsp. decora

Illustrations: J.W.Wrigley & M.Fagg, *Banksias Waratahs & Grevilleas* 243 (1989); D.J.McGillivray & R.O.Makinson, *Grevillea* 219 (1993); both as *G. goodii* subsp. *decora*; P.M.Olde & N.R.Marriott, *Grevillea Book* 2: 112 (bottom centre), 113 (90A, B) (1995), as *G. decora*.

Erect to spreading shrub 2–5 m tall. Branchlets with a mostly rusty-brown indumentum. Leaves 10–18 cm long, 15–40 (–50) mm wide, cuneate to broadly so at base; lower surface densely subsericeous, with biramous hairs mostly pale, sometimes rusty brown hairs present especially along veins. Ultimate floral rachis (30–) 40–66 mm long, with a rusty brown indumentum. Flower colour: buds rusty brown; perianth (after anthesis) and style pinkish red. Perianth subsericeous with mixed pale and rusty brown hairs outside. Pistil (40–) 50–55 mm long. Follicle 12–15 mm long, densely sericeous with alternating areas of pale silver-grey and ±extensive reddish blotching, with longer emergent pale hairs. Plate 14.

Occurs in the ranges of inland eastern Qld, where patchily widespread from Wandoan (NW of Miles) to the Burra Ra. (White Mtns) near Pentland, and reportedly almost to the coast near Townsville. Grows in open eucalypt and shrub associations in well-drained shallow soils on sandstone. Regenerates from seed only. Flowers Jan.–Sept. Map 85.

Qld: near Wandoan, *J.A.Dale 9* (BRI); 27.4 km W of Pentland on Flinders Hwy, Burra Ra., *R.O.Makinson 1527 & I.Telford* (CANB, MEL, NSW); 13.6 km W of Pentland, *R.A.Perry 3552* (AD, BRI, CANB, MEL, NSW).

Occasional small-flowered plants are known with pistils c. 40 mm long (e.g. Torrens Ck area), but it is uncertain whether these represent populations. On other features (rusty brown indumentum on branchlets and floral rachis and cuneate leaf base) they conform to subsp. *decora* rather than subsp. *telfordii*.

71b. Grevillea decora subsp. telfordii Makinson, *Fl. Australia* 17A: 494 (2000)

T: c. 14 km from Laura on Cairns road, 15°40'S, 144°30'E, Qld, 11 Apr. 1975, *L.A.Craven 3235*; holo: CANB; iso: BRI, DNA, L *n.v.*, NSW, PERTH *n.v.*, RSA *n.v.*

G. decora 'small-flowered form', of P.M.Olde & N.R.Marriott, *Grevillea Book* 2: 112 (1995).

Erect to spreading shrub 1–2 (–6?) m tall. Branchlets with a mostly silvery-grey indumentum. Leaves 5–8 (–10) cm long, 25–32 mm wide, obtuse-rounded at base; lower surface densely subsericeous with mostly pale (a few rusty) biramous hairs. Ultimate floral rachis 15–25 (–40) mm long, with a mostly silvery-grey indumentum. Flower colour: buds ?pinkish; perianth pink after anthesis; style pinkish red. Perianth subsericeous outside with mainly pale hairs. Pistil 35–40 mm long. Follicle 8–10 mm long, densely pale sericeous with a little reddish blotching towards apex and with a few longer emergent pale hairs.

Occurs in northern Qld in the Laura district, W of Cooktown. Grows in heath associations and open eucalypt forest on sandstone. Regeneration mode unknown. Flowers at least Mar.–June. Map 86.

Qld: 11 km S of Laura R. crossing on the Peninsula Development Rd, *J.R.Clarkson 6793 & W.J.F.McDonald* (BRI, MBA, NSW, QRS); near Hells Gate Ck, Cook District, *B.Hyland 8129* (CANB, NSW); *s. loc.* [Qld?] *L.Leichhardt NSW96949* (NSW); 16 km from Laura along road towards Lakeland Downs [Stn], *J.W.Wrigley & I.Telford NQ 1472* (CANB).

Similar to subsp. *decora*, which differs in its cuneate leaf base, rusty brown hairs on the young branchlets and floral rachises (both with pale hairs in *G. telfordii*), rachises 3–6 cm long, usually longer pistils ((40–) 50–55 mm long), and a larger fruit (12–15 mm long). *Grevillea decora* subsp. *telfordii* was noted by McGillivray (*in* McGillivray & Makinson, *Grevillea* 220 (1993)) as a small-flowered and small-fruited variant. He also stated that it rarely has 'pale hairs' on the outside of the perianth; this is a slip for 'rusty hairs'. Olde & Marriott (*Grevillea Book* 2: 112 (1995)) note it as the 'small-flowered form' of *G. decora*. They state that it has a 'low habit, rarely exceeding 1.5 m'; collection label data indicates a height of up to 6 m.

72. **Grevillea glossadenia** McGill., *Telopea* 1: 21 (1975)

T: near Barkerville, Qld, 17 Mar. 1972, *B.P.M.Hyland 5927*; holo: NSW; iso: CANB, QRS.

Illustrations: Society for Growing Australian Plants, *Hort. Guide Australia Pl.* 1: t. 4 (1975), as *G. sp. aff. decora*; W.R.Elliot & D.L.Jones, *Encycl. Austral. Pl.* 5: 58 (1990); P.M.Olde & N.R.Marriott, *Grevillea Book* 2: 180 (top left & 147) (1995).

Erect shrub 1–2 (–4) m tall. Leaves ±elliptic, 5–12 cm long, 20–35 mm wide, entire; margins slightly recurved, slightly undulate; lower surface sericeous. Conflorescence terminal or axillary, simple, short, a loose cluster, subsynchronous or flowers opening irregularly; floral rachis to 15 mm long. Flower colour: perianth deep yellow-orange to orange-red; style orange to reddish. Perianth subsericeous outside, pubescent inside. Pistil 26–32 mm long; stipe 7.5–9 mm long; ovary villous; style loosely villous near base becoming sparser above. Follicle obliquely obovoid, 10–14 mm long, subsericeous to subvillous; indumentum lacking coloured blotches.

Occurs in Qld, NW, W and SW of Atherton, mainly in the area from Walkamin to Irvinebank to Herberton to Watsonville, with a disjunct population W of Chillagoe. Grows in eucalypt woodland or low open forest, in shallow granitic soils. Probably regenerates from seed only. Flowers year round, mainly in Apr.–Aug. Map 87.

Qld: E of Walsh Bluff on headwaters of Granite Ck, 4 km SSW of Walkamin, *J.R.Clarkson 2853* (BRI, NSW); between Barkerville and Irvinebank, *B.P.M.Hyland 6712* (CANB, NSW, QRS); 5 km E of Irvinebank by road towards Herberton, *R.O.Makinson 1525 & I.Telford* (BRI, CANB, DNA, K, NSW, PERTH).

This species is occasionally confused with *G. decora*, which has the conflorescences longer, more floriferous, strongly secund and strongly basipetal, the flowers pinkish red, and the lateral veins of the leaves at a greater angle to the midvein. *Grevillea glossadenia* also has a conspicuous linguiform nectary, a 6–10-flowered conflorescence and rusty brown hairs on the style, characters which (together with flower colour) distinguish it from *G. venusta*.

This species is recognised as 'Vulnerable' in J.D.Briggs & J.H.Leigh, *Rare or Threatened Australian Plants* (1995).

73. **Grevillea venusta** R.Br., *Trans. Linn. Soc. London* 10: 175 (1810)

T: 'In Novae Hollandiae ora orientali; prope Cape Townsend' [protologue]; lecto: 'Bed of the Rivulet at the base of the Mountains opposite to strong tide passage east coast' [Shoalwater Bay, Cape Townshend, Qld], Aug. 1802, *R.Brown Iter Austral. 3347*; lecto: BM, *fide* D.J.McGillivray & R.O.Makinson, *Grevillea* 446 (1993); isolecto: K, MEL; probable isolecto: G.

Illustrations: J.W.Wrigley & M.Fagg, *Banksias Waratahs & Grevilleas* 332 (1989); D.J.McGillivray & R.O.Makinson, *Grevillea* 221 (1993); P.M.Olde & N.R.Marriott, *Grevillea Book* 3: 219 (top left & 180A–C) (1995).

Erect shrub 2–5 m tall. Leaves 3.5–19 cm long, entire and 5–20 (–30) mm wide and narrowly elliptic to narrowly ovate, and/or 2–7-partite with narrowly oblong to narrowly elliptic ascending lobes; lowermost lobes 4–11 cm long, 6–16 mm wide; margins flat or slightly recurved; lower surface sparsely to densely sericeous. Conflorescence terminal, erect, simple or basally 2-branched; unit conflorescence subcylindrical or a loose cluster, subsynchronous or weakly acropetal; ultimate floral rachis 50–90 mm long. Flower colour: perianth rich green in basal half, yellow to orange above; style deep maroon to purplish black, with conspicuous white hairs and a maroon tip. Perianth loosely to sparsely sericeous outside, pilose inside. Pistil 30–36 mm long; stipe 8.2–9.2 mm long; ovary villous; style loosely white-villous. Follicle ellipsoidal, 16–19 mm long, loosely to sparsely villous; indumentum lacking coloured blotches.

Occurs in Qld, in coastal areas from Many Peaks Ra. to Shoalwater Bay. Grows in sclerophyll forest or woodland, often in rocky situations or beside creeks, in sandy often granitic soils. Regenerates from seed. Flowers June–Sept. Map 88.

Qld: at foot of Mt Parnassus in Shoalwater Bay Army Camp, *C.H.Piedersen 1502* (BRI); 1 km W of Castletower peak, Many Peaks Ra., S of Calliope, *A.N.Rodd 1137* (NSW); Byfield near Keppel Bay, *C.T.White 8042* (AD, BRI).

Grevillea venusta differs from *G. glossadenia* in its arcuate, inconspicuous nectary, usually 10–20 flowers per conflorescence and its paler style hairs.

This species is recognised as 'Vulnerable' in J.D.Briggs & J.H.Leigh, *Rare or Threatened Australian Plants* (1995).

Heliosperma Group

Tall shrubs or small trees. Leaves entire or divided, dorsiventral; surfaces sometimes similar; margins flat to recurved or occasionally revolute. Conflorescence terminal, erect or decurved, simple or to 9-branched; unit conflorescence secund or occasionally conico-cylindrical, acropetal to basipetal. Flowers initially acroscopic but in some species twisting on pedicel and reversing orientation to basiscopic before anthesis, in some acroscopic throughout, rarely the flowers not always paired. Torus oblique to almost lateral. Perianth zygomorphic, glabrous outside or with simple glandular hairs and/or biramous hairs; inner surface with a narrow transverse beard or rarely glabrous; tepals held ventrally and loosely coherent after release of style-end except along dorsal suture; limb segments usually with a central ridge on outer surface and/or flanged along margins. Pistil 12–50 mm long; ovary stipitate, glabrous, sometimes developing glandular hairs later; style glabrous or with simple glandular hairs, usually ±exserted from late bud; pollen-presenter oblique to lateral, convex. Follicle glabrous; pericarp thin- or thick-walled, crustaceous to bony. Seed discoid to flat-obovoid, narrowly to broadly peripterous.

A group of seven species found in monsoon-tropical northern Australia, to fringes of the semi-arid zone. Related to the *Shiressii* and *Goodii* Groups. Bird-pollinated.

1 Pistil > 30 mm long; stipe > 8 mm long (including portion adnate to torus)

2 Limb segment of each tepal with a conspicuous central longitudinal keel or crest; leaf margins recurved to revolute; follicles thin-walled, crustaceous; seeds narrowly winged; leaf lobes 1–7 mm wide, ±linear; bushy shrubs, often < 1.5 m tall (rarely to 3 m) **76. G. dryandri**

2: Limb segment of each tepal smooth or with only a faint central longitudinal keel; leaf margins flat to shortly recurved; follicles thick-walled, bony-textured; seeds very broadly winged; leaf lobes sublinear to elliptic or obovate, 3–60 mm wide; erect slender shrubs or small trees

3 Erect shrub or tree to c. 8 m tall; primary leaf lobes usually again divided; ultimate leaf lobes 3–10 mm wide; lobe bases usually not decurrent; flowers red or deep pink, in dry season; leaves finely veined, with most veins similar size **74. G. heliosperma**

3: Erect shrub or tree usually < 4 m tall; primary leaf lobes usually entire, 10–40 mm wide; lobe bases decurrent; flowers pink and cream, in wet-season; leaves coarsely veined, with clearly descending orders of vein size **75. G. decurrens**

1: Pistil < 30 mm long; stipe ⩽ 6 mm long (including portion adnate to torus)

4 All leaves entire

5 Leaves round or broadly obovate to obcordate, 50–150 mm wide **80. G. latifolia**

5: Leaves elliptic or narrowly so to obovate or occasionally sublinear, < 35 mm wide **79. G. refracta**

4: Most or all leaves divided or toothed

6 Style glabrous; conflorescence with 2–6 spreading to recurved or refracted branches; pericarp > 1 mm thick **79. G. refracta**

6: Style with a sparse indumentum of minute erect hairs; conflorescence simple or if branched then branches ascending, not refracted; pericarp < 1 mm thick

7 Floral rachis ≤ 0.5 cm long; unit conflorescence ≤ 6-flowered; leaf indumentum of spreading hairs **78. G. versicolor**

7: Floral rachis 3–5 cm long; unit conflorescence many-flowered; leaf indumentum of appressed hairs **77. G. pungens**

74. Grevillea heliosperma R.Br., *Trans. Linn. Soc. London* 10: 176 (1810)

T: 'In Novae Hollandiae orâ septentrionali; Carpentaria; prope littora' [protologue]; lecto: North Coast [Qld or N.T.], *R.Brown Iter Austral. 3320*; lecto: K, *fide* D.J.McGillivray & R.O.Makinson, *Grevillea* 419 (1993); isolecto: BM, E.

Illustrations: J.Brock, *Top End Native Pl.* 203 (1988); W.R.Elliot & D.L.Jones, *Encycl. Austral. Pl.* 5: 62 (top right) (1990); P.M.Olde & N.R.Marriott, *Grevillea Book* 2: 192 (all), 193 (158C) (1995).

Spindly erect shrub or tree 2.5–8 m tall. Leaves 15–40 cm long, pinnatipartite with (1–) 2 (–3) orders of division; primary lobes 7–13, these often again 2–7-partite; ultimate lobes narrowly ovate, -elliptic or -obovate or falcate, rarely sublinear, 3–13 cm long, 3–10 mm wide, pliable, usually petiolulate or occasionally very narrowly decurrent, not pungent; margins shortly recurved to flat; lower surface glabrous to sericeous. Conflorescence simple or to 6-branched; unit conflorescence ascending to decurved, subsecund-conical, strongly acropetal; ultimate floral rachis 60–160 mm long. Flowers twisting to obliquely basiscopic orientation in bud stage; pedicels retrorse, 5–12 mm long. Flower colour: perianth red to pink in bud, greyish mauve at anthesis, red or deep pink afterwards; style red. Perianth glabrous outside, pilose or glabrous inside. Pistil 34–46.5 mm long, glabrous; stipe 9.8–14.0 mm long. Follicle subdiscoid to subglobose, 18.5–35 mm long, thick-walled, glabrous. *Red Grevillea, Rock Grevillea.* Plate 15.

Occurs in the monsoon tropics of northern Australia; in W.A. restricted to the northern and central Kimberley; in N.T. N of the Barkly Tableland and through the 'Top End' N of Katherine, and on coastal islands; in Qld W from c. 140° E. Grows in open eucalypt forest or woodland, often in rocky situations, in shallow sandy soils on quartzite, sandstone or conglomerate. Usually regenerates from seed, but sometimes weakly lignotuberous and with some ability to resprout from epicormic shoots. Flowers usually May–Sept. Map 89.

W.A.: Inglis Gap, King Leopold Ra., *J.R.Maconochie 1192* (DNA, K). N.T.: Nourlangie Rock, 17.5 km ENE of Jim Jim crossing, *L.G.Adams 2825* (CANB, K, L *n.v.*); Lawn Hill Stn, *P.K.Latz 1600* (DNA, K, NSW). Qld: 29 km NW of 'Corinda' [Stn], *S.Jacobs 1516* (CANB, NSW); 11.3 km W of Cliffdale Ck, *D.E.Symon 5036* (AD, BRI, K, NSW).

McGillivray & Makinson (*Grevillea* 215–217 (1993)) designated this as the 'red-flowered assortment' within a broad-concept *G. heliosperma* which included *G. decurrens*, here recognised as a separate species, albeit very closely related and with some tendencies in both species towards morphological intermediates. The delimitation attempted here remains somewhat unsatisfactory. *Grevillea decurrens* always has decurrent bases to the pinnae, and usually has a shorter habit (usually < 4 m tall), one order of leaf division, broader pinnae (to 4 cm wide) with coarser venation (veins clearly with several orders of size), cream to pink flowers, and flowers in the Wet season. *Grevillea heliosperma s. str.* has the pinnae usually petiolulate with non-decurrent bases (occasionally subdecurrent) and fine venation (most veins of similar size), often has two orders of leaf division, and darker flowers mainly in the Dry season. Some collections from the islands of the N.T. show trends towards *G. decurrens*, in decurrence of pinnae bases. Populations in the Mt Isa area of Qld are distinctive in having very globose fruits at the lower end of the size range, a glabrous inner surface of the perianth, and hairier leaves than most other populations.

75. **Grevillea decurrens** Ewart, *in* A.J.Ewart & O.B.Davies, *Fl. N. Terr.* 83 (1917)

T: N of 15°S, N.T., Sept. 1911, *W.S.Campbell*; holo: MEL; iso: NSW.

Illustrations: J.Brock, *Top End Native Pl.* 199 (1988); W.R.Elliot & D.L.Jones, *Encycl. Austral. Pl.* 5: 46 (top left) (1990); P.M.Olde & N.R.Marriott, *Grevillea Book* 2: 113 (bottom right), 114 (91A, B) (1995).

Shrub or tree 2–4 (–6?) m tall. Leaves 15–40 cm long, pinnatipartite with 7–13 primary lobes; lower lobes sometimes bipartite; ultimate lobes elliptic or broadly elliptic to obovate, 3–18 cm long, 10–40 mm wide, pliable, basally decurrent, not pungent; margins flat; lower surface glabrous to sericeous. Conflorescence simple or to 6-branched; unit conflorescence ascending to decurved, subsecund to conico-cylindrical, acropetal; ultimate floral rachis 50–150 mm long. Flowers twisting to obliquely basiscopic orientation in bud stage; pedicels spreading to retrorse, 5–7 mm long. Flower colour: perianth pink, or cream becoming flushed with pink, rarely reddish; style pink. Perianth glabrous outside, sparingly bearded inside. Pistil 32–35 mm long, glabrous; stipe 10–12 mm long. Follicle subglobose, 24–33 mm long, thick-walled, glabrous. Plate 16.

Occurs in the monsoon tropics of northern Australia; in W.A. throughout the Kimberley to Derby; in N.T. from Timber Ck N to Darwin and Melville Is., and E to eastern Arnhem Land. Grows in open woodland in sandy or gravelly soils. Some ability to resprout epicormically after fire. Flowers mainly Nov.–Mar. Map 90.

W.A.: 25 km SE of Kununurra, *K.Paijmans 2444* (CANB, K). N.T.: Timber Ck, road to Scarp, *M.Evans 3132* (CANB, DNA, K); 35 km N of 'Mainoru' old road to 'Bulman', *S.Jacobs 1741* (K, NSW).

Grevillea decurrens is a close sister-species to *G. heliosperma*, under which name it was included (as 'pink-flowered assortment') by McGillivray & Makinson (*Grevillea* 215–217 (1993)). See *G. heliosperma* for differences.

76. **Grevillea dryandri** R.Br., *Trans. Linn. Soc. London* 10: 175 (1810)

T: 'In Novae Hollandiae orâ septentrionali; Carpentaria, Arnhem's Land' [protologue]; lecto: islands in the Gulf of Carpentaria, N.T., Nov.–Dec. 1802, *R.Brown Iter Austral. 3345*; lecto: BM, *fide* D.J.McGillivray & R.O.Makinson, *Grevillea* 414 (1993); remaining syntypes: B *n.v.*, BRI, E, G, K, LE *n.v.*, MEL, NSW, NY *n.v.*, P.

Shrub 0.3–2 (–3) m tall. Leaves 4–28 cm long, usually all pinnatisect with (5–) 10–60 (–71) lobes, rarely lower leaves entire and linear; lobes linear to narrowly elliptic, entire or occasionally the lowermost bipartite; simple leaves and undivided lobes 4–21 cm long, 0.7–3 (–6.5) mm wide, pliable, sometimes pungent; margins recurved to revolute, occasionally enclosing most of lower surface except midveins of leaf and lobes; lower surface sericeous to (rarely) villous. Conflorescence erect-emergent, simple or to 9-branched; unit conflorescence many-flowered, secund, acropetal; ultimate floral rachis 100–600 mm long. Flowers twisting to basiscopic orientation in bud stage, also slumping laxly below rachis soon after anthesis; pedicels spreading, 4.5–10 mm long. Perianth glabrous or with a few erect glandular hairs outside, bearded inside near curve. Pistil 41–50 mm long, glabrous or with a few erect glandular hairs; stipe 8.5–10.5 mm long. Follicle obliquely ellipsoidal or subglobose, 7.5–15 mm long, glabrous or pilose with erect glandular hairs, viscid, thin-walled. *Dryander's Grevillea.*

Occurs in northern tropical Australia, from W.A. to Qld; two subspecies are recognised.

Grevillea dryandri is distinguished by (*inter alia*) the limb segments of the tepals having a conspicuous longitudinal median keel or crest.

Follicles glabrous, with viscid exudate; leaves mostly 15–29-lobed; lobes often irregularly curved and often not strongly mutually aligned (W.A., N.T., Qld) ..... **76a.** subsp. **dryandri**

Follicles with many erect simple glandular viscid hairs, not or scarcely viscid on surface; leaves mostly 39–59-lobed; lobes usually evenly curved to straight, neatly mutually aligned (N.T.) ..... **76b.** subsp. **dasycarpa**

76a. **Grevillea dryandri** R.Br. subsp. **dryandri**

G. rigens A.Cunn. ex R.Br., *Suppl. Prodr. Fl. Nov. Holl.* 23 (1830), as *G. ?rigens*. T: Cape Flinders [N.T.], 1819, *A.Cunningham*; holo: ?K, not found (purported type at K 'Cunningham, June 1821' is probably a subsequent collection); iso: A *n.v.*, NY *n.v.*

G. callipteris Meisn., *in* A.L.P.P. de Candolle, *Prodr.* 14: 375 (1856). T: Goulburn Islands and adjacent mainland coast [N.T.], 1818, *A.Cunningham 197*; holo: NY *n.v.*; iso: K.

G. callipteris Meisn., *Hookers J. Bot. Kew Gard. Misc.* 4: 187 (1852), *nom. nud.*

Illustrations: W.R.Elliot & D.L.Jones, *Encycl. Austral. Pl.* 5: 50 (bottom right) (1990); P.M.Olde & N.R.Marriott, *Grevillea Book* 2: 136 (all), 137 (110B, C) (1995).

Low shrub, usually < 1 m high, often lignotuberous. Leaf lobes (5–) 15–29 (–39), often irregularly curved and often not strongly mutually aligned. Flower colour: perianth white, red, orange-red, pink; limb of red flowers often purplish, limb of pink flowers often cream; style red, orange-red, pink or white, usually matching perianth. Follicle glabrous with a sticky exudate.

Occurs in northern Australia from the Kimberley, W.A., through N.T. N of about Tennant Creek, to Mt Isa and Normanton in Qld, with a possibly disjunct range also in the south-eastern part of Cape York (especially Herberton to Cooktown area). Grows in savanna woodland or open shrubland, in sandy or light soils on a variety of substrates, often in somewhat rocky situations. Regenerates from seed and lignotuber. Flowers Jan.–July, occasionally in other months. Map 91.

W.A.: 31 km SSW of Ord River Dam, *M.Lazarides 8503* (CANB, NSW, PERTH). N.T.: 5.6 km E of Wonarah Telegraph Stn, *G.Chippendale 1977* (BRI, CANB, DNA, MEL, NSW). Qld: 9.6 km from Petford on Herberton Rd, *H.S.McKee 9424* (BRI, CANB, NSW); 14.4 km E of Riversleigh Stn, *R.A.Perry 1441* (BRI, CANB, MEL, NSW).

In the area between Kununurra and Wyndham (and possibly extending E over the N.T. border) occurs a population with densely and distinctly subsericeous upper leaf surfaces; this population may deserve further investigation and formal recognition.

76b. **Grevillea dryandri** subsp. **dasycarpa** McGill., *New Names Grevillea* 5 (1986)

T: 10 km ESE of mouth of Deaf Adder Gorge on S side, N.T., 18 July 1978, *D.J.McGillivray 3937 & C.R.Dunlop*; holo: NSW; iso: DNA, K.

Illustrations: J.Brock, *Top End Native Pl.* 200, 201 (1988); D.J.McGillivray & R.O.Makinson, *Grevillea* 198, 199 (1993); P.M.Olde & N.R.Marriott, *Grevillea Book* 2: 137 (centre right), 138 (110A–C) (1995).

Shrub usually > 1 m tall. Leaf lobes (29–) 39–59 (–71), usually evenly curved to straight, neatly mutually aligned. Flower colour: perianth pink to bright red; style similar but darker. Follicle pilose with erect multicellular glandular hairs; tips of hairs with exudate, surface of fruit with very little.

Occurs in N.T. where widely distributed in the area between Mataranka, Daly River and Gove Peninsula. Grows in shrubby woodland associations in rocky situations in sandy soils over sandstone. Regenerates from seed and possibly lignotuber. Flowers mainly Feb.–July. Map 92.

N.T.: 70 km NE of Maranboy Police Stn, *L.G.Adams 111 & M.Lazarides* (BRI, CANB, NSW); 9.6 km NE of Goyder R. Crossing, *J.Must 1019* (CANB, DNA, K, L *n.v.*); 14 km W of East Alligator R. crossing (Oenpelli Rd), *D.E.Symon 7182* (AD, CANB, NSW).

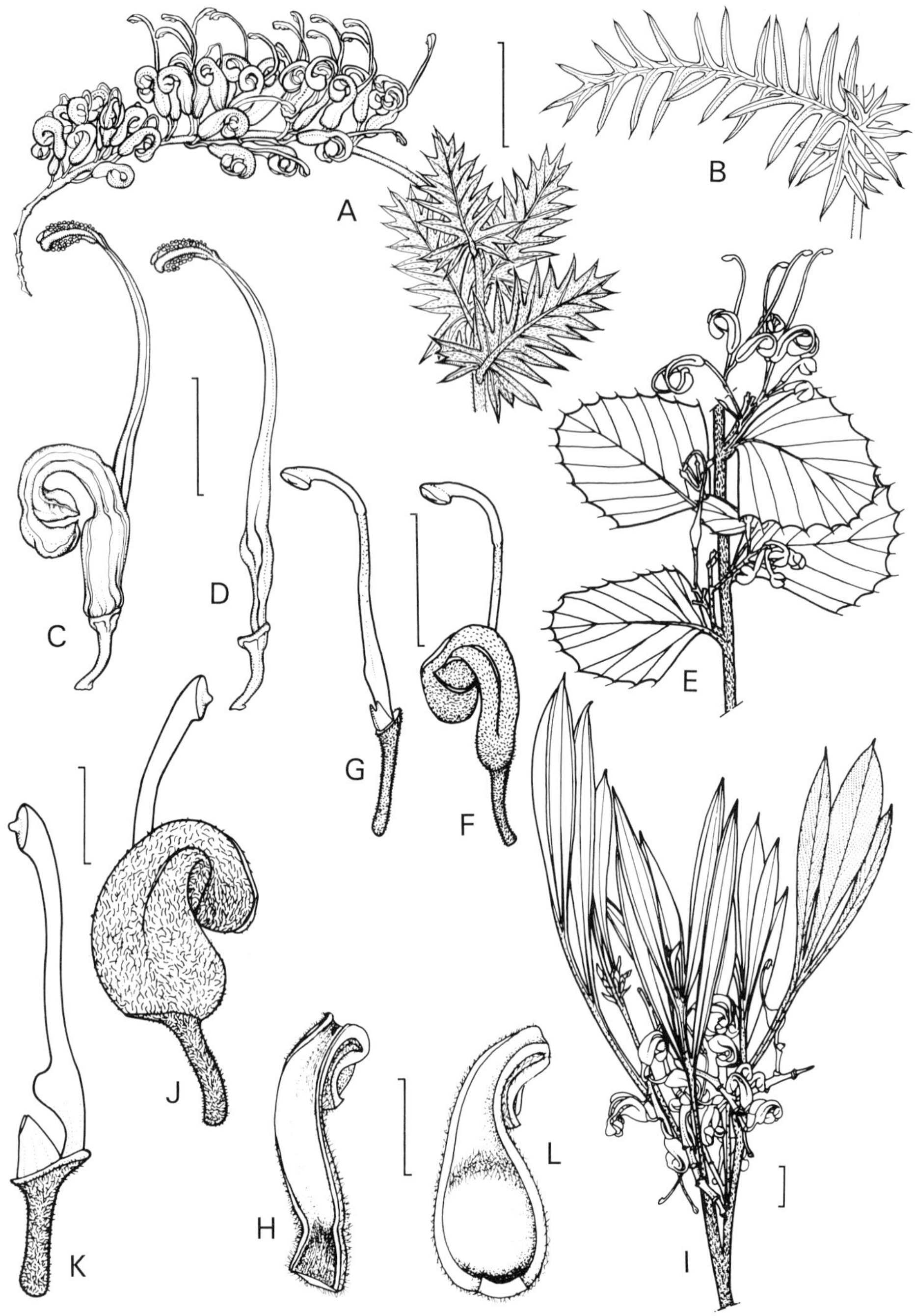

Figure 11. *Grevillea*. **A–D**, *G. pungens*. **A**, flowering branch; **B**, leaf; **C**, flower; **D**, pistil (**A–D**, J.Russell-Smith 3887 & D.Lucas, CANB). **E–H**, *G. versicolor*. **E**, flowering branch; **F**, flower; **G**, pistil; **H**, ventral tepal (**E–H**, D.J.McGillivray 3908 & C.R.Dunlop, NSW). **I**, *G. refracta* subsp. *glandulifera*, flowering branch (P.Ollerenshaw 1678, NSW); **J–L**, *G. refracta* subsp. *refracta*. **J**, flower; **K**, pistil (**J–K**, D.J.McGillivray 3855 & A.S.George, NSW); **L**, ventral tepal (D.J.McGillivray 3812, NSW). Scale bars: **A–B**, **E–G**, **I** = 1 cm; **C–D**, **H**, **J–L** = 5 mm. Drawn by: **A–D**, C.Payne; **E–G**, **I–K**, D.Fortescue; **H**, **L**, B.Chandler.

77. Grevillea pungens R.Br., *Trans. Linn. Soc. London* 10: 175 (1810)

T: Gulf of Carpentaria mainland opposite Groote Island, [N.T.], 4 Jan. 1803, *R.Brown Iter Austral. 3346*; lecto: BM, *fide* D.J.McGillivray & R.O.Makinson, *Grevillea* 436 (1993); isolecto: BM, BRI, E, G, G-DC, K, MEL, NSW, P *n.v.*

G. leichardtii S.Moore, *J. Linn. Soc., Bot.* 45: 211 (1920). T: west side of Gulf of Carpentaria, [N.T.], *L.Leichardt*; holo: BM; ?iso: K.

Illustrations: J.Brock, *Top End Native Pl.* 206 (1988); W.R.Elliot & D.L.Jones, *Encycl. Austral. Pl.* 5: 98 (top right) (1990); P.M.Olde & N.R.Marriott, *Grevillea Book* 3: 121 (centre right & 93A, B) (1995).

Shrub 1–3 m tall. Leaves 1.5–8.5 cm long, 15–30 mm wide, ovate to narrowly so in gross outline, dentate to pinnatipartite, with 6–30 (–50) triangular to linear, rigid, pungent teeth or lobes; lower lobes usually again toothed or divided; longest ultimate lobes 5–15 mm long; margins recurved; lower surface sericeous. Conflorescence usually simple, decurved to erect, secund, acropetal; floral rachis (30–) 40–110 mm long. Flowers acroscopic throughout or sometimes twisting to basiscopic orientation in late bud stage; pedicels retrorse, 2.5–4 mm long. Flower colour: perianth often yellow to light green in bud, becoming red, pink, white, purplish, or orange; style usually matching. Perianth glabrous or with scattered erect glandular hairs outside, glabrous or bearded inside in throat. Pistil 15.5–17.5 mm long; stipe 3.8–6.0 mm long, glabrous; style with scattered erect glandular hairs. Follicle ellipsoidal-obovoid with an obtuse apiculum, 16–25 mm long, thin-walled, glabrous or with glandular hairs. *Flame Grevillea.* Plate 17; Fig. 11A–D.

Occurs in N.T., in the 'Top End' from Oenpelli to Gove and the lower Roper R. Grows in open eucalypt woodland in sandy soils, often in rocky situations or near watercourses. Regenerates from seed. Flowers July–Nov. Map 93.

N.T.: N of Magela Ck, *N.Byrnes 1972* (CANB, DNA, NSW); Gove, *R.L.Specht 1015* (A *n.v.*, BRI, CANB, NSW, PERTH); Oenpelli, *R.L.Specht 1108* (A *n.v.*, AD, BRI, CANB, NSW).

This species varies greatly in depth of leaf division; variation on a single plant is not uncommon.

78. Grevillea versicolor McGill., *New Names Grevillea* 16 (1986)

T: SW of Nabarlek, N.T., 2 Oct. 1976, *C.R.Dunlop 4288*; holo: NSW; iso: BRI, CANB, DNA, K, L *n.v.*, LE *n.v.*, MEL, US *n.v.*

Illustrations: D.J.McGillivray & R.O.Makinson, *Grevillea* 193, 194, fig. 47, 196, fig. 49a (1993); P.M.Olde & N.R.Marriott, *Grevillea Book* 3: 220 (all) (1995).

Shrub 1.5–3 m tall. Leaves 2–12 cm long, 25–80 mm wide, flabelliform to ovate, serrato-dentate with 14–30 teeth, sometimes pungent; margins flat or slightly recurved; lower surface tomentose. Conflorescence simple or 2- or 3-branched; unit conflorescence erect to spreading, up to 6-flowered, sometimes subsecund, acropetal; ultimate floral rachis 2–5 mm long. Flowers acroscopic throughout; pedicels ascending, 5–8.5 mm long. Flower colour (may vary on and between plants): perianth yellow, white, or apricot, darkening to reddish after anthesis; style reddish, sometimes with a pale tip. Perianth tomentose outside with both biramous and erect simple glandular hairs, with a narrow transverse beard and papillae inside. Pistil 21–24.5 mm long; stipe 2.3–2.9 mm long; ovary glabrous; style with (sometimes very few) minute erect simple hairs. Follicle obliquely ellipsoidal, 12–19 mm long, thin-walled, glabrous or with scattered biramous and simple erect hairs. Fig. 11E–H.

Occurs in N.T., where known only from the Nabarlek area of the Arnhem Land escarpment. Grows in open sclerophyll forest in sandy soils over sandstone. Regenerates from seed. Flowers ?May–Nov. Map 94.

N.T.: WSW of Nabarlek, *D.J.McGillivray 3904, 3905, 3908 & C.R.Dunlop* (NSW); Nabarlek, 17 Nov. 1976, *R.Scott* (AD, BRI, CANB, DNA, MEL, NSW, PERTH).

The flowers are not always paired in the conflorescence in *G. versicolor* (a character shared with *G. latifolia*). This character is otherwise highly unusual in the genus.

This species is recognised as 'Rare' in J.D.Briggs & J.H.Leigh, *Rare or Threatened Australian Plants* (1995).

79. Grevillea refracta R.Br., *Trans. Linn. Soc. London* 10: 176 (1810)

T: Gulf of Carpentaria, [?Vanderlin's Island], [Qld], 14 Dec. 1802, *R.Brown Iter Austral. 3319*; holo: BM; iso: E, G *n.v.*, G-DC, K, MEL, P *n.v.*

Tree or shrub, 2–6 m tall. Leaves pinnatipartite to almost pinnatisect and 4–21 cm long, or sometimes entire and elliptic or narrowly so to obovate or occasionally sublinear, 7–19 cm long, 2–31 mm wide; lobes up to 27, entire, obovate to elliptic or linear, 4–13 cm long, 1–45 mm wide, decurrent, pliable, not pungent; margins recurved to flat; lower surface sericeous to tomentose or subvillous. Conflorescence with 2–6 spreading to refracted branches; unit conflorescence recurved to refracted, secund, basipetal; ultimate floral rachis 5–20 mm long. Flowers acroscopic; pedicels spreading to retrorse, 4–8 mm long. Perianth subsericeous to villous or glandular-tomentose outside, with a transverse beard of hairs and papillae inside opposite summit of ovary. Pistil 15–22.5 mm long, glabrous; stipe 2.0–4.2 mm long. Follicle obliquely broadly ellipsoidal to subglobose, 18–31 mm long, thick-walled, glabrous. *Silver Oak.*

Widespread in N.T. and W.A., north from about latitude 21°S, and with isolated occurrences in north-western Qld. Two subspecies are recognised.

Conflorescences and branchlets with indumentum of biramous hairs only; floral bracts < 2.5 mm long **79a.** subsp. **refracta**

Conflorescences and branchlets with indumentum of mainly erect simple glandular hairs, sometimes some biramous hairs mixed in, especially on perianth and pedicel; floral bracts ≥ 3 mm long **79b.** subsp. **glandulifera**

79a. Grevillea refracta R.Br. subsp. **refracta**

G. heterophylla A.Cunn. ex R.Br., *Suppl. Prodr. Fl. Nov. Holl.* 24 (1830); *Anadenia heterophylla* (A.Cunn. ex R.Br.) Ettingsh., *Denkschr. Kaiserl. Akad. Wiss., Math.-Naturwiss. Kl.* 15: t. 36, figs 7, 8 (1858). T: Cambridge Gulf, [W.A.], 1819, *A.Cunningham 410* (2nd Voyage of *Mermaid*); holo: BM; iso: CGE *n.v.*, E, G-DC, K, MANCH *n.v.*, NSW.

G. heterophylla var. *velutina* Meisn., *in* A.L.P.P. de Candolle, *Prodr.* 14: 382 (1856), as *β ?velutina*; *G. refracta* var. *velutina* (Meisn.) Meisn. ex Benth., *Fl. Austral.* 5: 458 (1870). T: Greville Island, Regents River, [W.A.], 1821, *A.Cunningham [352?]*; lecto: G, *fide* D.J.McGillivray & R.O.Makinson, *Grevillea* 437 (1993); isolecto: K, NY *n.v.*

Illustrations: D.J.McGillivray & R.O.Makinson, *Grevillea* 194, 195, fig. 48, 196, fig. 49b (1993); P.M.Olde & N.R.Marriott, *Grevillea Book* 3: 132 (top right & 104A), 133 (104B, C) (1995).

Branchlets and conflorescences with indumentum of biramous hairs only (sometimes also a few erect simple glandular hairs on flowers only). Leaves (4–) 8–17 (–21) cm long; divided leaves with 5–27 lobes each 2–30 (–45) mm wide. Floral bracts 1–2.2 mm long. Flower colour: perianth orange to cream on limb, red to orange or pinkish below curve, with white hairs, colour usually deepening with age; style cream to yellowish, with a yellow to reddish tip. Fig. 11J–L.

Occurs in W.A. and N.T. widespread N of about latitude 21°S, including coastal islands, and in north-western Qld between Mt Isa and Camooweal. Grows in open woodland or savanna or *Triodia* communities, usually in sandy soil on sandstone or quartzite. Regenerates from seed, with limited epicormic resprouting ability. Flowers mainly May–Aug., with sporadic and opportunistic flowering at other times. Map 95.

W.A.: 23 km SE of Mitchell Plateau Mining Camp, *A.S.George 14517* (CANB, NSW, PERTH). N.T.: 57 km W of Soudan, *G.Chippendale 7306* (AD, DNA, MEL, NSW, PERTH); Mt Tolmer, *J.McBean B755* (CANB, DNA, K). Qld: Settlement Ck, *L.G.Brass 360* (K).

The leaves, branchlets, and conflorescences normally have a closely appressed ±subsericeous indumentum. In the northern and eastern Kimberley (Mitchell Plateau, Karunjie Stn, W of Kununurra) occurs a 'subvelutinous form', which has a dense velvety indumentum of short soft spreading hairs on the leaves and branchlets. This variant has a significant populational basis and tends to have leaf lobes at the broad end of the range, but may not be entirely constant; Olde & Marriott (*op. cit.* 3: 132) report some reversion to an appressed indumentum in older plants at one site.

79b. Grevillea refracta subsp. **glandulifera** Olde & Marriott, *Grevillea Book* 1: 178 (1994)

T: near Lake Argyle, W.A., 14 July 1992, *P.M.Olde 92/66*; holo: NSW; iso: BRI, DNA, PERTH.

Illustrations: J.Brock, *Top End Native Pl.* 207 (1988); P.M.Olde & N.R.Marriott, *Grevillea Book* 3: 133 (bottom right), 134 (105) (1995).

Branchlets and conflorescences with an indumentum of erect simple glandular hairs, sometimes a few biramous hairs mixed in. Leaves 12–15 cm long; divided leaves with 3–17 lobes each 5–16 mm wide. Floral bracts 3–5 mm long. Flower colour: perianth orange to cream on limb, red to orange or pinkish below curve, with white hairs; style cream to yellowish, with a yellow to reddish tip. Fig. 11 I.

Occurs in W.A. on the Ord R. catchment between the Carr Boyd Ra., Spring Ck and Kununurra, and extending just into N.T. in the Pinkerton Ra. area. Grows in open woodland or savanna associations, in sandy soils on sandstone and possibly volcanic substrates. Regenerates mainly from seed. Flowers mainly May–Sept., possibly also at other times. Map 96.

W.A.: Dillens Springs, Oct. 1906, *W.V.Fitzgerald* (K, NSW); 4.8 km towards L. Argyle from turn-off on Kununurra road, *D.J.McGillivray 3780 & A.S.George* (K, NSW). N.T.: SW end of Pinkerton Ra., near Katherine to Kununurra road, *D.J.McGillivray 3791 & A.S.George* (K, NSW).

80. Grevillea latifolia C.A.Gardner, *Bull. Woods Forests Dept., W. Australia* 32: 43 (1923)

T: plateau between Lawley and King Edward Rivers, W.A., 30 July 1921, *C.A.Gardner 1498*; holo: PERTH; iso: B *n.v.*, PERTH.

Illustrations: D.J.McGillivray & R.O.Makinson, *Grevillea* 196, fig. 49c, 197, fig. 50 (1993); P.M.Olde & N.R.Marriott, *Grevillea Book* 2: 227 (bottom right), 228 (189A–C) (1995).

Shrub 1–2.5 m high, often multistemmed. Leaves entire, round or broadly obovate to obcordate, 3.5–15 cm long, 50–150 mm wide, not pungent; margins flat; lower surface with a sparse to dense appressed indumentum. Conflorescence erect, terminal or axillary, simple or few-branched; unit conflorescence 6–16-flowered, secund, acropetal; ultimate floral rachis 3–8 mm long; flowers acroscopic throughout, often not paired; pedicels ascending to spreading, 3.5–8 mm long. Flower colour: perianth pale to deep pink reddening from base with age; limb mauve, or rarely white with a mauve tinge; style white often with a pink to red tip. Perianth glabrous outside, transversely bearded inside. Pistil 12–19.5 mm long, glabrous; stipe 2–5 mm long. Follicle obliquely broadly ellipsoidal, 16–19 mm long, thin-walled, glabrous. Fig. 12A–D.

Occurs in W.A., where restricted to the Kimberley region in scattered populations on the Mitchell and Gardner Plateaus, the King Edward and Lawley R., and E almost to Wyndham. Grows in eucalypt woodland and grassland on plateaus, often on low rises, in sandy soils on sandstone, laterite or quartzite. Regenerates from seed and lignotuber. Flowers Mar.–Sept. Map 97.

W.A.: 45 km E from Karunjie turnoff on road to Wyndham, *D.J.McGillivray 3873, 3874* (NSW); Gauging Stn, Crystal Ck, N end Mitchell Plateau, *K.F.Kenneally 8155* (NSW, PERTH); 27 km NE of Karunjie Stn, *R.A.Perry 3099* (CANB, NSW, PERTH).

Unlike other members of the group, the style is scarcely (if at all) exserted from the late bud before the release of the style-end. The flowers are not always paired in the conflorescence in *G. latifolia* (a character shared with *G. versicolor q.v.*).

This species is recognised as 'Poorly Known' in J.D.Briggs & J.H.Leigh, *Rare or Threatened Australian Plants* (1995).

Shiressii Group

Shrubs. Leaves entire, dorsiventral; surfaces somewhat dissimilar; margins flat or undulate (not recurved). Conflorescence erect, terminal or axillary or rarely cauline, simple or occasionally basally 2- or 3-branched; unit conflorescence a 1–10-flowered irregular cluster. Flowers acroscopic to irregularly oriented. Torus lateral. Perianth markedly zygomorphic, pouched at base, glabrous to minutely hairy outside, inconspicuously hairy or tuberculate inside; limb segments of tepals remaining coherent and inverting ventrally after release of style-end. Pistil 23–32 mm long; ovary stipitate, narrow, glabrous or with minute simple hairs; style glabrous or with minute hairs, exserted from late bud; pollen-presenter lateral, convex. Follicle glabrous, faintly rugulose; pericarp crustaceous. Seed ellipsoidal, with a short waxy wing at apex and along anterior margin.

A group of two species occurring in eastern Australia. Related to the tropical *Heliosperma* Group and possibly also the *Goodii* Group. Bird pollinated. The complete coherence and post-anthesis eversion of the limb segments is unusual in the genus, and a somewhat similar pattern is seen elsewhere only in small-flowered, insect-pollinated species (*G. leptobotrys*, *G. cirsiifolia*).

Most leaves ≥ 10 cm long; pistil ≥ 27 mm long; perianth glabrous outside; erect shrub 1.5–5 m tall **81. G. shiressii**

Most leaves ≤ 3 cm long; pistil < 25 mm long; perianth glabrous or minutely and sparsely hairy outside; sprawling to spreading shrub 0.2–1.5 m tall **82. G. singuliflora**

81. Grevillea shiressii Blakely, *Proc. Linn. Soc. New South Wales* 50: 383 (1925)

T: Mullet Creek, Hawkesbury River, 1 mile [1.6 km] NW of Wondabyne Railway Station, N.S.W., 24 Dec. 1922, *D.W.C.Shiress & W.F.Blakely NSW20220*; holo: NSW; iso: K.

Illustrations: A.M.Blombery & B.Maloney, *Proteaceae Sydney Reg.* 123, t. 45 (1981); D.J.McGillivray & R.O.Makinson, *Grevillea* 217 (1993); P.M.Olde & N.Marriott, *Grevillea Book* 3: 170 (bottom right), 171 (138A–C) (1995).

Erect shrub 1.5–5 m high. Leaves oblong-elliptic to narrowly lanceolate, (6–) 8–19 (–21) cm long, (9–) 13–30 mm wide; margins sometimes undulate; surfaces glabrous except for appressed hairs when young. Conflorescence terminal or axillary or occasionally cauline, erect, usually simple, occasionally 2- or 3-branched, a loose 2–10-flowered cluster, acropetal to subsynchronous; floral rachis 1–8 mm long. Flowers ±acroscopic; pedicels 9–19 mm long. Flower colour: perianth green in bud, becoming translucent, basally and dorsally bluish to mauve, cream elsewhere; style reddish to brown-maroon, with bright green ovary and tip. Perianth glabrous outside, partly tuberculate inside. Pistil 27–32 mm long, glabrous or sometimes with scattered minute erect simple hairs on ovary and style; stipe 10.5–12.5 mm long; pollen-presenter lateral. Follicle ellipsoidal, 14–16 mm long, glabrous.

Occurs in N.S.W. where known from two populations on tributaries of the lower Hawkesbury R. N of Sydney (Mooney Mooney Ck and Mullet Ck). Grows in wet sclerophyll forest or shrub associations on lower slopes and beside streamlines, in alluvial sandy or loamy soils. Regenerates from seed only. Flowers mainly July–Dec. Map 98.

N.S.W.: Mooney Mooney Ck, between Calga and Kariong, 17 Oct. 1979, *H.Fallding & D.H.Benson* (NSW); NE of Wondabyne, *H.S.McKee 6687* (NSW, SYD); Mullet Ck, Wondabyne, *D.J.McGillivray 3096 & B.Payne* (NSW).

This species is recognised as 'Vulnerable' in J.D.Briggs & J.H.Leigh, *Rare or Threatened Australian Plants* (1995).

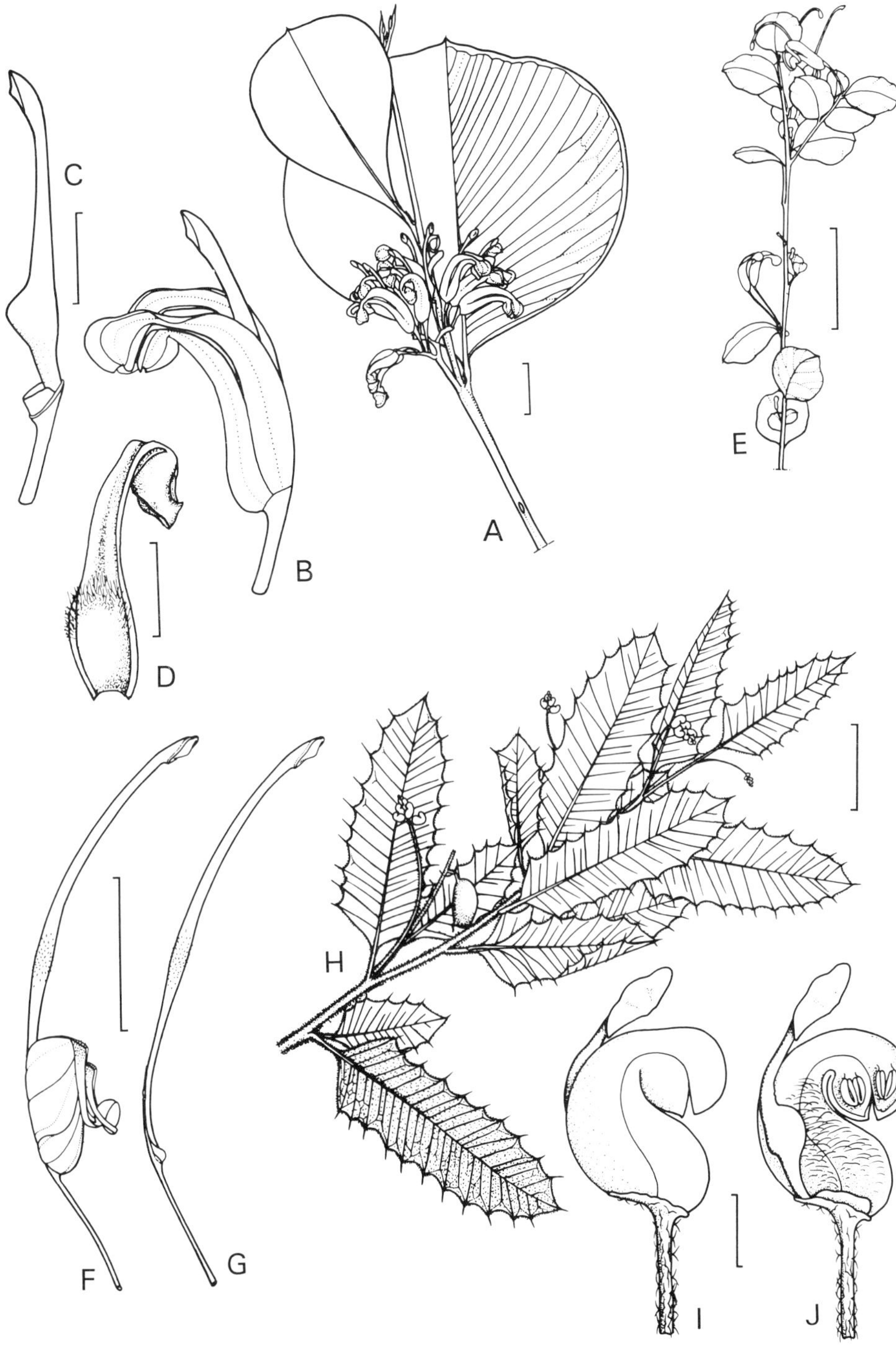

Figure 12. *Grevillea*. **A–D**, *G. latifolia*. **A**, flowering branch; **B**, flower; **C**, pistil; **D**, ventral tepal (**A–D**, D.J.McGillivray 3873 & A.S.George, NSW). **E–G**, *G. singuliflora*. **E**, flowering branch; **F**, flower; **G**, pistil (**E–G**, R?.Henderson 1162 *et al.*, NSW). **H–J**, *G. adenotricha*. **H**, flowering branch; **I**, flower; **J**, pistil and half perianth (**H–J**, May 1985, A.Turner, NSW). Scale bars: **A**, **F–G** = 1 cm; **B–D** = 5 mm; **E**, **H** = 2 cm; **I–J** = 1 mm. Drawn by: **A–C**, **E–G**, D.Fortescue; **D**, B.Chandler; **H–J**, D.Mackay.

82. Grevillea singuliflora F.Muell., *Fragm.* 6: 92 (1867)

T: Dogwood Creek, [Qld], *s.d.*, *L.Leichhardt*; holo: MEL; iso: BM.

Illustrations: B.A.Lebler, *Grevilleas SE Queensland*; *Dept Primary Industry Advis. Leaflet* 1453: 6 (1979), also issued in *Queensland Agric. J.* 105(2): 177–187 (1979); D.J.McGillivray & R.O.Makinson, *Grevillea* 218, fig. 59 & col. pl. (1993); P.M.Olde & N.R.Marriott, *Grevillea Book* 3: 175 (top right & 142A, B) (1995).

Sprawling to spreading shrub 0.2–1.0 (–1.5) m tall. Leaves broadly oblong-elliptic to ovate or almost round, (0.7–) 1.0–2.0 (–3.5) cm long, (5–) 10–16 (–23) mm wide; margins undulate; surfaces hairy when young; upper surface soon glabrous and often glossy; lower surface becoming glabrous or with persistent sparse to open appressed indumentum. Conflorescence terminal, 1- or 2-flowered, erect; floral rachis 1–3 mm long. Flowers irregularly oriented; pedicels 5–10 mm long. Flower colour: perianth translucent green or cream, blackish near base; style maroon with a green tip. Perianth glabrous or with scattered appressed minute hairs outside, pubescent inside near ovary. Pistil 23–24 mm long, with scattered minute simple erect hairs, or rarely glabrous; stipe 10 mm long; pollen-presenter lateral. Follicle ellipsoidal, 13–15 mm long, glabrous. Plate 19; Fig. 12E–G.

Occurs in Qld in scattered populations from Helidon near Toowoomba N to the Blackdown Tableland. Grows in open dry eucalypt forest in sandy soils, usually close to watercourses. Regenerates from seed. Flowers mainly Mar.–Sept. Map 99.

Qld: c. 11 km NNE of Helidon on Helidon to Hampton road, *F.E.Davies 1527 & M.M.Richardson* (BRI, CANB, NSW); Blackdown Tableland, c. 32 km SE of Blackwater, *R.J.Henderson 825 et al.* (BRI); c. 46 km (direct) just E of N of Chinchilla, Barakula State Forest, 0.4 km N of Hellhole Ck crossing of Auburn Rd, *R.O.Makinson 1377 et al.* (BRI, CANB, K, NSW, PERTH).

There is some variation in habit from near-prostrate to robust and upright; in leaf size and colour (light to dark green), and in details of flower colour.

This species is recognised as 'Rare' in J.D.Briggs & J.H.Leigh, *Rare or Threatened Australian Plants* (1995).

Agrifolia Group

Shrubs. Leaves shallowly lobed or dentate or entire, dorsiventral; surfaces often ±similar; margins flat (not recurved). Conflorescence usually axillary, decurved, simple or basally 2- or 3-branched; unit conflorescence conico-cylindrical to ovoid, acropetal. Flowers acroscopic. Torus oblique. Perianth zygomorphic, glabrous or hairy outside, usually bearded inside (rarely glabrous or papillose); tepals held ventrally and loosely coherent after release of style-end; dorsal tepals (except *G. aurea*) partially everting after anthesis, displaying inner surface with conspicuous white beard (sparse or absent in *G. glabrescens*). Pistil 8–24 mm long; ovary stipitate, glabrous or rarely with a few hairs; style minutely hairy with biramous and/or erect simple hairs or papillae (sometimes at style-end only), or glabrous, strongly exserted from late bud; pollen-presenter oblique to lateral, flat. Follicle rugulose to colliculose, glabrous; pericarp moderately to very thick, bony. Seed flat-ellipsoidal, narrowly peripterous.

Six species of the monsoon tropics, restricted to the Kimberley of W.A. and the 'Top End' of N.T. Pollination vectors uncertain, possibly both birds and insects. The group is most closely related to the *Wickhamii* Group.

1 Style (including apical 2–3 mm and back of style-end) completely glabrous and smooth

2 Leaves completely glabrous or with a few scattered minute appressed hairs only (denser along midvein) **84. G. prasina**

2: Leaves with a moderately dense to open (rarely very sparse), evenly distributed indumentum of minute appressed hairs

3 Leaves oblong in general outline or rarely narrowly elliptic, with teeth or lobes (if present) distributed ±evenly around margins; distance from base of leaf blade to apex of lowest lobe < 4 cm **85. G. angulata**

3: Leaves obovate to obovate-cuneate or obtrullate or rarely elliptic or almost round in general outline, with teeth or lobes (if present) mostly in distal half of leaf; distance from base of leaf blade to apex of lowest lobe usually > 4 cm **83. G. agrifolia**

1: Style with small biramous and/or minute simple erect hairs and/or papillae, sometimes confined to apical 1–3 mm and back of style-end

4 Pistil ⩽ 14 mm long

5 Lower surface of leaves completely glabrous or with a sparse sprinkling of minute coppery hairs; leaves narrowly elliptic to elliptic, < 2 cm wide, entire or with 1–3 triangular teeth or lobes; unit conflorescence 1–2 cm long; stylar hairs or papillae confined to apical 5 mm **87. G. brevis**

5: Lower surface of leaves with a dense but inconspicuous indumentum of minute coppery to translucent hairs; most or all leaves obovate to obovate-cuneate or obtrullate or rarely elliptic or almost round in general outline, ⩾ 3 cm wide, with 2–8 shallow teeth or lobes; unit conflorescence ⩾ 2.5 cm long; stylar hairs or papillae usually over most of length of style **83. G. agrifolia**

4: Pistil ⩾ 15 mm long

6 Leaves obovate to obovate-cuneate or obtrullate or rarely elliptic or almost round in general outline, mostly > 40 mm wide, teeth or lobes (if present) mostly in distal half; lower surface inconspicuously hairy; distance from base of leaf blade to apex of lowest tooth or lobe usually > 4 cm **83. G. agrifolia**

6: Leaves oblong to elliptic in general outline, most or all < 40 mm wide, with teeth or lobes ±evenly distributed around margins; lower surface glabrous or inconspicuously hairy; distance from base of leaf blade to apex of lowest tooth or lobe ⩽ 4 cm

7 Lower surface of leaves usually with an open, evenly distributed inconspicuous indumentum of minute hairs (lens feature); unit conflorescence dense; flowers light green in bud, becoming creamy yellow **85. G. angulata**

7: Lower surface of leaves completely glabrous or with minute hairs very thinly scattered (sometimes denser along midvein only); unit conflorescence loose; flowers green in bud, becoming cream or cream yellow, or coppery brown to red in bud becoming orange-red to bright yellow

8 Leaves with (3–) 5–11 teeth or lobes; inner surface of perianth glabrous and mealy or papillose, or with a sparse to dense beard of white hairs; tepals everting after anthesis and inner surface conspicuously displayed; floral rachis 2–4 cm long; unit conflorescence weakly acropetal (hence ovoid to subcylindrical or weakly conical); flowers green in bud, becoming white to cream or very pale yellow **86. G. glabrescens**

8: Leaves with (7–) 9–27 teeth or lobes; inner surface of perianth with a dense beard of white hairs, but tepals not everting after anthesis, so beard not conspicuously displayed; floral rachis 4–16 cm long; unit conflorescence strongly acropetal (hence markedly conical); flowers reddish in bud, becoming orange-red to bright yellow **88. G. aurea**

83. **Grevillea agrifolia** A.Cunn. ex R.Br., *Suppl. Prodr. Fl. Nov. Holl.* 24 (1830)

T: northern Australia [W.A. or N.T.], 1819, *A.Cunningham 411* (2nd Voyage of *Mermaid*); lecto: BM, *fide* D.J.McGillivray, *Telopea* 1: 24 (1975).

Shrub or tree, 1–4.5 m high. Leaves obovate to obovate-cuneate or obtrullate or rarely elliptic or almost round, 5–16 cm long, (10–) 20–55 mm wide, usually with 2–5 (–8) marginal teeth or shallow broadly triangular lobes mostly in apical half or occasionally entire; surfaces similar with an inconspicuous dense to open indumentum of appressed biramous hairs. Conflorescence terminal or axillary, decurved, simple or basally 2- or 3-branched; unit conflorescence dense, ovoid to shortly subcylindrical; ultimate floral rachis 20–50 mm long. Perianth glabrous outside or occasionally subsericeous with biramous hairs only, white-villous inside; tepals everting after anthesis with beard displayed. Pistil 13.5–20 mm long; ovary glabrous or with scattered erect hairs towards apex; style glabrous or with a sparse to open indumentum of minute biramous and/or erect simple glandular hairs over some or all of its length. Follicle obliquely ellipsoidal to obloid-ellipsoidal or subglobose, 11–24 mm long, glabrous.

Occurs in the Kimberley region of W.A. and adjacent offshore islands, and extends to the Victoria R. district of N.T.

Some populations in the northern Kimberley have smaller fruits and generally narrower leaves, and have been distinguished as a separate species, *G. microcarpa* Olde & Marriott (*Telopea* 5: 415–416 (1993)). The nominated differences in leaves, fruits, and stylar indumentum are however not absolute, and at various locations there is considerable intergradation, as well as (at others) apparent sympatry with no intergradation. Influences of microhabitat in these populations should be considered. In view of the relatively high frequency of intergrades, separate species are hard to maintain and two subspecies are recognised here.

Mature fruit (15–) 18–23 mm long; pericarp 2.8–3.1 (–4.6?) mm thick at centre-face; most leaves > 40 mm wide ..... **83a.** subsp. **agrifolia**

Mature fruit usually 11–15 mm long; pericarp ⩽ 1.5 mm thick at centre-face; most leaves usually < 40 mm wide ..... **83b.** subsp. **microcarpa**

83a. **Grevillea agrifolia** A.Cunn. ex R.Br. subsp. **agrifolia**

G. agrifolia var. *major* Ewart & B.Rees, *Proc. Roy. Soc. Victoria* 24: 68 (1911). T: Mission Station, Napier Broome Bay, W.A., 10 Dec. 1901, *G.F.Hill 22*; holo: MEL; iso: MEL, PERTH.

Illustrations: A.S.George, *Introd. Proteaceae W. Australia* 60, t. 84 (1984); D.J.McGillivray & R.O.Makinson, *Grevillea* 210 (1993); P.M.Olde & N.R.Marriott, *Grevillea Book* 2: 17 (bottom right), 18 (10A, B) (1995), as *G. agrifolia*.

Robust shrub 2–6 m tall, with ascending branches from above base. Leaves 5–16 cm long, (30–) 40–95 mm wide, with 2–8 teeth or lobes, grey-green to silvery blue-grey. Flower colour: perianth pale green in bud becoming cream at anthesis, with white erect hairs on inner surface conspicuously displayed after anthesis; style pale green to cream with green tip. Pistil 13.5–19 mm long; style sparsely pubescent with mixed biramous and simple-glandular hairs over its entire length or occasionally glabrous or almost so. Follicle (15–) 18–24 mm long at maturity; pericarp 2.8–3.1 (–4.6?) mm thick at centre-face. Plate 20.

Occurs in W.A and N.T., widespread in the south-western and eastern Kimberley, including offshore islands and as far S as Halls Creek, and in the N.T. in the north-western part of Victoria R. district (e.g. Timber Ck, Jasper Gorge). Grows in woodland or open shrub associations, often in rocky sites, in sandy or loamy soils, over sandstone, quartzite or laterite. Regenerates from seed, and also epicormic buds after low-intensity fire. Flowers mainly May–Sept. Map 100.

W.A.: near The Rocks, King Leopold Ra., *M.Lazarides 6454* (BRI, CANB, MEL, NSW, PERTH). N.T.: 192 km W of Timber Ck, *J.R.Maconochie 1126* (AD, CANB, DNA, NSW, PERTH); 162 km SE of Carlton Stn, *R.A.Perry 3003* (AD, BRI, CANB, MEL, NSW, PERTH).

Intergrades with subsp. *microcarpa* on leaf length, leaf width, and stylar indumentum, although in areas of co-occurrence plants assignable to subsp. *agrifolia* usually have the style inconspicuously hairy over the entire length. There is considerable independent variation in perianth and stylar indumentum. Subsp. *agrifolia* occurs in both rocky sites and on flat plateaus in deep sandy or lateritic loam soils.

Populations on the West Kimberley coast and islands (vicinity of Yampi Sound and Camden Sound) tend to have fruits at the very upper end of the size range; these populations are as yet poorly collected, but formal recognition may be warranted after further investigation.

83b. **Grevillea agrifolia** subsp. **microcarpa** (Olde & Marriott) Makinson, *Fl. Australia* 17A: 494 (2000)

G. microcarpa Olde & Marriott, *Telopea* 5: 415 (1993). T: Northern Kimberley: King Edward Crossing on road to Mitchell Plateau, W.A., July 1991, *P. & G.Keane s.n.*; holo: NSW; iso: CANB, DNA, PERTH.

Illustrations: P.M.Olde & N.R.Marriott, *Grevillea Book* 3: 25 (bottom right), 26 (15A–C) (1995), as *G. microcarpa*.

Bushy shrub 1–2.5 m tall, often with several ascending branches from ground level. Leaves 1–7.5 (–9) cm long, 10–40 (–50) mm wide, with 2–5 teeth or lobes, khaki to grey-green. Flower colour: perianth pale green in bud becoming cream or pale yellow at anthesis, with white erect hairs on inner surface conspicuously displayed after anthesis; style pale green to cream with green tip. Pistil 15–20 mm long; style often glabrous or sometimes sparsely pubescent with biramous and/or erect simple glandular hairs. Follicle 11–15 (–17?) mm long at maturity; pericarp 0.8–1.5 mm thick at centre-face.

Occurs in the northern Kimberley region of W.A., where confined to the area between Napier Broome Bay and the Drysdale and King Edward Rivers. Grows usually in open shrub associations, often around rocky sandstone outcrops in skeletal sandy soils. Regeneration mode unknown, possibly some lignotuber capability. Flowers mainly May–Aug. Map 101.

W.A.: Woppinbie Ck, 4 km SW from mouth of Napier Broome Bay, *S.Forbes 2136* (MEL, NSW, PERTH); Coucal Gorge, Carson Escarpment, *A.S.George 13838* (PERTH); 4 km NW of Kalumburu, *D.J.McGillivray 3860 & A.S.George* (K, NSW, PERTH, US).

Shows a more consistent preference for rocky sites (e.g. drop-offs) as compared to subsp. *agrifolia*, and some of the apparent differences may be ecotypic. Where the two co-occur, plants assignable to subsp. *microcarpa* usually have the style glabrous or nearly so, although at some sites there are plants apparently intergrading on the foliar, stylar and fructual characters.

This subspecies is recognised (under *G. microcarpa*) as 'Poorly Known' in J.D.Briggs & J.H.Leigh, *Rare or Threatened Australian Plants* (1995).

84. **Grevillea prasina** McGill., *New Names Grevillea* 12 (1986)

T: 39 miles [c. 63 km] WNW of Wave Hill Police Station, N.T., 27 June 1949, *R.A.Perry 2265*; holo: CANB; iso: AD, BRI, CANB, MEL, NSW, PERTH.

[*G. angulata auct. non* R.Br.: G.Bentham, *Fl. Austral.* 5: 455 (1870), *p.p.*; F.J.H. von Mueller, *Syst. Census* 69 (1882); J.S.Beard, *Descr. Cat. W. Austral. Pl.* 2nd edn, 38 (1970)]

Illustrations: D.J.McGillivray & R.O.Makinson, *Grevillea* 213 (1993); P.M.Olde & N.R.Marriott, *Grevillea Book* 3: 108 (top left & 82A, B), 109 (82C) (1995).

Shrub 1–4.5 m high. Leaves ovate or elliptic in general outline, 3–8 cm long, 20–50 mm wide, coarsely 5–11-dentate with teeth spaced ±evenly around margins; surfaces similar, bright yellow-green when live, usually glabrous when adult or sometimes with a few scattered minute appressed coppery hairs persisting, usually denser along midvein. Conflorescence terminal or axillary, decurved, simple or basally 2- or 3-branched; unit conflorescence dense, ovoid to shortly subcylindrical; ultimate floral rachis 10–30 (–45) mm long. Flower colour: perianth green becoming cream then pale yellow; style pale green to white with a green tip. Perianth glabrous or subsericeous outside, white-villous inside; tepals everting after anthesis with beard displayed. Pistil 18.5–22 mm long, glabrous. Follicle ellipsoidal, 8.5–16 mm long, glabrous.

Occurs in W.A. in the north-eastern Kimberley, between the Pentecost R. and the N.T. border, and in N.T. mainly in the north-western part of the Victoria R. district to Port Keats, and with remote easterly stations near the Gulf of Carpentaria (Cox R. and Nathan R. catchment). Grows in open woodland or shrubby associations, often rocky situations in skeletal sandy soils or beside watercourses, on sandstone, quartzite, laterite or limestone substrates. Regenerates from seed and lignotuber. Flowers fragrant, Mar.–Oct. Map 102.

W.A.: 59 km W on Wyndham–Gibb R. road, *P.M.Olde 92/65 & 65A* (CANB, DNA, NSW, PERTH). N.T.: Limbunya, *C.Dunlop 3519* (CANB, DNA, NSW, PERTH); Cox River Stn, *P.K.Latz 7160* (B *n.v.*, BRI, CANB, DNA, K, MO *n.v.*, NSW).

Grevillea prasina is similar to *G. angulata*, which almost always has inconspicuous papillae or short hairs near the apex of the style, leaves consistently and more densely minutely hairy, and branchlets round to slightly angular in cross-section (sharply angular in *G. prasina*). It is also similar to *G. glabrescens* which has the beard of the inner surface of the perianth sparser or absent, a usually larger fruit 14–16 mm long, and hairs or papillae on the style-end; and to *G. agrifolia* which has leaves with teeth or lobes mostly in the apical half and the lower surface inconspicuously but densely sericeous, and usually has minute hairs on the style.

85. Grevillea angulata R.Br., *Suppl. Prodr. Fl. Nov. Holl.* 24 (1830)

T: Sim's Island, [N.T.], Mermaids 1st Voyage, 1818, *A.Cunningham 163*; lecto: BM, *fide* D.J.McGillivray, *Telopea* 1: 24 (1975); isolecto: BM, E *n.v.*, G-DC (? *s.n.*), K, NSW, NY *n.v.*

Illustrations: J.Brock, *Top End Native Pl.* 198 (top) (1988); P.M.Olde & N.R.Marriott, *Grevillea Book* 2: 28 (top left & 18A–D) (1995).

Spreading to erect shrub 1–2.5 m tall, or rarely prostrate. Leaves usually oblong in general outline, 3.5–15 cm long, 15–35 (–50) mm wide, pinnatifid or coarsely dentate with (2–) 7–29 triangular ±evenly distributed teeth or lobes, or rarely oblong to narrowly elliptic and entire; surfaces similar, one or both usually with an open inconspicuous appressed indumentum, rarely almost glabrous. Conflorescence terminal or axillary, decurved, simple or basally 2- or 3-branched; unit conflorescence dense, ovoid to shortly subcylindrical; ultimate floral rachis 15–30 (–50) mm long. Flower colour: perianth green in bud becoming cream with white hairs of inner surface displayed after anthesis; style white to cream with a green tip. Perianth glabrous outside, densely white-villous inside; tepals everting after anthesis with beard displayed. Pistil 18–24 mm long; ovary glabrous; style glabrous except for short erect simple hairs or papillae in apical 1–3 mm, rarely completely smooth and glabrous. Follicle oblique-obovoid to -ellipsoidal, 10–14 mm long, glabrous. Plate 21.

Occurs in N.T. in north-western Arnhem Land, from the Oenpelli–Nabarlek area NE to the Murganella area, Sims Is. and Goulburn Is. Grows in shrub and woodland associations, along creeks or around rock platforms and rocky slopes, in sandy soils or (near coast) also on laterite. Regenerates from seed or epicormic buds. Flowers mainly Mar.–Sept. Map 103.

N.T.: 8 km E of East Alligator R. crossing, *C.Dunlop 3704* (DNA, K, MO *n.v.*, NSW); WSW of Nabarlek, *D.J.McGillivray 3910* (B *n.v.*, CANB, DAR *n.v.*, DNA, E, K, LE *n.v.* NSW, NY *n.v.*, PERTH, PRE *n.v.*, RSA *n.v.*, US *n.v.*); laterite cliffs E of Murganella on coast, *P.Olde 92/16* (CANB, DNA, NSW, PERTH).

There is considerable variation in habit and leaf division. In the N of the range (coastal strip and islands E of Murganella), plants vary in habit from ±erect and 1.5–2 m tall, to dense often low spreading or (near coast) completely prostrate shrubs. Leaves of plants from this area are usually at the shorter and broader end of the variance, with 2–11 very shallow teeth or lobes or occasionally with some or all leaves obtuse-oblong and entire, or very rarely narrowly acute-elliptic and entire (then similar to those of *G. brevis*, but with an evenly distributed mid-dense indumentum). The prostrate, entire-leaved form is illustrated in Plate 21. Some plants also have the style-end with very few papillae or hairs, or rarely completely smooth and glabrous. Plants from this area are referred to by Olde & Marriott (*loc. cit.*) as the 'Type form', although 'Northern form' may be a better appellation given the several dimensions of variance. At the southern (possibly disjunct) end of the range, around Nabarlek, Oenpelli and the lower East Alligator R., plants are taller (1.5–2.5 m), with longer, narrower leaves with deeper sinuses and 7–29 teeth or triangular lobes.

Grevillea angulata is very closely related to *G. glabrescens*, *G. brevis*, *G. aurea* and *G. prasina*; see under these taxa for differences. The five taxa are close sister species, and with the exception of *G. prasina* were grouped under a broad-concept *G. angulata* by McGillivray & Makinson (*Grevillea* 212 (1993)).

86. **Grevillea glabrescens** Olde & Marriott, *Telopea* 5: 406 (1993)

T: Kakadu National Park: c. 4 km S of El Sharana, N.T., 20 Apr. 1990, *A.V.Slee 2681 & L.Craven*; holo: CANB; iso: A *n.v.*, BRI, DNA, MEL, NSW.

[*G. angulata auct. non* R.Br.: K.Brennan, *Wildfl. Kakadu* 14, pl. 8 (1986)]

Illustrations: K.Brennan, *loc. cit.*, as *G. angulata*; P.M.Olde & N.R.Marriott, *Grevillea Book* 2: 176 (top right & 144A, B) (1995).

Open erect shrub 1–2.5 m tall. Leaves roughly oblong in general outline, 6.5–13 cm long, 10–40 mm wide, pinnatifid with (3–) 5–11 spreading triangular evenly distributed teeth or lobes; surfaces similar, glabrous or with scattered appressed hairs on midvein only. Conflorescence terminal or axillary, decurved, simple or basally 2- or 3-branched; unit conflorescence loose, conico-cylindrical; ultimate floral rachis (10–) 20–40 (–50) mm long. Flower colour: perianth green in bud, becoming white to cream to very pale yellow; style green becoming cream with a green tip. Perianth glabrous outside, inside glabrous and mealy or papillose or with a sparse to dense beard of short white hairs < 1 mm long; tepals everting after anthesis with beard displayed. Pistil 15–20 mm long; ovary glabrous; style glabrous except for papillae or erect hairs in apical 1–3 mm. Follicle ellipsoidal, 14–16 mm long, glabrous.

Occurs in N.T. in eastern Arnhem Land, in the area between Graveside Gorge and about 10 km S of El Sharana. Grows gregariously in shrub associations on sandstone escarpment and in nearby gullies, in shallow sandy soils and rock crevices. Regeneration from seed, possibly also has epicormic shoot capability. Flowers fragrant. Flowers most months, probably with a peak May–July. Map 104.

N.T.: above UDP Falls, *C.Gittins 2631* (DNA, K, NSW, PERTH); top of Gunlom Falls (formerly UDP Falls), *Johnstone 158* (B *n.v.*, NBG *n.v.*, NSW, PERTH, RSA *n.v.*); Kakadu Natl Park, Gunlom (UDP) Falls, 300 m above falls, *R.O.Makinson 1171 et al.* (AD, BRI, CANB, K, MEL, NSW, PERTH); tributary of Barramundi Ck, Djalkurumburr, *H.Thompson 513* (CANB, NSW, US *n.v.*).

Most closely related, and very similar, to *G. angulata* which has leaves with a denser inconspicuous appressed indumentum, markedly denser conflorescences, a longer ovarian stipe (2.5–) 3–5 mm long (2–2.5 mm in *G. glabrescens*), and a consistently dense and conspicuous beard on the inner perianth surface with the hairs mostly > 1 mm long. *Grevillea prasina* has a denser conflorescence, generally shorter leaves 3–8.5 cm long, and an entirely glabrous style. *Grevillea brevis* differs in having leaves entire and elliptic or with 1–3 teeth only, and pistils 8–12.5 mm long. *Grevillea aurea* has floral rachises usually 40–160 mm long, bright yellow to orange-red flowers, and the dorsal tepals not everting after anthesis (inner surface not displayed). The range of *G. glabrescens* marginally overlaps that of *G. aurea* and field observations and some specimens (e.g. *M.Lazarides 9040*, CANB) suggest occasional hybridisation.

This species is recognised as 'Rare' in J.D.Briggs & J.H.Leigh, *Rare or Threatened Australian Plants* (1995).

87. **Grevillea brevis** Olde & Marriott, *Telopea* 5: 410 (1993)

T: Kakadu National Park, 18.5 km S of Gimbat HS (below edge of Marawal Plateau), N.T., 21 Apr. 1990, *A.V.Slee 2689 & L.Craven*; holo: CANB; iso: A *n.v.*, AD, BRI, DNA, K, MEL, NSW.

G. angulata var. *lancifolia* F.Muell. ex Benth., *Fl. Austral.* 5: 455 (1870), as *?lancifolia*. T: Stony Ra., Central Australia, [loc. in error?], *coll. unknown*; holo: MEL; iso: AD.

Illustrations: P.M.Olde & N.R.Marriott, *Grevillea Book* 2: 71 (all) (1995).

Erect open shrub 1–2.5 m tall. Leaves narrowly elliptic to elliptic, 7–15 cm long, 6–15 mm wide, entire or some leaves with 1–3 spreading short triangular teeth or lobes (often well below apex); surfaces similar, glabrous or with scattered minute appressed coppery hairs.

Conflorescence terminal, decurved, simple or basally 2- or 3-branched; unit conflorescence dense to loose, conico-cylindrical to subovoid or subglobose; ultimate floral rachis 10–30 mm long. Flower colour: perianth white to yellow or cream-green; style pale green to cream. Perianth glabrous outside, bearded inside; tepals everting after anthesis with beard displayed. Pistil 8–12.5 mm long; ovary glabrous; style glabrous except for minute hairs or papillae in apical 2–3 (–5) mm. Follicle obliquely ellipsoidal, 13–15 mm long, glabrous.

Occurs in N.T. where known only from the Marawal Plateau, SE of El Sharana, in Kakadu Natl Park. Grows in heath or shrub associations, usually at the escarpment edge, in lateritic sands or sandy clays. Regeneration mode unknown. Flowers Mar.–July. Map 105.

N.T.: Mary R., Bloomfield Springs, *C.Dunlop 7078 & G.Wightman* (DNA); c. 29 km SE of El Sharana Mine, *M.Lazarides 7820* (CANB); Kakadu Natl Park, *C.Dunlop 8538 & P.F.Munns* (AD, BRI, CANB, DNA, MEL, NSW); escarpment above Douglas Springs, *P.Olde 92/27* (CANB, DNA, NSW, PERTH).

Closely related to *G. angulata* which has a longer pistil 18–24 mm long, and broader ±oblong leaves 15–50 mm wide, usually with > 6 teeth or lobes; *G. glabrescens* which has oblong leaves with 5–11 teeth and a pistil 15–20 mm long; *G. aurea* which has a floral rachis 40–160 mm long, bright yellow to orange-red flowers, and dorsal tepals not everting after anthesis; and *G. prasina* which has ovate to oblong-ovate leaves with 5–11 teeth, pistil 18.5–22 mm long, and a completely glabrous style.

The Type specimens of *G. angulata* var. *?lancifolia* F.Muell. ex Benth., labelled 'Stony Ranges – Centre', appear to be assignable to *G. brevis*, although the annotation 'dark red flowers' is discordant. The locality, if interpreted as 'Central Australia', is unlikely to be correct.

This species is recognised as 'Rare' in J.D.Briggs & J.H.Leigh, *Rare or Threatened Australian Plants* (1995).

88. **Grevillea aurea** Olde & Marriott, *Telopea* 5: 407 (1993)

T: Deaf Adder Gorge, along creek running N into the gorge, N.T., 17 July 1978, *D.J.McGillivray 3934 & C.Dunlop*; holo: NSW; iso: B *n.v.*, CANB, DNA, K, LE *n.v.*, PERTH, PRE *n.v.*, RSA *n.v.*, US *n.v.*

G. angulata 'Golden-flowered form', of D.J.McGillivray & R.O.Makinson, *Grevillea:* 212–213 (1993).

[*G. longicuspis auct. non* McGill.: W.R.Elliot & D.L.Jones, *Encycl. Austral. Pl.* 5: facing xvi (1990)]

Illustrations: W.R.Elliot & D.L.Jones, *loc. cit.*, as *G. longicuspis*; J.Brock, *Top End Native Pl.* 198 (lower) (1988); P.M.Olde & N.R.Marriott, *Grevillea Book* 2: 44 (31A), 45 (31B, C) (1995).

Tall open shrub 2–6 m tall. Leaves oblong in general outline, 7–16 cm long, 15–45 mm wide, dentate to shallowly lobed with (7–) 9–27 triangular teeth or lobes spaced evenly about margins; surfaces similar, minutely sericeous when young but soon glabrous, occasionally scattered hairs persisting. Conflorescence terminal, axillary, or cauline, decurved, simple or basally 2- or 3-branched; unit conflorescence conico-cylindrical, lax; ultimate floral rachis 40–160 mm long. Flower colour: perianth coppery brown in young bud, becoming brick red then orange-red to bright yellow; style orange to apricot or yellow. Perianth glabrous outside, white-bearded inside; tepals not everting after anthesis, with inner surface not displayed. Pistil 17–23 mm long; ovary glabrous; style with few to many papillae or short erect simple hairs in apical 2–3 mm or sometimes over apical half. Follicle ellipsoidal, 10–17 mm long, glabrous. Plate 18.

Occurs in N.T. where known from three areas in Kakadu Natl Park along the Arnhem Land Escarpment: NE of Jabiru; in the area from Graveside Gorge to Deaf Adder Gorge and Gilruth Ck; and NW of Barramundi Gorge. Grows in open heath and shrub associations and in shrubby forest understorey, on sandstone talus slopes or on plateaus, often near escarpment edge. Regenerates from seed. Flowers Apr.–Aug. Map 106.

N.T.: Graveside Gorge, *I.Cowie 275* (DNA); Djaljkurumburr, Kakadu Natl Park, *P.Olde 92/37, 37A* (CANB, DNA, NSW, PERTH); tributary of Barramundi Ck, *H.Thompson 511* (CANB, NSW, US *n.v.*); *loc. id.*, *H.Thompson 514* (CANB, KW *n.v.*, MEL, NSW, P *n.v.*, US *n.v.*).

Easily distinguished from other members of the *Agrifolia* Group by the combination of leaves glabrous (or almost so) and long conflorescences with bright yellow to orange-red flowers. The range of *G. aurea* abuts that of *G. glabrescens* in the Djalkurumburr area (at least), and field observations suggest occasional hybridisation.

Wickhamii Group

Shrubs or small trees. Leaves toothed or entire, dorsiventral; surfaces often similar; margins undulate, not recurved. Conflorescence axillary, terminal or cauline, decurved, simple or basally 2–5-branched; unit conflorescence conico-cylindrical to secund, acropetal. Flowers acroscopic. Torus moderately to very oblique. Perianth zygomorphic, glabrous or hairy outside, bearded inside; tepals flared open dorsally, otherwise ±loosely coherent and held ventrally. Pistil 5–14 mm long; ovary stipitate, glabrous or glandular-pubescent; style with biramous and/or simple hairs or tuberculate, visible but not exserted from late bud, scarcely longer than perianth after release of style-end; pollen-presenter lateral, flat to concave. Follicle rugulose or tuberculate, glabrous; pericarp thick, bony. Seed flat-ellipsoidal, narrowly peripterous.

A group of four species in the monsoon tropics of W.A. and N.T., with *G. wickhamii* extending through the northern Eremaean zone to Qld. Pollination vectors uncertain, possibly both birds and insects. Closely related to the *Adenotricha* Group, which has (except for *G. benthamiana*) shorter, subglobose to few- or 1-flowered conflorescences. Also related to the *Agrifolia* Group, which has longer styles. Most populations of most taxa in the *Wickhamii* Group have a slight extension of the dorsal edge of the toral rim out from the pedicels; this feature is shared with most taxa in the *Adenotricha* Group, where it is expressed more strongly.

1 Ovary glandular-pubescent; perianth limb glabrous outside; perianth yellow-green, becoming lemon to cream, sometimes with mauve tinges **91. G. velutinella**

1: Ovary glabrous; perianth limb hairy outside or rarely glabrous; perianth red, orange or yellow outside

2 Most leaves oblong, always spino-dentate with 9–25 spreading teeth spaced evenly around margins; blades of mature leaves usually at least twice as long as wide; leaf venation much more prominent on lower surface **92. G. miniata**

2: Most leaves broadly ovate, -obovate, -elliptic, or -rhombate, occasionally -oblong, most or all leaves with 3–7 teeth or shallow lobes mostly in apical half of leaf or rarely entire; leaf blades usually only slightly longer than wide, occasionally almost twice as long; leaf venation similar on both surfaces

3 Perianth limb either glabrous or with a sparse to dense, appressed indumentum of short sericeous biramous hairs, sometimes mixed with erect, glandular hairs **89. G. wickhamii**

3: Perianth limb with indumentum of short simple erect glandular hairs, with or without some ±appressed biramous hairs

4 Pedicels 6–10 mm long; perianth with simple erect hairs only on outer surface; pollen-presenter 2–2.5 mm wide; nectary long-U-shaped, relatively obscure; flowers orange **90. G. byrnesii**

4: Pedicels 2–6 (–9) mm long; perianth with simple erect hairs only or mixed biramous and simple-erect hairs outside; pollen-presenter < 2 mm wide; nectary cup-like; flowers red, sometimes orange or yellow or pink **89. G. wickhamii**

89. Grevillea wickhamii Meisn., *in* A.L.P.P. de Candolle, *Prodr.* 14: 380 (1856)

T: Usborne's Harbour, W.A., 1839, *J.C.Wickham*; holo: NY *n.v.*; ?iso: K.

G. wickhamii Meisn., *Hooker's J. Bot. & Kew Gard. Misc.* 4: 187 (1852), *nom. nud.*

Shrub or small spindly tree, 1–4 (–8) m tall. Leaves broadly ovate to -obovate or -rhombate in general outline, 3–9 cm long, 25–55 mm wide, usually pungent, coarsely and shallowly 2–7-dentate with most or all teeth in apical half, or very rarely mostly entire and then sometimes elliptical and 2–2.5 cm long, 8–15 mm wide; surfaces similar, inconspicuously tomentose to subsericeous or rarely glabrous. Conflorescence axillary or cauline, simple or 2–5-branched, decurved; unit conflorescence shortly and broadly secund to subcylindrical or subconical; ultimate floral rachis 10–70 (–120) mm long; pedicels 2–6 (–9) mm long. Perianth outer surface glabrous or with either appressed biramous hairs or short simple erect hairs; inner surface pilose. Pistil 5–10 mm long; stipe (0.5–) 1.4–3.2 mm long; ovary glabrous; style loosely pilose to densely tomentose with appressed-biramous and/or simple erect hairs, not exserted from perianth in late bud. Follicle obliquely obloid-ellipsoidal, 10–20 mm long, glabrous.

Widespread in northern W.A., and inland regions of N.T. and western Qld. Six subspecies are recognised here.

1 Outer surface of perianth glabrous; floral rachis and pedicels glabrous; follicle 10–15 mm long

2 Pollen-presenter elliptical to oblong-elliptical in face view, ⩽ 1 mm wide; floral rachis 10–20 (–40) mm long; nectary rising < 0.5 mm above the toral rim, its margin even to undulate or obscurely toothed ..... **89a.** subsp. **wickhamii**

2: Pollen-presenter circular to broadly elliptical or slightly obovate, 1.0–1.3 mm wide; floral rachis (15–) 30–60 (–70) mm long; nectary usually rising < 0.5 mm above the toral rim, its margin usually conspicuously dentate (a large triangular tooth at each end of the arc) ..... **89b.** subsp. **macrodonta**

1: Outer surface of perianth with hairs; floral rachis and pedicels usually with hairs, rarely glabrous; follicle (13–) 15–20 mm long

3 Outer surface of perianth with a very open indumentum of simple short erect hairs (sometimes inconspicuous and/or restricted to the limb segments), occasionally also with a few appressed biramous hairs; floral rachis similar or rarely glabrous or subsericeous with biramous hairs ..... **89c.** subsp. **hispidula**

3: Outer surface of perianth with an open to dense indumentum of ±appressed biramous hairs (sometimes restricted to the limb); floral rachis densely subsericeous with biramous hairs only

4 Pollen-presenter elliptic to oblong-elliptic in face view, ⩽ 1 mm wide; style with simple erect glandular hairs only; perianth pale creamy yellow ..... **89d.** subsp. **pallida**

4: Pollen-presenter circular to broadly elliptical or slightly obovate, 1.0–2.0 mm wide; style usually with mixed appressed-biramous and simple erect glandular hairs, rarely only one type or the other; perianth bright red to orange below a yellow to orange limb, or very rarely pink or golden-yellow throughout

5 Erect to spreading single-stemmed to multistemmed shrub 2–4 m tall; pedicels 3–5 mm long; floral rachis 15–40 (–50) mm long ..... **89e.** subsp. **aprica**

5: Erect single-stemmed shrub or small tree 3–8 m tall; pedicels 6–9 mm long; floral rachis 30–120 mm long ..... **89f.** subsp. **cratista**

89a. Grevillea wickhamii Meisn. subsp. wickhamii

Illustrations: P.M.Olde & N.R.Marriott, *Grevillea Book* 3: 227 (bottom centre & 186A), 228 (186B) (1995).

Erect to rounded shrub 1–3 m tall. Floral rachis 10–20 (–40) mm long, glabrous. Pedicels glabrous. Perianth glabrous outside. Nectary rising < 0.5 mm above the toral rim, its margin even to undulate or rarely obscurely toothed. Style with an open indumentum of simple erect glandular hairs only. Pollen-presenter elliptical to oblong-elliptical in face view, 0.7–0.9 mm wide. Follicle 10–13 (–15) mm long, not laterally compressed. Flower colour: perianth red, often with a yellow limb, or rarely yellowish throughout; style yellow with a greenish tip.

Occurs in the West Kimberley region of W.A., from the Mt Bell area to E of Gibb River Stn, and N to Bachsten Ck and the King Edward R., also on islands of the Buccaneer Archipelago (e.g. Powerful Is., MEL727855). Grows in open shrub associations on sandstone and quartzite outcrops and plateau edges in skeletal soils. Regenerates from seed. Flowers Apr.–Aug. Map 107.

W.A.: Bold Bluff, King Leopold Ra., *J.R.Maconochie 1204* (DNA, NSW, PERTH); 3 km NW from (new) Mount Elizabeth HS along track to Walcott Inlet, *R.O.Makinson 1107 et al.* (BM, CANB, L, MEL, MO, US); Manning R. Gorge, *B.Maloney 11/77* (CANB, K, NSW).

Subsp. *wickhamii* as here defined corresponds closely with the subsp. as circumscribed by McGillivray & Makinson (*Grevillea* 206 (1993)), but the reliance on fruit size is misplaced as there is overlap on this feature with the other subspecies.

One instance of hybridisation, apparently with *G. agrifolia*, has been reported from Bachsten Ck, in the West Kimberley (*W.Molyneux s.n.*, NSW). See also the note regarding the Edkins Ra. specimen, under subsp. *pallida*.

89b. **Grevillea wickhamii** subsp. **macrodonta** Makinson, *Fl. Australia* 17A: 494 (2000)

T: 65 km along Gibb River Rd from turnoff just E of Derby, c. 5 km W of Kimberley Downs Stn western boundary, W.A., 28 Apr. 1992, *R.O.Makinson 1088 et al.*; holo: CANB; iso: AD, BRI, DNA, K, NSW, PERTH.

G. wickhamii subsp. *aprica*, of D.J.McGillivray & R.O.Makinson, *Grevillea* 206 (1993), *p.p.*

G. wickhamii subsp. *aprica*, of P.M.Olde & N.R.Marriott, *Grevillea Book* 3: 228 (1995), *p.p.*

Illustration: K.F.Kenneally *et al.*, *Broome and Beyond* 170 (1996), as *G. wickhamii*.

Erect open shrub 1.5–2 (rarely to 4?) m tall. Floral rachis (15–) 30–60 (–70) mm long, glabrous. Pedicels glabrous. Perianth glabrous outside. Nectary rising 0.5–1.5 mm above the toral rim, its margin usually conspicuously dentate (a large triangular tooth at each end or the arc). Style with an open indumentum of simple erect glandular hairs only. Pollen-presenter circular to broadly elliptical or slightly obovate, 1.0–1.3 mm wide. Follicle 10–13 (–15) mm long, (usually somewhat?) laterally compressed. Flower colour: perianth red to deep orange or rarely bright yellow, often with a paler (to yellow) limb; style yellow, usually with a greenish tip.

Occurs in north-western W.A., from about Goldsworthy N along the coastal plain to Broome and E to Edgar Ra. and Derby and beyond (possibly to the Napier Ra.). Grows on low rises and rocky areas in lateritic gravelly loam soils in low open woodland and shrub associations. Regenerates probably from seed only. Flowers recorded for Dec. and Apr.–Aug. Map 108.

W.A.: Edgar Ra. area, 1.7 km NNW of Ardjorie HS on Mowla Bluff HS track, *S.J.Forbes 2487α, 2487β & K.F.Kenneally* (MEL, CANB); 12 km S of Thangoo [Stn] turnoff, *J.B.Martin KJM-088* (CANB); 33 km NW of Goldsworthy, c. 6 km along Shellborough track from Gt Northern Hwy, *I.R.Telford 6516* (CANB).

89c. **Grevillea wickhamii** subsp. **hispidula** Makinson, *Fl. Australia* 17A: 495 (2000)

T: Mt Windell Rd corridor, 10.6 km ENE of Mt Windell, 20.4 km ESE of Karijini Natl Park H.Q., W.A., 29 July 1991, *S. van Leeuwen 862*; holo: PERTH; iso: CANB, Pilbara Regional Herbarium (Karratha, W.A.).

Erect shrub 2–3 m tall. Floral rachis (10–) 20–40 mm long, with an open indumentum of simple erect hairs, occasionally also a few biramous hairs near the base, or rarely almost glabrous. Pedicels similar. Perianth minutely hispidulous outside with simple erect glandular hairs, occasionally also a few appressed biramous hairs. Nectary rising 0.5–1.0 mm above the toral rim, its margin even to weakly 3-toothed. Style with an open indumentum of erect simple glandular hairs (occasionally also some appressed to ascending biramous hairs). Pollen-presenter circular to slightly obovate in face view, 1.4–2.0 mm wide. Follicle 15 mm long (few seen), somewhat laterally compressed. Flower colour: perianth yellow or bright red (few data); style yellow.

Occurs in W.A. in the Hamersley Ra. and Pilbara areas, and E to Lake Disappointment and Windy Corner. Grows in open eucalypt and *Acacia* associations with *Triodia*, in red loamy to sandy soils, often near drainage lines. Regenerates probably from seed only. Flowers May–Aug. Map 109.

W.A.: 8 km N of Nullagine along Great Northern Hwy, *I.R.Telford 5949 & G.Butler* (CANB); Canning Stock Route, Durba Springs, *J.S.E.Jervis 006* (CANB); 64 km E of Windy Corner, *A.M.Ashby 5419* (AD, CANB).

Subsp. *hispidula* has leaves 2–5 (–7) cm long and 2–3.5 cm wide, and pedicels 3–6 mm long. There is a resemblance to *G. byrnesii*, which has a similar glandular-hispid indumentum on the floral parts and similarly shaped leaves. *Grevillea byrnesii* however has larger leaves 7.5–12 cm long and 3.5–7 cm wide, longer unit conflorescences (rachis 50–120 mm long), and longer pedicels 6–9 mm long.

The range of subsp. *hispidula* approaches that of subsp. *aprica* along the upper Fortescue R.

89d. **Grevillea wickhamii** subsp. **pallida** Makinson, *Fl. Australia* 17A: 495 (2000)

T: Prince Regent Nature Reserve, unnamed northern tributary of Prince Regent R., c. 800 m upstream of tidal limit, W Kimberley, W.A., 3 June 1998, *R.O.Makinson 1714*; holo: CANB; iso: AD, BRI, DNA, K, MEL, NSW, PERTH.

Erect open shrub 2–3 m tall. Floral rachis 10–25 (–30) mm long, densely subsericeous to subtomentose (biramous hairs only). Pedicels with an open indumentum of appressed biramous hairs. Perianth similar, the indumentum denser on the limb. Nectary rising 0.2–0.3 mm above the toral rim, its margin smooth and entire. Style with an open indumentum of simple erect glandular hairs. Pollen-presenter elliptical to slightly and narrowly obovate, 0.7–0.9 mm wide. Follicle (few seen) 10–13 mm long, not laterally compressed. Flower colour: perianth pale lemon-yellow, blackening from the limb on release of style-end; style pale yellow.

Occurs in north-western W.A., known as yet only from a small population at the Type locality in Prince Regent Nature Reserve. Possibly also on Koolan Is. (see below). Grows in *Eucalyptus–Terminalia–Livistona* woodland associations on lower slopes near creeks, in skeletal soils over sandstone. Probably regenerates from seed only. Flowers May–June (at least). Map 110.

Known only from the type locality.

Subsp. *pallida* resembles pale-flowered forms of subsp. *aprica*, but has a smaller, narrower pollen-presenter, generally shorter unit conflorescences and occurs in a different biotope. A budding specimen from Koolan Is. (*L.Vernon 001*, CANB, PERTH) with a note 'flowers white' may be assignable to subsp. *pallida*.

A population (e.g. *R.O.Makinson 1112 et al.*, CANB) in the Edkins Ra. area (W Kimberley), within the range of subsp. *wickhamii*, resembles subsp. *pallida* in pollen-presenter shape and in the indumenta of the perianth, style, pedicels and floral rachis; it differs in having a red perianth and perhaps a more bushy, multistemmed habit. It is here regarded as unassigned to subspecies.

89e. **Grevillea wickhamii** subsp. **aprica** McGill., *New Names Grevillea* 16 (1986)

T: Standley Chasm, c. 49 km W of Alice Springs, N.T., 14 July 1968, *A.E.Orchard 810*; holo: AD; iso: DNA, M *n.v.*, P *n.v.*

Illustrations: J.Brock, *Top End Native Pl.* 209 (1988), as *G. wickhamii*; D.J.McGillivray & R.O.Makinson, *Grevillea* 207 (1993); P.M.Olde & N.Marriott, *Grevillea Book* 3: 229 (187B–E) (1995).

Erect to spreading shrub 1.5–3 (–4) m tall. Floral rachis (15–) 20–40 (–50) mm long, densely to openly subsericeous with biramous hairs only. Pedicels 3–5 mm long, with appressed biramous hairs. Perianth with a dense to sparse indumentum of biramous hairs outside, sometimes confined to the limb. Nectary rising 0.2–0.4 (–0.6) mm above the toral rim, its margin smooth and entire or sometimes obscurely tridentate. Style with a dense indumentum, usually of mixed appressed biramous hairs and simple erect glandular hairs, rarely only one type or the other. Pollen-presenter circular to slightly obovate in face view, 1.0–1.3 mm wide. Follicle 13–17 (–19) mm long, laterally compressed. Flower colour: perianth red or orange, usually with a paler limb, or rarely golden yellow; style matching perianth or paler, yellow to orange or occasionally red. Plate 22.

Widespread in northern inland Australia; in W.A., S of the Kimberley from Margaret R. to the N.T. border, and disjunctly near Docker R. and S of Port Hedland; throughout N.T. S from about Larrimah, and into NW Qld in patchy occurrences from Lawn Hill area to Mt Isa, Cloncurry and Winton. Grows in open woodland or shrubby *Triodia* communities, often in rocky situations, on sandstone, quartzite, limestone, and granite in sand to clayey soil. Regenerates from seed. Flowers throughout the year, peaking in May–Aug. Map 111.

W.A.: 25.5 km SW of Lamboo Stn, *M.Lazarides 6321* (AD, CANB, MEL, NSW, PERTH). N.T.: 36.4 km N of Newcastle waters, *G.Chippendale 7003* (AD, CANB, DNA, MEL, NSW, PERTH). Qld: 4 km S of Mt Isa, *P.Ollerenshaw 1160* (BRI, CANB, NSW).

Previous work (McGillivray & Makinson *Grevillea* 206 (1993); Olde & Marriott *Grevillea Book* 3: 229 (1995)) has recognised a broad-concept subsp. *aprica*; this is here modified by the recognition of other subspecies. Subsp. *aprica* is generally a multi-stemmed shrub, rarely to small tree size with a single trunk, with ascending branches; it has appressed biramous hairs on the outer surface of the perianth, and a dense indumentum of (usually) mixed biramous and simple erect hairs on the style. Subsp. *hispidula* has an open indumentum of simple erect hairs only on these parts; subsp. *macrodonta* has a glabrous perianth outer surface, the style with simple erect hairs only, and a much more prominent nectary. Subsp. *cratista* is very similar to subsp. *aprica*; see under the former for differences.

In rare cases, plants of subsp. *aprica* may have most or all leaves entire and obovate to elliptic (*G.Chippendale s.n.*, 24 km W of Soudan HS, N.T., DNA, MEL).

89f. **Grevillea wickhamii** subsp. **cratista** Makinson, *Fl. Australia* 17A: 495 (2000)

T: 19.2 km SSW of Bungle Bungle outcamp, near Bellburn Ck, 8.2 km S of Tickalara Track en route to Piccaninny Creek, W.A., 7 July 1984, *S.J.Forbes 2557*; holo: PERTH; iso: CANB, MEL, NSW.

G. sp. 'A', B.L.Rye, *in* J.R.Wheeler (ed.) *et al.*, *Fl. Kimberley Region* 475 (1992).

Illustration: P.M.Olde & N.R.Marriott, *Grevillea Book* 3: 228 (187A) (1995).

Erect shrub or small tree 3–5 (–8) m tall. Floral rachis 30–70 (–120) mm long, densely subsericeous with biramous hairs only. Pedicels 6–9 mm long, with few to many appressed biramous hairs. Perianth with a mid-dense indumentum of ±appressed biramous hairs. Nectary rising 0.1–0.2 mm above the toral rim, margin smooth, entire. Style with a dense indumentum of mixed appressed-biramous hairs and simple erect hairs. Pollen-presenter circular to slightly obovate in face view, 1.5–1.9 mm wide. Follicle not seen. Flower colour: perianth red to deep pink, blackening from limb after anthesis; style more or less matching.

Occurs in northern W.A., known as yet only from the Bungle Bungle Ra. Grows in grassy woodland associations in rocky (sandstone) situations and on valley bottoms, in sandy loam soils. Regenerates probably from seed only. Flowers July. Map 112.

W.A.: track to Piccaninny Ck carpark, Bungle Bungle Natl Park, *P.C.Jobson 1396* (MEL, PERTH); 24 km E of Bungle Bungle Outcamp, *K.F.Kenneally 9273* (PERTH); 19.2 km SSW of Bungle Bungle outcamp, *N.H.Scarlett 84-298* (MEL, PERTH).

Subsp. *cratista* is distinctively robust in habit and conflorescences. It is very closely related and similar to subsp. *aprica*, but is a more robust plant to small tree size, with a single main trunk and the branches horizontal to widely ascending. It also has longer pedicels 6–9 mm long, longer unit conflorescences (ultimate floral rachises 30–100 mm long) and rather larger flowers; the conflorescences also tend to be more branched (and the branches less basal) and borne on older wood than in subsp. *aprica*. The torus structure in subsp. *cratista* is subtly different from the other subspecies, being a little wider and with a less-pronounced and later-developing dorsal extension. The development of the flower buds also appears different, with the limb developing faster relative to the lower portion of the perianth. Rye (*loc. cit.*) regarded this taxon as a distinct species based on the longer and hairy pedicels, gynoecium (pistil) length, and the blackening perianth limb. The last is not a unique feature in the species, and subsp. *aprica* shares the hairy pedicels; pistil length is not disjunct from other subspecies. Subspecies rank seems appropriate.

Subsp. *cratista* is similar to *G. byrnesii*, which has larger leaves 7.5–12 cm long, 35–70 mm wide, and the outer surface of the perianth with simple erect glandular hairs only.

90. Grevillea byrnesii McGill., *New Names Grevillea* 3 (1986)

T: between Gibbie Creek and Mt Sanford Station, N.T., 7 July 1978, *D.J.McGillivray 3902*; holo: NSW; iso: DNA, K, US *n.v.*

Illustrations: D.J.McGillivray & R.O.Makinson, *Grevillea* 205 (1993); P.M.Olde & N.R.Marriott, *Grevillea Book* 2: 76 (59A–C) (1995).

Shrub 4–5.5 m high. Leaves broadly ovate to -obovate or occasionally broadly elliptic, -rhombate or -oblong in general outline, 7.5–12 cm long, 35–70 mm wide, sub-pungent, entire or coarsely and shallowly 3–7-dentate with most or all teeth in apical half; surfaces very similar, minutely tomentose. Conflorescence axillary or terminal, simple or to 3-branched, decurved; unit conflorescence loosely conico-cylindrical; ultimate floral rachis (30–) 50–120 mm long; pedicels retrorse, 6–10 mm long. Flower colour: perianth orange with a paler limb; style orange (redder dorsally), with yellow tip. Perianth pilose to almost glabrous outside with erect simple glandular hairs especially on limb, pilose inside. Pistil 9–11 mm long; stipe 2.2–3.5 mm long; ovary glabrous; style minutely pilose with simple erect hairs, not exserted from perianth in late bud. Follicle subglobose to subreniform, 17–24 mm long, glabrous.

Occurs in the East Kimberley and Great Sandy Desert (Winnama Gorge, Dragon Tree Soak), W.A., and in the Victoria River District of the N.T. Grows in skeletal sandy soils or red gravelly lateritic soils on ridge tops or upper slopes in open shrubland or low open woodland or sand dunes. Regenerates from seed. Flowers May–Aug. Map 113.

W.A.: Winnama Gorge, Mabel Downs Stn, *E.A.Chesterfield 187* (MEL, NSW). N.T.: N side of Jasper Gorge, *D.J.McGillivray 3798 & A.S.George* (CANB, DNA, K, LE *n.v.*, NSW, PERTH, PRE *n.v.*, RSA *n.v.*, US *n.v.*); road to Halls Creek, c. 32 km SW of Wave Hill, *P.Ollerenshaw 1659* (CANB, NSW).

Grevillea byrnesii has a large torus 3–5 mm across, usually much wider than base of perianth; a long-U-shaped but relatively obscure nectary; and a pollen-presenter 2–2.5 mm wide. *Grevillea wickhamii* can be distinguished from *G. byrnesii* by its smaller torus (2–3 (–4) mm across) and pollen-presenter (< 2 mm wide) and its cup-like nectary. *Grevillea wickhamii* subsp. *wickhamii* also has smaller follicles and a shorter unit conflorescence than *G. byrnesii*.

This species is recognised as 'Poorly Known' in J.D.Briggs & J.H.Leigh, *Rare or Threatened Australian Plants* (1995).

91. Grevillea velutinella McGill., *New Names Grevillea* 16 (1986)

T: Spillway Creek area near outflow of Lake Argyle, W.A., 7 July 1974, *A.C.Beauglehole 46861 & G.W.Carr*; holo: NSW; iso: K, PERTH.

Illustrations: D.J.McGillivray & R.O.Makinson, *Grevillea* 208 (1993); P.M.Olde & N.R.Marriott, *Grevillea Book* 3: 217 (bottom right), 218 (179A–C) (1995).

Shrub 1.2–3.5 m tall. Leaves broadly obovate-cuneate or -rhombate in gross outline, 5–12 cm long, 35–80 mm wide, not pungent, entire or coarsely and shallowly 2–7-toothed with most or all teeth in apical half; surfaces similar, velvety. Conflorescence usually axillary, decurved, simple or usually 2–5-branched; unit conflorescence loosely conico-cylindrical; ultimate floral rachis 40–110 mm long; pedicels usually retrorse, 7–10 mm long. Flower colour: perianth yellow-green, becoming lemon to cream, sometimes with mauve tinges; style yellow-green to cream. Perianth glabrous outside, pilose to villous inside. Pistil 6.5–9 mm long; stipe 0.5–1.0 mm long; ovary and style coarsely pubescent with erect simple glandular hairs (these occasionally soon falling from ovary); style not exserted from perianth in late bud. Follicle laterally compressed, obliquely ellipsoidal, 18–19 mm long, glabrous.

Occurs in the Ord R. area of the Kimberley, W.A., from near Wyndham W to about El Questro and S to about Bedford Downs Stn. Grows in tall open shrubland or savanna woodland, usually in rocky situations in skeletal soils, usually on slopes of sandstone or quartzite. Regenerates from seed. Flowers Mar.–July. Map 114.

W.A.: The Grotto, c. 30 km SSE of Wyndham, *A.C.Beauglehole 54072* (NSW, PERTH); Ord R. Gorge, *C.A.Gardner 7338* (PERTH); 12.5 km SE of Bedford Downs Stn, King Leopold Ra., *M.Lazarides 6379* (AD,

BRI, CANB, MEL, NSW, PERTH); 6 km NE of Wyndham pumping stn, *J.R.Maconochie 125* (BRI, DNA); King R. near pumping stn, *E.M.Scrymgeour 1746* (PERTH).

Easily distinguished from close relatives such as *G. wickhamii* and *G. byrnesii* by the presence of erect glandular hairs on the ovary, and from *G. agrifolia* which has an ovary stipe 2.5–4 mm long, and a longer pistil (13.5–19 mm long) which loops outwards from the back of the perianth in late bud and projects well beyond the perianth after release of the style-end.

92. Grevillea miniata W.Fitzg., *Western Mail (Perth)* 21 (1066): 10, 28 incl. t. (1906)

T: Mount Leake, Lady Forrest Range, W.A., June 1905, *W.V.Fitzgerald 1198*; lecto: NSW, *fide* D.J.McGillivray & R.O.Makinson, *Grevillea* 429 (1993); isolecto: BM, K, PERTH.

Illustrations: D.J.McGillivray & R.O.Makinson, *Grevillea* 209 (1993); P.M.Olde & N.R.Marriott, *Grevillea Book* 3: 29 (top right & 18A), 30 (18B, C) (1995).

Shrub c. 2–4 m high. Leaves usually ±oblong, 4.5–15 cm long, 30–70 mm wide, shallowly sinuate to dentate with 9–25 evenly spaced teeth, usually pungent; surfaces dissimilar, with venation much more prominent below, subvelutinous. Conflorescence axillary, simple to few-branched, decurved; unit conflorescence broadly secund to subcylindrical; ultimate floral rachis (15–) 30–50 mm long; pedicels retrorse, 2–5 mm long. Flower colour: perianth bright yellow outside, becoming orange or deep yellow after anthesis; inner surface (partly exposed) very deep green becoming bright red after anthesis; style yellow. Perianth pilose outside with erect simple glandular hairs, bearded inside. Pistil 11–13.5 mm long; stipe 2.8–4 mm long; ovary glabrous; style glandular-pubescent, not exserted from perianth in late bud. Follicle obloid to ovoid-ellipsoidal, 14–18 mm long, glabrous.

Occurs in W.A., where restricted to an area in the Kimberley region from the southern end of the King Leopold Ra. to the Bungle Bungle Ra. and extending to the N.T. in the Victoria River District (Gregory Natl Park). Grows in tall open *Eucalyptus*, *Grevillea* and *Acacia* mixed shrubland or woodland, on low rises or rocky situations in yellow or red sandy soil over quartzite or sandstone. Regenerates from seed and also lignotuber (and/or basal suckers). Flowers Apr.–Aug. Map 115.

W.A.: 19.2 km SSW of Bungle Bungle Outcamp, *S.J.Forbes 2558* (MEL, NSW, PERTH); 17.5 km NE of Tableland Stn, *M.Lazarides 6400* (BRI, CANB, MEL, NSW, PERTH); c. 62 km from the Hann R. on road to Tableland Stn, *D.J.McGillivray 3830 & A.S.George* (E, NSW). N.T.: Gregory Natl Park, S of Wickham R., *I.D.Cowie 7820* (CANB, DNA, MEL, NSW, PERTH).

Easily distinguished from close relatives by its ±oblong leaves, with the teeth ±evenly spaced around the margins, and with the reticulate venation more pronounced on the lower surface.

This species is recognised as 'Rare' in J.D.Briggs & J.H.Leigh, *Rare or Threatened Australian Plants* (1995).

Adenotricha Group

Shrubs or small trees. Leaves toothed to pinnatifid or (*G. benthamiana*) deeply bi- or tri-pinnatipartite, dorsiventral; surfaces similar or not; margins flat or revolute. Conflorescence erect, terminal or axillary, simple or (*G. benthamiana*) up to 9-branched; unit conflorescence ovoid or a loose cluster or narrowly conical, acropetal. Flowers acroscopic. Torus oblique. Perianth zygomorphic with limb decurved, glabrous or with simple hairs outside, hairy inside; tepals remaining loosely coherent and held ventral to style. Pistil 4–10 or (*G. benthamiana*) 14–16 mm long; ovary stipitate, glabrous; style glabrous or with short simple hairs, exserted or not from late bud, scarcely longer than perianth; pollen-presenter lateral or (*G. benthamiana*) oblique, flat to convex. Follicle glabrous, rugulose; pericarp moderately thick-walled. Seed compressed-ellipsoidal and peripterous, or ellipsoidal with a pale revolute margin.

Five species from the wetter monsoon tropics of N.T. and W.A. The group (except *G. cunninghamii*) is characterised by the dorsal edge of the torus projecting well out from the pedicel, and the stipe being strongly reflexed from the line of the pedicel, features shared with some taxa in the *Wickhamii* group, which is very closely related. Pollination vector uncertain, probably birds. The placement of *G. benthamiana* in this group is tentative; it has features suggestive of a hybrid origin involving *G. pungens* (*Heliosperma* group) and *G. wickhamii*.

1 Pistil > 14 mm long; leaves deeply twice-divided; perianth with short glandular hairs outside **97. G. benthamiana**

1: Pistil < 10 mm long; leaves with shallow primary toothing only; perianth glabrous outside

2 Leaves sessile, cordate to amplexicaul **96. G. cunninghamii**

2: Leaves petiolate, ±truncate to cuneate at base

3 Adult leaves densely glandular-hispidulous on both surfaces (simple erect glandular hairs), mostly with 13–25 shallow teeth **93. G. adenotricha**

3: Adult leaves with a dense to sparse indumentum of appressed to ascending biramous hairs, sometimes also a few erect simple glandular hairs along margins and veins; leaves with 3–15 coarse teeth or shallow lobes

4 Pedicels glabrous or with simple erect glandular hairs only; pistil 7–9.5 mm long; floral rachis 3–18 mm long, the conflorescence ±ovoid **94. G. longicuspis**

4: Pedicels tomentose with biramous hairs only; pistil 5.5–7 mm long; floral rachis 5–8 mm long, the conflorescence dense, subglobose **95. G. microstyla**

93. Grevillea adenotricha McGill., *New Names Grevillea* 1 (1986)

T: Manning Gorge, W.A., 1 Aug. 1973, *B.Gill G20*; holo: QRS.

Illustrations: D.J.McGillivray & R.O.Makinson, *Grevillea* 203, fig. 54 (1993); P.M.Olde & N.R.Marriott, *Grevillea Book* 2: 16 (top centre & 8A, B) (1995).

Shrub 0.8–2 m tall. Leaves petiolate, narrowly oblong or narrowly ovate in general outline, 3–6.5 cm long, 15–25 mm wide, serrato-dentate with (6–) 13–25 pungent teeth spaced evenly around margins; base ±truncate to broadly cuneate; margins undulate, not recurved; surfaces similar, ±densely glandular-hispidulous. Conflorescence terminal or axillary, simple, subglobose tending subsecund; floral rachis 1–5? mm long. Pedicels 2–2.5 mm long. Torus extended dorsally well beyond pedicel. Flower colour: perianth red; style pale orange. Perianth glabrous outside, villous inside near base. Pistil 4–6 mm long, glabrous; stipe 1–3 mm long, almost perpendicular to pedicel; style not or scarcely exserted from late bud. Follicle obloid, 9.5–11 mm long, glabrous. Fig. 12H–J.

Occurs in the Kimberley district of W.A., where known only from Manning Gorge, Lushington Brook and the Prince Regent R. Fire response unknown. Flowers May–Aug. Map 116.

W.A.: Lushington Brook, *C.A.Gardner 9567* (PERTH); on plateau above Prince Regent R., May 1985, *A.Turner* (NSW).

Grevillea adenotricha is a distinctive species, characterised by the leaves with many evenly distributed marginal teeth, the very short pistil, and the dense indumentum of gland-tipped simple hairs occurring on most vegetative parts. This species also has the seed ellipsoidal with a recurved margin bearing a pale wavy rim, and on this feature is likely to be more closely related to *G. longicuspis* than to other members of this group.

This species is recognised as 'Rare' in J.D.Briggs & J.H.Leigh, *Rare or Threatened Australian Plants* (1995).

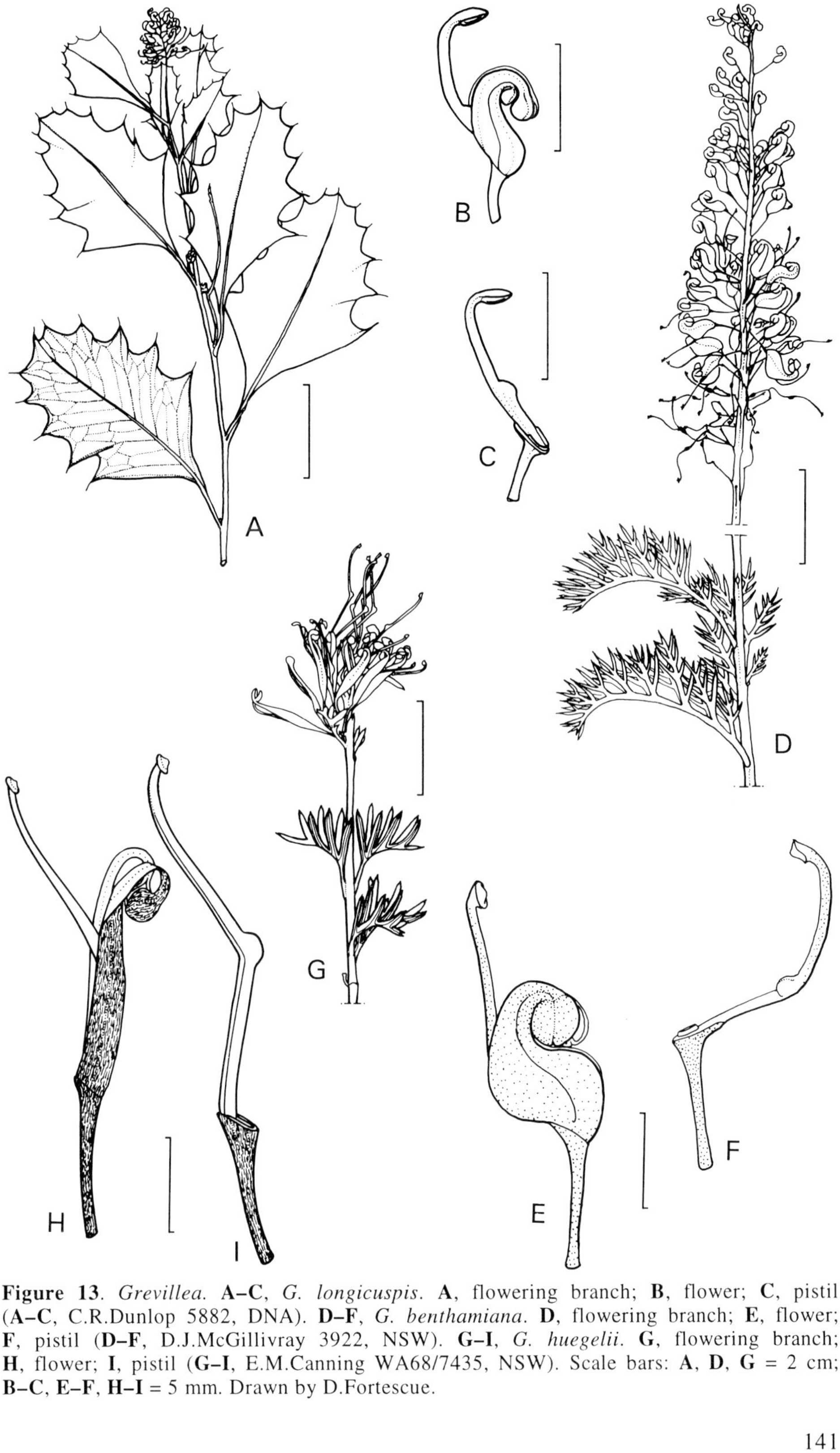

Figure 13. *Grevillea*. **A–C**, *G. longicuspis*. **A**, flowering branch; **B**, flower; **C**, pistil (**A–C**, C.R.Dunlop 5882, DNA). **D–F**, *G. benthamiana*. **D**, flowering branch; **E**, flower; **F**, pistil (**D–F**, D.J.McGillivray 3922, NSW). **G–I**, *G. huegelii*. **G**, flowering branch; **H**, flower; **I**, pistil (**G–I**, E.M.Canning WA68/7435, NSW). Scale bars: **A**, **D**, **G** = 2 cm; **B–C**, **E–F**, **H–I** = 5 mm. Drawn by D.Fortescue.

94. **Grevillea longicuspis** McGill., *New Names Grevillea* 9 (1986)

T: near Port Darwin, 'Camp Area', N.T., Nov. 1943, *T.H.Black*; holo: MEL; iso: MEL.

Illustrations: J.Brock, *Top End Native Pl.* 204 (1988); D.J.McGillivray & R.O.Makinson, *Grevillea* 202, fig. 53 (1993); P.M.Olde & N.R.Marriott, *Grevillea Book* 2: 244 (203A, B) (1995).

Shrub to 0.6 m high. Leaves petiolate, ovate in general outline, 2.5–8 cm long, 20–60 mm wide, pinnatifid with 4–8 (–13) pungent triangular teeth or shallow lobes, occasionally an entire leaf near inflorescence; base broadly cuneate; margins undulate, not recurved; surfaces similar; lower surface with an open to sparse indumentum of appressed biramous hairs, usually also with erect simple glandular hairs at least along veins and margins. Conflorescence terminal or subterminal-axillary, usually simple, subglobose to loosely ovoid; floral rachis 3–18 mm long. Pedicels 2–3.5 mm long. Torus extended dorsally well beyond pedicel. Flower colour: perianth red; style red or pinkish cream to lemon-cream. Perianth glabrous outside, villous inside. Pistil 6.5–9.5 mm long; stipe 1.5–2.2 mm long, adnate to dorsal wall of torus and almost perpendicular to pedicel; ovary glabrous; style glabrous or minutely glandular-pubescent or papilloid, not or only weakly exserted from late bud. Follicle oblong-ellipsoidal, 9.5–12.5 mm long, glabrous. Plate 23; Fig. 13A–C.

Occurs in N.T. where recorded only from Woolaning, Manton R. and Darwin. Grows in open *Eucalyptus* woodland on lateritic or quartz sand soil. Produces new stems from lignotubers and rhizomes in response to fire; also produces new annual stems from the base which form inflorescences and die off after fruiting. Flowers Aug.–Dec. Map 117.

N.T.: Woolaning, *B.Hyland 7840* (QRS); c. 3 km from Darwin R. Dam, *D.J.McGillivray 3932* (NSW); Manton R., 72 km S of Darwin, *J.McKean B160* (CANB, DNA, K); Port Darwin, 1874, *R.Schomburgk* (E).

Prior to being named this species was occasionally misidentified as *G. angulata* or *G. wickhamii*, both of which have decurved inflorescences.

This species is recognised as 'Rare' in J.D.Briggs & J.H.Leigh, *Rare or Threatened Australian Plants* (1995).

95. **Grevillea microstyla** M.D.Barrett & Makinson, *Fl. Australia* 17A: 495 (2000)

T: 4.6 km by road from Bachsten Ck (South Arm) campsite above falls, via side road to Wren Gorge ..., West Kimberley, W.A., 1 June 1998, *R.O.Makinson 1687 & H.Nicholson*; holo: CANB; iso: DNA, MEL, NSW, PERTH.

Shrub 0.3–0.8 m tall. Leaves petiolate, ovate to broadly oblong-elliptic or slightly obovate in general outline, (15–) 30–83 mm long, (8–) 20–43 mm wide, dentate to pinnatifid with (3–) 7–15 weakly pungent subtriangular teeth lobes; base broadly cuneate to truncate; margins flat; surfaces similar, with lower surface open-tomentose (biramous hairs only). Conflorescence terminal, usually simple, dense, subglobose; floral rachis 5–8 mm long. Pedicels 2 mm long. Torus markedly elongate at dorsal side. Flower colour: perianth crimson; style and style-end dull orange. Perianth glabrous outside, pilose inside at about level of ovary, glabrous below. Pistil 5.5–7 mm long; stipe 1.7–2.0 mm long, adnate to torus and perpendicular to pedicel; ovary glabrous; style glabrous or minutely glandular-pubescent, not or scarcely exserted from late bud. Follicle oblong-ellipsoidal, 11–16 mm long, glabrous. Plate 24.

Occurs in the western Kimberley region of W.A., known from near Bachsten Ck and in the Prince Regent Nature Reserve (southern foothills of the Princess May Ra.). Grows in grassy *Eucalyptus miniata* woodland in sandy loam soils on shallow valley bottoms below sandstone ridges. Regenerates from seed and lignotuber, probably also from rhizomes. Flowers Dec.–June. Map 118.

W.A.: Edkins Ra., c. 3.5 km WSW of Peter Lacy's Camp, along road to Wren Gorge, 8 Jan. 1995, *M.D. & R.L.Barrett s.n.* (PERTH); *loc. id.*, *R.L.Barrett 640* (KPBG, PERTH); W Kimberley, 4.6 km by road from Bachsten Ck (S arm) campsite above falls via side track to Wren Gorge, *R.O.Makinson 1686* (BRI, CANB, DNA, K, MEL, NSW); Prince Regent Nature Reserve, southern foothills of Princess May Ra., c. 5 km E of Purulba massif, *R.O.Makinson 1695 et al.* (AD, CANB, PERTH, QRS).

Grevillea microstyla is closely related to the Northern Territory endemic *G. longicuspis*, which has the pedicels glabrous or with an indumentum of inconspicuous simple erect glandular hairs; glandular hairs also occur, usually intermixed with biramous hairs, on the leaves, peduncles, and floral rachises. *Grevillea longicuspis* also has looser and usually longer unit conflorescences (dome-shaped to ovoid, rachis 3–18 mm long), a longer pistil 7–9.5 mm long and leaves drying khaki-green. *Grevillea microstyla* has only biramous hairs on the pedicels, lacks simple glandular hairs on all parts except (sometimes) the style, has very compact subglobose conflorescences, a shorter pistil and leaves drying grey-black.

Grevillea microstyla is also related to three other Kimberley endemics with partially overlapping ranges. *Grevillea adenotricha* has leaves and stems with a conspicuous indumentum of simple erect glandular hairs; *G. cunninghamii* lacks a dorsal elongation of the toral rim, lacks hairs on all parts of the plant except the floral bracts and inner surface of the perianth, and has stem-clasping leaf bases (free in *G. microstyla*). *Grevillea wickhamii* subsp. *wickhamii* grows in a distinct microhabitat on rocky ridges and platforms, is a larger shrub 1–4 m tall, and has more broadly cuneate leaves, a closely appressed leaf indumentum and glabrous pedicels.

96. **Grevillea cunninghamii** R.Br., *Suppl. Prodr. Fl. Nov. Holl.* 23 (1830)

T: Montague Sound, W.A., 7 Sept. 1820, *A.Cunningham 176*; lecto: BM, *fide* D.J.McGillivray & R.O.Makinson, *Grevillea* 412 (1993); isolecto: G-DC (*A.Cunningham s.n.*), K, MEL, NY *n.v.*, PH *n.v.*

Illustrations: D.J.McGillivray & R.O.Makinson, *Grevillea* 204, fig. 55 & col. pl. (1993); P.M.Olde & N.R.Marriott, *Grevillea Book* 2: 106 (all) (1995).

Shrub 1.2–3 m tall. Leaves sessile, ovate to narrowly so, 4–9 cm long (juveniles much longer), 30–55 mm wide, spino-dentate with 13–21 pungent teeth distributed evenly around margins; base cordate-amplexicaul; margins flat, not recurved; surfaces similar, glabrous. Conflorescence axillary, usually simple, loosely subglobose on a long peduncle; floral rachis 2–6 mm long. Pedicels 2–2.5 mm long. Torus extended a little beyond pedicel on dorsal side. Flower colour: perianth red with a paler (to yellow) limb; style red. Perianth glabrous outside, villous inside. Pistil 8–9.5 mm long, glabrous; stipe 1.2–2.2 mm long, adnate to torus and initially reclinate relative to pedicel; style not exserted from late bud. Follicle obloid, 9–10.5 mm long, glabrous.

Occurs in W.A. in coastal parts of the Kimberley district, including offshore islands, from Cape Londonderry area to King Sound area N of Derby. Grows in open shrub associations in sandy stony soils on sandstone and bauxite substrates. Regenerates from seed. Flowers (Jan.) May–Sept. Map 119.

W.A.: SE of Cape Londonderry, *A.S.George 13366* (NSW, PERTH); Planigale Ck, Drysdale River Natl Park, *K.F.Kenneally 4423* (CANB, PERTH); 4 km NW of Kalumburu, *D.J.McGillivray 3862 & A.S.George* (B *n.v.*, BRI, CANB, DNA, K, LE *n.v.*, MO *n.v.*, NSW, PERTH, RSA *n.v.*, US *n.v.*).

The sessile cordate-amplexicaul leaves and the lack of hairs on nearly all external surfaces of the plant make recognition of this species very easy.

97. **Grevillea benthamiana** McGill., *New Names Grevillea* 3 (1986)

T: 1 km N of Stuart Highway bridge over Fergusson River, N.T., 15 July 1978, *D.J.McGillivray 3922*; holo: NSW; iso: CANB, DNA, K, LE *n.v.*, PRE *n.v.*, US *n.v.*

Illustrations: J.Brock, *Top End Native Pl.* 199 (1988); D.J.McGillivray & R.O.Makinson, *Grevillea* 201, col. pl. & fig. 52 (1993); P.M.Olde & N.R.Marriott, *Grevillea Book* 2: 60 (all), 61 (44D) (1995).

Shrub 2–4 m high. Leaves petiolate, 2.5–5 cm long, 30–45 mm wide, bi- or partly tri-pinnatipartite, with 14–24 primary lobes; ultimate lobes linear to narrowly triangular, 3–10 mm long, 0.7–1.0 mm wide, rigid, pungent; base narrowly cuneate; margins revolute; surfaces dissimilar; lower surface mostly enclosed, subvillous. Conflorescence terminal or subterminal-axillary, simple or to 9-branched, unit conflorescence narrowly conico-cylindrical; ultimate floral rachis 80–100 mm long. Pedicels 6.5–7 mm long. Torus extended dorsally well beyond pedicel. Flower colour: perianth pink-red to purple-red, becoming black near or after anthesis, or rarely lemon in bud and cream with pink tinges at maturity (not blackening); style pinkish red or rarely (with perianth) cream. Perianth with very short glandular hairs outside, sparsely villous inside. Pistil 14.5–15.5 mm long; stipe 5.1–5.8 mm long, adnate to dorsal wall of torus and almost perpendicular to pedicel; ovary glabrous; style minutely glandular-pubescent, exserted from late bud. Follicle broadly oblique-elliptic to -obovate, 16–19 mm long, glabrous. Fig. 13D–F.

Occurs in N.T. where restricted to the Fergusson R. area. Grows in low, open *Eucalyptus* and *Terminalia* woodland on rocky slopes in loamy sand. Regenerates from lignotuber and seed. Flowers June–Aug. Map 120.

N.T.: 42 km NE of Claravale, 1 Dec. 1978, *C.Bosshart* (DNA); 32 km S of Pine Creek, *N.Byrnes 420* (DNA, NSW); 36 km SE of Pine Creek, *G.A.M.Chippendale NT11755* (DNA, MEL, NSW).

Grevillea benthamiana is an anomalous species in this group, differing from other members in many characters, but sharing a dorsally extended torus platform with an adnate and reclinate ovarian stipe. The seed is flat-ellipsoidal and peripterous. A hybrid origin involving *G. pungens* is possible. The rapid blackening of the perianth on reddish flowered plants is striking, and may be a cue to pollinators (probably birds); a number of other bird-pollinated species (usually with externally glabrous and glaucous perianths) also do this soon after anthesis. Cream-flowered plants of *G. benthamiana* appear to lack this blackening response.

This species is recognised as 'Rare' in J.D.Briggs & J.H.Leigh, *Rare or Threatened Australian Plants* (1995).

Huegelii Group

Shrubs. Leaves divided or entire, dorsiventral or dipleural; surfaces dissimilar; margins revolute against abaxial midvein. Conflorescence terminal, axillary or cauline, simple or to 3-branched; unit conflorescence subglobose to umbelloid or few-flowered and irregular, acropetal. Flowers acroscopic. Torus oblique. Perianth zygomorphic, glabrous or with biramous and/or simple glandular hairs outside, glabrous inside or with a few hairs only; tepals independently recoiled near apex, remaining coherent below. Pistil 16–30 mm long, glabrous; ovary stipitate; style partly exserted from late bud, afterwards refracted at ovary; pollen-presenter oblique, subconical. Follicle glabrous, apiculate, granulose; pericarp moderately thick, firmly crustaceous to weakly bony. Seed ellipsoidal, convex on outer face, with a narrow winged border.

A group of two species in southern semi-arid Australia. Bird-pollinated. The perianth is characteristically narrow and almost straight-sided below the limb; the pistil is reflexed sharply through the ovary. Affinities uncertain; a link with the *Acuaria* group is possible, as is a connection with tropical taxa (e.g. *G. robusta* and possibly the *Shiressii* group).

Most leaves pinnatipartite with at least 5 primary lobes; floral bracts usually absent at anthesis **98. G. huegelii**

Most leaves entire, rarely 2- or 3-partite; floral bracts usually persistent at anthesis **99. G. sarissa**

98. Grevillea huegelii Meisn., *in* J.G.C.Lehmann, *Pl. Preiss.* 1: 543 (1845)

T: ... ad fluvium Avon, 6 mill. ab urbicula York [Avon R., 6 miles [c. 10 km] from York, W.A.], 10 Sept. 1839, *L.Preiss 691*; lecto: NY *n.v.*, *fide* D.J.McGillivray & R.O.Makinson, *Grevillea* 420 (1993); isolecto: [*Preiss s.n.*] LD *n.v.*; ?isolecto: G, LE *n.v.*, MEL, P, TCD *n.v.* (some as *Preiss 274*, = *Preiss 691*).

G. rigidissima F.Muell. ex Meisn., *Linnaea* 26: 356 (1854). T: Pine forest, Gawlertown, S.A., 1851–52, *F.Mueller*; lecto: NY *n.v.*, *fide* D.J.McGillivray & R.O.Makinson, *Grevillea* 420 (1993); isolecto: MEL 47027 *p.p.*

Illustrations: L.Costermans, *Native Trees & Shrubs SE Australia* 164, 165 (1981); D.J.McGillivray & R.O.Makinson, *Grevillea* 112, fig. 17 & col. pl. (1993); P.M.Olde & N.R.Marriott, *Grevillea Book* 2: 201 (top right & 165A–D), 205 (165E–G) (1995).

Erect to decumbent shrub 0.3–3 m tall. Leaves dorsiventral, 1–6 cm long, 0.7–6.5 cm wide, usually pinnatipartite with (2–) 5–11 (–25) primary lobes, sometimes with secondary (rarely tertiary) division, rarely leaves pectinate or digitate; ultimate lobes linear to oblong,

0.2–4 cm long, 0.7–2.2 mm wide, divaricate or not; lower surface usually enclosed except for midveins, rarely slightly exposed. Conflorescence erect, simple or few-branched, terminal, axillary or sometimes cauline; unit conflorescence subglobose-umbelloid or few-flowered; rachis 2–15 mm long; bracts 0.4–1.2 mm long, falling early. Flower colour: perianth red to rich pink, orange or apricot, sometimes with limb pale, rarely entirely bright yellow to cream; style matching, usually with greenish apex. Perianth sericeous (hairs biramous) to glabrous outside, or rarely with simple glandular hairs, glabrous or sparsely pilose inside in throat. Pistil 19–29 mm long, glabrous, reflexed at ovary. Follicle ellipsoidal to subdiscoid, 10–12 mm long, granulose, apiculate. Plate 25; Fig. 13G–I.

Occurs in W.A., S.A., N.S.W. and Vic.; the distribution is disjunctly transcontinental south of 30°S. In W.A., widespread from Moora to Borden and E to Cundeelee and Balladonia (absent from the Nullarbor Plain). In S.A., widespread in drier parts of the southern half from Maralinga to Leigh Creek and Renmark, rare in the N (Musgrave Ra.). In N.S.W., scattered in W of State from Robinvale and Broken Hill N and E to Byrock. In Vic., far NW corner E to Swan Hill. Grows in varied habitats, often in mallee woodland, occasionally in heath-shrubland, in sandy to loamy (often alkaline) soils. Regenerates from seed. Flowers mainly July–Dec., sporadic in other months. Map 121.

W.A.: Gnarlbine, 12 Nov. 1891, *R.Helms* [*15*] (AD, MEL, PERTH). S.A.: Maralinga, *N.Forde 586* (CANB); Musgrave Ra., c. 15 km N of Mt Harriett, *D.J.E.Whibley 962* (AD, CANB). N.S.W.: 6.9 km S of intersection on Cobar to Roto road near Sandy (Crow) Ck, c. 6 km N of Bindi HS, *D.I.Wilson 35 & P.G.Wilson* (B *n.v.*, K, NSW, PERTH). Vic.: L. Boga, *H.I.Aston 947* (MEL).

There is considerable variation in degree of leaf division, spacing and divarication of leaf lobes, and perianth indumentum. Populations in W.A. usually have smaller flowers than those in eastern States. Secondary and tertiary leaf division is most common in W.A. in the NW of the range (Beverley to Moora to Kellerberrin area), and Olde & Marriott (*loc. cit.*) refer to populations from this area which have such leaf division, and maroon to burgundy flowers, as the 'maroon-flowered (typical) form'; this assemblage includes the species type, but there is still much variation within it. A 'glabrous-flowered form' occurs in W.A. mainly between Cape Arid, Esperance and Kumarl; it has short bluish grey pectinate leaves, an almost glabrous outer perianth surface, and a prostrate to sprawling habit. Some plants in the Kalgoorlie to Southern Cross to Lake Grace area of W.A. have glandular hairs on the outer perianth surface, and were referred to by McGillivray & Makinson (*loc. cit.*) as a 'glandular-haired form'; Olde & Marriott (*loc. cit.*) report that this may co-occur with plants with non-glandular hairs, and its taxonomic status is unclear. A specimen referred to by McGillivray & Makinson (*loc. cit.*) as *G.* cf. *huegelii* from near Wiluna, W.A., is assignable to *G. sarissa* subsp. *succincta*.

99. Grevillea sarissa S.Moore, *J. Linn. Soc. Bot.* 34: 222 (1899)

T: Gibraltar [Mt Gibraltar?, W of Coolgardie], W.A., Sept. 1895, *S.Moore*; holo: BM.

Shrub 0.6–3.5 m tall. Leaves dorsiventral to dipleural, 1.5–15 (–20) cm long, entire and linear to linear-subterete, or rarely 2- or 3-partite with ascending linear lobes 10–50 mm long; simple leaves and lobes 0.9–2.0 mm wide; lower surface enclosed except for midveins. Conflorescence ±erect, simple to 4-branched, terminal, axillary or often cauline; unit conflorescence wheel-like to subglobose; rachis 2.5–8 mm long; bracts 0.3–1.2 mm long, usually persistent to anthesis. Perianth villous to subsericeous (biramous hairs) to glabrous outside or rarely with simple glandular hairs, glabrous inside. Pistil (16–) 20–30 mm long, reflexed at ovary. Follicle ellipsoidal to slightly obovoid, 12–20 mm long, sometimes apiculate, granulose.

Grevillea sarissa occurs in drier parts of south-western W.A. and coastal S.A., often near salt lakes and often gregarious. Six subspecies are recognised.

While all six subspecies here recognised are intimately related, a case could be made, on comparative grounds with some other species complexes, for raising some or all to species rank. At least two, however, (subsp. *bicolor*, subsp. *succincta*) remain poorly known, and degrees of intergradation in the complex are unclear. McGillivray's (*New Names Grevillea* 13 (1986); *Grevillea* 115 (1993)) rankings are retained here, but further study is needed.

19 Leaves and lobes usually with lower surface on either side of midvein concealed by revolute margins (if lamina exposed then longest leaves ⩾ 4 cm long) **109. G. pinaster**

19: Leaves and lobes of some or all leaves with lower surface at least partly exposed on either side of midvein; leaves ⩽ 4.5 cm long

20 Branchlets with a dense short indumentum of ±spreading hairs **107. G. exposita**

20: Branchlets glabrous or appressed-sericeous, or with an open indumentum of weakly ascending hairs

21 All leaves entire; upper leaf surface with midvein usually clearly evident as a pale strip; leaves 1.5–4 mm wide; non-lignotuberous shrub **106. G. obtusifolia**

21: Usually at least a few leaves divided (occasionally all entire); upper leaf surface with midvein not or scarcely evident (sometimes a faint channel only); entire leaves 1–2 mm wide; lignotuberous shrub **101. G. thelemanniana**

101. Grevillea thelemanniana Hügel ex Endl., *Nov. Stirp. Dec.* 1: 6 (1 May 1839)

G. thelemanniana Hügel ex Lindl., *Edward's Bot. Reg. 1839, Suppl. Misc. Not.* 54 (? July 1839), as *G. Thelemaniana*, *nom. superfl.* T: Brook Rd, Wattle Grove, W.A., 6 Sept. 1976, *R.Coveny 8130*; neo: NSW, *fide* D.J.McGillivray & R.O.Makinson, *Grevillea* 444 (1993); isoneo: B *n.v.*, K, PERTH.

Illustrations: P.M.Olde & N.R.Marriott, *Grevillea Book* 3: 201 (bottom right), 202 (164A–C) (1995).

Spreading shrub 0.5–1 m high, 1–2 m wide. Branchlets angular to terete, glabrous or sparsely (rarely densely) sericeous. Leaves 1–2.5 (–3) cm long, obtuse, mucronate, entire and linear to very narrowly elliptic and/or 2–5-pinnatipartite or -pinnatisect (usually both states present); entire leaves 1–1.5 (–2) mm wide; primary lobes entire, 2–8 (–17) mm long, 0.5–1.5 mm wide; upper leaf surface smooth, not or scarcely pitted, soon glabrous; margins shortly recurved or revolute; lower surface usually narrowly exposed on either side of midvein, subsericeous. Conflorescence erect to decurved, secund, 6–14-flowered, basipetal; rachis 5–15 mm long, glabrous or with scattered appressed hairs. Flower colour: perianth pink-red with creamy limb; style red with green tip. Perianth usually with a few minute appressed hairs outside, especially on limb, otherwise glabrous or nearly so, pubescent inside. Pistil 24–28 mm long. Follicle obloid-acuminate to ellipsoidal, 12–13 mm long, faintly ridged at base, ribbed, wrinkled.

W.A., confined to the Cannington, Wattle Grove and Kenwick areas of Perth, in winter-wet heathland swamp. Regenerates from lignotuber and seed. Flowers July–Oct. Map 129.

W.A.: Kenwick, 1.1 km along Boundary Rd from intersection of Boundary and Bickley Rds, *M.Carter 25* (CANB, PERTH); Cannington, 17 Aug. 1898, *R.Helms* (NSW); University Reserve, Cannington Swamp, *R.Melville 8 & R.D.Royce* (K, NSW); Cannington, lower Cannington R., 13 June 1899, *A.Morrison* (CANB, E).

Grevillea thelemanniana has only primary leaf division (or leaves entire). *Grevillea preissii* subsp. *glabrilimba*, *G. delta*, *G. humifusa* and some plants of *G. fililoba* all have secondary leaf division on some or all leaves. *Grevillea fililoba* also has filiform leaf segments 0.5 mm wide and (like *G. preissii*) has the lower leaf surface enclosed except for the midvein. *Grevillea hirtella* and *G. exposita* both have spreading hairs on the branchlets (appressed in *G. thelemanniana*). *Grevillea evanescens* has all leaves entire, the widest > 5 mm wide (⩽ 2 mm wide in *G. thelemanniana*). *Grevillea obtusifolia* is very similar indeed, but also has all leaves entire, widest leaves 2–4 mm wide, the midvein visible on the upper leaf surface as a pale line, and a shorter pistil 20–24 mm long (in *G. thelemanniana* usually at least some leaves divided and the midvein not visible on the upper surface).

This species is recognised, under *G. thelemanniana* subsp. *thelemanniana*, as 'Rare' in J.D.Briggs & J.H.Leigh, *Rare or Threatened Australian Plants* (1995).

102. Grevillea hirtella (Benth.) Olde & Marriott, *Grevillea Book* 1: 182 (1994)

G. pinaster var. *hirtella* Benth., *Fl. Austral.* 5: 427 (1870); *G. thelemanniana* subsp. *hirtella* (Benth.) McGill., *New Names Grevillea* 15 (1986). T: Champion Bay, W.A., *P.Walcott 14*; holo: MEL.

Illustrations: D.J.McGillivray & R.O.Makinson, *Grevillea* 251 (bottom left) (1993); P.M.Olde & N.R.Marriott, *Grevillea Book* 2: 195 (top right), 196 (161A, B) (1995).

Shrub to c. 1 m high, spreading to 2.5 m wide. Branchlets terete, loosely or densely villous to pubescent. Leaves usually crowded, 0.6–2 cm long, obtuse, mucronate, entire, linear or bi- or trifid, occasionally pinnatipartite with up to 7 entire primary lobes, both entire and divided leaves usually present; leaf lobes 1.5–8 mm long, 1–1.5 mm wide, strongly ascending; upper surface glabrous to loosely to sparsely villous, conspicuously (rarely obscurely) pitted; margins smoothly to angularly revolute; lower surface obscured, sometimes villous in groove. Conflorescence decurved, secund, c. 10–16-flowered, weakly basipetal; floral rachis 5–13 mm long, densely to sparsely subsericeous to subtomentose or with scattered hairs. Flower colour: perianth pale to deep pink red; style red with green tip. Perianth glabrous outside or occasionally with a very few minute appressed hairs on limb, tomentose-pubescent inside. Pistil 20–25 mm long. Follicle obloid-acuminate to ellipsoidal, 12–13 mm long, with an obscure irregular slanting ridge or bulge just above base and a longitudinal subdorsal ridge on either side.

Occurs in south-western W.A., known only from scattered populations between Mingenew and Walkaway. Grows in open heathland associations. Regenerates probably only from seed. Flowers (Apr.–) Aug.–Nov. Map 130.

W.A.: 60 km (by road) E of Greenough HS on the Nangatta to Walkaway road, *B.J.Conn 2124* (MEL, NSW, PERTH, RSA); 39 km NW of Strawberry on Burma Rd, *M.G.Corrick 8285* (MEL, NSW); S of the Casuarinas, ESE of Geraldton, *A.S.George 11677* (CANB, NSW, PERTH); 10 km along Burma Rd from Strawberry turnoff, *D.J.McGillivray 3310 & A.S.George* (NSW); Allanooka Springs Rd, *S.Patrick & R.J.Cranfield 639* (NSW, PERTH).

Grevillea hirtella has a conspicuous keel along the midline of the limb-segment of each tepal, a usually pitted leaf surface and usually villous leaves and branchlets.

Grevillea thelemanniana, *G. evanescens* and *G. obtusifolia* can be distinguished from *G. hirtella* by their appressed indumentum on the leaves and branchlets. *Grevillea delta* and *G. humifusa* (and often *G. fililoba*) have secondary leaf division (primary division only in *G. hirtella*). *Grevillea exposita* can be distinguished by the appressed hairs on the leaves.

This species is recognised (under *G. thelemanniana* subsp. *hirtella*) as 'Poorly Known' in J.D.Briggs & J.H.Leigh, *Rare or Threatened Australian Plants* (1995).

103. Grevillea fililoba (McGill.) Olde & Marriott, *Grevillea Book* 1: 182 (1994)

G. thelemanniana subsp. *fililoba* McGill., *New Names Grevillea* 15 (1986). T: Ellendale Pool, Geraldton district, W.A., 29 July 1961, *R.D.Royce 6459*; holo: PERTH; iso: CANB.

Illustrations: D.J.McGillivray & R.O.Makinson, *Grevillea* 251 (top right) (1993); P.M.Olde & N.R.Marriott, *Grevillea Book* 2: 162 (bottom right), 163 (133A, B) (1995).

Spreading dense soft shrub c. 1.5 m high, to 3 m wide. Branchlets angular, sparsely subsericeous, soon ±glabrous. Leaves 2–4.5 cm long, pinnatipartite, sometimes basal lobes with secondary division; ultimate lobes very narrowly linear, sometimes curved, 2–20 mm long, 0.3–0.7 mm wide, becoming more spreading with age; upper surface smooth, not pitted, ±glabrous; margins tightly revolute; lower surface obscured except midvein, 2-grooved, subsericeous in grooves. Conflorescence erect to pendulous; unit conflorescence subsecund, lax, usually exceeding foliage, weakly basipetal, 12–30-flowered; rachis 18–30 mm long, sparsely silky. Flower colour: perianth pink to bright red with white limb; style pink with green tip. Perianth glabrous and ±glaucous outside, scantily bearded inside. Pistil 24–28 mm long. Follicles subtriangular, c. 13 mm long.

Occurs in south-western W.A. on catchments of the Greenough and Irwin Rivers E of Geraldton. Grows in a range of habitats, from creekside scrub to hilltops. Regenerates by seed. Flowers July–Sept. Map 131.

W.A.: Greenough Flat, 1869, *Gray s.n.* (MEL); Upper Irwin, *s.d.*, *M.Gwerin s.n.* (MEL); Greenough R., Nov. 1877, *F.Mueller* (MEL).

Among species with somewhat similar foliage (*G. preissii*, *G. humifusa*, *G. delta*) *G. fililoba* is usually easily distinguished by its glabrous filiform leaf segments c. 0.5 mm wide (usually wider and dorsally flat in the other species and often hairy on the upper surface).

This species is recognised (under *G. thelemanniana* subsp. *fililoba*) as 'Poorly Known' in J.D.Briggs & J.H.Leigh, *Rare or Threatened Australian Plants* (1995).

104. Grevillea humifusa Olde & Marriott, *Grevillea Book* 1: 182 (1994)

T: 26.5 km W of Brand Hwy on Cantabilling Rd (towards Jurien), W.A. [wording order differs in protologue], 15 Sept. 1991, *P.M.Olde 91/96*; holo: NSW; iso: PERTH *n.v.*

G. thelemanniana 'form e', of D.J.McGillivray & R.O.Makinson, *Grevillea* 251 (1993).

Illustrations: P.M.Olde & N.R.Marriott, *Grevillea Book* 2: 203 (top right), 204 (166A, B) (1995); A.Brown *et al.* (eds), *W. Australia's Threatened Flora* 49 (1998).

Prostrate shrub with trailing stems to 3 m long. Branchlets angular, openly villous. Leaves 1.5–2 cm long, bipinnatisect; ultimate lobes narrowly linear, ascending to spreading, 5–10 mm long, c. 0.5 mm wide; upper surface sparsely pilose, occasionally glabrous, smooth to finely granulose, not pitted; margins revolute; lower surface sometimes exposed on either side of midvein, pilose. Conflorescence erect or decurved, conico-secund, 12–30-flowered, acropetal; rachis 8–15 mm long, openly subsericeous to -tomentose. Flower colour: perianth pink to pale red with cream limb; style pink, red or orange red with yellow tip. Perianth glabrous outside except for scattered minute hairs on limb (these rarely absent), pubescent inside. Pistil 22–24 mm long. Follicle erect, obloid, 12–15 mm long, with a conspicuous transverse ridge just above base and 2–4 strong longitudinal ridges running from it.

Occurs in W.A., inland from Jurien, in gravelly loam soils in shrub/woodland associations. Regenerates from lignotuber and seed. Flowers (Mar.?–) July–Oct.? Map 132.

W.A.: c. 22.5 km Jurien Bay–Watheroo turnoff, *H.Demarz 107* (KPBG).

Grevillea delta and *G. preissii* are both taller more robust shrubs than *G. humifusa*, and both have longer pistils (c. 27 mm in *G. delta*, 25–28 mm in *G. preissii*). *Grevillea preissii* subsp. *preissii* usually has longer leaves (> 25 mm long). *Grevillea delta* also differs in its less conspicuously ridged follicles, and its narrowly triangular floral bracts (ovate in *G. humifusa*).

This species is recognised as 'Poorly Known' in J.D.Briggs & J.H.Leigh, *Rare or Threatened Australian Plants* (1995).

105. Grevillea delta (McGill.) Olde & Marriott, *Grevillea Book* 1: 181 (1994)

G. thelemanniana subsp. *delta* McGill., *New Names Grevillea* 15 (1986). T: NE of Mt Lesueur, W.A., 13 Oct. 1974, *A.S.George 12910*; holo: PERTH; iso: NSW.

Illustrations: P.M.Olde & N.R.Marriott, *Grevillea Book* 2: 116 (bottom right), 117 (93A, B) (1995).

Bushy, spreading shrub 1–2 m high. Branchlets terete, with open-villous (spreading) indumentum. Leaves 1–2.5 cm long, partly bi-pinnatisect; lower primary lobes often retrorse, always with secondary division, sometimes all primary lobes with secondary division; ultimate lobes linear, usually spreading, 1.5–12 mm long, 0.3–0.8 mm wide; upper surface not pitted, loosely villous becoming glabrous; margins revolute; lower surface enclosed except midvein, 2-grooved, loosely villous in grooves. Conflorescence erect or sometimes deflexed, 2–4 cm long, loosely oblong-secund, acropetal, 12–20-flowered, exceeding foliage; rachis 8–15 mm long, open-tomentose. Flower colour: perianth and style red; style-end green. Perianth sparsely subsericeous to glabrous outside, pubescent inside. Pistil 27 mm long. Follicles narrowly obloid-acuminate, 10–12 mm long, with a very conspicuous, sub-basal transverse ridge.

W.A., known only from the Mt Lesueur area. Grows mostly along creek-lines or in moist low lying areas in mallee heath. Regeneration mode not known, probably from seed only. Flowers July–Oct. Map 133.

W.A.: Lesueur Natl Park, *B.Evans WE579* (CANB, PERTH); vacant Crown Land, 8 km N of Mt Lesueur, *E.A.Griffin 2228* (NSW, PERTH); N of Mt Lesueur, *P.M.Olde 91/257* (NSW).

Grevillea delta can be distinguished from *G. humifusa* which has a prostrate habit and shorter pistils (22–24 mm long). *Grevillea preissii* differs in its more strongly ascending primary leaf lobes.

This species is recognised (under *G. thelemanniana* subsp. *delta*) as 'Poorly Known' in J.D.Briggs & J.H.Leigh, *Rare or Threatened Australian Plants* (1995).

106. Grevillea obtusifolia Meisn., *in* A.L.P.P. de Candolle, *Prodr.* 14: 356 (1856)

G. thelemanniana subsp. *obtusifolia* (Meisn.) McGill., *New Names Grevillea* 15 (1986). T: [Swan River, W.A.], 1848, *J.Drummond 4th coll. 278*; lecto: NY, photo NSW, *fide* D.J.McGillivray & R.O.Makinson, *Grevillea* 444 (1993); isolecto: BM, CGE *n.v.*, G, K, LE *n.v.*, MEL, NY *n.v.*, P *n.v.*, PERTH, TCD *n.v.*

Illustrations: P.M.Olde & N.R.Marriott, *Grevillea Book* 3: 60 (bottom centre), 61 (43A, B) (1995).

Spreading shrub, sometimes prostrate or decumbent, 0.5–1.5 m high, to 5 m wide. Branchlets angular or terete, densely to sparsely subsericeous, sometimes with an open indumentum of weakly ascending hairs. Leaves entire, oblong-linear to narrowly obovate or narrowly elliptic, 1–3.5 cm long, (1.5–) 2–4 mm wide, obtuse, mucronate; upper surface smooth to obscurely pitted, glabrous or with scattered appressed hairs; margins flat to shortly and smoothly recurved; lower surface sericeous. Conflorescence erect to decurved, a loose semi-secund cluster, mostly 8–12-flowered, weakly basipetal; rachis 8–15 mm long, sparsely subsericeous to tomentose. Flower colour: perianth and style pale to bright pink or red. Perianth glabrous outside or with a few minute appressed hairs on limb, tomentose-pubescent inside. Pistil 20–24 mm long. Follicle obloid-acuminate, 12–13 mm long, faintly ridged at base, rugulose.

Occurs in W.A. from Gingin to Muchea, with a disjunct occurrence at the Murray R., Pinjarra (1800s, no more recent collection). Grows in winter-wet deep dark sands in poorly drained areas. Regenerates from seed. Flowers (Apr.) Aug.–Nov. Map 134.

W.A.: Pinjarrah, c. 96.6 km S of Perth, Oct. 1897, *R.Helms s.n.* (PERTH); near Muchea, *D.J.McGillivray 3272 & A.S.George* (NSW); 20 km S of road to Gingin on Brand Hwy, *P.Olde 91/77* (NSW); Steer St., Muchea, 2.5 km from Brand Hwy, *P.Olde 92/129* (NSW).

Grevillea thelemanniana differs from *G. obtusifolia* in its lignotuberous habit, usually the presence of some divided leaves, lack of visible midvein on the upper leaf surface, and narrower leaf lamina (<+ 2 mm wide). *Grevillea exposita* is distinguished by the occasional presence of divided leaves, and by the short spreading hairs on the branchlets. *Grevillea evanescens* is a far more robust shrub 2–3.5 m tall, with leaves up to 9 mm wide. *Grevillea pinaster* has longer (to 8 cm) linear leaves.

Olde & Marriott (*Grevillea Book* 3: 60 (1995)) recognise 'shrubby' and 'prostrate' forms, but at least some populations are mixed.

This species is recognised as 'Endangered' in J.D.Briggs & J.H.Leigh, *Rare or Threatened Australian Plants* (1995).

107. Grevillea exposita Olde & Marriott, *Grevillea Book* 1: 181 (1994)

T: Arrowsmith R., Brand Hwy, N of Eneabba, W.A., 29 Sept. 1992, *P.M.Olde 92/161*; holo: NSW; iso: PERTH *n.v.*

Illustrations: P.M.Olde & N.R.Marriott, *Grevillea Book* 2: 156 (bottom right), 157 (127A, B) (1995).

Dense spreading shrub, 1–2 m high, to 3 m wide. Branchlets angular or terete, densely tomentose-pubescent (short ascending-spreading hairs). Leaves mostly entire, oblong-linear to very narrowly obovate or -elliptic, usually a few with 2 or 3 lobes 1.5–3 (–3.5) cm long, 1.2–4 (–5.5) mm wide, obtuse, mucronate; upper surface smooth or obscurely pitted, glabrous or with a few ±appressed hairs; margins shortly to strongly revolute; lower surface sometimes obscured except midvein, subsericeous. Conflorescence decurved or deflexed at base, secund to loosely subcylindrical, mostly 8–20-flowered, basipetal; rachis 7–17 mm long, openly subsericeous. Flower colour: perianth bright red with creamy white limb; style red with green tip. Perianth glabrous outside, sometimes a very few appressed hairs on limb only, openly tomentose-pubescent inside. Pistil 20–24 mm long. Follicles obloid-acuminate, 12–13 mm long, with an obscure slanting ridge at base, rugulose to almost smooth.

Occurs in south-western W.A. in the area bounded by Eneabba, Arrino and Arrowsmith. Grows in sandy loam (sometimes lateritic) soils near creeks. Reproduces by seed. Flowers winter–spring. Map 135.

W.A.: McQuins Caravan Park, 20 km N of Eneabba, *P.Armstrong 101* (PERTH); Drummonds Bridge over Arrowsmith R., 21 km ±N of Eneabba on Brand Hwy, *N.Hoyle 678* (CANB, PERTH); Arrowsmith R. c. 56.3 km from Dongara towards Eneabba, 22 Sept. 1968, *M.E.Phillips & D.Verdon CBG040397* (CANB, NSW); 4.6 km along Banney Rd from Slipper Rd (near Arrino), *P.Olde 86/588* (NSW); between Moore and Murchison Rivers, *E.Pritzel 381* (NSW).

See notes under *G. obtusifolia*, *G. hirtella* and *G. thelemanniana* for comparisons. *Grevillea evanescens* has appressed branchlet hairs and longer pistils (26–27 mm long).

108. Grevillea evanescens Olde & Marriott, *Grevillea Book* 1: 181 (1994)

T: Military Rd, N off Lancelin road towards Gingin, W.A., 29 Sept. 1991, *P.M.Olde 91/240*; holo: NSW; iso: PERTH *n.v.*

Illustrations: P.M.Olde & N.R.Marriott, *Grevillea Book* 2: 153 (124A–B), 154 (top left) (1995).

Erect to spreading single-stemmed shrub 2–3.5 m high. Branchlets angular or terete, glabrous or with scattered appressed hairs. Leaves entire, oblong to obovate to elliptic or narrowly so, 2–4.5 cm long, 2.5–9 mm wide; upper surface ±smoothly glabrous, not pitted; margins shortly recurved; lower surface sericeous on young leaves, soon becoming sparsely sericeous or glabrous. Conflorescence decurved, secund, mostly 8–12-flowered, weakly acropetal to subsynchronous; rachis 7–15 mm long, with scattered appressed hairs or glabrous. Flower colour: perianth pale to bright red with creamy limb; style red with prominent green apex. Perianth glabrous outside, tomentose-pubescent inside. Pistil 26–27 mm long. Follicles oblong-acuminate, 10–11 mm long, with an obscure slanting ridge at base, rugulose.

Occurs in south-western W.A., west of Gingin. Grows on ridges in sandy soil in *Banksia* woodland. Probably regenerating by seed only. Flowers winter–spring. Map 136.

W.A.: Gingin to Lancelin road, *H.Demarz 4847* (PERTH).

See under G. obtusifolia, *G. exposita* and *G. thelemanniana* for comparisons.

This species is recognised as 'Poorly Known' in J.D.Briggs & J.H.Leigh, *Rare or Threatened Australian Plants* (1995).

109. Grevillea pinaster Meisn., *Hooker's J. Bot. Kew Gard. Misc.* 7: 76 (1855)

G. thelemanniana subsp. *pinaster* (Meisn.) McGill., *New Names Grevillea* 15 (1986). T: W.A., Interior of Swan River, to the North A. 1850–51, *J.Drummond 6th coll., 182*; lecto: NY, *fide* D.J.McGillivray & R.O.Makinson, *Grevillea* 444 (1993); isolecto: BM, CGE *n.v.*, G-DC, K, LD *n.v.*, MEL, P *n.v.*

Illustrations: D.J.McGillivray & R.O.Makinson, *Grevillea* 252 (top left) (1993); P.M.Olde & N.R.Marriott, *Grevillea Book* 3: 97 (top right & 74A, B), 98 (74C, D) (1995).

Usually robust, erect shrub 2–3 m high or sometimes a low spreading shrub 0.2–1 m high and 2–3 m wide. Branchlets angular, subsericeous or sparsely so, rarely glabrous, sometimes densely pubescent. Leaves (1.5–) 2.5–8 cm long, entire and linear or rarely some or all 2–5-sect; lobes linear, 1–4 cm long, 0.5–2 (–2.8) mm wide, obtuse, mucronate; upper surface smooth, soon glabrous; margins revolute or sometimes recurved; lower surface 2-grooved or slightly exposed beside midvein and subsericeous. Conflorescence deflexed or decurved, shortly secund to subcylindrical, loose, 12–20-flowered, subsynchronous; rachis 5–20 mm long, subsericeous to sparsely so. Flower colour: perianth and style pink-red to red, with yellowish style-end. Perianth glabrous outside or rarely sparsely hairy on limb only, pubescent inside about ovary, elsewhere glabrous except along tepal margins. Pistil 20–25 mm long. Follicles obloid-acuminate to ellipsoidal, 10–13 mm long, usually faintly ribbed with an obscure slanting (rarely transverse) sub-basal ridge. Plate 27.

Occurs in south-western W.A., from Murchison R. S to Eneabba, and inland nearly to Mullewa, with isolated populations near Hamelin Pool, Wongan Hills and Pithara. Grows in shrubland and heath, often beside creek banks. Regenerates from seed, sometimes also from lignotuber. Flowers May–Sept., sporadically in other months. Map 137.

W.A.: c. 57 km from North West Coastal Hwy on road to Kalbarri, *D.J.McGillivray 3341* (NSW); 72.4 km N of Geraldton on highway, *M.E.Phillips CBG035562* (CANB, NSW); Moresby Ra., 12.9 km NE of Geraldton on road to Nabawa, *R.Coveny 3046* (NSW); c. 20.9 km N of Geraldton on Carnarvon road, *R.Melville 4149* (K, NSW); Oakabella Hills, 26 km N of outskirts of Geraldton on North West Coastal Hwy, *J.M.Fox 88/100* (CANB, MEL, NSW, PERTH).

Grevillea stenomera differs from *G. pinaster* in having all leaves divided, with the longest lobes usually > 40 mm long, and usually longer conflorescences (3–6 cm long). *Grevillea obtusifolia* and *G. exposita* have most simple leaves < 30 mm long, with the lower leaf surface exposed on either side of the midvein. *Grevillea thelemanniana* differs in its mostly shorter (mostly < 25 mm long), more spreading leaves with the lower surface usually narrowly exposed.

Olde & Marriott (*Grevillea Book* 3: 98 (1995)) distinguish three forms of *G. pinaster*. The 'typical form' has mid-green leaves, and upright habit, and leaves at the broad end of the variation; it occurs from Geraldton E to Mullewa and N beyond the Murchison R. The 'secund form' has grey-green leaves, often crowded to one side of the branchlets, and a lower more spreading habit; it occurs near Kalbarri and perhaps near Eneabba. Both these forms have occasional leaves with 2 or 3 linear lobes. A 'pinnate-leaved form', with most leaves having 3–5 linear lobes, occurs sporadically near Eneabba, Mullewa and Ogilvie (e.g. *D.J.McGillivray 3322 & A.S.George*, NSW, PERTH); its status is uncertain.

110. **Grevillea preissii** Meisn., *in* J.G.C.Lehmann, *Pl. Preiss.* 1: 543 (1845)

G. thelemanniana subsp. *preissii* (Meisn.) McGill., *New Names Grevillea* 15 (1986). T: Mt Eliza (Perth), W.A., 19 May 1839, *L.Preiss 709*; lecto: NY, photo NSW, *fide* D.J.McGillivray & R.O.Makinson, *Grevillea* 444 (1993); isolecto: B *n.v.*, G, G-DC, HBG *n.v.*, K, LE *n.v.*, MEL, P *n.v.*, TCD *n.v.*

Low mounded to spreading or erect dense shrub 0.3–1.2 m high. Branchlets terete, tomentose-pubescent to villous, with or without glabrous striation, sometimes almost glabrous. Leaves (1.5–) 2.5–5 cm long, (3–) 5–7-sect; primary lobes closely aligned, at least basal ones usually with secondary and sometimes tertiary division; ultimate lobes linear to narrowly linear to subterete, 5–20 mm long, 0.5–1.0 mm wide, weakly pungent, usually slightly incurved; upper surface pubescent to villous becoming glabrous, not pitted; margins revolute to midvein; lower surface tomentose to villous in grooves. Conflorescence decurved, secund, 12–30-flowered, acropetal; rachis 18–30 mm long, subsericeous to tomentose or glabrous. Perianth sparsely subsericeous or glabrous outside, densely pubescent inside. Pistil 25–28 mm long. Follicle obloid-triangular, c. 15 mm long, with a broad ±transverse ridge or ventral bulge just above base and an obscure longitudinal subdorsal ridge on each side, smooth and shiny or minutely rugose.

Occurs in W.A. from Greenhead to Bunbury, near the coast. There are two subspecies, with an apparent intergrade zone between them in the Moora–Jurien Bay–Moore R. area.

Perianth sparsely hairy outside, with hairs often restricted to limb; floral rachis ±densely subsericeous to tomentose; branchlets subsericeous to tomentose, rarely subvillous; leaves usually 2.5–5 cm long **110a.** subsp. **preissii**

Perianth glabrous outside; floral rachis glabrous or sparsely tomentose; branchlets usually villous, rarely tomentose; leaves usually 1.5–2.5 cm long **110b.** subsp. **glabrilimba**

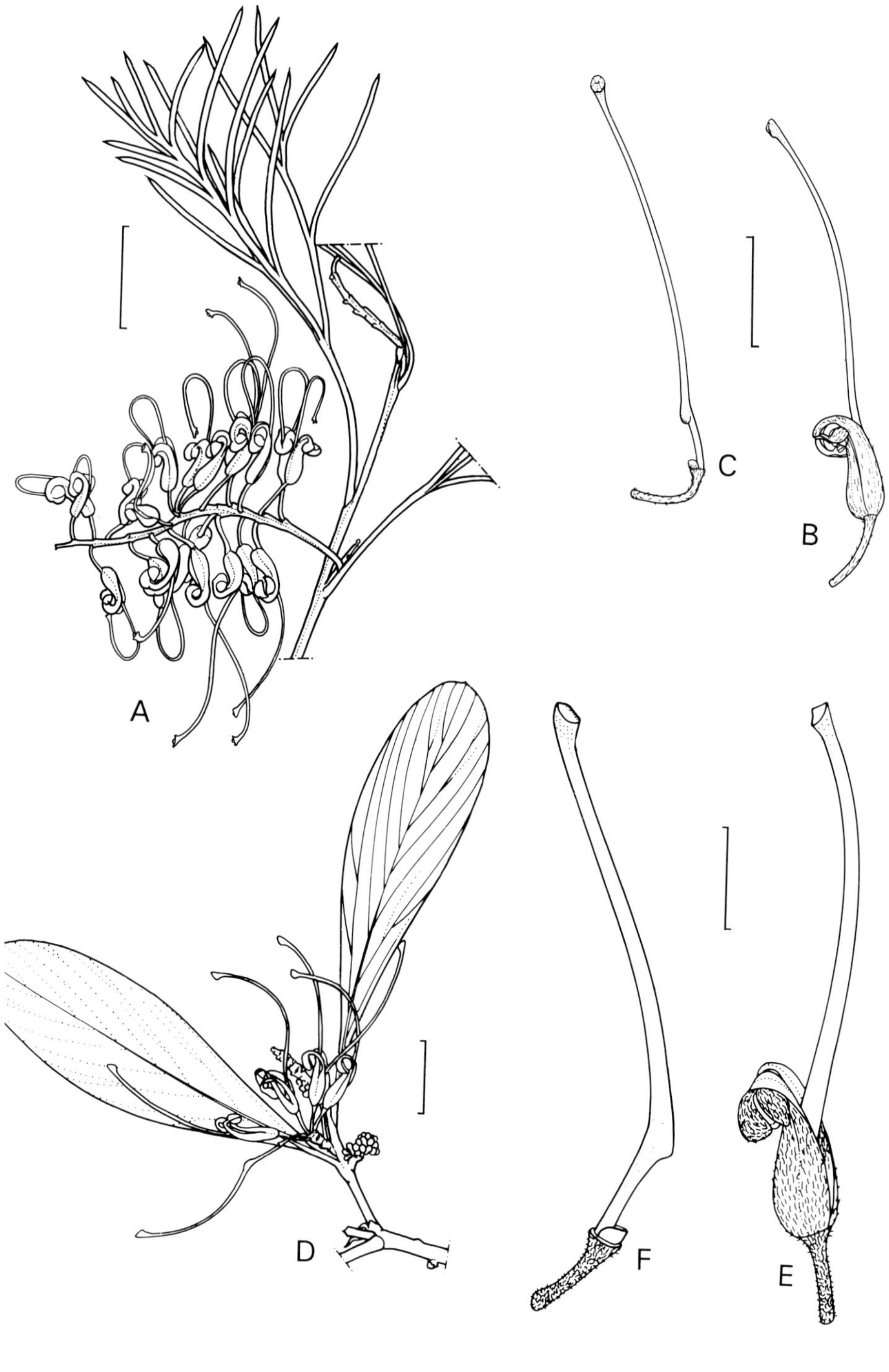

Figure 14. *Grevillea*. **A–C**, *G. preissii* subsp. *preissii*. **A**, flowering branch; **B**, flower; **C**, pistil (**A–C**, Aug. 1961, J.Burbidge, CANB). **D–F**, *G. olivacea*. **D**, flowering branch; **E**, flower; **F**, pistil (**D–F**, D.J.McGillivray 3294 & A.S.George, NSW). Scale bars: **A–D** = 1 cm; **E–F** = 5 mm. Drawn by: **A–C**, D.Boyer; **D–F**, D.Fortescue.

110a. Grevillea preissii Meisn. subsp. **preissii**

G. splendens Hort. ex de Noter, *Rev. Hort., n.s.*, 9: 33 (1909), *nom. illeg.* T: not designated.

Illustrations: P.M.Olde & N.R.Marriott, *Grevillea Book* 3: 109 (bottom right), 110 (83A, B) (1995).

Young leaves and branchlets subsericeous to tomentose, rarely subvillous. Leaves usually 2.5–5 cm long. Floral rachis ±densely subsericeous to tomentose. Flower colour: perianth and style red, orange-red or pinkish red with prominent yellow-green style-end. Perianth sparsely hairy outside, with hairs often restricted to limb. Fig. 14A–C.

Occurs in south-western W.A., in coastal areas from Lancelin to Bunbury, common in the Perth area. Grows in coastal heath. Probably reproducing from seed only. Flowers (Mar.–) May–Oct. Map 138.

W.A.: 36 km N of Yanchep on Lancelin road, *M.L.Clark 196* (NSW, PERTH); E of Medina, *A.S.George 21* (NSW, PERTH); Cottesloe, *R.Helms 41* (NSW).

Olde & Marriott (*loc. cit.*) recognise two forms. The 'prostrate green-leaf form', known mainly in cultivation, is a sprawling shrub with bright green leaves; MEL 1594142 from S of Mandurah may represent a wild population of this form. The 'superior form', occurring S of Perth, is more erect, growing to 1.2 m tall with darker green leaves.

Grevillea preissii subsp. *preissii* often has a markedly denser beard of hairs on the inner surface of the perianth as compared to the looser beard in subsp. *glabrilimba*; it also has more strongly ascending leaves, giving an upswept appearance to the foliage along a branch.

Subsp. *preissii* is distinguished from most other taxa in the group that have secondary leaf division by its longer leaves (those of *G. humifusa* and *G. delta* are generally ⩽ 2.5 cm long), and from *G. fililoba* which has very narrow (c. 0.5 mm wide) filiform leaf segments.

110b. Grevillea preissii subsp. **glabrilimba** Olde & Marriott, *Grevillea Book* 1: 182 (1994)

T: 26.1 km W of Brand Hwy, on road to Greenhead, W.A., 14 Sept. 1991, *P.M.Olde 91/89*; holo: NSW; iso: PERTH *n.v.*

Illustrations: P.M.Olde & N.R.Marriott, *Grevillea Book* 3: 110 (bottom right), 111 (84A, B) (1995).

Young leaves and branchlets usually villous, rarely tomentose. Leaves usually 1.5–2.5 cm long. Floral rachis glabrous or sparsely tomentose. Flower colour: perianth and style red to orange-red; style-end green. Perianth glabrous outside.

Occurs in south-western W.A., in near-coastal areas from Greenhead and Leeman to near Cervantes. Grows in low heath, in calcareous sands over limestone. Regenerates from lignotuber and seed. Flowers (Mar.–) May–Oct. Map 139.

W.A.: 10.5 km ENE of Jurien, *B.G.Briggs 6333* (NSW, PERTH); 3.5 km E of Jurien, *D.J.McGillivray 3292 & A.S.George* (CANB, NSW); 29.3 km W of Brand Hwy on road to Greenhead, *P.Olde 91/90* (NSW); 4.8 km E of Jurien Bay, *R.V.Smith 66/171* (MEL, NSW).

See under *G. humifusa*, *G. delta* and *G. fililoba* for differences from those taxa. Natural hybrids between this species and *G. olivacea* are known.

Olde & Marriott (*loc. cit.*) recognise two forms, one with conspicuously grey hairy foliage (the 'silver-leaf form') and the other (the 'fine-leaved form'), from the Leeman–Greenhead area, with less hairy, brighter green leaves and finer leaf segments.

A cultivated specimen (*CBG8703155*) with wild origin given as 1 km from Warren Natl Park on road from Pemberton to Northcliffe (*J.M.Powell 3089*, NSW?) is similar to *G. preissii* subsp. *glabrilimba*, but has the leaf lobes more divaricate and a very conspicuous bulbous nectary (similar to that of *G. hirtella*). This may represent a new subspecies.

111. Grevillea ripicola A.S.George, *Nuytsia* 1: 373 (1974)

T: Collie River, Collie, W.A., 15 Oct. 1965, *A.S.George 6848*; holo: PERTH; iso: AD, CANB, K, MEL, NSW, PERTH.

Illustrations: P.M.Olde & N.R.Marriott, *Grevillea Book* 3: 139 (top right), 140 (111A–C) (1995).

Dense, irregular, sprawling shrub c. 2–3 m high. Branchlets terete to slightly angular, ribbed, ±glabrous. Leaves 2.5–6 cm long, with rachis usually refracted at first node, divaricately pinnatipartite, once or twice divided into curved, elliptic to linear, rigid, pungent lobes 1–3 cm long and 1.5–5 mm wide; upper surface smooth to wrinkled, not pitted, glabrous; margins angularly but irregularly refracted; lower surface partly exposed, glabrous. Conflorescence deflexed or decurved or sometimes erect, secund to hemispherical, 12–20-flowered, acropetal; rachis 13–20 (?–30) mm long, ±glabrous. Flower colour: perianth yellowish orange, rarely yellow; style red, rarely yellow with green to yellowish tip. Perianth glabrous outside, sparingly bearded inside. Pistil 29–35 mm long. Follicle narrowly ovoid-ellipsoidal with truncated base, 13–17 mm long, with a broad basal transverse ridge or bulge, glabrous, slightly longitudinally ridged and faintly granular. Plate 26.

Occurs in south-western W.A., in the area bounded by Collie, Donnybrook and Bridgetown, possibly also near Northcliffe (R.Hearn, pers. comm.). Grows in loamy to clayey soils beside creeks in open Jarrah forest. Regenerates from seed. Flowers (Apr.–) Oct.–Nov. Map 140.

W.A.: Collie, *J.S.Beard 3949* (KPBG); Griffin Bridge, Collie, *A.R.Fairall 1720* (NSW, PERTH); creek line c. 6.4 km SE of Collie, on road to Cardiff, *A.S.George 7710* (PERTH); Collie, Oct. 1969, *P.N.Shedley* (PERTH).

See under *G. acropogon* for differences.

This species is recognised as 'Rare' in J.D.Briggs & J.H.Leigh, *Rare or Threatened Australian Plants* (1995).

112. Grevillea acropogon Makinson, *Fl. Australia* 17A: 496 (2000)

T: c. 6 km NNE of Lake Unicup, W.A., 7 July 1996, *E.Middleton ARA5755*; holo: PERTH.

Prostrate to erect shrub to 1.8 m high. Branchlets softly angular to subterete, loosely to sparsely sub-tomentose with straight hairs, becoming nearly glabrous. Leaves rigid, 1.5–2.5 cm long, divaricately pinnatisect with 5–7 primary lobes, the basal 1 or 2 pairs again divaricately 2- or 3-sect; ultimate lobes linear, 10–15 mm long, 0.8–1.1 mm wide, pungent; upper surface loosely subtomentose or -subsericeous, soon glabrous, not pitted; margins angularly revolute; lower surface mostly or wholly enclosed except for midveins, sometimes lamina narrowly exposed and then subsericeous. Conflorescence decurved, shortly and broadly secund, 18–24-flowered, acropetal; rachis 12–17 mm long (only c. 8–9 mm active), openly pubescent becoming almost glabrous. Perianth and style both red. Perianth glabrous outside except for a few inconspicuous appressed hairs near tip of limb segments, densely bearded inside. Pistil 20–22 mm long. Follicles and seeds not seen.

As yet known only from the Type population near Lake Unicup, about 30 km WNW of Frankland township in the SW of W.A. It is recorded as growing in shallow soils over ironstone on the margins of seasonally inundated areas, in a heath association. Flowering July, probably also June–Sept. Map 141.

Grevillea acropogon appears to belong to the *G. thelemanniana* group of species, although fruits are needed for an exact assignment. It is probably most closely related to *G. ripicola*, which has a glabrous leaf lower surface, longer leaves 2.5–6 cm long with longer and wider ultimate lobes 10–30 mm long and 1.5–5 mm wide, and a longer pistil 29–35 mm long. *Grevillea mccutcheonii* is also related but has cruciform leaf division with stubby rounded lobes and a usually longer pistil.

113. **Grevillea mccutcheonii** Keighery & Cranfield, *Nuytsia* 11: 33 (1996)

T: SE of Busselton, W.A., 6 Nov. 1993, *G.J.Keighery 13786*; holo: PERTH; iso: AD, CANB, K, MEL.

Illustrations: A.Brown, *Landscope* 11(2): 44 (1995); G.J.Keighery & R.J.Cranfield, *Nuytsia* 11: 35, fig. 1 (1996); A.Brown *et al.* (eds), *W. Australia's Threatened Flora* 143 (1998).

Erect and spreading shrub. Branchlets angular, ridged, glabrous. Leaves 1–2.5 (–3.5) cm long, 7–15 (–22) mm wide (across lobes), 3-fid to -partite and broadly cruciform with 3 spreading rounded apically spinescent lobes and sometimes a few secondary teeth, or occasionally (juvenile?) simple and panduriform (fiddle-shaped) and apically 3-toothed; leaves broadening towards base and with rounded basal auriculate stem-clasping lobes; upper surface with venation very conspicuous; margins undulate, sometimes slightly decurved; lower surface glabrous. Conflorescence decurved, secund, 26–28-flowered, acropetal; rachis 5–10 mm long, glabrous. Flower colour: perianth reddish green; style reddish. Perianth glabrous outside, densely bearded inside. Pistil (20–) 28–30 mm long. Follicles narrowly ovoid, 13–17 mm long, glabrous, faintly colliculose.

Occurs in south-western W.A., known only from a small area near the base of the Witcher Ra. SE of Busselton. Grows in tall shrubland in moist red sandy clay soil over ironstone. Regenerates apparently from seed only. Flowers most months, peaking July–Nov. Map 142.

W.A.: E of Ludlow to Hithergreen, *G.S.McCutcheon 2715–2718* (PERTH); Tutunup, *G.S.McCutcheon 2696* (PERTH); E of Ludlow to Hithergreen road, *J.A.Cochrane 1220* (PERTH).

Grevillea mccutcheonii is highly distinctive in leaf form, with stubby cruciform division and stem-clasping auricles at the base of the leaf. It is unlikely to be confused with any other species. It seems to be closely related to *G. ripicola*, which has oblong to subulate divaricate leaf lobes and lacks basal stem-clasping auriculate lobes, and to *G. acropogon*, which differs in its linear pungent leaf lobes c. 1 mm wide. *Grevillea manglesii* and *G. papillosa* grow in the same general area as *G. mccutcheonii*, and have some points of similarity, but both have much smaller flowers with pistils 6–11 mm long, and lack stem-clasping leaf lobes.

This species is recognised as 'Endangered' in J.D.Briggs & J.H.Leigh, *Rare or Threatened Australian Plants* (1995).

114. **Grevillea stenomera** F.Muell., *Fragm.* 4: 85 (1864)

T: Murchison River, W.A., *s.d.*, *A.Oldfield*; lecto: MEL, *fide* D.J.McGillivray & R.O.Makinson, *Grevillea* 442 (1993); isolecto: K, PERTH.

Illustrations: D.J.McGillivray & R.O.Makinson, *Grevillea* 248 (top right) (1993); P.M.Olde & N.R.Marriott, *Grevillea Book* 3: 184 (bottom right), 185 (150A–C) (1995).

Rounded silvery to bluish grey, sometimes lignotuberous, shrub to 2 m high. Branchlets conspicuously angular, silky. Leaves secund, 4–6 cm long, pinnatisect with 5–15 lobes; lobes long-linear, 4–16, 4–12.5 cm long, 0.8–2.2 mm wide, pliable; upper surface smooth, densely subsericeous becoming glabrous; margins angularly refracted; lower surface enclosed except prominent midvein, 2-grooved, subsericeous in grooves. Conflorescence decurved, secund to cylindrical, 20–30-flowered, subsynchronous or weakly basipetal; rachis 35–60 mm long, subsericeous to pubescent. Flower colour: perianth pale pink, orange-pink or reddish pink with a greenish yellow limb; style greenish pink with green tip. Perianth sparsely subsericeous outside, denser on limb, bearded inside at curve. Pistil 22–24 mm long, glabrous. Follicle 1–13 mm long, obloid-ellipsoidal, with obscure slanting basal ridge, rugulose, glabrous.

Occurs in south-western W.A., known only from scattered populations from the Kalbarri area N to Tamala. Grows in mixed shrubland or heath, in sandy soils over laterite or limestone. Regenerates from seed and/or lignotuber. Flowers (June–) Aug.–Oct. (–Dec.). Map 143.

W.A.: 56.2 km W of North West coastal Hwy along S side of State Barrier fence, *R.J.Cranfield 9219* (PERTH); c. 23 km direct N of Kalbarri on Murchison House Stn, c. 2 km direct W of Long Thicket Bore, *A.Lyne 910 et al.* (BRI, CANB, MEL, NSW, PERTH); c. 57 km from North West Coastal Hwy on road to Kalbarri, *D.J.McGillivray 3337 & A.S.George* (CANB, NSW); 63 mile tank – Junga tank NW of Ajana NE of Gantheaume Bay, 25 Sept. 1953, *N.H.Speck s.n.* (CANB).

Grevillea stenomera can be confused with the 'pinnate-leaved form' of *G. pinaster*, which differs in having shorter leaf lobes ⩽ 40 mm long, and shorter floral rachises ⩽ 25 mm long. Most forms of *G. pinaster* have most or all leaves entire. *Grevillea preissii* differs in its shorter, narrower leaf lobes (5–20 mm long, 0.5–1 mm wide) which are more strongly ascending than in *G. stenomera* and often incurved.

This species is recognised as 'Poorly Known' in J.D.Briggs & J.H.Leigh, *Rare or Threatened Australian Plants* (1995).

115. Grevillea variifolia C.A.Gardner & A.S.George, *J. Roy. Soc. W. Australia* 46: 129, 130, fig. 1 *p.p.* (1963)

T: Cape Range, near No. 3 Well, W.A., 2 June 1961, *A.S.George 2477*; holo: PERTH; iso: K, MEL, PERTH.

Spreading to sprawling shrub 0.3–1.3 m high. Branchlets angular, subsericeous. Leaves obovate-cuneate to obtrullate, occasionally entire and elliptic when young, 1–5.5 cm long, 3–25 mm wide, usually 3–7-dentate to -fid in upper half of leaf, sometimes 3–7-partite with narrowly triangular divaricate lobes, pungent or not; upper surface not pitted, ±glabrous, with reticulate venation usually prominent; margins recurved to shortly revolute; lower surface sericeous. Conflorescence often deflexed, ±secund, 12–30-flowered, acropetal to subsynchronous; rachis 15–50 mm long, ±densely subsericeous. Perianth with sprinkled appressed hairs outside below curve, subsericeous on limb, densely bearded inside. Pistil (18–) 25–28 mm long. Follicles ovoid-ellipsoidal, 12–16 mm long, glabrous, longitudinally ridged.

Occurs in W.A. from Cape Ra. to Lake McLeod. Two subspecies are recognised.

This species is recognised as 'Poorly Known' in J.D.Briggs & J.H.Leigh, *Rare or Threatened Australian Plants* (1995).

Petiole 3–6 mm long; leaf lamina usually > 25 mm long, with 3–7 teeth or shallow lobes, these not divaricate and not or scarcely rigid or pungent, and often obtuse **115a.** subsp. **variifolia**

Petiole c. 2 mm long; leaf lamina usually < 25 mm long, with 3–5 deep triangular lobes, these rigid, pungent, and often somewhat divaricate **115b.** subsp. **bundera**

115a. Grevillea variifolia C.A.Gardner & A.S.George subsp. variifolia

Illustrations: C.A.Gardner & A.S.George, *J. Roy. Soc. W. Australia* 46: 130, fig. 1A–J, L (1963); P.M.Olde & N.R.Marriott, *Grevillea Book* 3: 216 (178A, B), 217 (178C) (1995).

Mature leaves with petioles 3–6 mm long; lamina cuneate to narrowly so or occasionally oblanceolate, (17–) 25–43 mm long, with 3–7 subapical rounded teeth or shallow broadly triangular lobes; teeth and lobes not divaricate, obtuse or occasionally acute, weakly spinose, not or scarcely pungent. Flower colour: perianth red; style red with yellow tip.

Occurs in W.A. in the Cape Ra., northern part of North West Cape peninsula. Grows in rocky open shrub/*Triodia* associations on massive Tertiary limestones, in skeletal soils. Regenerates from seed. Flowers (Feb.) June–Oct. Map 144.

W.A.: Cape Range Natl Park, Charles Knife Rd, *Hj.Eichler 22581* (AD, CANB, DNA, MEL, PERTH); 1.6 km S of Vlaming Head, *A.S.George 2577* (PERTH); Mandu Mandu Gorge, *G.J.Keighery 12858* (PERTH).

115b. Grevillea variifolia subsp. bundera Keighery, *Nuytsia* 12: 294, fig. 1 (1998)

T: 15.6 km N of Coral Bay turnoff on Exmouth Rd, W.A., 25 Aug. 1992, *G.J.Keighery & N.Gibson 323*; holo: PERTH.

Illustrations: C.A.Gardner & A.S.George, *J. Roy. Soc. W. Australia* 46: 130, fig. 1K, M (1963); D.J.McGillivray & R.O.Makinson, *Grevillea* 253, fig. 68 (fruit only) (1993).

Mature leaves with petiole c. 2 mm long; lamina cuneate in basal half, 11–15 (–25) mm long, with 3–5 deep triangular lobes; lobes often somewhat divaricate, acute, rigid, conspicuously spinose, pungent. Flower colour: perianth red; style red with yellow tip.

Occurs in W.A., between Cape Cuvier (near S end of Lake McLeod) and Rough Ra. (inland from Ningaloo). Grows in open shrub/*Triodia* associations on Quaternary calcarenites and Pleistocene limestones with red sand soils. Regenerates probably from seed only. Flowers ?June–Oct. Map 145.

W.A.: 127.1 km S of Learmouth, *A.S.George 2402* (CANB, PERTH); Gnaraloo, *Gready 4* (PERTH *n.v.*); Rough Ra., *G.J.Keighery & N.Gibson 300* (PERTH *n.v.*).

116. Grevillea olivacea A.S.George, *Nuytsia* 1: 373 (1974)

T: on the mainland opposite Snag Is., S of Dongara, W.A., 6 Sept. 1966, *S.Chambers 88*; holo: PERTH.

Illustrations: A.S.George, *Introd. Proteaceae W. Australia* 51, 69 (1984); D.J.McGillivray & R.O.Makinson, *Grevillea* 254, 69a–c, 255, col. pl. (1993); P.M.Olde & N.R.Marriott, *Grevillea Book* 3: 67 (bottom right), 68 (49A–E) (1995).

Erect robust dense shrub to 4 m high. Branchlets terete to angular, white-sericeous with glabrous ribs. Leaves entire, elliptic to obovate, 3–6.8 cm long, 0.5 –1.7 cm wide, obtuse or acuminate, sometimes retuse, mucronate; upper surface ±smooth, not pitted, ±glabrous; margins flat to recurved or occasionally revolute; lower surface sericeous except for ±glabrous midvein. Conflorescence sometimes on old stem, erect, umbel-like, regular and dense, 14–28-flowered, subsynchronous; rachis 4–8 mm long, tomentose. Flower colour: perianth bright red, orange or yellow on inside, overlaid by white indumentum outside; style red to yellow with yellow tip. Perianth densely white-tomentose outside, bearded inside. Pistil 22–30 mm long. Follicle obloid-ovoid with a retrorse basal heel, 11–15 mm long, rugose, granular. Fig. 14D–F.

Occurs in south-western W.A., near the coast from Jurien Bay S to Coolimba. Grows in calcareous sands over limestone, in mixed shrub associations. Regenerates from seed. Flowers June–Oct. Map 146.

W.A.: Tunnel, Logue L., *C.A.Gardner 9104* (CANB, NSW, PERTH); Stockyard Gully, July 1963, *C.A.Gardner s.n.* (PERTH); Eneabba Reserve, 16 km NE of Leeman on seismic track near coast, *A.C.Kessell 631* (PERTH); Hastings Cave, Drovers Cave Natl Park, near Jurien, *D.J.McGillivray 3294 & A.S.George* (B, CANB, K, NSW, US).

Grevillea olivacea is potentially confusable with *G. evanescens*, which differs in having open to sparsely sericeous indumentum on the lower leaf surface, smaller leaves 20–45 mm long and 2.5–9 mm wide, a shorter floral rachis, a ridged fruit, and the outer surface of the perianth glabrous. *Grevillea obtusifolia* differs in similar respects, but has even smaller leaves. *Grevillea argyrophylla* (here placed in the Hakeoides group) has some similarities, and there may be a relationship, however *G. argyrophylla* has smaller and typically strongly obovate leaves, shorter pistils c. 10 mm long and white flowers.

McGillivray & Makinson (*Grevillea* 254 (1993)), left this species ungrouped, but its affinities lie with the taxa around *G. thelemanniana*; however, it lacks the ridging on the fruit typical of most other members (but see *G. mccutcheonii*). Natural hybrids between *G. olivacea* and *G. preissii* subsp. *glabrilimba* are known.

This species is recognised as 'Rare' in J.D.Briggs & J.H.Leigh, *Rare or Threatened Australian Plants* (1995).

Oncogyne Group

Shrubs. Leaves divided or entire, dorsiventral, surfaces dissimilar; margins recurved to revolute. Conflorescence terminal (sometimes conspicuously emergent) or axillary, erect or occasionally decurved to pendulous or decumbent, simple or irregularly and basally many-branched; unit conflorescence umbel-like to cylindrical or semi-secund, weakly basipetal to subsynchronous. Flowers basiscopic or abaxially oriented, rarely acroscopic or adaxially so. Torus reversely oblique to ±transverse. Perianth zygomorphic, mostly with erect, short glandular hairs outside, sometimes also with short biramous hairs, hairy inside; tepals remaining loosely coherent except along dorsal suture, and held ventrally. Pistil 15–52 mm long; ovary stipitate, usually glabrous, sometimes a few scattered biramous hairs on dorsal side of ovary and stipe; style glabrous, exserted from late bud; pollen-presenter lateral or very oblique, flat or convex. Follicle glabrous, conspicuously horned or ridged ventrally, elsewhere evenly granulate or smooth; pericarp moderately thick. Seeds ellipsoidal or almost so, with a central ridge surrounded by a membranous channel on inner face, the wing sometimes drawn to a short point at ends.

A group of eight species, endemic in south-western W.A. Bird-pollinated. This group corresponds to most of Bentham's (1870) series *Leiogyne* of section *Eugrevillea*. It is closely related to the *Thelemanniana* group. The *Oncogyne* group is also characterised by the ribbed tepals, with the ribs conspicuous on the limbs, and the ovary with prominent ridges or wrinkled protuberances on the ventral side. In several species the torus is reversely oblique to the pedicel, i.e. with the ventral rim higher (in side view) than the dorsal rim.

1 Pistil < 25 mm long

2 All leaves either entire or with primary division only

3 Leaves ⩽ 5 mm wide; ovarian stipe 1.3–2.2 mm long; inflorescences often emergent on long trailing branches **123. G. nudiflora**

3: Leaves 10–40 mm wide; ovarian stipe 0.5–1.2 mm long; inflorescences never on long emergent trailing branches **124. G. infundibularis**

2: Most leaves with secondary division **122. G. patentiloba**

1: Pistil > 25 mm long

4 All leaves entire and longest leaves ⩾ 5 cm long **117. G. oncogyne**

4: Some or all leaves either divided, or entire and then ⩽ 4 cm long

5 Some or all leaves pinnatisect (sometimes lower lobes with secondary division); leaf lobes not divaricate (primary lobes on either side of leaf rachis ±mutually aligned)

6 Most floral rachises ⩽ 5 mm long; conflorescence spreading to decurved; leaves dentate to pinnatipartite; longest leaf lobes < 25 mm long (and usually < 20 mm) **121. G. pectinata**

6: Most floral rachises > 5 mm long; conflorescence ±erect or spreading or decurved; leaves pinnatipartite; longest leaf lobes > 20 mm long (and usually > 25 mm)

7 Most leaves of non-flowering branches with < 10 primary lobes; conflorescence usually on short peduncle, borne within leaves **117. G. oncogyne**

7: Most leaves of non-flowering branches with > 10 primary lobes; conflorescence often on long peduncle projecting beyond foliage (sometimes trailing on ground) **120. G. plurijuga**

5: Leaves either not pinnatisect or if so, then leaf lobes divaricate

8 Leaves divaricately lobed

9 Leaf lobes ⩽ 1.4 mm wide; leaf margins smoothly revolute; leaves sometimes with secondary division; conflorescence usually pedunculate and branched **119. G. newbeyi**

9: Leaf lobes 1.5–3 mm wide; leaf margins refracted; leaves with primary division only; conflorescence usually sessile or almost so and unbranched **118. G. tripartita**

8: Leaves not divaricately lobed

10 Pistil > 45 mm long; conflorescence subsessile **118. G. tripartita**

10: Pistil < 40 mm long; conflorescence pedunculate **121. G. pectinata**

117. Grevillea oncogyne Diels, *Bot. Jahrb.* 35: 149 (1904)

T: Gilmores, W.A., 2 Nov. 1901, *L.Diels 5278*; lecto: B, *fide* D.J.McGillivray & R.O.Makinson, *Grevillea* 432 (1993); remaining syntypes: Boorabbin, Nov. 1901, *E.Pritzel 912*; syn: A *n.v.*, AD, E, G *n.v.*, K, NSW, P *n.v.*

Illustrations: P.M.Olde & N.R.Marriott, *Grevillea Book* 3: 69 (top right & 50A), 70 (50B, C) (1995).

Erect shrub 1–3 m high, or sometimes a low, multistemmed lignotuberous shrub to 1 m. Leaves entire, linear, (3–) 5–12 (–15) cm long, 1–2 mm wide, or some, rarely most, leaves 2–4 (–7)-lobed; lobes entire, linear, 5–9 cm long; margins smoothly revolute; lower surface obscured, 2-grooved. Conflorescence usually axillary or cauline on old stems, rarely terminal, erect, shortly pedunculate; unit conflorescence regular, shortly cylindrical, lax, basipetal; rachis 5–30 mm long, subsericeous. Flowers basiscopic. Flower colour: perianth and style (including tip) red or pinkish red. Perianth with both appressed biramous and erect simple glandular hairs outside, pilose inside. Pistil 36–40 (–47) mm long, glabrous; ovary with 2 prominent wrinkled outgrowths on either flank. Follicles 7–15 mm long, erect, with prominent ridge-like outgrowths on ventral side, the remainder irregularly tuberculate.

Occurs in W.A., E of the central and southern wheatbelt in an area bounded by Boorabbin, Coolgardie and Cave Hill, S to around Salmon Gums and W to Lake King and Hyden. Grows usually in well-drained light soils, rarely in heavy clays, often on lateritic, granitic or limestone outcrops, in mallee woodland or shrubland. Regenerates from lignotuber and seed. Flowers (Apr.–) Oct.–Dec. Map 147.

W.A.: Moorine Rock, *A.M.Ashby 1689* (AD, CANB); Southern Cross [to] Forrestiana road, 15.1 km N of Mt Holland T-junction, *F.H.Mollemans 3704* (CANB); Boorabbin, *s.d.*, *A.Morrison s.n.* (E, CANB); 11.9 km W of Mt Palmer, *B.H.Smith 1461* (CANB, HO, MEL, PERTH); 24 km SW of Norseman, *J.Taylor 675 et al.* (CANB, PERTH).

There are three forms. The 'type form' ('typical form' of Olde & Marriott, *loc. cit.*) is an upright, usually single-stemmed shrub 2–3 m tall, with entire leaves 1–1.2 mm wide; it occurs mainly in the northern part of the range (Boorabbin, Coolgardie, Kalgoorlie), with a disjunct occurrence S of Norseman. Lignotuberous reproduction has not been confirmed for this form. The 'coarse-leaved form' tends to have mixed entire and 2- or 3-lobed leaves, with leaves and lobes 1.5–2 mm wide; it is lower-growing (0.5–2 m), lignotuberous, and quite often multistemmed; it occurs in the Lake King–Hyden area and Frank Hann Natl Park. The 'pinnate-leaved form' occurs between Peak Charles and Salmon Gums; it has most or all leaves 3–7-lobed, the segments c. 1.0–1.2 mm wide, and a compact habit with lignotuberous reproduction.

Apparent hybrids with *G. pectinata* have been reported.

118. Grevillea tripartita Meisn., *in* A.L.P.P. de Candolle, *Prodr.* 14: 373 (1856)

T: Swan River Colony, W.A., c. 1848, *J.Drummond 6th coll: 285*; lecto: NY, *fide* D.J.McGillivray & R.O.Makinson, *Grevillea* 445 (1993); isolecto: A *n.v.*, BM, CGE *n.v.*, E, FI *n.v.*, G, K, LE *n.v.*, MEL, P *n.v.*, PERTH.

G. tripartita Meisn., *Hooker's J. Bot. Kew Gard. Misc.* 4: 186 (1852), *nom. nud.*

Erect to spreading, rarely decumbent, shrub (0.6–) 1–3 m high. Leaves 1–5 cm long, to 35 mm wide across lobes, usually apically 3 (–5)-fid to deeply 3 (–7)-partite, occasionally with some (rarely most or all) leaves entire and elliptic or narrowly so and 1–4 cm long by 2–5 (–9) mm wide; lobes ascending, sometimes divaricate, often rigid, broadly triangular to sublinear, 2–22 mm long, 1.5–3 mm wide, pungent; margins shortly refracted; lower surface sometimes enclosed, subsericeous to subvillous. Conflorescences terminal and in upper axils, often clustered, ±sessile, erect; unit conflorescence a loose 2–6-flowered cluster, basipetal; rachis 0.2–4 mm long, subsericeous or glandular-pubescent. Flowers basiscopic. Perianth with appressed biramous and simple-erect glandular hairs outside, pilose inside. Pistil 46–52 mm long, glabrous; ovary with a conspicuous rounded outgrowth on each flank. Follicles 12–18 mm long, glabrous, with a conspicuous transverse ventral ridge on each flank perpendicular to and across the ventral suture, colliculose elsewhere.

Occurs in southern W.A. near the coast from E of the Stirling Ra. to near Point Culver.

Olde & Marriott (*Nuytsia* 9: 298 (1993)) described *G. superba* as separate at specific rank from *G. plurijuga*, on the basis of the former lacking a lignotuber, and having a tendency to broader leaf lobes with more frequent secondary leaf division, the midvein of leaf and lobe lower surface level with the margins, the inflorescences emergent above the foliage, and the pistil tending to be longer. They acknowledged intermediacy on some features in some populations. The variability and partial intergradation on most of these characters is comparable to that of variant forms in *G. nudiflora*. The taxa do seem to be diverging, but in view of the degree of overlap *G. superba* is here regarded as a subspecies of *G. plurijuga*. Further field investigations are needed.

Leaf lobes 0.7–1.5 mm wide, mostly trigonous; peduncle decurved within or trailing at base of plant; pistil usually 35–40 mm long; usually a low shrub 0.5–1.5 (–2) m tall ..... **120a.** subsp. **plurijuga**

Leaf lobes 1–2.2 mm wide, usually ±oblong in cross-section; peduncle emergent above top of plant; pistil 40–47 mm long; robust shrub 2–4 m tall ..... **120b.** subsp. **superba**

120a. Grevillea plurijuga F.Muell. subsp. plurijuga

G. helmsiana Helms, *Trans. & Proc. Roy. Soc. S. Australia* 16: 325 (1895). T: Camp 71, 20 miles [32 km] NW from Fraser Range, W.A., 3 Nov. 1891, *R.Helms 12*; lecto: MEL, *fide* D.J.McGillivray & R.O.Makinson, *Grevillea* 435 (1993); isolecto: AD, K, MEL, NSW, PERTH.

Illustrations: D.J.McGillivray & R.O.Makinson, *Grevillea* 237, fig. 65b, 239, fig. 67b, 241, col. pl. (1993); P.M.Olde & N.R.Marriott, *Grevillea Book* 3: 102 (bottom centre), 103 (78A–D) (1995); all as *G. plurijuga*.

Prostrate to procumbent dense mounded shrub 0.3–2 m high. Leaves 4–13 cm long, pinnatisect with 9–25 closely aligned primary lobes, the lowermost lobes rarely again 2–4-partite, sometimes also entire linear leaves adjacent to inflorescences; ultimate lobes 2–7 (–10) cm long, 0.7–1.5 mm wide, pungent; lower surface 2-grooved with midvein prominent above revolute margins (lobe trigonous in cross-section). Conflorescences borne on long decurved or trailing peduncles beyond base of foliage at base of shrub, or rarely also a few emergent above the foliage. Flower colour: perianth pale grey to pinkish red with a cream limb; style pinkish red. Pistil 35–40 mm long. Fig. 15C–D.

Occurs in southern W.A., from Norseman, Peak Charles and Esperance E to Point Culver and Cundeelee area, with disjunct occurrences near Southern Cross. Grows in various habitats including open heath, eucalypt woodland, mallee scrub and mallee-*Triodia* associations, usually in sandy or loamy soils over laterite or limestone. Regenerates from seed and at least sometimes from lignotuber. Flowers Sept.–Jan. Map 151.

W.A.: Eyre Hwy, 77.2 km E of Norseman, *J.S.Beard 5215* (KPBG, PERTH); 115 km ENE of Esperance, *R.A.Saffrey 1263* (K, PERTH); 1.6 km from Ravensthorpe towards Hopetoun, 27 Oct. 1968, *J.W.Wrigley* (CANB, NSW).

Grevillea plurijuga subsp. *plurijuga* extends into considerably more arid habitats than does subsp. *superba*. Leaf lobe length is highly variable, as is habit. There is a strong tendency for the leaf rachis in this subspecies to be either straight or faintly incurved, very rarely slightly recurved (cf. subsp. *superba*). Olde and Marriott (*op. cit.* 3: 102 (1995)) recognise two forms. The 'typical form' (type form) is a dense mounded robust shrub 1–2 m high, usually very floriferous around the base, with leaf lobes 3–10 cm long; it occurs in the N of the species range, from Pontoon Ck near Cundeelee, W almost to Kalgoorlie, and S to Scaddan and Mt Ragged. The 'short-lobed form' is a dense, low mounded shrub eventually attaining 1 m in height, but prostrate when younger and often persistently so for some years; it has short fine leaf lobes 2–4 cm long and c. 0.7 mm wide, and occurs in the S of the subspecies range from near Esperance to Point Culver.

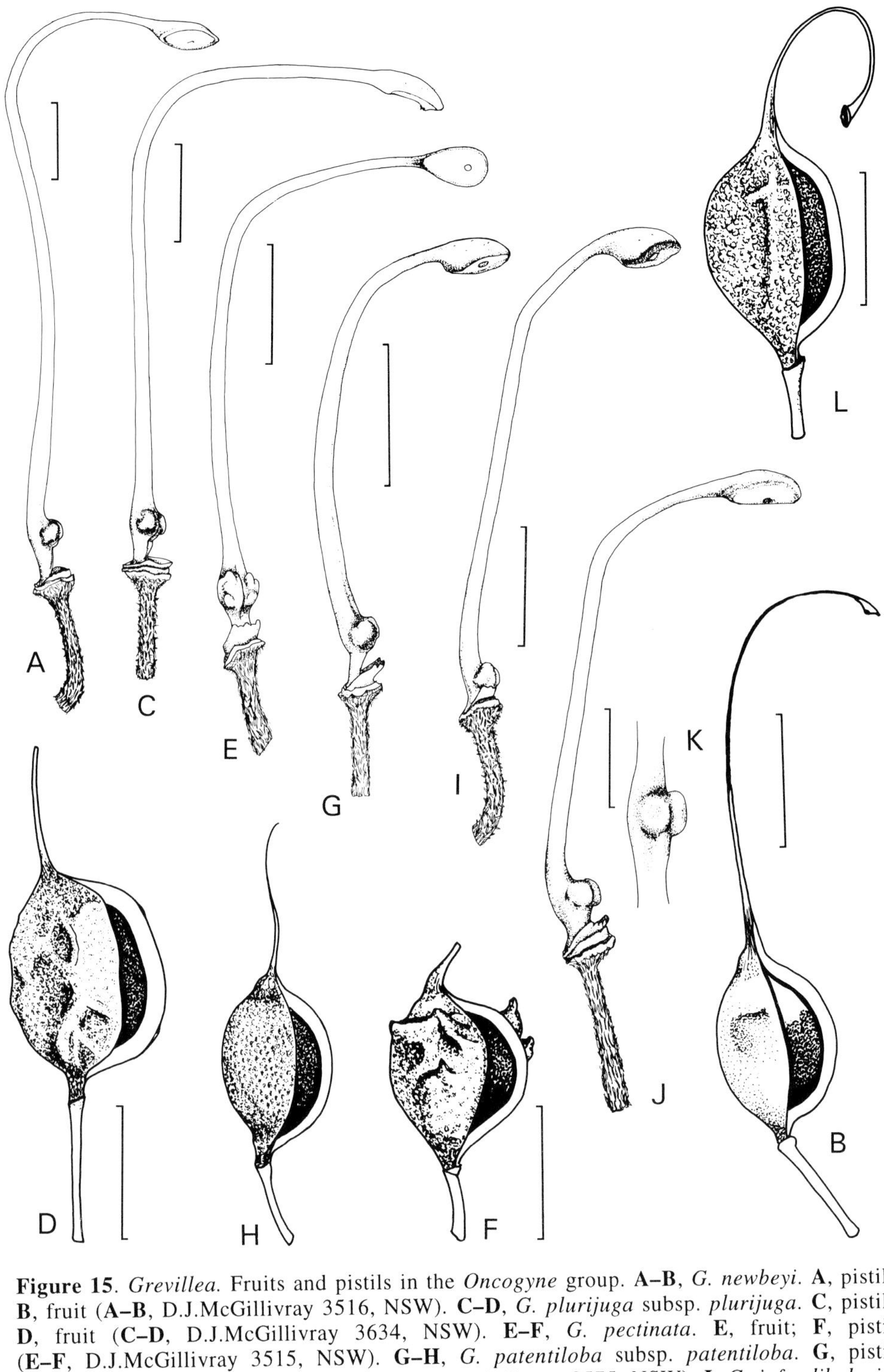

Figure 15. *Grevillea.* Fruits and pistils in the *Oncogyne* group. **A–B**, *G. newbeyi*. **A**, pistil; **B**, fruit (**A–B**, D.J.McGillivray 3516, NSW). **C–D**, *G. plurijuga* subsp. *plurijuga*. **C**, pistil; **D**, fruit (**C–D**, D.J.McGillivray 3634, NSW). **E–F**, *G. pectinata*. **E**, fruit; **F**, pistil (**E–F**, D.J.McGillivray 3515, NSW). **G–H**, *G. patentiloba* subsp. *patentiloba*. **G**, pistil (D.J.McGillivray 3589, NSW); **H**, fruit (D.J.McGillivray 3575, NSW). **I**, *G. infundibularis*, pistil (no voucher). **J–L**, *G. nudiflora*. **J**, pistil: **K**, ovary (**J–K**, D.J.McGillivray 3586, NSW) **L**, fruit (D.J.McGillivray 3598, NSW). Scale bars: **B**, **D**, **F**, **H**, **J**, **L** = 1 cm; **A**, **C**, **E**, **G**, **I**, **K** = 5 mm. Drawn by B.Chandler.

120b. Grevillea plurijuga subsp. **superba** (Olde & Marriott) Makinson, *Fl. Australia* 17A: 496 (2000)

G. superba Olde & Marriott, *Nuytsia* 9: 298 (1993). T: Norwood Rd, E of Scaddan, W.A., 13 Oct. 1991, *P.M.Olde 91/332*; holo: NSW; iso: NSW, PERTH.

Illustrations: P.M.Olde & N.R.Marriott, *Nuytsia* 9: 300, fig. 21 (1993), as *G. superba*; P.M.Olde & N.R.Marriott, *Grevillea Book* 3: 191 (bottom right), 192 (156A–D) (1995), as *G. superba*.

Dense robust shrub 2–4 m high. Leaves 2–7 cm long, pinnatisect, sometimes also entire, linear leaves adjacent to inflorescences; primary lobes 9–17, closely aligned, with lower lobes usually again 2–5-partite; ultimate lobes 0.5–2 cm long, 1.0–2.2 mm wide, pungent; lower surface 2-grooved, with midvein level with, to prominent above, revolute margins (lobes rectangular to trigonous in cross-section). Conflorescence borne on erect peduncles above foliage at top of shrub. Flower colour: perianth grey to pale or occasionally mid-pink with cream to pale yellow limb; style pale to deep pink or rarely cream with darker style-end. Pistil 40–47 mm long.

Occurs in south-western W.A. from Grasspatch and Scaddan E to the Wittenoom Hills and Mt Ney. Grows in eucalypt and mixed shrub associations in (calcareous?) loamy soils. Regenerates from seed; not known to be lignotuberous. Flowers Oct.–Dec. Map 152.

W.A.: N of Scaddan, *A.M.Ashby 2779* (PERTH); c. 6.5 km direct ENE of Scaddan, *A.M.Lyne 1118 et al.* (CANB, MEL, NSW, PERTH); Mt Burdett Rd near Mt Burdett, *P.Olde 86/1189* (NSW).

There is a high degree of overlap with subsp. *plurijuga* on leaf lobe width, and on prominence of the veins on the lower leaf surface. In addition to differences given in the key, subsp. *superba* tends to have the leaf rachis decurved (sometimes straight).

This species is recognised (under *G. superba*) as 'Poorly Known' in J.D.Briggs & J.H.Leigh, *Rare or Threatened Australian Plants* (1995).

121. Grevillea pectinata R.Br., *Suppl. Prodr. Fl. Nov. Holl.* 23 (1830)

T: between Cape Arid and Lucky Bay, [W.A.], 1824, *[W.]Baxter*; holo: BM *n.v.*, *fide* D.J.McGillivray & R.O.Makinson, *Grevillea* 433 (1993).

G. ctenophylla Meisn., *in* A.L.P.P. de Candolle, *Prodr.* 14: 372 (1856). T: Swan River Colony, W.A., *J.Drummond suppl. to 5th coll. 407*; holo: NY *n.v.*, *fide* D.J.McGillivray & R.O.Makinson, *Grevillea* 433 (1993); iso: BM, CGE *n.v.*, G *n.v.*, K, MEL, P *n.v.*

G. ctenophylla Meisn., *Hooker's J. Bot. Kew Gard. Misc.* 4: 186 (1852), as *G. chenophylla*, *nom. nud.*

Illustrations: D.J.McGillivray & R.O.Makinson, *Grevillea* 237, fig. 65c, 239, fig. 67c, 243, col. pl. (1993); P.M.Olde & N.R.Marriott, *Grevillea Book* 3: 87 (bottom centre), 88 (66A–E) (1995).

Spreading shrub 0.5–2.5 m high. Leaves oblong to obovate in general outline, 1.5–6 cm long, 7–25 mm wide (35 mm when flat), pinnatipartite to pinnatisect or sometimes flabelliform, with 8–24 ascending closely aligned teeth or lobes; lobes linear-oblong, 3–22 mm long, 0.8–1.6 mm wide, pungent, rarely lower lobes bifid; apices pungent; margins ±revolute; lower surface mostly or wholly enclosed except midvein, ±2-grooved. Conflorescence terminal, spreading to decurved; unit conflorescence a loose, sometimes subsecund, 1–10-flowered cluster, weakly acropetal; ultimate rachis 1–5 (–8) mm long, subsericeous. Flowers variably oriented, usually acroscopic. Flower colour: perianth mauve-pink to red, often cream to yellow at base and on limb, or occasionally blackish maroon or dull grey-red throughout; style red to deep pink. Perianth loosely hairy outside with appressed biramous and erect-simple glandular hairs, pilose inside. Pistil 27–37 mm long, glabrous; ovary irregularly rugose on each flank. Follicles erect, 8–11 mm long, glabrous with prominent subventral horns or ridges on each side. *Comb-leaf Grevillea.* Fig. 15E–F.

Occurs in south-western W.A., S coast and hinterland from the eastern end of the Stirling Ra. to E of Esperance, inland as far as Kulin and Lake King. Grows in mallee scrub, heath or low open woodland in sandy to clayey soils over laterite or limestone. Regenerates from seed and sometimes lignotuber. Flowers June–Jan. Map 153.

W.A.: c. 15 km by road N of Ongerup, *D.J.McGillivray 3515 & A.S.George* (K, NSW, PERTH, US); 12.9 km N of Gibson, *M.E.Phillips CBG016525* (CANB); Block 1099, Oldfield Location, c. 80 km W of Esperance, *P.G.Wilson 8014* (CANB, K, MEL, PERTH).

There is considerable variation in degree and depth of leaf division, plants with simple-flabelliform leaves being common N and NE of Ravensthorpe (the 'flabellate broad-leaf form' of Olde & Marriott, *op. cit.* 87), although sometimes in mixed stands with plants with deeply divided leaves. At least some populations in this area are lignotuberous. There is a trend for fewer and broader lobes in the NW of the range. A form with black-maroon perianths occurs in the W of the range (e.g. Ongerup area), and a narrow-leaved form in the Lake King area. Further analysis of the variation is required. Possible intermediates with *G. plurijuga*, *G. oncogyne* and *G. patentiloba* have been collected.

122. Grevillea patentiloba F.Muell., *Fragm.* 1: 137 (1859)

T: Phillips Range, W.A., [?*G.Maxwell*] *180*; holo: MEL; iso: B.

Prostrate to erect shrub 0.5–3 m high. Leaves 1.5–5 cm long, 15–65 mm wide, pinnatisect to pinnatipartite or pinnatifid with strongly divaricate lobing; primary lobes 3–7, these usually again 2- or 3-partite, occasionally with sparing tertiary division; ultimate lobes linear to oblong or triangular, 0.3–2 cm long, 1–9 mm wide, pungent; margins revolute or recurved; lower surface mostly or wholly enclosed. Conflorescences terminal, on decurved peduncles; unit conflorescence a loose, sometimes subsecund, 2–10-flowered cluster, ?acropetal; ultimate rachis 1–5 mm long, subsericeous, flattened, usually flexuouse. Flowers ±acroscopic. Perianth loosely hairy outside with appressed biramous and erect-simple glandular hairs, pilose inside. Pistil 15–20 mm long, glabrous or rarely with appressed hairs on dorsal side of ovary; ovary with a ridge on either flank. Follicles 10–14 mm long, glabrous, tuberculate, sometimes with a faint ridge on each flank next to the suture in the distal half.

Occurs in south-western W.A. from near Lake King S to Hopetoun. Two subspecies are recognised.

In this species the peduncles and primary floral rachises are characteristically and conspicuously flattened and usually flexuose.

Leaf lobes linear to narrowly triangular, usually divaricate, 1–4 mm wide; lower leaf surface usually fully enclosed except for midveins; prostrate to low shrub, usually ⩽ 1 m high	**122a.** subsp. **patentiloba**
Leaf lobes triangular to broadly so, not or scarcely divaricate, widest lobes 4–9 mm wide; lower leaf surface at least partly exposed on either side of midvein; robust shrubs 1.5–3 m high	**122b.** subsp. **platypoda**

122a. Grevillea patentiloba F.Muell. subsp. patentiloba

Illustrations: D.J.McGillivray & R.O.Makinson, *Grevillea* 237, fig. 65d, 239, fig. 67d (1993); P.M.Olde & N.R.Marriott, *Grevillea Book* 3: 81 (top right & 59A, B) (1995).

Prostrate to low spreading shrub, usually ⩽ 1 m high. Leaves pinnatipartite with 3–7 primary lobes, these simple or again divided, sometimes the basal ones with tertiary division; ultimate lobes linear to narrowly oblong or narrowly triangular, 1–2 cm long, 1–2 (–4) mm wide, usually divaricate; margins revolute; lower surface mostly or wholly enclosed except for midvein. Flower colour: perianth red to deep pink with a cream to bright yellow limb, sometimes with black at curve and tip; style and pollen-presenter red to deep pink. Plate 28. Fig. 15G–H.

Occurs in south-western W.A., in the Ravensthorpe area, from near Lake King in the N to the coast near Hopetoun, and from c. 30 km W of Ravensthorpe E to the Jerdacuttup R. Grows in heath and scrub-woodland associations in sand or loam soils. Regenerates from seed. Flowers June–Nov. Map 154.

W.A.: 25 km from Ravensthorpe along Lake King road, *B.Barnsley 477* (CANB, NSW, PERTH); 40 km ENE of Fitzgerald on Ravensthorpe to Ongerup road, *A.C.Beauglehole 49305* (NSW, PERTH); L. Varley turnoff,

c. 28 km W of crossroads E of Lake King, *D.J.McGillivray 3563 & A.S.George* (K, NSW, PERTH, US *n.v.*); N side of Mt Short, *D.J.McGillivray 3575 & A.S.George* (K, NSW); 2.4 km N of Ravensthorpe, 3 Sept. 1947, *J.H.Willis* (MEL).

Plants and populations trending towards or intermediate with subsp. *platypoda* on foliar features (i.e. with relatively flat leaves, and lobes 2–4 mm wide) occur in the Mt Desmond and Mt Short areas, where the ranges of the two taxa almost adjoin. Such intermediates are here provisionally assigned to subsp. *patentiloba*. Hybrids with *G. nudiflora* occur in the area E of Ravensthorpe. Hybrids with *G. pectinata* are also reported.

122b. Grevillea patentiloba subsp. platypoda (F.Muell.) Olde & Marriott, *Grevillea Book* 1: 180 (1994)

G. platypoda F.Muell., *Fragm.* 6: 205 (1868). T: Stirling Ra., W.A., *s.d.*, *F.Mueller*; holo: MEL; iso: B (as *Diels 7457*) *n.v.*, PERTH.

Illustrations: P.M.Olde & N.R.Marriott, *Grevillea Book* 3: 82 (bottom centre & 60A, B) (1995).

Robust ±upright shrub 1.5–3 m high. Leaves pinnatifid to pinnatipartite with 3–7 primary lobes, these simple or again 2- or 3-fid; ultimate lobes triangular to broadly so, 0.4–1.5 cm long, 4–9 mm wide, not or scarcely divaricate; margins shortly recurved to shortly revolute; lower surface mostly or wholly exposed. Flower colour: perianth red to deep pink with a cream to yellow limb; style and pollen-presenter red to deep pink.

Confined to the Ravensthorpe Ra. area in south-western W.A. The Type locality (Stirling Ra.) is probably in error; the species has not otherwise been collected there. Grows in lateritic soil in open eucalypt woodland and shrubland. Regenerates from seed. Flowers July–Nov. Map 155.

W.A.: 1.5 km E of Mt Desmond, *B.Barnsley 439* (CANB, PERTH); Ravensthorpe Ra., *C.A.Gardner 1846* (PERTH); Carlingup Rd, 1.8 km N of Highway 1, E of Ravensthorpe, *P.M.Olde 91/309* (NSW); 8 km E of Ravensthorpe, *C.Woolcock G119* (NSW).

123. Grevillea nudiflora Meisn., *in* A.L.P.P. de Candolle, *Prodr.* 14: 366 (1856)

T: Swan River Colony, W.A., c. 1849, *J.Drummond coll. 5: 406*; lecto: NY, *fide* D.J.McGillivray & R.O.Makinson, *Grevillea* 431 (1993); isolecto: CGE *n.v.*, G, K, LE *n.v.*, MEL, P *n.v.*, TCD *n.v.*

G. pedunculosa F.Muell., *Fragm.* 1: 135 (1859). T: in the vicinity of Salt R., Phillips R. and Fitzgerald R., W.A., *s.d.*, *coll. unknown*: syn: ?MEL *n.v.*, *fide* D.J.McGillivray & R.O.Makinson, *Grevillea* 431 (1993).

G. nudiflora Meisn., *Hooker's J. Bot. Kew Gard. Misc.* 4: 186 (1852), *nom. nud.*

Illustrations: D.J.McGillivray & R.O.Makinson, *Grevillea* 237, fig. 65e, 239, fig. 67e, 245, col. pl. (1993); P.M.Olde & N.R.Marriott, *Grevillea Book* 3: 53 (bottom centre), 54 (38A–C), 55 (38D) (1995).

Prostrate to low spreading clumping shrub 0.3 m high, to 2 m across, or sometimes erect to 2 m. Leaves entire, linear or rarely bifid, 4–25 cm long, 1–5 mm wide, often congested and clustered, leathery, straight or slightly curled, not pungent; margins revolute to refracted; lower surface on either side of midvein usually obscured, 2-grooved, sometimes exposed. Conflorescence erect, terminal, either (usually) much-branched and irregularly distant on long, secund trailing peduncles up to 2.5 m long, or (occasionally) simple and few-branched and shortly pedunculate and mostly within the foliage; unit conflorescence 2–6-flowered, secund, opening subsynchronously; ultimate rachis 3–5 mm long, subsericeous or sparsely so, sometimes flat. Flowers at least sometimes acroscopic, but possibly variable. Flower colour: perianth red or sometimes deep pink, yellow and/or grey at curve, with a blackish limb; style red. Perianth sparsely subsericeous with very short-armed biramous hairs and often also minute erect glandular hairs outside, villous inside in lower half. Pistil 19–24 mm long, glabrous; ovary with 2 rounded protuberances on ventral flank. Follicles erect, 10–18 mm long, when mature with faint ventral lobing or ridging, otherwise granulate. Fig. 15J–L.

Occurs in south-western W.A., from the Albany area E to Cape Arid, and inland to Mount Barker and Ravensthorpe. Grows in sandy to clay soils, in open woodland, open forest, mallee heath and heathland. Regenerates from rhizomes and seed. Flowers mainly July–Nov., but sporadically all year. Map 156.

W.A.: sandplain c. 2 km W of Mt Ragged, *R.H.Kuchel 1595* (AD, L, PERTH); Fitzgerald River Natl Park, c. 40 km by road from South Coast Hwy towards Point Anne, *A.Lyne 1050 et al.* (CANB, MEL, NSW, PERTH); ...c. 32 km NNE of coast at Stokes Inlet, *A.E.Orchard 1653* (AD, CANB); 31 km ESE of Ravensthorpe, *J.Taylor 773 et al.* (CANB, NSW, PERTH); 2 km NW of Middle Mt Barren, *P.G.Wilson 10131* (CANB, PERTH).

Very closely related to *G. patentiloba*. *Grevillea nudiflora* characteristically (but not always) has the inflorescences on the ground on long trailing branches; sometimes the branchlets are pilose with glandular hairs. *Grevillea infundibularis* differs in its flabelliform to almost round (rarely elliptic) funnel-like leaves 10–40 mm wide. *Grevillea patentiloba s. lat.* differs in its divaricately divided leaves and the flattened, zig-zag peduncle. *Grevillea plurijuga* has 4–24 primary leaf lobes and a longer pistil (35–47 mm long).

There is considerable variation in habit, mode of regeneration, leaf form, and exsertion of inflorescences. Olde and Marriott (*loc. cit.*) usefully recognise four forms, which, however, do not provide full resolution. The 'fine leaf (type) form' is usually prostrate (occasionally low clumping and rhizomatous), with straight linear leaves 0.8–2 mm wide, and inflorescences usually on long trailing peduncles (occasionally on short peduncles and scarcely exceeding the foliage); it often grows in moist depressions in low heath. The 'curled-leaf form' from the Point Ann area in Fitzgerald River Natl Park is a prostrate plant with laterally curved leaves c. 2 mm wide, and emergent or non-emergent inflorescences. The 'shrubby form' occurs on elevated ground in the Fitzgerald River Natl Park and towards the Stirling Ra.; plants may be prostrate or erect-spreading to almost 2 m tall, and again have variably exserted inflorescences. A 'broad-leaf form' with erect multistemmed habit to 30 cm tall and broad leaves (2–6.5 mm wide) with much of the lower surface exposed occurs on Middle Mt Barren; this form tends to have occasional divided leaves and flattened flexuose primary floral rachises similar to those of *G. patentiloba*, and may be of hybrid origin. Hybrids of these two species have been reported from other sites.

124. Grevillea infundibularis A.S.George, *Nuytsia* 1: 371 (1974)

T: W side of Middle Mt Barren, Fitzgerald River Natl Park, W.A., 16 July 1970, *A.S.George s.n.*; holo: PERTH; iso: AD, CANB, K, MEL, NSW, PERTH.

Illustrations: D.J.McGillivray & R.O.Makinson, *Grevillea* 239, fig. 67f, 246, col. pl. (1993); P.M.Olde & N.R.Marriott, *Grevillea Book* 2: 211 (top right & 174) (1995); A.Brown *et al.* (eds), *W. Australia's Threatened Flora* 196 (1998).

Spreading to decumbent shrub 0.2–1.0 m tall. Leaves obovate to hemispherical or rarely elliptic, 1–6 cm long, 1–4 cm wide, cuneate to amplexicaul at base, entire or with few to many broad, obtuse scarcely pungent teeth, glabrous or nearly so; margins shortly recurved; lower surface minutely pitted, with mid- and lateral veins prominent. Conflorescence terminal on short leafy branchlet or axillary, erect, shortly pedunculate; unit conflorescence umbel-like, 4–8-flowered, opening subsynchronously; peduncle glandular-pilose; rachis scarcely evident. Flowers adaxially acroscopic. Flower colour: perianth and style bright red. Perianth with both appressed biramous and simple erect glandular hairs outside (or sometimes the latter only), pilose inside. Pistil 18–20 mm long, glabrous; ovary with 2 ventral rounded protuberances. Follicles erect, 13–15 mm long, ventrally ridged or with 2 protruding triangular 'eyes' on ventral side, granulate. Figs 3G, 15 I.

Occurs in south-western W.A., confined to the Fitzgerald River Natl Park (Thumb Peak to Middle Mt Barren area). Grows in heath associations in quartzitic sand. Fire response not known. Flowers all year, mainly ?Oct.–?May. Map 157.

W.A.: coast SE of Middle Mt Barren, *A.S.George 10589* (PERTH); Twin Bays, old cark park, *C.R.Hart 35, 37* (PERTH); 2.4 km S of Thumb Peak, *K.Newbey 3121* (PERTH); 2 km S of Thumb Peak, *K.Newbey 4846* (PERTH); W slope of Thumb Peak, *R.D.Royce 9274* (PERTH)

Two forms may be distinguished (agreeing with those of Olde & Marriott, *loc. cit.*). The 'typical form' has very rotund, many-toothed, often stem-clasping leaves; it occurs inland from the sea, at the base and on the slopes of hills and peaks. The 'entire-leaved form', perhaps better named the 'beach-dune form', has the leaves entire or with a very few (to 6) broad obtuse lobes, a cuneate base (not stem-clasping) and a prostrate or nearly prostrate habit. It is apparently restricted to dunes and sand-drifts behind beaches.

Grevillea infundibularis is easily distinguished by the branchlets, which are densely pilose with simple and erect glandular hairs (long-seriate, up to 15 cells long) and by its glabrous and usually funnel-like, often amplexicaul leaves. *Grevillea tripartita* subsp. *macrostylis* sometimes has leaves similar to coastal variants of *G. infundibularis*, but they are hairy beneath, and the the former also has flowers with much longer pistils 36–40 mm long.

The name *G. asteriscosa* has been misapplied to this species in the past; *G. asteriscosa* has a hairy ovary, and smaller leaves with pungent lobes.

This species is recognised as 'Vulnerable' in J.D.Briggs & J.H.Leigh, *Rare or Threatened Australian Plants* (1995).

Hakeoides Group

Shrubs. Leaves entire or divided, dorsiventral with surfaces dissimilar, or sometimes almost dipleural (submarginal grooves almost lateral) and then the apparent surfaces similar; margins recurved to revolute or flat or obscure. Conflorescence terminal or axillary, simple or few-branched, rarely on emergent scape; unit conflorescence erect or rarely decurved to pendulous, regular-umbelloid and dome-shaped to ovoid or shortly subcylindrical, or secund, acropetal or rarely basipetal. Flowers adaxially acroscopic. Torus transverse to oblique. Perianth zygomorphic, glabrous or hairy outside, shortly bearded or papillose inside; tepals separating widely along dorsal suture and partially everting while limb segments are still coherent, tardily all fully separating. Pistil 4–20 mm long, glabrous or rarely a few hairs near ovary; ovary stipitate; style exserted from late bud; pollen-presenter oblique, convex to flat. Follicle rugose or colliculose, glabrous, sometimes obtusely apiculate; pericarp thin to moderately thick, crustaceous to bony. Seed ±ellipsoidal to obovate, with (or rarely without?) a short apical elaiosome and a waxy margin along one side.

Twelve species, all endemic to W.A., mostly in the SW of the State. All insect-pollinated. Possible affinities to the *Thelemanniana* Group. Placement of *G. murex*, *G. scapigera* and *G. bracteosa* in this group is tentative. The group is further subdivisible on the basis of fruit surface.

1 Adult leaves all entire and linear or subterete

2 Unit conflorescences decurved to pendulous; leaves linear (not subterete), 1.7–2.5 mm wide, appearing 5-veined below (lower surface with 3 parallel longitudinal veins flanked by the marginal rolls) **134. G. subterlineata**

2: Unit conflorescences erect; leaves linear to subterete and then 0.5–2 mm wide with the midvein on the lower surface flanked by 2 grooves, or leaves narrowly elliptic to narrowly oblanceolate, 2–11 mm wide, with lamina of lower surface exposed but with only a single prominent longitudinal vein

3 Floral bracts conspicuous, persistent to flower opening, broadly elliptic to obovate, 7–14 mm long, 5–13 mm wide; leaves 2-grooved underneath with no lamina exposed; style smoothly dilated for 2–4 mm below pollen-presenter **136. G. bracteosa**

3: Floral bracts inconspicuous, usually deciduous in early bud stage, linear to triangular, < 3 mm long and < 2 mm wide; leaves either 2-grooved underneath with lamina concealed, or with some or all lamina exposed; style not dilated until the very apex

4 Outer surface of perianth usually with simple erect glandular hairs (sometimes also biramous hairs); follicle 17–21 mm long with a rounded apiculum; pistil 12–17 mm long **127. G. brachystachya**

4: Outer surface of perianth glabrous or with biramous hairs only; follicle 7–13 mm long, lacking rounded apiculum; pistil 5–13 mm long

5 Leaf margins shortly recurved to loosely revolute; lower leaf surface exposed, sericeous; leaves pliable, dorsiventral and ±flat **130. G. diversifolia**

5: Leaf margins tightly revolute against abaxial midvein; lower leaf surface enclosed except for midvein or very rarely slightly exposed; leaves ±rigid, dorsiventral with 2 grooves on lower surface, or sometimes subterete-dipleural with a groove along each side **125. G. hakeoides**

1: Adult leaves (some or all) toothed or divided, or if all entire then not subterete or linear

6 Lower leaf surface enclosed by revolute margins, with only midvein visible; leaves divided with ±linear to oblong lobes

7 Floral bracts persistent to flower opening, conspicuous, broadly elliptic to obovate, 7–14 mm long, 5–13 mm wide; leaf segments linear, 1–3 mm wide, 2-grooved beneath; style smoothly dilated for 2–4 mm below pollen-presenter **136. G. bracteosa**

7: Floral bracts deciduous in early bud, inconspicuous, sublinear, < 1 mm long and < 1 mm wide; leaf segments linear to linear-obovate, 1–10 mm wide; style not dilated except at very apex

8 Leaf lobes strongly divaricate, rigid, pungent **133. G. lullfitzii**

8: Leaf lobes not divaricate, pliable, not or scarcely pungent

9 Leaves 3–12 cm long, irregularly 2–7-partite with lobes usually 30–60 mm long **126. G. commutata**

9: Leaves 0.4–1 cm long, regularly and almost palmately 3–6-partite with lobes ⩽ 8 mm long **129. G. murex**

6: Lower leaf surface exposed on either side of midvein; leaves entire or divided, with lobes or teeth (if present) linear or not linear

10 Lower leaf surface glabrous or with scattered (appressed) hairs

11 All leaves deeply divided with 5–11 primary lobes, these again with 3–7 ±triangular teeth or lobes; inflorescences usually borne on emergent erect scapes; fruits with viscid warts; prostrate shrub **135. G. scapigera**

11: Leaves entire or some or all toothed or divided, but lacking secondary division; inflorescences borne in or close to the foliage, not on erect scapes; fruits faintly colliculose to almost smooth, not viscid; erect to spreading shrubs

12 Perianth subsericeous to subtomentose outside, bearded and papillose inside; unit conflorescence regular **130. G. diversifolia**

12: Perianth glabrous outside, densely papillose inside but not bearded; unit conflorescence shortly secund **132. G. papillosa**

10: Lower leaf surface with a conspicuous open to dense indumentum (hairs appressed or not)

13 Unit conflorescences secund; indumentum of lower leaf surface open-spreading or appressed **131. G. manglesioides**

13: Unit conflorescences regular, umbelloid to ovoid or dome-shaped to hemispherical; indumentum of lower leaf surface appressed

14 Pistil 4–7 mm long; leaves entire and apically obtuse, to emarginate or apically bilobed **128. G. argyrophylla**

14: Pistil 6.5–15 mm long; leaves entire or divided, apically obtuse to acute (never emarginate)

15 Pistil 6.5–9.5 mm long; follicle faintly colliculose to almost smooth; branchlets glabrous or openly (rarely densely) subsericeous **130. G. diversifolia**

15: Pistil 10–15 mm long; follicle markedly rugose; branchlets densely subsericeous to tomentose **126. G. commutata**

125. **Grevillea hakeoides** Meisn., *in* J.G.C.Lehmann, *Pl. Preiss.* 2: 252 (1848)

T: W.A., 'Swan River Drummond. coll. II No. 325 et 326' [protologue]; lecto: *Drummond (Coll. II.) n. 325!* in hb Shuttl. Com. am. Shuttleworth 1849; lecto: NY *n.v.*, *fide* D.J.McGillivray & R.O.Makinson, *Grevillea* 419 (1993); isolecto: A *n.v.*, B *n.v.*, BM, CGE *n.v.*, E, G, K, LD *n.v.*, LE *n.v.*, MEL, NY *n.v.*, P; remaining syntypes: Swan River [W.A.], *[J.] Drummond (2nd coll.) 326*; syn: BM, CGE *n.v.*, G *n.v.*, K, LD *n.v.*, LE *n.v.*, MEL, P *n.v.*

Shrub 0.5–4 m tall. Branchlets subsericeous. Leaves ascending, 3–11 cm long, 0.5–2 mm wide; juvenile leaves divided; adult leaves entire, subterete to linear, dorsiventral to almost dipleural; margins revolute or obscure; lower surface 2-grooved, with grooves almost lateral, or abaxial and then rarely with a little subsericeous lamina exposed. Unit conflorescence erect, regular, dome-shaped subumbelloid, acropetal; floral rachis 2–5 mm long, tomentose. Flowers adaxially oriented; pedicels subsericeous to sparsely so. Perianth sparsely to densely subsericeous or glabrous outside, bearded or papillose inside. Pistil 5–12.5 mm long, glabrous or rarely with a few hairs near ovary; pollen-presenter oblique. Follicle obloid-ellipsoidal, 7–12 mm long, glabrous, mildly to strongly rugose.

Occurs in south-western W.A. from Shark Bay to Lake Grace, mainly inland; two subspecies are recognised.

Most leaves < 5 cm long, 0.5–0.8 mm wide, usually subterete, with submarginal grooves almost lateral; pistil < 8 mm long	**125a.** subsp. **hakeoides**
Most leaves 4–11 cm long, 0.7–2.0 mm wide, linear, with submarginal grooves usually clearly on topological lower surface; pistil 9–12.5 mm long	**125b.** subsp. **stenophylla**

125a. **Grevillea hakeoides** Meisn. subsp. **hakeoides**

Illustrations: D.J.McGillivray & R.O.Makinson, *Grevillea* 257 (1993); P.M.Olde & N.Marriott, *Grevillea Book* 2: 187 (top right & 153A–C) (1995).

Leaves usually 3–5 (rarely to 8) cm long, 0.5–0.8 mm wide, subterete, dipleural or almost so (grooves almost lateral). Unit conflorescence with rachis 2–3 mm long. Flower colour: perianth greenish white to cream-grey, sometimes tinged pink; style white to pale pink with a green tip. Perianth glabrous or sparsely to densely sericeous outside, bearded inside or with scattered hairs or papillae. Pistil 5–7 mm long. Follicle mildly to rarely strongly rugose. Plate 30.

Occurs in inland south-western W.A., in scattered populations in the southern part of the species range from Moora and Wongan Hills to Tammin and Lake Grace. Grows in open eucalypt woodland or tall mixed shrubland in sandy-loamy soils over clay. Regeneration mode unknown. Flowers July–Oct. Map 158.

W.A.: Cunderdin, *W.V.Fitzgerald, NSW100232* (BRI, NSW); 15 km NE of Calingiri, *A.S.George 14339* (CANB, PERTH); Bullaring, *A.S.George 14375* (PERTH); Cowcowing, *M.Koch 1100* (A *n.v.*, AD, E, K, MEL, NSW, P *n.v.*); 24 km from Lake Grace towards Newdegate, *J.W.Wrigley CBG038726* (CANB, NSW).

The type collections have the perianth glabrous or nearly so on the outer surface; most recent collections have an open to dense indumentum. *Grevillea hakeoides* subsp. *hakeoides* can be confused with *G. acacioides*, which has less dense unit conflorescences, the outer surface of the perianth glabrous, and subterete leaves with several longitudinal ridges. Some Wongan Hills plants have unusually long (to 8 cm) leaves.

125b. **Grevillea hakeoides** subsp. **stenophylla** (W.Fitzg.) McGill., *New Names Grevillea* 7 (1986)

G. stenophylla W.Fitzg., *J. W. Australia Nat. Hist. Soc.* 2: 30 (1905). T: Mingenew, W.A., Sept. 1903, *W.V.Fitzgerald*; holo: NSW.

Leaves 4–11 cm long, 0.7–2.0 mm wide, dorsiventral and linear, 2-grooved on lower surface, or sometimes linear-subterete almost dipleural (grooves almost lateral). Unit conflorescence with rachis 3–5 mm long. Flower colour: perianth silvery grey to white; style white to pale pink with a green tip. Perianth densely to sparsely subsericeous outside, bearded inside. Pistil 9–12.5 mm long. Follicle mildly rugose.

Occurs in south-western W.A., widespread from Dirk Hartog Is. and the lower Murchison R. inland to Paynes Find and S to Watheroo, Cowcowing and Tammin. Grows in heath, mallee heath or tall open shrubland, in sandy often poorly drained soils. Regenerates from seed. Flowers July–Oct. Map 159.

W.A.: 3.7 km ENE along Mt Gibson HS road from Wubin to Paynes Find road, *R.Coveny 7884* (K, NSW, PERTH); 5.5 km E of Tammin, *R.Coveny 8316* (K, NSW, PERTH); 17.4 km N of Sandy Point Outcamp, Dirk Hartog Is., *A.S.George 11564* (PERTH); Marchagee, 5 Aug. 1962, *F.Humphries* (PERTH); 37 km E of Kalli, *N.H.Speck 1065* (AD).

This subspecies has two forms. The 'stiff-leaved form', which includes the type of the subspecies, has leaves dorsiventral, 4–8 cm long, 1–2 mm wide, rigid, the lower surface bisulcate (sometimes in the Wubin area with the subsericeous lower lamina narrowly exposed), and pistil 9–11 mm long. It occurs in the S of the subspecies range from Indarra and Latham to about Tammin, its range partially overlapping with that of subsp. *hakeoides*, and some plants approach the latter in leaf and pistil length. The 'slender-leaved form' has leaves dorsiventral to dipleural, usually 5–11 cm long, 0.7–1.5 mm wide, pliable and often slightly wavy, and pistil 10–12.5 mm long. It occupies the N of the subspecies range from Dirk Hartog Is. and the Murchison R. to Paynes Find and Watheroo. This second form may deserve subspecies rank, although apparent intermediates occur in the Cowcowing area.

126. Grevillea commutata F.Muell., *Fragm.* 6: 207 (1868)

G. hakeoides subsp. *commutata* (F.Muell.) McGill., *New Names Grevillea* 7 (1986). T: Port Gregory, W.A., *A.Oldfield*; lecto: PERTH, *fide* D.J.McGillivray & R.O.Makinson, *Grevillea* 419 (1993); isolecto: MEL.

Spreading or rarely almost prostrate shrub (0.4–) 1.5–3 m tall. Branchlets subsericeous to tomentose. Leaves ascending, (1.5–) 3–12 cm long, 2–10 mm wide, entire and narrowly obovate to broadly linear-obovate, or some or all irregularly 2–7-partite, with lobes 3–6 cm long, 1–6 mm wide, broadly linear to narrowly obovate; margins shortly recurved to loosely revolute; lower surface usually exposed and subsericeous, rarely enclosed except for midvein. Unit conflorescence erect, regular, ovoid-subumbelloid, acropetal; floral rachis (5–) 8–20 mm long, subsericeous to tomentose. Flowers adaxially oriented; pedicels tomentose. Perianth subsericeous to subtomentose outside, bearded inside opposite ovary. Pistil 10–15 mm long, glabrous; pollen-presenter oblique. Follicle obloid-ellipsoidal, 10–14 mm long, glabrous, openly rugose-tuberculate.

Occurs in south-western W.A. from N of the Murchison R. to Greenough R. and inland to Yuna; two subspecies are recognised.

Most or all leaves entire, if divided leaves present then these 2–4-partite with ascending narrowly obovate to broadly linear lobes 2.5–6 mm wide; pistil 12–15 mm long	**126a.** subsp. **commutata**
Most or all leaves divided, 3–7-partite with ascending linear to broadly linear lobes; lobes and simple leaves 1–3 (–4) mm wide; pistil 10–13 mm long	**126b.** subsp. **pinnatisecta**

126a. Grevillea commutata F.Muell. subsp. commutata

Illustrations: P.M.Olde & N.R.Marriott, *Grevillea Book* 2: 94 (lower right), 95 (75A–C) (1995), all as *G. commutata*.

Leaves mostly entire and narrowly obovate, with widest leaves 3–10 mm wide, occasionally some leaves divided with 2–4 ascending narrowly obovate to broadly linear lobes 2.5–6 mm wide; margins recurved or sometimes revolute; lower surface usually mostly exposed. Unit conflorescence with rachis 10–20 mm long. Flower colour: perianth and style greenish white to cream or pale pink; style-end green to cream. Pistil 12–15 mm long.

Occurs in the south-west of W.A., in the N of the species range, from about Northampton to c. 70 km N of the Murchison R., and inland to about Yuna; also recorded from Houtman Abrolhos Is. Grows in low heath and tall shrub associations, in sandy to loamy soils, sometimes on sand-dunes. Regenerates from seed. Flowers Feb.–Oct. Map 160.

W.A.: 11.3 km from Ajana, *N.T.Burbidge 2197* (CANB, K); Red Bluff, Kalbarri, *R.Filson 8655* (MEL, PERTH); on road S of Red Bluff, between Rainbow Valley and Pot Alley Gorge, *R.J.Garraty 488* (CANB, K, PERTH); 35.4 km N of Galena, *C.H.Gittins 1578* (BRI, K); 7 km N from Whelarra on road to Balla, *D.J.McGillivray 3354 & A.S.George* (K, NSW).

126b. Grevillea commutata subsp. pinnatisecta (F.Muell.) Makinson, *Fl. Australia* 17A: 496 (2000)

G. commutata var. *pinnatisecta* F.Muell., *Fragm.* 6: 208 (1868); *G. pinnatisecta* (F.Muell.) Benth., *Fl. Austral.* 5: 473 (1870). T: between Moore and Murchison Rivers, W.A., *J.Drummond (6th coll.) 184*; lecto: K, *fide* D.J.McGillivray & R.O.Makinson, *Grevillea* 419 (1993); isolecto: BM, CGE *n.v.*, G *n.v.*, K, MEL, NY *n.v.*, PERTH.

G. hakeoides subsp. *commutata* 'divided-leaved form', of D.J.McGillivray & R.O.Makinson, *Grevillea* 258 (1993).

G. commutata 'divided-leaved form', of P.M.Olde & N.R.Marriott, *Grevillea Book* 2: 95 (1995).

Leaves mostly 3–5 (–7)-partite, with linear to broadly linear often slightly sinuous lobes 1–3 mm wide, sometimes also some leaves entire and broadly linear to c. 4 mm wide; margins loosely revolute, on narrower lobes sometimes enclosing lower surface except for midvein. Unit conflorescence with rachis (5–) 8–12 mm long. Flower colour: perianth and style greenish white to cream (?or pale pink); style-end green to cream. Pistil 10–13 mm long.

Occurs in south-western W.A., in the S of the species range (apparently narrowly disjunct from the range of subsp. *commutata*), from about Geraldton S to the Greenough R. Regeneration mode unknown. Flowers mainly June–Dec. Map 161.

W.A.: District Victoria, *E.Pritzel 421* (E, K); SW Australia, 1860, *Burgess s.n.* (K); Greenough, Champion Bay, Oct. 1898, *R.Helms* (E); Greenough R., *s.d.*, *F.Mueller* (MEL69957, 69961, 69962).

Differs from the type subspecies in its usually irregularly deeply divided leaves with narrow lobes, its marginally smaller unit conflorescence and slightly shorter pistils. Occasional specimens apparently intermediate between the subspecies are known from their nearest geographical approaches to each other.

127. Grevillea brachystachya Meisn., *in* J.G.C.Lehmann, *Pl. Preiss.* 2: 254 (1848)

T: Swan River, W.A., *J.Drummond [coll. II] 319*; holo: NY *n.v.*; iso: BM, CGE *n.v.*, G, K, LD *n.v.*, NY *n.v.*, P.

Illustrations: D.J.McGillivray & R.O.Makinson, *Grevillea* 117, fig. 19–20 (1993); P.M.Olde & N.R.Marriott, *Grevillea Book* 2: 66 (bottom centre), 67 (50A–C) (1995).

Bushy shrub 0.8–2.5 m high. Branchlets sericeous. Leaves ascending, entire, linear, 2–15 cm long, 1–2 mm wide; margins smoothly revolute; lower surface 2-grooved, enclosed except for midvein, subsericeous in grooves. Unit conflorescence erect, umbelloid to subglobose, acropetal; floral rachis 4–8 mm long, tomentose. Flowers adaxially oriented; pedicels with appressed biramous and erect glandular hairs. Flower colour: perianth cream to greenish cream; style cream tinged pink or green with a green style-end. Perianth with scattered simple-glandular and biramous hairs outside, glabrous inside or papillose or shortly bearded near base. Pistil 12–17 mm long, glabrous; pollen-presenter very oblique to lateral. Follicle ovoid with a rounded apiculum, 17–21 mm long, glabrous, faintly granulose. Fig. 16F–H.

Occurs in south-western W.A., at scattered localities from about Miling and Wongan Hills to a little N of the Murchison R. Grows in eucalypt woodland and tall shrubland, sometimes on sandplain, in sandy soils. Regeneration mode unknown. Flowers June–Nov. Map 162.

W.A.: 24 km N of Murchison R. on North West Coast Hwy, *A.C.Beauglehole 11908* (NSW); East Yuna, NE of Geraldton, *A.C.Burns 101* (PERTH); 1 km S of Wilroy Reserve, 16 km SSE of Mullewa, *B.G.Muir 416* (PERTH); N of Mt Curious, Murchison R., *A.Oldfield* (MEL); 18.7 km N of Murchison R. crossing on North West Coastal Hwy, *M.D.Tindale 2730* (NSW, PERTH).

Grevillea bracteosa is somewhat similar, see under that species for differences. *Grevillea gordoniana* is also similar, but is a more robust shrub to 7 m tall, with the conflorescences often emergent above the foliage on much-branched scapes, with short stiff ultimate peduncles and the bud-clusters enclosed by large brown bracts; the fruits are narrow and slightly sigmoid and viscid. In *G. brachystachya* the conflorescences are simple or few-branched, held within or near the foliage, and the bud-clusters are not enclosed by bracts.

Figure 16. *Grevillea*. **A–B**, *G. newbeyi*. **A**, flowering branch; **B**, flower (**A–B**, A.R.Fairall 1582, PERTH). **C–E**, *G. argyrophylla*. **C**, flowering branch; **D**, flower; **E**, pistil (**C–E**, M.E.Phillips *s.n.*, 22 Sept. 1968, CBG025872, NSW). **F–H**, *G. brachystachya*. **F**, flowering branch; **G**, flower; **H**, pistil (**F–H**, A.C.Beauglehole 11908, NSW). Scale bars: **A**, **F** = 2 cm; **B–C** = 1 cm; **D–E**, **G–H** = 5 mm. Drawn by D.Fortescue.

128. Grevillea argyrophylla Meisn., *Hooker's J. Bot. Kew Gard. Misc.* 7: 75 (1855)

T: W.A., Interior North of Swan River. A. 1850–51. legit. *[J.] Drummond Coll. VI: 179*; lecto: NY *n.v.*, *fide* D.J.McGillivray & R.O.Makinson, *Grevillea* 405 (1993); isolecto: BM, CGE *n.v.*, G-DC, K, LD *n.v.*, MEL, PERTH.

Illustrations: P.M.Olde & N.R.Marriott, *Grevillea Book* 2: 37 (bottom centre & 25A), 38 (25B, C) (1995).

Erect or rarely decumbent shrub 0.6–2.5 (–6) m high. Branchlets subsericeous to pubescent. Leaves ascending, usually entire, obovate to narrowly so, 1.5–6 cm long, (1.5–) 2.5–8 (–15) mm wide, often apically emarginate or rarely bilobed; margins recurved to shortly revolute; lower surface mostly exposed, subsericeous. Unit conflorescence erect, regular, ovoid to umbelloid, acropetal; floral rachis 1–4 (–8) mm long tomentose. Flowers adaxially oriented; pedicels sericeous. Flower colour: perianth and style white, sometimes tinged pink. Perianth loosely subsericeous outside, papillose inside near base. Pistil 4–7 mm long, usually glabrous, sometimes a few appressed hairs on ovary; pollen-presenter oblique. Follicle ovoid, 6–10 mm long, glabrous, rugose. Fig. 16C–E.

Occurs in south-western W.A., mainly near the coast from Dandaragan and near Jurien Bay N to the Murchison R. Grows in heath and tall shrubland, often in calcareous soils over limestone or sandstone. Regenerates from seed. Flowers July–Oct. Map 163.

W.A.: Cockleshell Gully, *C.A.Gardner 8419* (PERTH); Pot Alley Gorge, Kalbarri, *D.J.McGillivray 3346A & A.S.George* (NSW, PERTH); 14.5 km N of Geraldton, *K.Newbey 2168* (PERTH); 30.6 km from Dongara towards Eneabba, *M.E.Phillips CBG025872* (CANB, PERTH); Red Bluff, 4.8 km SW of Kalbarri; *R.V.Smith 66/355* (CANB, MEL).

Grevillea argyrophylla has a superficial similarity to entire-leaved plants of *G. commutata* subsp. *commutata*, which, however, have leaves usually 5–10 cm long and pistils 10–15 mm long. Natural hybrids between the two taxa occur in the lower Murchison R. area.

129. Grevillea murex McGill., *New Names Grevillea* 10 (1986)

T: 33.2 km SW of Morawa towards Three Springs, W.A., 1 Sept. 1976, *R.Coveny 7969 & B.R.Maslin*; holo: NSW; iso: B *n.v.*, K, PERTH, US *n.v.*

Illustrations: D.J.McGillivray & R.O.Makinson, *Grevillea* 260, fig. 71 (1993); P.M.Olde & N.R.Marriott, *Grevillea Book* 3: 42 (bottom right), 43 (29A, B) (1995); A.Brown *et al.* (eds), *W. Australia's Threatened Flora* 98 (1998).

Dense shrub 0.8–2.0 m tall. Branchlets subsericeous. Leaves ascending, 4–6 (–10) mm long, 3–6 mm wide across lobes, regularly 3–6-partite, almost palmate; lobes linear to oblong, 1–8 mm long, 1–1.5 mm wide; margins revolute; lower surface enclosed except for midveins, 2-grooved, subsericeous in grooves. Unit conflorescence erect, regular, dome-shaped or subumbelloid, acropetal; floral rachis 2–4 mm long subsericeous or sparsely so. Flowers adaxially oriented; pedicels glabrous or with a few appressed hairs. Flower colour: perianth greenish white to dull cream; style white to cream. Perianth glabrous outside, bearded inside. Pistil 8–10 mm long, glabrous; pollen-presenter oblique. Follicle obloid-ellipsoidal, 9–13 mm long, glabrous, very strongly muricate. Fig. 17A–C.

Occurs in south-western W.A., in the Morowa area. Grows in sclerophyll woodland or shrubland, in loamy soils usually over outcropping limestone. Regenerates from seed. Flowers July–Oct. Map 164.

W.A.: 19.3 km NE of Arrino, *J.S.Beard 7245* (NSW); 19 km NE of Arrino, Jan. 1975, *C.Chapman [as J.S.Beard 7383]* (NSW); 1.9 km E of Bligh Rd along Drew Rd, NE of Arrino, *R.J.Cranfield & P.J.Spencer 7842* (K, PERTH); 20 km NE of Arrino, *E.Wittwer 1612* (KPBG).

The foliage of *G. murex* is distinctive, the leaves being almost palmately divided, usually very small, and often ascending close to the stem, giving a semi-columnar appearance. *Grevillea murex* seems best placed in the *Hakeoides* group, although there are similarities suggestive of a relationship to *G. crithmifolia* (*Trifida* Group).

This species is recognised as 'Poorly Known' in J.D.Briggs & J.H.Leigh, *Rare or Threatened Australian Plants* (1995).

130. Grevillea diversifolia Meisn., *in* J.G.C.Lehmann, *Pl. Preiss.* 1: 547 (1845)

T: ad ripam fluvii Vasse (Sussex) [Vasse River, W.A.], d. 13. Decbr. 1839. *L.Preiss. No. 697*; lecto: NY *n.v.*, *fide* D.J.McGillivray & R.O.Makinson, *Grevillea* 414 (1993); isolecto: B *n.v.*, G-DC, HBG *n.v.*, LE *n.v.*, MEL, P.

Erect to spreading shrub 1–6 m tall, or rarely prostrate. Branchlets glabrous or openly (rarely densely) subsericeous. Leaves ascending, 1.5–8.5 cm long, entire and oblanceolate to narrowly elliptic or sublinear and 1.5–11 mm wide, and/or apically 2- or 3-fid or -partite and narrowly cuneate; lobes simple, subtriangular to oblong, 3–15 mm long, 2–4 mm wide; margins recurved or loosely revolute; lower surface exposed, glabrous or sparsely to densely sericeous. Unit conflorescence erect, subsessile or on wiry peduncles to 15 mm long, regular, umbelloid, acropetal; floral rachis 1.5–6 mm long, subsericeous to tomentose. Flowers adaxially oriented; pedicels sericeous to tomentose. Perianth 2–3 (–5) mm long, subsericeous to subtomentose outside, bearded and papillose inside. Pistil 6.5–9.5 mm long, glabrous; pollen-presenter oblique. Follicle obloid-ellipsoidal, 10–13 mm long, glabrous, faintly colliculose to almost smooth.

Occurs in south-western W.A. from the Darling Ra. E of Perth to Donnybrook and near the S coast from Broke Inlet to Albany. There are two subspecies.

Lower surface of leaves glabrous or with scattered appressed hairs **130a.** subsp. **diversifolia**

Lower surface of leaves with a dense appressed indumentum **130b.** subsp. **subtersericata**

130a. Grevillea diversifolia Meisn. subsp. diversifolia

Illustrations: P.M.Olde & N.R.Marriott, *Grevillea Book* 2: 131 (top right & 105A, B), 132 (105C) (1995).

Erect to prostrate shrub 0.2–6 m tall. Leaves soft and pliable, glabrous or with scattered appressed hairs on lower surface, entire and/or apically divided; entire leaves (1.5–) 3–11 mm wide. Flower colour: perianth white to cream; style dull red. Plate 31.

Occurs in south-western W.A., in the Darling Ra. from Mundaring Weir (Helena Valley) S to Donnybrook. Grows in shrubby forest or woodland associations in sandy to clayey soil, often in moist sites. Regenerates from seed. Flowers much of the year, peaking in Sept.–Oct. Map 165.

W.A.: Serpentine R., *W.V.Fitzgerald NSW100220* (NSW); Argyle, *J.W.Green 330* (PERTH); Dwellingup area, *P.C.Kimber 54* (PERTH); Midland Junction, Swan R., 8 Jan. 1898, *A.Morrison* (BRI, E, PERTH); Kelmscott, Canning R., 10 Sept. 1898, *A.Morrison* (BRI, PERTH).

130b. Grevillea diversifolia subsp. subtersericata McGill., *New Names Grevillea* 5 (1986)

T: Camfield, Broke Inlet, W.A., 10 May 1974, *A.S.George 11788*; holo: PERTH; iso: NSW.

G. manglesioides var. *angustissima* Benth., *Fl. Austral.* 5: 475 (1870), as var. *?angustissima*. T: Swan River, [W.A.], *s.d.*, *[J.] Drummond s.n.*; probable holo: K.

G. loboana Domin, *Vestn. Král. Ceské Spoledn. Nauk, Tr. Mat.-Prír* 1921–1922(2): 11 (1923). T: Slab Hut Creek to Cranbrook, W.A., 1910, *A.A.Dorrien-Smith*; lecto: K, *fide* R.O.Makinson, *Fl. Australia* 17A: 497 (2000); isolecto: PR.

Illustrations: P.M.Olde & N.R.Marriott, *Grevillea Book* 2: 132 (bottom left & 106A, B) (1995).

Compact to diffuse-erect shrub, 1.5–5 m tall. Leaves semi-rigid, with dense appressed indumentum on lower surface, entire, usually 1.5–6 mm wide. Flower colour: perianth white to cream; style dull red.

Occurs in south-western W.A., from N of Albany to about Broke Inlet. Grows in shrubby associations in sandy soil, often along water courses. Regenerates from seed. Flowers mainly July–Oct., sporadic in other months. Map 166.

W.A.: near Napier R., *E.M.Bennett 1044* (PERTH); Broke Inlet, *H.Demarz 1151* (KPBG); Kalgan, *G.Maxwell* (MEL); Stirling Ra. [area of?], 1870, *G.Maxwell per F.Mueller* (K); 14.5 km N of Albany, *K.Newbey 2437* (PERTH).

A variant with long narrow leaves (1.5–2 mm wide), named as '*G. manglesioides* var. ?*angustissima*' by Bentham, does not appear to represent a distinct taxon, but may represent a riparian ecotype.

The Maxwell record at Kew from 'Stirling Ranges' is not confirmed; the species probably does not occur there.

131. Grevillea manglesioides Meisn., *in* J.G.C.Lehmann, *Pl. Preiss.* 1: 547 (1845)

T: W.A., '... ad ripam flum. Vasse-River, Sussex, d. 13 Dec. 1839. Herb. Preiss. No. 720, et in distr. Plantagenet, m. Oct. 1840, No. 723' [protologue]; lecto: in districtu Plantagenet, [W.A.], Oct. 1840, *L.Preiss 723*; lecto: NY *n.v., fide* D.J.McGillivray & R.O.Makinson, *Grevillea* 428 (1993); isolecto: LD *n.v.*, LE? *n.v.*; excluding *Preiss 723* in DC. herb. at G (= *G. trifida* (R.Br.) Meisn.); remaining syntypes: *L.Preiss 720*; syn: B *n.v.*, G-DC, HBG *n.v.*, K, LD *n.v.*, LE *n.v.*, MEL.

Spreading shrub 0.3–4 m tall. Branchlets tomentose to subvillous. Leaves ascending, 2–6 cm long, usually cuneate to very narrowly cuneate, 2–25 mm wide, and apically 2–5 (–9)-fid or -partite, or sometimes deeply 3-partite with weakly divaricate narrowly triangular lobes, or occasionally few to many (rarely all?) leaves narrowly elliptic and entire and 2–6 mm wide; teeth or lobes entire, subtriangular to oblong, 1–20 mm long, 2–4 (–7?) mm wide; margins refracted; lower surface exposed, either sericeous to subsericeous (hairs appressed) or open-tomentose to -subvillous (hairs ascending to spreading). Unit conflorescence erect to decurved, broadly and shortly secund, acropetal; floral rachis 4–20 (–25) mm long, glabrous or sparsely to densely tomentose. Flowers acroscopic; pedicels glabrous or with a few hairs. Perianth outside with few to many appressed to ascending hairs on limb, hairs often sparse or absent below, bearded inside. Pistil 6.5–11 mm long, glabrous; pollen-presenter oblique. Follicle narrowly ovoid to narrowly ellipsoidal, 9.5–15 mm long, glabrous, faintly colliculose.

Grevillea papillosa was described by McGillivray (*New Names Grevillea* 9 (1986)) as a subspecies of *G. manglesioides*; it is here regarded as a distinct species, differing in its glabrous branchlets and lower leaf surfaces, and the papillose (non-bearded) inner surface of the perianth. Within *G. manglesioides*, as here circumscribed, three subspecies are recognised.

1 Leaf lower surface tomentose to subvillous (the hairs strongly ascending to spreading) **131a.** subsp. **manglesioides**

1: Leaf lower surface sericeous to subsericeous (the hairs closely appressed)

2 Leaves either narrowly cuneate and 2–5 mm wide and apically 2- or 3-fid, or deeply tripartite in the apical half with weakly divaricate narrowly subtriangular lobes, or sometimes few to many leaves entire and very narrowly elliptic, 2–4 mm wide; floral rachis glabrous or sparsely to openly (rarely densely) tomentose **131b.** subsp. **metaxa**

2: Leaves cuneate to narrowly so, 5–25 mm wide, apically 2–6 (–9)-toothed; floral rachis densely tomentose **131c.** subsp. **ferricola**

131a. Grevillea manglesioides Meisn. subsp. manglesioides

Illustrations: D.J.McGillivray & R.O.Makinson, *Grevillea* 262, col. pl. (1993); P.M.Olde & N.R.Marriott, *Grevillea Book* 3: 17 (bottom centre), 18 (9B, C) (1995), as *G. manglesioides*.

Spreading shrub 1–2.5 (–4) m tall. Branchlets tomentose to subvillous. Leaves 2–5.5 cm long, either entire and narrowly obovate or -oblanceolate to -elliptic, and then 2–9 mm wide, or apically 2–5-fid or -partite and cuneate; teeth or lobes entire, subtriangular to oblong, 1–20 mm long, 2–4 (–7?) mm wide; longer lobes weakly divaricate; lower surface open-tomentose to -subvillous, usually with ground tissue visible between hairs. Floral rachis 4–20 mm long, sparsely tomentose to glabrous. Flower colour: perianth greenish white to dull red; style dull red, or greenish white becoming red. Pistil 6.5–11 mm long. Follicle 13–15 mm long, faintly colliculose.

Occurs in south-western W.A., from Ludlow to Margaret River. The locality given on the lectotype is near Albany and may be in error; the species is not otherwise known to occur there. Grows in shrub associations or shrubby woodland, usually in moist, sometimes waterlogged sites in sandy soils. Regenerates from seed. Flowers year round, mainly July–Dec. Map 167.

W.A.: Ludlow R., Ludlow, *C.A.Gardner 1715* (PERTH); Ruabon, Busselton, *R.D.Royce 4871* (PERTH); Ruabon, 12 Oct. 1975, *D.Woolcock* (NSW).

131b. Grevillea manglesioides subsp. metaxa Makinson, *Fl. Australia* 17A: 497 (2000)

T: Rapids Crossing, Margaret River, W.A., 6 Dec. 1974, *R.Pullen 9884*; holo: PERTH; iso: BRI, CANB, NSW.

G. manglesioides 'small conflorescence form', of P.M.Olde & N.R.Marriott, *Grevillea Book* 3: 17 (1995).

Spreading shrub 1–2 m tall. Branchlets tomentose or occasionally subsericeous. Leaves 2.5–4 cm long, very narrowly cuneate, 2–5 mm wide and apically 2- or 3-fid to 2- or 3-partite, or deeply tripartite in the apical half with the lateral lobes weakly divaricate and the apical lobe often somewhat deflexed, occasionally few to many (rarely all?) leaves narrowly elliptical and entire and then 2–4 mm wide; teeth or lobes subtriangular to narrowly so, (1–) 5–17 mm long, (1–) 2–4 mm wide; lower surface densely sericeous or subsericeous. Floral rachis 3–7 (–15) mm long, sparsely to openly tomentose or sometimes glabrous. Flower colour: perianth cream or greenish yellow; style dark red. Pistil 7.5–9 mm long. Follicle not seen.

Occurs in south-western W.A., on the catchments of the Margaret and Blackwood Rivers, with one unconfirmed record from Cape Naturaliste (*J.C.Wiburd NSW100230*, NSW). Grows in streamside situations in shrub and forest associations. Probably regenerates from seed only. Flowers (Oct.–) Dec.–?Apr. Map 168.

W.A.: Stewart Rd, 17.7 km from junction of [sic] Nannup–Pemberton Rd, *J.W.Wrigley WA/684052* (CANB); The Rapids, Margaret R., *T.A.Halliday 241* (AK, CANB); 41.8 km from Augusta towards Nannup on Brockman Hwy, *M.E.Phillips CBG019058* (CANB, NSW); 16 km E of Augusta on Brockman Hwy, *U.Johnson 58* (K, NSW); 10 km from Cane Brake Rd E along Stewart Rd, WNW of Pemberton, *A.N.Rodd 4903 & G.Fensom* (NSW); Acton Park Rd, edge of Whicher Ra., *A.S.George 11744* (NSW, PERTH).

Grevillea manglesioides subsp. *metaxa* has secund unit conflorescences and usually apically lobed or toothed leaves, features that distinguish it from *G. diversifolia* subsp. *subtersericata*, which has regular (dome-shaped to hemispherical) conflorescences and entire leaves, and which occurs more than 100 km to the SE.

131c. Grevillea manglesioides subsp. ferricola Keighery, *Fl. Australia* 17A: 497 (2000)

T: Beenyup, Scott Plains, W.A., 5 Oct. 1997, *G.J.Keighery 15158*; holo: PERTH; iso: CANB, MEL, NSW.

G. manglesioides var. *sericea* Meisn., *in* J.G.C.Lehmann, *Pl. Preiss.* 2: 255 (1848), as β *sericea*. T: SW Australia, [W.A.], *s.d., [J.] Drummond [2nd coll.] 318*; lecto: K, *fide* R.O.Makinson, *Fl. Australia* 17A: 497 (2000); isolecto: A *n.v.*, BM, CGE *n.v.*, G, K, LD *n.v.*, LE *n.v.*, MEL, P *n.v.*

Spreading shrub to 0.5 (–1?) m tall. Branchlets densely subvillous to subsericeous. Leaves 3–6 cm long, usually cuneate and apically 2–6 (–9)-toothed or 2–5-fid, and then 5–25 mm wide across lobes, or sometimes some (rarely all?) leaves entire and narrowly oblanceolate and 2–6 mm wide; lower surface sericeous or subsericeous. Floral rachis (5–) 10–20 (–25) mm long, densely hairy. Flower colour: perianth greenish cream; style red or rarely blackish. Pistil 9–10 mm long. Follicle obloid-ellipsoidal, 9.5–10 mm long, glabrous, faintly colliculose.

Occurs in the south-west of W.A., where restricted to a few sites in the Scott R. area E of Augusta. Grows in heathy associations in peaty sand over ironstone, often streamside situations. Regenerates from seed. Flowers Oct. (–Dec.). Map 169.

W.A.: between Nannup and Margaret River, Dec. 1930, *W.E.Blackall s.n.* (PERTH); 2.7 km along Governor Broome Rd from Courtney Rd, Scott R. sandplain, *G.J.Keighery 4070* (PERTH); 1.5 km along Governor Broome Rd from junction with Dennis Rd, c. 15 km ENE of Augusta, *B.J.Lepschi 1895* (PERTH); Scott R., *G.S.McCutcheon 2826* (PERTH); Scott R., E of Augusta, *E.C.Nelson ANU17303* (CANB, PERTH).

Subsp. *ferricola* is very narrowly distinct from subsp. *metaxa*, which is a taller plant, has narrower and often more deeply divided leaves with only 3 apical teeth or lobes, a floral rachis that is glabrous or sparsely to openly hairy, and slightly smaller pistils 7.5–9 mm long. *Grevillea diversifolia* subsp. *subtersericata* differs in its regular (non-secund) inflorescences, and the appressed indumentum on the lower surface of the leaves.

Subsp. *ferricola* is sympatric (or very closely parapatric) with *G. papillosa* and *G. manglesioides* subsp. *metaxa* on two areas on the Scott Plains (Chester Block and Governor Broome Rd). There is no intergradation observed in these populations. The taxa are distinct in ecological preferences also. Subspecies *metaxa* is confined to sandy-clay based wetlands and depressions in the forest. Subspecies *ferricola* is always found on shallow red clays over massive ironstones where it is a dominant component of this shrubland and helps define the Scott Plain Ironstone community. Both subspecies also have different flowering periods on the Scott Plains: subsp. *ferricola* is strictly a spring-flowering taxon whereas subsp. *metaxa* flowers largely in late summer to early winter. *Grevillea papillosa* flowers over the same period as *G. manglesioides* subsp. *metaxa* on the Scott Plains.

A collection (*R.J.Cranfield 8337*, CANB, PERTH) from an ambiguous locality near Busselton (probably Scott Plains) has leaves and floral rachises consistent with subsp. *ferricola* but is recorded as 1 m tall; it is probably assignable to this subspecies.

This species is recognised (as *Grevillea sp. 15*) as 'Poorly Known' in J.D.Briggs & J.H.Leigh, *Rare or Threatened Australian Plants* (1995).

132. Grevillea papillosa (McGill.) Olde & Marriott, *Grevillea Book* 1: 183 (1994)

G. manglesioides subsp. *papillosa* McGill., *New Names Grevillea* 9 (1986). T: c. 57 km NW of Pemberton on Stewart Rd, 24 km NW of Nannup to Pemberton road, W.A., 2 Oct. 1976, *B.G.Briggs 6511*; holo: NSW; iso: PERTH.

G. diversifolia var. *lobata* Meisn., *in* J.G.C.Lehmann, *Pl. Preiss.* 2: 255 (1848), as β *lobata*. T: W.A., 'Swan River, Drummond. coll. II. No. 316'; lecto: NY *n.v.*, *fide* D.J.McGillivray & R.O.Makinson, *Grevillea* 428 (1993); isolecto: BM, CGE *n.v.*, G *n.v.*, K, LD *n.v.*, LE *n.v.*, MEL, P *n.v.*

Illustrations: D.J.McGillivray & R.O.Makinson, *Grevillea* 262, fig. 72 (1993), as *G. manglesioides* subsp. *papillosa*; P.M.Olde & N.R.Marriott, *Grevillea Book* 3: 73 (bottom centre), 74 (53A, B) (1995).

Spreading shrub to c. 1 m tall. Branchlets glabrous or with a few appressed hairs. Leaves ascending, 25–45 mm long, usually with some entire and linear to narrowly elliptic and 2–4 mm wide, sometimes deeply 3-fid to -partite; lobes entire, divaricate, 5–20 mm long, 1–2 mm wide, with terminal lobe slightly deflexed; margins loosely revolute; lower surface partially exposed, glabrous. Unit conflorescence erect to decurved, shortly secund, acropetal; floral rachis 3–7 mm long, glabrous. Flowers acroscopic; pedicels glabrous. Flower colour: perianth white with pinkish tinges; style pinkish red. Perianth glabrous outside, densely papillose inside. Pistil 6–7.5 mm long, glabrous; pollen-presenter oblique. Follicle (immature) ellipsoidal, glabrous, colliculose. Fig. 17D–F.

Occurs in the south-west of W.A., restricted to the area from Nannup to Scott R. Grows in heath associations in moist sites, in sandy to gravelly soils. Regenerates from seed. Flowers (Sept.–) Dec.–Apr. Map 170.

W.A.: Cape Beaufort, *A.J.Cough 59* (PERTH); Vasse and Augusta, *s.d.*, *J.Gilbert s.n.* (BM, K); 5.5 km S of Brockman Hwy on Scott River Rd, *P.M.Olde 91/282* (NSW); S of Blackwood, Nillup, *R.D.Royce 2972* (PERTH).

Grevillea papillosa differs from *G. manglesioides* in having the outside of the perianth and the lower leaf surface glabrous, and the inner surface of the perianth papillose (not bearded). It differs from *G. diversifolia* subsp. *diversifolia* in having the unit conflorescence secund, the outer surface of the perianth glabrous, and the inner surface of the perianth papillose (not bearded).

The name *G. diversifolia* var. *rigida* Meisn. is a synonym of *G. curviloba* McGill.

This species is recognised (as *G. manglesioides* subsp. *papillosa*) as 'Poorly Known' in J.D.Briggs & J.H.Leigh, *Rare or Threatened Australian Plants* (1995).

Figure 17. *Grevillea.* **A–C**, *G, murex.* **A**, flowering branch; **B**, flower; **C**, pistil (**A–C**, J.S.Beard 7245, NSW). **D–F**, *G. papillosa.* **D**, flowering branch; **E**, flower; **F**, pistil (**D–F**, B.G.Briggs 6511, NSW). **G–J**, *G. gordoniana.* **G**, leaves; **H**, inflorescences showing bracts; **I**, flower; **J**, pistil (**G–J**, J.S.Beard 7055, PERTH). Scale bars: **A**, **D**, **H** = 1 cm; **B–C** = 2 mm; **E–F** = 1 mm; **G** = 2 cm; **I–J** = 5 mm. Drawn by D.Fortescue.

133. Grevillea lullfitzii McGill., *New Names Grevillea* 9 (1986)

T: Diggers Rock, W.A., 10 Dec. 1964, *E.Lullfitz 4001*; holo: KPBG.

Illustrations: P.M.Olde & N.R.Marriott, *Grevillea Book* 2: 247 (top right), 248 (206A–C) (1995).

Spreading shrub to 1.5 m tall. Branchlets loosely subsericeous. Leaves ascending, 3–5 cm long, deeply 3–6-partite; lobes divaricate, entire, linear, 1.5–3.5 cm long, 1.2–1.5 (–2.0) mm wide, rigid, pungent; margins revolute; lower surface enclosed except for midveins, 2-grooved, sericeous in grooves. Unit conflorescence erect, regular and umbelloid dome-shaped, or sometimes subsecund, acropetal; floral rachis 2–3 mm long, subsericeous. Flowers adaxially oriented; pedicels loosely subsericeous. Flower colour: perianth grey-white outside, cream inside; style off-white becoming pinkish brown, with a green tip. Perianth loosely subsericeous outside, bearded inside opposite ovary. Pistil 8–10.5 mm long, glabrous; pollen-presenter very oblique to almost lateral. Follicle ellipsoidal to obovoid, 14–17 mm long, obtusely attenuate at apex, glabrous, colliculose.

Occurs in south-western W.A., known only from a few collections at Digger Rocks and Hatters Hill, NE of Lake King. Grows in eucalypt woodland, usually on rises, in lateritic loam. Regenerates from seed, possibly also rhizomes. Flowers ?Sept.–Dec. Map 171.

W.A.: locality not given, *C.A.Gardner 14621* (PERTH).

Unlikely to be confused with other species. *Grevillea lullfitzii* is probably most closely related to *G. brachystachya*, which has entire linear leaves and glandular hairs on the outer surface of the perianth.

This species is recognised as 'Poorly Known' in J.D.Briggs & J.H.Leigh, *Rare or Threatened Australian Plants* (1995).

134. Grevillea subterlineata Makinson, *in* D.J.McGillivray & R.O.Makinson, *Grevillea* 443 (1993)

T: Dairy Creek Station, W.A., 22 Aug. 1965, *J.S.Beard 4356*; holo: PERTH; iso: KPBG.

Illustrations: P.M.Olde & N.R.Marriott, *Grevillea Book* 3: 188 (top right & 153A), 189 (153B, C) (1995).

Open shrub 1–2.5 m tall. Branchlets subsericeous. Leaves ascending, entire, linear, 6–13 cm long, 1.7–2.5 mm wide; margins smoothly revolute; lower surface appearing 5-nerved but actually with 3 longitudinal veins flanked by the similar marginal rolls, subsericeous. Unit conflorescence decurved to pendulous, ovoid-subsecund to shortly cylindrical, acropetal; floral rachis 7–20 mm long, tomentose. Flowers acroscopic; pedicels subsericeous to tomentose. Flower colour: perianth white, tinged pink; style white with a green tip. Perianth subsericeous to tomentose outside, bearded inside just above ovary. Pistil 14–17 mm long, glabrous; pollen-presenter oblique. Follicle not seen.

Occurs in western central W.A., known only from an area E of Gascoyne Junction. Grows in open shrub associations and mulga woodland. Regeneration mode unknown. Flowers Aug.–? Map 172.

W.A.: 75 km E of Gascoyne Junction on highway to Meekatharra, 20 Aug. 1987, *F. & N.Johnston* (NSW).

135. Grevillea scapigera A.S.George, *Nuytsia* 1: 374 (1974)

T: between Corrigin and Quairading, W.A., 14 Feb. 1960, *C.V.Malcolm s.n.*; holo: PERTH.

Illustrations: D.J.McGillivray & R.O.Makinson, *Grevillea* 378 (1993); P.M.Olde & N.R.Marriott, *Grevillea Book* 3: 162–163 (131A–D) (1995); A.Brown *et al.* (eds), *W. Australia's Threatened Flora* 99 (1998).

Prostrate shrub to 2 m across, with emergent floral scapes to 0.4 m high. Leaves pinnatipartite to pinnatisect, 3–7.5 (–20?) cm long; margins slightly thickened, flat or occasionally shortly recurved; primary lobes 5–11, ovate-cuneate, 3–7-fid; ultimate lobes or teeth broadly triangular, ±pungent; both surfaces glabrous, sometimes glaucous. Conflorescence terminal or rarely axillary, usually on erect emergent scapes, simple, umbelloid or hemispherical, acropetal, many-flowered; floral rachis 5–8 mm long. Flowers adaxially acroscopic. Flower colour: perianth green in bud, white to cream at anthesis; style white with greenish pollen-presenter. Perianth glabrous outside, pilose inside near ovary, glabrous above; tepals remaining loosely coherent except along dorsal suture. Pistil 15–19 mm long; ovary stipitate, glabrous; style papillose above ovary, otherwise glabrous, exserted in late bud; pollen-presenter strongly oblique. Follicles ±ellipsoidal, 10–13 mm long, with viscid warts. *Corrigin Grevillea.* Plate 32.

A very rare species, found in south-western W.A. between Quairading, Lake Mears, Corrigin and Bullaring, in heathland in sand or lateritic soils. There is also an uncorroborated record from the Hyden area (*O'Donnell 51334*). Probably regenerates mostly by seed, rhizomes have also been recorded. Flowers recorded Feb., Oct. & Nov. Map 173.

W.A.: 4.2 km W of Corrigin towards Brookton, at entrance to Corrigin airfield, *J.D.Briggs 658* (CANB, NSW, PERTH); WNW of Corrigin, *A.S.George 12913* (NSW, PERTH); 4.5 km W of Bullaring, *A.S.George 14376* (NSW, PERTH); Hyden area, *O'Donnell 51334* (PERTH).

Olde and Marriott (*Grevillea Book* 3: 162 (1995)) distinguish three forms in this species; the 'Corrigin (type) form' with glaucous leaves to 7.5 cm long and white flowers on scapes to 10–20 cm long; the 'Lake Mears form' with generally larger glaucous leaves 5–10 cm long and cream flowers on scapes to 30 cm long; and the 'Quairading form' (not seen) with non-glaucous leaves reportedly to 20 cm long (flowers not seen).

This species is recognised as 'Endangered' in J.D.Briggs & J.H.Leigh, *Rare or Threatened Australian Plants* (1995).

136. Grevillea bracteosa Meisn., *in* J.G.C.Lehmann, *Pl. Preiss.* 2: 254 (1848)

T: Swan River, W.A., *J.Drummond coll. 3: 269*; holo: NY *n.v.*; iso: BM, K, LE *n.v.*, MEL, P, PERTH, NY *n.v.*

Illustrations: D.J.McGillivray & R.O.Makinson, *Grevillea* 375 (1993); P.M.Olde & N.R.Marriott, *Grevillea Book* 2: 70 (53A–C) (1995).

Erect to spreading shrub 1–2 m high. Leaves usually simple and linear, rarely divided near base into 2 or 3 linear segments, 5–14 (–25) cm long, 1–2 (–3) mm wide, acute to obtuse; margins revolute or sometimes refracted; lower surface 2-grooved, glabrous except for hairs on midvein. Conflorescence terminal or axillary, erect, simple or few-branched, pedunculate; unit conflorescence shortly obovoid to subglobose, basipetal; floral rachis 8–15 mm long. Flowers adaxially acroscopic. Flower colour: perianth pale green to greenish pink, rarely white; style deep rose pink to pale pink or white. Perianth glabrous outside, pubescent inside to c. level of ovary, glabrous above; tepals partially everted, remaining coherent at limb segments. Pistil 16–23 mm long; ovary stipitate, glabrous; style minutely papillose above ovary, swollen below pollen-presenter, exserted in late bud; pollen-presenter ±lateral. Follicles narrowly obovoid, apically attenuate, 12–15 (–20?) mm long, glabrous, very finely tuberculate to smooth.

Occurs in south-western W.A., from E of Geraldton to near Mogumber; found in shrubland or sometimes in sandplain heath. Regenerates from seed. Flowers Aug.–Oct. Map 174.

W.A.: Northern Gully, c. 30 km E of Geraldton, *A.M.Ashby 1673* (AD, CANB); between Pithara and Milling, Sept. 1958, *C.A.Gardner s.n.* (PERTH); McFarland's, Howatharra Hill Reserve, *A.S.George 14867* (PERTH); c. 6 km S of Bindi Bindi railway telephone pole 110–08, *B.H.Smith 1448* (CANB, HO, MEL, S *n.v.*); Babilion Hills, Mogumber, Oct. 1934, *H.Steedman* (PERTH).

Grevillea bracteosa has persistent prominent floral bracts which are broadly spathulate to broadly elliptic, 7–14 mm long and membranous, and a smooth conspicuous dilation of the style in the apical 2–4 mm (style-end club-like). *Grevillea eryngioides* also has large less-persistent bracts, but is easily distinguished by its habit (ground-level foliage with long erect flowering scapes), leaves (pinnatipartite with lobes 10–20 mm wide) and obovoid-cylindrical conflorescence. *Grevillea brachystachya* is also somewhat similar, but has smaller linear to elliptic floral bracts, 1–3.5 mm long, which are inconspicuous and fall before anthesis, a ±discoid style-end, and (usually) cream to yellow flowers.

Olde and Marriott (*op. cit.* 2: 70) recognise a 'small-conflorescence form' in the southern part of the range (Mogumber–Pithara–Miling) which also has a greater range of flower colour (white to deep pink) than the more northerly (pink-flowered) populations.

This species is recognised as 'Poorly Known' in J.D.Briggs & J.H.Leigh, *Rare or Threatened Australian Plants* (1995).

Acacioides Group

Robust shrubs. Leaves entire or divided, unifacial, usually terete, rarely dorsiventral; surfaces similar; margins flat (not recurved). Conflorescence erect, terminal or axillary, simple or many-branched; unit conflorescence regular-umbelloid, dome-shaped or subglobose to shortly subcylindrical, strongly to weakly basipetal. Flowers acroscopic or adaxially oriented. Torus straight to slightly oblique. Perianth zygomorphic, glabrous outside, hairy inside; tepals variably separating, widely splayed or independently apically recoiled or everting at the dorsal suture. Pistil 7–17 mm long, glabrous; ovary stipitate; style strongly to weakly exserted from late bud; pollen-presenter oblique to lateral, flat to convex. Follicle glabrous, usually oblong-ellipsoidal (in *G. gordoniana* elongate and sigmoid), smooth, sometimes viscid; pericarp thick (c. 0.6–1.0 mm) relative to small size of fruit. Seed ellipsoidal to obovoid, peripterous or marginate.

A group of three species, endemic to south-western W.A. Insect pollinated. *Grevillea endlicheriana* and *G. gordoniana* are highly derived in some features; the phylogenetic coherence of this group remains to be tested. The affinities of the species involved are however almost certainly with the *Hakeoides* group.

1 Longest leaves > 15 cm long; ovarian stipe > 2 mm long; follicle > 20 mm long **139. G. gordoniana**

1: Longest leaves < 15 cm long; ovarian stipe < 2 mm long; follicle < 15 mm long

2 Leaves usually terete (rarely flat), with several fine (sometimes obscure) longitudinal ridges; fruit often with a transverse ridge near middle; floral bracts > 4 mm long; base of ovary bulging retrorsely; unit conflorescences usually enclosed within foliage with peduncles short to absent **137. G. acacioides**

2: Leaves usually flat (rarely terete), lacking longitudinal ridges; follicles smooth to faintly warty, lacking a transverse ridge; floral bracts < 4 mm long; base of ovary spreading from stipe, lacking retrorse bulge; unit conflorescences usually exceeding foliage, on short to long primary or secondary peduncles **138. G. endlicheriana**

137. Grevillea acacioides C.A.Gardner ex McGill., *New Names Grevillea* 1 (1986)

T: east of Sandstone, W.A., 16 Aug. 1931, *C.A.Gardner 2486*; holo: PERTH; iso: PERTH.

Illustrations: P.M.Olde & N.R.Marriott, *Grevillea Book* 2: 7 (top right & 1A, B) (1995).

Erect shrub 1–2.2 m tall. Leaves entire, terete or rarely flat-linear to very narrowly obovate, 3–7 (–8) cm long, 0.8–1.1 mm diam. or to 1.5 mm wide if flat, with several fine longitudinal ridges (sometimes obscure), acute, pungent, either sericeous and soon glabrous, or pubescent-sericeous. Conflorescence erect, terminal on short lateral branches or axillary, enclosed within foliage, simple, umbel-like, pedunculate, obscurely basipetal to subsynchronous; floral rachis < 2 mm long. Flower colour: perianth green in bud, becoming greenish white to cream [greener towards base?]; style white, sometimes pinkish with age. Perianth bearded inside; tepals independently recoiled after anthesis. Pistil 10–12.5 mm long; style exserted from late bud; pollen-presenter oblique, convex. Follicles obloid-ellipsoidal, (6.5–) 9–13 mm long, smooth to granular-rugulose, often with a blunt transverse ridge on each flank, glabrous, sometimes glaucous, sometimes slightly viscid.

Widespread in inland W.A., from near Cosmo Newbery and Wiluna to Perenjori, S to Ravensthorpe, and E to Queen Victoria Spring; usually in shrubland; regenerates mostly from seed, sometimes by rhizomes. Flowers (May–) July–Sept. Map 175.

W.A.: 10 km SSE of Perrinvale HS, *R.J.Cranfield 7146* (CANB, PERTH); 2.5 km NE of Comet Vale townsite and 2 km N of edge of L. Goongarrie, *A.V.Milewski 1087* (CANB, PERTH); 42.7 km E of Sandstone, *B.H.Smith 401* (CANB, MEL, NSW, PERTH).

Superficially similar to *G. hakeoides* which has leaves with two grooves only (either on the lower surface or almost lateral), and markedly rugose fruits with a thin wall c. 0.5 mm thick (0.6–3 mm thick in *G. acacioides*). *Grevillea brachystachya* also has some similarity, but has much more robust and larger flowers (pistils 12–17 mm long) with hairs on the outer surface of the perianth.

138. Grevillea endlicheriana Meisn., *in* J.G.C.Lehmann, *Pl. Preiss.* 1: 546 (1845)

T: Darling Ra., W.A., 25 July 1839, *L.Preiss 698*; lecto: LD *n.v.*; isolecto: B *n.v.*, G, G-DC, HBG *n.v.*, MEL, NY *n.v.*, P.

G. filifolia Meisn., *in* J.G.C.Lehmann, *Pl. Preiss.* 1: 547 (1845). T: interior of W.A., *L.Preiss 699*; holo: NY *n.v.*; iso: B *n.v.*, G-DC, LD *n.v.*, MEL.

Illustrations: P.M.Olde & N.R.Marriott, *Grevillea Book* 2: 143 (centre right), 144 (116A–D) (1995).

Erect shrub 1–2.5 m high. Leaves entire, linear to very narrowly obovate or rarely subterete, 3–13 cm long, 0.5–3 (–5) mm wide/diam.; margins flat or incurved; both surfaces subsericeous to sericeous. Conflorescence on long flowering branches above foliage, terminal or axillary, usually branched; unit conflorescence subglobose to obovoid or subcylindrical, basipetal; floral rachis 5–20 mm long. Flowers adaxially acroscopic. Flower colour: perianth and style white tinged with pink, or pale pink (rarely perianth reddish?). Perianth villous inside; tepals independently recurved after anthesis. Pistil 7.5–12 mm long; style exserted in late bud; pollen-presenter very oblique to lateral, almost flat. Follicles broadly ellipsoidal to subglobose, 8–8.5 mm long, glabrous, faintly warty to almost smooth.

Occurs in south-western W.A., between about Mogumber and Kelmscott, in the Darling Ra., with a disjunct occurrence at Wongan Hills; mostly in granitic or lateritic soils in woodland or shrubland. Regenerates from lignotuber and sometimes epicormic shoots after fire. Flowers July–Nov. Map 176.

W.A.: 20 km N of Gingin–Bindoon road toward Mogumber, *J.D.Briggs 576* (CANB, NSW); Red Hill, Upper Swan, *R.J.Cranfield 417* (CANB, PERTH); Kalamunda, 19 km E of Perth, *R. & M.Hamilton 161* (CANB, MEL, NSW); Bindoon–Dewars Pool road, 6.2 km E of Great Northern Hwy, *N.Hoyle 860* (CANB, PERTH).

A 'fine-leaved form', with narrow leaves 0.5–1.5 mm wide, subterete or with strongly incurved margins, occurs in the NE of the range (Wongan Hills, Wannamal and Bindoon areas). The more typical, flatter leaved form occurs from Mogumber S along the Darling Escarpment.

139. Grevillea gordoniana C.A.Gardner, *J. Roy. Soc. W. Australia* 47: 56 (1964)

T: [near No 1 Tank, N of Murchison R. on the Carnarvon road (North West Coastal Hwy)], W.A., 21 Dec. 1962, *C.A.Gardner 14273*; holo: PERTH; iso: PERTH.

Illustrations: P.M.Olde & N.R.Marriott, *Grevillea Book* 2: 182 (top centre & 149A, B) (1995).

Erect shrub or small tree 2.5–7 m high, with almost leafless, emergent, flowering branches. Leaves ascending, simple or rarely bi- or tripartite, terete, (8–) 15–36 cm long, 0.9–1.5 mm diam.; surface initially subsericeous, soon glabrous. Conflorescence erect, many-branched, pedunculate, often emergent above foliage; unit conflorescence umbelloid, subglobose, subsynchronous to weakly basipetal; floral rachis (1–) 2–7 (–9) mm long. Flowers with sutures adaxial. Flower colour: perianth yellow to orange; style similar, becoming red on dorsal side. Perianth densely bearded inside; tepals mostly everted below limb before anthesis, independently lax afterwards. Pistil 15–17 mm long; style exserted in late bud; pollen-presenter oblique, flat to convex, obovate. Follicles erect, slightly sigmoid, pod-like, narrowly obovoid to narrowly obloid, 23–28 mm long, rugulose to tuberculose, viscid. Fig. 17G–J.

Occurs near the W coast of W.A., from near Exmouth Gulf S to Yuna, growing in sand in shrubland or steppe. Regeneration strategy unknown. Flowers (July) Sept.–Dec. Map 177.

W.A.: 17.7 km SW of Barradale Crossing on sandhill, *J.S.Beard 6179* (NSW, PERTH, SYD); near Number 1 Tank, North West Coastal Hwy, *F.Lullfitz 1958* (CANB, PERTH); c. 15 km SW of Yannarick on North West Coastal Hwy, *J.Z.Weber 4876* (AD, CANB); Shark Bay, c. 20 km SW of Hamelin Pool HS, c. 5 km W of crossing to Tamala, *J.Z.Weber 5046* (AD, CANB).

The robust, umbelloid inflorescences on emergent branchlets, the enclosure of young flower-bud clusters in large brown papery bracts, and the pod-like sigmoid follicles with a viscid, exfoliating exocarp are highly distinctive. An affinity to other members of this group is clear, but there may also be a relationship between this species and *G. stenostachya*. *Grevillea brachystachya* (*Hakeoides* group) is somewhat similar, but is a smaller shrub to 2.5 m tall, with the conflorescences always held within or close to the foliage, the young flower-bud clusters are not bract-enclosed, and the fruits are ellipsoidal-apiculate, not sigmoid and not viscid.

Acuaria Group

Shrubs. Leaves entire, dorsiventral; surfaces dissimilar; margins recurved to revolute. Conflorescence erect, axillary or terminal, simple or basally 2- or 3-branched; unit conflorescence usually a few-flowered (many-flowered in *G. kennedyana*) irregular cluster, usually synchronous. Flowers abaxiallly or rarely adaxially oriented. Torus oblique. Perianth zygomorphic, glabrous or hairy outside, bearded inside; tepals remaining basally coherent, recoiled above, independently or in 2 pairs. Pistil 14–31 mm long, glabrous; ovary stipitate; style exserted from late bud; pollen-presenter very oblique to lateral, convex to slightly concave. Follicle glabrous, colliculose to smooth, or sometimes faintly ribbed; pericarp thin, crustaceous. Seed ellipsoidal, with a short apical wing and a waxy margin along one edge.

A group of seven species, from south-western W.A. except for one species in far north-western N.S.W./far south-western Qld. Bird pollinated. Affinities are unclear, probably close to the *Linearifolia* group, perhaps also to the *Thelemanniana* group.

1 Outer perianth surface glabrous

2 Torus > 4 mm across **146. G. kennedyana**

2: Torus < 3 mm across

3 Perianth limb 2.5–3 mm wide; fruit 6–7 mm wide; upper surface of leaves punctate with prominent midvein; nectary very conspicuous, rising > 0.5 mm above toral rim

4 Leaves oblong-elliptic or very slightly obovate, obtuse, mucronate; upper surface convex, lacking ribs parallel to midvein; lamina of lower surface exposed **141. G. punctata**

4: Leaves linear, acute; upper surface flat with prominent ribs parallel to midvein; lamina of lower surface obscured by margins **142. G. sulcata**

3: Perianth limb 2–2.2 (–2.5) mm wide; fruit 3–5 mm wide; upper surface of leaves rarely punctate, with midvein usually obscure (rarely evident to raised); nectary obscure to prominent, rising 0.2–0.5 mm above toral rim **140. G. acuaria**

1: Outer perianth surface hairy

5 Leaf lower surface exposed on either side of midvein; widest leaves usually > 2 mm wide, narrowly obovate or narrowly oblanceolate to sublinear **144. G. oligantha**

5: Leaf lower surface enclosed by the revolute margins except for midvein; widest leaves ⩽ 2 mm wide, linear

6 Outer perianth surface densely brown-silky; upper leaf surface prominently veined; most leaves > 5 cm long; leaf margins angularly revolute **143. G. decipiens**

6: Outer perianth surface sparsely white-silky; upper leaf surface smooth or with a few discontinuous ridges; most leaves < 5 cm long; leaf margins smoothly revolute **145. G. sparsiflora**

140. Grevillea acuaria F.Muell. ex Benth., *Fl. Austral.* 5: 452 (1870)

T: W. Australia, *s.d.*, *J.Drummond s.n.*; holo: K; iso: K, MEL.

G. aculeolata S.Moore, *J. Linn. Soc. Bot.* 34: 222 (1899). T: between Wilson's Creek and Lake Darlot, W.A., *s.d.*, *S.Moore s.n.*; holo: BM; ?iso: K.

G. aculeolata var. *longifolia* S.Moore, *J. Linn. Soc. Bot.* 34: 222 (1899). T: between Wilson's Pool and Lake Darlot, W.A., May, *S.Moore s.n.*; holo: BM; ?iso: K.

G. arida C.A.Gardner, *J. & Proc. Roy. Soc. W. Australia* 9: 39 (1923). T: Widgiemooltha, W.A., 25 Sept. 1922, *C.A.Gardner 1272*; holo: PERTH; iso: MEL, PERTH.

Illustrations: D.J.McGillivray & R.O.Makinson, *Grevillea* 362, fig. 88, 363, col. pl. (1993); P.M.Olde & N.R.Marriott, *Grevillea Book* 2: 13 (bottom right), 14–15 (7A–I) (1995).

Rounded bushy to erect shrub 0.2–1.5 m tall. Leaves entire, spreading, subterete-trigonous to linear-subulate or linear, rarely narrowly elliptic, 0.5–3 cm long, 0.5–3.5 (–4.5) mm wide, rigid to pliable, pungent or not; upper surface glabrous or subsericeous to pubescent, dull or shiny, rarely punctate or with prominent ribs; margins smoothly revolute or recurved; lower surface usually obscured except midvein and then 2-grooved, occasionally 1-grooved when dry. Conflorescence terminal or axillary, erect, sessile or almost so, simple, usually (2–) 4–6 flowers in a loose cluster, sometimes solitary; rachis < 1 mm long, subsericeous or pubescent. Flowers abaxially oriented. Flower colour: perianth scarlet, deep burgundy or cherry-red; style green or red, often with green tip. Perianth glabrous outside and slightly glaucous, bearded inside; tepals rolled back to curve after anthesis. Pistil 14–21 mm long; stipe 0.5–2 mm long; style exserted in late bud; pollen-presenter lateral to very oblique. Follicles ±erect on stipe, 8–11 mm long, 3–5 mm wide, usually ovoid to obloid-ovoid with attenuated apex, colliculose. Plate 34.

Widespread in south-western W.A. between Lake Darlot, Brookton, Mt Ragged and Balladonia. Habitat variable, often in winter-wet situations in woodland, heath or mallee communities in clay to sand soils. Regenerates from seed or lignotuber. Flowers mainly May–Oct., sporadically in other months. Map 178.

W.A.: 46 km from Coolgardie along Eyre Hwy towards Norseman, *B.Barnsley 1051* (CANB); near to McPherson Rock, by Coolgardie–Esperance Hwy, c. 30 km S of Norseman, *M.D.Crisp 973* (CANB); c. 29 km NE of Southern Cross, *K.Newbey 2515* (CANB, PERTH); Beacon Hill, Norseman, *M.E.Phillips CBG025389* (CANB, PERTH); 28 km from Corrigin along road to Kondinin, *J.Taylor 2441* & *P.Ollerenshaw* (CANB, NSW, PERTH).

Grevillea acuaria is similar to *G. punctata* and *G. sulcata*, which differ in their slightly broader perianth limb (2–2.2 (–2.5) mm wide in *G. acuaria*), leaves with a prominent midvein on the upper surface, usually refracted leaf margins, very prominent nectary, and follicles 6–7 mm wide; *G. punctata* also has leaves with a punctate upper surface. *Grevillea decipiens*, *G. oligantha* and *G. sparsiflora* differ in having an open to dense appressed indumentum on the outer perianth surface. *Grevillea kennedyana* differs in its larger flowers with torus > 4 mm across (< 3 mm across in *G. acuaria*), ovary stipe 4–5.5 mm long, and 8–12-flowered unit conflorescence.

Olde & Marriott (*Grevillea Book* 2: 13–15 (1995)) recognise nine forms; some of these are noted here and further investigation of variation in the species is required. The 'needle-leaf form' (which includes the type material) has leaves spreading to ascending, subterete-trigonous, 1–3 cm long, with an aristate point. It occurs roughly from Brookton to Coolgardie. The 'dagger-leaf form' (which includes the type of *G. aculeolata*) is very similar but the leaves are broader, linear, and the flowers are often larger. It is found from Coolgardie and Norseman to Balladonia. The 'Parker Range form', found only in the Parker Ra., where it occurs with the 'soft-leaf form', has obtuse pungent leaves with margins not completely obscuring the lower leaf surface beside the midvein, and obloid-ellipsoidal follicles. The 'soft-leaf form' has leaves that are soft, bright green (or sometimes glaucous), ascending to erect, crowded, c. 1 cm long, not pungent. It occurs mostly in the Lake King to Bremer Ra. area, and has also been collected from Norseman and Mt Madden. The 'glaucous leaf form' has broader leaves which are obovate to elliptic, 1.5–2 (–4.5) mm wide, obtuse-mucronate, with margins obscuring lower leaf surface. This form is found between Lake Cronin and Bremer Ra., with isolated collections from Mt Ragged and SE of Norseman.

141. Grevillea punctata Olde & Marriott, *Grevillea Book* 1: 182 (1994)

T: 21 km east of Ravensthorpe, W.A., July 1978, *W.Molyneux s.n.* [information differs in protologue]; holo: NSW; iso: PERTH.

Illustrations: P.M.Olde & N.R.Marriott, *Grevillea Book* 3: 120 (bottom left & 92A), 121 (92B) (1995).

Erect open shrub 1.5 m high. Leaves spreading to ascending, entire, oblong-elliptic or very slightly obovate, 5–18 mm long, 1.5–3 mm wide, obtuse with a short recurved mucro, leathery; upper surface densely granulate with scattered dark pits, with midvein evident to prominent; margins ±smoothly revolute about obscure longitudinal vein on each side; lower surface partly exposed, subsericeous. Conflorescence terminal on short lateral branchlet or sometimes in upper axil, almost sessile, 2–4-flowered; rachis c. 1 mm long, tomentose. Flowers abaxially oriented. Flower colour: perianth and style scarlet; limb of bud sometimes yellowish. Perianth glabrous outside, lightly bearded inside to about halfway above and below level of ovary, sparsely sericeous above beard, glabrous below; tepals shortly recurved in 2 lateral pairs after anthesis. Pistil 14–16 mm long; stipe 1–1.5 mm long; style exserted in late bud; pollen-presenter lateral to very oblique. Follicles erect but ±transverse on stipe, ovoid, 12–15 mm long, 6–7 mm wide, apically attenuate, somewhat distended at base, colliculose, glabrous.

Occurs in south-western W.A., near the Jerdacuttup R. E of Ravensthorpe. In open low shrubland in loamy soil over 'blue stone' substrate. Probably regenerates from seed only. Flowers June–Nov. Map 179.

W.A.: Jerdacuttup R., E of Kundip, *C.A.Gardner 16220* (PERTH); 20 km E of Ravensthorpe on Esperance road, 1978, *W.M.Molyneux s.n.* (CANB, PERTH); 20.8 km E of Ravensthorpe, just E of Jerdacuttup R., *P.Olde 91/304* (NSW, PERTH).

Closely related to *G. acuaria*, which has a smooth to shiny upper leaf surface (lacking granules and pits), and smoothly revolute leaf margins, smaller perianth limb and fruit (limb 2.5–3 mm wide in *G. punctata*), less conspicuous nectary and generally longer pistil (14–21 mm long). *Grevillea sulcata* differs in its linear leaves with flat upper surface (convex in *G. punctata*), prominent, raised venation on upper surface and enclosed lower surface.

This species is recognised as 'Poorly Known' in J.D.Briggs & J.H.Leigh, *Rare or Threatened Australian Plants* (1995).

142. Grevillea sulcata C.A.Gardner ex Olde & Marriott, *Grevillea Book* 1: 183 (1994)

T: Cocanarup, near Ravensthorpe, W.A., 14 May 1924, *C.A.Gardner 2169*; holo: PERTH.

Illustrations: P.M.Olde & N.R.Marriott, *Grevillea Book* 3: 190 (centre bottom & 155A), 191 (155B, C) (1995).

Dense robust shrub 1–1.5 m high. Leaves ±spreading, linear, 1–3 cm long, 1–1.8 mm wide, acute, not or scarcely pungent; upper surface punctate, glabrous, with 2–5 ±prominent longitudinal ribs; margins refracted; lower surface enclosed except midvein, 2-grooved, subsericeous in grooves. Conflorescence terminal or axillary, erect, 2–14-flowered, subsessile; rachis 1–1.5 mm long, subsericeous. Flowers abaxially oriented. Flower colour: perianth and style scarlet; pollen-presenter orange. Perianth glabrous outside, bearded inside opposite ovary, glabrous at base, sparsely subsericeous above beard; tepals shortly recurved in 2 lateral pairs after anthesis. Pistil 15–16 mm long; stipe 1.5 mm long; style exserted in late bud; pollen-presenter lateral. Follicles oblique to stipe, ovoid, 13 mm long, 6 mm wide, acuminate, somewhat distended at base.

Occurs in W.A., confined to a small area W of Ravensthorpe. Grows in eucalypt woodland in rich brown loam with bluestone with *Dodonaea*. Regenerates probably from seed only. Flowers winter to early spring. Map 180.

W.A.: 11 km W of Ravensthorpe, Cocanarup Timber Reserve, *P.Olde 92/260* (NSW); 10 km W of Ravensthorpe, *C.Woolcock G230* (NSW).

Very narrowly distinct from *G. acuaria* which has a smooth to shiny upper leaf surface and smoothly revolute leaf margins, less conspicuous nectary, usually longer pistil (14–21 mm long), smaller floral bracts 0.25–0.5 mm long (0.8–1 mm long in *G. sulcata*), somewhat smaller perianth limb and follicles (limb 2.5–3 mm wide in *G. sulcata*). *Grevillea punctata* differs in its elliptic to obovate leaves with convex granular and pitted upper surface, and shortly revolute margins with the lower surface exposed.

Collections seen of *G. acuaria* from the Mt Jackson area (e.g. *R.A.Saffrey 947*) closely approach *G. sulcata* in leaf form; the taxon may be better reduced to a subspecies of *G. acuaria*.

143. **Grevillea decipiens** McGill., *New Names Grevillea* 4 (1986)

T: 1 mile [1.6 km] S of Mt Gibbs, W.A., 31 July 1969, *A.S.George 9454*; holo: PERTH; iso: distribution uncertain.

Illustrations: P.M.Olde & N.R.Marriott, *Grevillea Book* 2: 111 (top centre & 89A, B), 112 (89C) (1995).

Dense compact shrub to 1.5 m high, with ascending to erect branches. Leaves entire, ascending, linear, 2.5–9 cm long, 1–1.6 mm wide, obtuse-mucronate; upper surface prominently 7–9-veined, sometimes sparsely pubescent between ridges; margins angularly revolute; lower surface enclosed except midvein, 2-grooved, sometimes subsericeous in grooves. Conflorescence erect, usually axillary, or sometimes terminal on short lateral branchlets, or cauline, sessile, usually simple, 1–6-flowered; rachis < 1 mm long, tomentose. Flowers adaxially oriented. Flower colour: perianth red; style pale orange to red with green tip. Perianth subsericeous outside with brown hairs, bearded inside opposite ovary; tepals shortly recurved in 2 lateral pairs after anthesis. Pistil 17–20.5 mm long; stipe 1.3–2.2 mm long; style exserted in late bud; pollen-presenter very oblique. Follicles 10–11.5 mm long, 3–4.5 mm wide, erect, narrowly subtriangular-ovoid, apically attenuate, colliculose-rugulose, glaucous and with a slight ridge along dorsal side, ?rarely smooth.

Confined to southern inland W.A. from Ongerup to Frank Hann Natl Park. Grows in low to tall shrubland and mallee woodland, often in depressions, in sandy to clayey soils. Regenerates from seed. Flowers May–Aug. Map 181.

W.A.: 77.3 km S of crossroads Hyden–Norseman, Mt Holland–Lake King, *H.Demarz 128* (KPBG, PERTH); c. 20 km by road N of Ongerup, *D.J.McGillivray 3519 & A.S.George* (NSW, PERTH); 18.5 km E of crossroads which is just E of Lake King, *D.J.McGillivray 3570 & A.S.George* (NSW); 5 km S of Baanga Hill and 18 km SE of Lake King, *P.G.Wilson 7009* (PERTH).

Grevillea oligantha is similar to *G. decipiens* but has thinner, generally broader obovate-cuneate leaves 1–12 mm wide, with shortly recurved margins that usually do not conceal the lower surface and without multiple ridges on the upper surface. *Grevillea sparsiflora* has an open white-subsericeous outer perianth surface and generally shorter leaves (2–6 cm long) without clear ridging on the upper surface.

Grevillea decipiens is sometimes sympatric with *G. oligantha* (e.g. in the Ongerup area) in disturbed roadside habitats, and hybrids between the two species have been reported.

This species is recognised as 'Poorly Known' in J.D.Briggs & J.H.Leigh, *Rare or Threatened Australian Plants* (1995).

144. **Grevillea oligantha** F.Muell., *Fragm.* 6: 206 (1868)

T: near the Phillips R., W.A., *s.d.*, [*G.Maxwell*]; lecto: MEL, *fide* D.J.McGillivray & R.O.Makinson, *Grevillea* 432 (1993); isolecto: MEL.

[*G. pauciflora auct. non* R.Br.: G.Bentham, *Fl. Austral.* 5: 453 (1870), *p.p.*]

Illustrations: D.J.McGillivray & R.O.Makinson, *Grevillea* 357, col. pl. (1993); P.M.Olde & N.R.Marriott, *Grevillea Book* 3: 65 (bottom centre & 47A), 66 (47B, C) (1995).

Dense shrub 0.5–2.5 m high with erect branches. Leaves ascending, entire, obovate-cuneate or oblanceolate to rarely narrowly elliptic or linear, 1.2–8 cm long, 1–8 (–12) mm wide, obtuse to emarginate, mucronate; upper surface sprinkled with appressed hairs, soon glabrous and slightly glaucous, minutely pitted, with several fine longitudinal ribs; margins smoothly recurved or rarely revolute; lower surface exposed, subsericeous. Conflorescence erect, sessile or pedunculate, axillary or terminal on short lateral branchlet or cauline, usually simple, 1–6-flowered; rachis 0.4–2.5 mm long, subvillous. Flowers abaxially oriented. Flower colour: perianth brownish yellow, orange or reddish brown; style pale yellow to reddish with green tip. Perianth subsericeous to sparsely so outside, bearded inside above ovary; tepals independently recoiled after anthesis. Pistil 18–22.5 mm long; stipe 1.9–3.5 mm long; style exserted in late bud; pollen-presenter very oblique. Follicles erect, ovoid-ellipsoidal, 10.5–12.5 (?–20) mm long, 3.5–5 mm wide, with faint longitudinal dorsal ridges, irregularly warted, glabrous, glaucous.

146. Grevillea kennedyana F.Muell., *Trans. & Proc. Roy. Soc. Victoria* 24: 172 (1888)

T: Grey Ra., N.S.W., c. 1887, *W.Baeuerlen 358*; lecto: MEL, *fide* D.J.McGillivray & R.O.Makinson, *Grevillea* 424 (1993); isolecto: K, MEL, ?NSW.

Illustrations: D.J.McGillivray & R.O.Makinson, *Grevillea* 361, col. pl. & fig. 87 (1993); P.M.Olde & N.R.Marriott, *Grevillea Book* 2: 224 (bottom right), 225 (187A–C) (1995).

Diffuse shrub 0.7–2 m high. Leaves often fasciculate, ascending to spreading, entire, linear, 0.7–3.3 cm long, 1.2–2 mm wide, rigid, pungent with long aristate point; upper surface subsericeous becoming sometimes almost glabrous, with midvein evident only as a faint groove; margins angularly revolute against prominent midvein; lower surface 2-grooved. Conflorescence erect, ±sessile, terminal, simple, a loose cluster of c. 8–20-flowers; rachis 2–5 mm long, subsericeous. Flowers abaxially oriented. Flower colour: perianth pink, red or orange-red with limb shiny green in bud; style red. Perianth glabrous or with few to many appressed hairs outside, subvillous inside; tepals shortly and independently recurved after anthesis. Pistil 27–31 mm long; stipe 4–5.5 mm long; style exserted in late bud; pollen-presenter very oblique to lateral. Follicles erect, obovoid-ellipsoidal, 15–19 mm long, 5.5–7 mm wide, finely tuberculate to rugulose. Plate 35; Fig. 18 I–K.

Occurs in far south-western Qld, at Naryilco (Bygrave Ra.), and in N.S.W. in the far NW corner (Grey Ra., McDonald Peak, Olive Downs escarpment, Mt Wood, Mt Wood Hills, Onepah Stn). Grows on slopes, steep jump-ups and drainage lines; often gregarious, in loamy soils on weathered silcrete. Regenerates mainly from sucker and lignotuber, occasionally from seed. Flowers mainly July–Nov. Map 184.

N.S.W.: Olive Downs to Mt King road, *F.Armitage NSW156423* (NSW); Yandama Stn, *C.R.Collier NSW93160* (NSW); Onepah Stn, *W.E.Mulham 1230* (NSW); 1 km E of Jumpup Loop Rd, 1 km before turnoff to Olive Downs lookout, c. 3.5 km S of Olive Downs, Sturt Natl Park, *M.F.Porteners 65 & J.E.Plaza* (BRI, CANB, K, NSW, PERTH).

Grevillea kennedyana is distinctive in this group for its broad torus (c. 4 mm across), broad perianth, long pistil (27–31 mm) and more floriferous unit conflorescence. It is similar to *G. acuaria*, which differs in its prominent ovary (ovary scarcely wider than style in *G. kennedyana*), its fewer-flowered unit conflorescence, smaller torus and perianth and in its fruit (8–12 mm long). There is superficial similarity to taxa allied to *G. juniperina* and *G. victoriae*, but these species have minute hairs near the apex of the style and a much less oblique torus.

This species is recognised as 'Vulnerable' in J.D.Briggs & J.H.Leigh, *Rare or Threatened Australian Plants* (1995).

Linearifolia Group

Shrubs. Leaves entire, dorsiventral; surfaces ±dissimilar; margins recurved to revolute. Conflorescence terminal or axillary or occasionally cauline, erect to decurved or pendent, sessile or pedunculate, usually simple, regular-umbelloid or shortly cylindrical or secund, acropetal or rarely subsynchronous. Flowers acroscopic or rarely abaxially oriented. Torus straight to oblique. Perianth zygomorphic, hairy outside (rarely almost glabrous), bearded inside or rarely glabrous with a small swelling (pulvinus) in the beard position; tepals either coherent over at least the basal third and independently recoiled above (ventral pair more strongly so), or separating almost to base and widely splayed. Pistil 4–35 mm long; ovary glabrous, stipitate; style usually appearing ±glabrous, but always minutely hairy or papilloid at least on rear of style-end (see Fig. 3H), weakly to strongly exserted from late bud; pollen-presenter oblique to lateral, slightly concave to convex. Follicle glabrous, smooth or densely colliculose or rarely ribbed; pericarp usually thin-walled, crustaceous. Seeds ellipsoidal; margins revolute, a waxy strip on one side extending into a short apical elaiosome.

Forty five species, predominantly in moister areas of south-eastern Australia, a few species in south-western W.A., S.A., and the southern half of Qld. Pollinated by birds and insects. Corresponds to Section *Lissostylis* Benth. (1870).

1 Leaves linear to terete, and most or all with lower surface of leaf enclosed except for midvein by the revolute margins and 2-grooved, or midvein also enclosed and then 1-grooved

2 Pistils ⩾ 14 mm long

3 Leaves 40–120 mm long, pliable, not pungent, not clustered on short lateral branchlets; flowers deep purplish pink **158. G. sericea**

3: Leaves ⩽ 45 mm long, usually rigid, pungent, often clustered on short lateral branchlets; flowers red, yellow, orange or rarely pink

4 Branchlets angular in cross-section; leaves planar, the lower surface 2-grooved with a prominent midvein **150. G. molyneuxii**

4: Branchlets terete; leaves in cross-section planar to angularly deltoid or trigonous **151. G. juniperina**

2: Pistils < 14 mm long

5 Leaf lower surface 1-grooved; pistils < 8 mm long **173. G. australis**

5: Leaf lower surface 2-grooved; pistils 4.5–13 mm long

6 Follicle with faint to very conspicuous longitudinal ridges (W.A.)

7 Branchlets subterete; leaves basally deflexed, often slightly twisted **176. G. inconspicua**

7: Branchlets angular in cross-section; leaves ascending to spreading, not deflexed, not twisted

8 Follicle with several very prominent longitudinal flange-like ridges **175. G. costata**

8: Follicle with faint longitudinal ridges **177. G. christinae**

6: Follicle surface colliculose to almost smooth, lacking longitudinal ridges (south-eastern Qld to S.A.)

9 Unit conflorescence secund (sometimes shortly and broadly so, but all styles held within a 180° arc in end view)

10 Unit conflorescence usually very dense and if terminal then usually > 30-flowered; style with papilloid hairs usually over the apical 1–2 (–6) mm (western Vic.) **156. G. confertifolia**

10: Unit conflorescence loose to dense, < 26-flowered; style with papilloid hairs confined to apical 1 mm or less (sometimes only on rear of pollen-presenter) (south-eastern Qld, N.S.W., far eastern Vic.)

11 Leaves 5–17 cm long; prostrate to scandent or rarely weakly erect sub-shrub to 0.3 m tall; branches usually scandent or arching, pliable **163. G. reptans**

11: Leaves 1–5 (–7) cm long; erect or low spreading shrubs 0.3–2 m tall; branches ascending, ±rigid

12 Pistil 6.5–10.5 mm long; leaves ±rigid, pungent, usually somewhat clustered (often in threes) **165. G. patulifolia**

12: Pistil 4.5–7.5 mm long; leaves pliable, not pungent, usually not clustered **164. G. parviflora**

9: Unit conflorescence regular, subumbelloid or arrangement obscure

13 Inner surface of perianth glabrous, or with a few hairs only opposite the ovary (then the hairs on a pulvinus and usually < 0.2 mm long); leaves often gently wavy or incurved **168. G. micrantha**

13: Inner surface of perianth with a dense to scanty beard (hairs 0.2–1.0 mm long) at least on the ventral tepals; leaves usually straight, occasionally gently incurved

14 Leaf margins smoothly revolute; submarginal veins not evident **172. G. halmaturina**

14: Leaf margins angularly revolute or refracted about submarginal veins

15 Leaves ±rigid, pungent, usually widely ascending to spreading

16 Conflorescences often terminal on short lateral branchlets, often many aggregated in short columns below branch tips; leaves solitary; leaf upper surface with 3–7 prominent longitudinal ridges (S.A.) **172. G. halmaturina**

16: Conflorescences terminal, if on short lateral branchlets then not heavily aggregated below branch tips; leaves in clusters of 3 or occasionally solitary; leaf upper surface with midvein and 2 edge veins evident, rarely 3-ridged (Vic. & N.S.W.)

17 Longest leaves usually > 2.5 cm long, > 1.5 mm wide; pistils 6.5–10.5 mm long; inner surface of perianth with a profuse beard; leaf cross-section obliquely elliptic, with margins refracted at least 90° relative to upper surface (N.S.W., north-eastern Vic.) **165. G. patulifolia**

17: Longest leaves usually < 2.5 cm long and < 1.5 mm wide; pistils 6.5–7 mm long; inner surface of perianth with a scanty beard ±opposite ovary; leaf cross-section usually oblong, with margins refracted ±vertically downwards at 90° to plane of upper surface (eastern Vic.) **167. G. alpivaga**

15: Leaves pliable, not or scarcely pungent, closely ascending to spreading

18 Inner surface of perianth profusely bearded, the beard usually > 1 mm in extent and the hairs ⩾ 0.5 mm long; unit conflorescence sessile or subsessile (peduncle to 1 mm long); very young conflorescences densely clad with rusty-brown hairs **164. G. parviflora**

18: Inner surface of perianth with a short dense to scanty beard, < 1 mm in extent, the hairs < 0.5 mm long; unit conflorescence sessile or variably pedunculate (if present then peduncles 1–15 (–30) mm long); very young conflorescences with pale, fawn or sometimes a few rusty-brown hairs

19 Leaves very widely ascending to spreading, very straight, 0.7–1.3 (–2) mm wide, longest leaves > 30 (and usually > 40) mm long **170. G. wiradjuri**

19: Leaves closely to widely ascending, straight or wavy or incurved, if the latter then usually either > 1.5 mm wide or < 30 mm long

20 Leaves crowded and ⩽ 30 mm long **167. G. alpivaga**

20: Leaves either well-spaced, or if crowded then longest leaves > 30 mm long

21 Most leaves > 3 cm long, ±straight; shrub 1–2.5 m tall; branches long, pliable **166. G. neurophylla**

21: Most leaves < 2.5 cm long; often gently incurved; shrub 0.3–1 m tall; branches short, rigid **169. G. gariwerdensis**

1: Leaves either not linear or terete, or linear but then with lower surface of leaf narrowly exposed between the midvein and the recurved to revolute margins

22 Pistil > 23 mm long

23 Apex of flower bud acute (limb subpyramidal) **185. G. oxyantha**

23: Apex of flower bud obtuse (limb rounded)

24 Longest leaves ⩽ 40 mm long

25 Leaves narrowly obovate to subulate or linear, usually rigid and pungent, 1–6 mm wide; midvein of lower leaf surface densely hairy and diminishing markedly in prominence towards apex, often scarcely or not evident in apical half of leaf **151. G. juniperina**

25: Leaves narrowly elliptic to elliptic or lanceolate or almost round, usually pliable (occasionally rigid), not pungent, (4–) 7–45 mm wide; midvein of lower leaf surface densely to sparsely hairy or glabrous, but always evident over full length of leaf

26 Pistils 25–35 mm long; longest floral rachises 5–12 mm long; midvein on lower leaf surface glabrous or nearly so **147. G. speciosa**

26: Pistils ⩽ 26 mm long; longest floral rachises 10–90 mm long; midvein on lower leaf surface densely hairy or rarely sparsely hairy to almost glabrous

27 Leaf lower surface with an open to sparse appressed indumentum (ground tissue visible between hairs) or glabrous **182. G. monslacana**

27: Leaf lower surface with a dense indumentum (ground tissue obscured by hairs) except on midvein

24: Longest leaves ⩾ 50 mm long

28 Floral rachis 1–8 mm long; conflorescence terminal, axillary or cauline; midvein on leaf lower surface glabrous or almost so

29 Flowers adaxially acroscopic (ventral sutures facing towards axis of floral rachis); conflorescence terminal or occasionally axillary; pistil 27–36 mm long (N.S.W.) **148. G. oleoides**

29: Flowers abaxially acroscopic (ventral sutures facing away from axis of floral rachis); conflorescence axillary or cauline; pistil ⩽ 26 mm long (Vic.) **149. G. dimorpha**

28: Floral rachis 10–90 mm long; conflorescence terminal (sometimes on short lateral branchlets); midvein on lower leaf surface densely hairy or rarely sparsely hairy to almost glabrous

30 Leaf lower surface with an open to sparse appressed indumentum (ground tissue visible between hairs) or glabrous **182. G. monslacana**

30: Leaf lower surface with a dense indumentum (ground tissue obscured by hairs) except on midvein **178. G. victoriae**

22: Pistil ⩽ 23 mm long

31 Pistil 14–23 mm long

32 Leaf upper surface finely granulose to coarsely asperous

33 Leaves oblong to obovate or narrowly so, or occasionally elliptic 5–10 times as long as wide; pistil 14–17 mm long; lateral veins on lower leaf surface usually not evident (SE of N.S.W.) **184. G. irrasa**

33: Leaves ovate to elliptic or broadly so, or slightly obovate, 1.5–4 times as long as wide; pistil ⩾ 17 mm long; lateral veins evident on lower leaf surface (Vic.) **183. G. miqueliana**

32: Leaf upper surface smooth (sometimes hairy but not granulose or asperous)

34 Leaves with a tangible velvety indumentum on upper surface; conflorescence 2–8-flowered

35 Pistil to 16 mm long; perianth bichromatic, green in basal half, red above **189. G. linsmithii**

35: Pistil ⩾ 16 mm long; perianth bright red throughout **190. G. mollis**

34: Leaves with upper surface glabrous or hairy but then the hairs usually appressed, rarely ascending, never velvety; conflorescence usually with > 8 flowers, rarely 2–8

36 Apex of flower bud acute (limb subpyramidal) **185. G. oxyantha**

36: Apex of flower bud obtuse (limb subglobose)

37 Most or all conflorescences axillary or cauline

38 Pistil ⩾ 21 mm long (Vic.) **149. G. dimorpha**

38: Pistil ⩽ 18 mm long (Qld) **188. G. hockingsii**

37: Most or all conflorescences terminal (sometimes on short lateral branchlets)

39 Widest leaves ⩽ 6 mm wide

40 Leaves usually rigid, pungent, ⩽ 35 mm long, often many clustered together on short lateral branchlets; midvein on lower surface of leaf as densely hairy as the adjacent lamina, and declining markedly in prominence in apical half of leaf; branchlets ±terete **151. G. juniperina**

40: Leaves rigid or pliable, usually not pungent, 1–5.5 cm long, solitary or in clusters of 3; midvein on lower surface of leaf glabrous or markedly less hairy than the adjacent lamina, maintaining prominence over full length of leaf; branchlets angular or terete

41 Branchlets terete; pollen-presenter lateral on style; perianth and style green; unit conflorescence usually with ⩽ 6 flowers **191. G. cyranostigma**

41: Branchlets angular in cross-section; pollen-presenter oblique on style; perianth and style white, pink or red; unit conflorescence usually with > 10 flowers

42 Pedicels 2.5–3.5 mm long; perianth and style bright red; low almost prostrate to weakly erect shrub 0.2–1 m tall **150. G. molyneuxii**

42: Pedicels 4–10 mm long; perianth and style white to cream or pink, or rarely mauve; prostrate to robust-erect shrubs 0.2–2 m tall

43 Hairs on lower leaf surface highly reflective, sparkling under strong light; leaves narrowly oblong-elliptic to sublinear, ⩽ 40 mm long, < 4 mm wide (south-eastern Qld & far north-eastern N.S.W.) **159. G. humilis**

43: Hairs on lower leaf surface matt, not sparkling; leaves variable, obovate or elliptic or narrowly so and then 10–50 mm long and 2–9 mm wide, or rarely linear and then 60–120 mm long and 1–3 mm wide (Sydney, Blue Mtns & Hunter R. area, N.S.W.) **158. G. sericea**

39: Widest leaves > 6 mm wide

44 Tepals all separating to level of ovary or below, and widely and evenly arching and splayed, all held with a 180° arc ventral to the pistil; hairs of the inner perianth surface displayed **158. G. sericea**

44: Tepals coherent over basal third to half, independently recoiled or recurved above on either side of pistil, the ventral pair more strongly recoiled than the dorsal pair; hairs of the inner perianth surface not displayed

45 Limb of bud (and limb segments of tepals) tomentose to villous (±spreading hairs)

46 Lower leaf surface densely to loosely velvety or villous (ground tissue usually obscured by hairs); peduncle, floral rachis and outer surface of perianth all with biramous nonglandular hairs only; leaf apices usually obtuse **183. G. miqueliana**

46: Lower leaf surface very open-tomentose (ground tissue visible between hairs); peduncle, floral rachis and outer surface of perianth with both biramous hairs and simple erect glandular hairs (the latter sometimes few and inconspicuous); leaf apices usually ±acute **186. G. rhyolitica**

45: Limb of bud (and limb segments of tepals) subsericeous to sericeous or sparsely so (appressed hairs)

47 Lower leaf surface with an open to sparse indumentum (ground tissue visible between hairs)

48 Longest leaves ⩽ 4 cm long; new growth usually flushed pink or purplish pink **180. G. parvula**

48: Longest leaves > 4 cm long; new growth green or very briefly flushed pink or tan

49 Limb of bud subcubic, distinctly square in face view; pollen-presenter ventrally concave in cross-section **181. G. epicroca**

49: Limb of bud subglobose, rounded in face view; pollen-presenter ventrally convex in cross-section **182. G. monslacana**

47: Lower leaf surface with a dense indumentum (ground tissue concealed by hairs)

50 Longest floral rachises 20–90 mm long; longest leaves 6–12 (–20) cm long; flower buds rich rusty brown; upper surface of leaf with lateral veins pale, strongly evident **178. G. victoriae**

50: Longest floral rachises 6–25 mm long; longest leaves 2–5 (–7) cm long; flower buds rusty brown, red to pink, or green; upper surface of leaf with lateral veins obscure or weakly evident

51 Leaves narrowly oblong to narrowly oblong-ovate, with widest leaves 5–11 mm wide; unit conflorescence usually with ⩽ 6 flowers; perianth and style green; longest floral rachises 6–12 mm long; stigma a very prominent beak c. 1 mm long **191. G. cyranostigma**

51: Leaves elliptic to ovate or narrowly so, with widest leaves 8–30 mm wide; unit conflorescence usually with > 6 flowers; perianth and style brown, lilac, yellow, white, apricot, pink or red; longest floral rachises 10–35 mm long; stigma a convex boss

52 New vegetative growth usually strongly flushed pink or purplish pink (rarely green); pollen-presenter usually strongly ventrally concave in cross-section; branchlets tomentose to subvillous (spreading hairs) **180. G. parvula**

52: New vegetative growth green; pollen-presenter usually flat in cross-section; branchlets subsericeous (appressed hairs) or occasionally weakly tomentose (ascending hairs) **179. G. brevifolia**

31: Pistil < 14 mm long

53 Inner surface of perianth glabrous or with a few hairs only ±opposite the ovary

54 Lower surface of leaf with hairs of a similar density on midvein and lamina; midvein either not evident between hairs, or prominent only in lower half of leaf **173. G. australis**

54: Lower surface of leaf with hairs less dense on midvein than on lamina, or midvein glabrous; midvein evident and prominent over full length of leaf

55 Leaves mostly elliptic to narrowly so, straight, (1.2–) 2–6.5 mm wide; prostrate or very low spreading shrub ⩽ 0.4 m tall **171. G. imberbis**

55: Leaves ±linear, often gently curved, 0.6–2 (–3) mm wide; erect to prostrate shrub 0.3–1.0 m tall

56 Upper leaf surface often minutely scabrid on and between veins; leaves 0.6–1.0 (–1.5) mm wide **168. G. micrantha**

56: Upper leaf surface ±smooth, usually faintly graniculate on veins only; leaves 1–3 mm wide **169. G. gariwerdensis**

53: Inner surface of perianth with a distinct, scanty to profuse beard ±opposite the ovary

57 Tepals remaining coherent over basal 1/3–1/2 after release of style-end, independently recoiled or recurved above on either side of pistil, with ventral pair more strongly recoiled than dorsal pair; hairs on inner surface of perianth not displayed; pollen-presenter lateral on style or almost so

58 Upper surface of leaf either granulose to scabrid, or with a tangible velvety indumentum; lower leaf surface velutinous to villous (long ±spreading hairs)

59 Leaves granulose to scabrid above; perianth monochromatic, red to pinkish red (southern N.S.W., north-eastern Vic.) **184. G. irrasa**

59: Leaves velvety above, not granulose to scabrid beneath hairs; perianth bichromatic, green at base, red above (north-eastern N.S.W., south-eastern Qld) **189. G. linsmithii**

58: Upper surface of leaf smooth, either glabrous or with ±appressed hairs, lacking a velvety feel; lower leaf surface sericeous or subsericeous or occasionally shortly tomentose

60 Longest leaves ⩽ 2 cm long; conflorescence on strongly decurved to pendulous peduncle; flower buds with a dense rusty brown indumentum all over; pistils 10–11 mm long **187. G. diminuta**

60: Longest leaves 2.5–14 cm long; conflorescence erect to decurved or pendulous; flower buds pink to red, lacking dense brown indumentum except sometimes on the limb; pistils 12–20 mm long

61 Branchlets tomentose to subvillous (spreading hairs); conflorescence terminal or axillary; floral rachis 5–25 mm long (southern N.S.W.) **180. G. parvula**

61: Branchlets sericeous (closely appressed hairs); conflorescence axillary or cauline, rarely a few terminal; floral rachis 2–5 mm long (Qld) **188. G. hockingsi**

57: Tepals all separating to level of ovary or below after release of style-end, all held with a ±180° arc ventral to the pistil; hairs on inner surface of perianth usually displayed; pollen-presenter oblique on style

62 Unit conflorescence ±regular-umbelloid (open flowers arranged in full 360° about rachis), or a loose irregular cluster (sometimes 4–6-flowered) cluster of indeterminate structure

63 Leaf lower surface glabrous or openly to sparsely hairy on either side of midvein (ground tissue visible between hairs)

64 Follicles with faint longitudinal ridges (south-western W.A.) **177. G. christinae**

64: Follicles lacking longitudinal ridges, colliculose (south-eastern Qld) **162. G. leiophylla**

63: Leaf lower surface with a ±dense indumentum on either side of midvein (ground tissue not visible between hairs)

65 Leaves narrowly obovate with an obtuse or emarginate apex

66 Leaves with upper surface minutely pitted, 5–7-veined, 1.5–6 cm long; conflorescences on very short axillary branchlets (appearing axillary), aggregated subterminally on major branches (S.A.) **174. G. quinquinervis**

66: Leaves with upper surface not pitted, 1–3-veined, 1.5–3.5 cm long; conflorescences terminal on main or lateral branchlets, not appearing axillary, not subterminally aggregated (N.S.W.) **159. G. humilis**

65: Leaves linear, elliptic or lanceolate, with a ±acute apex

67 Leaves with upper surface minutely pitted (S.A.) **174. G. quinquinervis**

67: Leaves with upper surface not pitted (N.S.W., Vic.)

68 Inner surface of perianth with a scanty beard of very short hairs; leaves often gently curved (central Vic.)

69 Pistil 7–8.5 mm long; leaves not crowded; conflorescence often pedunculate, sometimes subsessile; leaves 1–5 cm long, 1–3 mm wide **169. G. gariwerdensis**

69: Pistil 6.5–7 mm long; leaves usually very crowded; conflorescence sessile; leaves 0.8–2.5 (–3) cm long, 0.8–1.4 mm wide **167. G. alpivaga**

68: Inner surface of perianth with a profuse beard; leaves ±straight (N.S.W., far eastern Vic.)

70 Leaves ±rigid, pungent **165. G. patulifolia**

70: Leaves pliable, not pungent

71 Leaves 0.2–1.3 mm wide; hairs of branchlets, leaves, and perianth dull, not sparkling under strong light; young flower buds deep rusty brown in colour (hairs) **164. G. parviflora**

71: Leaves 1.5–10 mm wide; hairs of branchlets, leaves, and perianth sparkling under strong light; young flower buds silver and fawn (hairs) **159. G. humilis**

62: Unit conflorescence secund, sometimes shortly and broadly so and appearing subglobose, but all styles of open flowers held within a 180° arc when viewed along rachis (bud sometimes held below this arc)

72 Late flower buds subvillous to villous, especially on limb (hairs ascending to spreading); unit conflorescence often very dense and appearing almost globose, actually usually broadly secund, sometimes pendulous, sometimes held on or near the ground

73 Leaves pungent, conspicuously 3-veined, often twisting on drying; perianth with both biramous hairs and simple erect glandular hairs on outer surface **155. G. oldei**

73: Leaves not or scarcely pungent, not conspicuously 3-veined, remaining straight on drying; perianth with biramous hairs only on outer surface

74 Leaves linear, 6–17 cm long; unit conflorescence secund, not dense or subglobose; perianth and style mauve-pink **163. G. reptans**

74: Leaves narrowly elliptic to oblong or slightly obovate, or ovate to triangular, 0.5–9 cm long; unit conflorescence densely and broadly secund, often appearing subglobose; perianth deep maroon or blackish red, rarely white; style red, rarely greenish white

75 Limb of bud with a dark rusty brown indumentum; nectary shortly linguiform, projecting clearly over the toral rim **154. G. evansiana**

75: Limb of bud with a pale indumentum (sometimes a few light brown hairs); nectary arcuate, scarcely emergent above toral rim **153. G. capitellata**

72: Late flower buds sericeous including on limb (hairs appressed); unit conflorescences various (dense to loose, secund to broadly so or appearing subglobose, pendulous or not, close to ground or not)

76 Unit conflorescence dense, broadly secund (often appearing subglobose), often pendulous or sometimes held on or near the ground; flowers red to deep maroon **152. G. diffusa**

76: Unit conflorescence loose to dense, secund or broadly so (but not appearing subglobose), not pendulous; flowers white to cream or greenish yellow, or pink

77 Leaf lower surface on either side of midvein glabrous or with an open to sparse indumentum, ground tissue visible between hairs

78 Hairs of leaves and branchlets matt, not sparkling under strong light

79 Low erect to spreading shrub to 1 m tall; leaves ascending, usually solitary, occasionally a few clusters of 3; branchlets brownish to green; flowers pink or white **159. G. humilis**

79: Erect open shrub 0.5–2 m tall; leaves ±spreading, usually in well-spaced clusters of 3–5; branchlets reddish; flowers white (style sometimes pink with age) **161. G. virgata**

78: Hairs of leaves and branchlets sparkling under strong light

80 Weakly erect to almost prostrate shrub 0.2–1 m tall; branchlets short, stiff, brownish; leaves often crowded but not in clusters; perianth and style pale to deep pink (south-eastern Qld) **162. G. leiophylla**

80: Erect open virgate shrub 0.5–2 m tall; branchlets long, flexible, reddish; leaves usually in well-spaced clusters of 3–5; perianth and style white (style sometimes pink with age) (mid-coastal N.S.W.) **161. G. virgata**

77: Leaf lower surface on either side of midvein with a ±dense indumentum, ground tissue not visible between hairs

81 Largest conflorescences with > 30 flowers (subterminal axillary 2-flowered conflorescences sometimes also present), very dense; flowers reddish purple or rich mauve; papilloid hairs of the style-end extending 1–6 mm down style (Grampians area, Vic.) **156. G. confertifolia**

81: Largest conflorescences with < 24 flowers, dense to loose; flowers (after anthesis) white or pale to deep pink or mauve, rarely styles yellowish or mauve; papilloid hairs of the style-end extending no more than 1 mm below tip

82 Styles (fresh) lemony yellow or yellow-green; perianth greenish becoming white after anthesis (montane habitats of northern N.S.W. & south-eastern Qld on granite) **160. G. viridiflava**

82: Styles (fresh) white or pink; perianth white or pink to mauve after anthesis (coastal to montane N.S.W. & far north-eastern Vic.)

83 Leaves pungent, usually ±rigid **165. G. patulifolia**

83: Leaves not or scarcely pungent, pliable

84 Erect shrubs 1–3 m tall; flowers white or very pale pink; leaves not crowded

85 Pistils 7–13 mm long; early buds pale with a fawn-brown tip; leaves 3.5–11 cm long, 1–5 mm wide, light green; stipe of ovary 1.3–1.5 mm long; shrub, not rhizomatous **157. G. linearifolia**

85: Pistils 4–8 mm long; early buds with dense rusty brown hairs all over; leaves 1.5–5 cm long, 0.8–1.3 mm wide, dark green; stipe of ovary ⩽ 1.2 mm long; rhizomatous shrub **164. G. parviflora**

84: Low spreading to weakly erect shrubs 0.2–1.3 m tall; flowers white or pale to deep pink; leaves crowded or not

86 Leaves ⩾ 1.5 mm wide, variously sublinear or narrowly elliptic or oblanceolate; hairs of branchlets and leaves sparkling under strong light or matt; conflorescence with 10–24 flowers; early buds usually silver-grey with a fawn limb, rarely brown all over (coastal, from L. Macquarie, N.S.W., to south-eastern Qld) **159. G. humilis**

86: Leaves ⩽ 1.3 mm wide, linear or very slightly oblanceolate; hairs of branchlets and leaves matt; conflorescence usually with 6–14 flowers; early buds rusty brown all over (Mossvale to lower Hunter, N.S.W.) **164. G. parviflora**

Speciosa Subgroup

Unit conflorescence a loose broadly secund to irregular cluster, erect to decurved. Floral rachis 1–17 mm long. Tepals remaining coherent over at least the basal third, independently recoiled above (usually the ventral pair more strongly so). Pistil (13–) 18–36 mm long. Pollen-presenter oblique. Follicle colliculose to smooth, not ridged.

Five species, Sydney Basin (N.S.W.) and Grampians (Vic.). Bird pollinated.

147. Grevillea speciosa (Knight) McGill., *Telopea* 1: 24 (1975)

Lysanthe speciosa Knight, *Cult. Prot.* 118 (1809). T: Port Jackson [N.S.W.], *s.d.*, [*W.*]*Paterson*; holo: BM; ?iso: G *n.v.*

G. speciosa (Knight) McGill. subsp. *speciosa*, *sensu* D.J.McGillivray & R.O.Makinson, *Grevillea* 329–332 (1993).

Embothrium sericeum var. *major* Sm., *Spec. Bot. New Holland* 3: 27, t. 9, fig. 5 (1794), as β *major*; *G. punicea* R.Br., *Trans. Linn. Soc. London* 10: 169 (1810). T: New South Wales, 1793, *Mr* [*J.*] *White*; lecto: LINN, *fide* D.J.McGillivray & R.O.Makinson, *Grevillea* 442 (1993); probable isolecto: G, PH *n.v.*

G. dubia R.Br., *Trans. Linn. Soc. London* 10: 169 (1810). T: Port Jackson towards the entrance of Lane Cove, [N.S.W.], *R.Brown Iter Austral. 3332*; lecto: BM, *fide* D.J.McGillivray & R.O.Makinson, *Grevillea* 442 (1993); isolecto: K, ?NSW.

G. punicea var. *crassifolia* A.A.Hamilton, *Proc. Linn. Soc. New South Wales* 45: 261 (1920). T: Gosford, on road to Wiseman's Ferry, N.S.W., Jan. 1916, *A.A.Hamilton*; lecto: NSW 94360 *p.p.*, *fide* D.J.McGillivray & R.O.Makinson, *Grevillea* 442 (1993); isolecto: K, NSW94360 p.p., NSW117348; remaining syntypes: K, NSW94354, NSW94356.

Illustrations: N.C.W.Beadle *et al.*, *Fl. Sydney Reg.* t. 10 (1972), as *G. punicea*; D.J.McGillivray & R.O.Makinson, *Grevillea* 329 (1993), as *G. speciosa* subsp. *speciosa*; P.M.Olde & N.R.Marriott, *Grevillea Book* 3: 177 (bottom centre & 144A), 178 (144B, C) (1995).

Erect shrub 0.4–3 m tall. Branchlets angular, subsericeous to villous. Leaves obovate to almost round or elliptic to narrowly so, 1–4 cm long, 4–12 (–19) mm wide; upper surface punctate; margins shortly refracted; lower surface exposed, subsericeous or pubescent. Unit conflorescence terminal on a short lateral branchlet or rarely axillary, shortly pedunculate, decurved, hemispherical-subsecund to loosely globose, many-flowered. Flowers acroscopic. Flower colour: perianth and style bright red or rarely pink, very rarely cream. Perianth subsericeous to tomentose outside, bearded inside. Pistil 25–35 mm long; style gently curved, with scattered minute erect simple hairs over apical 3–15 mm; pollen-presenter oblique. Follicle narrowly ellipsoidal to narrowly ovoid, 12–20 mm long, colliculose. *Red Spider Flower.*

Occurs in N.S.W. in the northern half of the Sydney Basin, from just S of Port Jackson N to Gosford, Kulnura and Bucketty; one early record from Port Stephens is unconfirmed. Grows in moister areas of heath and low sclerophyll woodland or forest, in sandy soils on sandstone. Regenerates from seed. Flowers mainly July–Oct., sporadic in other months. Map 185.

N.S.W.: Mangrove Mtn, *E.F.Constable NSW19368* (NSW); Pearl Beach, *M.Craig CBG012604* (CANB); 2.5 km N of Wisemans Ferry on road to Ten Mile Hollow, *L.A.S.Johnson NSW96980* (NSW); Park Hill Reserve, North Head, *J.Pulley 56* (CANB); Flat Rock Ck gully, *J.Seur 418* (NSW).

Similar and very closely related to *G. oleoides*, which has longer leaves 5–14 mm long, a subsessile to shortly pedunculate conflorescence with peduncle to 5 mm long (2–10 mm long in *G. speciosa*), and the inflorescence usually held within the foliage. Specimens apparently intermediate between *G. speciosa* and *G. oleoides* are known from what are now the eastern and southern suburbs of Sydney. *Grevillea dimorpha* is also very similar, but has consistently axillary conflorescences held very close to the stems, often with fewer flowers and these abaxially oriented. Some variants of *G. victoriae* are also similar, but have pistils < 25 mm long, usually larger leaves, and the midvein on the lower surface of the leaf densely hairy (glabrous or nearly so in *G. speciosa*). Occasional hybrid swarms with *G. sericea* subsp. *sericea* occur in areas of sympatric occurrence.

The variant named as var. *crassifolia* A.A.Hamilton occurs in the Gosford area and sporadically elsewhere; it has rotund leaves with the margins refracted about a thick, prominent intramarginal vein; such plants occur in mixed stands with plants with a less thick-leaved appearance, and there is little justification for formal recognition except as a phenotype. The 'Maroota form', occurring from north of Dural to Wisemans Ferry, has relatively short and narrow elliptic leaves, with pistils also at the lower end of the range of length; these features are suggestive of possible past gene-flow with the geographically disjunct *G. oldei* or a related taxon, and resemble a known hybrid swarm of *G. speciosa* × *G. oldei* from near Kulnura.

148. Grevillea oleoides Sieber ex Schult. & Schult.f., *Mant.* 3: 277 (1827)

G. speciosa subsp. *oleoides* (Sieber ex Schult. & Schult.f.) McGill., *New Names Grevillea* 14 (1986). T: 'Sieber Herb. nov. Holl. Nro. 35' [protologue], lecto: Fl. Novae Holl. [N.S.W.], *s.d.*, *F.W.Sieber 35*; lecto: NSW, *fide* R.O.Makinson, *Fl. Australia* 17A: 497 (2000); isolecto: A *n.v.*, B *n.v.*, BM, G, G-DC, K *p.p.?*, LE *n.v.*, NY *n.v.*, P.

G. seymouriae Sweet ex Meisn., *in* A.L.P.P. de Candolle, *Prodr.* 14: 354 (1856). T: 'circa George's River (Mac Arthur, n. 214!)' [protologue]; lecto: George's R. [N.S.W.], *s.d.*, [*J.*] *McArthur 214*; lecto: NY *n.v.*, *fide* D.J.McGillivray, *Telopea* 1: 28 (1975).

Illustrations: E.R.Rotherham *et al.*, *Fl. & Pl. New South Wales & S Queensland* 43, t. 96 (1975); D.J.McGillivray & R.O.Makinson, *Grevillea* 330 (1993), as *G. speciosa* subsp. *oleoides*; P.M.Olde & N.R.Marriott, *Grevillea Book* 3: 63 (top right), 64 (46A–C) (1995).

Erect shrub 0.5–3 m tall. Branchlets angular, glabrous to pubescent or subsericeous. Leaves obovate to narrowly elliptic or sublinear, (4–) 5–14 cm long, 2–20 mm wide; upper surface faintly punctate, wrinkled; margins recurved to revolute; lower surface subsericeous to tomentose. Unit conflorescence terminal on very short lateral branchlets or occasionally axillary, usually enclosed within foliage, subsessile or with peduncles < 5 mm long, umbelloid-subsecund, (5–) 12–16-flowered. Flowers acroscopic. Flower colour: perianth and style (including style-end) red or deep reddish pink, rarely pale pink. Perianth openly subsericeous to -tomentose outside, bearded inside. Pistil 27–36 mm long; style gently curved, with scattered minute erect hairs or papillae in apical half and behind pollen-presenter; pollen-presenter oblique. Follicle ellipsoidal, 10–15 mm long, colliculose. Plate 36.

Occurs in N.S.W. in the southern half of the Sydney Basin, from Botany Bay and Georges R. S to the northern Illawarra, with isolated populations to the W on the ranges near Wentworth Falls and Robertson. Grows in open heath and in shrubby eucalypt woodland and forest, in skeletal sandy soils; frequently in streamside sites. Regenerates from seed and sometimes rhizomes. Flowers mainly Aug.–Nov., sporadically in other months. Map 186.

N.S.W.: The Pheasants Nest, Nepean R., 16 km S of Camden, *E.F.Constable 6190* (AD, K, NSW); Georges R. crossing on the Wedderburn to Campbelltown road, *R.G.Coveny 15527 & J.Dalby* (B, CHR, HO, K, NBG, PERTH, RSA); Georges R., Woolwash, Campbelltown, *E.McBarron 9137* (NSW); near Avon Dam via Bargo, *R.Pullen 2163* (CANB); Robertson, c. 16 km E of Moss Vale, *J.Pulley 1280* (NSW).

Narrowly distinct from *G. speciosa* (see under that species for differences); the two taxa (with *G. dimorpha*) were treated as subspecies of *G. speciosa s. lat.* by McGillivray (*New Names Grevillea* 14 (1986)). *Grevillea dimorpha* has similar leaves but has the conflorescences almost always axillary or cauline, with a shorter pistil 21–26 mm long, and a more prominent ovary (ovary scarcely wider than style in *G. oleoides*). Some variants of *G. victoriae* are similar, but have a longer floral rachis 5–65 mm long, shorter pistil < 25 mm long, and abaxial leaf midrib densely hairy (in *G. oleoides* the rachis is 2–5 mm long, and the midrib glabrous or nearly so).

Grevillea oleoides shows great variation in leaf form and size. The 'riparian form' phenotype occurs in streamside (often flood-zone) habitats, and has narrowly elliptic to sublinear leaves usually 1–5 mm wide with a subsericeous lower surface; habit varies from compact and c. 0.5 m tall (in some populations genetically fixed) to spindly and 2–3 m tall. The 'non-riparian form' phenotype occurs in heathy associations on flat ground and in ridge-top woodland and forest; its has generally broader leaves than the previous form but of variable size and shape, sometimes with a tomentose lower surface; habit is variable from low and compact to erect and spindly to 3 m, often with the leaves crowded at the branch apices.

149. **Grevillea dimorpha** F.Muell., *Trans. Philos. Soc. Victoria* 1: 21 (1855)

G. oleoides subsp. *dimorpha* (F.Muell.) Benth., *Fl. Austral.* 5: 469 (1870); *G. speciosa* subsp. *dimorpha* (F.Muell.) McGill., *New Names Grevillea* 14 (1986). T: 'In the Grampians, Serra- & Victoria Ranges' [protologue], [Vic., *s.d., F.Mueller*]; lecto: MEL, *fide* D.J.McGillivray & R.O.Makinson, *Grevillea* 442 (1993); isolecto: MEL; remaining syntypes: K, MEL, NSW.

G. dimorpha var. *latifolia* F.Muell., *Trans. Philos. Soc. Victoria* 1: 21 (1855), as *a, latifolia, nom. illeg.*; *G. dimorpha* var. *lanceolata* Meisn., *in* A.L.P.P. de Candolle, *Prodr.* 14: 353 (1856), as α *lanceolata, nom. illeg.* T: 'In the Grampians, Serra & Victoria Ranges' [protologue], [Vic., *s.d., F.Mueller*]; lecto: MEL (same specimen as lectotype of *G. dimorpha*), *fide* D.J.McGillivray & R.O.Makinson, *Grevillea* 442 (1993); isolecto: MEL; remaining syntypes: K, MEL, NSW.

G. dimorpha var. *angustifolia* F.Muell., *Trans. Philos. Soc. Victoria* 1: 21 (1855), as *b, angustifolia*; *G. dimorpha* var. *linearis* Meisn., *in* A.L.P.P. de Candolle, *Prodr.* 14: 353 (1856), as β *linearis, nom. illeg.* T: Mt Abrupt [Vic.], *s.d.*, Dr M [*F.Mueller*]; lecto: MEL, *fide* D.J.McGillivray & R.O.Makinson, *Grevillea* 442 (1993); isolecto: MEL; remaining syntypes: K, MEL.

Illustrations: L.F.Costermans, *Native Trees & Shrubs of SE Australia* 160 (1981); D.J.McGillivray & R.O.Makinson, *Grevillea* 331, col. pl. (1993), as *G. speciosa* subsp. *dimorpha*; P.M.Olde & N.R.Marriott, *Grevillea Book* 2: 127 (top right & 102A), 128 (102B, C) (1995).

Erect to spreading shrub 0.4–3 m tall, rarely decumbent to prostrate. Leaves linear or narrowly to broadly elliptic or obovate, 5–15 cm long, 1.5–20 (–40) mm wide; upper surface obscurely punctate; margins recurved to revolute; lower surface subsericeous, occasionally mostly enclosed. Conflorescence usually axillary or occasionally cauline, or rarely terminal on short lateral branchlets, usually decurved, sessile or very shortly pedunculate, a loose irregular to subsecund cluster, 2–16-flowered, opening indeterminate. Flowers abaxially oriented on twisted or retrorse pedicels. Flower colour: perianth and style bright red.

Perianth subsericeous to openly tomentose outside, bearded inside. Pistil 21–26 mm long; style gently curved, usually with scattered minute erect hairs near tip; pollen-presenter strongly oblique. Follicle ellipsoidal, 12–16 mm long, colliculose. *Flame Grevillea, Olive Grevillea.* Plate 33.

Occurs in Vic., where endemic to the Grampians Range area. Grows in heathy woodland and forest, in well-drained sites in sandy or light loam soils. Regenerates from seed. Flowers mainly Aug.–Dec. Map 187.

Vic.: Victoria Valley, Grampians, *A.C.Beauglehole 7767 & L.G.Dale* (NSW); Mt Abrupt, S end of Serra Ra., *A.C.Beauglehole 30200 & B.A.Fuhrer* (MEL); Mirranatwa Gap, Grampians, 29 Aug. 1958, *L.A.S.Johnson NSW47218* (NSW); Teddy Bear Gap, Serra Ra., *R.Melville 1841 et al.* (MEL); foot of Mt Sturgeon, *A.N.Rodd NSW130792* (NSW).

Grevillea dimorpha was reduced by McGillivray (*New Names Grevillea* 14 (1986), see also McGillivray & Makinson *Grevillea* 330–332 (1993)) to one of three subspecies of *G. speciosa*, with *G. oleoides* similarly treated (both the latter are Sydney Basin endemics). *Grevillea dimorpha* possesses a narrow suite of distinctive character-states and is geographically very disjunct from its sister taxa.

Grevillea dimorpha is easily distinguished from *G. speciosa* on leaf length and inflorescence position; *G. dimorpha* also has the ovary conspicuously wider than the stipe and style (scarcely so in *G. speciosa*), and a more oblique pollen-presenter. *Grevillea oleoides* is more easily confused, but has a longer pistil 28–36 mm long, the ovary scarcely broader than the style, and usually more flowers (5–16) per unit conflorescence.

Grevillea dimorpha shows considerable variation in habit and leaf form, and phenotypes with varying degrees of populational integrity can be distinguished. A 'narrow-leaved form' with sublinear leaves 1–3 mm wide occurs in the south of the range from Yarram Gap to Mt Abrupt; the leaf form is not associated with riparian habitat. Broad-leaved plants with leaves 15–40 mm wide are widespread in southern and central parts of the range. A 'robust form' growing 2–4 m tall occurs in the Fyans Ck area; it has greyish leaves (dark green in all other populations). A 'decumbent form' occurs in the Jimmy Ck area.

150. Grevillea molyneuxii McGill., *New Names Grevillea* 10 (1986)

T: on old timber track c. 7 km SE of Wingello, N.S.W., 23 Sept. 1973, *T. & J.Whaite 3541*; holo: NSW.

Illustrations: A.Fairley & P.Moore, *Nat. Pl. Sydney District* 169, pl. 561 (1989); P.M.Olde & N.R.Marriott, *Grevillea Book* 3: 34 (top right & 22A), 35 (22B) (1995).

Low spreading to weakly erect shrub 0.2–0.8 (–1.0) m tall. Branchlets angular, subsericeous. Leaves linear or narrowly elliptic, 1.5–4.5 cm long, 1–3 (–5) mm wide, pungent or not; upper surface smooth, not punctate; margins angularly recurved to refracted; lower surface subsericeous, sometimes enclosed except for midvein. Conflorescence terminal, erect to deflexed, pedunculate (peduncles 3–15 mm long), shortly secund, 5–15-flowered. Flowers acroscopic. Flower colour: perianth and style red. Perianth loosely subsericeous outside, bearded inside. Pistil 18–21 mm long; style glabrous except for scattered minute erect hairs in the apical 3–4 mm; pollen-presenter oblique. Follicle ovoid-ellipsoidal, 12–14 mm long, smooth to colliculose.

Occurs in N.S.W., where restricted to a few sites near Wingello ENE of Goulburn. Grows in moist heath and shrub associations in skeletal sandy soils over sandstone pavement. Regenerates from seed. Bird-pollinated. Flowers mainly Aug.–Nov. Map 188.

N.S.W.: Tallong, *E.Cheel NSW94420* (NSW); between Wingello State Forest & Moreton Natl Park, end of Bull Ridge Rd, Tallowa Gully, *P.Gilmore 5240* (CANB); off Gap Ridge Rd, Wingello, 2 Oct. 1984, *W.Molyneux* (NSW); 10 km ESE of Marulan, *S.N.Panter & G.Fulton 1* (CANB); Teudts Lookout, at end of Teudts Rd, c. 7 km SW of Bundanoon, *K.E.Thiele 2512* (CANB).

Grevillea molyneuxii has the tepals more deeply separated and more evenly recurved than other species in the subgroup, and in this feature resembles taxa of the *Linearifolia* subgroup. It has some similarity to *G. juniperina*, which has terete branchlets and the leaf abaxial midvein evident only at the base of the leaf or not at all. *Grevillea sericea* has pink or white flowers, pedicels 4–9 mm long (2.5–3.5 mm long in *G. molyneuxii*), usually more flowers per unit conflorescence, pistils 14–19 mm long, and leaves either broader or longer.

This species is recognised as 'Poorly Known' in J.D.Briggs & J.H.Leigh, *Rare or Threatened Australian Plants* (1995).

151. Grevillea juniperina R.Br., *Trans. Linn. Soc. Bot.* 10: 171 (1810)

T: about 7 miles [11.2 km] NW. ½ N from Prospect, N.S.W., Oct. 1803, *G.Caley*; lecto: BM, *fide* D.J.McGillivray & R.O.Makinson, *Grevillea* 423 (1993); isolecto: A *n.v.*, E *n.v.*; remaining syntype: Port Jackson, *s.d.*, *A.Gordon*; syn: ?BM *n.v.*

Prostrate to erect shrub 0.2–2 (–4) m high, to 3 m across; branches sometimes appearing columnar. Branchlets terete, tomentose to villous. Leaves spreading to ascending, often crowded on short lateral branchlets, usually rigid, narrowly ovate to subulate or linear, 5–35 mm long, 0.5–6 mm wide, usually pungent; upper surface usually with 3–5 longitudinal veins, glabrous or nearly so; margins angularly or rarely smoothly recurved to revolute, rarely ±flat; lower surface often enclosed or almost so, usually densely sericeous or occasionally openly so, rarely glabrous, or open-tomentose. Conflorescence terminal, occasionally also axillary and subterminal, usually simple or occasionally 2 (–4)-branched; unit conflorescence erect or slightly decurved, acropetal, subsecund. Perianth densely to openly subsericeous to tomentose outside, rarely subvillous especially on limb, bearded inside between 2.5 and 9 mm above base. Pistil (15–) 18–27 mm long; style glabrous except for minute scattered erect simple hairs extending from back of style-end down at least 3 mm and sometimes almost to ovary; pollen-presenter usually oblique or occasionally lateral. Follicles narrowly ovoid or oblong-ellipsoidal, 10–18 mm long.

Occurs from southern Qld through eastern N.S.W. to the Southern Tablelands; seven subspecies are recognised.

Despite the number of subspecies and their variation, *G. juniperina* is not easily confused with other taxa. See under *G. molyneuxii* for distinctions from that species. *Grevillea juniperina* has a distinctive leaf lower surface (shared with *G. australis*): the midvein is equally as hairy as the adjacent lamina, and usually decreases markedly in prominence from the base towards the apex of the leaf.

Grevillea juniperina has been an important plant in ornamental horticulture, as selections and as a parent of many hybrids.

1 Limb at apex of flower-bud tomentose or shaggy-subvillous (±spreading hairs); outer surface of perianth tomentose (ascending hairs); lower leaf surface glabrescent or with only a sparse indumentum of strongly ascending hairs, concentrated along midvein **151f.** subsp. **villosa**

1: Limb at apex of flower-bud subsericeous (appressed hairs); outer surface of perianth subsericeous to open-appressed; lower leaf surface densely to openly subsericeous (to almost glabrous)

2 Outer surface of perianth with few to many minute erect simple hairs (c. 0.1–0.2 mm long; microscope feature) mixed in with much larger biramous (2-armed) hairs

3 Prostrate to spreading shrub 0.3–0.8 (–1.2) m tall; adult and juvenile leaves both with a dense appressed indumentum over the whole lower surface; juvenile leaves ⅓–½ as long as adult leaves (grows on slopes and ridges usually away from streams) **151d.** subsp. **amphitricha**

3: Ascending to erect shrub with erect columnar branches to 2 m tall, or rarely prostrate; adult leaves with a sparse to dense indumentum on lower surface; juvenile leaves as long as adult leaves, with a sparse to open indumentum beneath (grows in moist creek-side and swamp habitats) **151g.** subsp. **sulphurea**

2: Outer surface of perianth lacking minute erect hairs (biramous hairs only present)

4 Adult leaves elliptic to narrowly elliptic or narrowly ovate; margins angularly and very shortly recurved, with most of lower leaf surface exposed; upper surface ±planar except for 3 (or 5) ±prominent pale veins

5 Adult leaves 10–20 mm long, 1.5–2.5 mm wide; spreading shrub 0.5–2 m high; flowers usually yellow or orange, rarely red; innermost pair of lateral veins on upper leaf surface joining the midvein abruptly at ±90° at extreme base of leaf **151b.** subsp. **trinervis**

5: Adult leaves (10–) 15–35 mm long, 2–4.5 (–6) mm wide; prostrate shrub or rarely low spreading shrub to 1.2 m tall; flowers usually red, rarely yellow; innermost pair of lateral veins on upper leaf surface joining midvein at acute angle just above leaf base **151c.** subsp. **allojohnsonii**

4: Adult leaves linear, sublinear, or acicular-subulate; margins ±strongly recurved to revolute, concealing much or all of lower leaf surface; upper surface angularly ridged (leaf trigonous or deltoid in cross-section) or smoothly convex, rarely planar on adult foliage, with venation variably conspicuous to obscure; weakly ascending to spreading or erect shrubs to 3 m tall

6 Widest adult leaves < 1 mm broad

7 Upper leaf surface with three prominent longitudinal veins; adult leaf angularly deltoid to trigonous in cross-section, not adaxially convex-rounded; leaf margins angularly revolute or refracted; low spreading shrub to 1 (rarely 1.5) m tall **151a.** subsp. **juniperina**

7: Upper leaf surface with only the midvein evident or if intramarginal veins evident then the latter not prominent; adult leaf usually rounded (adaxially convex) in cross-section or sometimes angularly deltoid; leaf margins smoothly to angularly revolute; ascending to erect shrub with columnar branches to 2 m tall, or rarely prostrate **151g.** subsp. **sulphurea**

6: Widest adult leaves ⩾ 1 mm broad

8 Robust shrub 1–3 m tall, with strong erect central stem at base and many spreading to ascending lateral branches; leaves angularly deltoid to trigonous in cross-section (not smoothly convex); upper leaf surface with 1–3 prominent longitudinal veins; lower surface with a dense appressed indumentum; flowers red or rarely pink **151e.** subsp. **fortis**

8: Sprawling or weakly erect shrub with main branches spreading at ground level and then ascending or erect-columnar, to 2 m tall; leaves usually markedly rounded-convex in cross-section; upper leaf surface with 1–3 veins visible but usually scarcely prominent; lower surface with a dense to sparse appressed indumentum; flowers dull yellow or orange **151g.** subsp. **sulphurea**

151a. Grevillea juniperina R.Br. subsp. juniperina

More or less erect to spreading dense divaricate shrub 0.5–1.5 m tall; major branches appearing subcolumnar (leaves clustered on short lateral branchlets); foliage dense. Adult leaves often dark green with paler veins, usually narrow, needle-like, 10–22 mm long, 0.6–0.8 mm wide, angularly deltoid to trigonous in cross-section; midvein and intramarginal veins usually very prominent; upper surface with appressed hairs; margins strongly and angularly revolute; lower surface usually fully enclosed; juvenile leaves scarcely broader than adults. Flower colour: perianth red, yellow, pale orange, or rarely greenish; style similar to perianth or a little paler. Perianth subsericeous outside with biramous hairs only. Pistil (13–) 20–25 mm long.

Occurs in the N.S.W. Central Coast botanical district, in the far western suburbs of Sydney, from Marsden Park, Riverstone, Blacktown and Penrith and N to the Pitt Town area. Grows in open dry sclerophyll (eucalypt-dominated) forest or woodland, at altitudes of less than about 50 m, in sandy to clay-loam soils and red pseudolateritic gravels. Regenerates from seed only. Flowers mostly Aug.–Sept., sporadically in other months. Map 189.

N.S.W.: Pitt Town area, c. 7.9 km (direct) ENE of Windsor, *R.O.Makinson 1286 & D.J.McGillivray* (CANB, MEL, NSW); 1.6 km E of St Marys, 22 Dec. 1955, *L.A.S.Johnson & E.F.Constable* (K, NSW); St Marys, 43.2 km W of Sydney on the Great Western Hwy, *R.Melville 622 & L.A.S.Johnson* (K, MEL, NSW); between Windsor & Kingswood, S of Blacktown road, 3 May 1967, *M.E.Phillips* (CANB, NSW).

This subspecies shows some ability to colonise mechanically disturbed areas where open ground surface persists; repeated disturbance seems to eliminate it. Populations are often restricted to infrequently managed road verges or ungrazed semi-cleared land. In the Pitt Town area ENE of Windsor, an apparent hybrid swarm of *G. juniperina* × *G. mucronulata* is known.

151b. **Grevillea juniperina** subsp. **trinervis** (R.Br.) Makinson, *Fl. Australia* 17A: 497 (2000)

G. trinervis R.Br., *Suppl. Prodr. Fl. Nov. Holl.* 18 (1830). T: near Port Jackson, N.S.W., [?]*Fraser*; lecto: BM, *fide* D.J.McGillivray & R.O.Makinson, *Grevillea* 424 (1993): isolecto: BM; remaining syntypes: 'Country west from Bathurst in the vale at the source of Bell's River Nov. 1822 C. Fraser spcn from J. Smith'; syn: K; 'Bed of the Fish river near old fish R. bridge and in the vale at the source of the Bell's river N.W. from Batht [Bathurst] very rare Fraser'; syn: K.

Spreading low shrub 0.5–1.2 (–2) m tall, or rarely prostrate; branches spreading, not columnar, the leaves usually clustered on short lateral branchlets. Adult leaves often dark green with contrasting pale veins, narrowly ovate to narrowly elliptic, 10–20 mm long, 1.5–2.5 mm wide, ±flat with the upper surface venation prominent and conspicuous with 3 (–5) longitudinal veins; margins shortly recurved or shortly refracted; lower surface exposed, densely subsericeous; juvenile leaves broader (to 4 mm wide) and flatter than adult leaves, with scattered hairs below. Flower colour: perianth and style dull orange to yellow or sometimes red; style-end and pollen-presenter straw-coloured to green. Perianth subsericeous outside, with biramous hairs only. Pistil 21–26 mm long.

Occurs in the Central Tablelands of N.S.W., western Blue Mtns at higher altitudes from Hampton and Bindo Ck area S to Jenolan State Forest and Boyd Plateau, NW historically perhaps to Orange area. Flowers mostly Aug.–Dec., sporadically later. Map 190.

N.S.W.: Whalania Heights, c. 600 m NNW of head of Belarah Swamp, Boyd Plateau *D.H.Benson 2245 & D.Keith* (NSW); Jenolan Caves, Dec. 1899, *W.F.Blakely* (NSW); near Hampton on the Oberon road, 26 Nov. 1948, *L.Fraser NSW\94478* (NSW); Merrong Falls, Kanangra Boyd Natl Park, *R.O.Makinson 1761* (CANB, K, MEL, NSW); Bindo Ck, E of Oberon *R.O.Makinson 1578* (CANB, MEL, NSW).

Subsp. *trinervis* as here defined is restricted to the N.S.W. Central Tablelands. Populations are sporadic and it appears to be quite rare. It intergrades with subsp. *sulphurea*, and specimens are not always clearly assignable. In general, subsp. *trinervis* has less densely leafy and less columnar branches compared to subsp. *sulphurea*, and adopts a more spreading habit; the leaves have 3–5 usually distinct veins on the upper (adaxial) surface, and are narrowly elliptic or narrowly ovate with scarcely recurved margins (cf. subsp. *sulphurea* with leaves linear or subulate, venation obscure and margins markedly refracted to revolute). The epithet '*trinervis*' has been widely applied (at species or varietal rank) to the taxon recognised here as subsp. *allojohnsonii*. *Grevillea oldei* has in the past also been misidentified as *G. trinervis*.

151c. **Grevillea juniperina** subsp. **allojohnsonii** Makinson, *Fl. Australia* 17A: 497 (2000)

T: c. 30 km (direct) NE of Guyra, Pheasant Mtn (near Backwater), N.S.W., 17 Nov. 1993, *R.O.Makinson 1449, I.R.Telford & J.Nightingale*; holo: CANB; iso: K, NSW.

Prostrate or decumbent shrub to 0.3 m high, or rarely a weakly erect-spreading shrub to 1.2 m high; branches spreading, with leaves in short clusters or not. Adult leaves dark green with contrasting pale veins, narrowly ovate to narrowly elliptic, (10–) 15–35 mm long, 2–4.5 (–6) mm wide, ±flat with the upper surface venation prominent and conspicuous with 3–5 longitudinal veins; margins shortly refracted; lower surface exposed, densely subsericeous; juvenile leaves not seen. Flower colour: perianth and style scarlet (rarely salmon pink or yellow), with a green style-end. Perianth openly subsericeous outside, with biramous hairs only. Pistil 21–27 mm long.

Occurs on the Northern Tablelands and North West Slopes of N.S.W., widespread but uncommon from Walcha area to Tingha and N to Tenterfield, extending to SE Qld in Girraween Natl Park and near Stanhope. Regenerates from seed only. Flowers mainly Sept.–Feb. Map 191.

Qld: 5 km SE of Stanthorpe on Quart Pot Ck, *C.Hockings*, 20 Feb. 1989 (BRI, NSW). N.S.W.: Glen Elgin, 3 Feb. 1930, *J.W.H[aney] s.n.* (CANB); c. 14 km W along fire trail from forestry hut in Carrai State Forest, *R.Coveny 408 & D.J.McGillivray* (NSW); c. 10 km from turn-off road Inverell to Tingha, 25 km SSE from Inverell, *J.Seur 378* (NSW); Road to Boonoo Boonoo Falls, NE of Tenterfield, *I.R.Telford 2543* (CANB).

This subspecies is very similar to subsp. *trinervis*, but most populations have a consistently prostrate habit and red flowers (subsp. *trinervis* is usually a spreading shrub with yellow flowers). Subsp. *allojohnsonii* also has rather broader and often longer leaves, and the inner pair of lateral veins on the adaxial leaf surface are usually more obviously curved, and tend to meet the midvein at an acute angle slightly above the extreme base of the lamina. In subsp. *trinervis* these veins are straighter and join the midvein almost at 90° at the very base of the leaf lamina. The venation characters are difficult to observe, and the two taxa are clearly very closely related.

151d. Grevillea juniperina subsp. amphitricha Makinson, *Fl. Australia* 17A: 498 (2000)

T: Southern Tablelands, Windellama Cemetery, N.S.W., 30 Nov. 1954, *C.W.E.Moore 3004*; holo: CANB.

Prostrate to sprawling shrub 0.2–1.2 m tall, to 3 m wide; branches ascending to spreading, not divaricate, not appearing columnar, the leaves occasionally in short clusters. Adult leaves light- to mid-green, sublinear to narrowly elliptic, 8–18 mm long, 1.0–2.5 (–3.5) mm wide; upper surface convex, with venation often obscure with only the midvein evident and scarcely prominent; margins smoothly recurved to revolute; lower surface usually partially exposed, densely subsericeous; juvenile leaves shorter (1/3 to 1/2 length), proportionately wider, and flatter than adult leaves. Flower colour: perianth dull pale yellow to apricot-orange (rarely red), with gold or brownish hairs especially on limb; style dull yellow (rarely red) with greenish style-end. Perianth densely to openly subsericeous with biramous hairs, and a few to many minute erect simple multicellular hairs scattered through. Pistil 20–25 mm long.

Occurs in the N.S.W. Southern Tablelands in the area between Braidwood and Nerriga, with isolated populations in the Windellama area, and (one doubtful record) Tianjara Falls. All populations are within the Shoalhaven R. catchment. Grows in woodland and grassland habitats on slopes and ridges (rarely along creeks), at altitudes of 550–600 m. Regenerates from seed only. Flowers mostly Aug.–Sept., sporadically in other months. Map 192.

N.S.W.: ... near Charleyong (24 km from Braidwood at turnoff to Tarago). *M.E.Phillips 842 & Moore* (CANB); c. 38 km NE of Braidwood along Nerriga road; between Black Bobs Ck and Corang R., *Taylor 428 et al.* (K, NSW); 37 km NE along Nerriga road from Braidwood, 250 m SW of Black Bobs Ck crossing, *R.O.Makinson 1531 et al.* (CANB); Braidwood to Nerriga, 21 km from Braidwood, 28 Oct. 1962, *D.W.Shoobridge s.n.* (A *n.v.*, CANB, L *n.v.*, NSW); Durran Durra [trig.], 13 Oct 1962, *C.Burgess s.n.* (CANB).

There are some indications that this subspecies may occasionally intergrade with subsp. *villosa*, and it definitely intergrades with subsp. *sulphurea* (see under each of these taxa). Red-flowered variants (in predominantly yellow-flowered populations) are known only from the Windellana area in the far N of the range.

151e. Grevillea juniperina subsp. **fortis** Makinson, *Fl. Australia* 17A: 498 (2000)

T: Pine Island, Lanyon district, A.C.T., 28 Oct. 1961, *R.D.Hoogland 8422*; holo: CANB; iso: A *n.v.*, AD, B *n.v.*, BH *n.v.*, BM, BO *n.v.*, BRI, CANB, E *n.v.*, FI *n.v.*, G *n.v.*, K, L *n.v.*, MEL, NE *n.v.*, NSW, P *n.v.*, UC *n.v.*, US *n.v.*

Robust shrub 1–3 m tall; branches ascending to spreading (often arching upwards), appearing columnar (leaves mostly clustered on short lateral branchlets); foliage often in pseudo-whorls. Adult leaves dark green, linear to subulate, 10–20 (–26) mm long, (0.7–) 1.0–1.4 (–2.0) mm wide, angularly deltoid to trigonous (rarely convex-upwards) in cross-section; upper surface with 1–3 prominent longitudinal veins; margins angularly to smoothly revolute; lower surface densely subsericeous, often mostly or wholly obscured by margins; juvenile leaves sometimes slightly broader than adults (1.5–3 mm wide), with strongly revolute margins. Perianth densely to openly subsericeous with biramous hairs only. Flower colour: perianth red or rarely pink (very rarely yellow); style red (rarely pink or yellow). Pistil 19–27 mm long.

Known from the A.C.T. area, where it occurs along Ginninderra Ck, lower reaches of the Molonglo and Cotter R., and the Murrumbidgee R. from the Molonglo confluence upstream to Pine Is. There is also a single record from much further upstream on the Murrumbidgee (11.5 km N of Cooma, *Forbes 597*), and the taxon may occur sporadically in rugged country along the river between the two areas. Grows in open dry sclerophyll forest, woodland, and shrubland, often in rocky situations, along banks or slopes above permanent watercourses, at altitudes of 450–750 m. Regenerates from seed only. Flowers mostly Aug.–Sept., sporadically in other months. Map 193.

N.S.W.: bank of Murrumbidgee R., 11.5 km N of Cooma P.O., *S.J.Forbes 597* (CANB, MEL, NSW). A.C.T.: Pine Is., 16 km S of Canberra, *Evans 2638* (A *n.v.*, AD, CANB, CHR *n.v.*, K, L *n.v.*, NSW); Ginninderra Gorge, Kilby Bros. property, 11 Oct. 1967, *I.R.Telford* (CANB); Pine Is., Murrumbidgee R., below picnic area, *J.E.Ward 29 & A.Hughes* (CANB, MEL).

151f. Grevillea juniperina subsp. **villosa** Makinson, *Fl. Australia* 17A: 498 (2000)

T: Southern Tablelands, banks of Corang River, at crossing on Braidwood to Nerriga road, c. 35 km (direct) NNE of Braidwood, N.S.W., 31 Aug. 1994, *R.O.Makinson 1533 et al.*; holo: CANB; iso: NSW.

Robust erect shrub to 2 m high; branches arching to erect, columnar; foliage often in pseudo-whorls with leaves clustered. Adult leaves dark green with paler veins, sublinear to narrowly elliptic, 11–21 mm long, (0.7–) 0.9–1.8 (–2.6) mm wide; upper surface convex, with midvein and intramarginal veins variably conspicuous; margins smoothly recurved to refracted; lower surface usually partially exposed, openly to sparsely subvillous, especially along midvein, or sometimes glabrous; juvenile leaves with margins scarcely recurved. Flower colour: perianth and style red or yellow; style-end and pollen-presenter sometimes paler red or yellow, or green. Perianth tomentose to subvillous outside; limb of bud conspicuously shaggy-subvillous or tomentose. Pistil 18–25 mm long.

Occurs in N.S.W. in the Southern Tablelands, known only from an area E and NE of Braidwood, and one old record from Currockbilly. Grows beside permanent watercourses in riparian shrub associations in remnant eucalypt forest, at altitudes of 550–600 m. Regenerates from seed only. Flowers mostly Aug.–Sept., sporadically in other months. Map 194.

N.S.W.: Mongarlowe R. near Charlies Forest, Sept. 1915, *J.Boorman s.n.* (NSW); Mongarlowe R. 3.3 km S of Mongarlowe towards Clyde Mtn to Braidwood road, *R.Coveny 11032 & T.James* (K, NSW); banks of Corang R., at crossing on Braidwood to Nerriga road, *R.O.Makinson 1534 et al.* (CANB, NSW); 3.2 km S of Mongarlowe on road from Braidwood, *J. de Nardi 538* (NSW); Nettletons Ck, 8.5 km NE of Mongarlowe, *I.R.Telford 9623 & H.Hadobas* (CANB).

Normally a single plant bears flowers of only one colour, but *Fox & Cowley 88/009* (CANB) records flowers as 'creamish yellow although flowers with red perianths occur on same bush'. Another collection (*Boorman*, Sept. 1915, NSW) from Mongarlowe R. near Charlies Forest, records flowers as 'pink and a few white', but these colour variants have not been confirmed. This subspecies may rarely intergrade with subsp. *amphitricha*.

crimson to burgundy; perianth beard white and conspicuously displayed at anthesis. Style with papilloid hairs confined to apical c. 1 mm.

Occurs in N.S.W., south of Sydney in the area bounded by Menai, Heathcote, Appin, and Cordeaux Dam. Grows in dry heath or shrubby woodland or open forest, usually on low ridges and slopes in sandy to loamy soils. Regenerates from seed. Flowers July–Dec. Map 196.

N.S.W.: Woronora Dam catchment area, c. 6 km NNW of Darkes Forest, *R.Coveny 9517* (K, NSW); 5.1 km N of Menai turnoff on Heathcote Rd, *R.Coveny 11154 & J.Thomas* (K, NSW); Woronora R., Sept. 1897, *W.Forsyth* (K, NSW); 5 km E of Leumeah, 6 Nov. 1948, *L.A.S.Johnson NSW94408* (NSW); [Royal] Natl Park at top of falls, 2 Nov. 1954, *M.Tindale NSW29863* (K, NSW).

Subsp. *diffusa* as here defined corresponds to *G. diffusa* subsp. *diffusa* 'short-leaved form' of McGillivray & Makinson (*Grevillea* 334 (1993)) and Olde & Marriott (*op. cit.* 123). In the east of the range, around Waterfall, subsp. *diffusa* intergrades to a limited degree with subsp. *constablei*, with leaf lengths overlapping. Distinguished from *G. sericea* which has secund conflorescences with (in the area of sympatry) pink flowers, and leaves often obovate and arising in clusters of 3 (solitary in *G. diffusa*). Several instances of hybridisation of subsp. *diffusa* with *G. sericea* are known; the resulting plants usually have loose, weakly secund conflorescences, and solitary elliptic leaves < 4 cm long.

152b. **Grevillea diffusa** subsp. **constablei** Makinson, *Fl. Australia* 17A: 499 (2000)

T: c. 1.5 miles [2.4 km] from waterfall on E side of line, N.S.W., 2 July 1950, *T.M.Whaite 722*; holo: NSW.

Illustration: P.M.Olde & N.R.Marriott, *Grevillea Book* 2: 124 (98C) (1995).

Open spreading to erect shrub 1–2 m tall. Branchlets angular in cross-section and with longitudinal ridges, glabrous or sericeous (rarely pubescent with short upright hairs) between ridges, rarely secund. Leaves narrowly elliptic to sublinear, 4.5–7 (–10) cm long, 2.5–4 (–7) mm wide, green, margins ±smoothly recurved to irregularly revolute. Conflorescence dense to loose, often pendulous below the foliage, usually pedunculate (peduncles angular, 10–15 mm long, usually 0.6–1 mm diam., sericeous to almost glabrous, often flexuose). Flower colour: perianth and style deep crimson to burgundy; perianth beard white and conspicuously displayed at anthesis. Style with papilloid hairs scattered over apical 1–4 mm.

Occurs in N.S.W., south of Sydney in the area from Waterfall to Helensburgh and the Georges R.; fairly widespread in Royal Natl Park. Grows in shady eucalypt forest or woodland, often near creeks, in sandy loam soils. Regenerates from seed. Probably insect-pollinated. Flowers Aug.–?Nov. Map 197.

N.S.W.: [Royal] Natl Park, c. 29 km SW of Sydney, 31 Aug. 1948, *E.F.Constable NSW27335* (K, NSW); *loc. id.*, 22 Sept. 1955, *E.F.Constable NSW39916* (K, NSW); *s. loc.?*, *L.Fraser NSW94006* (NSW); Waterfall, 7 Sept. 1949, *P.K.Macnichol s.n.* (CANB); [Royal] Natl Park, S of Sydney, 22 July 1933, *G.Rodway 1133* (K).

In the west of the range, subsp. *constablei* apparently tends towards subsp. *diffusa*, having somewhat shorter leaves and sometimes pubescent rather than sericeous branchlet indumentum. Plants with longer leaves and longer, nearly glabrous peduncles approach very closely the phenotype of the geographically disjunct subsp. *filipendula*. The floral rachis is sometimes curved and the inflorescence noticeably subsecund, suggestive of possible genetic influence from (e.g.) *G. linearifolia* or *G. sericea*; the taxon however seems consistent in most features (apart from a partial intergrade towards subsp. *diffusa*).

152c. **Grevillea diffusa** subsp. **filipendula** McGill., *New Names Grevillea* 4 (1986)

T: between Calga and Mt White, N.S.W., 14 Sept. 1968, *D.J.McGillivray 3097*; holo: NSW; iso: CANB, K *n.v.*

Illustrations: A.Fairley & P.Moore, *Nat. Pl. Sydney District* 170 (1989); D.J.McGillivray & R.O.Makinson, *Grevillea* 335 (1993); P.M.Olde & N.R.Marriott, *Grevillea Book* 2: 124 (lower left & 99A, B) (1995).

Low spreading shrub to 1 m tall. Branchlets angular in cross-section and with longitudinal ridges, glabrous or nearly so (scattered appressed hairs between ridges), usually strongly secund. Leaves very narrowly elliptic to broadly linear, 5–15 cm long, 2–4 mm wide, green, margins refracted. Conflorescence moderately dense, usually pendulous below the foliage, pedunculate (peduncles angular, (10–) 15–40 mm long, c. 0.6 mm diam., glabrous, flexuose, thin and wiry). Flower colour: perianth and style scarlet to light crimson or light burgundy; perianth beard white and conspicuously displayed at anthesis. Style with papilloid hairs usually scattered over apical half.

Occurs in N.S.W., north of Sydney in the area between Calga and Mt White, N of Brooklyn. Grows in open heath or shrubby eucalypt woodland, often in rocky sites, in shallow sandy soils over sandstone. Regenerates from seed. Flowers mainly July–Nov., sporadic in other months. Map 198.

N.S.W.: W of Mt White at Greenmans Valley and Neverfail Rd junction, *R.Coveny 11217 et al.* (K, NSW); Mt White, 15 Oct. 1945, *D.O.Cross NSW3258* (K); Mt White, W of Gosford, *D.O.Cross NSW3259* (NSW); Calga, *L.A.S.Johnson NSW94361* (NSW); 3.2 km W of Mooney Mooney Ck Bridge on Pacific Hwy, *M.Tindale NSW34371* (NSW).

In addition to the key characters, subsp. *filipendula* has reddish branchlets; other subspecies have the branchlets neutral in colour (rarely reddish in subsp. *constablei*) or concealed by the indumentum. Subsp. *filipendula* is reported (Olde & Marriott, *op. cit.* 2: 124) as occasionally hybridising with *G. linearifolia.*

153. Grevillea capitellata Meisn., *in* A.L.P.P. de Candolle, *Prodr.* 14: 356 (1856)

T: 'Circa Port Jackson (A. Cunningh.!) G. diffusa A. Cunn.! mss. in herb. (non Sieber) ... v.s. in herb. DC. et Mart.' [protologue]; lecto: 'barren country near Port Jackson [N.S.W.] Octr. 1818 AC' [A. Cunningham]; lecto: G, *fide* D.J.McGillivray & R.O.Makinson, *Grevillea* 413 (1993); isolecto: BR; remaining syntype: 'barren country nr Illawarra Mountain 1818/Octr in swampy heath, A.Cunningham 53/1818'; syn: K.

G. capitellata Meisn., *Hooker's J. Bot. Kew Gard. Misc.* 5: 187 (1852), *nom. nud.*

G. diffusa subsp. *diffusa* "form 'b', robust form", of D.J.McGillivray & R.O.Makinson, *Grevillea* 334 (1993).

Illustrations: E.R.Rotherham *et al., Fl. & Pl. New South Wales & S Queensland* 58, t. 148 (1975); P.M.Olde & N.R.Marriott, *Grevillea Book* 2: 86 (67A, B) (1995).

Low dense mounded or prostrate shrub to 50 cm tall, with branches often arching. Branchlets angular and ribbed, tomentose to villous or loosely so, usually secund. Leaves narrowly elliptic to oblong-lanceolate, 2–9 cm long, 2–8 mm wide; upper surface sparingly punctate; margins shortly refracted vertically about intramarginal vein; lower surface appearing subvillous (subsericeous with numerous longer emergent ascending often dark hairs, these occasionally lacking). Conflorescence erect, terminal, subsessile, dense, subglobose to broadly subsecund, often held at ground level at edge of or underneath foliage. Flowers acroscopic. Flower colour: perianth dull deep crimson to black-maroon, with white inner beard displayed; style maroon. Perianth subsericeous outside, with a conspicuously tomentose to villous limb, or rarely subsericeous throughout, bearded inside. Pistil 10–12 mm long; style with curve stronger near apex, and with minute papilloid hairs scattered over apical 3–5 mm; pollen-presenter oblique. Follicle narrowly ovoid, c. 18 mm long, glabrous.

Occurs in N.S.W., restricted to the far S of the Sydney Basin and northern Illawarra, between Cordeaux Dam, Cataract Dam, Bulli and Mt Ousley. Grows in poorly drained depressions, swamp margins, or moist rock pockets in pavement, in sandy soils over sandstone or pebbly ironstone. Regenerates from seed or (sometimes?) lignotuber or basal suckers. Pollinator unknown. Flowers July–Dec. Map 199.

N.S.W.: Cordeaux Dam, *H.K.Mair & E.F.Constable NSW16389* (G, NSW); Bulli to Appin road, 9.6 km by road NW of Lodden R. crossing, *R.O.Makinson 1295* (CANB, K, MEL, NSW); 1.6 km S of Bulli on Princes Hwy, *D.J.McGillivray 1501* (NSW).

Grevillea capitellata is distinguished from *G. diffusa* which has the limb of the bud subsericeous (closely appressed hairs), the peduncles likewise, the branchlets sericeous or almost glabrous or shortly pubescent, usually narrower leaves, and the new growth lacking a rusty indumentum (in *G. capitellata* with a distinctly rusty brown indumentum). Also, all populations of *G. diffusa* have the leaf lower surface with a very closely appressed silky indumentum; that of *G. capitellata* has, over most of the range, an underlying ±appressed layer of hairs with emergent longer often darker hairs giving a subvillous appearance (these sometimes absent in the far south of the range, e.g. Mt Keira, Mt Ousley). The range almost abuts that of *G. diffusa* subsp. *diffusa*, but the preferred habitats are different, with the latter occurring mainly in well-drained sites on slopes and low ridges. The degree of intergradation is uncertain. Some material from Woronora Dam area seems intermediate.

Grevillea capitellata is very similar to the geographically disjunct *G. evansiana*, and the two could reasonably be regarded as conspecific subspecies. *Grevillea evansiana* has the limb of the bud always with a rusty-brown indumentum (pale or with a few light brown hairs scattered through in *G. capitellata*), the nectary much more emergent over the toral rim and shortly linguiform (scarcely emergent and arcuate in *G. capitellata* and *G. diffusa*), the conflorescences usually noticeably pedunculate (peduncles 5–15 mm long), and the intramarginal veins smooth (usually minutely scabrid in *G. capitellata*).

154. **Grevillea evansiana** MacKee, *Proc. Linn. Soc. New South Wales* 78: 49 (1953)

G. diffusa subsp. *evansiana* (MacKee) McGill., *New Names Grevillea* 4 (1986). T: Currant Mtn Gap, 5 ml [8 km] E of Olinda, N.S.W., 2 Sept. 1951, *L.A.S.Johnson & H.S.McKee NSW22651*; lecto: NSW, *fide* R.O.Makinson, *Fl. Australia* 17A: 499 (2000); isolecto: BM, K; ?isolecto: BRI, MEL, SYD, US *n.v.*

Illustrations: A.M.Blombery & B.Maloney, *Prot. Sydney Reg.* 100, 101 (1992), as *G. diffusa* subsp. *evansiana*; P.M.Olde & N.R.Marriott, *Grevillea Book* 2: 154 (bottom centre), 155 (125A–C) (1995).

Low dense spreading shrub to 50 (–80) cm tall. Branchlets angular to subterete, subtomentose. Leaves elliptic to slightly obovate or occasionally sublinear, 2.5–6 cm long, 3–10 mm wide; upper surface sparsely punctate; margins sharply and shortly refracted about the intramarginal vein; lower surface subsericeous, sometimes with emergent untidy longer hairs. Conflorescence terminal, usually on short pendulous branchlets, usually pedunculate (peduncles subterete, 5–15 mm long, subsericeous) or sometimes subsessile, dense, subglobose. Flowers acroscopic. Flower colour: buds with rusty indumentum; flower with perianth usually blackish red and style burgundy, or rarely perianth white and style pale creamy green. Perianth subsericeous outside below a conspicuously ferruginous villous limb, bearded inside. Pistil 9–10 mm long; style curved, with minute papilloid hairs in apical 1–3 mm; pollen-presenter oblique. Follicle obloid, 10–12 mm long. Plate 38.

Occurs in N.S.W., where known only from an area E of Rylstone mostly on the western catchment but just extending into the Colo R. catchment. Grows in dry sclerophyll forest or woodland, occasionally in swampy heath, in sandy soils over sandstone. Regenerates from seed. Pollinator unknown. Flowers Aug.–Dec. Map 200.

N.S.W.: Currant Mtn Gap [Khyber Pass], 24 km E of Rylstone, *R.Coveny 9550* (K, NSW); 1.5 km S of Kandos Weir, c. 22 km E of Rylstone, *M.D.Crisp 1274* (PERTH); Currans Mtn Gap [Khyber Pass], 8 km E of Olinda, *L.A.S.Johnson NSW22649* (K, NSW); Currant Mtn Gap [Khyber Pass], *H.S.MacKee 469* (BRI, MEL, NSW); 22 km SE of Rylstone, *H.Streimann 841* (AD, BRI, CANB).

The rusty brown indumentum on the bud limb, and the emergent shortly linguiform nectary, serve to distinguish *G. evansiana* from most related species (*G. diffusa*, *G. oldei* and *G. capitellata* all have a pale indumentum on the limb). See under *G. capitellata* for additional differences.

This species is recognised as 'Vulnerable' in J.D.Briggs & J.H.Leigh, *Rare or Threatened Australian Plants* (1995).

155. **Grevillea oldei** McGill., *New Names Grevillea* 11 (1986)

T: Dog Trap Rd, about 3 miles [4.8 km] from Ourimbah, N.S.W., 12 Apr. 1972, *C.W.E.Moore 6169*; holo: CANB; iso: CANB.

Illustrations: J.W.Wrigley & M.Fagg, *Banksias, Waratahs & Grevilleas* 262 (1989); D.J.McGillivray & R.O.Makinson, *Grevillea* 336, fig. 84, 337, col. pl. (1993); P.M.Olde & N.R.Marriott, *Grevillea Book* 3: 62 (bottom right), 63 (45A, B) (1995).

Diffuse shrub 0.4–1.2 m tall; branches often arching and subcolumnar. Branchlets angular, villous with spreading hairs. Leaves narrowly ovate to subtriangular, 0.5–3.5 cm long, (1.5–) 3–6 mm wide, (juveniles leaves to 5 cm long and 8 mm wide); upper surface faintly foveolate; margins shortly refracted; lower surface loosely and untidily villous. Conflorescence terminal, usually pedunculate (peduncles pendulous, often conspicuous, wiry, to 5 cm long), sometimes 2-branched; unit conflorescence subglobose. Flowers usually adaxially oriented. Flower colour: perianth and style bright to dark red, white inner beard displayed. Perianth outer surface with loose appressed biramous hairs mixed with erect simple glandular hairs and villous on limb; inner surface bearded. Pistil 9.5–15 mm long; style curved, with scattered papilloid hairs over most of its length; pollen-presenter oblique. Follicle obloid, 14–15 mm long. Fig. 18F–H.

Occurs in N.S.W., north of Sydney, restricted to a small area from Mangrove Mtn to Woy Woy and Gosford. Grows in heath or woodland in well-drained sites in sandy soils over sandstone. Regenerates from seed. Pollinator not known. Flowers June–Feb. Map 201.

N.S.W.: Mangrove Mtn near road to Somersby Falls, *W.F.Blakely & P.Murphy NSW94344* (K, NSW); Penang, *J.H.Maiden & J.L.Boorman NSW94350* (NSW); Narara, May 1927, *A.Murphy et al.* (NSW); Woy Woy, *H.M.R.Rupp NSW94351* (NSW).

Grevillea oldei has conspicuously 3-nerved leaves, and has in the past been misidentified as *G. trinervis* (see *G. juniperina* subsp. *trinervis*, which has pistils ⩾ 20 mm long and a subsericeous lower leaf surface). The leaves often twist distinctively on drying. *Grevillea diffusa* subsp. *filipendula* occurs at a distance of a few kilometres, and differs in its much longer, sublinear leaves, and lack of erect glandular hairs on the outer surface of the perianth. Just beyond the west of the range (Kulnura area), there is an extended zone of apparent hybridity with *G. speciosa* (see comments under that species).

This species is recognised as 'Rare' in J.D.Briggs & J.H.Leigh, *Rare or Threatened Australian Plants* (1995).

156. **Grevillea confertifolia** F.Muell., *Trans. Philos. Soc. Victoria* 1: 22 (1855)

T: Vic., 'On the subalpine summit of Mt William, and on rocky ridges towards Mt Zero' [protologue], *s.d.*, *?F.Mueller*; lecto: MEL 47014 *p.p.*, *fide* D.J.McGillivray & R.O.Makinson, *Grevillea* 411 (1993); isolecto: E, K, MEL 47218; remaining syntypes: MEL 64004, MEL 47219.

Illustrations: J.W.Wrigley & M.Fagg, *Banksias, Waratahs & Grevilleas* 224 (1989); D.J.McGillivray & R.O.Makinson, *Grevillea* 338 (1993); P.M.Olde & N.R.Marriott, *Grevillea Book* 2: 98 (top centre & 78A), 99 (78B, C) (1995).

Low spreading (sometimes prostrate) to erect shrub to 1 m tall. Leaves linear to narrowly oblong-elliptic, 1–4.5 cm long, 0.7–2.0 (–3.3) mm wide; upper surface convex; margins refracted; lower surface usually enclosed except for midvein, 2-grooved, sericeous beside midvein when exposed. Conflorescence terminal and many-flowered (usually > 30), often also subterminal-axillary and 2-flowered, erect to decurved, subsessile, dense, broadly secund. Flowers adaxially acroscopic. Flower colour: perianth reddish purple, with white beard on inner surface conspicuously displayed; style pink to reddish mauve. Perianth loosely subsericeous outside, bearded inside. Pistil 10.5–12.5 mm long; style strongly curved near apex, with papilloid hairs on apical 1–2 (–6) mm; pollen-presenter oblique. Follicle obloid-ovoid, 11–12 mm long, glabrous. Plate 40; Fig. 18A–B.

Occurs in Vic., where endemic in the Grampians Ra., with one doubtful record from the Mt Bolangum area SW of St Arnaud. Grows in woodland or shrubby streamside or rock drop-off associations, in sandy soils over sandstone. Regenerates from seed. Pollinator unknown. Flowers Aug.–Dec. Map 202.

Vic.: Balangum [Bolangum] Ra., Nov. 1920, *J.W.Audas* (MEL); Billywing Ck near Glen Isla Camp, Victoria Ra., Grampians, *D.N.Kraehenbuehl 219* (AD); between Moora Moora Ck and Sandy Ck on Henham Track, Grampians, *D.J.McGillivray 3229 & C.Bartlett* (MEL, NSW); Major Mitchell Plateau, 22 km SSE of Halls Gap, *H.Streimann 3079* (CANB); Mt Difficult, Grampians, Jan. 1882, [*D.?*]*Sullivan* (MEL).

This species is recognised as 'Rare' in J.D.Briggs & J.H.Leigh, *Rare or Threatened Australian Plants* (1995).

Linearifolia Subgroup

Unit conflorescence dense to loose, broadly secund or regular-umbelloid, erect to decurved. Floral rachis 1–15 (–27) mm long. Tepals separating to ovary level or base and widely arching or splayed within 90° of ventral suture, with apices recoiled or recurved. Pistils 4.5–19 mm long. Pollen-presenter oblique. Follicle colliculate to ±smooth, not ridged.

Eighteen species in south-eastern Australia (Qld, N.S.W., Vic., Tas., S.A.). Insect pollinated.

McGillivray & Makinson (*Grevillea* 341–349 (1993)) recognised a broad species based on the N.S.W. endemic *G. linearifolia*, with some 14 'forms' distributed from south-eastern Qld to S.A., with some intermediate and unassigned populations; most of these are here recognised as separate species. Members of the alliance are certainly closely related, and diagnosis can be difficult, sometimes resting on suites of non-exclusive character states, with intermediate populations and intergrades occurring in some cases. Genetic studies to test the various morphological classifications that have been proposed are desirable.

Two broad alliances seem to be defined on inflorescence form and geographical distribution. Species with weakly to strongly secund unit conflorescences (*G. linearifolia* to *G. parviflora*) form a northern alliance, ranging from south-eastern Qld S to Nowra (one species extending S to Mallacoota near the N.S.W./Vic. border). Species with regular-umbelloid unit conflorescences form a southern alliance, from southern N.S.W. to Kangaroo Is., S.A.

157. Grevillea linearifolia (Cav.) Druce, *Bot. Soc. Exch. Club Brit. Isles*, suppl. 2: 625 (1917)

Embothrium linearifolium Cav., *Icon.* 4: 59, t. 386, fig. 1 (1798), as *E. linearefolium*; *Lysanthe linearifolia* (Cav.) Knight, *Cult. Prot.* 119 (1809); *G. linearis* R.Br., *Trans. Linn. Soc. London* 10: 170 (1810), *nom. illeg.* T: prope oppidum Jackson [near Port Jackson, N.S.W., 1799?, *coll. unknown*]; ?holo: MA *n.v.*; ?iso: G [attrib. J.White], P-JU.

Embothrium sericeum var. *angustifolium* Sm., *Spec. Bot. New Holland* 3: 27, t. 9, fig. 6 (1794), as γ *angustifolia.* T: LINN *n.v.*, *fide* D.J.McGillivray & R.O.Makinson, *Grevillea* 426 (1993).

Embothrium lineare J.Kennedy, *Bot. Repos.* 4: t. 272 (1803). T: plate in J.Kennedy, *Bot. Repos.* 4: t. 272 (1803), *fide* D.J.McGillivray & R.O.Makinson, *Grevillea* 426 (1993).

G. linearis var. *incarnata* Sims, *Bot. Mag.* 53: t. 2661 (1826). T: 'cult. by Loddiges, Mar. 1824, propagated from collection in Australia by J. Banks' [protologue]; lecto: *op. cit.*, t. 2661, *fide* R.O.Makinson, *Fl. Australia* 17A: 499 (2000). [See note in Appendix, p. 499, re identification.]

Illustrations: A.M.Blombery & B.Maloney, *Prot. Sydney Reg.* 108, 109 (1992); D.J.McGillivray & R.O.Makinson, *Grevillea* 345 (1993), as "form 'e', broader leaved Sydney form"; P.M.Olde & N.R.Marriott, *Grevillea Book* 2: 240 (top right & 200B) (1995), as 'Typical form'.

Open erect shrub 1–2 (–3) m tall, not rhizomatous. Branchlets angular, ridged, subsericeous. Leaves spreading to widely ascending, straight, pliable, not crowded, single or occasionally in loose clusters of 3, linear to very narrowly elliptic, (3.5–) 5–9 (–11) cm long, 1–3 (–5) mm wide; upper surface usually punctate, midvein and intramarginal veins prominent and faintly granular; margins shortly refracted; lower surface partly exposed, subsericeous. Conflorescence terminal on longer ultimate branchlets, pedunculate (peduncle 1–20 mm long), usually erect and borne clear of the foliage, usually simple, shortly and broadly secund, usually many-flowered, acropetal. Flowers acroscopic. Flower colour: perianth dead white with brown hairs on limb, or rarely tinged faintly pink; style dead white becoming reddish with age after tepal drop. Perianth subsericeous outside, profusely bearded inside. Pistil 7–13 mm long; stipe 1.3–1.5 mm long; style strongly hooked in apical 1–3 mm, glabrous except for minute hairs or papillae in apical 1 mm; pollen-presenter oblique. Follicle narrowly ovoid, (11–) 13–15 mm long, colliculose. Fig. 18C–E.

Occurs in eastern N.S.W., from Gosford and Putty area S to Parramatta R. and Port Jackson, with disjunct populations, apparently assignable to this species, in the lower Blue Mtns (Lawson area) and to the S in the ranges just inland from Ulladulla and near Nowra (Cambewarra, Parma Ck, Falls Ck). Grows in shrubby eucalypt woodland in moist but well-drained semi-shaded to open situations, usually in skeletal sandy soils over sandstone, occasionally in light sandy clay or pebbly pseudolateritic soils. Regenerates only from seed. Insect-pollinated. Flowers mainly Aug.–Dec., sporadic Jan.–May. Map 203.

N.S.W.: St Ives, *C.Burgess CBG033073* (AD, CANB, DNA, NE, SYD); Pennant Hills Park, Cheltenham, *R.Coveny 11231 et al.* (K, NSW); Wheelbarrow Forest Rd, 4 km W of Princes Hwy, c. 5 km direct W of Burrill L., *R.O.Makinson 416* (CANB, MEL, NE, NSW); Cambewarra near Nowra, Sept. 1925, *F.A.Rodway* (K); Lane Cove R., *C.T.White 5548* (A, BRI, K).

Grevillea linearifolia as here defined corresponds to *G. linearifolia* 'form e, broader-leaved Sydney form' of McGillivray & Makinson (*loc. cit.*), and to the 'type form' of Olde & Marriott (*loc. cit.*). *Grevillea linearifolia* is similar in leaf size and shape to *G. sericea* subsp. *riparia*, which has strongly ascending leaves (spreading to weakly ascending in *G. linearifolia*), deep mauve-pink flowers, and pistils ⩾ 14 mm long.

Some populations in the Blue Mtns (Lawson area) and in the Nowra area have shorter and narrower leaves (1–1.5 mm wide) than is usual, and their assignation here to *G. linearifolia* is provisional; in the latter case there may be intergradation with *G. patulifolia*. Other populations in the same areas correspond well to the Sydney populations except in having the conflorescences more enclosed by the foliage. Some early collections at BM from the Parramatta R. valley have shorter, broader leaves than is usual in this species; it is unclear whether these represent a distinct form (now possibly extinct) similar to *G. humilis*, or a hybrid swarm with *G. sericea*. Several extant swarms of the latter hybrid are known, the flowers usually being very pale pink in colour and the leaves being sublinear to narrowly elliptic and tending to occur in clusters of three. Occasional hybrids with *G. diffusa* subsp. *filipendula* have also been reported.

A capability for rhizomatous reproduction is widespread in the *Linearifolia* subgroup, but is apparently absent in *G. linearifolia s. str.*, which while often very gregarious appears to reproduce from seed only.

158. Grevillea sericea (Sm.) R.Br., *Trans. Linn. Soc. London* 10: 170 (1810)

Embothrium sericeum Sm., *Spec. Bot. New Holland* 3: 25 & t. 9 (1–6) (1794); *Embothrium sericeum* var. *minor* Sm., *Spec. Bot. New Holland* 3: 25–6 & t. 9 (1–4) (1794), *nom. illeg.*; *Lysanthe sericea* (Sm.) Knight, *Cult. Prot.* 118 (1809). T: New South Wales, *s.d.*, [*J.White*]; lecto: LINN, *fide* D.J.McGillivray & R.O.Makinson, *Grevillea* 441 (1993); isolecto: G, MANCH (photo seen), P-JU *n.v.*

Shrub 0.5–2 m tall. Branchlets angular and ridged, subsericeous. Leaves either obovate or elliptic to sublinear, and 1–3 (–5) cm long, (2–) 3–9 mm wide, or linear, and 6–12 cm long, 1–3 mm wide; upper surface sparingly punctate, granulose along veins; margins refracted about intramarginal vein; lower surface at least partly exposed, sericeous. Conflorescence terminal, deflexed, sessile or shortly pedunculate, loosely secund, many-flowered. Flowers acroscopic. Perianth subsericeous to loosely so outside with biramous hairs and often also scattered inconspicuous minute erect hairs, bearded inside. Pistil 14–19 mm long; style gently curved or sharply decurved in apical 4 mm, glabrous except for minute erect hairs and papillae in apical 1–4 mm; pollen-presenter oblique. Follicle narrowly ellipsoidal to narrowly ovoid, 9–16 mm long, colliculose.

Occurs in N.S.W. from S of Sydney inland almost to Mudgee and N to the Toronto and Wyee area. Two subspecies are recognised.

Leaves narrowly obovate to elliptic or narrowly so, usually ⩽ 4 cm long (very rarely linear and 6–9 cm long), (2–) 3–9 mm wide; some or all conflorescences standing clear of the foliage; flowers deep to pale pink or white; habitat non-riparian ..... **158a.** subsp. **sericea**

Leaves linear, 6–12 cm long, 1–3 mm wide; conflorescences usually enclosed within the foliage; flowers purplish pink; habitat riparian ..... **158b.** subsp. **riparia**

Plants in the north and inland parts of the range usually have pink flowers and leaves ⩽ 4 mm wide, and are very like subsp. *humilis*, differing mainly in their neater and more reflective indumentum, less strongly refracted leaf margins, and preference for moist heath associations. In the Angourie area there is a strong cline from such 'inland' populations (2–3 km from the sea) towards the very distinctive headland Shelley Beach/Bald Knob population, the 'type form', which has a denser, much-branched habit, broader (often narrowly obovate) leaves with scarcely recurved margins, a dense white indumentum on the lower leaf surface, hairs more persistent on the upper leaf surface, shorter and broader (rarely almost regular-umbelloid) unit conflorescences, and usually white flowers (styles reddening later).

Broader-leaved plants of subsp. *maritima* are outwardly very similar to *G. imberbis*; the latter species however lacks a dense beard on the inner surface of the ventral tepals, having at most a few hairs opposite the ovary.

160. Grevillea viridiflava Makinson, *Fl. Australia* 17A: 500 (2000)

T: 12.3 km N of Torrington on road to Silent Grove, N.S.W., 21 Oct. 1988, *R.O.Makinson 576 & S.Krauss*; holo: CANB; iso: AD, BRI, K, MEL, NE, NSW.

G. linearifolia form 'c' (montane elements), of D.J.McGillivray & R.O.Makinson, *Grevillea* 343 (1993).

G. linearifolia 'Torrington form' and 'Darling Downs form', of P.M.Olde & N.R.Marriott, *Grevillea Book* 2: 240 (1995).

G. linearifolia 'Northern form' *p.p.*, of R.O.Makinson, *in* G.J.Harden (ed.), *Fl. New South Wales* 2: 51 (1991).

Illustration: P.M.Olde & N.R.Marriott, *Grevillea Book* 2: 240 (200A) (1995), as '*G. linearifolia* Torrington form'.

Erect spindly shrub 0.5–1 (–2) m tall, often suckering. Branchlets angular to rounded, weakly ridged, subsericeous. Leaves ascending, broadly linear to very narrowly oblong-elliptic, (2–) 3–6 cm long, 2–3 mm wide (rarely to 7 mm wide on juveniles); upper surface soon glabrous, not punctate; margins refracted; lower surface partly exposed, subsericeous except for glabrous midvein. Conflorescence terminal, erect, subsessile, not exceeding the foliage, sometimes 2-branched; unit conflorescence shortly and broadly secund, acropetal, 12–24-flowered. Flowers acroscopic. Flower colour: perianth greenish or yellowish cream outside in very late bud, becoming white later, with brown hairs on limb; inner surface white; style lemony yellow-green or occasionally yellowish cream. Perianth subsericeous outside, scantily bearded inside. Pistil 7–9.5 mm long; stipe 0.6–1.0 mm long; style glabrous except for papillae in apical c. 0.5 mm; pollen-presenter oblique. Follicle 9–11 mm long, colliculose. Fig. 19H–J.

Occurs in montane areas of south-eastern Qld (Wyberba to Girraween areas) and far north-eastern N.S.W. (Torrington, Emmaville and Bald Rock areas), at altitudes of 850–1000 m. Grows in low open eucalypt forest with shrubby understorey, moister well-drained sites in granitic gravelly-loam soils over granite. Regenerates from seed or rhizomes. Insect-pollinated. Flowers Aug.–Jan. Map 209.

Qld: The Junction, Girraween Natl Park, *R.O.Makinson 560* (BRI, CANB, NSW); Wyberba, Nov. 1964, *G.Ward s.n.* (PERTH). N.S.W.: Torrington, *W.T.Jones 4088* (CANB); Tungsten via Deepwater, *T.D.Lynch 7326/13* (NSW); Bald Rock Ck, 10 km N of Wallangarra, *I.R.Telford 3192* (A, CANB, K, L, NSW).

Grevillea viridiflava differs from *G. linearifolia* in its shorter leaves, usually lower habit and rhizomatous propagation, green- or yellow-tinged flower colour, and generally smaller fruits and shorter pistils. It is very closely related to *G. humilis*, which occurs at coastal altitudes in heavier clay soils, and which has either pink or white styles.

There are two barely distinguishable forms. The 'type form' or 'Torrington form' has greyish green strongly ascending leaves 4–6 cm long, pedicels 5–6 (–7.5) mm long, and pistils 8–9.5 mm long; it occurs in the tin-fields around Tungsten, Torrington and Emmaville. The 'Girraween form' ('Darling Downs form' of Olde & Marriott, *loc. cit.*) has slightly more spreading, shorter and greener leaves (2–) 3.5–5 cm long, pedicels 6.5–8 mm long, and pistils 7–8.5 mm long; the yellow-green colour of the style is sometimes less pronounced in this form, which occurs in the Wyberba and Girraween area of Qld and near Bald Rock (N.S.W.) S of Wallangarra.

A population NW of Tabulam (NSW, W of Casino) may be assignable to this species; it has plain white styles, but in other respects conforms well to the circumscription.

161. **Grevillea virgata** Makinson, *Fl. Australia* 17A: 500 (2000)

T: 5.3 km S of Bulahdelah on Pacific Hwy, N.S.W., 4 July 1997, *R.O.Makinson 1627*; holo: CANB; iso: BRI, K, NSW.

G. linearifolia, 'form c' *p.p.* and 'Unassigned 2', of D.J.McGillivary & R.O.Makinson, *Grevillea* 343–348 (1993).

G. linearifolia 'northern form', of R.O.Makinson, *Fl. New South Wales* 2: 51(1991), *p.p.*

G. linearifolia 'Coastal form', of P.M.Olde & N.R.Marriott, *Grevillea Book* 2: 241(1995), *p.p.*

Erect open shrub 0.5–2 m tall. Branchlets angular, ribbed, soon nearly glabrous. Leaves ±spreading, narrowly elliptic, 1–3 cm long, 1.5–3.0 mm wide, usually in well-spaced clusters of 3–5; upper surface punctate or not, with midvein and intramarginal veins prominent, smooth or faintly granular; margins shortly refracted; lower surface exposed, loosely to sparsely subsericeous. Conflorescence terminal or leaf-opposed, erect to decurved, subsessile or on peduncles to 20 mm long, simple or rarely branched, broadly secund, 10–24-flowered, acropetal. Flowers acroscopic. Flower colour: perianth and style white, with style becoming pink after tepal drop. Perianth sparsely sericeous outside (densely so on limb), sparingly bearded inside. Pistil 7–8 mm long; style hooked in apical 1–2 mm, glabrous except for minute subapical papillae; pollen-presenter oblique. Follicle narrowly ovoid, 10–11 mm long, colliculose. Fig. 21A–B.

Occurs on the N.S.W. mid-North Coast, known only from flood plain and hinterland of lower Myall R. between Bulahdelah and Nerong. Grows in eucalypt/*Allocasuarina* forest and on margins of *Melaleuca* swamps, in heavy clay-loam soils. Regenerates from rhizomes and seed. Insect-pollinated. Flowers at least May–Oct. Map 210.

N.S.W.: 5.3 km S of Bulahdelah on Pacific Hwy, *R.O.Makinson 1626* (AD, CANB, COFF, MEL, NE, NSW, PERTH); S of Myall R., 3 May 1975, *J.Mowatt s.n.* (NSW); 8 km S of Bulahdelah, *J.Garden NSW17740* (K, NSW); Bulladelah [sic], 21 Aug. 1923, *H.M.Rupp* (NSW).

Grevillea virgata has reddish young branchlets, with the leaves ±spreading and usually in well-spaced clusters of three. It is closely related to *G. humilis* which has a lower, weaker habit, shorter internodes, ascending leaves, and (in subsp. *humilis*) the leaves not usually in conspicuous clusters of three. *Grevillea linearifolia* is distinguished by its non-rhizomatous habit, longer leaves usually 5–11 cm long, usually longer pistils (7–13 mm long), and larger fruit (11–15 mm long). *Grevillea patulifolia* is similar but has usually denser foliage with more rigid, pungent leaves, usually pink perianths and styles (rarely white), and young branchlets brownish green.

162. **Grevillea leiophylla** F.Muell. ex Benth., *Fl. Austral.* 5: 471 (1870)

T: 'Glasshouse ranges, Moreton Bay, F.Mueller, and probably from the same neighbourhood, Leichhardt' [protologue]; lecto: Glass houses Moreton Bay [Glass House Mtns, Qld], *s.d.*, *F.Mueller*; lecto: K, *fide* R.O.Makinson, *Fl. Australia* 17A: 500 (2000); remaining syntypes: Moreton Bay [Qld], *s.d.*, *Hill & Mueller*; syn: K; round Tibburrocan [Qld], *s.d.*, *Leichhardt* [*per*] *Mueller*; syn: MEL 65617, ?65618, 65619, 65620.

G. linearifolia form 'b' (smooth-leaved Queensland form), of D.J.McGillivray & R.O.Makinson, *Grevillea* 343 (1993).

Illustrations: T.D.Stanley & E.M.Ross, *Fl. SE Queensland* 2: 26, fig. 3E (1986); P.M.Olde & N.R.Marriott, *Grevillea Book* 2: 231 (top right), 232 (192A, B) (1995).

Weakly erect to low-spreading or decumbent shrub 0.2–0.6 (–1) m tall. Branchlets angular and ridged, glabrous or nearly so. Leaves narrowly oblong-elliptic or -obovate to sublinear, 2–6 (–8) cm long, 1–3 (–4.5) mm wide, often crowded but not in clusters of 3; upper surface glabrous, with midvein evident but scarcely prominent; margins very shortly recurved or refracted; lower surface exposed with an open to sparse indumentum of appressed, short-

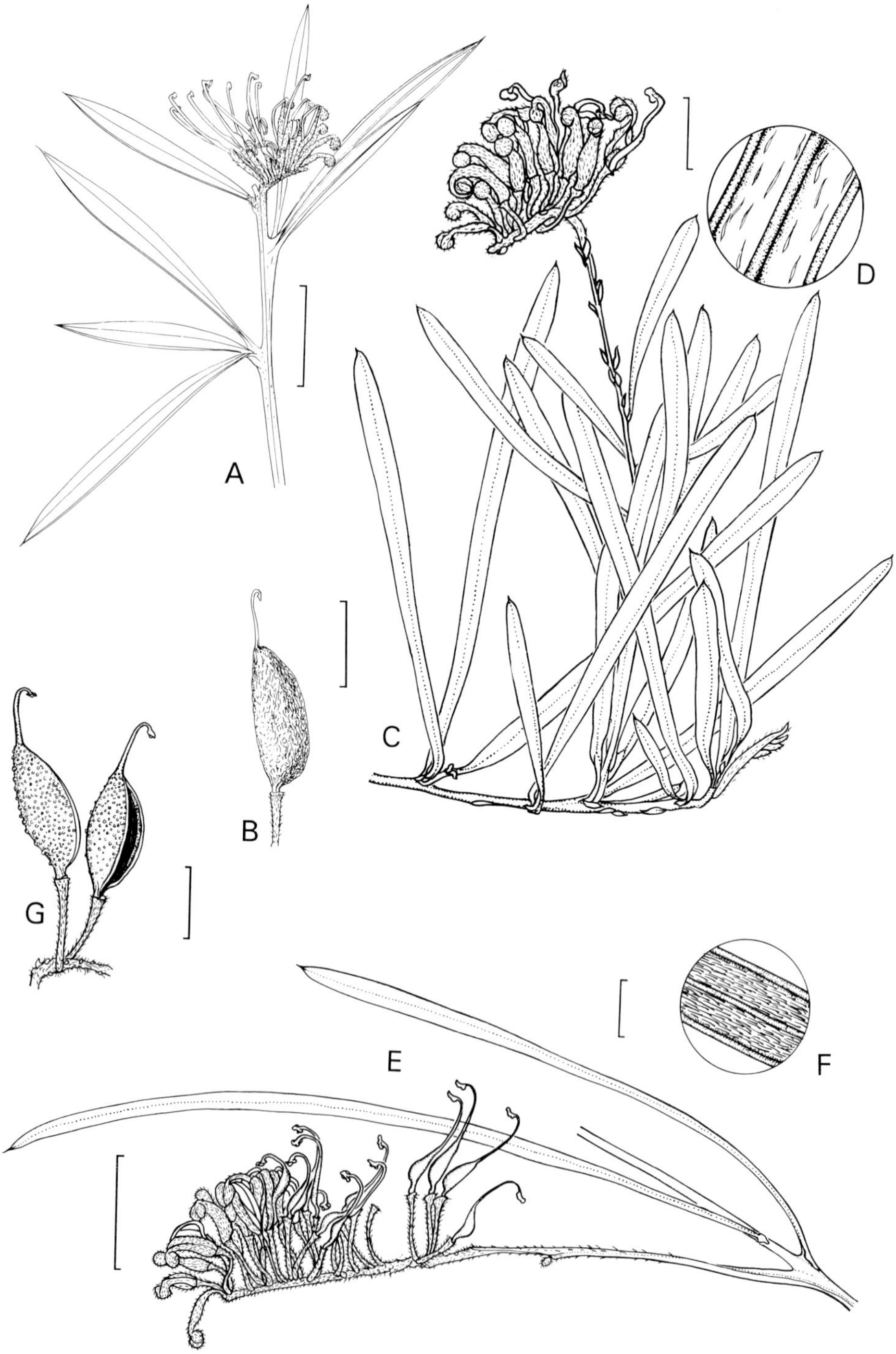

Figure 21. *Grevillea*. **A–B**, *G. virgata*. **A**, flowering branch; **B**, fruit (**A–B**, R.O.Makinson 1626, CANB); **C–D**, *G. leiophylla*. **C**, flowering branch; **D**, detail of lower side of leaf (**C–D**, R.O.Makinson 1415, CANB). **E–G**, *G. reptans*. **E**, flowering branch; **F**, detail of lower side of leaf; **G**, fruit (**E–G**, R.O.Makinson 1410, CANB). Scale bars: **A–B**, **E** = 1 cm; **C**, **G** = 5 mm; **D** = 1 mm; **F** = 2 mm. Drawn by: **A–B**, C.Wardrop; **C–G**, C.Payne.

armed sparkling biramous hairs (ground tissue visible between hairs) or glabrous. Conflorescence terminal, often decurved, scarcely exceeding foliage, sessile to shortly pedunculate (peduncles to 5 mm long), simple, broadly secund (sometimes obscurely so), 8–24-flowered, acropetal. Flowers acroscopic. Flower colour: perianth pale to deep pink; style pale pink, becoming red with age. Perianth openly to sparsely subsericeous outside below a densely subsericeous (rarely tomentose) limb, bearded inside. Pistil 6–10 (–13) mm long; style hooked in apical c. 1 mm, glabrous except for minute subapical papillae; pollen-presenter oblique. Follicle c. 9 mm long, colliculose. Fig. 21C–D.

Endemic in south-eastern Qld, occurring in scattered populations N of Brisbane from Deception Bay and North Stradbroke Is. to Caloundra and Caboolture area; possibly also Yeppoon area. Grows in shrubby woodland or open forest, open drier wallum heath associations, and *Themeda* grassland, usually on level well-drained sites in sandy soils. Regenerates from seed, rhizomes, and basal suckers, sometimes with a weakly developed lignotuber. Presumed insect-pollinated. Flowers mainly Aug.–Nov., sporadic in other months. Map 211.

Qld: N of Brisbane, between Narangba and Burpengary, *S.T.Blake 3151* (BRI, K); Caloundra, *S.T.Blake 4194* (BRI, K); c. 35 km (direct) NNW of Brisbane centre, Freshwater Creek Natl Park, Deception Bay road, opposite entrance to Caboolture Shire Lawn Cemetery, *R.O.Makinson 1415 et al.* (BRI, CANB, K, NSW); 5 km S of Caboolture on the Bruce Hwy, *R.Melville 3523 et al.* (AD, BRI, K, MEL, NSW, PERTH); between Caboolture and coast, *M.E.Phillips CBG022052* (CANB).

Grevillea leiophylla is variable in habit, leaf form, and pistil length, but is usually a dense-foliaged dwarf shrub with short stiff branches, narrowly oblong-elliptic leaves, scarcely refracted leaf margins, and the lower leaf surface with scattered appressed hairs. *Grevillea leiophylla* is closely related to *G. reptans*, which has longer linear leaves 6–17 cm long, longer, more gracile and pliable branches with more open foliage, and the outer surface of the perianth tomentose to subvillous, especially on the limb. The ranges overlap in the Caloundra area, and some specimens (including one from Deception Bay, outside the known range of *G. reptans*) suggest intergradation. In the Glass House Mtns area, the range of *G. leiophylla* abuts that of *G. humilis* subsp. *lucens*; see under that taxon for differences.

A fragmentary specimen at BRI (*H.S.McKee 10272*) from Jim Crow Mtn near Yeppoon (c. 450 km NW of the confirmed range), is very similar to *G. leiophylla*.

163. Grevillea reptans Makinson, *Fl. Australia* 17A: 500 (2000)

T: Burrum Heads road, 18 miles [c. 29 km] NNE of Howard, Qld, Sept. 1966, *C.H.Gittins 1215*; holo: BRI; iso: K.

G. linearifolia 'form a, villous-flowered Queensland form', of D.J.McGillivray & R.O.Makinson, *Grevillea* 342–343 (1993).

G. sp. aff. leiophylla, P.M.Olde & N.R.Marriott, *Grevillea Book* 2: 232 (1995).

Illustrations: P.M.Olde & N.R.Marriott, *Grevillea Book* 2: 232 (centre right & 193A) (1995), as *G. sp. aff. leiophylla*.

Prostrate to scandent or occasionally weakly erect sub-shrub, to 0.3 m tall. Branches usually long, scandent or arching, pliable. Branchlets angular, ridged, usually some hairs in grooves. Leaves linear or sublinear, (5–) 6–12 (–17) cm long, 1–2 mm wide (to 4 mm on juvenile growth), usually solitary (occasionally crowded but not in clusters of 3); upper surface not punctate, with midvein and intramarginal veins scarcely prominent; margins strongly refracted; lower surface sometimes concealed except for midvein, when partly exposed densely to sparsely subsericeous or glabrous. Conflorescence terminal, erect, usually pedunculate (peduncles 10–40 mm long), sometimes 2- or 3-branched; unit conflorescence secund, 16–24-flowered, acropetal. Flower colour: perianth mid to deep mauve-pink; style similar but paler, becoming red with age. Flowers acroscopic. Perianth tomentose to subvillous outside, especially on limb, bearded inside. Pistil 7–9 mm long; style sharply hooked in apical c. 3 mm, glabrous except for a few subapical papillae (sometimes these only on rear of pollen-presenter); pollen-presenter oblique. Follicle 7–9 mm long, colliculose. *Tin Can Bay Grevillea*. Fig. 21E–G.

Low open to erect shrub 0.3–1 m tall. Major branches ascending to erect; branchlets not secund with leaves not all held skywards. Leaves mostly 0.8–1.3 mm wide. Flower colour: perianth white with rusty brown hairs on limb; style white, sometimes becoming red with age. Stipe of ovary 1.0–1.2 mm long. Fig. 22A–B.

Endemic to N.S.W., occurring W and S of Sydney from W of Prospect (where now almost certainly extinct), Kemps Ck and lower Georges R. S to Camden, Appin and Cordeaux Dam, with disjunct northern populations S of Putty and near Cessnock and Cooranbong, possibly also S of Moss Vale. Grows in heathy associations or shrubby woodland, in sandy or light clay soils usually over shale substrates. Regenerates from seed and at least sometimes from rhizomes. Insect-pollinated. Flowers July–Dec. Map 213.

N.S.W.: Wedderburn, 18 Oct. 1959, *C.E.Chadwick* (NSW); Tahmoor, S of Picton, 18 Jan. 1956, *E.F.Constable* (K, NSW); Kemps Ck, *R.Coveny 11931 et al.* (K, NSW); 1.6 km N of Bargo R., Bargo, *C.L.Wilson 565* (NSW); Cooranbong, Freemans Drive, 1.0 km by road NW of Martinsville Rd intersection, *R.O.Makinson 1629* (CANB, MEL, NSW).

Subsp. *parviflora* varies in habit from almost prostrate (e.g. at Voyager Point), to an erect spindly shrub to 2 m tall at Wedderburn; populations in the latter area are less rhizomatous. This taxon is very similar to *G. micrantha*, which differs in having the leaves often gently curved and the unit conflorescence ±regular (not secund).

164b. Grevillea parviflora subsp. supplicans Makinson, *Fl. Australia* 17A: 501 (2000)

T: Berrilee, c. 32 km (direct) NNW of Sydney GPO, 1.9 km by road from Berowra Waters Ferry (W bank terminus) towards Glenorie, N.S.W., 24 Sept. 1993, *R.O.Makinson 1294*; holo: NSW; iso: AD, BRI, CANB, G, HO, K, MEL, MO, NE, PERTH.

G. linearifolia 'form d, narrower-leaved Sydney form', of D.J.McGillivray & R.O.Makinson, *Grevillea* 343, 348 (1993).

G. parviflora 'Maroota-Berrilee form', of P.M.Olde & N.R.Marriott, *Grevillea Book* 3: 80 (1995).

Illustration: P.M.Olde & N.R.Marriott, *Grevillea Book* 3: 80 (58B) (1995), as *G. parviflora.*

Low shrub 0.3–0.6 (–1.0) m tall. Major branches spreading; branchlets usually strongly secund with leaves held skywards. Leaves 0.6–2 (–3) mm wide. Flower colour: perianth white with rusty brown hairs on limb; style white, sometimes becoming red with age. Stipe of ovary 0.5–0.6 mm long. Fig. 22C.

Endemic to N.S.W., occurring NW of Sydney at Berrilee near Arcadia, and in the Maroota to Marramarra Ck area. Grows in heathy woodland associations in skeletal sandy soils over sandstones. Regenerates from seed and possibly also suckers. Insect-pollinated. Flowers Aug.–Nov. Map 214.

N.S.W.: N Maroota, 14 Oct. 1961, *L.H.Williams* (CANB, NSW); Berrilee, Arcadia, *H.S.McKee 819* (K, MEL, NSW); Coba Ridge, Marramarra Natl Pk, *L.McDougall 75 & B.Bishop* (K, NSW); Maroota [State] Forest, W of Old Northern Rd, *R.G.Coveny 15492 et al.* (B *n.v.*, K, MEL, NBG *n.v.*, NSW, PERTH, RSA *n.v.*); Glenorie, Nov. 1929, *A.H.Winter* (NSW).

Subsp. *supplicans* sometimes grows sympatrically with *G. linearifolia*, with no known indications of interbreeding.

165. Grevillea patulifolia Gand., *Bull. Soc. Bot. France* 66: 231 (1919)

T: Barbers Ck, N.S.W., Nov. 1899, '*J.H.Maiden*' [probably *H.J.Rumsey*]; holo: LY; ?iso: NSW (Barbers Ck, Nov. 1899, *H.J.Rumsey*).

G. linearifolia Southern Sandstone form, of R.O.Makinson, *in* G.J.Harden (ed.), *Fl. New South Wales* 2: 52 (1991).

G. linearifolia 'form f - southern sandstone form', of D.J.McGillivray & R.O.Makinson, *Grevillea* 344–345 (1993).

[*G. parviflora auct. non* R.Br.: A.Fairley & P.Moore, *Nat. Pl. Sydney District* 169 (1989)]

Illustrations: A.Fairley & P.Moore, *Native Pl. Sydney District* 169 (1989), as *G. parviflora*; R.O.Makinson, *in* G.J.Harden (ed.), *Fl. New South Wales* 2: 52, fig. adj. 39d (1991), as *G. linearifolia* Southern Sandstone form; P.M.Olde & N.R.Marriott, *Grevillea Book* 3: 83 (top centre & 61) (1995).

Erect or rarely low spreading shrub (0.2–) 0.8–2.0 m tall. Branchlets angular, subsericeous between ridges. Leaves sublinear to narrowly elliptic or slightly oblanceolate, (0.8–) 2–5 (–7) cm long, 1–3 mm wide (juveniles to 7 mm), straight, usually rigid, pungent, sometimes in clusters of 3; upper surface punctate with veins prominent and granulose to scabrid; margins refracted, usually at more than 90° to plane of upper surface; lower surface exposed and subsericeous or mostly enclosed and 2-grooved. Conflorescence terminal, often aggregated, erect to decurved, sessile to shortly pedunculate, regular and umbelloid or occasionally subsecund, scarcely exceeding adjacent leaves, usually many-flowered. Flowers adaxially oriented to acroscopic. Flower colour: perianth pale pink to dark mauve-pink or rarely white; style similar. Perianth subsericeous outside, bearded inside. Pistil (6.5–) 8–8.5 (–10.5) mm long; style gently curved with a sharp hook in the apical 2 mm, minutely hairy or papillose in apical 1 mm; pollen-presenter oblique. Follicle narrowly ellipsoid, c. 11 mm long, faintly warty. Fig. 22G–H.

Occurs in N.S.W. and Vic. In N.S.W., from southern outskirts of Sydney (Heathcote, Yerrinbool) NW to Kanimbla Valley and Sunny Corner, S to Tabourie and Braidwood areas and inland nearly to Marulan, then disjunct in Vic. where restricted to far eastern Gippsland (Mallacoota area), with one doubtful record from Bemm R. Doubtfully naturalised in the Canberra area. Grows usually in moist heath or woodland/heath associations in sandy soils. Regenerates from seed and (most populations?) basal suckers and/or rhizomes. Probably insect-pollinated. Flowers July–Feb. Map 215.

N.S.W.: Heathcote Ck, *J.G.McKern NSW94107* (NSW); Flat Rock Ck to Yalwal Ck, [W of] Nowra, *M.McMillan 72/216* (CANB); 3.2 km S Fitzroy Falls, *K.Mowle 71* (CANB, NSW). Vic.: N of Howe Ck, E of Mallacoota Inlet, *A.C.Beauglehole 34381* (MEL, NSW); SE sector of Mallacoota Inlet Natl Park, 4 Nov. 1969, *J.H.Willis s.n.* (MEL, NSW).

The leaves of *G. patulifolia* are often spreading to widely ascending, rigid, and pungent, as compared with most similar species. Leaves are usually 1–5 cm long, with a rare riparian variant with strongly ascending linear leaves 5–7 cm long known from the Endrick R. near Nerriga. A rare form, apparently assignable to this species, with strongly scabrid leaf veins is known from a few collections from the Kanimbla Valley and Sunny Corner area; elsewhere the veins are obscurely granulose. Over much of the range the habit is a low shrub to 60 cm tall; in the Heathcote area (N.S.W.) it forms an erect shrub to 2 m tall, often with subsecund inflorescences (regular-umbelloid elsewhere).

A tiny, probably uniclonal, white-flowered population occurs in disturbed bushland on the southern footslopes of Mt Ainslie, A.C.T. While natural origin cannot be ruled out, it is likely that this is introduced; other *Grevillea* species and cultivars have naturalised in the same area apparently from plantings in the 1920–1940 period.

Grevillea patulifolia can be confused with some variants of *G. linearifolia* and *G. parviflora* subsp. *parviflora* (which have pliable non-pungent leaves and consistently secund to subsecund unit conflorescences), and with *G. neurophylla* (which has often longer and narrower, pliable, non-pungent, more strongly ascending leaves and shorter pedicels 4.5–5 mm long, and a usually shorter pistil 6–7.5 mm long). There are some indications of an intergrade between *G. patulifolia* and *G. linearifolia* in the Nowra/Conjola area.

166. Grevillea neurophylla Gand., *Bull. Soc. Bot. France* 66: 231 (1919)

T: Australia merid. [Vic.], 1902, *coll. unknown*; holo: LY.

G. linearifolia "form 'g' (southern submontane form)", of D.J.McGillivray & R.O.Makinson, *Grevillea* 345 (1993).

Spreading to erect shrub 1–2.5 m tall. Branchlets angular, subsericeous between ridges. Leaves linear or occasionally very narrowly elliptic, 2.5–5.5 (–8) cm long, 0.8–2 mm wide, crowded or not but not in clusters of 3; upper surface sparingly punctate, with veins slightly granular; margins refracted (sometimes forming a short vertical 'wall'); lower surface enclosed except for midvein, 2-grooved. Conflorescence erect, usually terminal, sessile or pedunculate (peduncles to 6 (–30) mm long), exceeded by adjacent leaves, regular and umbelloid, 8–20-flowered. Flowers adaxially oriented. Flower colour: perianth white to pale pink, with brownish hairs on limb; style white to pink. Perianth subsericeous or sparsely so outside, bearded inside. Pistil 6–7.5 mm long; style strongly hooked just below apex, glabrous except for minute hairs or papillae in apical 1 mm; pollen-presenter oblique. Follicle narrowly ovoid-ellipsoidal, 6.5–10 mm long, faintly colliculose.

Occurs in south-eastern N.S.W. and eastern Vic. Two subspecies are recognised.

In addition to features above, *G. neurophylla* has the leaves spreading to strongly ascending, pliable, not or scarcely pungent, often slightly curved or wavy.

Grevillea neurophylla is closely related to *G. wiradjuri* (which is usually a more open shrub with more space between the consistently widely ascending to spreading leaves, and the leaf margins always twice-refracted forming a short vertical 'wall' between the two refractions, giving a box-like cross-section). *Grevillea gariwerdensis* is also very similar, but has the leaves 1–3 (–5) cm long and the leaf margins shortly refracted, usually with some subsericeous lamina exposed on either side of the abaxial midvein. *Grevillea patulifolia* has the leaves almost spreading, straight, rigid, pungent, and often in clusters of three, and sessile conflorescences. Some western populations of *G. patulifolia* (Nerriga area, N.S.W.) are barely distinct from *G. neurophylla*, differing mainly in robustness and length of peduncles. These other three species all have fruits 9–11 mm long. *Grevillea alpivaga* has generally shorter leaves to 3 cm long, which are crowded in clusters on suppressed lateral branchlets, and more rigid, with the lower surface on either side of the midvein usually partly exposed.

Leaves spreading to ascending, not or only moderately crowded; longest leaves ⩽ 4 cm long; newer branches not columnar; fruits 8–10 mm long	**166a.** subsp. **neurophylla**
Leaves strongly ascending, crowded; longest leaves usually > 4 cm long; newer branches subcolumnar; fruits 6.5–8 mm long	**166b.** subsp. **fluviatilis**

166a. Grevillea neurophylla Gand. subsp. neurophylla

Illustrations: P.M.Olde & N.R.Marriott, *Grevillea Book* 3: 50 (top right & 35B) (1995), as *G. neurophylla.*

Erect to spreading shrub. Branches not columnar, the foliage usually open. Leaves irregularly ascending or occasionally spreading, not or only moderately crowded, sometimes irregularly curved or wavy; longest leaves ⩽ 4 cm long. Flower colour: perianth and style white to very pale pink; limb with brown hairs. Fruits 8–10 mm long. Fig. 22D–F.

Occurs in montane and subcoastal areas in south-eastern N.S.W. and montane eastern Vic. In N.S.W., S from the Brindabella Valley W of Canberra, along the western edge of the Snowy Mtns (Talbingo, Lobs Holse, Tumut area), and a longer-leaved variant on the S coast near Eden (Yambulla State Forest). In Vic., Omeo, Nunniong Plateau, Mitta Mitta, upper Genoa R. and Mt Buffalo. Grows in eucalypt woodland or open forest, often in gullies or on rocky slopes near creeks but not usually in flood-zones, in sandy granite-derived or sometimes sandstone soils. Regenerates from seed and (at least sometimes) rhizomes. Presumed insect-pollinated. Flowers Sept.–Feb. Map 216.

N.S.W.: Kosciuszko Natl Park, c. 15 km SSE of Tumut, top of waterfall at end of Blowering Cliffs Walk, *N.Taws 302* (CANB, MEL, NSW); Yumbulla [Yambulla] State Forest, Newtons Crossing Picnic area near junction of Imlay Ck, *M.J.Taylor 206 & T.James* (K, NSW). Vic.: sandy-rocky banks of the Snowy R., *s.d.*, *Anon.* (MEL 68654); Nunniong Plateau, Mundy Plain Track, *A.C.Beauglehole 41425* (MEL, NSW); Bundarra R. Bridge on road from Omeo to Bogong High Plains, 27 km from Benambra turnoff, *R.O.Makinson 910 & P.Carmen* (AD, BRI, CANB, MEL, NE, NSW).

Some collections of *G. neurophylla* subsp. *neurophylla* from the Nunniong Plateau area have relatively short curved leaves and could be confused with *G. alpivaga*, which however, has leaves with the lower surface partly exposed. One collection from Cobungra lacks a beard inside the perianth and has a pulvinus in the beard position, and is similar in this respect to *Grevillea micrantha*; however, it lacks the leaf-vein scabridity of the latter species. Occasional collections from the Mitta Mitta R. and the upper Genoa R. (the latter sympatric with subsp. *fluviatilis*), closely approach subsp. *fluviatilis* in leaf length and crowding; fruit size for these plants is not known. Populations in N.S.W. tend to have straighter leaves and denser foliage than those in Vic.

166b. **Grevillea neurophylla** subsp. **fluviatilis** Makinson, *Fl. Australia* 17A: 501 (2000)

T: upper Genoa R., Vic., *A.C.Beauglehole 35004 & K.C.Rogers*; holo: MEL.

Illustrations: P.M.Olde & N.R.Marriott, *Grevillea Book* 3: 50 (top right & 35A) (1995), as *G. neurophylla* form from Bemm River.

Erect shrub. Newer branches columnar, the foliage dense. Leaves strongly ascending, straight, crowded; longest leaves usually > 4 cm long. Flower colour; perianth white to very pale pink; limb with brown hairs. Fruits 6.5–8 mm long. Fig. 23A.

Occurs in Vic., restricted to a few rivers in East Gippsland (Genoa R., Bemm R., Cann R., Wingan R. and possibly Snowy R.), at generally much lower altitudes than the type subspecies. Also possibly occurs in N.S.W. (Wallagaraugh R.). Grows in open riparian shrub associations, usually on rocky outcrops in the flood-zone of permanent rivers. Regenerates from seed and basal suckers after floods. Presumed insect-pollinated. Flowers ?Aug.–Nov. Map 217.

Vic.: Wingan Inlet Natl Park, The Rapids, W side of Wingan R., 13 Nov. 1969, *A.C.Beauglehole 31674* (MEL); Bemm R., c. 3.2 km NNW of Sydneham Inlet, *A.C.Beauglehole 34010* (MEL); c. 4 km S of Cann River P.O. Gauge Track at Cann R., *S.J.Forbes 2947* (MEL, NSW); East Gippsland, Wingan R., *s.d.*, ?*F.Mueller* (MEL 68691).

Very narrowly distinct from the type subspecies, but consistent enough in habit, longer leaf length, dense foliage, slightly smaller fruits, and ecological preference to warrant some recognition. Subsp. *fluviatilis* occurs at much lower elevations (usually less than 150 m alt.) than most N.S.W. and all Vic. populations of subsp. *neurophylla* (mostly 600–1500 m alt.). There is variation in leaf width, with populations on the Wingan, Bemm and Cann Rivers having the narrowest leaves, 0.8–1.0 mm wide, and those on the Genoa R. having coarser and wider leaves 1.2–1.6 mm wide (comparable to those of subsp. *neurophylla*).

167. **Grevillea alpivaga** Gand., *Bull. Soc. Bot. France* 66: 231 (1919)

T: Victoria, Vict. Alps, Nov. 1903, ex Herb. [& leg.?] [C.]*Walter*; holo: LY.

G. linearifolia 'form h (Mt Buffalo form)', of D.J.McGillivray & R.O.Makinson, *Grevillea* 345 (1993).

Illustrations: P.M.Olde & N.R.Marriott, *Grevillea Book* 2: 22 (bottom right), 23 (13A, B) (1995).

Erect to prostrate shrub (0.3–) 0.7–1 m tall. Branchlets angular, ridged, subsericeous between ridges. Leaves linear, often slightly curved, 0.8–2.5 (–3) cm long, 0.8–1.4 mm wide, crowded on suppressed or very short lateral branchlets in clusters of 3; upper surface with longitudinal veins prominent and usually granular; margins shortly twice-refracted about intramarginal vein, with the first refraction forming a short vertical wall perpendicular to the topological upper surface, the second at 90° again (leaf oblong in cross-section); lower surface usually partly exposed and subsericeous, occasionally enclosed. Conflorescence erect, terminal, sessile, regular, umbelloid, c. 8–20-flowered. Flowers adaxially acroscopic. Flower colour: perianth pale green to cream or white, with brown hairs on limb; style white to pale pink, becoming reddish later. Perianth subsericeous outside, with a scanty beard inside opposite ovary. Pistil 6.5–7 mm long; style strongly hooked in apical 3 mm, with minute hairs or papillae in apical 1 mm; pollen-presenter oblique. Follicle narrowly ellipsoidal, c. 9 mm long, faintly warty. Fig. 23B–D.

Occurs in Vic. on the Mt Buffalo massif and towards Porepunkah. Recorded for granite/slate contact zone and on slate, in *Eucalyptus piperita* woodland at c. 900 m alt. Regeneration mode not known. Presumed insect-pollinated. Flowering recorded for Oct.–Feb. Map 218.

Vic.: Mt Buffalo Plateau on flat by Ranger's cottage, *R.Melville 2692* (K); c. 16 km from Porepunkah on road to Mt Buffalo, *D.J.McGillivray 3188 & C.Bartlett* (K, L, MEL, NSW, PERTH); Mt Buffalo at Mackies Lookout, *M.E.Phillips CBG012565* (CANB); Buffalo, *G.Weindorfer 585* (MEL); Mt Buffalo Natl Park, at E extremity of Hospice Plain, 21 Feb. 1963, *J.H.Willis s.n.* (MEL).

Grevillea alpivaga is very similar, and very closely related, to *G. gariwerdensis* (which has a slightly longer pistil 7–8.5 mm long, less crowded leaves with the margins once-refracted, often secund branchlets, and often pedunculate conflorescences), and to *G. neurophylla* subsp. *neurophylla* (which has longer leaves, the leaf upper surface usually convex, the lower leaf surface usually fully enclosed on either side of the midvein, and shorter pedicels 3–5 mm long).

Figure 23. *Grevillea*. **A**, *G. neurophylla* subsp. *fluviatilis*. **A**, flowering branch (**A**, A.C.Beauglehole 34010, CANB). **B–D**, *G. alpivaga*. **B**, flowering branch; **C**, detail of lower side of leaf; **D**, fruit (**B–D**, R.H.Cambage 3716, CANB). **E–F**, *G. gariwerdensis*. **E**, flowering branch; **F**, detail of lower side of leaf (**E–F**, M.E.Phillips *s.n.*, 29 Oct. 1960, CBG012825, CANB). **G–I**, *G. micrantha*. **G**, flowering branch; **H**, detail of lower side of leaf; **I**, fruit (**G–I**, T.B.Muir 4904, MEL). Scale bars: **A–B**, **D–E**, **G**, **I** = 5 mm; **C**, **F**, **H** = 1 mm. Drawn by C.Payne.

168. Grevillea micrantha Meisn., *Linnaea* 26: 358 (1854)

T: 'Australia felix. [Vic.] 1851–52. legit Dr Ferd. Mueller. Grevillea micrantha nob. (5 Sept. 53.) ex Herb. Sonder'; holo: NY *n.v., fide* D.J.McGillivray & R.O.Makinson, *Grevillea* 426 (1993); ?iso: G-DC.

G. linearifolia 'form j' (Western Victorian form) and 'Unassigned 7', of D.J.McGillivray & R.O.Makinson, *Grevillea* 345–347 (1993).

Illustrations: P.M.Olde & N.R.Marriott, *Grevillea Book* 3: 23 (bottom right), 24 (14A, B) (1995).

Spreading shrub 0.3–0.6 m tall. Branchlets angular and ridged, subsericeous between ridges. Leaves linear to narrowly elliptic or oblanceolate, 1–4 cm long, 0.6–1.0 (–1.5) mm wide, often gently incurved or wavy; upper surface prominently to obscurely 3-ribbed, minutely scabrid at least on ribs; margins vertically refracted about intramarginal vein (leaf oblong in cross-section) and usually abutting midvein; lower surface usually enclosed except for midvein and 2-grooved, occasionally partly exposed and subsericeous. Conflorescence usually terminal, erect, sessile or very shortly pedunculate, regular, umbelloid, c. 6–14-flowered. Flowers adaxially oriented. Flower colour: perianth and style white to pale pink. Perianth subsericeous to sparsely so outside, usually glabrous inside with a discolorous cushion-like pulvinus on each ventral tepal opposite ovary, occasionally pulvinus with a few minute hairs. Pistil 5–6.5 (–8?) mm long; style sharply hooked in apical 2 mm, glabrous except for minute hairs or papillae on back of style-end; pollen-presenter oblique. Follicle narrowly ovoid, 10–11 mm long, glabrous, faintly colliculose. Fig. 23G–I.

Endemic in south-central to south-western Vic.: sporadic occurrences in the area bounded by Brisbane Ra., Mt Macedon, Wedderburn, the eastern fall of the Victoria Ra., and Portland area. Grows in poor stony soils in ironbark or mallee woodland. Regenerates from seed and often rhizomes. Presumed insect-pollinated. Flowers Aug.–Jan. Map 219.

Vic.: Mt Clay, N of Portland, *E.Finck & A.C.Beauglehole 5878* (MEL); Victoria Ra., E side, Big Cord, Glenelg R., *A.C.Beauglehole 22243* (MEL, NSW); Midlands, 6.5 km SW of Wedderburn, Nine Mile South Rd, *J.Connock 368* (MEL); on the Wychitella to Wedderburn road, 4 km N of Calder Hwy, *T.B.Muir 4904* (K, MEL); Brisbane to Anakie Ra. area, c. 1.6 km SE of Durdiwarrah Reservoir, 30 Sept. 1945, *R.V.Smith* (K, MEL).

Most populations of *G. micrantha* have linear leaves 0.8–1.0 mm wide, with the margins fully refracted against the abaxial midvein. There is considerable variation in the coarseness of foliage. Populations in the Portland area, the 'Mt Clay form', have the leaves sublinear to narrowly elliptic or narrowly oblanceolate, to 1.5 mm wide, the margins less refracted with the lower surface often exposed on either side of the midvein, and the veins of the upper surface less granular-scabrid. Some Brisbane Ra. plants also have the leaf lower surface partly exposed; these populations also tend to have fewer-flowered conflorescences.

169. Grevillea gariwerdensis Makinson, *Fl. Australia* 17A: 501 (2000)

T: 2 miles [3.2 km] NE of Halls Gap on Pomonal Rd, Vic., 14 Nov. 1966, *A.C.Beauglehole 22204*; holo: MEL; iso: NSW.

G. linearifolia form 'i' (Grampians form), of D.J.McGillivray & R.O.Makinson, *Grevillea* 345 (1993).

G. sp. aff. micrantha Grampians, of P.M.Olde & N.R.Marriott, *Grevillea Book* 3: 25 (1995).

G. sp. 1, of R.O.Makinson, *Fl. Victoria* 3: 846, 856, fig. 173h (1996).

[*G. linearifolia auct. non* (Cav.) Druce: I.R.McCann, *Grampians in Flower* 92 (1994)]

Illustrations: I.R.McCann, *Grampians in Flower* 92 (1994), as *G. linearifolia*; P.M.Olde & N.R.Marriott, *Grevillea Book* 3: 24 (14C), 25 (top centre) (1995), as *G. sp. aff. micrantha* Grampians.

Shrub 0.3–1 m tall. Branchlets angular and ridged, subsericeous between ridges, often secund. Leaves sublinear to narrowly oblong-elliptic or -obovate, often slightly incurved, 1–3 (–5) cm long, 1–3 mm wide; upper surface with midvein and intramarginal veins usually prominent and faintly granulose; margins shortly refracted; lower surface usually partly exposed, subsericeous. Conflorescence usually terminal, erect, level with or scarcely exceeded by leaf apices, usually pedunculate (peduncles to 15 mm long), regular, umbelloid, 6–16-flowered. Flowers adaxially oriented. Flower colour: perianth white to pink with brownish limb (hairs); style white to pink. Perianth subsericeous to tomentose outside, scantily bearded (rarely a few hairs only) inside opposite ovary. Pistil 7–8.5 mm long; style sharply geniculate 1–2 mm below apex, with minute hairs or papillae within apical 1 mm; pollen-presenter oblique. Follicle narrowly ovoid, 9–10 mm long, faintly warty. Fig. 23E–F.

Occurs in Vic. in the Grampians Ra. Grows usually in low moist heath in sandy soils. Regenerates from seed or rhizomes. Presumed insect-pollinated. Flowers Oct.–Jan. Map 220.

Vic.: Black Rd, Oct. 1929, *J.W.Audas* (MEL); 16 km N of Halls Gap, Deep Ck, on Mt Zero road, Oct. 1955, *N.F.Learmonth* [*A.C.Beauglehole 7463*] (MEL); Plantation Rd, 12.8 km N of Halls Gap, *T.B.Muir 2228* (K, MEL).

Grevillea gariwerdensis is narrowly distinct from *G. micrantha*, which has the upper leaf surface often densely granulose, and lacks a beard on the inner surface of the perianth or at most has a few hairs surmounting a cushion-like pulvinus opposite the ovary. *Grevillea gariwerdensis* usually has a more conspicuous (albeit still scanty and short-haired) beard; it also has coarser, broader leaves, and a tendency to pedunculate or pseudo-pedunculate conflorescences (sometimes some sessile). *Grevillea gariwerdensis* is very similar indeed to *G. parviflora*, a N.S.W. endemic, which has a profuse beard of longer hairs on the inner surface of the perianth.

170. **Grevillea wiradjuri** Makinson, *Fl. Australia* 17A: 501 (2000)

T: 19 km W of Temora [towards] Griffith, N.S.W., 13 Nov. 1975, *M.D.Crisp 1529*; holo: CANB; iso: A *n.v.*, AD, NSW.

G. linearifolia '39f Western Slopes form', of R.O.Makinson, *Fl. New South Wales* 2: 52 (1991).

G. linearifolia 'form k' (Western Slopes form), of D.J.McGillivray & R.O.Makinson, *Grevillea* 346 (1993).

G. sp. aff. neurophylla Temora-Barmedman, of P.M.Olde & N.R.Marriott, *Grevillea Book* 3: 51 (1995).

Illustrations: P.M.Olde & N.R.Marriott, *Grevillea Book* 3: 51 (top right & 36) (1995), as *G. sp. aff. neurophylla* Temora-Barmedman.

Open erect or dwarf shrub 0.2–1.8 m tall. Branchlets angular and ridged, subsericeous. Leaves linear, widely ascending to spreading, not crowded, very straight, (2–) 3–7.5 cm long, 0.7–1.3 (–2.0) mm wide, not pungent, pliable; upper surface with midvein and intramarginal veins obscure to slightly prominent and scabrid to faintly granular; margins vertically refracted (the angles soft); lower surface usually completely enclosed except glabrous midvein, 2-grooved. Conflorescence terminal, erect, sessile, regular and umbelloid, exceeded by the ascending foliage, c. 8–20-flowered. Flowers adaxially oriented. Flower colour: perianth white with a brown limb (hairs); style white sometimes becoming tinged pink or red with age. Perianth subsericeous to sparsely so outside, conspicuously bearded inside opposite ovary. Pistil 7–9 mm long; style strongly curved near apex, with minute hairs or papillae in apical c. 0.5 mm; pollen-presenter oblique. Follicle narrowly ovoid-ellipsoidal, 9–10 mm long, glabrous, faintly warty. Fig. 24G–I.

Occurs in N.S.W., known from the area bounded by Temora, Barmedman and Ariah Park, disjunctly to the NE at Bumberry near Parkes, and in Goobang Natl Park (Harvey Ra.). Grows in eucalypt woodland or tall open ironbark forest in dry sites (usually on low ridges), in stony clay-loam soils. Regenerates from seed or (Bumberry populations, not confirmed elsewhere) seed and rhizomes. Presumed insect-pollinated. Flowers Aug.–Dec. Map 221.

N.S.W.: Ingalba Nature Reserve, 10 km W of Temora, *J.Brickhill 605–4* (NSW); Griffith district, *T. van den Brock 671* & *E.Richards* (CANB, DNA); 9.8 km by road from Temora P.O. on main road to Griffith, Ingalba Nature Reserve, S side of road, *R.O.Makinson 1303* & *D.J.Mallinson* (AD, BRI, CANB, HO, K, L, MEL, NSW, PERTH); Bumberry, *M.Tindale NSW94150* (NSW).

Grevillea wiradjuri is very similar to non-riparian forms of *G. neurophylla*, which differ in the leaves being usually shorter (2–4 cm long) and broader (1.2–2 mm wide), more ascending, and often curved or wavy, the conflorescence usually pedunculate, the pedicels slightly longer at 4.5–5 mm long (c. 3.5 mm long in *G. wiradjuri*) and the pistil usually shorter (6–7.5 mm long), and occur in wetter, montane habitats. Also, in *G. neurophylla* the leaf margins is usually more angularly once-refracted at more than 90° (in *G. wiradjuri*, twice-refracted, with a short wall-like section roughly perpendicular to the (topographically) upper leaf surface, before a smoother secondary refraction towards the abaxial midrib).

Figure 24. *Grevillea*. **A–C**, *G. halmaturina* subsp. *halmaturina*. **A**, flowering branch; **B**, detail of lower surface of leaf (**A–B**, E.J.Carroll, 29 Sept. 1965, CBG020897, CANB). **C–E**, *G. halmaturina* subsp. *laevis*. **C**, flowering and fruiting branch; **D**, cross-section and detail of upper surface of leaf; **E**, detail of lower surface of leaf (**C–E**, J.D.Briggs 1319, CANB). **F–H**, *G. wiradjuri*. **F**, flowering branch; **G**, detail of lower surface of leaf; **H**, fruit (**F–H**, M.D.Crisp 1529, CANB). **I–J**, *G. imberbis*. **I**, flowering branch; **J**, detail of lower surface of leaf (**G–H**, I.R.Telford 10106 & M.D.Crisp, CANB). Scale bars: **A**, **C**, **F**, **H**, **I** = 5 mm; **B**, **D–E**, **G** = 1 mm; **J** = 2 mm. Drawn by C.Payne.

173. Grevillea australis R.Br., *Trans. Linn. Soc. London* 10: 171 (1810)

T: 'In Insula Diemen; plagis australioribus; ad fluviorum ripas' [protologue]; lecto: [verso] Ad ripas fluv: Derwent rarius [Derwent R., Tas.], *s.d.*, *R.Brown Iter Austral. 3338*; lecto: BM, *fide* D.J.McGillivray & R.O.Makinson, *Grevillea* 406 (1993); isolecto: K; remaining syntypes: BM (as for lecto, rest of sheet), E, ?K, ?NSW 94049.

G. tenuifolia R.Br., *Trans. Linn. Soc. London* 10: 171 (1810); *G. australis* var. *tenuifolia* (R.Br.) Meisn., *in* A.L.P.P. de Candolle, *Prodr.* 14: 360 (1856), as η *tenuifolia*. T: Cataract R., Port Dalrymple, [Tas.], *s.d.*, *R.Brown Iter Austral. 3339*; lecto: K, *fide* D.J.McGillivray & R.O.Makinson, *Grevillea* 406 (1993); remaining syntypes: BM, K.

G. australis var. *erecta* Hook.f., *in* W.J.Hooker, *Hooker's London J. Bot.* 6: 282 (1847), as var. α *erecta*. T: Launceston, [Tas.], 5 Oct. 1841, [*R.C.Gunn*] *535/1842*; lecto: K, *fide* D.J.McGillivray & R.O.Makinson, *Grevillea* 406 (1993); isolecto?: A *n.v.*

G. australis var. *brevifolia* Hook.f., *in* W.J.Hooker, *Hooker's London J. Bot.* 6: 282 (1847), as var. ε *brevifolia*. T: Westn Mts, [Tas.], 18 Feb. [18]43, [*R.C.Gunn*] *1260/1842*; holo: K.

G. australis var. *linearifolia* Hook.f., *in* W.J.Hooker, *Hooker's London J. Bot.* 6: 282 (1847), as var. β *linearifolia*. T: South Esk [R.], Launceston, [Tas.], 14 Oct. [18]43, [*R.C.Gunn*] *534*; lecto: K, *fide* D.J.McGillivray & R.O.Makinson, *Grevillea* 406 (1993); isolecto?: A *n.v.*, K, NSW 94056.

G. australis var. *montana* Hook.f., *in* W.J.Hooker, *Hooker's London J. Bot.* 6: 282 (1847), as var. δ *montana*. T: Hampshire Hills, [Tas.], *R.C.Gunn 199* (=*J.Milligan 543*); lecto: K, *fide* D.J.McGillivray & R.O.Makinson, *Grevillea* 407 (1993); isolecto: CGE *n.v.*, FI *n.v.*, HO, K, LE *n.v.*, MEL, NSW; remaining syntypes: Summit Western Mts 20 Dec. [18]43, [*R.C.Gunn*] *730*; syn: CGE *n.v.*, K, NSW, TCD *n.v.*; Tasmania, *J.D.Hooker*; syn: MEL.

G. australis var. *planifolia* Hook.f., *in* W.J.Hooker, *Hooker's London J. Bot.* 6: 282 (1847), as var. γ *planifolia*. T: common on the North and South Esk Rivers, near Launceston, [Tas.], 19 Oct. [18]39, [*R.C.Gunn*] *535/1842*; lecto: K, *fide* D.J.McGillivray & R.O.Makinson, *Grevillea* 406 (1993); remaining syntypes: BRI, CGE *n.v.*, K, NSW.

G. australis var. *subulata* Hook.f., *in* W.J.Hooker, *Hooker's London J. Bot.* 6: 282 (1847), as var. ζ *subulata*. T: South Esk [R.], [Tas.], *R.C.Gunn 1240/1842*; lecto: K, *fide* D.J.McGillivray & R.O.Makinson, *Grevillea* 407 (1993); ?isolecto: K, NSW.

G. stuartii Meisn., *Linnaea* 26: 357 (1854). T: Tasmania, *s.d.*, *C.Stuart*; holo: NY.

G. amplifica F.Muell., manuscript name sub *G. stuartii* Meisn., *Linnaea* 26: 358 (1854), *nom. nud.*

Illustrations: Anon, *Guide Fl. & Pl. Tasmania* 57, t. 119 (1981); D.J.McGillivray & R.O.Makinson, *Grevillea* 349 (1993); P.M.Olde & N.R.Marriott, *Grevillea Book* 2: 45 (bottom right), 46 (32A, B) (1995).

Open to dense, prostrate to erect shrub 0.2–1 (–3) m tall. Branchlets angular to terete, sericeous to tomentose. Leaves narrowly elliptic to narrowly obovate to linear or subterete, 0.3–5.0 cm long, 0.5–5.5 mm wide; upper surface flat to convex, often glossy; margins (often unevenly) revolute to recurved; lower surface enclosed and 1-grooved, or partly to wholly exposed and usually sericeous. Conflorescence terminal or subterminal-axillary, erect, sessile or shortly pedunculate, regular, umbelloid, 2–16-flowered. Flowers adaxially acroscopic. Flower colour: perianth and style white to cream; style sometimes becoming pink to dull red. Perianth subsericeous or loosely so outside, glabrous or nearly so inside with a near-basal cushion-like pulvinus (this sometimes bearing a few erect hairs). Pistil 6–7.5 mm long; style strongly incurved, often hooked near apex, with papilloid or short erect hairs in apical 1 mm; pollen-presenter oblique. Follicle narrowly ovoid, 8.5–9.5 mm long, with a longitudinal line or shallow furrow along either side. *Alpine Grevillea.* Fig. 25A–D.

Occurs in montane south-eastern mainland Australia and Tas.: at alpine and subalpine altitudes in the western mountains of the A.C.T., Snowy Mtns of N.S.W. (and one old record at G from Marulan), and eastern highlands of Vic. and in Tas. widespread from near sea-level to c. 1400 m. Grows in heath, woodland, or herbfield, often in moist and/or rocky situations. Regenerates from seed. Insect-pollinated. Flowers Sept.–Jan. Map 225.

A.C.T.: Snowy Flats, Bimberi Ra., *P.J.Darbyshire 84* (CANB, NSW). N.S.W.: Kiandra, Dec. 1901, *W.Forsyth* (B *n.v.*, MEL, NSW). Vic.: Mt Buller, slopes of Baldy, *R.Melville 3216 et al.* (BRI, K, NSW, PERTH); tributary of Nigothoruk Ck, 2.5 km SW of Mt Wellington, *T.B.Muir 3089* (MEL). Tas.: 3 km from Breona on Great L. near Pine L., *N.T.Burbidge 3452* (CANB, HO).

Grevillea australis is the only *Grevillea* occurring naturally in Tas. On the mainland, it is best distinguished from related and similar species in the *G. linearifolia* alliance by a combination of pistil length and (when exposed) the lower surface of the leaf. In *G. australis* the abaxial midvein is covered by hairs of a similar density to those on the adjacent lamina, and the midvein is either not prominent beneath the hairs or prominent only in the lower half of the leaf. See Fig. 25B–D. This feature is shared by taxa in the *G. juniperina* alliance, but these all have much longer pistils (≥ 15 mm long); all species likely to be confused with *G. australis* have the abaxial midvein evident over the full length of the leaf and nearly always either glabrous or with a markedly less dense indumentum than the adjacent lamina.

Within *G. australis* there is little floral variation, but much diversity in habit and leaf form, especially in Tas. Many populations and individuals are intermediate between Hooker's varietal circumscriptions. Infraspecific taxa do seem to exist, but formal recognition is at this stage premature, as further investigation of variation (especially in the field) is needed. Prostrate forms generally have broader leaves with the lower surface more exposed. Mainland populations often have the leaves more or less obovate with obtuse apices and with 3–7 nerves evident on the upper leaf surface, whereas those from Tas. are usually subterete or linear to narrowly elliptic, with acute apices, and generally have only the midvein evident. A particularly unusual Tasmanian population occurs in riparian habitat on O'Connors Rivulet, N of Swansea; this is a robust shrub, 2 m tall, with scarcely revolute leaves.

174. Grevillea quinquinervis J.M.Black, *Trans. & Proc. Roy. Soc. S. Australia* 33: 225, t. XIV *p.p.* (1909).

T: 'Collected by Mr Griffith at Snug Cove, Harvey's Return, and Ravine Creek, K.I., Oct. 1908, and by Dr R.S. Rogers near the same localities about a month later' [protologue]; lecto: Kangaroo Is., [S.A.], Harvey's Return, Oct. 1908, *Mr Griffith*; lecto: AD *p.p.*, *fide* D.J.McGillivray & R.O.Makinson, *Grevillea* 437 (1993); remaining syntypes: AD *p.p.*, MEL, NSW.

Illustrations: J.P.Jessop & H.R.Toelken (eds), *Fl. S. Australia*, 4th edn, 1: 134, fig. 71E (1986); P.M.Olde & N.R.Marriott, *Grevillea Book* 3: 127 (top right & 99A); 128 (99B, C) (1995).

Erect dense shrub 1–1.5 m tall. Branchlets angular and ridged, sericeous to tomentose. Leaves sublinear to oblong-elliptic or narrowly obovate, 1.5–6 cm long, 1.5–8 mm wide; upper surface minutely punctate, with (3–) 5 (–7) prominent longitudinal parallel veins; margins smoothly and strongly recurved to revolute; lower surface sometimes partly enclosed, sericeous including on midvein. Conflorescence terminal on short lateral branchlets, erect, sessile or rarely pedunculate, regular, umbelloid, 6–20-flowered. Flowers adaxially oriented. Flower colour: perianth and style pale to deep pink. Perianth subsericeous to tomentose outside, bearded inside. Pistil 9.5–12 mm long; style sharply curved, with papilloid hairs in apical 1 mm; pollen-presenter oblique. Follicle ovoid-ellipsoidal or obloid, 13–15 mm long. *Five-nerved Grevillea.* Plate 39.

Occurs in S.A., where restricted to the western half of Kangaroo Is. Grows in heath or shrubby woodland in sandy soils, over ironstone and ?limestone. Regenerates from seed. Probably insect-pollinated. Flowers Aug.–Feb. Map 226.

S.A.: W boundary of Kelly Hill reserve c. 12 km ENE of Cape du Couedic, *P.G.Wilson 722* (AD, B *n.v.*, UC *n.v.*); Cape Borda, Dec. 1909, *J.M.Black* (MEL); 8 km from Rocky R. on Western Hwy, *M.E.Phillips CBG020661* (AD, CANB); 14.4 km from Rocky R., at Breakneck Ck on track around coast to Cape Borda, *M.E.Phillips CBG016860* (CANB, NSW).

Rarely confused with any other species; the leaves are distinctively and prominently (3–) 5 (–7)-nerved above, sometimes with a reticulum evident, and the leaf apex is usually very obtuse (and sometimes slightly emarginate).

Costata Subgroup

Unit conflorescence mid-dense, ±umbelloid, erect. Floral rachis 0.5–5 mm long. Tepals separating at least to ovary, usually widely independently arching or splaying. Pistils 7–12 mm long. Pollen-presenter oblique. Follicle with faint to conspicuous longitudinal ridges.

Three species in south-western W.A. Insect pollinated.

175. Grevillea costata A.S.George, *Nuytsia* 1: 370 (1974)

T: Murchison R., W.A., 30 Aug. 1931, *C.A.Gardner 2597*; holo: PERTH; iso: K, NSW.

Illustrations: D.J.McGillivray & R.O.Makinson, *Grevillea* 355, fig. 86a (1993); P.M.Olde & N.R.Marriott, *Grevillea Book* 2: 101 (bottom right), 102 (81A–C) (1995).

Dense shrub 0.5–1.5 m tall. Branchlets angular and ridged, sericeous between ridges. Leaves linear, 1.5–4.5 cm long, 0.8–1.3 mm wide; upper surface flat to convex; margins revolute; lower surface enclosed except for midvein, 2-grooved. Conflorescence terminal or axillary, erect, pedunculate, loosely regular, subumbelloid, c. 4–10-flowered. Flowers adaxially oriented. Flower colour: perianth white; style white to cream, sometimes tinged pink. Perianth subsericeous outside, bearded inside. Pistil 7–9.5 mm long; style incurved, with minute papilloid or erect hairs in apical 1 mm; pollen-presenter oblique. Follicle ellipsoidal to ovoid, 9.5–11 mm long, with several very prominent regular longitudinal flange-like ridges.

Occurs in south-western W.A., restricted to lower reaches of the Murchison R. Grows in or near the river-bed in alluvial sands and among rocks. Regenerates from basal shoots and possibly lignotuber after floods. Flowers May–Sept. Map 227.

W.A.: North West Coastal Hwy at Murchison R. Bridge, *R.Filson 8626* (MEL); Murchison R., *C.A.Gardner 13206* (PERTH); 589 km N of Perth, near Murchison R. on North West Coastal Hwy, 10 Sept. 1965, *F.W.Humphreys* (PERTH); 200 m W of bridge over Murchison R. on North West Coastal Hwy, *D.J.McGillivray 3330 & A.S.George* (NSW); Murchison R. at base of Ross Graham Lookout, Kalbarri Natl Park, *P.G.Wilson 6616* (PERTH).

The heavily ridged fruits are highly diagnostic, and unusual in the group. *Grevillea costata* is otherwise similar to the closely related *G. christinae*, which has the leaf lower surface usually partly exposed and glabrous. *Grevillea inconspicua* has grey-green foliage with vertically oriented, basally deflexed leaves.

This species is recognised as 'Poorly Known' in J.D.Briggs & J.H.Leigh, *Rare or Threatened Australian Plants* (1995).

176. Grevillea inconspicua Diels, *Bot. Jahrb. Syst.* 35: 153 (1904)

T: Hab. in distr Austin prope Cue ..., W.A., June–July 1904, *L.Diels 3277*; holo: B.

G. brachyclada W.Fitzg., *J. W. Australia Nat. Hist. Soc.* 2: 30 (1905). T: Greenough R., W.A., July 1903, *C.R.P.Andrews*; holo: NSW.

Illustrations: B.L.Rye & S.D.Hopper, *Guide Rare Fl. W. Australia* 14 (1981); D.J.McGillivray & R.O.Makinson, *Grevillea* 355, fig. 85 (1993); P.M.Olde & N.R.Marriott, *Grevillea Book* 2: 207 (bottom right), 208 (170A, 170A bis) (1995).

Dense prickly shrub 0.6–1.0 (–1.8) m tall. Branchlets wiry, subterete, slightly flexuose, subsericeous to glabrous. Leaves linear, 1–5 cm long, 0.8–1.3 (–1.8) mm wide, deflexed at base and often slightly twisted; upper surface convex to peaked along midline; margins refracted; lower surface enclosed except for midvein, 2-grooved. Conflorescence terminal, erect, pedunculate, loosely regular and subumbelloid, 6–8-flowered. Flowers adaxially oriented. Flower colour: perianth off-white to silvery grey; style white to pale pink. Perianth subsericeous outside, bearded inside. Pistil 9.5–11.5 mm long; style sharply hooked near apex, with minute papilloid hairs in apical 1 mm; pollen-presenter oblique. Follicle subellipsoidal, 7–14 mm long, with faint longitudinal ridges. *Cue Grevillea*. Fig. 25E–G.

Occurs in the central west of W.A., in a few localities near Meekatharra, Cue, Weld Ra., and NE of Sandstone. Grows in open to sparse shrub associations, usually along drainage lines and gullies, in red clay-loam over green-stone. Regeneration mode unknown, probably regenerates from seed only. Flowers July–Sept. Map 228.

W.A.: Cue, *W.E.Blackall 119* (PERTH); Cue, 27 July 1927, *C.A.Gardner s.n.* (MEL, PERTH); NE side of Cue Hill, Cue, *D.J.McGillivray 3373 & A.S.George* (B, CANB, K, NSW, US); Meekatharra, *A.A.Mitchell 990* (NSW); Weld Ra., *E.Wittwer 1259* (PERTH).

The silvery grey-green foliage consists of distinctively basally deflexed leaves, usually oriented more or less vertically and descending. The leaf deflexion is diagnostic as compared to similar species, including *G. christinae* which has the straight leaves usually broader (1–6.5 mm wide).

This species is recognised as 'Vulnerable' in J.D.Briggs & J.H.Leigh, *Rare or Threatened Australian Plants* (1995).

177. Grevillea christinae McGill., *New Names Grevillea* 4 (1986), as *G. christiniae*

T: near Mortlock R., 9 km SW of Goomalling, W.A., 1 Sept. 1979, *A.S.George 15738*; holo: PERTH; iso: NSW, PERTH.

Illustrations: P.M.Olde & N.R.Marriott, *Grevillea Book* 2: 88 (bottom right), 89 (70A, B) (1995); A.Brown *et al.* (eds), *W. Australia's Threatened Flora* 96 (1998).

Dense shrub 0.5–1.0 m tall. Branchlets angular, slightly flexuose, often secund, subsericeous. Leaves linear to narrowly elliptic or slightly oblanceolate, 2–6 cm long, 1–6.5 mm wide; upper surface smooth with midvein and flanking intramarginal veins evident; margins revolute; lower surface exposed and glabrous or enclosed except for midvein and then 2-grooved. Conflorescence terminal or axillary, erect, shortly pedunculate, regular and umbelloid, 8–12-flowered. Flowers adaxially oriented. Flower colour: perianth white; style white, sometimes becoming pink or red. Perianth sericeous outside, bearded inside. Pistil 7–8.5 mm long; style sharply hooked just below apex and with minute papilloid hairs in apical 1 mm; pollen-presenter oblique. Follicle narrowly ovoid-ellipsoidal, 10–15 mm long, with faint longitudinal ridges.

Occurs in south-western W.A., north of Perth, where restricted to scattered populations between Watheroo and Goomalling. Grows in eucalypt woodland and shrubland in clay loam or sandy clay soils. Regenerates apparently from seed only. Flowers ?July–Sept. Map 229.

W.A.: *loc. incert.* [W.A.], *J.Drummond* (MEL); 6.5 km E of Watheroo, *s.d.*, *A.R.Main s.n.* (PERTH); 45 km N of Moora on Geraldton highway, *M.D.Tindale 1271* (NSW).

Narrower-leaved plants have some similarity to *G. hakeoides* subsp. *stenophylla*; the latter taxon however has denser ovoid conflorescences and pistils completely glabrous (lacking papillae near apex).

This species is recognised as 'Endangered' in J.D.Briggs & J.H.Leigh, *Rare or Threatened Australian Plants* (1995).

Victoriae Subgroup

Unit conflorescence dense to loose, acropetal, broadly secund or a loose cluster or regular and then conico-cylindrical, decurved to pendulous or sometimes erect. Floral rachis (2–) 5–60 (–90) mm long. Tepals mostly coherent in basal half, independently recurved or recoiled above. Pistil 12–26 mm long. Pollen-presenter lateral or very oblique. Follicle smooth, colliculate or with faint longitudinal subdorsal ridging along either side.

Fourteen species, mostly south-eastern Australia (south-eastern Qld, N.S.W., Vic.) with two species in central Qld. Primarily bird-pollinated. Many of the taxa in this subgroup have a rather narrow, straight erect perianth with the limb abruptly nodding or deflexed; in these taxa the style in late bud stage is sometimes weakly exserted to the lower rear of the perianth rather than arching above the perianth curve as in related subgroups (e.g. *G. speciosa* and relatives).

McGillivray & Makinson (*Grevillea* 320–324 (1993)) adopted a superspecies concept of *G. victoriae* with an informal interim structure of 11 races and several unassigned populations. Olde & Marriott (*Grevillea Book* 3: 223–225 (1995)) reinstated three named varieties and Makinson (*Telopea* 7: 129–138 (1997)) segregated some as species and subspecies. Most of McGillivray and Makinson's races are here recognised at formal ranks (equivalences with earlier informal designations are noted in the text).

One new character introduced here is *cross-sectional shape of the pollen-presenter*; this may require magnification and actual sectioning of fresh or wet-pickled material; it cannot usually be reliably determined from dried specimens.

Delimitation and reliable keying in the alliance remains problematic, especially for the 'core' taxa of the alliance (*G. victoriae* subsp. *victoriae* and subsp. *nivalis*, *G. parvula*, *G. brevifolia* subsp. *brevifolia* and subsp. *polychroma*, *G. monslacana* and *G. epicroca*), which represent a radiation of morphologically very similar micro-species. V.Stajsic and W.Molyneux are developing (at MEL) a synoptic review of the subgroup. Several new names for taxa in the 'core' of the alliance are here published by them to facilitate availability of names, and they here reinstate *G. brevifolia* F.Muell. ex Benth. Data for descriptions of these taxa derive from both their work and Makinson's. Comments on diagnosis and character reliabilty in notes after the descriptions are Makinson's responsibility.

Some populations remain unassigned. One, near Bemboka, N.S.W., is mentioned under *G. parvula* below, and one in the NW corner of the A.C.T. under *G. victoriae* subsp. *nivalis*.

178. Grevillea victoriae F.Muell., *Trans. Philos. Soc. Victoria* 1: 107 (1855)

T: Vic., 'Along the waters of the Buffalo Range, on the summits of Mt Buller and Mt Tambo, on the sources of the Mitta Mitta, at Mt Hotham and Mt Latrobe.' [protologue]; lecto: MEL, *fide* D.J.McGillivray & R.O.Makinson, *Grevillea* 447 (1993); remaining syntypes: ?A *n.v.*, ?B *n.v.*, K, ?K, ?LE *n.v.*, MEL (several sheets), ?MEL, ?NY *n.v.*, NSW. [some of these collections may belong to taxa here recognised as distinct, e.g. the Mt Tambo syntype locality almost certainly relates to *G. brevifolia*, and to the Type collection of that species].

G. victoriae 'race c', of D.J.McGillivray & R.O.Makinson, *Grevillea* 321 (1993).

G. victoriae var. *victoriae*, of P.M.Olde & N.R.Marriott, *Grevillea Book* 3: 234 (1995), *p.p.*

G. victoriae 'race c', of R.O.Makinson, *Fl. Victoria* 3: 852 (1996).

Erect to spreading shrub 1–3 m tall. Branchlets terete, subsericeous or rarely tomentose. Leaves elliptic or narrowly elliptic to lanceolate, rarely oblanceolate, (2–) 6–12 (–20) cm long, 10–45 mm wide; upper surface smooth, with lateral veins conspicuous; margins flat or shortly recurved; lower surface densely sericeous to subsericeous or subtomentose. Conflorescence terminal or axillary, decurved to pendulous, pedunculate, simple or few-branched; unit conflorescence regular, conico-cylindrical or a loose ovoid cluster, many-flowered; floral rachis (10–) 20–60 (–90) mm long. Perianth densely subsericeous or very shortly tomentose outside, bearded inside. Pistil 17–26 mm long, minutely pubescent near the apex. Pollen-presenter lateral or almost so, the base not decurrent to the style, the face flat to convex. Follicle 17–20 mm long. *Royal Grevillea.*

Occurs in south-eastern N.S.W. and montane Vic. Two subspecies are recognised. References in McGillivray & Makinson (1993) to occurrences of *G. victoriae* in Qld refer to the taxon now recognised as *G. hockingsii*.

The floral rachis indumentum of *G. victoriae* is often dead white; contrasting strongly with the (usually rusty-brown) flower buds. Other taxa sharing this feature are *G. diminuta*, which has much smaller leaves ⩽ 2 cm long and a shorter pistil 10–11 mm long, and *G. oxyantha* subsp. *ecarinata*, which has a subpyramidal limb of the flower bud (in *G. victoriae* and *G. diminuta* the limb is subglobose and obtuse). Other related taxa have the rachis hairs reddish or brownish to fawn, or rarely pale but the with the buds usually red, not brown.

New vegetative growth in *G. victoriae* is green or rusty brown, never pink (as it is in *G. parvula* and *G. epicroca*) and the pollen-presenter cross-section shape also differs from those species. The dense indumentum of the adult leaf lower surfaces also distinguish it from *G. epicroca* and from *G. monslacana*, which both have an open indumentum of appressed hairs on the lower surface of adult leaves, with ground tissue clearly visible between the hairs.

Grevillea victoriae is distinguished from *G. brevifolia* subsp. *brevifolia* which has shorter, more obtuse leaves 1.5–3 (–4.5) cm long, obscure venation on the leaf upper surface , and shorter floral rachises ≤ 25 mm long. *Grevillea brevifolia* subsp. *polychroma*, *G. parvula* and *G. epicroca* have floral rachises usually ≤ 20 mm long, bearing subsecund unit conflorescences, and the perianth outer surface lacks the dense rusty brown indumentum of *G. victoriae*. *Grevillea parvula* also often has much smaller leaves.

Leaves usually narrowly elliptic to lanceolate (rarely oblanceolate), 6–12 cm long, 10–45 mm wide, with length/width ratio of (3:1–) 4:1–5:1 (–6:1); upper surface dull; leaf margins shortly recurved; floral rachis 40–90 mm long **178a.** subsp. **victoriae**

Leaves elliptic to narrowly so or rarely ovate, usually 3.5–10 (–13.5) cm long, 10–37 mm wide, with length/width ratio of 2.5:1–4:1 (–5:1); upper surface usually glossy; leaf margins flat or very shortly recurved; floral rachis 10–50 (–60) mm long **178b.** subsp. **nivalis**

178a. Grevillea victoriae F.Muell. subsp. **victoriae**, *Trans. Philos. Soc. Victoria* 1: 107 (1855)

G. victoriae 'race c', of D.J.McGillivray & R.O.Makinson, *Grevillea* 321 (1993).

G. victoriae var. *victoriae*, of P.M.Olde & N.R.Marriott, *Grevillea Book* 3: 234 (1995), *p.p.*

G. victoriae 'race c', of R.O.Makinson, *Fl. Victoria* 3: 852 (1996).

Illustrations: P.M.Olde & N.R.Marriott, *Grevillea Book* 3: 224 (184B) (1995); N.G.Walsh & T.J.Entwisle (eds), *Fl. Victoria* 3: 853 (fig. 172e, f) (1996); G.R.Cochrane *et al.*, *Fls & Pls Victoria* 161 (506) (1968).

Erect to spreading shrub 1–3 m tall. Branchlets subsericeous. Leaves narrowly elliptic to lanceolate, rarely oblanceolate, (4–) 6–12 (–20) cm long, 10–25 (–45) mm wide, length/width ratio of (3:1–) 4:1–5:1 (–6:1); upper surface smooth, dull, with the lateral veins conspicuous; margins shortly recurved; lower surface densely sericeous to subsericeous. Unit conflorescence decurved to pendulous, conico-cylindrical or a loose ovoid cluster; floral rachis 40–90 mm long. Flower colour: buds rusty brown or red-brown (hairs), open flower red with brown hairs outside perianth, reddish inside; style reddish. Perianth densely subsericeous or very shortly tomentose outside. Pistil 17–24 mm long. Pollen-presenter face convex in cross-section. Follicle (17–) 19–20 mm long.

Endemic in Vic., primarily on the Mt Buller, Mt Buffalo and Mt Bogong massifs, also on Mt Torbreck, Mt St Bernard, Mt Hotham and Mt Howitt. Grows in high montane forest (e.g. *Eucalyptus delegatensis*) and snow-gum woodland, in rocky sites usually on granite. Regenerates from seed. Flowers mainly Oct.–Jan. Map 230.

Vic.: Mt McKay to Howmans Gap Rd, 3 km W of Falls Creek, *L.G.Adams 2647* (CANB, K, L, MEL, NSW); Mt Buffalo plateau, *R.Hill 1267* (AD, CANB); Macalister Springs near Mt Howitt, *A.C.Beauglehole 40788* (CANB, MEL, NSW).

Grevillea victoriae subsp. *victoriae* has green new vegetative growth. See under subsp. *nivalis* for differences.

178b. Grevillea victoriae subsp. **nivalis** Stajsic & Molyneux, *Fl. Australia* 17A: 502 (2000)

T: near Tooma Reservoir, N.S.W., 29 Apr. 1973, *R.J.Chinnock 298*; holo: CANB; iso: AD.

G. victoriae 'race d', of D.J.McGillivray & R.O.Makinson, *Grevillea* 321 (1993).

G. victoriae var. *victoriae*, of P.M.Olde & N.R.Marriott, *Grevillea Book* 3: 234 (1995), *p.p.*

Illustration: P.M.Olde & N.R.Marriott, *Grevillea Book* 3: 224 (184A) (1995), a cultivated form, as *G. victoriae* 'Murray Queen'; A.Costin *et al.*, *Kosciusko Alpine Flora* 151, 152 (1982).

Erect to spreading shrub 1–2 m tall. Branchlets subsericeous or rarely tomentose. Leaves elliptic or occasionally narrowly so, rarely ovate, (2–) 3.5–10 (–13.5) cm long, 10–37 mm wide, length/width ratio of 2.5:1 to 4:1 (rarely 5:1); upper surface smooth, usually glossy, with conspicuous lateral veins; margins flat or occasionally very shortly recurved; lower surface densely subsericeous to subtomentose. Unit conflorescence decurved to pendulous, pedunculate, simple or basally few-branched; unit conflorescence regular, shortly conico-cylindrical or a loose ovoid cluster; floral rachis 10–50 (–60) mm long. Flower colour: perianth (indumentum) rusty brown in bud, becoming red-brown or red outside, reddish inside; style reddish. Perianth densely and very shortly tomentose outside. Pistil (18–) 20–23 (–26) mm long; pollen-presenter face flat to convex in cross-section. Follicle 18–20 mm long. Plate 37.

Occurs in N.S.W. on the western flank of the Kosciuszko massif, from The Pilot N to Talbingo and towards Tumut; possibly also just in Vic. near Tom Groggin. Grows in rocky situations in high montane forest and snow-gum woodland, in shallow soils on granites and slates. Regenerates from seed. Flowers Sept.–Dec., sometimes to Apr. Map 231.

N.S.W.: near Junction Shaft camp site, on track to gauging station in Happy Jacks R., *M.E.Phillips CBG024635* (CANB, NSW); between Pilot Lookout and Dead Horse Gap, Snowy Mtns, *M.E.Phillips & E.J.Carroll CBG017439* (CANB, NSW); Tumut Pond to Tooma road at Burns Ck turnoff, *M.E.Phillips & E.J.Carroll CBG039711* (AD, CANB, DNA).

A geographically isolated and poorly collected population occurring in the far NW of the A.C.T. (Baldy Range area) appears to be closely related to (if not part of) subsp. *nivalis*, despite a disjunction of some 60 km; it is similar except in having short floral rachises c. 15 mm long, and a somewhat shorter pistil (c. 16 mm long). It is suggested that this taxon should be known by the designations "*G.* aff. *victoriae* 'Baldy Range' or '*G.* aff. *victoriae M.Richardson 9* (CBG8601975)' ", to avoid confusion with *G. oxyantha* which was for many years known as *G. victoriae* 'A.C.T. form'.

179. Grevillea brevifolia F.Muell. ex Benth., *Fl. Austral.* 5: 467 (1870)

G. victoriae var. *brevifolia* (F.Muell. ex Benth.) F.Muell. ex Maiden & Betche, *Census New South Wales Pl.* 60 (1916). T: Mount Tambo, 5000', Vic., *Dr Ferd. Mueller*; holo: K; iso: MEL (2 sheets).

Shrub 0.5–2.5 m tall. Branchlets subsericeous or occasionally tomentose. Leaves elliptic or very slightly ovate, 1.1–7 cm long, 6–20 (–30) mm wide; upper surface smooth, usually glossy, lateral veins obscure to conspicuous; margins shortly recurved; lower surface sericeous. Conflorescence terminal, decurved to pendulous, pedunculate, usually simple or up to 3 (4)-branched; unit conflorescence a loose subregular cluster or shortly conico-cylindrical, usually many-flowered; floral rachis 10–15 (–32) mm long. Perianth tomentose outside, bearded inside. Pistil 17–22 mm long; style minutely pubescent towards apex; pollen-presenter flat or occasionally very slightly concave in cross-section. Follicle 18–23 mm long, with faint longitudinal ridges.

Occurs in montane areas of north-eastern Vic. and (subsp. *brevifolia* only) just extending to N.S.W. S of Mt Kosciuszko. There are two subspecies.

Grevillea brevifolia has greenish new vegetative growth. The floral rachises are subtomentose with a usually pale indumentum. The limb of the flower bud is subglobose and obtuse.

Leaves usually 1.5–3 cm long, usually 7–15 mm wide; venation of leaf upper surface consistently obscure; floral bracts (deciduous when buds c. 1 mm long) 1.2–1.4 mm long, c. 0.3 mm wide, linear-crescentic; flowers red **179a.** subsp. **brevifolia**

Leaves usually 3–5 cm long, 10–20 mm wide; venation of leaf upper surface obscure to evident; floral bracts (deciduous when buds c. 1 mm long), 3.2–3.5 mm long, 1 mm wide at widest, spathulate; flowers red, pink, orange, yellow, or cream **179b.** subsp. **polychroma**

179a. Grevillea brevifolia F.Muell. ex Benth. subsp. **brevifolia**

G. victoriae var. *brevifolia* (F.Muell. ex Benth.) F.Muell. ex J.H.Maiden & E.Betche.

G. victoriae 'race e', and 'Unassigned 3', of D.J.McGillivray & R.O.Makinson, *Grevillea* 321 (1993).

G. victoriae 'race e', R.O.Makinson, *Fl. Victoria* 3: 852 (1996).

Low compact to tall open shrub 0.5–3 m tall. Leaves (1.2–) 1.5–3.2 (–4.5) cm long, (6–) 7–14 (–20) mm wide, length/width ratio 2:1 to 2.5:1; venation of upper surface consistently obscure. Floral bracts linear-crescentic, widest at base, 1.2–1.4 mm long, 0.3 mm wide. Pistils 17–21 mm long. Flower colour: perianth and style red.

Occurs in N.S.W. on The Pilot S of Mt Kosciuszko and in north-eastern Vic. in Cobberas-Tingaringy Natl Park, and towards Omeo and Benambra (including Mt Seldom Seen). Grows at subalpine altitudes on exposed rocky sites in open heathy situations, or occasionally on lower slopes in open forest. Regenerates by seed. Flowers Nov.–Dec.(–Feb.?) Map 232.

Vic.: Mt Tingaringy summit, *H. van Rees 051* (CANB, MEL); Wombargo Ra, c. 13 km NNW of Black Mountain Stn, Wulgulmerang, 4 Dec. 1962, *J.H.Willis s.n.* (MEL); Alpine Natl Park, Mt Tingaringy summit, *A.M.Lyne 1216 & J.Lyne* (CANB, MEL).

179b. Grevillea brevifolia subsp. **polychroma** Molyneux & Stajsic, *Fl. Australia* 17A: 502 (2000)

T: Betts Creek Track, 1.2 km E of Tulloch Ard Rd, 9.5 km E of Mt Murrindal, Vic., 9 Feb. 1980, *S.J.Forbes 264*; holo: MEL.

G. victoriae 'race f' *p.p.*, and 'Unassigned 3', of D.J.McGillivray & R.O.Makinson, *Grevillea* 321, 322 (1993).

G. victoriae , 'race f' [Victorian populations only], R.O.Makinson, *Fl. Victoria* 3: 852 (1996).

Open shrub 1–3 m tall. Leaves 3–5 (–7) cm long, (6–) 10–20 (–30) mm wide, length/width ratio 2.1:1 to 3.6:1; venation of upper surface of leaf obscure to conspicuous. Floral bracts spathulate, widest below tip, 3.2–3.5 mm long, c. 1 mm wide. Pistil 18–22 mm long. Flower colour: perianth red to pale lilac or apricot pink or yellow, rarely creamy white; style similar shades but often different to perianth colour except in the yellow form.

Occurs in Vic. in the area N of Buchan, from W Tree Falls N to Tulloch Ard and Butchers Ridge, E to Mt Elizabeth, and W to Wentworth R. near Dargo. Grows in open forest and rocky shrub associations. Regeneration probably from seed only. Flowers Oct.–Feb. Map 233.

Vic.: 'Seldom Seen' [property] at southern end of Jones' Rd, S of Wentworth R., *N.G.Walsh 142* (MEL); on Rodger R., 2.4 km @ 90° E Rodger R./Snowy R. junction, *S.J.Forbes 1777* (MEL); 'W Tree', near Buchan, *[Mrs] L.Hodge s.n.*, Oct. 1947 (MEL).

The difference in size and shape of the floral bracts given above is based on a very few specimens, and the character is rarely available as the bracts are generally caducous when the floral buds are c. 1 mm long; its diagnostic use requires validation.

Subsp. *polychroma* is congruent with those populations of McGillivray & Makinson's 'race f' of *G. victoriae* that lie in Victoria on the catchments of the Buchan and Snowy Rivers, i.e. those mapped as W of 149° longitude (McGillivray & Makinson, *Grevillea* 322, map 90 (1993)). Note that the 'Seldom Seen' property cited above is near Dargo; this subspecies does not occur at Mt Seldom Seen, which is about 75 km to the NE and is a locality for subsp. *brevifolia*.

Olde & Marriott (*Grevillea Book* 3: 225 (1995)), make reference under *G. victoriae* var. *leptoneura* to plants from W Tree Ck, Mt Elizabeth and Mt Kaye appearing to constitute a separate taxon; these are congruent with subsp. *polychroma* as here circumscribed. However, the colour illustration of '*G. victoriae*, form from W Tree Falls' (Olde & Marriott, *loc. cit.*), is unlikely to be subsp. *polychroma*. The plant depicted has acute apices to the flower buds, a feature of *G. oxyantha*; the apparently pale flower colour shown is not characteristic of that species and may be an artefact. *Grevillea oxyantha* is not known to occur naturally there.

Grevillea enthusiast Leo Hodge (1904–199?) bred grevilleas on his property 'Poorinda' at W Tree; the photo may be of a cultivated plant.

Grevillea brevifolia subsp. *polychroma* differs from *G. victoriae* in its smaller and thinner leaves, less conspicuous leaf venation, shorter floral rachises and shorter less regular (often subsecund) unit conflorescences, and possibly in the size and shape of the floral bracts, although floral bracts in *G. victoriae* are poorly sampled and variable, ranging in shape from subulate to subspathulate and in length from 1–2.5 mm.

Grevillea parvula differs from *G. brevifolia* subsp *polychroma* in its consistently tomentose to subvillous branchlets, usually pink-flushed new foliage, strongly concave pollen-presenter, and narrower range of flower colour. Some specimens of *G. brevifolia* subsp. *polychroma* approach some populations (Mt Kaye area) of *G. parvula* very closely in most other features; there may be some intergradation, and the delimitations given here may require adjustment.

180. Grevillea parvula Molyneux & Stajsic, *Fl. Australia* 17A: 502 (2000)

G. victoriae var. *leptoneura* Benth., *Fl. Austral.* 5: 468 (1870), as var. *?leptoneura*. T: sources of the Genoa river, Vic., *F.Mueller* recd 1870; lecto: K, *fide* D.J.McGillivray & R.O.Makinson, *Grevillea* 447 (1993); remaining syntypes: K, MEL (2 sheets).

G. victoriae 'race f' *p.p.*, of D.J.McGillivray & R.O.Makinson, *Grevillea* 321 (1993).

G. victoriae, unassigned to race *p.p.* ('Mallacoota Inlet'), of R.O.Makinson, *Fl. Victoria* 3: 852 (1996).

Illustrations: J.W.Wrigley & M.Fagg, *Banksias, Waratahs & Grevilleas* 332 (1991), as *G. victoriae* var. *leptoneura*; D.J.McGillivray & R.O.Makinson, *Grevillea* 323 (1993), as *G. victoriae*; P.M.Olde & N.R.Marriott, *Grevillea Book* 3: 225 (184G) (1995), as *G. victoriae* var. *leptoneura*.

Shrub, widely spreading or rarely compact, 1–2 m tall. Branchlets tomentose to subvillous. Leaves elliptic to narrowly so or occasionally obovate to oblanceolate, (1.5–) 2.5–4 cm long, (5–) 8–18 mm wide, upper surface smooth, dull, with lateral veins obscure to weakly evident; margins shortly recurved; lower surface densely (very rarely openly) subsericeous to subtomentose. Conflorescence terminal or axillary, usually decurved to pendulous, pedunculate, simple or to 3-branched; unit conflorescence a loose irregular often subsecund cluster, 8–36-flowered; floral rachis 5–16 (–25) mm long. Flower colour: perianth red outside with red hairs, style red or reddish pink. Perianth densely to openly subsericeous outside, bearded inside. Pistil (12–) 17–20 mm long; style with minute hairs towards apex; pollen-presenter lateral or almost so. Follicle (14.5–) 17–19.5 mm long, with faint longitudinal ridges. Fig. 3F.

Occurs in southern N.S.W. (isolated mountains between Eden and the Main Range, and on the Wallagaraugh R. and historically on the Towamba R.) and in Vic. on the upper Genoa R. and lower Wallagaraugh R., also (Molyneux & Stajsic, pers. comm., vouchers not seen) Thurra R. headwaters on E side of Mt Kaye, and NW of Buldah on E side of Errinundra Natl Park. Grows usually near watercourses in forest communities, on granitic and slatey substrates. Regenerates from seed. Flowers Oct.–Mar. (–July). Map 234.

N.S.W.: Mt Kaye, *A.C.Beauglehole 34095* (CANB, MEL); Nalbaugh Natl Park, White Rock Mtn, *P.Beesley 321 & D.Binns* (AD, CANB, MEL, NSW); Letts Mtn, 7 km ESE of Mt Wog Wog, *I.R.Telford 6741* (CANB, NSW); spur on E face of Wog Wog Mtn, *D.E.Albrecht 965* (CANB, MEL, NSW). Vic.: Wallagaraugh R., upper tidal reaches c. 2 km upstream from Johnstons Bridge, W bank, *R.O.Makinson 1648 & B.J.Harrison* (AD, BRI, CANB, HO, K, MEL, NE, NSW, PERTH).

A latinised version of the former varietal name, 'var. *tenuinervis*', has been applied in the horticultural trade, but has no formal standing.

Grevillea parvula is congruent with those populations of McGillivray & Makinson's 'race f' of *G. victoriae* that lie in N.S.W., and in Victoria on the catchment of the Wallagaraugh and Genoa Rivers, i.e. those mapped as E of 149° longitude (McGillivray & Makinson, *Grevillea* 322, map 90 (1993)).

Grevillea parvula has new growth usually pink or purplish pink (rarely green), a leaf length/width ratio 2.5:1 to 4.5:1, and floral rachises with a pale to brownish indumentum. Foliage of plants from higher altitudes tend to have a light green leaf colour, less pronounced pink flushing of the new growth, and a tendency to a denser beard on the inner surface of the perianth.

Grevillea parvula has usually smaller leaves, flowers and follicles as compared with most of its close relatives, and the pollen-presenter strongly concave or rarely flat in cross-section. It has a subglobose, obtuse limb of the flower bud, which is thus round in face view; this is a strong distinguishing character state from *G. oxyantha* (limb acute, subpyramidal) and *G. epicroca* (limb of flower bud obtuse but distinctly square in face view).

It is easily distinguished by its tomentose to subvillous branchlets (the hairs spreading) from *G. victoriae* and *G. epicroca*, both of which have the branchlets hairs appressed. Other differences from *G. victoriae* are noted under that species. The strongly pink-flushed new leaf growth of *G. parvula* also distinguishes it from other taxa in the subgroup (except *G. epicroca*).

The lower leaf surface is usually densely hairy, but occasional specimens from Mt Kaye (Vic.) and the upper Genoa R. have an open indumentum resembling that of *G. epicroca* (see differences above) and *G. monslacana*.

Other taxa in the subgroup which resemble *G. parvula* in having spreading branchlet hairs include *G. rhyolitica* and *G. irrasa*.

A population near Bemboka, N.S.W., (*Telford 11565*) resembles *G. parvula* but has a less shaggy indumentum on the branchlets, a tendency towards a squarish limb on the flower bud (due to a median keel on each tepal's limb-segment), and thicker peduncles. It may constitute either a distinct taxon allied to *G. parvula*, or an intergrade between it and *G. oxyantha*. See also note under *G. epicroca* re Minuma Ra. population.

181. Grevillea epicroca Stajsic & Molyneux, *Fl. Australia* 17A: 502 (2000)

T: Merricumbene Fire Trail, 18.4 km S of Batemans Bay road, c. 19 km SW of Braidwood, N.S.W., 30 Mar. 1976, *M.D.Crisp 2008*; holo: CANB; iso: NSW.

G. victoriae 'Unassigned 5', of D.J.McGillivray & R.O.Makinson, *Grevillea* 322 (1995).

Shrub 1–2.5 m tall. Branchlets subsericeous to almost glabrous, softly angular in cross-section. Leaves elliptic to narrowly so or lanceolate, (3.5–) 4–7 cm long, 8–12 (–20) mm wide, upper surface smooth, ?dull, with the lateral veins evident; margins shortly recurved; lower surface with an open to sparse indumentum of appressed hairs (ground tissue easily visible between hairs). Conflorescence terminal or subterminal-axillary, pedunculate, decurved to pendulous, simple or basally few-branched; unit conflorescence ovoid or a loose sometimes subsecund cluster, few- to many-flowered; floral rachis 10–25 mm long. Flower colour: perianth red with reddish hairs; style red or reddish pink. Perianth densely subsericeous outside (biramous hairs, with a few scattered very inconspicuous erect glandular hairs intermixed), bearded inside. Pistil 16–20 mm long; style with minute hairs towards apex; pollen-presenter lateral. Follicle 18–24 mm long, with faint longitudinal ridges.

Occurs in south-eastern N.S.W., known only from a small area of the tablelands escarpment country W of Moruya (upper Deua R. and Bendethera Mtn area). Grows at 700–1000 m alt. in tall moist *Eucalyptus* forest, on steep rocky slopes on substrates variously reported as sandstone, quartzite, phyllite, or volcanic. Regenerates probably from seed only. Flowers Nov.–May. Map 235.

N.S.W.: 1 km NE along boundary fire trail from junction of River Forest Rd and Gollaribee Fire Trail, 14 km S of Monga, *P.Ollerenshaw 1806* (CANB, MEL, NSW); Deua Natl Park, knoll c. 1.3 km directly N of Bendethera Mtn, *N.Taws 329* (CANB, NSW); downstream on Deua R., 400 m from 'Alpine' property above junction of Deua R. and Curmullee Ck, *T.Whaite 3008 & J.Whaite* (NSW).

Grevillea epicroca has new vegetative growth green or briefly pink, and with a leaf length/width ratio of (3:1–) 4:1 to 6:1 or more, floral rachises apparently with mixed tan and rusty brown hairs, the limb of the bud obtuse but distinctly square in face view, and the pollen-presenter ventrally very concave in cross-section. The open appressed indumentum of the leaf lower surface is distinctive relative to *G. victoriae*, *G. oxyantha*, *G. brevifolia* and most populations of *G. parvula* (for exceptions, see notes under that species). Taxa with a similar lower leaf surface indumentum include *G. monslacana*, which differs in having thicker, more leathery leaves that are proportionately a little wider, the bud limb subglobose and obtuse and rounded in face view, and a pollen-presenter that is ventrally convex in cross-section. *Grevillea rhyolitica* may have an open indumentum on the leaf lower surface, but either (subsp. *rhyolitica*) has the hairs spreading, or (subsp. *semivestita*) appressed but very short and almost rhombic in shape, and in either case has the outer perianth surface tomentose (±spreading hairs).

A population in the Big Badja to Minuma Ra. area of the N.S.W. tablelands escarpment is here unassigned to species. It has some points of similarity to both *G. epicroca* and *G. parvula* and to '*G. aff victoriae* (Bemboka)'; see also note under *G. parvula*.

182. Grevillea monslacana Molyneux & Stajsic, *Fl. Australia* 17A: 502 (2000)

T: Rubicon State Forest, NE junction of Ruoaks Rd and Boundary Trail West, Blue Ra., Vic., 10 Dec. 1995, *N.H.Sinnot 3136*; holo: MEL.

G. victoriae 'race h' of D.J.McGillivray & R.O.Makinson, *Grevillea* 321 (1993).

G. victoriae 'Lake Mountain form' of P.M.Olde & N.R.Marriott, *Grevillea Book* 3: 225 (1995).

G. victoriae 'race h', of R.O.Makinson, *Fl. Victoria* 3: 852 (1996).

Shrub 1–2 m tall. Branchlets tomentose or rarely subsericeous, terete. Leaves oblong-elliptic to broadly lanceolate, (2.0) 3–7 (–11?) cm long, 8–13 (–25?) mm wide; upper surface smooth, variably dull or glossy, the lateral veins obscure to weakly evident; margins shortly recurved; lower surface with an open appressed indumentum. Conflorescence terminal or axillary, pedunculate, decurved to pendulous, simple or basally few-branched; unit conflorescence ovoid or a loose sometimes subsecund cluster, usually many-flowered; floral rachis (10–) 15–30 mm long. Flower colour: perianth solid pink to pinkish red or rarely white with pink highlights; style pink to pinkish red or white. Perianth open-subsericeous outside, bearded inside. Pistil (16–) 18–23 mm long, style with minute hairs towards apex. Pollen-presenter strongly oblique on style. Follicle c. 18 mm long, faintly longitudinally ridged.

Occurs in Vic. limited to a small area from Lake Mountain to Mt Margaret, N and NE of Marysville. Grows at high altitudes in wet sclerophyll forest and snow-gum woodland. Regenerates probably from seed only. Flowers Oct.–Apr. Map 236.

Vic.: 8 km from Mt Margaret Gap towards Lake Mountain Playground, *E.J.Carroll CBG014405* (CANB, NSW); Lake Mountain, *M.D.Tindale 763* (CANB, K, NSW); Lake Mountain, *D.B.Foreman 1044 & R.Filson* (CANB, MEL, NSW).

Grevillea monslacana has fairly leathery leaves with a leaf length/width ratio of 3:1 to 4:1, the limb of the flower bud subglobose and obtuse (rounded in face view); and the pollen-presenter ventrally convex in cross-section. The colour of new vegetative growth is briefly tan, with whitish hairs. It is most similar to *G. epicroca*; see under that species for differences.

183. Grevillea miqueliana F.Muell., *Trans. & Proc. Victorian Inst.* 1: 132 (1855)

T: Vic., '... near Mt McMillan, and along the upper valleys of the Avon in Gipp's Land.' [protologue]; lecto: ... inter Castle Hill & Mt Angus ... Nov. 1854 [in error? probably 1853], *Dr M.* [*F.Mueller*]; lecto: MEL 65738, *fide* D.J.McGillivray & R.O.Makinson, *Grevillea* 447 (1993); remaining syntypes: Avon Ranges, Vic., *s.d.*, [*F.*]*Mueller*; syn: A *n.v.*, K, MEL (several sheets); Avon R., Vic., *s.d.*, [*F.*]*Mueller*; syn: TCD; Mt Angus, Vic., *s.d.*, [*F.*]*Mueller*; syn: NSW.

Illustrations: P.M.Olde & N.R.Marriott, *Grevillea Book* 3: 31 (bottom right), 32 (20A) (1995); N.G.Walsh & T.J.Entwisle, *Fl. Victoria* 3: 853 (fig. 172g, h) (1996).

Erect to spreading shrub 1.5–2.5 m tall. Branchlets angular to terete, villous to tomentose. Leaves ovate to elliptic or slightly obovate, 0.9–8.5 cm long, (10–) 15–30 mm wide; upper surface dull, granulose to almost smooth; margins shortly recurved to revolute; lower surface exposed, velutinous or villous to loosely so. Conflorescence terminal or axillary, decurved, pedunculate, sometimes basally branched; unit conflorescence a loose ovoid to shortly cylindrical cluster, 6–20-flowered; floral rachis 11–22 (–36) mm long. Perianth loosely tomentose to subvillous outside, bearded inside. Pistil 17–22 mm long; style sparsely pubescent in apical 2–5 mm; pollen-presenter lateral. Follicle 15–20 mm long, with faint longitudinal ridges.

Occurs in montane areas of eastern Vic. Two subspecies are recognised.

Grevillea miqueliana is distinguished from *G. victoriae*, *G. brevifolia*, *G. parvula*, *G. monslacana* and *G. epicroca* by the very spreading hairs on branchlets, leaf lower surfaces, and perianth outer surface; the upper surface of the leaf is usually also faintly to strongly granulose (smooth in the other species named) and the pollen-presenter is ventrally convex in cross-section. *Grevillea irrasa* from south-eastern N.S.W. is similar and probably most closely related to *G. miqueliana*, but the former is easily distinguished by its usually oblong to narrowly obovate leaves, which have the lateral veins obscure on both surfaces (evident at least on the lower surface in *G. miqueliana*), and shorter pistil 14–17 mm long. *Grevillea oxyantha* has leaves of similar rotundity to *G. miqueliana*, but has a subpyramidal, pointed limb of the flower bud, which also bears appressed hairs; the tomentose to villous limb in *G. miqueliana* is subglobose and obtuse.

Largest leaves 3–8.5 cm long, 15–30 mm wide, soft-textured, with shortly recurved margins; leaf upper surface finely granulose to almost smooth; peduncles slender, 0.6–0.9 mm thick **183a.** subsp. **miqueliana**

Largest leaves 1–2.5 (–5) cm long, 8–20 (–25) mm wide, leathery, with shortly revolute margins; leaf upper surface usually coarsely granulose; peduncles stout, c. 1 mm thick **183b.** subsp. **moroka**

183a. Grevillea miqueliana F.Muell. subsp. miqueliana

G. victoriae 'race j' of D.J.McGillivray & R.O.Makinson, *Grevillea* 321 (1993).

G. miqueliana 'Typical form' of P.M.Olde & N.R.Marriott, *Grevillea Book* 3: 31 (1995).

Grevillea victoriae 'race j', of R.O.Makinson, *Fl. Victoria* 3: 852 (1996).

Tall open shrub to 2.5 m tall. Leaves 3–8.5 cm long, 15–30 mm wide, soft-textured; margins shortly recurved; upper surface finely granulose to almost smooth. Conflorescences borne on slender peduncles. Flowers dull red.

Occurs in eastern Vic., in scattered localities from Walhalla N to Mt Useful and E to Valencia Ck and the Moroka R. area. Grows in high-altitude wet sclerophyll forest or rarely in snow-gum woodland, usually in rocky sites, ridges and slopes, in stony loam soils. Regenerates from seed. Flowers mainly Aug.–Jan., second flush in Mar.–Apr. Map 237.

Vic.: 7 km N of Esteppy yards, Valencia Ck Rd c. 16 km SE of Mt Wellington, *A.C.Beauglehole 43464* (MEL, NSW); 1.5 km from Walhalla on road to Moe, *D.J.McGillivray 3199 & C.Bartlett* (G, K, MEL, NSW); Moroka R. crossing of track from Millers Gap to the Moroka road, *D.E.Albrecht 2979 & N.G.Walsh* (MEL).

The ranges of the two subspecies abut in the Moroka R. area.

183b. Grevillea miqueliana subsp. moroka Molyneux & Stajsic, *Fl. Australia* 17A: 502 (2000)

T: The Sentinels, 4 mls [6.4 km] SW of Mt Wellington, Vic., 2 Jan. 1964, *T.B.Muir 3044*; holo: MEL.

G. victoriae 'race k' of D.J.McGillivray & R.O.Makinson, *Grevillea* 321 (1993).

G. miqueliana 'Mt Wellington form' of P.M.Olde & N.R.Marriott, *Grevillea Book* 3: 31–32 (1995).

G. victoriae 'race k', R.O.Makinson, *Fl. Victoria* 3: 852 (1996).

Illustrations: P.M.Olde & N.R.Marriott, *Grevillea Book* 3: 32 (20A) (1995), as '*G. miqueliana* Mt Wellington'; N.G.Walsh & T.J.Entwisle (eds), *Fl. Victoria* 3: 853 (fig. 172a, g, h).

Compact to dense shrub, 1–2 m tall. Leaves 0.9–2.5 cm long, 8–20 (–25) mm wide, leathery; margins shortly revolute; upper surface coarsely granulose. Conflorescences borne on stout peduncles. Flowers orange to pink.

Occurs in montane eastern Vic. in the Mt Wellington–Moroka R. area NE of Licola. Grows in high-altitude wet sclerophyll forest or rarely in snow-gum woodland, usually in rocky sites on ridges and slopes, in stony loam soils. Regenerates from seed. Flowers mainly Aug.–Jan., second flush in Mar.–Apr. Map 238.

Vic.: Mt Wellington, *W. & J.Molyneux NSW98771* (NSW); Mt Hump, *T.M.Whaite NSW93344* (NSW); The Razorback between Mt Hump and The Gable End, *E.A.Chesterfield 3115* (MEL).

184. Grevillea irrasa Makinson, *Fl. Australia* 17A: 503 (2000)

T: Nullica State Forest, Yowaka R. catchment, Mine Rd, 500 m uphill along side track from pyrophyllite mine entrance, N.S.W., 23 Nov. 1997, *R.O.Makinson 1649, W.Molyneux, S.Forrester & B.J.Harrison*; holo: CANB; iso: AD, BRI, HO, K, MEL, NE, NSW.

G. victoriae 'race I', of D.J.McGillivray & R.O.Makinson, *Grevillea* 321 (1993).

G. sp. aff. miqueliana, of P.M.Olde & N.R.Marriott, *Grevillea Book* 3: 32 (1995).

Erect spreading shrub 1.5–3 m tall. Branchlets subterete, tomentose to subvillous. Leaves oblong to obovate or narrowly so or occasionally elliptic, 3–6 (–9) cm long, 5–22 mm wide; upper surface granulose or asperous; margins shortly recurved; lower surface exposed, velutinous or villous. Conflorescence usually terminal, decurved to pendulous, pedunculate (peduncle 10–40 mm long), sometimes branched; unit conflorescence a loose cluster, 6–16-flowered; floral rachis 8–20 (–30) mm long. Flowers acroscopic. Perianth subvillous outside, bearded inside. Pistil 14–17 mm long; style straight to gently curved, sparsely pubescent in apical 2–5 mm; pollen-presenter lateral. Follicle ovoid to ellipsoidal, (?12–) 15–18 mm long, with faint longitudinal ridges.

Occurs in south-eastern N.S.W., in two separate ranges, separated by an apparent gap of c. 80 km. The two metapopulations are here regarded as subspecies.

Grevillea irrasa is closely related to *G. miqueliana*, for differences see notes under the latter species. The pinkish new growth and the fine wiry peduncles suggest some relationship also to *G. victoriae* var. *leptoneura*.

Leaves narrowly oblong or narrowly obovate, l:w ratio 5:1 to 10:1; indumentum of the leaf lower surface with all hairs more or less spreading, not obviously 2-layered ..... **184a.** subsp. **irrasa**

Leaves obovate to narrowly so or elliptic or oblong, l:w ratio 5:1 to 7:1; indumentum of the leaf lower surface (×20–30 magnification) with spreading hairs emergent from a denser layer of appressed hairs ..... **184b.** subsp. **didymochiton**

184a. Grevillea irrasa Makinson subsp. **irrasa**

Illustration: P.M.Olde & N.R.Marriott, *Grevillea Book* 3: 32 (20C) (1995), as '*G. sp. aff. miqueliana* Form from Yowaka'.

Leaves narrowly oblong or narrowly obovate, 3.5–6 (–9) cm long, 5–8 mm wide, l:w ratio 5:1 to 10:1; lower surface with all hairs ±spreading, lacking an underlying appressed layer; apices obtuse-mucronate. Floral bracts sublinear, c. 2 mm long. Flower colour: perianth and style red or reddish pink, with brown hairs on bud limb.

Occurs in south-eastern N.S.W., in the area inland from Pambula on the catchments of the Yowaka R. and Nullica R. Grows in dry sclerophyll forest on steep stony hillsides, often on rhyolitic or related volcanic substrates. Regenerates probably by seed only. Flowers at least Nov.–Jan., also June–July. Map 239.

N.S.W.: 3 km WSW of the Old Hut creek crossing of the Nethercote Rd, Nullica State Forest, *D.E.Albrecht 2068* (CANB, MEL, NSW); in the vicinity of the Back Creek pyrophyllite mine, Nullica SF, *D.E.Albrecht 2394* (CANB, MEL, NSW); 14.2 km SW of Pambula, Sugarloaf Fire Rd, near crossing of Old Hut Ck, *M.Parris 8930* (CANB).

Subsp. *irrasa* has been variously known as *G. sp. nov. aff miqueliana*, *G. sp. nov.* 'Yowaka', and *G. sp. nov.* 'Nullica'.

184b. Grevillea irrasa subsp. **didymochiton** Makinson, *Fl. Australia* 17A: 503 (2000)

T: 22.1 km W from Nerrigundah on road to Belowra, N.S.W., 21 Nov. 1997, *R.O.Makinson 1646 & B.J.Harrison*; holo: CANB; iso: AD, BRI, HO, K, MEL, NE, NSW.

Illustration: P.M.Olde & N.R.Marriott, *Grevillea Book* 3: 32 (20B) (1995), as '*G. sp. aff. miqueliana* Form from near Yowrie'.

Leaves obovate to narrowly so or elliptic or oblong, 3–6 cm long, 8–15 (–22) mm wide, l:w ratio 5:1 to 7:1; lower surface (×20–30 magnification) with spreading hairs emergent from a denser layer of appressed hairs; apices obtuse-mucronate or often some emarginate. Floral bracts not seen. Flower colour: perianth and style red or reddish pink, with brown hairs on bud limb.

Occurs in south-eastern N.S.W., W and SW from Moruya on the catchment of the Tuross R. in the area bounded by Nerrigundah, Belowra and Yowrie. Grows in dry sclerophyll forest associations, on steep rocky slopes, often near watercourses or drainage lines, on rhyolitic or slatey substrates. Regenerates from seed, possibly also from rhizomes. Flowers (June–) Sept.–Nov. Map 240.

N.S.W.: Tuross R., Belowra, c. 40 km W of Narooma, *E.F.Constable NSW53915* (A, AD, BRI, CANB, CHR, COLO, HO, K, L, LE, MEL, MO, NBG, NY, PE, PRE, RSA, W); 9 km along Bourkes Rd from Yowrie River entrance to Wadbilliga Natl Park, *P.Carmen 130* (CANB); 14 km N of Yowrie, *A. & C.Tyrrel 148* (CANB).

Subsp. *didymochiton* has been variously known as *G. sp. nov. aff. miqueliana*, *G.* sp. 'Yowrie', and *G. sp. nov.* 'Belowra'. The pattern of plants at some sites suggests the possibility of rhizomatous propagation, but this has not been confirmed.

A collection from the upper Wadbilliga R., near the range for subsp. *didymochiton*, has the rough upper leaf surface of *G. irrasa* but has subacute leaf apices and an appressed indumentum on the leaf lower surface. It may represent an intergrade or hybrid with *G. epicroca*.

185. Grevillea oxyantha Makinson, *Telopea* 7: 130 (1997)

T: Southern Tablelands (A.C.T. boundary): Brindabella Range c. 3 miles [4.8 km] N of Mt Franklin, N.S.W., 6 Dec. 1961, *R.D.Hoogland 8433*; holo: CANB; iso: A *n.v.*, AD, B *n.v.*, BISH *n.v.*, BM, BRI, C *n.v.*, E, FI *n.v.*, G, K, L *n.v.*, M *n.v.*, MEL, NE, NSW, NY *n.v.*, P, PERTH, S *n.v.*, UC *n.v.*, US *n.v.*, W *n.v.*, Z *n.v.*

Erect to spreading shrub (0.5–) 1–3 m tall. Branchlets softly angular becoming terete, subsericeous. Leaves ovate to obovate or broadly so, occasionally suborbicular, 2–6 cm long, 15–35 mm wide; upper surface smooth, minutely foveolate; margins very shortly recurved; lower surface densely tomentose to subtomentose with short curled to wavy hairs, or (subsp. *ecarinata*) subsericeous. Conflorescence terminal (often on short axillary branchlets), decurved to pendulous, pedunculate, usually a short loose cluster or sometimes loosely cylindrical, few- to many-flowered; floral rachis 8–45 mm long. Flowers acroscopic. Perianth densely subsericeous to shortly subtomentose outside, bearded inside. Pistil 18–24 mm long; style nearly straight, minutely and sparsely pubescent or papillose in apical 3–10 mm; pollen-presenter lateral. Follicle narrowly and obliquely ovoid to -ellipsoidal, 15–20 mm long, glabrous, smooth except for 1 or 2 longitudinal ridges on each side.

Occurs in southern montane areas of N.S.W. (including north-western A.C.T.), in the general area bounded by Wee Jasper, Tumut, Batlow, Mt Franklin, Brown Mtn and Mongarlowe. Two subspecies are recognised.

Grevillea oxyantha is characterised by the pointed, subpyramidal limb of the bud, a feature unique in the group.

Limb-segments of tepals each with a pronounced longitudinal keel along midline; lower leaf surface usually densely tomentose to subtomentose with short tightly curled to wavy or curved hairs, rarely almost subsericeous; floral rachis 8–13 mm long ..... **185a.** subsp. **oxyantha**

Limb-segments of tepals not or scarcely keeled along midline; lower leaf surface densely subsericeous to subtomentose with appressed to weakly ascending, ±straight mutually-aligned hairs; floral rachis (12–) 20–45 mm long ..... **185b.** subsp. **ecarinata**

185a. Grevillea oxyantha Makinson subsp. oxyantha

G. victoriae 'race a', of D.J.McGillivray & R.O.Makinson, *Grevillea* 321 (1993).

G. sp. aff. victoriae 'A' *p.p.*, of P.M.Olde & N.R.Marriott, *Grevillea Book* 3: 225 (1995).

Illustrations: P.M.Olde & N.R.Marriott, *Grevillea Book* 3: 224 (184C), 225 (184E) (1995), as *G. sp. aff. victoriae* Big Badja.

Spreading to erect shrub 0.5–3 m tall. Leaves broadly ovate or broadly elliptic or sometimes almost orbicular, obtuse with a small mucro, or rarely subacute; lower surface densely tomentose with short curled to wavy hairs or sometimes subtomentose with wavy to curved ascending hairs, rarely almost subsericeous. Floral rachis 8–13 mm long. Flower colour: perianth scarlet to crimson outside (the indumentum red to brownish), with inner surface (partly displayed) pinkish red with pale hairs; style deep to pale red. Limb-segments of tepals each strongly keeled along external midline. Fig. 3B.

Occurs in southern montane areas of N.S.W. (including north-western A.C.T.) mainly in the Brindabella, Tinderry and Kybean Ranges S to Brown Mtn area, in a variety of habitats, usually in or on the margins of open eucalypt forest in rocky sites, sometimes near creeks, usually in skeletal soils over granite or quartzite. Regenerating from seed only. Flowers mainly Aug.–Dec., occasional in other months. Map 241.

A.C.T.: c. 2 km N of Mt Aggie, Brindabella Ra., *L.A.S.Johnson 7262 & B.G.Briggs* (NSW); Tidbinbilla Ra., E of Fishing Gap, *L.G.Adams 607* (CANB, E, K, L *n.v.*, MEL, NSW, US *n.v.*). N.S.W.: Tinderry Ra. on Michelago to Anembo road, *L.A.Craven 1971* (A *n.v.*, CANB, G, L *n.v.*, NSW, P, PERTH, RSA *n.v.*, SI *n.v.*, UC *n.v.*); between Snowball and Boggy Plain, c. 4.8 km W of Big Badja Mtn, 5 Jan. 1968, *L.A.S.Johnson* (NSW); Bumberry Ck on Wadbilliga Fire Trail, *R.Coveny 6598 et al.* (CANB, NSW).

Grevillea oxyantha subsp. *oxyantha* has also been widely known in the botanical and horticultural literature as the 'A.C.T. form' or 'Canberra form' of *G. victoriae*. It typically has broadly elliptic or broadly ovate leaves, with the venation very conspicuous on the dark green upper surface. The flower buds are densely covered with rusty brown, red or purplish crimson hairs with an almost metallic sheen.

There is a cline in leaf form and indumentum from W to E of the range, with plants from the Brindabella Ra. having very round leaves with tightly curled hairs, and those from the E having narrower, more acute leaves with only mildly wavy hairs.

185b. Grevillea oxyantha subsp. ecarinata Makinson, *Telopea* 7: 133 (1997)

T: Bulls Flat Ck, 10 miles [16 km] N of Snowy Mts Hwy on Yarrangobilly to Goobarragandra road, 10 miles [16 km] ENE of Talbingo, N.S.W., 31 Mar. 1970, *A.N.Rodd 1021*; holo: NSW; iso: PERTH.

G. victoriae 'race b', of D.J.McGillivray & R.O.Makinson, *Grevillea* 321 (1993).

G. sp. aff. victoriae 'A', of P.M.Olde & N.R.Marriott, *Grevillea Book* 3: 225 (1995), *p.p.*

Erect to spreading shrub to 2 m tall. Leaves ovate or ovate-elliptic to broadly so, subacute; lower surface densely subsericeous to subtomentose with loosely appressed to ascending hairs. Floral rachis (12–) 20–45 mm long. Flower colour: perianth scarlet to crimson outside (the indumentum red to brownish), with inner surface (partly displayed) pinkish red with pale hairs; style deep to pale red. Limb-segments of tepals not or scarcely keeled along external midline.

Occurs in N.S.W. from Micalong Ck area near Wee Jasper S to near Tumut and Bago State Forest near Batlow. Poorly recorded, but appears to grow in small localised populations in similar sites to the type subspecies. Regeneration mode is uncertain, but probably from seed only. Flowers mainly ?Aug.–Dec. Map 242.

N.S.W.: Micalong Ck (near Wee Jasper via Tumut), W.*Wingate Hall NSW93294* (NSW); 24 km SE of Batlow, Bago State Forest, *R.J.Allen NSW58417* (NSW); Bago State Forest, S of Batlow, 10 Aug. 1961, '*Z.M.*' *[Z.Mazanec]* (CANB); Kosciuszko Natl Park, c. 25 km directly SE of Tumut, Peak R., 1 km upstream of junction with Waterfall Ck, *N.Taws 236* (CANB, MEL, NSW); Jounama Ra., 11 Nov. 1952, *E.Gauba s.n.* (CANB).

This subspecies remains poorly known and is rarely collected. In the Tumut area it occurs within a few kilometres of populations assignable to *G. victoriae*. It resembles *G. victoriae* in its long floral rachis with pale hairs, and in leaf indumentum, but can be distinguished by its acute flower buds.

186. **Grevillea rhyolitica** Makinson, *Telopea* 7: 134 (1997)

T: Donald Creek near crossing of the Coondella Trig Road (new alignment), c. 4.5 km E of Coondella Trig, N.S.W., 24 Mar. 1990, *D.Albrecht 3993*; holo: MEL; iso: CANB.

G. sp. aff. victoriae 'B', of P.M.Olde & N.R.Marriott, *Grevillea Book* 3: 325 (1995).

More or less erect, often gregarious shrub 0.5–2 m tall. Branchlets angular to subterete. Leaves narrowly to broadly elliptic, 4–11 cm long, 10–25 mm wide; upper surface glabrous or with a few hairs especially near midvein; margins very shortly recurved; lower surface with an open to sparse indumentum (ground tissue clearly visible between hairs). Conflorescence terminal or subterminal axillary, simple or basally 2-branched; unit conflorescence decurved to pendulous, pedunculate, acropetal to subsynchronous, a loose ±regular ovoid to subglobose cluster, (2–) 5–18-flowered; floral rachis 10–20 (–50) mm long. Flowers acroscopic. Perianth loosely tomentose outside (becoming densely so on limb segments), glabrous near base inside, densely bearded above. Pistil 16–20 mm long; style with scattered minute erect simple hairs near apex, or occasionally (subsp. *semivestita* only?) over most of length; pollen-presenter ±lateral. Follicle (subsp. *rhyolitica* only seen) slightly sigmoid when young, at maturity narrowly and obliquely ellipsoidal, 18–22 mm long, with several longitudinal ridges especially near apex.

Occurs in subcoastal montane areas of southern N.S.W., in the Moruya area in Dampier State Forest and Deua Natl Park. Two subspecies are recognised.

An apparent intergrade with *G. oxyantha* subsp. *oxyantha* is discussed under that taxon, *q.v.*

Leaf lower surface with an open indumentum of ascending biramous hairs (often also with simple erect glandular hairs); outer surface of perianth with both ascending biramous hairs and simple erect glandular hairs, the latter usually confined to basal third, occasionally more extensive	**186a.** subsp. **rhyolitica**
Leaf lower surface with scattered minute closely appressed biramous hairs (surface sometimes appearing glabrous); outer surface of perianth with both ascending biramous and simple erect glandular hairs, the latter distributed over the whole surface, including limb segments	**186b.** subsp. **semivestita**

186a. **Grevillea rhyolitica** Makinson subsp. **rhyolitica**

G. sp. aff. victoriae 'B', of P.M.Olde & N.R.Marriott, *Grevillea Book* 3: 225 (1995).

Illustration: P.M.Olde & N.R.Marriott, *Grevillea Book* 3: 224 (184D) (1995), as *G. sp. aff. victoriae* form from Deua Natl Park.

Shrub to 2 m tall. Branchlets open-tomentose, usually with a mixture of non-glandular brownish biramous hairs, and simple erect pale ?glandular hairs (sometimes only the latter). Lower leaf surface with an open indumentum of ascending biramous hairs usually denser along midvein, often also with few to many scattered minute erect simple hairs; hairs mostly pale. Peduncle and floral rachis tomentose with biramous hairs and usually also numerous simple erect hairs. Flower colour: perianth and style red. Perianth loosely tomentose outside, with mainly biramous ascending hairs and usually also relatively few simple erect hairs, the latter usually confined to basal third and inconspicuous, or rarely extending over whole outer surface. Fig. 3E.

Occurs in southern N.S.W., in montane areas W and SW of Moruya, in Dampier State Forest and Deua Natl Park, on the catchments of Donalds Ck and the Coondella Ck–Burra Ck system (all tributaries from the S to the Deua R.). Grows in both moist gully and steep rocky ridge situations, in skeletal soils on rhyolite. Probably regenerates from seed only. Flowers Sept.–Dec. Map 243.

N.S.W.: 'Mountains of the Moon', Deua Natl Park, *G.Moran 110 & L.Thomson* (CANB); Deua Natl Park, c. 2 km N of Coondella Trig, *P.Gilmour 4320* (CANB); Dampier State Forest, c. 22 km W of Moruya, *R.Pullen 4758* (CANB); Deua Natl Park, Prominence 1.9 km N of Coondella Trig., *N.Walsh 1875* (CANB, MEL).

186b. Grevillea rhyolitica subsp. **semivestita** Makinson, *Telopea* 7: 137 (1997)

T: Araluen Valley, N.S.W., 3 Aug. 1953, *E.Gauba CBG005044*; holo: CANB; iso: NSW.

G. victoriae 'race g', of D.J.McGillivray & R.O.Makinson, *Grevillea* 321 (1993).

G. sp. aff. victoriae 'B', of P.M.Olde & N.R.Marriott, *Grevillea Book* 3: 225 (1995).

Shrub to 1 m tall. Branchlets loosely subsericeous with appressed short-armed biramous hairs only. Lower leaf surface often appearing glabrous but with an open inconspicuous indumentum of very short-armed appressed biramous hairs only; hairs mostly ferruginous, some pale. Peduncle and floral rachis loosely to sparsely pubescent with mainly simple erect hairs and often also a few ascending biramous hairs. Flower colour: perianth and style red. Perianth loosely tomentose outside with brownish ascending biramous hairs mixed with numerous weakly erect simple ?glandular pale hairs, the latter distributed ±evenly from base to limb.

Occurs in N.S.W. in the northern part of the species range, in coastal mountain ranges NW of Moruya, recorded from the NE slopes of Mt Donovan (Oulla Ck catchment) and from Araluen Valley; possibly also Donalds Ck catchment. Grows in broken escarpment country, in open *Eucalyptus sieberi* forest. Regenerates from seed only. Flowers recorded for Oct.–Dec. Map 244.

N.S.W.: Deua Natl Park, ridge running off Mt Donovan, 1.8 km NE of summit, c. 20 km WNW of Moruya, *P.Beesley 391 & D.Binns* (CANB, NSW); Araluen Valley, *Mrs Shoobridge, CBG006729* (CANB, NSW).

187. Grevillea diminuta L.A.S.Johnson, *Contr. New South Wales Natl. Herb.* 3: 95 (1962)

T: Brindabella Ra. near Mt Franklin, A.C.T., 11 Nov. 1956, *R.D.Hoogland 6279*; holo: NSW; iso: A *n.v.*, BM, BRI, G, K, MEL, US *n.v.*

Illustrations: J.W.Wrigley & M.Fagg, *Banksias, Waratahs & Grevilleas* 225 (1989); P.M.Olde & N.R.Marriott, *Grevillea Book* 2: 126 (bottom left & 101A–C) (1995).

Spreading shrub 0.3–1 m tall. Branchlets angular, subsericeous. Leaves narrowly elliptic to ovate, 0.5–2 cm long, 3–9 mm wide; upper surface foveolate, glossy, with midvein scarcely apparent; margins slightly recurved; lower surface subsericeous. Conflorescence terminal and axillary, decurved to pendulous, pedunculate, regular and cylindrical or shortly so, many-flowered; floral rachis 7–30 mm long. Flowers acroscopic. Flower colour: perianth rusty brown outside, with inner surface (displayed) bright red; style reddish pink to red. Perianth subsericeous outside, bearded inside. Pistil 10–11 mm long; style gently curved, with minute scattered hairs in apical 4–6 mm; pollen-presenter strongly oblique. Follicle ellipsoidal, 10–15 mm long, finely wrinkled. Plate 42.

Occurs in south-eastern Australia (A.C.T. and N.S.W.), endemic to the Brindabella and Bimberi Ranges SW of Canberra. Grows in sub-alpine sclerophyll woodland, on rocky slopes (often in scree) at c. 1400–1800 m alt. Regenerates from seed. Bird-pollinated. Flowers mainly Sept.–Dec. Map 245.

A.C.T.: between Mt Franklin and Mt Ginini, *L.A.S.Johnson 7264 & B.G.Briggs* (NSW); Mt Ginini, *H.S.McKee 7634* (NSW); Mt Gingera, *C.W.E.Moore 2516* (NSW); Mt Kelley, *I.Olsen 682* (NSW). N.S.W.: Mt Gingera, NW slope, *I.R.Telford 2872* (AD, BRI, CANB, NE).

This species is distinguished from *G. victoriae*, *G. oxyantha* and *G. miqueliana* which have pistils > 15 mm long, more oblique pollen-presenters, and more enclosed nectaries; *G. oxyantha* also has an acute, pyramidal apex to the floral bud.

This species is recognised as 'Rare' in J.D.Briggs & J.H.Leigh, *Rare or Threatened Australian Plants* (1995).

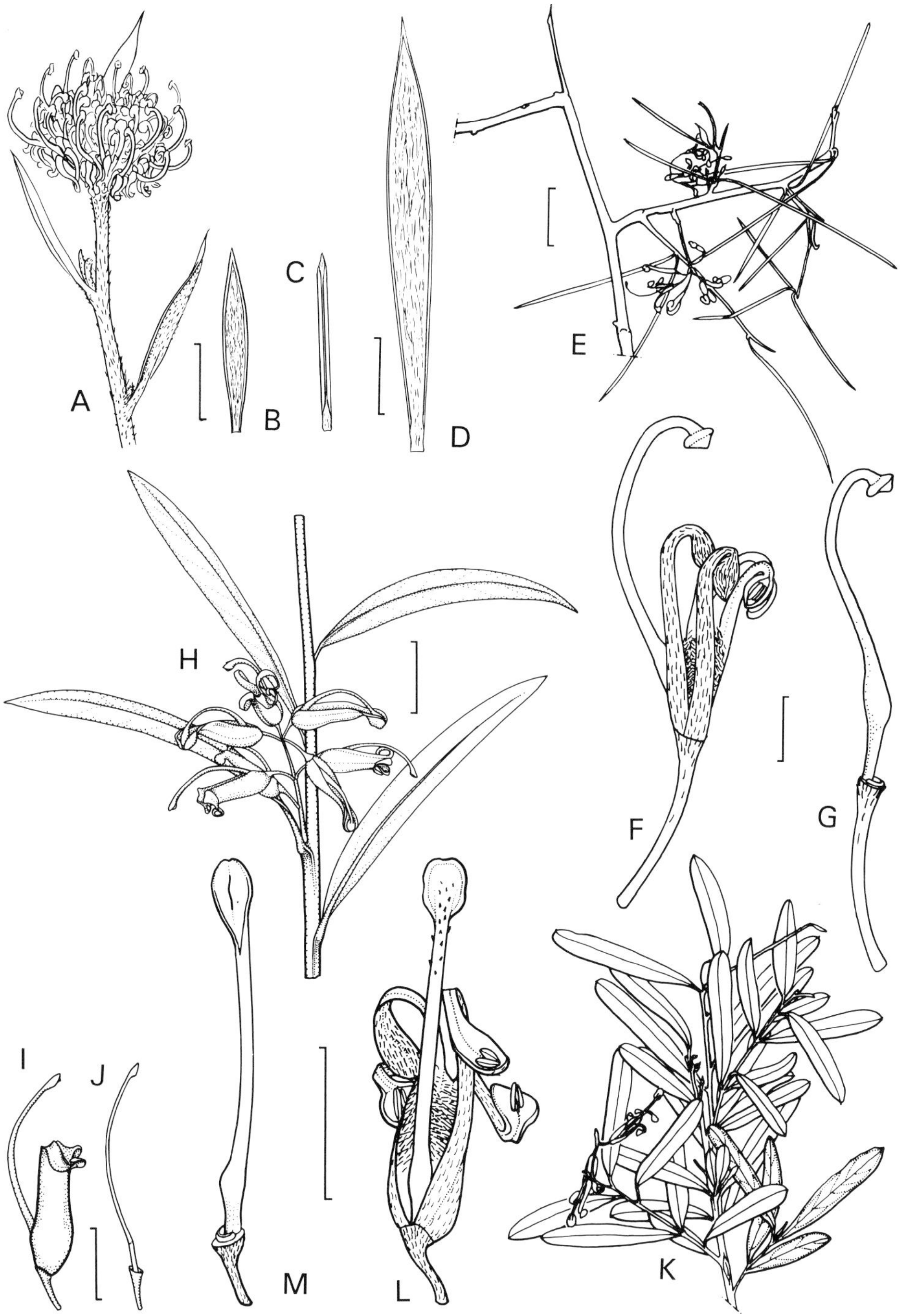

Figure 25. *Grevillea*. **A–D**, *G. australis*. **A**, flowering branch; **B–D**, leaves (**A–B**, M.E.Phillips CBG005098, CANB; **C–D**, R.Burns 217, CANB). **E–G**, *G. inconspicua*. **E**, flowering branch; **F**, flower; **G**, pistil (**E–G**, C.A.Gardner 2265, PERTH). **H–J**, *G. hockingsii*. **H**, flowering branch; **I**, flower; **J**, pistil (**H–J**, R.O.Makinson 1380 *et al.*, CANB). **K–M**, *G. linsmithii*. **K**, flowering branch; **L**, flower; **M**, pistil (**K–M**, I.R.Telford 3226, CANB). Scale bars: **A–D**, **I–J**, **L–M** = 5 mm; **E**, **H**, **K** = 1 cm; **F–G** = 2 mm. Drawn by: **A–D**, **H–J**, C.Wardrop; **E–G**, **K–M**, D.Fortescue.

BISH *n.v.*, BRI, CANB, NSW); N of Injune on Rolleston Rd at crossing of the Great Divide, *D.F.Blaxell 1507 & J.Armstrong* (BRI, NSW).

This species is recognised as 'Rare' in J.D.Briggs & J.H.Leigh, *Rare or Threatened Australian Plants* (1995).

Aspera Group

Shrubs. Leaves entire, dorsiventral; surfaces dissimilar; margins recurved to revolute. Conflorescence terminal, axillary or rarely cauline, simple or basally few-branched; unit conflorescence semi-secund to subcylindrical or subglobose or 1–few-flowered, weakly acropetal to subsynchronous. Flowers acroscopic. Torus oblique to almost lateral. Perianth zygomorphic, hairy outside and inside; tepals either remaining loosely coherent, apically somewhat everted and held ventrally, or (*G. pauciflora*) independently recoiled. Pistil 7–11 mm long; ovary stipitate, glabrous; style glabrous or with scattered minute hairs at least near apex, exposed but not or scarcely exserted from late bud; pollen-presenter lateral, slightly concave to flat. Follicle glabrous, smooth to rugulose; pericarp thin, crustaceous. Seed narrowly ellipsoidal, revolutely marginate, with an apical waxy elaiosome.

A group of three species, in S.A. and the south coast of W.A. Bird pollinated. Placement of *G. pauciflora* in this group is tentative; it has points of similarity to the *Acuaria* Group. *Grevillea aspera* and *G. parallelinervis* have some similarities to the *G. victoriae* alliance, albeit with shorter, non-exserted styles and different tepal posture after anthesis.

1 Tepals independently recoiled in apical third after anthesis, 2 held dorsally to style and 2 ventrally; conflorescences 1–4 (–10)-flowered, erect ..... **194. G. pauciflora**

1: Tepals remaining loosely coherent almost to tip after anthesis, all held ventrally to style; conflorescences many-flowered, deflexed to decurved

2 Leaves oblong to obovate or narrowly so or sublinear, (1.5–) 3–12 mm wide, with margins shortly recurved; lower leaf surface exposed, sericeous to tomentose; conflorescences terminal and axillary ..... **192. G. aspera**

2: Leaves linear, 0.7–1.5 mm wide, with margins angularly revolute; lower leaf surface ±enclosed except midvein, 2-grooved; conflorescences axillary or on older wood ..... **193. G. parallelinervis**

192. Grevillea aspera R.Br., *Trans. Linn. Soc. London* 10: 172 (1810)

T: 'In Novae Hollandiae ora australi; Flinder's Land' [protologue]; lecto: South Coast Bay X [Port Lincoln, S.A.], 1802, *R.Brown Iter Austral. 3341*; lecto: BM, *fide* D.J.McGillivray & R.O.Makinson, *Grevillea* 406 (1993); isolecto: E, K.

Illustrations: J.Carrick, *Contr. Herb. Austral.* 15: 3, fig. 2(6) (1976); D.J.McGillivray & R.O.Makinson, *Grevillea* 327, fig. 83 & col. pl. (1993); P.M.Olde & N.R.Marriott, *Grevillea Book* 2: 40 (bottom right), 41 (28A–C) (1995).

Low spreading to erect shrub, 0.6–2.5 m tall. Branchlets terete, tomentose. Leaves oblong to obovate or narrowly so or sublinear, 1.5–8 cm long, (1.5–) 3–12 mm wide; margins shortly recurved; lower surface sericeous to tomentose. Conflorescences terminal and axillary, deflexed to decurved, pedunculate, cylindrical or subsecund, many-flowered; floral rachis 15–50 mm long. Flowers acroscopic; pedicels 2.5–5 mm long. Flower colour: perianth pinkish red to bright red in basal half, with apical half (or at least the limb) cream, green, yellow or white; style pale green, becoming reddish pink. Perianth villous to subsericeous outside, bearded inside. Pistil 7.5–10.5 mm long; ovary glabrous; style dorsally exposed but not exserted before release of style-end, afterwards straight to barely incurved, scarcely exceeding tepals, glabrous or with scattered minute hairs; pollen-presenter lateral. Follicle narrowly ovoid to ellipsoidal, 13–17 mm long, glabrous, smooth to rugulose. Fig. 26E–G.

Occurs in W.A. (known only from the Rawlinson Ra) and in S.A. (Gawler Ra., northern and central Flinders Ra., and Eyre Peninsula). In S.A., grows in open dry heath, mixed scrub and eucalypt woodland, often on rocky quartzitic slopes, in loam, clay-loam or lateritic soils. Regenerates from seed and (at least in Gawler Ra.) from basal-suckers and rhizomes. Bird-pollinated. Flowers May–Nov. Map 250.

W.A.: Rawlinson Ra., S side, below Circus Rock Holes, *J.M.Bechervaise & J.Kelso 125* (MEL). S.A.: Carappee Hill, *L.Haegi 1315* (AD, K, L, NA *n.v.*, NSW); Heyson Ra. near Mt Hayward, Oraparinna Natl Park, *D.E.Symon 7310* (AD, B *n.v.*, CANB, K, L, NSW); Mt Nott, S of Thurlga, Gawler Ra., *D.E.Symon 8038* B (AD, CANB, NSW); Kirton Point Reserve, Port Lincoln, *D.J.E.Whibley 353* (AD, B *n.v.*, UC *n.v.*).

The Rawlinson Range [W.A.] form occurs over 1000 km from the main range, and is very poorly known; it has slightly smaller flowers than populations in S.A., a smoother leaf upper surface, and branched conflorescences. It is probably a distinct taxon. It is listed, under *Grevillea sp. 12* (Rawlinson Range; *J.M.Bechervaise & J.Kelso 125*), as 'Poorly Known' in J.D.Briggs & J.H.Leigh, *Rare or Threatened Australian Plants* (1995).

In S.A. there is variation within and between the three areas of occurrence. The 'Flinders Range form' has pliable olive-green leaves and relatively inconspicuous conflorescences and small flowers. The 'Gawler Ranges form' has grey-green leaves usually at the broad end of the variance, flowers prolifically and sets ramets vigorously. The 'Eyre Peninsula form' is somewhat intermediate, but often has narrow leaves < 4 mm wide.

193. Grevillea parallelinervis Carrick, *Contr. Herb. Austral.* 15: 1, figs 1–3 (1976)

T: c. 7 km W of Yardea Homestead [Gawler Ra.], S.A., 31 Aug. 1968, *B.J.Copley 2051*; holo: AD; iso: CANB, K.

Illustrations: J.Carrick, *loc. cit.*; J.P.Jessop & H.R.Toelken (eds), *Fl. S. Australia*, 4th edn, 1: 134, fig. 71B (1986); P.M.Olde & N.R.Marriott, *Grevillea Book* 3: 78 (bottom right), 79 (57A–C) (1995).

Spreading shrub 0.5–2 m tall. Branchlets terete, subsericeous to tomentose. Leaves linear, 2.5–9 cm long, 0.7–1.5 mm wide, rigid, pungent; margins angularly revolute; lower surface mostly or wholly enclosed except for midvein, 2-grooved. Conflorescences axillary or cauline, decurved, shortly pedunculate, broadly secund, many-flowered; floral rachis 20–50 mm long. Flowers acroscopic; pedicels 3.5–5.5 mm long. Flower colour: perianth red with limb paler; style red or reddish pink with green to yellow tip. Perianth loosely subsericeous outside (denser and to subvillous on limb), bearded inside. Pistil 8–10 mm long; ovary glabrous; style dorsally exposed but not exserted before release of style-end, afterwards nearly straight, scarcely exceeding tepals, glabrous except for scattered hairs and papillae in apical 2–4 mm; pollen-presenter lateral. Follicle narrowly ovoid-ellipsoidal, 13.5–17 mm long, glabrous, smooth. Plate 41.

Occurs in S.A., where restricted to the western end of the Gawler Ra., from Yardea Stn to Mt Wallaby. Grows in mixed open shrub associations, usually in rocky sites, in shallow soils. Regenerates from seed or rhizomes. Bird-pollinated. Flowers Aug.–Oct. Map 251.

S.A.: Mt St Mungo, c. 40 km NW of Yardea, *J.Carrick 2439* (AD); Yardea, 23 Sept. 1928, *J.B.Cleland* (AD); 7.4 km W of Yardea HS, *B.Copley 2052* (AD); Hiltaba Stn, c. 80 km W of S end of L. Gairdner, Sept. 1962, *E.Newman* (AD); 8 km N of Hiltaba HS, *A.G.Spooner 2363* (AD).

Closely related to *G. aspera*, which has leaves usually > 3 mm broad with the lower surface on either side of the midvein exposed and the upper surface lacking longitudinal ridges.

194. Grevillea pauciflora R.Br., *Trans. Linn. Soc. London* 10: 171 (1810)

T: Flinders' Land [S.A.], 1802, *R.Brown*; lecto: *R.Brown Iter Austral 3340*; BM, *fide* R.O.Makinson, *Fl. Australia* 17A: 503 (2000); isolecto: BM, E, FI *n.v.*, G *n.v.*, G-DC, K, LE *n.v.*, MEL.

Shrub to 2 m high, usually spreading. Leaves linear or narrowly cuneate to obovate, rarely filiform, (0.7–) 2–5 (–9) cm long, (0.6–) 2.5–8 (–12) mm wide; margins recurved to revolute, sometimes ±flat and thickened; lower surface sericeous or glabrous. Indumentum usually terminal on axillary branchlets, sometimes axillary, rarely cauline, erect, simple, usually ±sessile, basipetal, 1–4 (–10)-flowered; floral rachis 0.2–1.5 mm long. Flowers abaxially oriented; pedicels 3–5.5 mm long. Perianth with scattered appressed hairs outside, denser on limb where sometimes subsericeous, bearded inside almost to limb. Pistil 7–10.5 mm long, glabrous; ovary stipitate; style not exserted in late bud; pollen-presenter lateral. Follicles obliquely ellipsoidal, 10–14 (–16) mm long, colliculate-rugulose, glabrous.

Occurs in south-western W.A. between Esperance and Point Malcolm, and in S.A. on the Eyre and Yorke Peninsulas and on Kangaroo Is. There are four subspecies.

Grevillea pauciflora can be confused with *G. sparsiflora* and *G. oligantha*, both of which are distinguished by their much longer pistils (> 17 mm long).

1 Branchlets terete or slightly angular, not ridged; distance between perianth beard and base of ventral limb at least 1 mm (S.A.)

2 Leaves obovate to narrowly linear, sericeous or sometimes glabrous on lower surface; margins ±flat to loosely revolute **194a.** subsp. **pauciflora**

2: Leaves filiform, sericeous on lower surface; margins revolute to form a single groove on lower surface **194b.** subsp. **leptophylla**

1: Branchlets angular, ridged; distance between perianth beard and base of ventral limb < 1 mm (W.A.)

3 Leaves narrowly obovate or narrowly cuneate, glabrous on lower surface; margins ±flat **194c.** subsp. **psilophylla**

3: Leaves vary narrowly obovate to sub-linear, sericeous on lower surface; margins revolute **194d.** subsp. **saxatilis**

194a. Grevillea pauciflora R.Br. subsp. pauciflora

Illustrations: P.M.Olde & N.R.Marriott, *Grevillea Book* 3: 85 (bottom left & 63A, C) (1993).

Shrub to 1 m high. Branchlets terete or slightly angular (not ridged), sericeous. Leaves obovate to narrowly linear, mostly 1.2–5.5 cm long, relatively rigid; margins ±flat to loosely revolute; lower surface sericeous or sometimes glabrous. Flower colour: perianth orange to bright red with limb yellow; style red with yellow tip. Perianth beard not reaching within 1–2 mm of base of ventral limb.

Occurs in S.A., widespread on Eyre and Yorke Peninsulas and on Kangaroo Is., on calcareous substrates. Regenerates from seed. Flowers (Apr.) June–Oct. (Dec.). Map 252.

S.A.: between Streaky Bay and Talia, 23 Sept. 1957, *J.B.Cleland* (AD); turnoff to Memory Cove, from road to Spalding Cove S of Port Lincoln, *A.S.George 13107* (CANB, PERTH); Kangaroo Is., Flinders Chase Natl Park, Cape du Couedic, *E.N.S.Jackson 4383* (AD, CANB); between Warooka and Stenhouse Bay, c. 40 km from Stenhouse Bay, *K.B.Warnes 45* (AD); Kelly Hill, c. 21 km ENE of Cape du Couedic, *P.G.Wilson 738* (AD, B *n.v.*, K, UC *n.v.*).

Subsp. *pauciflora* shows variation in leaf width from obovate to narrowly linear, and some specimens appear to be intergrades between subsp. *pauciflora* and subsp. *leptophylla*. Mixed populations of broad- and narrow-leaved plants, the latter approaching subsp. *leptophylla*, are known. Leaf indumentum is also variable; some plants from drier areas of Eyre Peninsula retain the sericeous leaf indumentum which is usually confined to juvenile foliage, while other plants have glabrous adult leaves.

194b. Grevillea pauciflora subsp. leptophylla W.R.Barker, *J. Adelaide Bot. Gard.* 17: 211–215 (1996)

T: Eyre Peninsula, Koolidie Stn, NW of Yeelana, S.A., 16 Sept. 1990, *T.Hall 412*; holo: AD; iso: CANB, NSW.

G. pauciflora R.Br. subsp. 'Narrow-leaved' (*R.Bates 6936*), W.R.Barker, *List. Vasc. Pl. S. Australia* 4th edn, 8 (1993).

[*G. pauciflora* subsp. *pauciflora auct. non* R.Br.: D.J.McGillivray & R.O.Makinson, *Grevillea* 364 (1993), *p.p.*; P.M.Olde & N.R.Marriott, *Grevillea Book* 3: 84 (63B) (1995)]

Illustration: P.M.Olde & N.R.Marriott, *Grevillea Book* 3: 85 (63B) (1995), as *G. pauciflora* subsp. *pauciflora*.

Shrub to 1 m high. Branchlets terete or slightly angular (not ridged), sericeous. Leaves filiform, mostly 2–4 cm long, relatively flexible; margins revolute; lower surface enclosed including midvein, 1-grooved, sericeous in groove and at base. Flower colour: perianth red; style red with yellow tip. Perianth beard not reaching within 1.5–2 mm of base of ventral limb.

Occurs in S.A., only on the Eyre Peninsula, N and NW of Cummins, S.A., on calcareous substrates. Regenerates from seed. Flowers ?Aug.–?Oct. Map 253.

S.A.: E side of Section 51, Hundred of Shannon, *C.R.Alcock 2673* (AD, CANB, DNA, LSU, MA, MO, SI, TAI, TUR; all *n.v.* except AD & CANB); 33 km S of Lock, Section 53, Hundred of Shannon, 13 Sept. 1991, *R.Barratt s.n.* (AD); 10 km N of Cummins, *R.Bates 6936* (AD, CAF *n.v.*); Section 45, Hundred of Mitchell, *s.d., coll. unknown* (AD99050229).

There are morphological intergrades between subsp. *pauciflora* and subsp. *leptophylla*. *Alcock 2673*, here assigned to subsp. *leptophylla*, approaches this intergrade.

194c. **Grevillea pauciflora** subsp. **psilophylla** McGill., *New Names Grevillea* 11 (1986)

T: Balladonia track, 19.5 km NE of junction with Fisheries Rd, SW of Mt Ragged, W.A., 30 June 1976, *D.J.McGillivray 3608 & A.S.George*; holo: NSW; iso: PERTH.

Illustrations: P.M.Olde & N.R.Marriott, *Grevillea Book* 3: 86 (bottom left & 64A, B) (1993).

Shrub 1.5–2 m high. Branchlets conspicuously angular (ridged), sericeous, sometimes sparsely so. Leaves narrowly obovate or narrowly cuneate, 2–5 cm long, relatively rigid; margins ±flat; lower surface glabrous. Flower colour: perianth red; style red with yellow tip. Perianth beard not reaching within c. 0.6 mm of base of ventral limb.

Occurs near the south coast of W.A. between Aroona Stn (E of Esperance) and Point Malcolm, growing in sand over laterite. Regenerates from seed. Flowers (June) Sept.–Dec. Map 254.

W.A.: between Mt Ragged and Israelite Bay, *J.S.Beard 5292* (KPBG); on Fisheries Rd, opposite Aroona Stn, E of Esperance *D.J.McGillivray 3601 & A.S.George* (NSW); Cape Arid Natl Park, 7 km W of E boundary on Fisheries Rd, *J.W.Green 5149* (CANB, PERTH).

194d. **Grevillea pauciflora** subsp. **saxatilis** McGill., *New Names Grevillea* 11 (1986)

T: S end of Mt Ragged on W side, W.A., 1 July 1976, *D.J.McGillivray 3608 & A.S.George*; holo: NSW; iso: CANB, K, PERTH, US *n.v.*

Illustrations: P.M.Olde & N.R.Marriott, *Grevillea Book* 3: 86 (bottom right), 87 (65) (1993).

Spindly shrub to 0.7 m high. Branchlets angular (ridged), sericeous to tomentose. Leaves very narrowly obovate to sublinear, mostly 4–8 cm long, relatively rigid; margins revolute; lower surface usually enclosed including midvein, 1-grooved, sericeous in groove. Flower colour: perianth red with paler limb; style red with orange tip. Perianth beard not reaching within c. 0.6 mm of base of ventral limb.

Confined to Mt Ragged, N of Cape Arid, W.A., near the western end of the Great Australian Bight, in sand over quartzite. Regenerates from seed. Flowers recorded for Jan., July and Aug. Map 255.

W.A.: Mt Ragged Ra., 2.5 km S of Tower Peak, isolated range, lower W slopes, *M.D.Crisp 4816* (CANB); S end of Mt Ragged, *A.S.George 7428* (PERTH); below summit of Tower Peak, *A.S.George 16130* (PERTH); S end of Mt Ragged on W side, *D.J.McGillivray 3617 & A.S.George* (CANB, K, NSW, PERTH, US *n.v.*).

Floribunda Group

Shrubs. Leaves entire, dorsiventral or rarely dipleural; surfaces dissimilar; margins recurved to revolute or flat. Conflorescence terminal, axillary or cauline, erect to pendent, usually simple, regular-subglobose or a loose 1–20-flowered cluster, or occasionally secund or conico-cylindrical, acropetal or basipetal or subsynchronous. Flowers acroscopic or ad- or abaxially oriented. Torus transverse to oblique or rarely lateral. Perianth zygomorphic, hairy or glabrous outside, hairy inside; tepals remaining loosely coherent and held ventrally, sometimes everted, or all independently recurling; limb-segments sometimes keeled, horned or apiculate. Pistil 4–38 mm long; ovary sessile or stipitate, hairy or rarely glabrous and then stipe swollen and hairy; style exserted from late bud or not, hairy or rarely glabrous; pollen-presenter oblique to lateral, flat to convex. Follicle with a few longitudinal ridges, hairy (lacking indumental markings) or rarely glabrous; pericarp thin, crustaceous. Seed ellipsoidal, with a waxy margin along one side and a short apical wing or elaiosome.

A group of 47 species mainly in southern temperate zones (SE and SW), and margins of the southern Eremaean zone. Primarily bird-pollinated. Most closely related to the *Buxifolia* group; less certain affinities to the *Longistyla* group.

1 Perianth completely glabrous on outer surface (including on limb)

2 Pistil < 12 mm long

3 Pistil 4–6 mm long, scarcely exceeding perianth after release of style-end (south-western Australia.) **235. G. drummondii**

3: Pistil 9–11 mm long, clearly exceeding perianth after release of style-end (south-eastern Australia) **214. G. jephcottii**

2: Pistil > 12 mm long

4 Leaf lower surface with lamina on either side of midvein exposed and glabrous or with scattered hairs along the midvein only

5 Style loosely villous over at least the lower two-thirds; ovary villous to openly so including on upper half **216. G. baueri**

5: Style glabrous or with scattered hairs in lower half only; ovary glabrous or with a tuft of hairs on lower half (and stipe) only

6 Leaves narrowly elliptic, acute, ⩽ 3 mm wide; inflorescence usually simple **211. G. rosmarinifolia**

6: Leaves narrowly oblong-elliptic, obtuse, usually > 4 mm wide; inflorescence usually 2–10-branched **213. G. iaspicula**

4: Leaf lower surface with lamina on either side of midvein either enclosed by margins, or if exposed then with hairs on the lamina to either side of midvein

7 Style loosely tomentose to villous over at least two-thirds of length; ovary villous over whole surface **215. G. lanigera**

7: Style glabrous or with scattered erect hairs in basal half; ovary glabrous or with hairs on ventral side near base only

8 Leaves strongly ascending, 0.7–3.0 mm wide, smooth or faintly granulose on upper surface; plant not lignotuberous; pistil 15–23 mm long **211. G. rosmarinifolia**

8: Leaves spreading, 0.5–0.6 mm wide, scabrid on upper surface; plant sometimes lignotuberous; pistil c. 16 mm long **212. G. divaricata**

1: Perianth with hairs on outer surface (sometimes few and/or restricted to limb)

9 Pistil ⩽ 12 mm long

10 Leaves dipleural (linear-subterete with a groove along either side) **239. G. pinifolia**

10: Leaves dorsiventral (with upper and lower surfaces dissimilar)

11 Torus lateral or almost so (toral rim almost co-linear with pedicel), 2–5 mm long from dorsal edge to ventral, and pistil 7–16 mm long; stipe 2–5 mm long; style-end sometimes bluish (SW of W.A.)

12 Nectary spreading, sometimes tongue-like, projecting 1–2 mm beyond toral rim; perianth ±ovoid with distinct pouch at base; stipe of ovary usually free for much of length; leaves mostly 1–3 (–6) cm long; style-end never bluish **238. G. depauperata**

12: Nectary long-U-shaped, scarcely emergent above toral rim; perianth narrowly ovoid or ellipsoidal to cuneate, broadest towards middle; stipe of ovary adnate to dorsal wall of toral cavity for most of length; leaves 1–16 cm long; style-end sometimes bluish

13 Perianth 7–9 mm long; pedicels usually 6–8 mm long **240. G. brachystylis**

13: Perianth 10–14 mm long; pedicels usually 3.5–4.5 mm long **241. G. bronwenae**

11: Torus either transverse on pedicel to oblique at ⩽ c. 60° (toral rim clearly not co-linear with pedicel), 1–3.5 mm long from dorsal edge to ventral, or almost lateral (colinear with pedicel) and then the pistil ⩽ 8 mm long; stipe absent or 0.5–2.5 mm long; style-end never bluish (south-western & south-eastern Australia)

14 Longest ultimate floral rachises 10–45 mm long; unit conflorescences decurved to deflexed **195. G. floribunda**

14: Longest ultimate floral rachises ⩽ 10 mm long; unit conflorescences erect to decurved

15 Young conflorescences with conspicuous bracts concealing buds; floral bracts 2.5–7 mm long, usually persistent at anthesis **196. G. polybractea**

15: Young conflorescences with buds not concealed by bracts; floral bracts inconspicuous, < 2 mm long, usually falling in bud stage

16 Style clearly exceeding perianth when released, projecting > 3 mm beyond tepals

17 Ovary stipitate; torus oblique at > 45°; pollen-presenter concave (south-western Australia) **238. G. depauperata**

17: Ovary sessile; torus oblique at < 45°; pollen-presenter flat to convex (south-eastern Australia)

18 Nectary tongue-like, spreading laterally beyond toral rim; lower surface of leaf usually villous (ascending to spreading hairs), rarely subsericeous or glabrous **199. G. alpina**

18: Nectary arcuate to reniform, not or scarcely spreading beyond toral rim; lower surface of leaf subsericeous (hairs straight and appressed) **201. G. kedumbensis**

16: Style scarcely exceeding perianth when released, projecting < 2 mm beyond tepals

19 Pistil > 10 mm long **238. G. depauperata**

19: Pistil < 10 mm long

20 Perianth grossly saccate, 4–6 mm wide **232. G. saccata**

20: Perianth ventrally dilated at base or not, but not grossly saccate, ⩽ 3.5 mm wide

21 Pedicels 1–4 mm long; outer surface of perianth with an open indumentum; leaf margins usually conspicuously ciliate with hairs 2–4 mm long; most or all leaves with margins shortly recurved and most or all of lower surface exposed; leaf lower surface glabrous or hairy

22 Lower leaf surface glabrous; stigma located almost basally on pollen-presenter; unit conflorescences usually 13–18-flowered **236. G. pimeleoides**

22: Lower leaf surface with ascending to spreading hairs; stigma usually located near centre of pollen-presenter; unit conflorescences usually 5–8-flowered **237. G. centristigma**

21: Pedicels 3–10 mm long; outer surface of perianth with a dense to open indumentum; leaf margins glabrous, not ciliate; leaves with margins shortly recurved to strongly revolute, sometimes enclosing most or all of leaf lower surface; leaf lower surface always hairy

23 Hairs of leaf lower surface appressed, straight and mutually aligned **230. G. fasciculata**

23: Hairs on leaf lower surface ascending to erect, straight or wavy to curled or irregularly matted

24 Perianth ±ovoid below limb, broadest at base

25 Leaves 0.5–3.5 cm long, elliptic to suboblong, often plump; most unit conflorescences 2–4-flowered; perianth bearded inside near base; style-end red **231. G. crassifolia**

25: Leaves mostly 4–9 cm long, narrowly obovate to broadly linear, thin and leathery; most unit conflorescences 8–15-flowered; perianth not bearded inside (narrow line of hairs on dorsal tepals only) or with a very scanty beard in throat; style-end yellow **233. G. fistulosa**

24: Perianth ±ellipsoidal below limb, broadest at about middle

26 Limb of bud sericeous to shortly tomentose (hairs appressed to weakly ascending, < c. 0.5 mm long); inner surface of perianth with a conspicuous dense beard extending well below constricted 'throat' (especially on dorsal tepals); leaves usually < 5 cm long, occasionally to 9 cm; fruit lacking longitudinal ridges **230. G. fasciculata**

26: Limb of bud villous (hairs spreading, 0.6–1.0 mm long); inner surface of perianth glabrous except for either a narrow line of hairs near dorsal suture, or a wispy beard in throat; leaves (1.5–) 4–9 cm long; fruit with longitudinal ridges

27 Inner surface of perianth glabrous except for a narrow line of hairs on dorsal tepals adjacent to dorsal suture; perianth red; style-end yellow **233. G. fistulosa**

27: Inner surface of perianth glabrous except for a silky beard in constricted throat; perianth dull yellow-orange; style-end yellow **234. G. fuscolutea**

9: Pistil > 12 mm long

28 Torus almost lateral (its rim nearly co-linear with the pedicel), and style apex scarcely exceeding perianth (projecting by < 3 mm); pollen-presenter concave; stipe of ovary adnate to dorsal wall of toral cavity for most of length (SW of W.A.)

29 Perianth 7–9 mm long; pedicels usually 6–8 mm long **240. G. brachystylis**

29: Perianth 11–14 mm long; pedicels usually 3.5–4.5 mm long **241. G. bronwenae**

28: Torus transverse to oblique (its rim clearly not co-linear with pedicel); style clearly exceeding perianth (projecting by ⩾ 3 mm); pollen-presenter flat to convex; stipe (if present) usually mostly free (south-western & south-eastern Australia)

30 Leaf lower surface mostly or wholly exposed, and either glabrous or with only a few hairs along midvein

31 Outer surface of perianth with hairs restricted to the limb; nectary arcuate, projecting a little above the toral rim but not linguiform, not projecting laterally into perianth pouch **216. G. baueri**

31: Outer surface of perianth tomentose to villous, with hairs not confined to the limb; nectary linguiform, projecting outwards into perianth pouch **199. G. alpina**

30: Leaf lower surface either concealed by the revolute margins, or exposed and then hairy (hairs not confined to midvein)

32 Ovary sessile or subsessile

33 Leaves linear, the upper surface usually striate with 3–6 longitudinal ridges (sometimes faint); torus strongly oblique and sometimes markedly elongate on dorsal side; all conflorescences axillary or cauline (W.A.)

34 Leaf lower surface 2-grooved (margins revolute against sides of midvein) **223. G. haplantha**

34: Leaf lower surface 1-grooved (revolute margins fully enclosing whole lower surface, or occasionally with a narrow recessed strip of lamina visible)

35 Leaf upper surface not scabrous; torus c. 1.5 mm across; branchlets not secund; longest leaves usually < 2.5 cm long **224. G. disjuncta**

35: Leaf upper surface usually scabrous; torus 2–3 mm across; branchlets usually secund; longest leaves usually 3–5 cm long (rarely only 2.5 cm long) **225. G. dolichopoda**

33: Leaves linear or not, the upper surface lacking longitudinal ridges; torus transverse to oblique, not elongate on dorsal side; conflorescences terminal or axillary (W.A., Qld, N.S.W., Vic.)

36 Floral bracts persistent at anthesis and mostly ⩾ 2 mm wide; perianth persistent to fruiting stage; conflorescence dense, subglobose **196. G. polybractea**

36: Either floral bracts caducous or persistent and < 2 mm wide; perianth falling early or (*G. floribunda p.p.*) persistent to fruiting stage; conflorescence never dense and subglobose

37 Nectary very conspicuous when perianth removed, linguiform or shovel-like, projecting > 1 mm laterally beyond toral rim into perianth pouch

38 Nectary spreading beyond toral rim, ±perpendicular to pedicel; leaves 5–20 (–30) mm long, 1.5–4 (–10) mm wide; floral bracts 0.3–0.8 mm wide, often persistent to late bud; pistil 8–21 mm long; shrub, not usually rhizomatous **199. G. alpina**

38: Nectary angled at c. 45° to pedicel; leaves 20–60 mm long, (4–) 7–12 (–18) mm wide; floral bracts 1.2–1.4 mm wide, falling when buds c. 3 mm long; pistil 18–25 mm long; shrub, usually rhizomatous **198. G. celata**

37: Nectary not spreading into perianth-pouch, variously shaped but not linguiform or shovel-like, usually inconspicuous

39 Apex of bud (limb) pointed (acute to acuminate or caudate)

40 Inner surface of perianth with area above beard bearing a dense to open spreading indumentum (especially on ventral tepals) all the way to the limb segments (N.S.W. south from Blue Mtns) **205. G. arenaria**

40: Inner surface of perianth with area above beard glabrous or bearing a sparse appressed indumentum (at least on ventral tepals) (SE Qld and N.S.W. north from Hunter R.)

41 Leaf lower surface with a dense, matted indumentum; fruit not developing **207. G. rhizomatosa**

41: Leaf lower surface with sparse to dense indumentum of straight mutually aligned hairs; fruit developing

42 Inner perianth beard situated just below curve; perianth prominently saccate **210. G. masonii**

42: Inner perianth beard situated near base; perianth oblong with slight basal dilation

43 Leaf lower surface on either side of midvein obscured by a dense even appressed indumentum (ground tissue not visible between hairs); branchlets with ±appressed hairs **206. G. montana**

43: Leaf lower surface with ground tissue clearly visible between hairs; branchlets with ±spreading hairs

44 Leaves ⩽ 2 cm long, obtuse; fruit transverse to pedicel with the style sharply inflexed at base; conflorescence 1–4-flowered **208. G. quadricauda**

44: Leaves > 2 cm long, acute; fruit erect with erect style; conflorescence ⩾ 6-flowered **211. G. banyabba**

39: Apex of bud (limb) rounded-obtuse, depressed or emarginate

45 Perianth outer surface with an open to dense indumentum of rusty brown hairs; longest floral rachis usually > 15 mm long **195. G. floribunda**

45: Perianth outer surface with hairs mostly or all pale in colour (off-white to fawn or pinkish); longest floral rachis < 15 mm long

46 Broadest leaves ⩽ 2.5 mm wide, linear; lower surface of lamina (sometimes including midvein) usually enclosed by revolute margins

47 Pistil 14–18 mm long (south-eastern Australia) **204. G. obtusiflora**

47: Pistil 18–25 mm long (south-western Australia)

48 Leaf lower surface 2-grooved (margins revolute against sides of prominent abaxial midrib); robust shrub to c. 2 m tall **223. G. haplantha**

48: Leaf lower surface 1-grooved (margins enclosing leaf lower surface including midvein, or rarely a narrow strip of recessed lamina visible); low procumbent shrub to 0.6 m tall **225. G. dolichopoda**

46: Broadest leaves > 2.5 mm wide, broadly linear to elliptic, ovate, oblong, obovate or almost round; lower surface of lamina usually partly or wholly exposed

49 Inner surface of perianth with beard situated in upper half near throat

50 Longest floral rachis (5–) 10–20 mm long; leaves 40–50 mm long **195. G. floribunda**

50: Longest floral rachis 1–5 mm long; leaves 5–20 (–30) mm long **210. G. masonii**

49: Inner surface of perianth with beard situated in basal third

51 Hairs of style reddish or brownish coloured (rarely white); nectary ±oblong, thin, with a narrow upper margin, ascending to spreading above or outwards from toral rim; tepal limb segments sometimes with prominent median blunt 'horn' or ridge, or unornamented; lower leaf surface with ±spreading hairs (Vic.)

52 Nectary extending not more than 1 mm above toral rim; habit not rhizomatous; perianth dull to bright yellow **197. G. chrysophaea**

52: Nectary extending more than 1 mm above or outward from toral rim; habit rhizomatous; perianth usually bichromatic red and yellow, or red and white, or apricot and white, rarely monochromatic yellow **198. G. celata**

51: Hairs of style white; nectary cushion-like, thick, lacking a clearly defined upper margin; tepal segments sometimes with median ridge, never an erect blunt 'horn', often unornamented; lower leaf surface with ±appressed hairs (N.S.W.)

53 Most leaves with apical mucro 1–2 mm long; style of fruit often inflexed; limb segment of each tepal with a prominent median keel **200. G. mucronulata**

53: Most leaves with apical mucro not more than 1 mm long; style of fruit erect or deflexed; limb segment of each tepal with a faint to strong median ridge or keel

54 Pollen-presenter ⩽ 2 mm long, spherical to obovate-elliptic

55 Pollen-presenter convex with inconspicuous stigma; pistil (14–) 18–23 mm long; perianth pink to pinkish red with cream limb, persistent; habit rhizomatous **204. G. obtusiflora**

55: Pollen-presenter flat with prominent stigma; pistil 12.5–17.5 mm long; perianth green to cream, caducous; habit lignotuberous **201. G. kedumbensis**

54: Pollen-presenter > 2 mm long, obovate, elliptic or oblong, with short basal attenuation

56 Leaf lower surface appressed-sericeous, with upper surface finely granulose (rarely coarsely so); peduncle densely hairy; pistil 21–24 mm long; fruit < 15 mm long **204. G. granulifera**

56: Leaf lower surface subvillous, with upper surface coarsely granulose; peduncle glabrous or sparsely hairy; pistil (21–) 25–26 mm long; fruit c. 20 mm long **203. G. guthrieana**

32: Ovary stipitate

57 Perianth segments after anthesis cohering except at limb and along dorsal suture, limb segments held loosely together just ventral to style, the tepals not curled back in 2 opposing pairs

58 Pistil ⩽ 16 mm long **238. G. depauperata**

58: Pistil ⩾ 18 mm long

59 Most conflorescences terminal (sometimes on short lateral branchlets)

60 Upper leaf surface longitudinally ridged (SW of W.A.) **220. G. phillipsiana**

60: Upper leaf surface not longitudinally ridged (S.A.)

61 Style loosely villous over most of length; leaves plump-textured with a recurved callous point, spreading, solitary along longer branchlets **218. G. muricata**

61: Style glabrous or nearly so in distal half; leaves ±flattened and leathery-textured, usually with a straight pungent tip, ascending, usually clustered on short lateral branchlets **217. G. lavandulacea**

59: All conflorescences axillary or cauline

62 Ovarian stipe > 2 mm long **219. G. extorris**

62: Ovarian stipe < 2 mm long

63 Leaf lower surface 2-grooved (margins revolute against sides of prominent midvein) **223. G. haplantha**

63: Leaf lower surface 1-grooved (margins fully enclosing whole lower surface, or occasionally with a narrow recessed strip of lamina visible)

64 Upper leaf surface not scabrous; torus c. 1.5 mm across; branchlets not secund; longest leaves usually < 2.5 cm long **224. G. disjuncta**

64: Upper leaf surface usually scabrous; torus 2–3 mm across; branchlets usually secund; longest leaves (2.5–) 3–7 cm long **225. G. dolichopoda**

57: Perianth segments separating after anthesis over their apical $^{2}/_{3}$; dorsal and ventral tepals curled back as independent opposed pairs (ventral tepals curled back further)

65 Upper leaf surface with c. 4 longitudinal ridges or grooves **221. G. tetrapleura**

65: Upper leaf surface without longitudinal ridging or grooves (except raised midvein)

66 Floral rachis and peduncle together ⩾ 20 mm long **222. G. deflexa**

66: Floral rachis and peduncle together < 10 mm long

67 Most leaves ⩾ 4 cm long

68 Outer perianth surface with a spreading indumentum (often densely villous) **226. G. pityophylla**

68: Outer perianth surface with an appressed indumentum, sometimes tomentose on limb only **227. G. granulosa**

67: Most leaves < 4 cm long

69 Upper leaf surface faintly granulose; leaves 1–4 cm long; pistil 20–23 mm long **228. G. rosieri**

69: Upper leaf surface smooth; leaves 0.3–1.8 cm long; pistil 17–22 mm long **229. G. yorkrakinensis**

Plate 33. *Grevillea dimorpha*
Photograph — D.Foreman.

Plate 34. *Grevillea acuaria*
Photograph — D.Foreman.

Plate 35. *Grevillea kennedyana*
Photograph — D.L.Jones.

Plate 36. *Grevillea oleoides*
Photograph — J.Plaza.

Plate 37. *Grevillea victoriae* subsp. *nivalis*
Photograph — D.Foreman.

Plate 38. *Grevillea evansiana*
Photograph — T.Low.

Plate 39. *Grevillea quinquinervis*
Photograph — T.Low.

Plate 40. *Grevillea confertifolia*
Photograph — P.Abell.

Plate 41. *Grevillea parallelinervis*
Photograph — D.L.Jones.

Plate 42. *Grevillea diminuta*
Photograph — R.Boden (ANBG).

Plate 43. *Grevillea mucronulata*
Photograph — D.Greig (ANBG).

Plate 44. *Grevillea rosmarinifolia* subsp. *glabella*
Photograph — J.Plaza.

Plate 45. *Grevillea iaspicula*
Photograph — D.L.Jones.

Plate 46. *Grevillea lavandulacea*
Photograph — T.Low.

Plate 47. *Grevillea buxifolia* subsp. *buxifolia*
Photograph — M.Fagg (ANBG).

Plate 48. *Grevillea haplantha* subsp. *haplantha*
Photograph — D.Foreman.

Plate 49. *Grevillea scabra*
Photograph — P.Abell.

Plate 50. *Grevillea longistyla*
Photograph — M.Fagg (ANBG).

Plate 51. *Grevillea uncinulata*
Photograph — M.Fagg (ANBG).

Plate 52. *Grevillea insignis* subsp. *insignis*
Photograph — M.Fagg (ANBG).

Plate 53. *Grevillea pythara*
Photograph — A.Lyne (ANBG).

Plate 54. *Grevillea baileyana*
Photograph — D.L.Jones.

Plate 55. *Grevillea paradoxa*
Photograph — D.L.Jones.

Plate 56. *Grevillea petrophiloides* subsp. *petrophiloides*
Photograph — D.Foreman.

Plate 57. *Grevillea parallela*
Photograph — B.Gray.

Plate 58. *Grevillea glauca*
Photograph — T.Low.

Plate 59. *Grevillea leucopteris*
Photograph — D.L.Jones.

Plate 60. *Grevillea teretifolia*
Photograph — D.Foreman.

Plate 61. *Grevillea obliquistigma* subsp. *obliquistigma*
Photograph — D.Foreman.

Plate 62. *Grevillea triloba*
Photograph — D.Greig (ANBG).

Plate 63. *Grevillea pulchella* subsp. *ascendens*
Photograph — D.L.Jones.

Plate 64. *Grevillea quercifolia*
Photograph — A.S.George.

Floribunda Subgroup

Leaf lower surface usually exposed, hairy. Flowers with perianth hairy (at least sparsely so) on both surfaces; late bud with limb obtuse; tepals not apiculate, usually with a strong median ridge or keel in upper part (sometimes concealed by indumentum). Nectary inconspicuous and arcuate to U-shaped or occasionally (*G. alpina*, *G. celata*) conspicuous and linguiform to shovel-like and projecting into perianth pouch; margin smooth to undulate. Pistil 8–28 mm long; ovary sessile or subsessile, villous; style usually exserted strongly to weakly from late bud; style-end not apiculate. Follicle with 2–5 longitudinal ridges (sometimes faint); style persistent, remaining ±erect.

A group of 10 species occurring in south-eastern Australia, from inland central Qld to Vic. Primarily bird-pollinated.

195. Grevillea floribunda R.Br., *Suppl. Prodr. Fl. Nov. Holl.* 1: 19 (1830)

T: 'Ora orient., mont. prope Port Jackson, 1817. D. Cunningham' [protologue]; lecto: [on reverse] New South Wales, *s.d.*, [A.] Cunningham Oxley's 1st Expedition, *s.n.*; lecto: BM, *fide* D.J.McGillivray & R.O.Makinson, *Grevillea* 417 (1993); isolecto: [most as 1817, *A.Cunningham 25*] BM, CGE *n.v.*, G-DC, NSW.

Spreading shrub, 0.2–1.8 m high. Leaves entire, oblong to narrowly elliptic or narrowly obovate or rarely sublinear, 1.5–8 cm long, 3.5–15 (–35) mm wide; margins recurved to loosely revolute; upper surface soon glabrous and granular; lower surface subsericeous, tomentose, or subvelutinous. Conflorescence usually terminal and simple, basally decurved, loosely hemispherical to shortly subcylindrical or subsecund, acropetal; floral rachis 5–45 mm long, tomentose. Flowers acroscopic. Perianth pubescent to tomentose or villous outside, bearded inside in lower third. Pistil 9–19.5 mm long; ovary sessile, white-villous; style villous, scarcely exserted from bud before release of style-end; pollen-presenter very oblique to lateral. Follicle ovoid to ellipsoidal, 10.5–17 mm long, loosely villous, ridged.

Occurs from central Qld down the western side of the Great Dividing Ra. to about Albury in N.S.W., doubtfully in Vic. Two subspecies are recognised.

Perianth with a dense to open (rarely sparse) indumentum of rusty brown hairs outside; perianth usually persistent to fruiting stage; torus 2.2–3.5 mm across from dorsal to ventral edges, transverse on pedicel; peduncle and rachis usually > 1 mm thick, usually villous ..... **195a.** subsp. **floribunda**

Perianth with mainly whitish hairs outside, sometimes with brownish hairs intermixed; perianth falling soon after anthesis; torus 1.2–1.8 mm across, slightly oblique on pedicel; peduncle and rachis c. 0.7 mm thick, tomentose ..... **195b.** subsp. **tenella**

195a. Grevillea floribunda R.Br. subsp. floribunda

Illustrations: G.M.Cunningham *et al.*, *Pl. W New South Wales* 213 (1981), as *G. floribunda*; W.R.Elliot & D.L.Jones, *Encycl. Austral. Pl.* 5: 54 (1990); P.M.Olde & N.R.Marriott, *Grevillea Book* 2: 166 (bottom right), 167 (136A, B) (1995).

Shrub 0.4–2 m tall. Leaves narrowly oblong-elliptic or -obovate or rarely linear, 1.5–8 cm long, 3–15 (–35) mm wide; lower surface tomentose to velvety or sometimes subsericeous. Conflorescence 6–20-flowered; floral rachis 10–45 mm long, usually > 1 mm thick, villous to tomentose; floral bracts 0.8–2.0 mm long, often persistent at anthesis. Torus 2.2–3.5 mm across from dorsal to ventral edges, transverse on pedicel. Flower colour: perianth greenish yellow outside beneath a rusty brown indumentum; inner surface greenish and strongly displayed; style pinkish yellow to brownish pink. Perianth 3–8 mm wide, villous to tomentose outside, with indumentum usually rusty brown; perianth usually persistent to fruiting stage. Pistil 9–19.5 mm long. *Seven Dwarfs Grevillea, Rusty Spider Flower.* Fig. 26H–J.

Widespread on the western slopes of the Great Dividing Ra. and adjacent plains of N.S.W., with disjunct occurrences in central Qld in the area bounded by Blackwater, Augathella and Chinchilla; doubtfully present in Vic. where known from a single record from Killawarra

Forest, at the northern end of the Warby Ra.; this may represent an introduction. Grows in dry sclerophyll forest or woodland or sometimes open shrub associations, often in rocky sites, usually in sandy soils over granite or sandstone. Regenerates from seed or (some populations) lignotuber or rhizome. Flowers May–Nov. Map 256.

Qld: 27.4 km WSW of Rolleston township, *M.Lazarides 33 & R.Story* (AD, BRI, CANB, NSW); 33.8 km SE of Bedourie, *N.H.Speck 1851* (BRI, CANB, NSW, PERTH). N.S.W.: 14.5 km from Coonabarabran towards Gilgandra, 13 Sept. 1963, *M.E.Phillips CBG026129* (AD, CANB); Mt Bingar, Cocoparra Ra., 25 km NE of Griffith, *I.R.Telford 3894 & P.Ollerenshaw* (CANB, NSW). Vic.: Killawarra Forest at N end of Warby Ra., 10 Oct. 1976, *W.B.McDonald* (MEL).

The dense rusty floral indumentum is diagnostically useful, as is the tendency of the perianth to persist in the fruiting stage (but cf. *G. polybractea* below). A 'small-flowered form', with pistils usually 9–12 mm long, small fruits, and a low-growing habit to about 1 m tall, occurs mainly on granite outcrops in the Tingha area of northern N.S.W., with occasional florally similar specimens from the Pilliga Scrub area and Sandy Hollow (upper Hunter system) and on granites in the Cracow–Mundubbera area of SE Qld; in flower size and slenderness of peduncles these approach subsp. *tenella*, but they have the rusty indumentum and persistent perianths characteristic of subsp. *floribunda*. A similarly small-flowered variant, with an open golden brown indumentum over a yellow-green perianth on rhizomatous plants, is known from near Wee Jasper NW of Canberra, where it hybridises with *G. lanigera*.

195b. Grevillea floribunda subsp. **tenella** Olde & Marriott, *Grevillea Book* 1: 184 (1994)

T: Crows Nest Falls National Park, Qld, 3 Sept. 1993, *P.M.Olde 93/50*; holo: NSW; iso: BRI, CANB.

G. floribunda 'delicate form', of D.J.McGillivray & R.O.Makinson, *Grevillea* 274 (1993).

Illustrations: D.J.McGillivray & R.O.Makinson, *op. cit.* 275 (1993), as *G. floribunda* delicate form; P.M.Olde & N.R.Marriott, *Grevillea Book* 2: 167 (bottom right), 168 (137A–C) (1995).

Erect shrub 0.3–2 m tall. Leaves narrowly elliptic or narrowly oblong, 4–5 cm long, 2–5 mm wide; lower surface subsericeous. Conflorescence 4–10 (–20)-flowered; floral rachis 5–20 mm long, c. 0.7 mm thick, tomentose; floral bracts c. 0.8 mm long, falling before anthesis. Torus 1.2–1.8 mm across, slightly oblique on pedicel. Flower colour: perianth variable, often yellowish green becoming reddish from upper half, occasionally all greenish yellow; style pinkish yellow to brownish pink. Perianth 3–4 mm wide, basally saccate, sparsely pubescent to villous outside, with hairs mostly pale; perianth usually falling soon after floral maturity, rarely persistent on young fruits only. Pistil 9–12 mm long.

Occurs in Qld, mainly in the Darling Downs area from SE of Toowoomba to Durong, and extending patchily NE to near Eidsvold. Grows in open to closed low dry eucalypt forest, in sandy soils. Fire response variable: seed, rhizomes, possibly lignotuber. Flowers July–Oct. Map 257.

Qld: 35 km by road W of Tingoora on the Durong South road, *R.Coveny 6762 & P.Hind* (K, NSW); 8 km W of Manar HS, Boondooma, *P.I.Forster PIF4646* (BRI, CANB, GUAM *n.v.*, NSW, PNH *n.v.*); Crows Nest Falls Natl Park, on foot-track to falls, *R.O.Makinson 1365, 1366* (both: BRI, CANB, K, MEL, NSW); Tobacco Rd, 17 km ESE of Inglewood, *A.N.Rodd 4116* (BRI, NSW); Crows Nest, N Darling Downs, Oct. 1921, *C.T.White* (BRI, K).

Olde & Marriott (*op. cit.* 2: 167) erect three forms within subsp. *tenella*, although there is considerable minor variation within these. The 'Crows Nest (type) form' occurs between Crows Nest and Inglewood, and is a robust shrub to 2 m (not lignotuberous or rhizomatous) with the flowers bright green becoming red-brown. The 'Durong form' occurs near the township of Durong, and is a small (to 50 cm tall) compact, possibly lignotuberous shrub with bright greenish yellow flowers. The 'northern form' occurs on the Burnett R. catchment near Eidsvold and near Boondooma; it has been reported as a spindly shrub to 1 m tall with dull green flowers becoming dusky pink-brown.

Figure 26. *Grevillea*. **A–D**, *G. mucronulata*. **A**, flowering branch; **B**, leaves; **C**, flower; **D**, pistil and half perianth (**A–D**, J.McLuckie 1635, CANB). **E–G**, *G. aspera*. **E**, flowering branch; **F**, flower; **G**, pistil (**E–G**, R.D.Pearce 356, CANB). **H–J**, *G. floribunda* subsp. *floribunda*. **H**, flowering branch; **I**, flower; **J**, pistil (**H–J**, R.Pullen 2135, CANB). Scale bars: **A–B**, **E**, **H** = 1 cm; **C–D**, **F–G**, **I–J** = 5 mm. Drawn by D.Boyer.

196. Grevillea polybractea H.B.Will., *Victorian Naturalist* 44: 139, 141, fig. B, D (1927)

T: Granya Gap, Tallangatta, Vic., Oct. 1925, *H.B.Williamson*; lecto: MEL, *fide* D.J.McGillivray & R.O.Makinson, *Grevillea* 435 (1993); remaining syntypes: Mt Granya, Vic., Oct. 1891, *C.Walter*; syn: MEL; Corryong, Vic., Oct. 1917, *E.W.Curtis*; syn: MEL; near Dubbo, N.S.W., 1882, *M.Curran*; syn: MEL.

Illustrations: H.B.Williamson, *op. cit.* 44: 141 fig. B, D; J.W.Wrigley & M.Fagg, *Banksias, Waratahs & Grevilleas* 279 (1989); P.M.Olde & N.R.Marriott, *Grevillea Book* 3: 106 (bottom right), 107 (81A–C) (1995).

Spreading shrub 0.3–1.8 m high. Leaves entire, linear to narrowly oblong-elliptic, 3–7.5 cm long, 1–6 mm wide; margins loosely to tightly revolute; upper surface becoming glabrous and granular; lower surface sometimes obscured, subvillous. Conflorescence terminal, simple, decurved, very dense, subglobose and often subsecund, with 10–20 (–34) flowers, acropetal; floral rachis 3–10 mm long, villous. Flowers adaxially acroscopic. Flower colour: perianth yellow and orange or yellow and red (hairs reddish); style dull pink to reddish orange. Perianth villous outside, bearded inside, usually persistent to fruiting stage. Pistil 9.5–14 mm long; ovary sessile, villous; style villous at base, becoming glabrous towards apex and ventrally, weakly exserted in late bud; pollen-presenter lateral. Follicles obloid-ellipsoidal, 12–15 mm long, villous and subvelutinous, with 3 faint longitudinal dorsal ridges. *Crimson Grevillea.*

Occurs in south-eastern N.S.W. (W of the Divide from the Khancoban area N sporadically as far as Dubbo; possibly extinct or endangered in the far northern part of the range) and in north-eastern Vic. (Corryong to Mt Granya area). Grows in well-drained sites in open eucalypt forest with sclerophyllous shrubby understorey, in stony soils often over granite. Regenerates from seed. Flowers Sept.–Dec. Map 258.

N.S.W.: 8 km WSW of Carabost, *A.E.Logan NSW84408* (NSW); E side of Warrumba Ra. NE of Grenfell, *D.J.McGillivray 3149 & R.Coveny* (K, MEL, NSW, PERTH); Yambira State Forest, between Cowra and Grenfell, *R.J.Turner NSW117300* (NSW). Vic.: 29 km from Walwa on road to Tallangatta, *D.J.McGillivray 3185 & C.Bartlett* (K, NSW); Murray Valley Hwy, 7 km W of Burrowye, 18 Nov. 1964, *J.H.Willis* (AD, MEL).

The floral bracts of *G. polybractea* are conspicuous in late bud stage and diagnostic; they are usually ovate, 2.5–7 mm long and 2–5.5 mm wide, villous on the outer surface, and are usually present at anthesis and more persistent towards the base of the very dense conflorescence. The persistence of the perianth is also unusual and somewhat diagnostic. *Grevillea floribunda* subsp. *floribunda* also has persistent perianths, but has smaller floral bracts 0.8–2.0 mm long, and rusty brown (rather than reddish) perianth hairs and a looser conflorescence.

The illustration in McGillivray & Makinson (*Grevillea* 276 (1993)) given as '*G. polybractea*', is of a hybrid between that species and *G. lanigera*; swarms of these hybrids occur in the Mt Granya area of Vic.

197. Grevillea chrysophaea F.Muell. ex Meisn., *Linnaea* 26: 357 (1854)

T: Australia, Vic., 1850–52, *F.Mueller*; holo: NY *n.v.*; ?iso: BM, G-DC, MEL.

G. chrysophaea var. *canescens* H.B.Will., *Victorian Naturalist* 44: 141 (1927). T: Sperm Whale Head, Vic., Oct. 1926, *F.Barton*; holo: MEL; ?iso: MEL.

[*G. floribunda auct. non* R.Br.: G.Bentham, *Fl. Austral.* 5: 440 (1870), *p.p.*]

Illustrations: J.W.Wrigley & M.Fagg, *Banksias, Waratahs & Grevilleas* 224 (1989); D.J.McGillivray & R.O.Makinson, *Grevillea* 271 (1993); P.M.Olde & N.R.Marriott, *Grevillea Book* 2: 89 (bottom right), 90 (71A, B) (1995).

Spreading or rarely prostrate shrub 0.3–2.5 m tall. Leaves entire, oblong-elliptic (or broadly or narrowly so), to almost linear, 1.5–6 cm long, 3–15 (–23) mm wide; margins recurved or revolute; upper surface pubescent to glabrous, granulose or rarely almost smooth; lower surface usually exposed, tomentose to velvety. Conflorescence terminal on short lateral apically decurved branchlets, usually simple, 2–8 (–12)-flowered, subsecund; floral rachis 1–6.5 mm long, tomentose to subvillous. Flowers acroscopic. Flower colour: perianth dull yellow to golden yellow with a sprinkling of white and brown hairs; style red or orange-red with reddish or brownish, rarely white hairs. Perianth basally dilated, deciduous soon after anthesis, sparsely tomentose to villous outside, bearded inside in lower third. Pistil 15–21.5 mm long; ovary sessile, villous; style villous, scarcely exserted in late bud; pollen-presenter lateral. Follicle ellipsoidal, 11–13 mm long, villous, longitudinally ridged. *Golden Grevillea.*

Occurs in southern Vic. in the Brisbane Ra. in the Anakie to Steiglitz area (near Tallarook and Merton) and in Gippsland (Woodside to Licola area). Grows in eucalypt or *Banksia* woodland or heath in silty sand to sandy loam. Regenerates from seed. Flowers mainly June–Dec. Map 259.

Vic.: Brisbane Ra., along the Mt Anakie to Durdidwarrah road, c. 5 km SE of the reservoirs, *H.I.Aston 62* (MEL); c. 5.5 km NNE of Licola, c. 1.5 km E of Tamboritha Rd, *A.C.Beauglehole 43373 et al.* (MEL, NSW); 16.5 km SE of Gormandale P.O., *A.C.Beauglehole 50268* (MEL, NSW); 4 km ESE of Steiglitz, *D.J.McGillivray 3205 & C.Bartlett* (MEL, NSW); Sperm Whale Head, *M.E.Phillips CBG11326* (CANB, NSW).

Grevillea chrysophaea has a moderately conspicuous nectary that is ±oblong and upright, projecting up to 1 mm above the toral rim (cf. *G. alpina*). In some populations there is also a prominent median ridge or blunt horn on the limb segments of the tepals.

Olde & Marriott (*loc. cit.*) usefully distinguish three forms. The 'Brisbane Ranges form' has an open-straggly to compact habit, and flowers at the upper end of the size range. The 'Holey Plains form' occurs SW of Sale and has a low spreading to prostrate habit, narrow leaves up to c. 5 mm wide, and is very floriferous with bright yellow flowers. The 'Gippsland form' is variable in habit, ranging from open to dense, and in flower colour.

McGillivray & Makinson's (*Grevillea* 270–271 (1993)) circumscription included the since-published *G. celata* (see below) within *G. chrysophaea.* References in that treatment to 'root-suckering' habit, red coloration on the perianth, and occurrence in the Buchan area refer to *G. celata.* McGillivray & Makinson (*loc. cit.*) mentions Tallarook and Merton as localities for *G. chrysophaea*; these records appear to relate to *G. alpina.*

198. Grevillea celata Molyneux, *Muelleria* 8: 311 (1995)

T: Colquhoun State Forest ... c. 5.5 km south east of Bruthen, Vic., 13 Oct. 1993, *W.M.Molyneux*; holo: MEL; iso: AD, BRI, CANB, K, NSW, PERTH.

Illustrations: W.M.Molyneux, *op. cit.* 8: 313, fig. 1 (1995); P.M.Olde & N.R.Marriott, *Grevillea Book* 2: 91 (bottom left & 72) (1995), as *G. sp. aff. chrysophaea.*

Erect and open to low and dense shrub, 0.4–1.8 m tall. Leaves entire, oblong-elliptic to broadly elliptic or -linear, usually 20–45 (–60) mm long, 4–18 mm wide; margins recurved to loosely revolute; upper surface granulose; lower surface tomentose. Inflorescences terminal or axillary, usually simple, usually decurved, a short subsecund 2–8-flowered cluster, acropetal; floral rachis 1–5 mm long, pubescent. Flowers adaxially acroscopic. Flower colour: perianth usually red in basal half, shading to yellow above, or sometimes red and white, apricot and white, or rarely all yellow; style green at base, shading to pink or cherry red above with a green tip, rarely light green throughout. Perianth basally dilated, pubescent to tomentose outside, bearded inside in lower third. Pistil 18–25 mm long; ovary sessile, villous; style dorsally villous in lower half, glabrescent above, scarcely exserted in late bud; pollen-presenter lateral. Follicle ellipsoidal, 14–16 mm long, tomentose, longitudinally ridged. *Nowa Nowa Grevillea.*

Occurs in south-eastern Vic., where known only from the Colquhoun State Forest E of Bruthen. Grows on red siliceous or pale granitic sands in shrubby patches in dry sclerophyll forest. Regenerates from seed and rhizomes. Flowers July–Feb. Map 260.

Vic.: Reformatory Rd, c. 1.55 km N from junction with Bruthen to Buchan road, c. 14.5 km E of Bruthen, 13 Oct. 1989, *W.Molyneux* (MEL); W side of Stony Ck, W of Nowa Nowa, 13 Dec. 1991, *W.Molyneux* (MEL); W of Nowa Nowa, 15 Nov. 1992, *W.Molyneux* (MEL, NSW 117361); 'Foggy' [Boggy?] Ck, c. 13 km E of Bruthen, c. 1937, *F.Robbins* (MEL); Stony Ck on Bruthen to Buchan road, 23 Sept. 1969, *K.C.Rogers* (MEL).

Grevillea celata has a prominent, ascending linguiform or shovel-like thin-edged nectary extending clearly above the toral rim. It is generally similar to the closely related species *G. chrysophaea* (which is not rhizomatous and lacks red colouration on the perianth), and *G. alpina* (which is not usually rhizomatous, has a more prominent and spreading tongue-like nectary, and usually a shorter pistil). *Grevillea celata* is only narrowly distinct from, and shares features of, these two relatives and is unstable in several character states (e.g. flower colour, perianth width); it may be of recent hybrid origin.

199. Grevillea alpina Lindl., *in* T.L.Mitchell, *Three Exped. Australia* 2: 178 (1838)

T: Mt. William, [Vic.], Major Mitchell's Expedition, 1836, [*coll. not known*]; lecto: CGE *n.v.*, *fide* D.J.McGillivray & R.O.Makinson, *Grevillea* 403 (1993); ?isolecto: BM, CGE *n.v.*, K, P.

G. alpestris Meisn., *Hooker's J. Bot. Kew Gard. Misc.* 4: 187 (1852), *nom. nud.*

G. alpestris Meisn., *in* A.L.P.P. de Candolle, *Prodr.* 14: 361 (1856). T: 'v.s. in herb. Lindl., Cunn. et comm. a cl. Sonder' [protologue]; lecto: Australia felix. Sept. [18]52. legit Dr Ferd. Müller. ex Herbar. W.Sonder. acc. m. Apr. 1853; lecto: NY *n.v.*, *fide* D.J.McGillivray & R.O.Makinson, *Grevillea* 403 (1993); remaining syntypes: CGE *n.v.*, K, MEL, NY *n.v.*

G. alpestris var. *helianthemifolia* Meisn., *Hooker's J. Bot. Kew Gard. Misc.* 4: 187 (1852), *nom. nud.*

G. alpestris var. *helianthemifolia* Meisn., *in* A.L.P.P. de Candolle, *Prodr.* 14: 361 (1856), as β *helianthemifolia*. T: Port Phillip ['Phillipp'], [Vic.], *s.d.*, [*C.J.*] *Latrobe*; holo: NY *n.v.*

G. alpina var. *aurea* Guilf., *Austral. Pl. Suitable for Gard., Parks, Timb. Reserves, etc.* 193 (1911). T: 'Golden-flowered Alpine Grevillea ... Vic.' [protologue]; type not found.

G. alpina var. *dallachiana* Benth. ex Guilf., *Austral. Pl. Suitable for Gard., Parks, Timb. Reserves, etc.* 193 (1911). T: 'Dallachy's Grevillea ... Vic. & Qld' [protologue]; type not found. Qld locality is erroneous.

G. dallachiana F.Muell. ex Hazlewood, *Favourite Roses, Trees & Shrubs* 1927: 78 (1927). T: none cited.

Illustrations: W.R.Elliot & D.L.Jones, *Encycl. Austral. Pl.* 5: 20 (1990); D.J.McGillivray & R.O.Makinson, *Grevillea* 272, fig. 73 & col. pl. (1993); P.M.Olde & N.R.Marriott, *Grevillea Book* 2: 20 (top left & 12A, B), 21 (12C–F), 22 (12G–H) (1995).

Spreading to decumbent shrub, 0.3–2 m tall. Leaves entire, oblong to elliptic to almost round or linear, 0.5–2 (–3) cm long, 1.5–4 (–10) mm wide; margins recurved to revolute; upper surface villous to glabrous and granulose; lower surface villous or rarely subsericeous or glabrous. Conflorescence terminal or axillary, usually simple, decurved to erect, subsecund to subglobose and then 4–26-flowered, or 2-flowered, basipetal; floral rachis 2–10 (–18) mm long, tomentose to villous. Flowers adaxially acroscopic. Flower colour: perianth colour variable, usually orange, pink or red, sometimes yellow or cream, often paler (to cream) in apical or dorsal half; style usually reddish, occasionally greenish yellow to pure yellow. Perianth tomentose to villous outside, bearded inside. Pistil 8.5–20.5 mm long; ovary sessile, white-villous; style villous near base, tomentose above, weakly exserted from late bud; pollen-presenter lateral. Follicle ovoid, 8.5–12 mm long, loosely villous, ridged. Chromosome number: $n = 10$. H.P.Ramsay, *Austral. J. Bot.* 11: 5 (1963). *Mountain Grevillea, Cat's Claws Grevillea.*

Widely distributed in Vic. from the Grampians and Black Ra., through the 'Goldfields' area to Mt Buffalo area, S to the Dandenongs N to near Albury; also in N.S.W. where occasional from Albury to Canberra. Grows in a variety of well-drained sites, including in dry sclerophyll forest or woodland, heath or mallee; usually in sandy soils, sometimes on granitic hills or slopes. Despite the name, this species does not extend to alpine altitudes. Regenerates from seed or (very rarely?) rhizomes. Flowers July–Dec. (–May). Map 261.

N.S.W.: Monument Hill, Albury, *E.J.McBarron 4243* (NSW, SYD). A.C.T.: Black Mtn, *R.D.Hoogland 8417* (BRI, CANB, MEL, NSW). Vic.: W of Huntly to Kamarooka road., 8.3 km N of junction with Bendigo to Echuca highway, *H.I.Aston 28* (MEL); near Kelvin View, Strathbogie Ra., *H.I.Aston 652* (MEL, NSW); N foot of Mt Stapylton, Grampians Mtns, *T.B.Muir 2118* (AD, MEL).

In *G. alpina* the nectary is characteristically very conspicuous and linguiform, often almost perpendicular to the pedicel, and protruding into the perianth pouch. *Grevillea alpina* is a variable taxon, in habit, leaf details, inflorescence posture and flower number, and floral characters including colour. Many local forms have informal names in botany and horticulture, but the natural genetic and phylogenetic status of these remains unclear. Olde &

Marriott (*loc. cit.*) group them into five major forms, although these remain variable and do not accommodate all populations. The 'Grampians (type) form' occurs in the Grampians Ra. area, and is large-leaved with a few large bright flowers in erect conflorescences. The 'small-flowered form', congruent with the '2-flowered race' of McGillivray & Makinson (*loc. cit.*) occurs from Canberra to Albury and S to Beechworth and Chiltern; it has usually 2-flowered conflorescences which are often subterminally aggregated, and short styles usually 8.5–9.5 mm long. The 'northern Victorian form' occurs in the Warby and Strathbogie Ranges; it has an open habit and larger leaves and flowers than the previous forms, with more flowers per conflorescence. The 'Goldfields form' occurs in the central and western goldfields of Vic., and has a dense habit to 1 m tall, crowded foliage, small grey hairy leaves, and conspicuous pendent conflorescences of often monotonal flowers. The 'Southern Hills form' occurs N and E of Melbourne, from Mt Slide to Kinglake and the Dandenongs, and is normally an open shrub to 2 m tall, with rounded leaves, short conflorescences and red and cream perianths.

Grevillea alpina has been reported as hybridising with *G. dryophylla* near Bendigo, with *G. obtecta* near Daylesford and with *G. lavandulacea* in the Black Ra. (Grampians). Rhizomatous regeneration is not confirmed in *G. alpina*; references in literature appear to relate to either *G. celata* or to hybrids between *G. alpina* and *G. lanigera.*

200. Grevillea mucronulata R.Br., *Trans. Linn. Soc. London* 10: 173 (1810)

T: Port Jackson, N.S.W., *R.Brown*; lecto: BM, *fide* D.J.McGillivray & R.O.Makinson, *Grevillea* 429 (1993); isolecto: BM; probable isolecto: BM, E, G-DC, K *n.v.*, P *n.v.*

G. cinerea R.Br., *Trans. Linn. Soc. London* 10: 173 (1810). T: '...prope Port Jackson; in montosis ad ripas saxosas fluviorum' [protologue]; lecto: 17 Grevillea cinerea prodr. 378 Port Jackson ad ripas fl: Grose [Grose R., N.S.W.], *R.Brown, Iter Austral. 3328*; lecto: BM, *fide* D.J.McGillivray & R.O.Makinson, *Grevillea* 429 (1993); isolecto: BM, E, K.

G. acuminata R.Br., *Trans. Linn. Soc. London* 10: 173 (1810); *Embothrium acuminatum* (R.Br.) Dum.Cours., *Bot. Cult.* 2nd edn, 7: 112 (1814); *G. mucronulata* var. *angustifolia* Benth., *Fl. Austral.* 5: 443 (1870). T: 'In Novae Hollandiae ... prope Port Jackson; in montosis' [protologue]; lecto: 16 Grevillea acuminata prodr. 378 Hunters River [N.S.W., 1802-5], *R.Brown, Iter Austral. 3327*; lecto: BM, *fide* D.J.McGillivray & R.O.Makinson, *Grevillea* 430 (1993); isolecto: BM.

G. myrtacea Sieber ex Spreng., *Syst. Veg.* 4(2), *Cur. Post.* 46 (1827); *G. myrtacea* Sieber ex Schult. & Schult.f., *Mant.* 3: 280 (1827), *nom. illeg. non* Spreng.; *G. cinerea* var. *myrtacea* Meisn., *in* A.L.P.P. de Candolle, *Prodr.* 14: 358 (1856), as β *myrtacea, nom. illeg. non* Spreng. T: Australia, N.S.W., 1823, *F.W.Sieber 39*; syn: A *n.v.*, BM, G, K, LE *n.v.*, MEL, NY *n.v.*, PR.

G. attenuata A.Cunn. ex Meisn., *in* J.G.C.Lehmann, *Pl. Preiss.* 1: 538 (1845); *G. cinerea* var. *angustifolia* Benth., *Fl. Austral.* 5: 441 (1870). T: Bulga Mtns, Hunters R., N.S.W., 1825, *A.Cunningham*; holo: G-DC; iso: BM, NY *n.v.*

Illustrations: J.W.Wrigley & M.Fagg, *Banksias, Waratahs & Grevilleas* 262 (1989); D.J.McGillivray & R.O.Makinson, *Grevillea* 269 (1993); P.M.Olde & N.R.Marriott, *Grevillea Book* 3: 39 (bottom right), 40 (27A–D) (1995).

Spreading to erect shrub, 0.3–3.5 m tall. Leaves entire, elliptic or narrowly so to obovate or round, 0.4–4 cm long, 2–18 mm wide; margins recurved; upper surface granulose; lower surface loosely villous to subsericeous or sparsely so. Conflorescence terminal and axillary, usually simple, erect to decurved, a loose semi-secund cluster, (2–) 6–18-flowered, basipetal; floral rachis 3–9 mm long, tomentose to villous. Flowers acroscopic. Flower colour: perianth mid to pale green, sometimes with red or brown markings, ageing to almost black; style red or purplish with a green tip. Perianth hairy (usually tomentose) outside with mostly pale hairs, bearded inside in lower third. Pistil 18–28 mm long; ovary sessile, villous; style loosely villous, moderately exserted from late bud; pollen-presenter very oblique to lateral. Follicle ovoid, 13–20 mm long, loosely villous; style persistent, usually deflexed, sometimes decurved or erect. Plate 43; Fig. 26A–D.

Occurs in eastern N.S.W., from Singleton and the Rylstone to Denman area S to Mittagong, with disjunct populations on the S coast at Mossy Point and between Moruya and Eden. Grows in dry sclerophyll forest, in sand to clay soils over sandstone and shale or rarely granite. Regenerates from seed. Flowers mainly May–Oct. Map 262.

N.S.W.: Castlereagh State Forest, *D.Benson 287* (NSW); Baerami, 24 km W of Denman, *R.H.Cambage 2634* (NSW); Green Gully 1.7 km S of Glen Davis, *M.D.Crisp 2200 & I.R.Telford* (CANB, NSW); Kurnell Penin., *J.Pulley 212* (CANB); 5 km N of Merimbula on Tathra road, *A.N.Rodd 4282* (NSW).

Grevillea mucronulata is characterised by the often slightly undulate leaf margins and the subcubic limb of the flower bud (each tepal with a prominent medial keel on the limb segment); the latter feature occurs in some related species but is usually concealed by a denser indumentum. *Grevillea mucronulata* shows considerable variation in leaf form, habit, regeneration mode, and flower colour, with populations often fairly consistent on a geographic basis. While not all populations conform, the following variants can be identified. The 'Lower Blue Mountains form' occurs from about Warragamba N to Singleton, is non-lignotuberous, and tends to have ovate-acuminate leaves; the type of the name *G. cinerea* R.Br. belongs to this form. The 'Cumberland Plain form' occurs in the western Sydney area (e.g. Richmond to Blacktown) has broadly ovate to almost orbicular leaves and is often a low-growing lignotuberous shrub to about 0.5 m tall. A 'large-leaved form' occurs from Botany Bay and the Georges R. to Mittagong and perhaps to Springwood, with leaves and flowers at the large end of the size ranges. The 'Picton form' occurs in the Picton Lakes to Thirlmere area and has narrowly elliptic to sublinear leaves with small flowers (pistils c. 18 mm long); in the Couridjah area this form becomes very similar to *G. kedumbensis*, which has a shorter pistil 12.5–17.5 mm long, style pubescent rather than villous, and leaves more strongly granulose; this population is provisionally assigned to *G. mucronulata*. Further research into variation and relationships is needed.

201. Grevillea kedumbensis (McGill.) Olde & Marriott, *Telopea* 5: 727 (1994)

Grevillea obtusiflora subsp. *kedumbensis* McGill., *New Names Grevillea* 11 (1986). T: Kedumba Valley, 8 km beyond homestead, N.S.W., 7 Oct. 1977, *A.M.Blombery NSW 117349*; holo: NSW.

[*G. obtusiflora auct. non* R.Br.: A.Fairley & P.Moore, *Native Pl. Sydney District* 168, t. 555 (1989)]

Illustrations: A.Fairley & P.Moore, *Native Pl. Sydney District* 168, t. 555 (1989), as *G. obtusiflora*; D.J.McGillivray & R.O.Makinson, *Grevillea* 265 (1993), as *G. obtusiflora* subsp. *kedumbensis*; P.M.Olde & N.R.Marriott, *Grevillea Book* 2: 222 (top right & 185A), 223 (185B) (1995).

Multistemmed shrub 0.2–1 m tall. Leaves entire, narrowly elliptic to narrowly obovate, 1–3 cm long, 1–5 mm wide, often varying much in size on a plant; margins revolute to recurved; upper surface granulose; lower surface often enclosed on smaller leaves, subsericeous. Conflorescence terminal or upper-axillary, ±erect, simple to 3-branched; unit conflorescence a loose subglobose to shortly subcylindrical or semi-secund cluster, 6–20-flowered, basipetal; floral rachis 2–5 mm long, sericeous to sparsely so. Flowers acroscopic. Flower colour: perianth green to cream; style reddish. Perianth loosely tomentose to almost glabrous outside, bearded inside in lower third. Pistil 12.5–17.5 mm long; ovary sessile, villous; style tomentose to subvillous (white biramous hairs) also with simple erect hairs, scarcely to moderately exserted from late bud; pollen-presenter almost lateral. Follicle ovoid, 13–15 mm long, loosely tomentose, faintly ridged; style persistent, ?erect.

Occurs in N.S.W. in the Blue Mtns W of Sydney, in the Kedumba Valley and Yerranderie areas. Grows in dry sclerophyll forest in sandy soils over quartzite or sandstone. Regenerates from seed and lignotuber or ?rhizomes. Flowers most months, peaking June–Oct. Map 263.

N.S.W.: The Cedar Track near Kiaromba Ridge to Devitts Ra., [and?] Coxs R. to Kowmung R., *L.A.S.Johnson NSW15640* (NSW); E slope of Mt Solitary, Kedumba Valley, *A.Willows NSW92639* (NSW); 3.6 km S of Mt Cookem, Scotts Main Ra., *M.Kennedy 22 et al.* (NSW); Kedumba Valley, 1.6 km S of Reedy Ck, *L.A.S.Johnson & D.Benson* (K, NSW, US *n.v.*).

Plants very similar to *G. kedumbensis* but with less strongly granulose leaves and slightly longer pistils occur in the Couridjah area; these are here assigned tentatively to *G. mucronulata*, but may represent an intergrade between the two species.

Grevillea kedumbensis was first recognised as a subspecies of *G. obtusiflora*, but the latter species differs in its red perianth, fewer flowers per unit conflorescence, and fairly consistent leaf size on any one plant.

This species is recognised as 'Poorly Known' in J.D.Briggs & J.H.Leigh, *Rare or Threatened Australian Plants* (1995).

202. Grevillea granulifera (McGill.) Olde & Marriott, *Telopea* 5: 729 (1994)

G. obtusiflora subsp. *granulifera* McGill., *New Names Grevillea* 11 (1986). T: 1–2 km W of Mt George, North Coast, N.S.W., 26 Jan. 1980, *L.A.S.Johnson 8518*; holo: NSW; iso: K.

Illustrations: W.R.Elliot & D.L.Jones, *Encycl. Austral. Pl.* 5: 86 (1990), as *G. obtusiflora* subsp. *granulifera*; P.M.Olde & N.R.Marriott, *Grevillea Book* 2: 183 (bottom left & 150A), 184 (150B, C) (1995).

Compact to open-erect shrub 0.7–5 m tall. Leaves entire, narrowly elliptic, 1.5–6 cm long, 2–11 mm wide; margins recurved to loosely revolute; upper surface obscurely (rarely coarsely) granulose; lower surface usually exposed, subsericeous. Inflorescences terminal or axillary or cauline, simple to 3-branched, on decurved peduncles; unit conflorescence a subsecund cluster, 6–16 (–20)-flowered, basipetal; floral rachis 2–13 mm long, tomentose. Flowers acroscopic. Flower colour: perianth pink to reddish at base, cream above; style reddish brown to burgundy with white hairs and a green style-end. Perianth open-tomentose outside, densely bearded inside in lower third. Pistil 21–24 mm long; ovary sessile, white-appressed-villous; style openly white-tomentose, strongly exserted from late bud; pollen-presenter very oblique to almost lateral. Follicle narrowly ellipsoidal, c. 14 mm long, open-tomentose, with longitudinal ridges; style persistent, ±erect.

Occurs in N.S.W. on the N coast and adjacent ranges, from near Wingham to Barrington Tops and Wollomombi Falls. The 'Gloucester Tops' locality cited in McGillivray (*loc. cit.*) is a collector's labelling error and is incorrect. Grows in sclerophyll forest or woodland, in stony soils over serpentine or rarely granite or schists. Regenerates from seed, not known to be rhizomatous. Flowers Sept.–Nov. (–Jan.). Map 264.

N.S.W.: Watchim [Watchimbark?] Ck, NW of Gloucester, *D.F.Blaxell 3* (NSW); Burrell Ck area near Wingham, 14 Sept. 1968, *R.Gray* (NSW); 4.8 km N of Curricabark Ck, Curricabark, 15 Oct. 1953, *L.A.S.Johnson* (NSW); near lookout at Wollomombi Falls, *D.J.McGillivray 3168 & R.Coveny* (NSW); Polblue Ck, Barrington Tops State Forest, *P.Olde 93/75 et al.* (BRI, CANB, NSW).

This species occurs in small disjunct populations, and is recognised as 'Poorly Known' in J.D.Briggs & J.H.Leigh, *Rare or Threatened Australian Plants* (1995). It is similar to and very narrowly distinct from *G. guthrieana*, which has leaf upper surface coarsely granulose, slightly longer pistils (25–26 mm long), leaf lower surface loosely subvillous, and the perianth green. In addition, *G. guthrieana* has longer peduncles (7–22 mm long), which are glabrous to sparsely hairy, and rather slender and wiry (sometimes weakly flexuose); the peduncles of *G. granulifera* are 3–10 mm long, subsericeous to tomentose, relatively stout, and not at all flexuose.

A population below Wollomombi Falls has the leaf lower surface with wavy, weakly ascending hairs, pistils 25–26 mm long and peduncles < 12 mm long and loosely pubescent. This population differs from those upstream and its assignation as *G. granulifera* is provisional.

203. Grevillea guthrieana Olde & Marriott, *Telopea* 5: 731 (1994)

T: 3 km E of Booral, N.S.W., 20 Sept. 1992, *P.M.Olde 92/96*; holo: NSW; iso: CANB.

G. obtusiflora subsp. *granulifera* McGill. *p.p.*, in D.J.McGillivray & R.O.Makinson, *Grevillea* 264–266 (1993).

Illustrations: D.J.McGillivray & R.O.Makinson, *Grevillea* 264 (1993), as *G. obtusiflora* subsp. *granulifera*; P.M.Olde & N.R.Marriott, *Telopea* 5: 732, fig. 4a–e (1994); P.M.Olde & N.R.Marriott, *Grevillea Book* 2: 186 (bottom left & 152A, B) (1995).

Shrub 1–4.5 m tall. Leaves entire, oblong-elliptic, 2–6 cm long, 4–9 mm wide; margins loosely revolute; upper surface coarsely granulose; lower surface loosely appressed-subvillous. Conflorescence terminal, simple or 2-branched, on decurved to stiffly pendulous peduncles; unit conflorescence a loose cluster, 2–6 (–10)-flowered, basipetal; floral rachis 2–6 mm long, glabrous or sparsely tomentose. Flowers acroscopic. Flower colour: perianth light green; style maroon with green tip. Perianth sparsely tomentose outside, bearded inside near base and openly subsericeous to glabrous above. Pistil 25–26 mm long; ovary subsessile, pubescent to villous; style loosely tomentose in basal half, becoming glabrous above, strongly exserted from late bud; pollen-presenter very oblique. Follicle narrowly ellipsoidal, attenuate, c. 20 mm long, sparsely tomentose, longitudinally ridged; style persistent, ±erect.

Occurs in N.S.W., in scattered localities on the N coast (mainly near Booral and Carrai Plateau). Grows in sclerophyll associations near creeks-lines in sandy-loam soils, sometimes over sandstone. Regenerates from seed and basal shoots. Flowers Aug.–Oct. Map 265.

N.S.W.: W edge of Carrai Plateau overlooking Kunderang Brook, Aug. 1982, *J.Benson s.n.* (NSW); 3 km by road E of Booral towards Bulahdelah, bridge over Booral Ck, *R.O.Makinson 1465 & J.Nightingale* (BRI, CANB, K, MEL, NE, NSW); Booral Ck, Booral to Bulahdelah road, *A.N.Rodd 2355* (K, NSW, PERTH).

Narrowly distinct from *G. granulifera*, which has a more finely granulose upper leaf surface, a neater indumentum on the lower leaf surface, densely hairy peduncles and rachises, shorter peduncles, a pink to reddish and cream perianth and slightly shorter pistils 21–24 mm long.

This species is recognised as 'Vulnerable' in J.D.Briggs & J.H.Leigh, *Rare or Threatened Australian Plants* (1995).

204. Grevillea obtusiflora R.Br., *Suppl. Prodr. Fl. Nov. Holl.* 19 (1830)

T: '... mont. Port Jackson, 1822. D.Cunningham' [protologue]; lecto: Brushy hills north of Bathurst, N.S.W., [Dec.] 1822, *[A.]Cunningham 197*; BM, lecto: *fide* R.O.Makinson, *Fl. Australia* 17A: 503 (2000); isolecto: K; ?remaining syntype: on the hills north of the settlement of Bathurst, N.S.W., 1822, *[A.]Cunningham 157*; syn: BM.

Low multistemmed or spreading to erect shrub 0.2–1 (–2) m tall. Leaves narrowly elliptic to oblong, or linear to very narrowly obovate (15–) 20–30 (–50) mm long, 1–4 mm wide; margins recurved to revolute; upper surface soon glabrous, granulose; lower surface sometimes enclosed, sparsely to densely subsericeous or rarely loosely villous. Inflorescences terminal, simple to 3-branched; unit conflorescence erect, subumbelloid or subsecund, 1–6-flowered, opening uncertain; floral rachis 1.5–3 mm long, tomentose. Flowers acroscopic. Perianth tomentose outside, bearded inside in lower third. Pistil (14–) 18–23 mm long; ovary subsessile, villous; style loosely villous on flanks and dorsal side (biramous hairs), sometimes also with erect simple multicellular ?glandular hairs in apical 3–4 mm or glabrous, scarcely exserted from late bud; pollen-presenter very oblique. Follicle (subsp. *fecunda*) narrowly ellipsoidal or slightly ovoid, 11–13.5 mm long, loosely subsericeous to -subvillous, with faint dorsal ridges; style persistent, erect.

Occurs in the central tablelands of N.S.W., NW of Lithgow. Two subspecies are recognised. Prior to recognition of subsp. *fecunda*, the species was recognised as 'Endangered' in J.D.Briggs & J.H.Leigh, *Rare or Threatened Australian Plants* (1995).

Leaves 1.5–5 mm wide; margins shortly recurved or revolute, usually leaving most of lower surface exposed; leaf lower surface densely subsericeous (ground-tissue not visible between hairs), or occasionally open-villous; perianth 2.5–3 mm across ..... **204a.** subsp. **obtusiflora**

Leaves 1.0–1.2 (–1.8) mm wide; margins strongly revolute, usually completely enclosing lower surface (occasionally some exposed); leaf lower surface, when visible, with an open appressed indumentum (ground-tissue clearly visible between the hairs at ×10 magnification); perianth 1.5–2 mm across ..... **204b.** subsp. **fecunda**

204a. Grevillea obtusiflora R.Br. subsp. obtusiflora

Illustrations: D.J.McGillivray & R.O.Makinson, *Grevillea* 264 (1993); P.M.Olde & N.R.Marriott, *Grevillea Book* 3: 59 (top right & 42A), 60 (42B, C) (1995).

Low multistemmed shrub 0.2–0.4 m tall. Leaves narrowly obtuse-elliptic to oblong, 2–5 cm long, 2–4 mm wide; margins shortly recurved or shortly revolute; upper surface openly and moderately granulose; lower surface usually exposed on most leaves, densely subsericeous (ground tissue completely obscured) or sometimes open-villous. Flower colour: perianth pink to pinkish red with a cream limb; style red. Perianth 2.5–3 mm across from dorsal edge to ventral. Pistil 18–23 mm; style with biramous hairs and sometimes also minute erect simple hairs ventrally. Not known to set fruit or seed.

Occurs in N.S.W. where known only from the Clandulla area NW of Lithgow; the type collection, from 'hills N of Bathurst' and a contemporaneous collection from 'Capity' (Capertee) may refer to a separate occurrence. Grows in open low understorey of eucalypt forest. Regenerates from rhizomes only; fruits, seeds and seedlings have never been recorded. Flowers July–Oct. Map 266.

N.S.W.: 0.4 km from Clandulla, Clandulla State Forest, *R.Coveny 9563* (NSW); Clandulla, *R.O.Makinson 1576 & R.Johnstone* (CANB).

Subsp. *obtusiflora* has broader leaves than subsp. *fecunda*, with the lower surface more exposed and with a denser indumentum; the perianth is broader, and the pistil longer.

204b. Grevillea obtusiflora subsp. fecunda Makinson, *Telopea* 7: 144 (1997)

T: c. 18 km (direct) NNE of Capertee ... along road running NE from 'Port Macquarie' property, N.S.W., 10 Mar. 1995, *R.O.Makinson 1574 & R.Johnstone*; holo: CANB; iso: NSW.

Low spreading to erect dense shrub 0.5–1 (–2) m tall. Leaves linear to very narrowly obovate, 15–40 mm long, 1.0–1.2 (–1.8) mm wide; margins smoothly revolute; upper surface densely granulose; lower surface usually enclosed, including midvein, sometimes narrowly exposed and then with an open appressed indumentum (ground-tissue visible between hairs). Flower colour: perianth pale to deep pink to crimson, paling to pink or cream along dorsal side and with a cream limb, or occasionally cream with a weak red tinge along dorsal side, or rarely yellow; style deep pink to red (rarely yellow) with white hairs; style-end sometimes yellowish. Perianth 1.5–2 mm across. Pistil 14–18 mm long; style with biramous hairs and sometimes also minute erect simple hairs ventrally. Follicles setting freely.

Occurs in N.S.W., NW of Lithgow on the catchment of the Capertee R. Grows in low open dry sclerophyll forest in orange loamy soil with sandstone boulders at c. 570 m alt. Regenerates from seed and also from rhizomes; establishes well on disturbed ground. Flowers Aug.–Dec. (–Mar.). Map 267.

N.S.W.: 6.7 km along road towards 'Port Macquarie' property from Glen Alice to Rylstone road, *R.Johnstone 485 et al.* (NSW); c. 17 km (direct) NNE of Capertee, 4 km along road running NE from 'Port Macquarie' property, *R.Makinson 1571 & R.Johnstone* (CANB, MEL, NSW); c. 14 km SSE of Kandos, 2.2 km S along road from 'Kooringle' Woolshed, from Rylstone to Glen Alice road, *R.Makinson 1575 & R.Johnstone* (CANB).

Subsp. *fecunda* differs from subsp. *obtusiflora* in foliar and some floral characters. Subspecies *obtusiflora* grows to c. 40 cm tall, reproduces from rhizomes only, has leaves 1.5–5 mm wide, the leaf upper surface with granules less prominent than in subsp. *fecunda*, the margins only shortly recurved or -revolute (usually most leaves on a plant with most of the lower surface visible), and the lower surface usually densely subsericeous or sometimes open-villous; the perianth significantly broader (2.5–3 mm across), and the pistil longer (18–23 mm). Distribution of biramous hairs on the style, and presence or absence of simple erect ?glandular hairs near the style apex, are not helpful for diagnosis of the two subspecies. Subsp. *fecunda* does reproduce from rhizomes, but also sets copious numbers of fruits.

Arenaria Subgroup

Leaf lower surface exposed, hairy. Flowers with perianth hairy on both surfaces; late bud with limb acute to caudate; tepals usually apiculate, drawn out into an acute erect appendage 0.2–5 mm long; apical limb-segments of tepals with or without a median ridge or keel. Nectary arcuate to U-shaped, usually inconspicuous; margin smooth to undulate. Pistil 18–32 mm long; ovary sessile or subsessile, villous; style exserted strongly from late bud; style-end not or scarcely apiculate. Follicle with 2–5 longitudinal ridges; style persistent, remaining erect or basally sharply deflexed.

A group of six species; south-eastern Australia (south-eastern Qld to south-eastern N.S.W.). Primarily bird-pollinated. Differences between the northern species (*G. montana* to *G. masonii*) are very fine.

205. Grevillea arenaria R.Br., *Trans. Linn. Soc. London* 10: 172 (1810)

Embothrium arenarium Dum.Cours., *Bot. Cult.* 2, 7: 112 (1814). T: 14 ... low sandy island of the Nepean [R.], a little above its junction with Grose [R.], [N.S.W.], 1803–04, *R.Brown Iter Austral. 3325*; lecto: BM, *fide* D.J.McGillivray & R.O.Makinson, *Grevillea* 405 (1993); isolecto: E, K, ?MEL, ?NSW (NSW92786), NY *n.v.*

Erect to spreading shrub, 0.3–4 m tall. Leaves entire, oblong to elliptic or obovate, rarely linear, 1–7.5 cm long, 3–15 mm wide; margin recurved to revolute; upper surface velvety to pubescent or silky, smooth or granulose; lower surface villous, subsericeous, tomentose or velutinous. Conflorescence terminal on short lateral branchlets, usually simple, erect to deflexed, 2–6 (–10)-flowered, a loose cluster or rarely subcylindrical, opening irregularly; floral rachis 1–10 mm long, subsericeous to tomentose or velvety. Flowers adaxially acroscopic. Perianth subsericeous or pubescent to subvillous or velutinous outside, bearded inside above ovary with spreading hairs above (especially on ventral tepals); tepals apiculate to caudate, with apiculum 0.2–2 (–5) mm long; limb of bud acute to caudate. Pistil 24–32 mm long; ovary sessile, villous; style pubescent to tomentose; pollen-presenter very oblique to lateral. Follicle obloid-ellipsoidal, apically attenuate, 12–22 mm long, sparsely tomentose, longitudinally ridged; style erect.

Occurs in N.S.W. along the Great Dividing Ra. and adjacent slopes and ranges from a line between Sydney and Dubbo S to about the line between Narooma and Kiandra. Two subspecies are recognised.

Leaf lower surface subsericeous, tomentose, or villous (sometimes loosely so); pistil 22–27 mm long; leaves usually 3–5 times as long as wide	**205a.** subsp. **arenaria**
Leaf lower surface with a dense velvety indumentum; pistil 26–32 mm long; leaves usually 2–3 times as long as wide	**205b.** subsp. **canescens**

205a. Grevillea arenaria R.Br. subsp. arenaria

G. ferruginea Sieber ex Spreng., *Syst. Veg.* 4(2), *Cur. Post.* 46 (1827); *G. ferruginea* Sieber ex Schult. & Schult.f., *Mant.* 3: 280 (1827), *nom. superfl.* T: Australia, N.S.W., 1823, *F.W.Sieber 27*; lecto: BM, fide R.O.Makinson, *Fl. Australia* 17A: 503 (2000); isolecto: B *n.v.*, G, G-DC, K, LE *n.v.*, NSW, NY *n.v.*, P, PR, TCD *n.v.*

Illustrations: A.Fairley & P.Moore, *Native Pl. Sydney District* 168 (t. 556) (1989), as *G. arenaria*; D.J.McGillivray & R.O.Makinson, *Grevillea* 267 (1993); P.M.Olde & N.R.Marriott, *Grevillea Book* 2: 35 (top right & 23A, B) (1995).

Single-trunked shrub 1–4 m tall. Branchlets subsericeous, pubescent to tomentose, or subvillous. Leaves 1–7.5 cm long, 3–10 (–15) mm wide, oblong-elliptic to narrowly obovate or rarely obovate or sublinear; upper surface subsericeous to glabrous, granulose; margins recurved to loosely revolute; lower surface subsericeous, tomentose, or villous. Conflorescence erect to deflexed; peduncle and floral rachis subsericeous to villous; floral bracts 1–2 mm long; pedicels 2–8 mm long, subsericeous to villous. Flower colour: perianth basally green to yellow, pink to reddish above; style green. Perianth with tepal appendages 0.2–5.0 mm long. Pistil 22–27 mm long; style pubescent to subvillous; pollen-presenter 2–3 mm long (dorsal edge to ventral).

Occurs in N.S.W., mainly on the eastern fall of the Great Dividing Ra. and adjacent subcoastal ranges from Richmond W of Sydney S to the Deua R. area W of Narooma. Grows in woodland associations, often around rocky drop-offs or near stream lines, in a range of shallow stony soils over various substrates including sandstone, granite, rhyolite and occasionally limestone. Regenerates from seed. Flowers most months, peaking in July–Nov. Map 268.

N.S.W.: Endrick R. near Nerriga, *L.A.Craven 643* (AD, B *n.v.*, BH *n.v.*, BM, BRI, C *n.v.*, CANB, CHR *n.v.*, E, FI *n.v.*, G, HO *n.v.*, KYO *n.v.*, K, L *n.v.*, MEL, NSW, P, US *n.v.*, W *n.v.*, WELT *n.v.*); Bungonia

Lookdown, *A.N.Rodd 3636* (NSW); Goulburn, *C.W.E.Moore 2621* (CANB, NSW); Wingello, *G.D'Aubert 770* (NSW); Moruya, June 1890, *W.Bauerlen* (NSW).

A loose division into two forms is possible, following Olde & Marriott (*loc. cit.* 2: 35), although these are variable and other assortments are possible. The 'silky form' occurs from the middle reaches of the Nepean R. (e.g. Bents Basin) to Yerranderie and S to Nowra and Moruya; it has the lower leaf surface subsericeous to subvillous, conflorescences usually pedunculate, and pedicels 2–5 mm long; the limb of the late bud is strongly revolute and shortly acute (tepal apiculations usually < 2 mm long). The 'villous form' occurs from Bungonia and Wombeyan Caves S to Braidwood and the Deua R. area; it has the lower leaf surface villous, the conflorescences usually sessile, and pedicels 5–8 mm long; the limb of the late bud is usually less strongly recurved and is long-acute to caudate (tepal apiculations 3–5 mm long).

Subsp. *arenaria* may occasionally hybridise with *G. rosmarinifolia*, *G. lanigera* and *G. mucronulata*. It occurs sympatrically with subsp. *canescens* in the Mt Werong area.

205b. Grevillea arenaria subsp. canescens (R.Br.) Olde & Marriott, *Telopea* 5: 711 (1994)

G. canescens R.Br., *Suppl. Prodr. Fl. Nov. Holl.* 18 (1830); *G. arenaria* var. *canescens* Benth., *Fl. Austral.* 5: 443 (1870). T: N.S.W., '... mont. prope Port Jackson, 1817. D. Cunningham' [protologue]; lecto: '*Grevillea cinerea* Cunningh.' in *Field Mem. New South Wales* [specimen marked 1, attrib. on reverse to [C.] Frazer, *sic.*]; lecto: BM, *fide* D.J.McGillivray & R.O.Makinson, *Grevillea* 405 (1993); remaining syntype: New South Wales, 1817, A.Cunningham, Mr Oxley's 1st Expedition, No. 26 [sheet 2]; syn: BM, G-DC.

Illustrations: P.M.Olde & N.R.Marriott, *Grevillea Book* 2: 36 (top right & 24A, B) (1995).

Single-trunked shrub 1–3 m tall, or occasionally suckering and 0.3–1.0 m tall. Branchlets velvety to villous. Leaves 1–4 cm long, 4–10 (–17) mm wide, obovate or sometimes oblong-elliptic; margins shortly recurved; upper surface pubescent; lower surface velvety. Conflorescence usually decurved; peduncle and floral rachis velvety; floral bracts 1.2–3.0 mm long; pedicels 3–5 mm long, velvety. Flower colour: perianth usually bichromatic, green, yellow, or orange to red at base, and pink to reddish above, or occasionally monochromatic; style green. Perianth with tepal appendages 0.2–3.0 mm long. Pistil 26–32 (–35) mm long; style pubescent to subvillous; pollen-presenter usually 3–4 mm long (dorsal edge to ventral).

Occurs in N.S.W. in the western, drier part of the species range, along the western fall of the Great Dividing Ra. and adjacent slopes, from Gilgandra and near Tamworth S to Bathurst and western Blue Mtns, Kiandra and Tumut area. Grows in dry sclerophyll forest, often in rocky situations, in various soils and substrates, often on granite. Regenerates usually from seed, occasionally (e.g. Gilgandra area) from rhizomes. Flowers Aug.–Nov. Map 269.

N.S.W.: Goonoo [-Goonoo] Forest, NE of Dubbo, *Chapman 1244, 1245* (NSW); Gilgandra Flora Reserve, *R.Coveny 2354* (NSW); Winburndale Nature Reserve (20 km E of Bathurst), *R.Coveny 9631* (B, CANB, K, NSW, PER, RSA); Ruby Creek Falls, near Mt Werong, *R.O.Makinson 102 et al.* (NSW); upper Tumut R., SW of Yarrangobilly, Sept. 1911, *Trainor* (NSW).

The dense velvety indumentum of the leaf lower surface (the hairs short, erect, and soft), is characteristic of this subspecies; it also has a tendency to shorter, more obovate and less revolute leaves with a bluish cast to the upper surface. Over most of the range, plants are apparently non-rhizomatous and robust (2–3 m tall), with usually bichromatic perianths (occasionally monochromatic and red to orange or yellow). In the Gilgandra area, populations occur with a low (to 1.5 m tall) rhizomatous habit and bright red monochromatic perianths. In the Kiandra and Tumut areas the flowers often have a very short tepal apiculation ⩽ 0.5 mm long.

Subsp. *canescens* occurs sympatrically with subsp. *arenaria* in the Mt Werong area; the extent of intergradation is uncertain, and it is possible that the two taxa are effectively reproductively isolated; treatment at species rank would also be appropriate.

206. Grevillea montana R.Br., *Trans. Linn. Soc. London* 10: 172 (1810)

G. arenaria subsp. *montana* (R.Br.) McGill., *New Names Grevillea* 1 (1986). T: '... prope Port Jackson; in montosis' [protologue]; lecto: 15 Grevillea montana prodr. 378 Gathered in Mr Barralier's Journey [1802, N.S.W.], [prob. leg. R.Brown], *Iter. Austral. 3326*; lecto: BM, *fide* D.J.McGillivray & R.O.Makinson, *Grevillea* 405 (1993); isolecto: BM.

Illustrations: W.R.Elliot & D.L.Jones, *Encycl. Austral. Pl.* 5: 26 (1990), as *G. arenaria* subsp. *montana*; A.M.Blombery & B.Maloney, *Prot. Sydney Reg.* 80, 81 (1992); P.M.Olde & N.R.Marriott, *Grevillea Book* 3: 35 (bottom right), 36 (23A, B) (1995).

Dense shrub 0.3–1.5 m tall. Leaves entire, narrowly elliptic to oblanceolate, 1–2 (–3) cm long, 1–7 mm wide; margins revolute, sometimes obscuring whole lower surface including midvein; upper surface granulose; lower surface sericeous. Conflorescence erect, terminal, simple, sessile, a 1–4-flowered cluster, opening uncertain; floral rachis 1–4 mm long, subsericeous. Flowers irregularly oriented. Flower colour: perianth bi- or often trichromatic, bright green at base, pinkish red apically, often becoming blackish medially and dorsally; style green. Perianth sparsely tomentose outside, bearded inside above ovary, then with scattered hairs above; tepals with apiculum 1–2 mm long; limb of bud acute-caudate. Pistil (19–) 25–28 mm long; ovary sessile, villous; style loosely villous; pollen-presenter lateral. Follicle ovoid to ellipsoidal, c. 12 mm long, loosely and patchily villous, longitudinally ridged; style persistent, erect.

Occurs in N.S.W., restricted to the southern margin of the Hunter R. catchment between Denman and Kurri Kurri. Grows in open eucalypt forest in sandy loam soils over sandstone or shallow shales. Regenerates from seed and also lignotuber and rhizomes. Flowers mainly Sept.–Oct. Map 270.

N.S.W.: Greta, Nov. 1904, *J.L.Boorman NSW 92649* (NSW); 6.4 km SE of Bulga on Milbrodale to Broke road, *R.Coveny 5592 & S.W.L.Jacobs* (BRI, NSW); 0.8 km S of Pelaw Main P.O. towards Cooranbong, *R.Coveny 7383* (NSW, PERTH); Denman, Sept. 1908, *H.Heron NSW 92642* (NSW); North Rothbury, near Branxton, *M.E.Phillips 361* (CANB, NSW).

Grevillea montana was regarded as a subspecies of *G. arenaria* by McGillivray (*loc. cit.* and in McGillivray & Makinson (*loc. cit.*)); the two are very closely related and morphologically similar, with *G. arenaria* usually having broader more obtuse leaves, usually with a less (or not-) appressed indumentum on the lower surface, 2–6 (–10)-flowered unit conflorescences, and the style pubescent to tomentose with shorter hairs. *Grevillea masonii* is similar but has the style tomentose or pubescent (shorter hairs than *G. montana*) and leaves usually broader (to 11 mm wide), with an obtuse apex (acute or subacute in *G. montana*).

This species is recognised as 'Poorly Known' in J.D.Briggs & J.H.Leigh, *Rare or Threatened Australian Plants* (1995).

The locality given in the protologue, while vague, is somewhat discordant with a taxon from the Hunter R. area. Moreover, Barralier is not known to have visited the Hunter region, but did visit the Blue Mtns area. The protologue refers to 'glabrous peduncles more or less equal in length to the sub-glabrous perianth'; the conflorescence in *G. montana* is sessile. The lectotype designated by McGillivray (*in* McGillivray & Makinson (*loc. cit.*)) is consistent with the concept of *G. montana* applied for the last century or so to the Hunter R. taxon, but it is possible that the (Bennett) label has been misplaced onto the lectotype and originally related to a specimen of another taxon (possibly a form of *G. mucronulata*, although this does not conform fully to the protologue either). The lectotype is in any event a poor match for the protologue. Conservation of the name *G. montana* on a neotype belonging to the Hunter River taxon may be desirable if the situation can be clarified.

207. Grevillea rhizomatosa Olde & Marriott, *Telopea* 5: 724 (1994)

T: Mulligan's Hut, Dandahra Creek, Gibraltar Ra. Natl Park, N.S.W., 21 Sept. 1992, *P.M.Olde 92/101*; holo: NSW; iso: BRI, CANB.

Illustrations: P.M.Olde & N.R.Marriott, *op. cit.* 5: 725, fig. 3a–f (1994); P.M.Olde & N.R.Marriott, *Grevillea Book* 3: 136 (bottom right), 137 (108) (1995).

Spreading shrub 0.3–1.0 m tall. Leaves entire, obovate to narrowly so or almost orbicular, 0.7–3.0 cm long, 7–12 mm wide; margins shortly recurved to almost flat; upper surface smooth to faintly granulose; lower surface exposed, with a dense matted indumentum with longer emergent hairs. Inflorescences terminal, simple, erect, a loose 1–4-flowered cluster, opening uncertain; floral rachis 2–3 mm long, villous. Flowers adaxially acroscopic. Flower colour: perianth bi- or trichromatic, green basally, pinkish red apically, often purplish black about middle; style green. Perianth sparsely tomentose outside, inconspicuously bearded inside near base, glabrous or with a few appressed hairs above; tepals with apiculum 1–2.5 mm long; limb of bud acute-caudate. Pistil 23–25 mm long; ovary sessile, villous; style sparsely tomentose, usually sharply inflexed in the apical half; pollen-presenter lateral. Not known to set fruit or seed.

Occurs in north-eastern N.S.W. where known only from the Gibraltar Ra. area E of Glen Innes. Grows in densely shrubby eucalyptus forest, usually near creek lines, in rocky sandy loam soils. Regenerates apparently from rhizomes only; searches have not found any mature fruits. Flowers (Aug.–) Sept.–Nov. Map 271.

N.S.W.: Gibraltar Range Natl Park, 27 Oct. 1968, *Grieves* (NSW); Gibraltar Range Natl Park, Little Dandahra Ck, bank opposite Mulligans Hut picnic area, *R.O.Makinson 1447* (AD, BRI, CANB, K, MEL, NE, NSW); Dandarah Ck near Mulligans Hut, *K.Wilson 819 & L.A.S.Johnson* (NSW).

The leaf lower surface has a characteristic 2-layered indumentum, with a dense matted appressed indumentum of short hairs overlain by an open layer of longer emergent weakly ascending hairs. *Grevillea rhizomatosa* is closely related to *G. montana*, which has a sericeous indumentum on the lower leaf surface, and to *G. arenaria* which has the inner surface of the perianth with a dense spreading indumentum continuing above the near-basal beard.

This species is recognised as 'Vulnerable' in J.D.Briggs & J.H.Leigh, *Rare or Threatened Australian Plants* (1995).

208. Grevillea quadricauda Olde & Marriott, *Telopea* 5: 722 (1994)

T: Helidon Hills, Qld, 16 Aug. 1992, *N.McCarthy s.n.*; holo: NSW; iso: BRI, CANB, K, MEL, US *n.v.*

Illustrations: P.M.Olde & N.R.Marriott, *op. cit.* 5: 723, fig. 2g–m (1994); P.M.Olde & N.R.Marriott, *Grevillea Book* 3: 125 (centre & 97A, B) (1995).

Shrub ±erect, 1–2 m tall. Leaves entire, narrowly obovate to oblong-elliptic, 1–2 cm long, 3–6 mm wide; margins shortly recurved to almost flat; upper surface pubescent to villous, granulose to scabrous; lower surface loosely tomentose to loosely villous. Conflorescence usually terminal, erect, usually simple, a loose 1–4-flowered cluster, basipetal; floral rachis 1–2 mm long, tomentose. Flowers acroscopic. Flower colour: perianth bichromatic, green basally, pinkish red or occasionally white to cream apically; style green. Perianth openly tomentose outside, bearded inside at or below level of ovary and glabrous or with a few appressed hairs above; tepals with apiculum 1.5–2 mm long; limb of bud caudate. Pistil 25–27 mm long; ovary sessile, villous; style loosely villous; pollen-presenter lateral. Follicle ovoid to obloid-ellipsoidal, 15–18 mm long, pubescent to loosely villous, faintly ridged; style persistent, sharply basally inflexed. Fig. 27A–C.

Occurs in south-eastern Qld (Helidon Hills to Murphys Ck area near Toowoomba) and in north-eastern N.S.W. (near Whiporie SW of Grafton). Grows in dry sclerophyll forest or woodland, usually near creek lines, in sandy or loam soils over deep sands or sandstone. Regenerates from seed. Flowers mainly July–Nov. Map 272.

Qld: Helidon Hills, 6.4 km NW of Helidon, *P.Baxter 1086 & B.Lebler* (BRI); Murphys Ck, [*s.d.*?], *Lord 4/44* (BRI); c. 5 km (direct) NE of Murphys Ck, 0.9 km along White Mountain Rd, from Murphys Ck to Hampton road, *R.O.Makinson 1364 et al.* (BRI, CANB, K, MEL, NSW). N.S.W.: c. 32 km SSW of Casino, c. 5 km from Lawrence Rd along Brewers Rd, near junction of Cabbage Tree Ck with Fullers Arm, *R.O.Makinson 1438 et al.* (CANB, K, MEL, NE, NSW); Cabbage Tree Ck, Mt Neville Nature Reserve, SW of Rappville, *P.Olde 93/49 & D.Mason* (BRI, NSW).

Similar to *G. masonii*, which has a low lignotuberous habit, a shorter tepal apiculation < 1 mm long, and the inside of the perianth with the beard positioned higher, and a slightly

211a. Grevillea rosmarinifolia A.Cunn. subsp. rosmarinifolia

G. latrobei Meisn., *in* J.G.C.Lehmann, *Pl. Preiss.* 1: 539 (1845), as *G. Latrobii*; *G. latrobei* var. *leiostylis* Meisn., *in* A.L.P.P. de Candolle, *Prodr.* 14: 364 (1856), as α *leiostylis*, *nom. illeg.* T: e, Nov. Holl. Port Phillipe, [Vic.], 1842, [*C.J.*] *Latrobe*; holo: NEU *n.v.*; ?iso: G-DC (as '1845').

G. latrobei var. *dasystylis* Meisn., *in* A.L.P.P. de Candolle, *Prodr.* 14: 364 (1856), as β *?dasystylis*. T: Vic., 'circa Port Phillip alibique in Australia Felici (Ferd. Müller!)'; holo: not located.

G. nutans Meisn., *in* A.L.P.P. de Candolle, *Prodr.* 14: 364 (1856). T: interior of New Holland, [?Vic.], Mitchell's expedition, 5 July 1836, *T.L.Mitchell 219*; holo: NY *n.v.*; iso: CGE *n.v.*, K.

Illustrations: L.F.Costermans, *Native Trees Shrubs SE Australia* 160, 161 (1981); W.R.Elliot & D.L.Jones, *Encycl. Austral. Pl.* 103 (1990); P.M.Olde & N.R.Marriott, *Grevillea Book* 3: 147 (top right & 117A), 148 (117B, C), 149 (117F) (1995); all as *G. rosmarinifolia*.

Dense to open, erect to low spreading shrub, 0.2–2.0 m tall. Leaves usually well-spaced along branchlets or occasionally tightly clustered on short lateral branchlets, strongly ascending to almost spreading, narrowly elliptic or narrowly oblong to linear or occasionally linear-subterete, 0.8–4.0 cm long, 0.8–3.0 mm wide, pungent or not; margins loosely recurved to tightly revolute; lower surface exposed and subsericeous to tomentose or rarely almost glabrous, or mostly or wholly enclosed by revolute margins and then 1-grooved. Flower colour: perianth dichromatic and basally pink and cream above, or monochromatic and pink to pinkish red or rarely yellow, pale green, cream or orange; style pink or matching perianth.

Occurs in montane areas in south-eastern N.S.W., and patchily through much of inland Vic., from Gippsland SW to Melbourne area and W to Skipton and the Brisbane Ra.; also reportedly naturalised in S.A. (Lofty Ra.). Grows in open eucalypt forest or woodland or in riparian shrub associations, on rocky slopes or near creeks. Regenerates usually from seed, in some populations also from rhizomes. Flowers July–Dec. Map 275.

N.S.W.: Germanton [now Holbrook], Oct. 1900, *W.Forsyth NSW93009* (NSW); 2.3 km along Evens Rd from Oberon road (bear left at 1 km), Bindo Ck, *R.O.Makinson 1579* (CANB); Lobs Hole, 9 Nov. 1961, *M.E.Phillips CBG010690* (CANB). Vic.: 8 km NE of Lara, in railway reserve, *G.W.Carr 7161* (NSW); Moroka R. below Mt Kent (c. 110 km NE of Traralgon), 13 Mar. 1966, *J.H.Willis* (MEL).

Subsp. *rosmarinifolia* is often very sporadic in occurrence, but where it does occur it is often gregarious. Even with the excision of subsp. *glabella* and *G. divaricata*, much variation in habit, foliage and flower colour remains. The 'type form' is a dense erect shrub with tomentose branchlets and greyish foliage, leaves 2–3 mm wide, recurved to loosely revolute leaf margins, and the subsericeous leaf lower surface mostly exposed on most leaves; it is still extant near the Type locality, on a tributary of the Coxs R. E of Bathurst, N.S.W. Very similar populations survive in the Hampton and Tuglow R. areas. Forms with soft green leaves occur sporadically from the Blue Mtns S to Blowering. Horticultural selections made from some of these populations have naturalised in some areas (e.g. N of Goulburn, Canberra area). An 'Upper Lachlan River' form is a low shrub to c. 40 cm tall, with narrowly elliptic leaves, lower leaf surface variably exposed or enclosed, sparsely hairy branchlets and leaf lower surfaces, and flowers cream with touches of pink. The 'Lobs Hole form' occurs S of Tumut and has very strongly ascending crowded leaves 1.5–3 cm long, with fully revolute margins; except in leaf size it is morphologically similar to subsp. *glabella*. In eastern Vic., the 'Crooked River form' is a dense shrub to 2 m tall, with green linear leaves and dull pink and cream flowers. In central Vic., the 'Lara form' is known only from a small and now extinct population on basalt plains at Lara near Geelong; it is a low rhizomatous shrub to c. 40 cm tall, with bluish grey leaves, and may survive in cultivation. Similar, though more robust, populations occur elsewhere in central Vic. (Craigieburn, South Mandurang, Anakie area). The 'Whipstick form' from near Bendigo has a virgate habit with the leaves crowded on short lateral branchlets and also on longer branchlet internodes, short (c. 1 cm long), strongly ascending, and with the lower surface slightly exposed; flower colour varies from red or pink and cream, to pale green or cream or yellow; a similar variant with orange flowers occurs near Rushworth. The 'Whipstick form' is close morphologically and geographically to subsp. *glabella*.

Grevillea rosmarinifolia apparently occasionally hybridises with *G. lanigera*, which normally has thicker obtuse hairy leaves with a villous to lanate lower surface, a rather linguiform nectary (cushion-like in *G. rosmarinifolia*), a densely villous ovary, and much hairier styles. Apparent morphological intermediates between *G. rosmarinifolia* and *G. lanigera* also occur in areas of eastern and central Victoria beyond the known range of *G. lanigera* (cf. Willis, 1973: 46, where a 'widespread riparian population' of such intermediates is referred to *G. lanigera*). These populations, where they still survive, need study; they may reflect a former distribution of *G. lanigera*, or deserve inclusion in a more broadly circumscribed *G. rosmarinifolia*. See note under *G. lanigera*, below.

211b. Grevillea rosmarinifolia subsp. **glabella** (R.Br.) Makinson, *Fl. Australia* 17A: 503 (2000)

G. glabella R.Br., *Suppl. Prodr. Fl. Nov. Holl.* 20 (1830). T: N.S.W., '... mon. prope Port Jackson 1817. D. Cunningh.' [protologue]; lecto: Oxley's 1st Expedition [R.Brown script; leg. prob. A.Cunningham]; lecto: BM, *fide* D.J.McGillivray, *Telopea* 1: 28 (1975); isolecto: K, NY *n.v.*, both as '31 May 1817, *A.Cunningham 28*'.

Illustrations: G.M.Cunningham *et al.*, *Pl. W New South Wales* 213 (1981), as *G. glabella*; P.M.Olde & N.R.Marriott, *Grevillea Book* 3: 148 (117D, E) (1995), as *G. rosmarinifolia*, Narrow-leaf form.

Much branched rounded shrub 0.8–2.0 m tall. Leaves strongly ascending, usually tightly clustered on short lateral branchlets or occasionally solitary on larger branchlets, linear-subterete, 0.8–1.2 (–1.5) cm long, 0.7–0.8 mm wide, pungent; margins tightly revolute, enclosing lower surface including midvein, very rarely a little subsericeous lamina exposed at base of leaf. Flower colour: perianth dichromatic and basally pink and cream above, or monochromatic and pink to pinkish red or rarely yellow, pale green, cream or orange; style pink or matching perianth. Plate 44.

Occurs in the Rankins Springs to Griffith area of central-southern N.S.W. and in western Vic. (Little Desert area). Grows in mallee or shrub associations in sandy soils. Regenerates from seed. Flowers Aug.–Nov. Map 276.

N.S.W.: 24 km E of Rankins Springs by road, 21 Mar. 1959, *E.F.Constable NSW92859* (NSW); 4 km W of Kamarah, c. 50 km E of Griffith, *M.D.Crisp 1504* (NSW); 20.1 km W of Temora P.O. (by road) towards Griffith, *R.O.Makinson 1304 & D.J.Mallinson* (AD, K, MEL, NSW). Vic.: Little Desert, 25 Nov. 1973, *D.J.McGillivray & C.Bartlett* (NSW); Serviceton, S.A. border, 1887, *Turner s.n.* (MEL).

Grevillea rosmarinifolia is very variable, grading morphologically and ecologically from the N.S.W. Southern Tablelands (where it has a bushy to erect habit, relatively broad leaves, open foliage, and a mesic habitat) through northern and central Victoria (sporadic occurrences, variable shrubby to near prostrate habit, and variable leaves), to western Victoria, and north into inland N.S.W. (where it is a more gregarious plant characterised by a consistently upright and densely foliaged habit with the leaves ≤ 1 mm across). The ends of the cline are distinct but the transition zone, involving most Victorian populations, grades in both directions. The inland N.S.W. variant has intermittently been recognised as *G. glabella* R.Br., and some populations from W Vic. (Little Desert area) are very similar to it. Recent treatments (McGillivray & Makinson (1993); Makinson *in* Harden (1991); Makinson *in* Walsh & Entwistle (1996); Olde & Marriott (1995)), have synonymised *G. glabella* because of the difficulty of delimitation and the ambiguous placement of many populations from central and western Vic. and one or two populations in montane N.S.W. (e.g. Lobs Hole).

The foliar and ecological distinctions of the ends of the cline are such that some degree of formal recognition of the inland populations is required, pending genetic analysis of the complex. Subsp. *glabella* as here recognised is restricted to those populations with a dense foliage of very fine leaves, occurring in sandy soils in mallee or mallee-shrub communities.

212. Grevillea divaricata R.Br., *Suppl. Prodr. Fl. Nov. Holl.* 1: 20 (1830)

G. rosmarinifolia var. *divaricata* (R.Br.) Maiden & Betche, *Census New South Wales Pl.* 60 (1916). T: '... mont. prop. Port Jackson, 1823, D. Cunningh.' [protologue]; lecto: 30 Interior [N.S.W.] Apr. 1823 [*A.Cunningham*]; lecto: BM, *fide* D.J.McGillivray & R.O.Makinson, *Grevillea* 439 (1993); isolecto: A *n.v.*, B *n.v.*, BM, CGE *n.v.*, G-DC, K, NY *n.v.*

Low shrub to ?40 cm tall. Leaves entire, well-spaced along branchlets, spreading, linear, often gently curved, 0.8–1.3 cm long, 0.5–0.6 mm wide; margins revolute; upper surface scabrid; lower surface enclosed including midvein and 1-grooved, or rarely slightly exposed near leaf base. Conflorescence terminal, simple to 3-branched; unit conflorescence a decurved 1–4-flowered loose cluster, opening uncertain; floral rachis 2–6 mm long, glabrous. Flowers acroscopic. Flower colour: not known, probably red or red and cream. Perianth glabrous outside, bearded inside. Pistil c. 16 mm long; ovary shortly stipitate, glabrous or with a few ascending hairs ventrally on basal half; stipe swollen, c. 0.5 mm long, ventrally tomentose; style glabrous, slightly exserted from late bud; pollen-presenter lateral. Fruit and seed not known.

Occurs in N.S.W., known only from the Type collection made N of Bathurst. Recorded as growing in dry open forest. Fire response unknown, but type material appears to have been rhizomatous or lignotuberous, and capable of stem-basal suckering. Flowering recorded for Apr., probably also spring months. Map 277.

Recent treatments have considered *G. divaricata* as representing a depauperate plant of *G. rosmarinifolia.* Re-examination of the types indicates that while very closely related, it should be recognised as distinct. *Grevillea rosmarinifolia* is, with the exception of a few populations in Vic., non-lignotuberous and non-rhizomatous, and has pistils usually 17–23 mm long; most montane populations (subsp. *rosmarinifolia*) have much longer, broader leaves (1.5–4 cm long, 1–3 mm wide) than *G. divaricata*, with a non-scabrid upper surface and often with the lower surface partly exposed. Narrower-leaved populations of *G. rosmarinifolia* (in either subspecies) have the leaves strongly ascending and smooth or faintly granular.

213. Grevillea iaspicula McGill., *New Names Grevillea* 7 (1986)

T: Thermal Paddock, Mr A.Howard's property, Wee Jasper, N.S.W., 16 May 1980, *D.J.McGillivray 3962*; holo: NSW; iso: K.

Illustrations: W.R.Elliot & D.L.Jones, *Encycl. Austral. Pl.* 5: 64 (1990); J.W.Wrigley & M.Fagg, *Banksias, Waratahs & Grevilleas* 244 (1989); D.J.McGillivray & R.O.Makinson, *Grevillea* 282, fig. 74 & col. pl. (1993).

Spindly shrub, 1–1.5 m high. Leaves entire, narrowly oblong-elliptic, 0.6–3.5 cm long, (2–) 4–10 mm wide, obtuse; margins flat to shortly recurved; upper surface smooth; lower surface exposed, glabrous. Conflorescence terminal or upper-axillary, 2–10-branched or rarely simple, often deflexed near base to pendulous or on downward-directed branchlets; unit conflorescence a loose irregular to semi-secund cluster, 8–20-flowered, weakly basipetal to subsynchronous; floral rachis 3–14 mm long, glabrous. Flowers acroscopic. Flower colour: perianth pale green to cream, often tinged pinkish near curve; style and pollen-presenter pink to red. Perianth glabrous outside, bearded inside. Pistil 16–18 mm long; ovary shortly stipitate, glabrous or with a few hairs only; stipe 0.5–0.7 mm long, swollen and ventrally pilose to almost glabrous; style glabrous, weakly exserted from late bud; pollen-presenter lateral. Follicle narrowly ellipsoidal, 12–17 mm long, glabrous, longitudinally ridged. Plate 45; Fig. 27H–J.

Occurs in N.S.W., where known only from the Wee Jasper area NW of Canberra. Grows on limestone exposures where inaccessible to sheep grazing. Regenerates from seed. Flowers May–Nov. Map 278.

N.S.W.: on old Goodradigbee R. channel of L. Burrinjuck, *G.Butler 1738 & D.Mallinson* (CANB); Wee Jasper, *A.Howard NSW128280* (NSW); L. Burrinjuck, 1.5 km S of Wade Is., *M.M.Richardson 359a et al., 360 et al.*, (both CANB); Macphersons Ck, Wee Jasper, 22 Jan. 1966, *J.Webb* (CANB); Wee Jasper Caves, *T. & J.Whaite 2836* (NSW).

This species is recognised as 'Endangered' in J.D.Briggs & J.H.Leigh, *Rare or Threatened Australian Plants* (1995).

214. Grevillea jephcottii J.H.Willis, *Muelleria* 1: 117 (1967)

T: SW slopes of Pine Mtn, c. 7 miles [11.3 km] SE of Walwa, Vic., 17 Nov. 1964, *J.H.Willis*; holo: MEL; iso: AD, BRI, CANB, NSW.

Illustrations: W.R.Elliot & D.L.Jones, *Encycl. Austral. Pl.* 5: 68 (1990); D.J.McGillivray & R.O.Makinson, *Grevillea* 281 (1993); P.M.Olde & N.R.Marriott, *Grevillea Book* 2: 216 (bottom right), 217 (180) (1995).

Dense to open shrub 1–3 m tall. Leaves entire, usually narrowly oblong-elliptic or sometimes tending lanceolate or oblanceolate, 10–35 mm long, 1.5–6 (–8) mm wide; margins shortly recurved; upper surface scabrid, tuberculate; lower surface glabrous (juvenile leaves sparsely hairy along midvein). Conflorescence terminal, erect, usually clustered, simple, a dense 3–8-flowered semi-secund cluster; floral rachis 2–3 mm long, sparsely tomentose. Flowers acroscopic. Flower colour: perianth pale green becoming pale lemon or cream, then blackening with age; style purplish with white hairs. Perianth glabrous outside, bearded inside. Pistil 9–10.5 mm long; ovary sessile, villous; style dorsally villous in lower half becoming glabrous above, not or scarcely exserted from late bud; pollen-presenter very oblique to almost lateral. Follicle narrowly ovoid, 10–15 mm long, sparsely villous to pilose, weakly to obscurely ridged. *Pine Mountain Grevillea.*

Occurs in north-eastern Vic., where restricted to a small area between Walwa and Corryong, centring on Burrowa and Pine Mountain Natl Park. Grows in dry sclerophyll forest, in rocky situations over granite, at altitudes of c. 550–650 m. Regenerates from seed. Flowers July–Nov. Map 279.

Vic.: Cudgewa Bluff, *R.V.Smith 73/35* (AD, BRI, CANB, HO, K, MEL, NSW, PERTH); mid-tops of Mt Burrowa, 17 Oct. 1971, *J.H.Willis* (MEL, NSW); Pine Mtn, higher declivities, 8 Dec. 1974, *J.H.Willis* (MEL).

Reported to hybridise in the wild with *G. lanigera.*

This species is recognised as 'Rare' in J.D.Briggs & J.H.Leigh, *Rare or Threatened Australian Plants* (1995).

215. Grevillea lanigera A.Cunn. ex R.Br., *Suppl. Prodr. Fl. Nov. Holl.* 1: 20 (1830)

G. lanigera var. *revoluta* Meisn., *in* A.L.P.P. de Candolle, *Prodr.* 14: 364 (1856), as β *revoluta*, *nom. illeg.* T: '... ad rip. fl. Lachlan, 1824. D. Cunningham.' [protologue]; lecto: Rocky bed of the Murrumbidgee [R.] or Lachlan [R.], SW from Lake George [N.S.W.], Apr. 1824, *A.Cunningham 42*; lecto: BM, *fide* D.J.McGillivray & R.O.Makinson, *Grevillea* 424 (1993); isolecto: B *n.v.*, BM, G-DC, K, NSW, NY *n.v.*

G. lanigera var. *planifolia* Meisn., *in* A.L.P.P. de Candolle, *Prodr.* 14: 364 (1856), as α *planifolia.* T: Yass, [N.S.W.], *s.d.*, *W.MacArthur 137*; holo: NY *n.v.*; iso: CGE *n.v.*, FI *n.v.*

G. ericifolia R.Br., *Suppl. Prodr. Fl. Nov. Holl.* 1: 20 (1830). T: '... mont. prope Port Jackson, 1817, D. Cunningh.' [protologue]; lecto: [N.S.W.] Grevillea Pt. ericifolia nob. Grevillea phylicoides Cunn. herb et 1 Oxley; lecto: BM, *fide* D.J.McGillivray & R.O.Makinson, *Grevillea* 424 (1993); isolecto: ?E, K, NSW, NY *n.v.* (all as Limestone Ck, Wn country prope Bathurst [N.S.W.], 23 Apr. 1817, A.Cunningham, Oxley's Expedition).

G. ericifolia var. *muelleri* Meisn., *in* A.L.P.P. de Candolle, *Prodr.* 14: 365 (1856), as β *muelleri.* T: 'Ad flum. Delalite [sic] et inter Ovens et Maday Hills Australiae merid. (Ferd. Müller)'; lecto: on stony hills between the Ovens [R.] and May day hills (Australia Felix) [Vic.], d. 23 Febr. 1853, *F.Mueller*; NY *n.v.*, *fide* D.J.McGillivray & R.O.Makinson, *Grevillea* 424 (1993); isolecto: MEL; remaining syntypes: MEL.

G. scabrella Meisn., *in* A.L.P.P. de Candolle, *Prodr.* 14: 365 (1856); *G. ericifolia* var. *scabrella* (Meisn.) Benth., *Fl. Austral.* 5: 445 (1870). T: 'Circa Nargus in Nov. Holl. orient. australi interiore (Mac Arthur, n. 134!)' [protologue]; lecto: Nangus [near Gundagai, N.S.W.], *s.d.*, *W.MacArthur 134*; lecto: NY *n.v.*, *fide* D.J.McGillivray & R.O.Makinson, *Grevillea* 424 (1993); isolecto: CGE *n.v.*, FI *n.v.*

G. baueri var. *pubescens* Benth., *Fl. Austral.* 5: 444 (1870), *nom. illeg. non* (Hook.) Meisn. (1856). T: Shoalhaven, N.S.W., *s.d. W.Woolls*; holo: MEL.

Illustrations: W.R.Elliot & D.L.Jones, *Encycl. Austral. Pl.* 5: 71 (1990); D.J.McGillivray & R.O.Makinson, *Grevillea* 277 (1993); P.M.Olde & N.R.Marriott, *Grevillea Book* 2: 226 (top left & 188A, B), 227 (188C, D) (1995).

Prostrate to mounded or dense erect shrub 0.2–2 m tall. Leaves entire, narrowly oblong-elliptic to narrowly obovate or sublinear, sometimes plumply subterete, 0.5–4 cm long,

1–5 mm wide; margins revolute to recurved; upper surface scabrid and sometimes villous; lower surface enclosed and then 1-grooved, or exposed and then villous to sublanate. Conflorescence terminal and/or upper-axillary, usually simple, decurved or erect, usually a subsecund cluster, 2–10-flowered, basipetal; floral rachis 2.5–10 (–35) mm long, almost glabrous. Flowers acroscopic. Flower colour: perianth pale pink to red at base and cream above, or red or cream throughout, rarely cream-green or pale yellow; style pink to red or rarely yellow-green. Perianth glabrous outside, bearded inside. Pistil 13.5–19.5 mm long; ovary subsessile (stipe < 1 mm long), villous; style dorsally loosely tomentose to villous except near apex, weakly to moderately exserted from late bud; pollen-presenter lateral. Follicle ellipsoidal to narrowly ovoid, 10–15 mm long, loosely villous, faintly ridged.

Occurs in mountains (rarely coast) of south-eastern N.S.W., S from Bathurst, and in eastern Vic., E and N from Wilsons Promontory. Grows in various habitats, usually moist sites near water, including riparian scrub associations, open snow-gum woodland, and tall eucalypt forest, in sandy, gravelly or light clay soils over granite, sandstone, serpentine or limestone. Regenerates from seed or (some populations only) from rhizomes. Flowers mainly July–Dec., sporadic in other months. Map 280.

N.S.W.: Snowy R., Mt Kosciuszko, *S.Helms 810* (A *n.v.*, C *n.v.*, NSW, US *n.v.*); E side of Warrumba Ra., NE of Grenfell, *D.J.McGillivray 3155 & R.Coveny* (NSW); Green Cape, *M.E.Phillips CBG005392* (CANB); Goodradigbee Gorge, Cooleman Plain, *D.J.Walker ANU1076* (CANB, NSW). Vic.: near mine on Dolodrook R., S of Mt Wellington, 20 Oct. 1973, *J.H.Willis* (MEL).

Grevillea lanigera is very variable in foliage and habit. Data are insufficient and too inconsistent for infraspecific taxa to be formalised, and on a whole-species basis the variable characters form a continuum of states; nevertheless, adjacent populations are often different from each other, and occasionally some populations have two morphs. The extent of genetic versus micro-environmental influence in these cases needs to be assessed. Some forms are very distinctive. Olde and Marriott (*Grevillea Book* 2: 236–237 (1995)) nominate several populations or horticultural selections. To these may be added a 'fine-leaved form' which occurs from the Goodradigbee R. W of Canberra, N to Boorowa and beyond; the type of *G. ericifolia* is similar to this form. Olde and Marriott's 'coastal form' (*loc. cit.*) occurs, at Green Cape, within 1–2 km of a glabrous-leaved, rhizomatous population with some resemblance to *G. baueri*. Two forms occur together near Tumut, one of which is a diffuse open shrub to 2 m tall with short, scarcely revolute hairy leaves and very small flowers.

Grevillea lanigera sometimes intergrades with *G. rosmarinifolia* subsp. *rosmarinifolia*. The former has a prominent nectary with the exposed portion recurved and extending down over the rim of the torus, a loosely villous style, and a prominent style-end with an oblong-elliptic pollen-presenter. In contrast, *G. rosmarinifolia* has an inconspicuous cushion-like nectary (not emergent above the toral rim), a glabrous or sparsely hairy style, and a slightly-expanded style-end with a usually obovate pollen-presenter. In addition, many specimens of *G. lanigera* have a denser and more spreading indumentum on the branchlets, and broader more obtuse leaves than *G. rosmarinifolia*. The leaves of *G. rosmarinifolia* are never plumply revolute and sausage-like (sometimes the case in *G. lanigera*).

There are numerous collections from the range overlap, sometimes with apparently low pollen fertility, which share features of both species and cannot be placed in either with certainty. Apparent intermediates of this sort also occur in central Vic., well beyond the current range of *G. lanigera* (see McGillivray & Makinson, *Grevillea* 278 (1993), and discussion under *G. rosmarinifolia* subsp. *rosmarinifolia*).

Grevillea lanigera is also recorded as hybridising with *G. polybractea* at Granya Gap (Vic.), with *G. jephcottii* in Pine Mountain and Burrowa Natl Park area (Vic.), and with *G. floribunda* near Wee Jasper (N.S.W.).

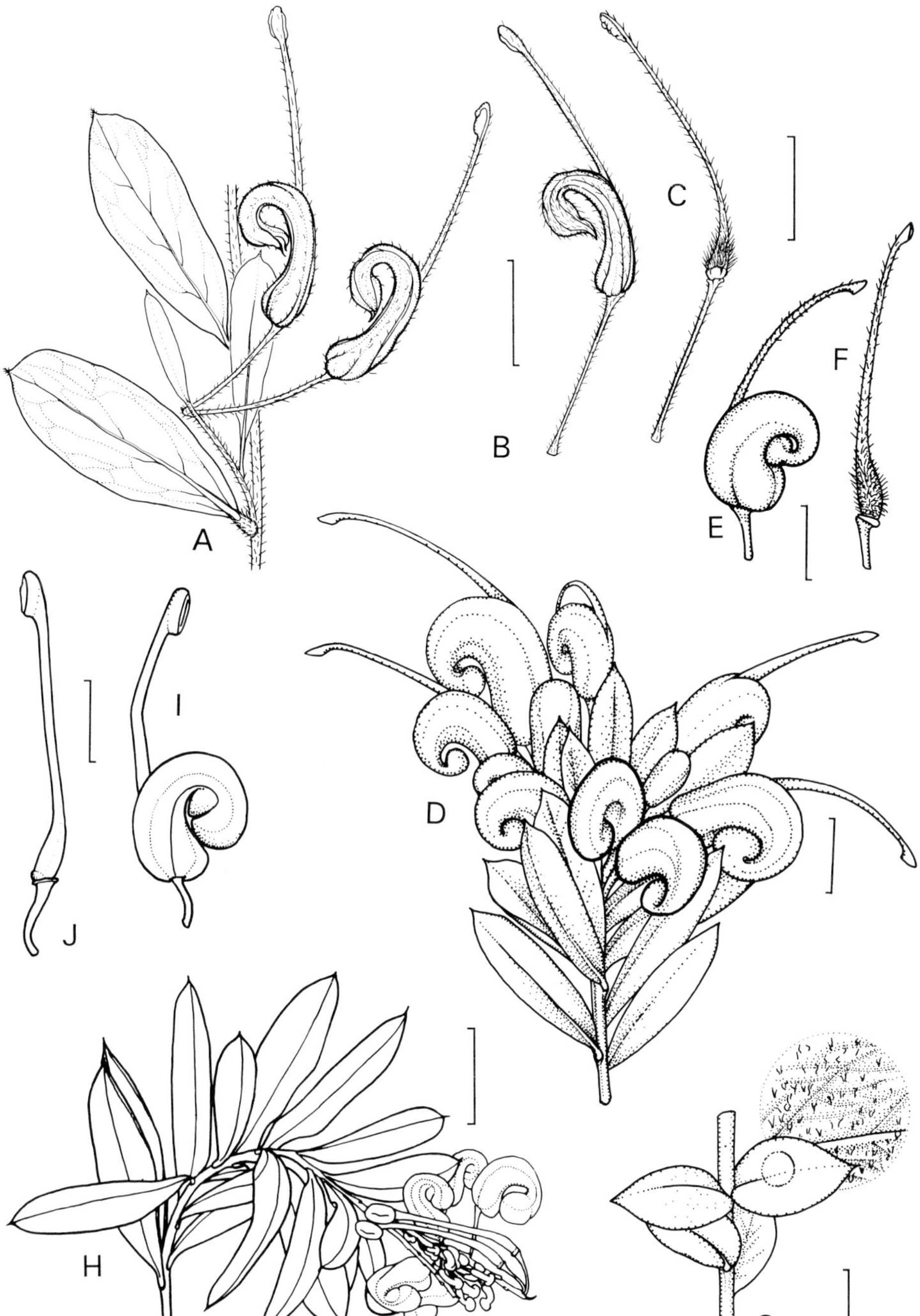

Figure 27. *Grevillea*. **A–C**, *G. quadricauda*. **A**, flowering branch; **B**, flower; **C**, pistil (**A–C**, R.O.Makinson 1438, CANB). **D–F**, *G. baueri* subsp. *baueri*. **D**, flowering branch; **E**, flower; **F**, pistil (**D–F**, J.D.Briggs 698, CANB). **G**, *G. baueri* subsp. *asperula*, leaves and detail of leaf surface (cult., coll. unknown, CBG037853, CANB). **H–J**, *G. iaspicula*. **H**, flowering branch; **I**, flower; **J**, pistil (**H–J**, D.J.McGillivray 3962, NSW). Scale bars: **A–C**, **G–H** = 1 cm; **D–F** = 2 cm; **I–J** = 5 mm. Drawn by: **A–C**, C.Payne; **D–G**, L.Spindler; **H–J**, D.Fortescue.

216. Grevillea baueri R.Br., *Trans. Linn. Soc. London* 10: 173 (1810)

T: '... prope Port Jackson; in depressis a littore remotis' [protologue]; lecto: Cowpastures prope fluv. Nepean [Camden area, N.S.W.], ?19 Oct. 1803, *R.Brown Iter Austral. 3330*; lecto: BM, *fide* D.J.McGillivray & R.O.Makinson, *Grevillea* 408 (1993); isolecto: BM, E; ?isolecto: G-DC, K, LE *n.v.*, MEL, P.

Low spreading shrub 0.3–1.5 m tall. Leaves oblong-elliptic to narrowly ovate, 0.5–3 cm long, (1–) 3–15 mm wide; margins recurved, often more revolute near apex; upper surface smooth and glossy or matt and strongly granulose; lower surface usually exposed, glabrous or with scattered hairs along midvein only. Conflorescence terminal and axillary, simple to 4-branched, erect to decurved; unit conflorescence a dense irregular to semi-secund cluster, (4–) 8–18-flowered, basipetal; floral rachis 2–5 mm long, loosely to sparsely tomentose. Flowers adaxially acroscopic. Perianth usually glabrous outside or sometimes with a few erect hairs on limb, bearded inside. Pistil 16–23 (–28.5) mm long; ovary subsessile to shortly stipitate, loosely villous; stipe 0.5–1.4 mm long; style loosely villous becoming glabrous at or near apex, weakly to moderately exserted from late bud; pollen-presenter lateral. Follicle narrowly ovoid, 13–14 mm long, sparsely tomentose, longitudinally ridged.

Occurs in coastal ranges of south-eastern N.S.W. Two subspecies are recognised.

Upper surface of leaves almost smooth (granules absent or very few) and glabrous; conflorescence usually simple, erect; leaves usually oblong-elliptic to narrowly oblong-ovate, 3–7 mm wide ..... **216a.** subsp. **baueri**

Upper surface of leaves rough, with numerous granules and sometimes also with ascending hairs; conflorescence usually branched and decurved; leaves usually oblong-ovate to ovate, 5–10 mm wide ..... **216b.** subsp. **asperula**

216a. Grevillea baueri R.Br. subsp. baueri

G. pubescens Hook., *Exot. Fl.* 3: t. 216 (June 1826); *G. pubescens* Graham, *Edinburgh New Philos. J.* 1: 172 (post Oct. 1826), *nom. illeg. non* Hook.; *G. baueri* var. *pubescens* (Hook.) Meisn., *in* A.L.P.P. de Candolle, *Prodr.* 14: 357 (1856), as α *pubescens*. T: holo: t. 216 in *Exot. Fl.* 3 (1826).

G. baueri β *sieberi* Endl., *Gen. Pl. Suppl.* 4(2): 84 (1848). T: *Grevillea daphnoides* Sieber Herb. Nov. Holl. n. 25 [N.S.W., 1823, *F.W.Sieber 25*]; syn: BM, G, G-DC, K, NY *n.v.*, P, PR.

G. baueri var. *glabra* Meisn., *in* A.L.P.P. de Candolle, *Prodr.* 14: 357 (1856), as β *glabra*, *nom. illeg.* T: [*F.W.*] *Sieber 25* [N.S.W., 1823]; lecto: NY *n.v.*, *fide* D.J.McGillivray & R.O.Makinson, *Grevillea* 408 (1993); isolecto: BM, G, G-DC, K, P, PR; remaining syntype: Country of Argyll, N.S.W., 1824, *A.Cunningham 43*; syn: MB, P.

Illustrations: W.R.Elliot & D.L.Jones, *Encycl. Austral. Pl.* 5: 32 (1990); A.M.Blombery & B.Maloney, *Prot. Sydney Reg.* 84, 85 (1992); P.M.Olde & N.R.Marriott, *Grevillea Book* 2: 53 (centre right), 54 (38A–C) (1995).

Low usually bushy shrub 0.5–1 m tall. Leaves usually oblong-elliptic to narrowly oblong-ovate, 3–7 mm wide; upper surface smooth and glossy or with occasional granules only. Conflorescence usually simple and erect. Flower colour: perianth red to pink, with cream to lemon on limb and ventrally, or rarely cream or pink throughout, variably blackening after anthesis; style pinkish red (rarely lemon yellow) with a green tip. Fig. 27D–F.

Occurs in south-eastern N.S.W., from the Camden to Picton area S to Mittagong and towards Bundanoon, with one very disjunct record from Eden. Grows in dry sclerophyll woodland or heath in sandy soils on sandstone. Regenerates from seed. Flowers mainly July–Nov. Map 281.

N.S.W.: 19 km WNW of Mittagong on road to Wombeyan Caves, *B.G.Briggs NSW92843* (NSW); Buxton, *C.Burgess NSW124589* (NSW); Eden, *H.Deane NSW84353* (NSW); Belanglo State Forest near Moss Vale, *P.Snowdon CBG021558* (CANB).

One instance of possible wild hybridisation with *G. juniperina* has been recorded.

216b. Grevillea baueri subsp. **asperula** McGill., *New Names Grevillea* 2 (1986)

T: Round Hill, 3 miles [4.8 km] S of Sassafras, N.S.W., 20 Sept. 1961, *E.F.Constable NSW 57888*; holo: NSW.

Illustrations: P.M.Olde & N.R.Marriott, *Grevillea Book* 2: 55 (top left & 39A, B) (1995).

Dense low domed shrub to 50 cm tall or open spreading to erect shrub 0.5–1.2 m tall. Leaves usually oblong-ovate to ovate, 5–10 mm wide; upper surface markedly granulose-scabrid, sometimes also with an open indumentum of ascending hairs. Conflorescence usually branched and decurved. Flower colour: perianth deep reddish pink, with cream on the limb and ventrally, blackening soon after anthesis; style pinkish red with a green tip. Fig. 27G.

Occurs in south-eastern N.S.W. from E of Nerriga to near Nowra and S to the northern edge of the Budawang Ra. Grows in heath or open eucalypt woodland in shallow sandy soils over sandstone or conglomerate. Fire response uncertain, probably regenerates from seed only. Flowers mainly June–Oct. Map 282.

N.S.W.: The Jumps, c. 10.75 km due NNE of Nerriga at head of Bullfrog Ck, *J.M.Fox & K.J.Cowley 88/027* (CANB); Budawang Ra., near head of Wog Wog Ck, NE of Mongarlowe, *J.Pulley & I.Telford 121* (CANB); Naval College Rd, Jervis Bay, *F.A.Rodway 7640* (NSW); Nerriga, between Nerriga and Tonga, above The Jumps, *D.Walker ANU1132* (CANB); road to Tolwong Stn, 48 km N of Nerriga, 1.5 km past Tonga turnoff, *T.S.Whaite 3102* (NSW).

Lavandulacea Subgroup

Leaf lower surface exposed and hairy or enclosed by revolute margins. Flowers with perianth hairy on both surfaces (hairs on outer surface sometimes restricted to limb); late bud with limb rounded, obtuse; tepals not ridged or apiculate, apical limb-segments mostly without a median ridge or keel. Torus oblique. Nectary arcuate to U-shaped or occasionally linguiform, margin smooth to undulate. Pistil 16–38 mm long; ovary stipitate or occasionally subsessile, villous to sericeous; style exserted strongly from late bud; style-end not apiculate. Follicle smooth or occasionally with faint longitudinal ridges; style persistent, remaining ±erect.

A group of 13 species in W.A. (south-west and southern interior), southern S.A., and the western half of Vic. Primarily bird-pollinated.

Within this subgroup the pattern of tepal recurvature is diagnostically and taxonomically important. The perianth segments either (e.g. *G. lavandulacea*) remain mostly coherent except along the dorsal suture, with the individual tepals held in two lateral pairs barely sparated ventral to the style, or (e.g. *G. deflexa*) the tepals all separating for much of their length and coiled back in two opposing (ventral and dorsal) pairs; the ventral pair are usually rolled back further than the dorsal pair. For brevity in descriptions, these conditions are referred to respectively as 'tepals remaining coherent' and 'tepals recoiled independently'.

217. Grevillea lavandulacea Schltd., *Linnaea* 20: 586 (1847)

T: '... im Walde Pine forest zwischen dem Gawler- und Lightriver. Die andere Form auf kalkig-sandigem Boden (sandeplaine) bei Bethanien. Decbr.' [protologue]; lecto: circa Port Adelaide [S.A.] .... *s.d.*, *Dr. Behr*; lecto: HAL 40153, *n.v.*, *fide* D.J.McGillivray & R.O.Makinson, *Grevillea* 425 (1993); remaining syntype: In solo calcareo-arenosa (Sandeplaine) pr[ope] Bethaniam [Bethany, S.A.], *s.d.*, *Dr.* [*H.H.*] *Behr*; syn: HAL 40154, *n.v.*

Spreading or procumbent, usually low mounded shrub 0.2–1.5 m tall. Leaves ascending, sometimes clustered on short lateral branchlets, entire, narrowly elliptic to -obovate or linear, 0.5–4 cm long, 0.5–10 mm wide; margins shortly recurved to loosely revolute; upper surface glabrous and granular or scabrid, or subsericeous; lower surface enclosed including midvein and 1-grooved, or exposed and tomentose to lanate. Conflorescence usually terminal on short lateral branchlets, sometimes also subterminal axillary and 2-flowered, simple, a loose 2–10 (–16)-flowered cluster, irregular to subsecund, opening uncertain; floral rachis 0.5–5 (–10) mm long, tomentose to villous. Flowers acroscopic. Perianth subsericeous to loosely so outside, or glabrous below a subsericeous limb, bearded inside; tepals remaining coherent. Pistil 21.5–28.5 mm long; ovary stipitate, sericeous; style with a loose appressed to ascending indumentum in basal half becoming sparser or glabrous above; pollen-presenter oblique. Follicle narrowly ovoid, 11–15 mm long, pubescent to subvillous. *Lavender Grevillea.* Plate 46.

Occurs in drier parts of the south-eastern quarter of S.A., including Kangaroo Is., and in central and western Vic. Two subspecies are recognised.

Grevillea muricata is similar; see under that species for differences.

Most unit conflorescences > 4-flowered; longest leaves usually > 10 mm long; upper surface of leaf faintly granular when glabrous; perianth 2–5 mm across from dorsal to ventral suture **217a.** subsp. **lavandulacea**

Most unit conflorescences 1–4-flowered; longest leaves usually < 10 mm long; upper surface of leaf scabrid when glabrous; perianth 2–3 mm across from dorsal to ventral suture **217b.** subsp. **rogersii**

217a. Grevillea lavandulacea Schltd. subsp. lavandulacea

G. rosea Lindl., *Paxton's Fl. Gard.* 2: 91, 92, t. 56 (1851). T: S.A. *s.d.*, *Mr Hutt*; holo: CGE *n.v.*

G. ramulosa F.Muell. ex Meisn., *Linnaea* 26: 357 (1854). T: Encounter Bay, [S.A.], *s.d.*, [*J.D.?*] *Stuart*; lecto: NY *n.v.*, *fide* D.J.McGillivray & R.O.Makinson, *Grevillea* 425 (1993); ?isolecto: MEL.

G. lavandulacea var. *latifolia* Meisn., *in* A.L.P.P. de Candolle, *Prodr.* 14: 362 (1856). T: Lofty Ranges prope Adelaide, S.A., *s.d.*, *F.Mueller*; holo: NY *n.v.*; iso: MEL.

G. lavandulacea var. *latifolia* Meisn. *Linnaea* 26: 354 (1854), as α *latifolia*, *nom. nud.*

G. lavandulacea var. *lanceolata* Meisn., *in* A.L.P.P. de Candolle, *Prodr.* 14: 362 (1856). T: Lovethal prope Adelaide, [Lobethal, S.A.], *s.d.*, *F.Mueller*; holo: NY *n.v.*; iso: G-DC.

G. lavandulacea var. *lanceolata* Meisn., *Linnaea* 26: 354 (1854), as β *lanceolata*, *nom. nud.*

G. lavandulacea var. *angustifolia* Meisn., in A.L.P.P. de Candolle, *Prodr.* 14: 362 (1856). T: Salt Creek prope Adelaide, [S.A.], July [no year], *F.Mueller*; holo: NY *n.v.*

G. lavandulacea var. *angustifolia* Meisn., *Linnaea* 26: 354 (1854), as γ *angustifolia*, *nom. nud.*

G. lavandulacea var. *sericea* Benth., *Fl. Austral.* 5: 448 (1870). T: '... Montem Barkeri, [Mt Barker, S.A.] & fl Murray...', Oct. 1848, *F.Mueller*; syn: MEL; Wimmera, [Vic.], *s.d.*, [*J.*] *Dallachy*; syn: MEL.

Illustrations: W.R.Elliot & D.L.Jones, *Encycl. Austral. Pl.* 5: 74 (1990); D.J.McGillivray & R.O.Makinson, *Grevillea* 307 (1993); P.M.Olde & N.R.Marriott, *Grevillea Book* 2: 229 (top right & 191A), 230 (191A–D) (1995).

Low often mounded to straggling erect shrub, 0.6–1 (–1.5) m high. Leaves 5–40 mm long, pungent, sometimes crowded on short lateral branchlets (higher order branches roughly columnar), and/or more widely spaced along main branchlets; upper surface of leaf faintly granular when glabrous. Most unit conflorescences > 4-flowered. Flower colour: perianth usually red (sometimes pink to mauve, rarely white), often paler (to white) on dorsal side or limb; style pink or red or rarely white. Perianth 2–5 mm across from dorsal to ventral suture, subsericeous to almost glabrous outside (hairs sometimes on limb only).

Occurs in drier parts of the south-eastern quarter of S.A., extending N to the Flinders Ra., and in western Vic. W of the Grampians, with an isolated occurrence near St Arnaud. A record from Clarendon, Coromandel in S.A. (*Tepper 280*, MEL) was misinterpreted in McGillivray & Makinson (*Grevillea* 307 (1993)) as Clarendon near Ballarat in Vic. Grows in various habitats including open heath, mallee heath, open woodland and dense shrubland, in sandy or light-loam acidic to alkaline soils on various substrates, occasional on limestone. Regenerates mainly from seed, in some populations also from rhizomes. Flowers mainly July–Dec. Map 283.

S.A.: 1 km SW of Tanderra Saddle, Wilpena Pound Ra., *A.J.A.Sikkes 758 & P.Ollerenshaw* (CANB, NSW); Clarendon, Coromandel, *O.Tepper 280* (MEL); Menglers Hill near Tanunda, *D.J.E.Whibley 3804* (AD, AK *n.v.*, C *n.v.*, CANB, PERTH). Vic.: Big Desert, Nhill to Murrayville track, 23 km N of Yanac, *J.J.Ackland 73* (CANB, MEL); NE of Mt Zero, *J.C.Anway 394* (AD, MEL, NSW, PERTH).

This subspecies is variable between populations in habit, leaf dimensions and colour, and tendency to rhizomatous reproduction. Olde & Marriott (*loc. cit.*) characterise a number of these as forms, though the delimitations are not rigid and not all populations are consistent. The 'Victor Harbour form' grows to 1 m tall and has glabrous green leaves, most other forms having grey or bluish foliage. In S.A., the 'Adelaide Hills form' grows to 50 cm tall and has broad, soft, grey-green leaves 5–10 mm wide with a pubescent lower surface; the flowers are pink. The 'Flinders Ranges form' grows to 60 cm and has small grey rigid leaves and pink, red, or cream flowers. The 'Aldinga form' grows to about 80 cm tall, has silvery grey elliptic leaves, and conspicuous clusters of large pink flowers. The 'Tanunda form' has silvery grey hairy leaves and mauve-pink flowers. In Vic., the 'Black Range form' is an open shrub to 1 m tall with narrow, linear, rigid grey leaves and bright red flowers. Also in the Grampians, the 'Billywing form' is a rhizomatous shrublet to 30 cm tall, with leaves and flowers similar to the previous form. The 'Desert form' occurs in mallee heath in the Big Desert area and adjacent portions of S.A., grows from 30–100 cm tall with rhizomes, and has small silvery leaves with very revolute margins, and pink to red flowers. More research is desirable.

217b. Grevillea lavandulacea subsp. **rogersii** (Maiden) Makinson, *Fl. Australia* 17A: 503 (2000)

G. rogersii Maiden, *Trans. Proc. & Rep. Roy. Soc. S. Australia* 32: 257 (1908), as *G. Rogersi*. T: Cape Borda, Kangaroo Is., S.A., Sept. 1907, *R.S.Rogers*; lecto: NSW, *fide* D.J.McGillivray & R.O.Makinson, *Grevillea* 425 (1993).

G. lavandulacea 'small-leaved form', of D.J.McGillivray & R.O.Makinson, *Grevillea* 306 (1993).

Illustrations: I.Holliday *et al., Kangaroo Island's Native Pl.* 6 (1994); P.M.Olde & N.R.Marriott, *Grevillea Book* 3: 144 (top left & 114A–C) (1995); both as *G. rogersii*.

Low often mounded shrub to 60 cm high. Leaves 2–10 mm long, not pungent, often crowded on short lateral branchlets (higher order branches roughly columnar); upper surface of leaf scabrid when glabrous. Most unit conflorescences 1–4-flowered. Flower colour: perianth rose-pink with a cream limb, or pink throughout; style reddish pink. Perianth 2–3 mm across from dorsal to ventral suture, sparsely subsericeous outside (hairs sometimes on limb only).

Occurs in S.A., apparently restricted to the central northern and far western parts of Kangaroo Is. Similar populations (not seen) are reported from the lower Glenelg R. in Vic.; their placement between the subspecies is uncertain. Grows in shrubland or open eucalypt forest and woodland, in calcareous or ?lateritic sandy soils, often in association with limestone. Fire response unknown. Flowers Sept.–Dec. Map 284.

S.A.: Ravine des Casoars, 20 Oct. 1951, *G.F.Gross* (AD); 9.7–11.3 km from Rocky R. towards Cape Borda, 29 Sept. 1965, *M.E.Phillips CBG020894* (AD, CANB); Harveys Return, Kangaroo Is., Sept. 1908, *R.S.Rogers* (K); Flinders Chase c. 5 km W of Rocky River HS, *J.R.Wheeler 1291* (AD).

Subsp. *rogersii* is narrowly distinct, although visually distinctive, from subsp. *lavandulacea* and was recognised at species rank by Olde & Marriott (*loc. cit.*). It is characterised by a strongly scabrid-granulate leaf upper surface, a strong tendency for unit conflorescences to have no more than 1–4 flowers, and small leaves (2–10 mm long) which are usually crowded on short lateral branchlets. It differs from subsp. *lavandulacea* which has a more finely granulose leaf upper surface, usually has > 4 flowers per unit conflorescence (at least when terminal; axillary unit conflorescences may have only two flowers), and longest leaves usually > 10 mm long, crowded on short lateral branchlets or not. The differences are of degree, and various populations of the highly variable *G. lavandulacea* approach the appearance of subsp. *rogersii*. It is recognised, under *G. rogersii*, as 'Rare' in J.D.Briggs & J.H.Leigh, *Rare or Threatened Australian Plants* (1995).

218. Grevillea muricata J.M.Black, *Trans. Roy. Soc. S. Australia* 63: 244, 247 fig. 4 (1939)

T: between Kingscote & Vivonne Bay, Kangaroo Island, S.A., 16 Nov. 1924, *J.B.Cleland*; lecto: AD, *fide* D.J.McGillivray & R.O.Makinson, *Grevillea* 430 (1993); remaining syntypes: between Kingscote & Vivonne Bay, Kangaroo Island, S.A., 10 Nov. 1924, *J.B.Cleland*; syn: AD; Birchmoore [Birchmore], Kangaroo Island, 7 Nov. 1886, and W of Western Cove, Kangaroo Island, 19 Nov. 1886, *J.G.O.Tepper*; syn: AD, MEL, MO *n.v.*

[*G. rogersii auct. non* Maiden: J.M.Black, *Fl. S. Australia* 2nd edn, 2: 271, fig. 329 (1948), *p.p.*]

Illustrations: J.P.Jessop & H.R.Toelken (eds), *Fl. S. Australia* 4th edn, 1: 136, fig. 72B (1986); I.Holliday *et al.*, *Kangaroo Island's Native Pl.* 6 (1994); P.M.Olde & N.R.Marriott, *Grevillea Book* 3: 44 (top left & 30A–C) (1995).

Low spreading shrub 0.3–1.0 m tall. Leaves entire, narrowly oblong to sublinear-subterete, usually plump, 0.2–1.7 cm long, 1.0–2.1 mm wide; margins revolute; upper surface muricate; lower surface completely enclosed including midvein, 1-grooved, sublanate. Conflorescence terminal, simple, erect, subumbelloid, 1–6-flowered, opening uncertain; floral rachis 0.5–0.7 mm long, tomentose. Flowers mostly adaxially oriented. Flower colour: perianth and style bright orange-red; style-end yellow-green. Perianth pubescent to loosely tomentose outside, glabrous inside near base and bearded about middle, subsericeous above; tepals remaining coherent. Pistil 24–25 mm long; ovary stipitate, villous; style loosely tomentose to villous, with hairs over most of its length; pollen-presenter oblique. Follicle narrowly ellipsoidal or narrowly ovoid, 12.5–14.5 mm long, pubescent to subvillous.

Occurs in S.A., where endemic to the central eastern part of Kangaroo Is. Grows in open woodland and dense shrub associations in sandy to loamy soil over ironstone. Fire response unknown. Flowers Aug.–Nov. Map 285.

S.A.: American R. turnoff, D'Estrees Bay Rd, *U.Johnson 91* (NSW); c. 10 km from Kingscote Aerodrome and 1.5 km NW of Ballast Head turnoff on the road between Kingscote and American R., Kangaroo Is., *D.J.McGillivray 3951* (K, NSW); near Destrees Bay, 30 Aug. 1964, *M.E.Phillips CBG011147* (AD, CANB, NSW); 6.4 km from American R. towards Kingscote, Sept. 1965, *M.E.Phillips CBG016858* (CANB, NSW).

Grevillea muricata, *G. rogersii* and *G. lavandulacea* have been variously interpreted taxonomically. Here, *G. muricata* is treated as a separate species on the basis of its plumply subterete (sausage-like), highly revolute leaves, which have a recurved callous point, are muricate on the upper surface and usually spreading and evenly distributed along the branchlets, and the strongly ascending to spreading stylar hairs. In *G. lavandulacea* the leaves are usually flatter, narrowly elliptic to narrowly obovate or linear or oblong, sometimes scabrous, often ascending, and are mostly borne on short lateral branchlets or, rarely, singly along the branches; the hairs on the pistil are more appressed.

This species is recognised as 'Rare' in J.D.Briggs & J.H.Leigh, *Rare or Threatened Australian Plants* (1995).

219. Grevillea extorris S.Moore, *J. Linn. Soc., Bot.* 34: 221 (1899)

T: track between Wilson's Pool and Lake Darlôt, W.A., May 1895, *S. le M.Moore*; lecto: BM, *fide* D.J.McGillivray & R.O.Makinson, *Grevillea* 416 (1993); remaining syntype: BM.

Illustrations: P.M.Olde & N.R.Marriott, *Grevillea Book* 2: 158 (top centre & 128A, B) (1995).

Erect shrub 1–2.5 m tall. Leaves linear or narrowly oblong, (2.5–) 4–12 cm long, 1–3 mm wide; margins recurved to angularly revolute; upper surface sericeous, becoming glabrous, 7–11-grooved; lower surface scarcely to mostly enclosed except for midvein, if exposed then subsericeous with several ridges. Conflorescence cauline and/or axillary, never terminal, simple or few-branched; unit conflorescence a loose irregular to subsecund 3–12-flowered cluster, opening uncertain; floral rachis 0.5–4 mm long, subsericeous to subvillous. Flowers basiscopic. Flower colour: perianth pink to red or yellow; style matching. Perianth subsericeous outside, bearded inside; tepals remaining coherent. Pistil 28–38 mm long; ovary stipitate, appressed-villous; style subsericeous near base, glabrous above except for papillae in apical 1–3 mm; pollen-presenter oblique. Follicle oblong ellipsoidal to ovoid, 11–15 mm long, sericeous.

Occurs in southern inland W.A., in two broad belts from Mullewa area E to Lake Darlot and from Cue S to Mt Jackson. Grows in low woodland or shrubland, often in rocky situations, on laterite, granite or conglomerate. Regenerates from seed and possibly lignotuber. Flowers Apr.–Sept. Map 286.

W.A.: near Lennonville, Mt Magnet, *C.A.Gardner 2250* (PERTH); 31 km S of Malcolm, *A.S.George 3026* (PERTH); breakaways between Thundelarra and Yalgoo, *E.McCrumm* (PERTH); 10 km E of Mullewa towards Yalgoo, *B.R.Maslin 3630* (PERTH); 22.5 km W of Paynes Find, *R.A.Saffrey 857* (CANB, PERTH).

Olde & Marriott (*loc. cit.*) identify a 'broad-leaved form', occurring from Leonora to Comet Vale and W to Sandstone, with oblong leaves 2–3 mm wide and recurved rather than revolute margins.

220. **Grevillea phillipsiana** McGill., *New Names Grevillea* 12 (1986)

T: 5 miles [8 km] from Norseman towards Coolgardie, W.A., 4 Sept. 1968, *M.E.Phillips WA/68 397* [*CBG025390*]; holo: CANB; iso: L *n.v.*

Illustrations: J.W.Wrigley & M.Fagg, *Banksias, Waratahs & Grevilleas* 279 (1989); P.M.Olde & N.R.Marriott, *Grevillea Book* 3: 91 (bottom right), 92 (69A, B) (1995).

Spreading to erect shrub 1–1.5 m tall. Leaves entire, linear to subulate, 1–3.5 cm long, 0.8–1.4 mm wide; margins angularly revolute; upper surface longitudinally ridged; lower surface enclosed except for midvein, 2-grooved. Conflorescence terminal or rarely axillary, usually simple, erect or decurved, umbelloid and semi-secund, 2–14-flowered, acropetal; floral rachis 1–4 mm long, subsericeous. Flowers acroscopic. Flower colour: perianth and style red. Perianth loosely subsericeous outside, bearded inside; tepals remaining coherent. Pistil 22–28 mm long; ovary stipitate, villous; style glabrous; pollen-presenter very oblique. Follicle narrowly ovoid-ellipsoidal, 15 mm long, sparsely tomentose.

Occurs in southern inland W.A., restricted to an area between Norseman and Zanthus. Grows in shrubland, woodland and mallee scrub in rocky soils on granite. Regenerates from seed. Flowers July–Sept. Map 287.

W.A.: Cardunia Rocks, 5.5 km N of Karonie Siding, *D.F.Blaxell 1748* (NSW, PERTH); 23 km SE of Sinclair Soak, c. 70 km NE of Norseman, *K.Newbey 6905* (PERTH); 7.5 km by road from Norseman P.O. on W side of Lake Cowan, *D.J.McGillivray 3650 & A.S.George* (CANB, K, NSW, PERTH).

This species is recognised as 'Poorly Known' in J.D.Briggs & J.H.Leigh, *Rare or Threatened Australian Plants* (1995).

221. **Grevillea tetrapleura** McGill., *New Names Grevillea* 15 (1986)

T: Duladgin Rock, c. 14 km N of Yellowdine, W.A., 3 July 1976, *D.J.McGillivray 3667 & A.S.George*; holo: NSW; iso: B *n.v.*, CANB, K, NY *n.v.*, PERTH.

Illustrations: D.J.McGillivray & R.O.Makinson, *Grevillea* 305, fig. 79 & col. pl. (1993); P.M.Olde & N.R.Marriott, *Grevillea Book* 3: 200 (bottom right), 201 (163A–C) (1995).

Low spreading shrub 0.1–0.9 m tall. Leaves entire, linear-subterete to subulate, rigid, pungent, 1–6 cm long, 0.9–1.1 mm wide; margins obscure, tightly revolute; upper surface 4-ridged; lower surface enclosed except for prominent midvein, obscurely 2-grooved. Conflorescence axillary and cauline, aggregated, simple, erect, a loose 1–4-flowered cluster; floral rachis 0.5–1 mm long, subsericeous. Flowers with orientation uncertain. Flower colour: perianth and style pinkish red. Perianth loosely subsericeous outside, bearded inside; tepals recoiled independently. Pistil 16.5–20.5 mm long; ovary stipitate, villous; style glabrous or with a few hairs and papillae in apical 1–2 mm; pollen-presenter very oblique. Follicle ellipsoidal to obovoid, 8–11 mm long, sparsely pubescent. Fig. 28A–C.

Occurs in southern inland W.A., in the area bounded by Bullfinch, Mt Jackson, and Duladgin Rock. Grows in open shrub associations in shallow sandy soil over granite. Regenerates from seed and rhizomes. Flowers July–Sept. Map 288.

W.A.: 10 km N of Weowanie Rock, *J.S.Beard 5943* (KPBG); Warralakin to Bullfinch soil survey area, *K.Hunter* [*J.S.Beard 5058*] (KPBG); 12 km NNE of Mt Jackson on road to Die Hardy Ra., *D.J.McGillivray 3676 & A.S.George* (B *n.v.*, CANB, MO *n.v.*, NSW, PERTH, US *n.v.*).

This species is recognised as 'Rare' in J.D.Briggs & J.H.Leigh, *Rare or Threatened Australian Plants* (1995).

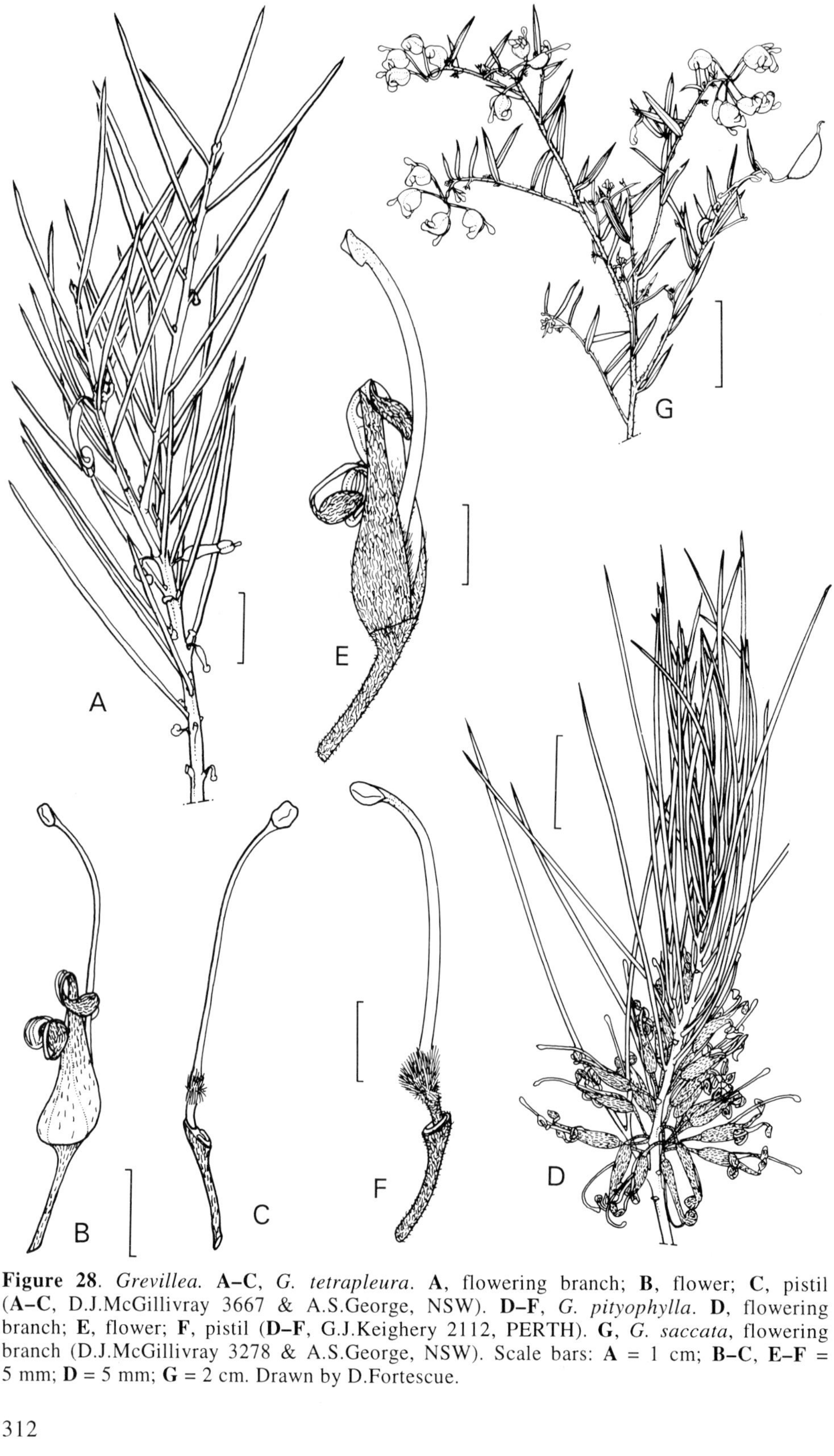

Figure 28. *Grevillea*. **A–C**, *G. tetrapleura*. **A**, flowering branch; **B**, flower; **C**, pistil (**A–C**, D.J.McGillivray 3667 & A.S.George, NSW). **D–F**, *G. pityophylla*. **D**, flowering branch; **E**, flower; **F**, pistil (**D–F**, G.J.Keighery 2112, PERTH). **G**, *G. saccata*, flowering branch (D.J.McGillivray 3278 & A.S.George, NSW). Scale bars: **A** = 1 cm; **B–C**, **E–F** = 5 mm; **D** = 5 mm; **G** = 2 cm. Drawn by D.Fortescue.

222. Grevillea deflexa F.Muell., *Australas. Chem. Druggist* 5: 72 (1883)

T: Gascoyne R., [W.A.], 1882, *J.Forrest*; lecto: MEL, *fide* D.J.McGillivray & R.O.Makinson, *Grevillea* 412 (1993); ?isolecto: B *n.v.*, K, MEL, PERTH; remaining syntype: Gascoyne R., [W.A.], 1882, *Pollack*; syn: NSW.

G. deflexa F.Muell., *Syst. Census* 140 (1882), *nom. nud.*

G. ninghanensis C.A.Gardner, *J. Roy. Soc. W. Australia* 47: 55 (1964). T: 'Hab. in distr. Austin prope Ninghan ... Aug. Gardner n. 12502' [protologue]; neo: Wanarra [W.A.], Aug. 1960, *C.A.Gardner 12502*; PERTH, *fide* D.J.McGillivray & R.O.Makinson, *Grevillea* 412 (1993); ?isoneo: MEL.

Illustrations: R.Erickson *et al.*, *Fl. & Pl. W. Australia* 137, col. pl. 427 (1973), as *G. ninghanensis*; A.A.Mitchell & D.G.W.Wilcox, *Arid Shrubland Pl. W. Australia* 258 (1994); P.M.Olde & N.R.Marriott, *Grevillea Book* 2: 115 (bottom left & 92A, B), 116 (92C, D) (1995).

Low domed to open erect shrub 0.2–2 m tall. Leaves linear or elliptic to narrowly ovate or oblanceolate, 1–9 cm long, 0.7–10 mm wide; margins smoothly revolute; upper surface hairy or glabrous, smooth; lower surface exposed or enclosed and then 2-grooved, sericeous. Conflorescence axillary or cauline, rarely terminal, simple, deflexed to spreading, a loose subsecund 2–5-flowered cluster, opening uncertain; floral rachis 2–5 cm long, subsericeous to villous. Flowers acroscopic, lower ones sometimes abaxial. Flower colour: perianth red with a yellow limb, or all yellow; style red or yellow. Perianth loosely to densely subsericeous to apically subvillous, bearded inside above ovary; tepals recoiled independently. Pistil 17–24 mm long; ovary stipitate, villous; style glabrous except for small erect hairs in apical 2 mm; pollen-presenter very oblique to lateral. Follicle ellipsoidal to obovoid, 11–16 mm long, loosely villous.

Occurs in inland central W.A. from Laverton to the upper Gascoyne R., and from the Thomas R. south to the Mt Singleton area. Grows in open mulga associations, often along drainage lines in red loam or sand. Regenerates from seed and probably lignotubers. Flowers May to Oct. Map 289.

W.A.: Wooleen HS, *A.M.Ashby 2519* (AD, CANB, PERTH); 41.8 km S of Nookawarra HS, *J.S.Beard 6627* (NSW, PERTH); 4 km S of Meekatharra P.O., *A.C.Beauglehole 49041* (NSW, PERTH); Aermotor Mill, Glenorn Stn, *N.T.Burbidge 189* (PERTH); 225 km from Wiluna, *A.R.Fairall 1932* (PERTH).

A 'short-leaved form', including the type of *G. ninghanensis*, occurs between Mongers Lake and Sandstone and N to Cue; it has leaves 1–2.5 cm long, and up to 1.5 mm wide, with a sparsely sericeous perianth. The 'long-leaved form' has leaves 2–9 cm long, 2–10 mm wide, with a tomentose perianth. *Grevillea yorkrakinensis* is similar but has more crowded foliage with leaves < 2 cm long and always linear and about 1 mm wide, and fewer-flowered conflorescences which are always spreading, never deflexed.

223. Grevillea haplantha F.Muell. ex Benth., *Fl. Austral.* 5: 451 (1870)

T: W.A., *J.Drummond*; lecto: MEL, *fide* D.J.McGillivray & R.O.Makinson, *Grevillea* 419 (1993). The other syntype nominated by Bentham (East Mt. Barren, [W.A.], *s.d.*, *Maxwell*, at K) is a specimen of *G. dolichopoda* (McGill.) Olde & Marriott.

Dense rounded shrub 0.6–2.2 m tall. Leaves linear, 1.5–8 cm long, 1.0–1.7 mm wide; margins angularly revolute; upper surface finely and densely granulose; lower surface enclosed except for midvein, 2-grooved, subsericeous in grooves. Conflorescence axillary or cauline, usually simple, 1–6-flowered in a loose cluster, opening uncertain; floral rachis 0.7–1.5 mm long, villous. Flowers abaxially oriented. Perianth subvillous outside, bearded inside; tepals remaining coherent. Pistil 18–25 mm long; ovary subsessile to obscurely stipitate, villous; style villous, sometimes glabrous to minutely pubescent in apical 2–3 mm; pollen-presenter very oblique. Follicle ovoid to ellipsoidal, 10–13 mm long, villous, with longitudinal ridging.

Occurs in the south-west of W.A. Two geographically disjunct subspecies are recognised.

Style with indumentum extending onto back of pollen-presenter, the hairs not reducing in size; pedicels 7–9 mm long; pistil (20–) 24–25 mm long **223a.** subsp. **haplantha**

Style with indumentum either lacking in apical 2–3 mm, or hairs much reduced in size; pedicels 5–7 mm long; pistil 18–20 (–22) mm long **223b.** subsp. **recedens**

223a. **Grevillea haplantha** F.Muell. ex Benth. subsp. **haplantha**

G. haplantha 'Coolgardie District race', of D.J.McGillivray & R.O.Makinson, *Grevillea* 301 (1993).

Illustrations: D.J.McGillivray & R.O.Makinson, *op. cit.* 300, as *G. haplantha*; P.M.Olde & N.R.Marriott, *Nuytsia* 9: 288, fig. 17H–J (1993); P.M.Olde & N.R.Marriott, *Grevillea Book* 2: 190 (lower left & 156A, B) (1995).

Dense shrub 1–2.2 tall, to 3 m wide. Branchlets ascending, not or scarcely divaricate. Pedicels 7–9 mm long. Flower colour: perianth dull deep pink to red with whitish hairs, often cream to orange on limb; style red with white hairs. Pistil (21–) 24–25 mm long; stipe c. 1 mm long; style villous throughout, with indumentum extending onto back of pollen-presenter. Plate 48.

Occurs in inland southern W.A., in the Coolgardie area (Koolyanobbing to W of Goongarrie and to c. 50 km S of Coolgardie). Grows in mallee heath or shrubland associations in sandy, sometimes lateritic soils. Regenerates probably from seed only. Flowers May–Dec. Map 290.

W.A.: S of Coolgardie, *T.E.H.Aplin 1874* (PERTH); between Callion and Musson Soak, W of Goongarrie, *J.S.Beard 6251* (NSW, PERTH); Queen Victoria Rock, 50 km S of Coolgardie, *R.Filson 8890* (MEL, PERTH).

223b. **Grevillea haplantha** subsp. **recedens** Olde & Marriott, *Nuytsia* 9: 290 (1993)

T: near Manmanning, W.A., 6 July 1986, *B.H.Smith 658*; holo: NSW; iso: CANB, HO, PERTH.

G. haplantha 'Avon District race', of D.J.McGillivray & R.O.Makinson, *Grevillea* 301 (1993).

Illustrations: P.M.Olde & N.R.Marriott, *Nuytsia* 9: 288, fig. 17A–G (1993); P.M.Olde & N.R.Marriott, *Grevillea Book* 2: 190 (top right & 157) (1995).

Shrub 0.6–1 m high, to 1 m wide. Branchlets divaricate. Pedicel 5–7 mm long. Flower colour: perianth pink to red, with brown hairs giving a fawn or rusty tinge; style pinkish red with a yellow-green tip. Pistil 18–20 (–22) mm long; stipe c. 0.5 mm long; style villous, becoming glabrous or sparsely and minutely pubescent in apical 2–3 mm.

Occurs in inland south-western W.A. from Mollerin to near Ballidu, extending S to Cunderdin and Merredin. Grows in open shrubland or woodland in heavy clay loam (often strongly lateritic) soils. Regenerates from seed. Flowers June–Sept. Map 291.

W.A.: Yorkrakine, *C.A.Gardner 228* (PERTH); Cunderdin, Aug. 1903, *W.V.Fitzgerald* (NSW, PERTH); 3 km SE of Manmanning, *D.J.McGillivray 3419 & A.S.George* (K, NSW, PERTH, US *n.v.*); Kodjkodjin Nature Reserve, *E.Mattiske HLA38* (PERTH); 20.9 km E of Ballidu, *R.D.Royce 1262* (PERTH).

224. **Grevillea disjuncta** F.Muell., *Fragm.* 6: 206 (1868)

G. disjuncta F.Muell. subsp. *disjuncta*, *sensu* D.J.McGillivray & R.O.Makinson, *Grevillea* 302 (1993). T: 'Ad rivum Salt-River Australiae occidentalis. Mx. In tractu montano Stirling's Range.' [protologue]; lecto: Salt R. [Pallinup R., W.A.], *s.d.*, *G.Maxwell*; lecto: MEL, *fide* D.J.McGillivray & R.O.Makinson, *Grevillea* 414 (1993); remaining syntypes: rocky ridges E of Stirling Ra., [W.A.], *s.d.*, *G.Maxwell*; syn: MEL, K; N of Stirling Ra., [W.A.], *s.d.*, *G.Maxwell*; syn: K.

Illustrations: J.W.Wrigley & M.Fagg, *Banksias, Waratahs & Grevilleas* 225 (1989); P.M.Olde & N.R.Marriott, *Grevillea Book* 2: 129 (top right & 103A, B) (1995).

Low mounded shrub 0.3–1.0 m tall. Leaves entire, linear to subterete, 0.8–2 cm long, 1–3 mm wide; margins angularly and tightly revolute; upper surface longitudinally ridged, smooth; lower surface enclosed including midvein, 1-grooved. Conflorescence axillary and cauline, usually simple, spreading, a loose 1–4 (–12?)-flowered cluster, opening uncertain; floral rachis 1–2 (–4) mm long, villous. Flowers abaxially oriented. Flower colour: perianth pale orange to bright red, with limb green or yellow to red; style pale to reddish pink with a green tip. Perianth subsericeous to loosely so (biramous hairs) outside, sometimes also with short simple glandular hairs, bearded inside; tepals remaining coherent. Pistil 19.5–28 mm long; ovary subsessile, villous; style loosely villous at base (biramous hairs), these becoming sparse or absent above, usually also short erect hairs in apical third; pollen-presenter very oblique. Follicle ovoid, 9.5–14 mm long, sparsely villous.

Occurs in south-western W.A., mostly in the area bounded by Dumbleyung, Nyabing and Pingrup, with isolated occurrences near York, Kellerberrin and Stokes Inlet. Grows in heath, low shrubland and open woodland in coarse-grained granitic soils. Regenerates probably from seed only. Flowers Apr.–Sept. Map 292.

W.A.: c. 8 km NW of Young R. crossing on Ravensthorpe to Esperance road, *N.N.Donner 2787* (AD, PERTH); Greenhills, E of York, Sept. 1934, *C.A.Gardner* (PERTH); 1.3 km NNW of Kukerin, *D.J.McGillivray 3535 & A.S.George* (CANB, LE *n.v.*, NSW, PERTH, US *n.v.*).

Grevillea disjuncta is very closely related to *G. dolichopoda*, which has the leaf upper surfaces usually strongly granulose to scabrid, the torus 2–3 mm across (c. 1.5 mm across in *G. disjuncta*), and the branchlets usually secund (not secund in *G. disjuncta*).

Olde & Marriott (*op. cit.* 129) designate a 'longer-leaved form', equivalent to McGillivray & Makinson's (*Grevillea* 302 (1993)) *G. disjuncta* subsp. *dolichopoda* 'smooth-leaf form'. This element is here retained as a form within *G. dolichopoda*, *q.v.*

225. Grevillea dolichopoda (McGill.) Olde & Marriott, *Nuytsia* 9: 291 (1993)

G. disjuncta subsp. *dolichopoda* McGill., *New Names Grevillea* 5 (1986). T: 21 km by road N of Ongerup, W.A., 26 June 1976, *D.J.McGillivray 3521 & A.S.George*; holo: NSW; iso: PERTH.

Illustrations: W.R.Elliot & D.L.Jones, *Encycl. Austral. Pl.* 5: 48 (1990), as *G. disjuncta* subsp. *dolichopoda*; P.M.Olde & N.R.Marriott, *Grevillea Book* 2: 133 (top left & 107A, B) (1995).

Low procumbent shrub 0.3–0.6 m tall. Leaves entire, linear to subterete, (1.5–) 2.5–7 cm long, (0.5–) 1–2.5 mm wide; margins smoothly or angularly revolute; upper surface granulose to scabrid, or rarely smooth, with faint longitudinal ridges; lower surface usually completely enclosed including midvein, usually 1-grooved, rarely slightly exposed and sericeous. Conflorescence axillary or cauline, ±spreading, simple, 1–4-flowered, opening uncertain; floral rachis c. 1 mm long, villous. Flowers abaxially oriented. Flower colour: perianth red with an orange limb; style red with a green tip. Perianth sparsely sericeous to almost glabrous outside, rarely also with short erect hairs, bearded inside; tepals remaining coherent. Pistil 23–24 mm long; ovary subsessile, villous; style villous at base, sparser above, usually with erect simple hairs in apical third; pollen-presenter lateral. Follicle ovoid, c. 12 mm long, sparsely villous.

Occurs in south-western W.A. in three slightly disjunct areas, from Nyabing to the Gairdner R., between Lake Varley and Ravensthorpe, and in the East Mt Barren to lower Hamersley R. area. Grows in heath or mallee shrubland in granitic, quartzitic or lateritic soils. Regenerates probably from seed only. Flowers Apr.–Nov. Map 293.

W.A.: 3.4 km N of Hopetoun turnoff from Hamersley R. mouth track, SW of Eyre Ra., *A.S.George 11690* (PERTH); 1 km W of Needilup, *D.J.McGillivray 3512 & A.S.George* (CANB, K, LE *n.v.*, NSW, NY *n.v.*, PERTH, PRE *n.v.*, US *n.v.*); 45 km S of Ravensthorpe, E of Hamersley R., *E.Wittwer 405* (PERTH).

Grevillea dolichopoda is narrowly distinct from *G. disjuncta*: see that species for differences.

Two forms are currently recognised. The 'rough-leaved (type) form' is more widespread, occurring in the Nyabing to Borden and Gairdner R. area and the Bottle Rock to Lake Varley and Ravensthorpe area; it has a very granular upper leaf surface. Rough-leaved populations with short leaves 1.5–2 cm long occur in the Newdegate area; these may constitute a distinct form. The 'smooth-leaved form' is restricted to the vicinity of East Mt Barren and the Lower Hamersley R., and has the upper leaf surface smooth (without granules). Olde & Marriott (*loc. cit.*) placed this as *G. disjuncta* 'long-leaved form', but it has the long floral torus, stylar indumentum, and usually longer leaves of *G. dolichopoda*.

226. Grevillea pityophylla F.Muell., *Fragm.* 6: 208 (1868)

T: Western Australia, *s.d.*, *J.Drummond*; holo: MEL.

G. blackallii C.A.Gardner, *J. Roy. Soc. W. Australia* 22: 121 (1936). T: Paynes Find, W.A., 10 July 1931, *C.A.Gardner 2224*; lecto: PERTH, *fide* D.J.McGillivray, *Telopea* 1: 30 (1975).

Illustrations: D.J.McGillivray & R.O.Makinson, *Grevillea* 298, fig. 78 (1993); P.M.Olde & N.R.Marriott, *Grevillea Book* 3: 100 (top right & 76A–D) (1995).

Shrub 0.3–1.5 m tall. Leaves entire, linear to subterete, sometimes curved or twisted, 4–15 cm long, 0.9–1.8 mm wide; margins tightly revolute; upper surface villous to glabrous, sometimes granulose; lower surface obscure, with marginal grooves abaxial or lateral on midrib. Conflorescence axillary and cauline, simple, spreading to ascending, 3–6-flowered, umbelloid, opening uncertain; floral rachis c. 0.5 mm long, villous. Flowers abaxially oriented. Flower colour: perianth and style pinkish red to bright red. Perianth densely to sparsely villous outside, bearded inside; tepals independently recoiled. Pistil 18–23 mm long; ovary stipitate, villous; style either with an open indumentum of erect simple hairs, or glabrous except for scattered biramous hairs and with simple erect hairs in apical c. 4 mm; pollen-presenter lateral. Follicle ovoid-ellipsoidal, 11–13.5 mm long, villous. Fig. 28D–F.

Occurs in inland central W.A. from N of Mullewa to Mt Magnet and Wanarra Stn. Grows in open *Acacia* shrubland in red loamy or sandy soils over granite or quartzite. Regenerates from seed or lignotuber. Flowers July–Oct. Map 294.

W.A.: c. 75 km N of Mullewa, *A.M.Ashby 2882* (AD, CANB, FI *n.v.*, L *n.v.*, SI *n.v.*); White Wells, Ninghan, *C.A.Gardner 12049* (PERTH); c. 75 km N of Paynes Find and c. 5 km S of Kirkalocka on Great Northern Hwy, *D.J.McGillivray 3379 & A.S.George* (CANB, NSW, PERTH, US *n.v.*); Wanarra Stn, Sept. 1960, *D.Mason* (PERTH); 5 km S of Paynes Find, *P.G.Wilson 8628* (AD, PERTH, SYD).

Most specimens are assignable to one of two forms. The 'dipleural-leaved form' occurs from SW of Paynes Find to Mt Magnet and Kirkalocka, and has the leaves subterete, 0.9–1.2 mm wide, with the submarginal grooves lateral on the very prominent abaxial midvein; the perianth outer surface has a fairly dense, spreading indumentum, and the styles have a uniform, open indumentum of short simple hairs. The 'dorsiventral-leaved form' occurs mainly in the area between Wurarga and Mullewa, and has the leaves linear, 1.1–1.8 mm wide, and with 2 marginal grooves on the lower (abaxial) surface; the outer surface of the perianth has a softer, more open indumentum with ground tissue visible between the hairs, and the style has simple erect hairs restricted to the apical 5 mm. These variants warrant subspecies status but investigation of some unassigned populations is needed prior to formal recognition.

227. **Grevillea granulosa** McGill., *New Names Grevillea* 7 (1986)

T: half way between Mullewa to Pindar, W.A., 2 Aug. 1965, *A.M.Ashby 1574*; holo: PERTH; iso: AD, NSW.

Illustration: W.R.Elliot & D.L.Jones, *Encycl. Austral. Pl.* 5: 62 (1990).

Low spreading shrub 0.6–1.5 m high; branches often secund. Leaves entire, linear, 3.5–16 cm long, 1.2–2.5 (–3.5) mm wide; margins smoothly revolute; upper surface sericeous, becoming glabrous and granulose; lower surface mostly or wholly enclosed, sometimes including midvein and then 1- or 2-grooved, or exposed and sericeous on broader leaves. Conflorescence axillary and cauline, simple or few-branched; unit conflorescence decurved, a subsecund umbelloid cluster, 3–8-flowered, opening uncertain; floral rachis 0.5–6 mm long, tomentose. Flowers acroscopic. Flower colour: perianth and style red to orange or rarely yellow. Perianth subsericeous to sparsely so outside, sometimes tomentose on limb; inner surface loosely villous to openly pilose; tepals independently recoiled. Pistil 20–23 mm long; ovary stipitate, villous; style glabrous except for simple hairs and papillae in apical c. 2 mm; pollen-presenter very oblique to lateral. Follicle narrowly ovoid or narrowly ellipsoidal, 10–14 mm long, villous, longitudinally ridged.

Occurs in W.A. between Wubin, Lake Moore and Yuna. Grows in shrubland, mallee scrub, or eucalypt woodland in sandy or clayey soils. Fire response unknown. Flowers July–Oct. Map 295.

W.A.: Eradu sand plain, *A.M.Ashby 1493* (PERTH); E Yuna Reserve, NE of Geraldton, *A.C.Burns 26* (PERTH); Jibberding, 2 Oct. 1959, *W.H.Butler* (PERTH); White Wells, Ninghan, *C.A.Gardner 12049 bis* (PERTH); 8 km NE of Wubin, Great Northern Hwy, *A.S.George 5672* (PERTH).

This species is recognised as 'Poorly Known' in J.D.Briggs & J.H.Leigh, *Rare or Threatened Australian Plants* (1995).

228. Grevillea rosieri McGill., *New Names Grevillea* 13 (1986)

T: Doodarding, W.A., July 1959, *B.Rosier 51*; holo: PERTH.

Illustrations: J.W.Wrigley & M.Fagg, *Banksias, Waratahs & Grevilleas* 314 (1989); P.M.Olde & N.R.Marriott, *Grevillea Book* 3: 146 (116A–C), 147 (top left) (1995).

Low spreading to erect shrub to 0.5 m high. Leaves linear, 1–4 cm long, 0.8–1.5 mm wide; margins smoothly and tightly revolute; upper surface soon glabrous, faintly granulose; lower surface mostly or wholly enclosed including midvein, 1-grooved, with appressed hairs. Conflorescence axillary, simple, spreading, 1–3-flowered, opening uncertain; floral rachis 0.2–0.5 mm long, tomentose. Flowers abaxially oriented. Flower colour: perianth red to rusty red; style red or rarely cream with a red tip. Perianth villous outside, bearded inside; tepals independently recoiled. Pistil 17.5–19 mm long; ovary stipitate, villous; style glabrous except for short hairs in apical 3 mm; pollen-presenter lateral. Follicle obloid-ellipsoidal, 6–10 mm long, villous.

Occurs in south-western W.A., from near Wubin to near Kirwan, and reportedly (not seen) also at Edah Stn SW of Mt Magnet. Grows in shrubland in sandy or loamy sometimes granitic soils. Regenerates from seed. Flowers July–Sept. Map 296.

W.A.: 20 km from Kirwan NNW along the Rabbit Proof Fence road, *R.W.Purdie 5285* (CANB, NSW); Mollerin, *B.Rosier 254* (PERTH); c. 19 km N of Kirwan, *B.H.Smith 1043* (CANB, MEL, NSW).

This species is known from few collections and is recognised as 'Poorly Known' in J.D.Briggs & J.H.Leigh, *Rare or Threatened Australian Plants* (1995). It has been previously been identified as *G. disjuncta*, which has hairy styles and a less conspicuous, appressed indumentum on the outside of the perianth; *G. rosieri* has the style villous only at the very base and then glabrous except for short hairs in the apical 3 mm.

229. Grevillea yorkrakinensis C.A.Gardner, *J. & Proc. Roy. Soc. W. Australia* 9: 34 (1923)

T: Yorkrakine, W.A., 4 Sept. 1922, *C.A.Gardner 1726*; lecto: PERTH, *fide* D.J.McGillivray & R.O.Makinson, *Grevillea* 449 (1993); isolecto: (*Gardner 1226*) PERTH.

Illustrations: D.J.McGillivray & R.O.Makinson, *Grevillea* 296 (1993); P.M.Olde & N.R.Marriott, *Grevillea Book* 3: 237 (top right & 194B, C), 238 (194A) (1995).

Small compact shrub to 0.4 m tall. Leaves linear, 0.3–1.8 cm long, 0.6–1.2 mm wide; margins smoothly revolute; upper surface glabrous or with scattered hairs, often glaucous; lower surface enclosed except midvein, 2-grooved. Conflorescence axillary and cauline, simple, spreading, 2–5-flowered, sometimes subsecund, opening uncertain; floral rachis 0.5–3 mm long, subvillous. Flowers adaxially oriented. Flower colour: perianth red to yellow-orange, sometimes red with purplish base or yellow limb; style red to yellow or greenish. Perianth sparsely pubescent to subvillous outside, bearded inside; tepals independently recoiled. Pistil 17–22 mm long; ovary stipitate, villous; style open-subvillous in lower third, otherwise glabrous except for minute erect hairs and papillae in apical 2 mm; pollen-presenter very oblique. Follicle narrowly ovoid to narrowly ellipsoidal, 7–10.5 mm long, sparsely villous, faintly ridged.

Occurs in south-western W.A. from Mt Gibson and Wubin SE and S to Hyden and near Southern Cross. Grows in shrubland or mallee scrub in coarse sandy or gravelly soils often over granite. Regenerates from seed. Flowers May–Oct. Map 297.

W.A.: 48 km from Wubin towards Paynes Find on Great Northern Hwy, *D.J.McGillivray 3403 & A.S.George* (K, NSW, PERTH, US *n.v.*); 17 km W of Mukinbudin on road E of Waddouring, *D.J.McGillivray 3436 & A.S.George* (NSW, PERTH); 40 km S of Merredin [and?] Kondinin, *D.J.McGillivray 3701 & A.S.George* (NSW, PERTH); W of Moorine Rock towards Nulla Nulla, 10 Sept. 1968, *M.E.Phillips CBG032211* (CANB, PERTH); 10 km S of Bruce Rock, *R.D.Royce 7881* (PERTH).

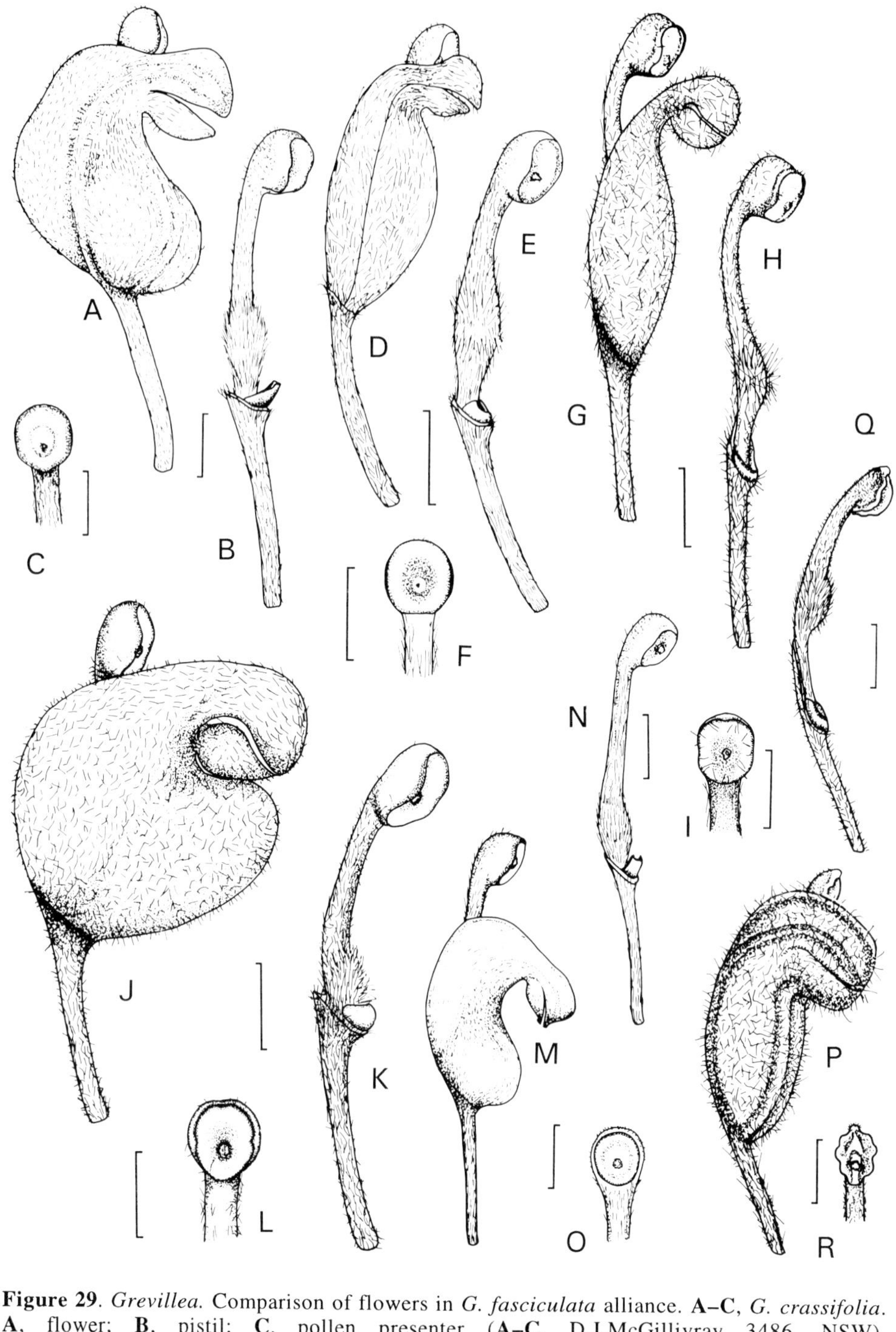

Figure 29. *Grevillea.* Comparison of flowers in *G. fasciculata* alliance. **A–C**, *G. crassifolia*. **A**, flower; **B**, pistil; **C**, pollen presenter (**A–C**, D.J.McGillivray 3486, NSW). **D–F**, *G. fasciculata*. **D**, flower; **E**, pistil; **F**, pollen presenter (**D–F**, D.J.McGillivray 3493, NSW). **G–I**, *G. centristigma*. **G**, flower; **H**, pistil; **I**, pollen presenter (**G–I**, D.J.McGillivray 3464, NSW). **J–L**, *G. saccata*. **J**, flower; **K**, pistil; **L**, pollen presenter (**J–L**, D.J.McGillivray 3278, NSW). **M–O**, *G. depauperata*. **M**, flower; **N**, pistil; **O**, pollen presenter (**M–O**, no voucher). **P–R**, *G. bronwenae*. **P**, flower; **Q**, pistil; **R**, pollen presenter (**P–R**, no voucher). Scale bars: **A–L** = 2 mm; **M–R** = 3 mm. Drawn by B.Chandler.

232. Grevillea saccata Benth., *Fl. Austral.* 5: 450 (1870)

T: W.A., *s.d.*, [*J.*] *Drummond*; lecto: MEL 74715, *fide* D.J.McGillivray & R.O.Makinson, *Grevillea* 439 (1993); ?isolecto: B *n.v.* [as *Diels 7473*], K, MEL, NSW, PERTH.

Illustrations: P.M.Olde, *Austral. Pl.* 13 (108): 357 (1986); W.R.Elliot & D.L.Jones, *Encycl. Austral. Pl.* 5: 107 (1990); D.J.McGillivray & R.O.Makinson, *Grevillea* 287, fig. 75e, 292, fig. 76 & col. pl. (1993).

Low spreading shrub 0.3–0.5 m high. Leaves entire, sublinear to oblong-elliptic or narrowly obovate, 1–5 cm long, (0.6–) 1–4 (–6.5) mm wide; margins recurved to shortly revolute; upper surface hairy, later scabrous; lower surface partly exposed, loosely subvillous or with an open indumentum of ascending to spreading hairs. Conflorescence axillary or terminal, usually simple, erect to decurved, few-flowered, subumbelloid, sometimes subsecund, opening uncertain; floral rachis 1–2 mm long, tomentose. Flowers acroscopic. Flower colour: perianth red with an orange limb; style lime-green. Perianth loosely tomentose outside (denser on limb) with mixed biramous and simple glandular hairs, bearded inside; tepals lacking median ridge. Pistil 7–9 mm long; ovary tomentose, subsessile to shortly stipitate; style pubescent, scarcely exceeding perianth after release of style-end; pollen-presenter lateral. Follicle ovoid, 13–16 mm long, pubescent. Figs 28G, 29J–L.

Occurs in the south-west of W.A., restricted to an area from just N of the Hill R. to Mullering Brook and Dandaragan. Grows in eucalypt woodland or open heath in sandy to clayey soils on laterite, sometimes in swampy sites. Regenerates from seed. Flowers June–Nov. Map 300.

W.A.: New Geraldton road, 3.2 km N of Watheroo/Jurien Bay crossroads, *R.Filson 8477* (MEL, PERTH); 18 km from Dandaragan on road to Badgingarra, *D.J.McGillivray 3277 & A.S.George* (B *n.v.*, CANB, K, NSW, US *n.v.*); 1.6 km W of Dinner Hill, *K.Newbey 2863* (PERTH); Badgingarra, 10 Nov. 1960, *L.Steenbohm* (NSW); 16 km S of Badgingarra, *P.G.Wilson 3868* (K, PERTH).

Grevillea saccata has a distinctively broadly ovoid saccate perianth (4.5–6 mm across, almost as broad as long); very oblique torus; leaves variable in shape, size and colour (greyish to green); and often secund branchlets. It is closely related to *G. fasciculata*, which has a narrow-based perianth that is widest at the middle, and to *G. crassifolia*, which has a pouched perianth 2–4 mm wide abruptly narrowed to a short tube above, and a red style. *Grevillea depauperata* is also somewhat similar, but has a pistil 11–16 mm long.

This species is recognised as 'Rare' in J.D.Briggs & J.H.Leigh, *Rare or Threatened Australian Plants* (1995).

233. Grevillea fistulosa A.S.George, *Nuytsia* 1: 371 (1974)

T: Middle Mt Barren, Fitzgerald River National Park, W.A., 23 Sept. 1925, *C.A.Gardner 1861 & W.E.Blackall*; holo: PERTH; iso: CANB, K, MEL, NSW.

Illustrations: P.M.Olde, *Austral. Pl.* 13(108): 353 (1986); P.M.Olde & N.R.Marriott, *Grevillea Book* 2: 164 (top right & 134A, B) (1995).

Shrub 0.6–2 m high. Leaves entire, narrowly obovate or oblong-elliptic to broadly linear when strongly revolute, 2.5–9 cm long, 5–15 mm wide; margins recurved to revolute; upper surface granulose; lower surface sometimes enclosed, felty-tomentose with wavy to curled hairs. Conflorescence erect, terminal or axillary, usually simple, umbelloid or a subsecund 10–14-flowered cluster, opening uncertain; floral rachis 1–2 mm long, villous. Flowers acroscopic. Flower colour: perianth orange-red to scarlet; style-end yellow. Perianth loosely tomentose to villous outside, with an open appressed indumentum inside near dorsal suture and otherwise glabrous or with a very scanty beard in throat; tepals with slight median ridge. Pistil 6–8 mm long; ovary villous, shortly stipitate; style appressed-villous, scarcely exceeding perianth after release of style-end; pollen-presenter lateral. Follicle narrowly ovoid, 12–18 mm long, tomentose, longitudinally ridged. *Barrens Grevillea, Mount Barren Grevillea.*

Occurs in the south-west of W.A., where restricted to Middle Mt Barren and the Whoogarup Ra. in the Fitzgerald River Natl Park. Grows in heath and dense shrubland, in quartzitic or ?granitic soils in rocky situations. Regenerates from seed. Flowers July–Dec. Map 301.

W.A.: top of Mt Woolbernup, Fitzgerald R., *M.I.H.Brooker 2727* (PERTH); Whoogarup Ra. near Middle Mt Barren, *C.A.Gardner 2967* (PERTH); NE side of Whoogarup Ra., *A.S.George 1912* (PERTH); Thumb Peak Ra., Fitzgerald River Natl Park, *A.S.George 7119* (PERTH); plateau N of Hamersley Ra., 21 km from Phillips R. crossing, *E.Wittwer 433* (PERTH).

Grevillea fistulosa is closely related to *G. fuscolutea*, which has a more villous lower leaf surface with the hairs longer and more strongly ascending, a narrower torus < 1.5 mm wide (c. 2 mm wide in *G. fistulosa*), shorter pedicels 3–6 mm long (5–10 mm long in *G. fistulosa*), the inside of the perianth with a beard of silky hairs in the throat, purplish new growth (pale fawn in *G. fistulosa*) and a dull yellow perianth colour. *Grevillea fasciculata* is also similar, but also has a narrower torus, usually shorter leaves with usually appressed hairs on the lower surface, and the perianth below the curve more usually ellipsoidal than narrowly ovoid.

This species is recognised as 'Poorly Known' in J.D.Briggs & J.H.Leigh, *Rare or Threatened Australian Plants* (1995).

234. Grevillea fuscolutea Keighery, *Nuytsia* 8: 228 (1992)

T: Mt Lindesay, south-western W.A., 10 Apr. 1989, *G.J.Keighery 11271*; holo: PERTH; iso: CANB, K, MEL.

Illustrations: P.M.Olde & N.R.Marriott, *Grevillea Book* 2: 172 (bottom centre), 173 (141A, B) (1995).

Erect shrub to 2 m tall. Leaves entire, obovate to sublinear, 1.5–8 cm long, 5–12 mm wide; margins recurved to revolute; upper surface finely granulose; lower surface densely villous. Conflorescence axillary or terminal on lateral branchlets, erect or decurved, simple or 2-branched; unit conflorescence umbelloid, secund, 4–10-flowered, opening variable; floral rachis 1–3 mm long, villous. Flowers acroscopic. Flower colour: perianth dull yellow-orange; style dull yellow. Perianth villous outside, glabrous inside except for silky to wispy beard in throat; tepals with a median ridge. Pistil 6–8 mm long; ovary shortly stipitate to subsessile, villous; style villous, scarcely exceeding perianth after release of style-end; pollen-presenter lateral. Follicle narrowly ovoid, c. 17 mm long, longitudinally ridged, sparsely villous.

Occurs in the south-west of W.A., restricted to upper slopes of Mt Lindesay, NE of Denmark. Grows in low heath in shallow granitic loam soil over granite. Regenerates from seed. Flowers May–Dec. Map 302.

W.A.: Mt Lindesay, *A.P.Brown 269* (PERTH): Mt Lindesay, Apr. 1980, *Forests Dept 38* (PERTH); Mt Lindesay, *Survey Team, DW80* (PERTH); side of Mt Lindesay, *W.Webb 33* (MEL).

The population now recognised as *G. fuscolutea* was included within *G. fistulosa* by McGillivray & Makinson (*Grevillea* 291–292 (1993)); see under *G. fistulosa* for differences.

This species is recognised as 'Poorly Known' in J.D.Briggs & J.H.Leigh, *Rare or Threatened Australian Plants* (1995).

235. Grevillea drummondii Meisn., *in* J.G.C.Lehmann, *Pl. Preiss.* 1: 536 (1845)

T: [W.A.], *s.d.*, [J.] Drummond [Coll. III] 335 Shuttl. 1849; lecto: NY *n.v.*, *fide* D.J.McGillivray & R.O.Makinson, *Grevillea* 414 (1993); isolecto: BM, G, K, MEL, P, PERTH; remaining syntype: ad flumen Cygnorum [Swan R., W.A.], *L.Preiss 2623*; syn: LD? [not found].

Illustrations: J.W.Wrigley & M.Fagg, *Banksias, Waratahs & Grevilleas* 225 (1989); D.J.McGillivray & R.O.Makinson, *Grevillea* 289 (1993); P.M.Olde & N.R.Marriott, *Grevillea Book* 2: 135 (top centre & 109A–C) (1995).

Low spreading shrub 0.2–1.0 m tall. Leaves entire, narrowly elliptic to narrowly obovate, 1–3 (–4) cm long, 1.5–5 (–10) mm wide; margins shortly recurved, often ciliate; upper surface loosely villous, smooth; lower surface loosely villous or rarely glabrous. Conflorescence terminal, usually simple, erect to decurved, dense, 6–8-flowered, umbelloid, subsecund, subsynchronous; floral rachis 1–3 mm long, villous. Flowers acroscopic. Flower colour: perianth and style cream in bud (limb greenish), ageing to pink or red. Perianth glabrous outside, bearded inside; tepals with a prominent median ridge. Pistil 4–6 mm long; ovary shortly stipitate, villous; style sparsely tomentose, scarcely exceeding perianth after release of style-end; pollen-presenter lateral. Follicle ovoid, 12–14 mm long, sparsely villous. *Drummond's Grevillea.*

Occurs in south-western W.A., N of Perth between Bindoon and Bolgart. Grows in eucalypt woodland and shrubland in gravelly loam soils on lateritic rises. Regenerates from seed. Flowers June–Dec. Map 303.

W.A.: between Bindoon and Moore R., Sept. 1966, *C.A.Gardner s.n.* (PERTH); c. 118.8 km NE along Geraldton Hwy, *R.J.Garraty 157* (PERTH); Bolgart, Aug. 1963, *B.Gorey* (PERTH); Bindoon to Gillingarra, *F.Lullfitz 4456* (KPBG).

Grevillea drummondii is fairly uniform, usually having leaves acutely elliptic and ≤ 5 mm broad with a loosely villous lower surface; one collection of unknown locality (*Drummond coll. 3, 327*) has larger obtuse leaves to 10 mm broad with an almost glabrous lower surface. *Grevillea pimeleoides* and *G. centristigma* are similar, but both have hairs on the outer surface of the perianth, larger leaves, and the perianth initially yellow becoming reddish.

This species is recognised as 'Rare' in J.D.Briggs & J.H.Leigh, *Rare or Threatened Australian Plants* (1995).

236. Grevillea pimeleoides W.Fitzg., *J. Proc. Mueller Bot. Soc. W. Australia* 1(10): 37 (1902)

G. drummondii subsp. *pimeleoides* (W.Fitzg.) McGill., *New Names Grevillea* 5 (1986). T: between Smiths Mill and Helena R., W.A., July 1901, *W.V.Fitzgerald*; lecto: NSW, *fide* D.J.McGillivray & R.O.Makinson, *Grevillea* 414 (1993); isolecto: PERTH.

Illustrations: W.R.Elliot & D.L.Jones, *Encycl. Austral. Pl.* 5: 50 (1990), as *G. drummondii* subsp. *pimeleoides*; D.J.McGillivray & R.O.Makinson, *Grevillea* 291 (1993), as *G. drummondii* subsp. *pimeleoides*; P.M.Olde & N.R.Marriott, *Grevillea Book* 3: 96 (top right & 73A–C) (1995).

Robust erect shrub 0.6–2.5 m tall. Leaves entire, elliptic to obovate, 2–6.5 cm long, 7–20 mm wide; margins flat to recurved, ciliate; upper surface with a sparse indumentum, otherwise smooth; lower surface glabrous. Conflorescence terminal on lateral branchlets, simple, erect to decurved, secund, umbelloid, usually 13–18-flowered, subsynchronous; floral rachis 1–2 mm long, villous. Flowers adaxially acroscopic. Flower colour: perianth light orange in bud, yellow at anthesis, then deepening to orange again; style-end yellow becoming red. Perianth sparsely pubescent to pilose outside, bearded inside; tepals with a prominent medial ridge. Pistil (6–) 7–8 mm long; ovary shortly stipitate villous; style sparsely villous, scarcely exceeding perianth after release of style-end; pollen-presenter lateral. Follicle ovoid, c. 15 mm long, pubescent, longitudinally ridged.

Occurs in south-western W.A., in the Helena and Canning R. areas E and SE of Perth. Grows in open shrubby eucalypt woodland or forest in gravelly or clay-loam soils. Regenerates from seed. Flowers July–Nov. Map 304.

W.A.: S side of Mundaring Weir, *A.S.George 11656* (CANB, PERTH); bank of Canning R. near Gleneagle, 30 May 1963, *A.H.Larner 3* (PERTH); [loc. variously as 'montium Darling Range' or 'Glen Forrest'], Aug. 1901, *E.Pritzel 517* [also as *L.Diels 3697*] (A *n.v.*, AD, B *n.v.*, E, G, GH *n.v.*, K, NSW, P *n.v.*, PERTH).

Grevillea pimeleoides differs from *G. centristigma* in its usually 13–18-flowered unit conflorescences, the obovate pollen-presenter with its base concurrent with the style or nearly so, and the stigma being near-basal on the pollen-presenter face. *Grevillea centristigma* typically has unit conflorescences 5–8-flowered, the pollen-presenter nearly circular, its base clearly not concurrent with the style, and the stigma central or nearly so (occasionally tending basal). There is a tendency for *G. pimeleoides* to have larger leaves, and a slightly longer pistil, although there is some overlap on both features among specimens in the NW of the range of *G. centristigma*.

This species is recognised as 'Rare' in J.D.Briggs & J.H.Leigh, *Rare or Threatened Australian Plants* (1995).

237. Grevillea centristigma (McGill.) Keighery, *Nuytsia* 8: 227 (1992)

G. drummondii subsp. *centristigma* McGill., *New Names Grevillea* 5 (1986). T: 24 km SE of Pemberton on road to Northcliffe, W.A., 1 Oct. 1967, *P.G.Wilson 6289*; holo: PERTH.

Illustrations: D.J.McGillivray & R.O.Makinson, *Grevillea* 287, fig. 75d, 290, col. pl. (1993), as *G. drummondii* subsp. *centristigma*; P.M.Olde & N.R.Marriott, *Grevillea Book* 2: 86 (centre right), 87 (68A–C) (1995).

Low compact to erect rounded shrub, 0.3–1.0 m tall. Leaves entire, narrowly elliptic to narrowly obovate, 1.5–4 cm long, 2–9 mm wide; margins recurved, usually long-ciliate; upper surface sparsely villous; lower surface loosely to sparsely villous. Conflorescence terminal, simple, erect to decurved, subsecund-umbelloid, usually 5–8-flowered; floral rachis 0.5–1 mm long, sparsely villous. Flowers adaxially oriented. Flower colour: perianth deep yellow becoming orange; style yellow becoming brick red with age. Perianth loosely pilose outside, bearded inside in throat; tepals with a median rib. Pistil 5–6 (–7) mm long; ovary stipitate, villous; style loosely tomentose, scarcely exceeding perianth after release of style-end; pollen-presenter lateral. Follicle ovoid, 8–12 mm long, tomentose. Fig. 29G–I.

Occurs in south-western W.A., where widespread between Armadale, Shannon and the Whicher Ra. Grows in moister areas of Karri or Marri-Jarrah forest, usually in gravelly granitic soils. Regenerates mainly from seed, possibly also lignotuber and rhizome. Flowers June–Nov. Map 305.

W.A.: 34 km by road SE from Armadale on the Albany Hwy, *D.J.McGillivray 3464 & A.S.George* (CANB, K, NSW, PERTH, US *n.v.*); Whicher, *G.McCutcheon 820* (PERTH); Last Bottle Rock, W of Shannon, *N.G.Marchant 79/58* (PERTH); 64 km from Perth towards Williams, 30 Sept. 1968, *M.E.Phillips CBG027721* (CANB, NSW); 24 km SE of Pemberton on road to Northcliffe, *P.G.Wilson 6289* (BH *n.v.*, MEL, NSW, PERTH).

McGillivray & Makinson (*Grevillea* 289–290 (1993)) treated this taxon as a subspecies within a broadly defined *G. drummondii.* It is narrowly distinct in some features from *G. pimeleoides* (see under that species for differences), and in the NW of its range sometimes appears to approach *G. pimeleoides* in stigma position, pistil length and pollen-presenter shape.

238. Grevillea depauperata R.Br., *Suppl. Prodr. Fl. Nov. Holl.* 1: 20 (1830)

T: inland from King George III'd Sound [W.A.], 1828–9, *W.Baxter*; lecto: BM, *fide* D.J.McGillivray & R.O.Makinson, *Grevillea* 413 (1993); isolecto: NSW, PERTH; possible isolecto: BM.

G. brownii Meisn., *in* J.G.C.Lehmann, *Pl. Preiss.* 1: 537 (1845). T: ... montis Wuljenup [also as 'Wiljenup'], Plantagenet District [W.A.], 14 Oct. 1840, *L.Preiss 719*; holo: LD *n.v.*; iso: LE *n.v.*, NY (photo seen).

Illustrations: A.S.George, *Introd. Prot. W. Australia* 50, pl. 67 (1984), as *G. brownii*; D.J.McGillivray & R.O.Makinson, *Grevillea* 286, fig. 75c, 288, col. pl. (1993); P.M.Olde & N.R.Marriott, *Grevillea Book* 2: 117 (bottom right), 118 (94A, B) (1995).

Low dense or sometimes prostrate shrub 0.2–0.8 m tall. Leaves elliptic to oblong-elliptic or narrowly so or ovate, 0.6–3.0 (–6.0) cm long, 1.5–10 mm wide; margins smoothly recurved or revolute; upper surface granulose to smooth and glossy; lower surface with a dense matted indumentum of wavy to curled hairs. Conflorescence terminal and axillary, simple or few-branched; unit conflorescence erect, umbelloid-subsecund, 2–6 (–8)-flowered, opening uncertain; floral rachis 0.5–1 mm long, villous. Flowers acroscopic. Flower colour: perianth and style red to dull or bright orange. Perianth tomentose outside, bearded inside; tepals with a faint median ridge. Pistil 11–15.5 mm long; ovary stipitate, villous; style tomentose, weakly exserted before release of style-end, ±erect and exceeding perianth by > 3 mm after release of style-end; pollen-presenter lateral. Follicle ovoid, c. 15 mm long, glabrous, faintly ribbed. Fig. 29M–O.

Occurs in south-western W.A., in the area bounded by Albany, Cranbrook and Manjimup. Grows in eucalypt woodland in sandy or gravelly soils often over laterite. Regenerates from seed. Flowers mainly May–Oct. Map 306.

W.A.: on Porongorup road 0.8 km W of Many Peak turnoff, Porongorup Ra., *R.Filson 9074* (MEL); S of Mt Barker, *A.S.George 11692* (PERTH); 3 km E of Mt Barker, *R.H.Kuchel 1989* (AD, PERTH); Tenterden, 23 Sept. 1902, *A.Morrison* (BRI); c. 74.7 km SE of Manjimup, Muirs Hwy, *M.D.Tindale 3908* (NSW, PERTH).

Grevillea depauperata shows considerable variation in habit and flower colour. It has a strongly oblique torus (angled at 50°–60° to the pedicel), 1.4–2.3 mm across. The nectary is linguiform and usually projects strongly into the 'pouch' of the perianth. The stipe of the ovary is 1.5–2 mm long and free for most of its length (not or scarcely adnate to the wall of the torus cavity). It is somewhat similar to *G. fasciculata* and *G. crassifolia*; both these species have the pistil < 10 mm long.

239. **Grevillea pinifolia** Meisn., *in* A.L.P.P. de Candolle, *Prodr.* 14: 350 (1856)

T: Swan River, W.A., 1848, [*J.*] *Drummond* [*4th coll.*] *281*; lecto: NY *n.v.*, *fide* D.J.McGillivray & R.O.Makinson, *Grevillea* 434 (1993); isolecto: A *n.v.*, BM, CGE *n.v.*, E, G, K, LE *n.v.*, MEL, NY *n.v.*, P, TCD *n.v.*

G. pinifolia Meisn., *Hooker's J. Bot. Kew Gard. Misc.* 4: 186, 1852, *nom. nud.*

Illustrations: W.R.Elliot & D.L.Jones, *Encycl. Austral. Pl.* 5: 94 (1990); P.M.Olde & N.R.Marriott, *Grevillea Book* 3: 99 (top right & 75A, B) (1995).

Low mounded shrub 0.3–0.6 (–1.0) m tall. Leaves entire, linear-subterete, dipleural with a submarginal groove along each side, 2.5–5 cm long, 0.5–0.7 mm wide; margins obscure, tightly revolute against prominent abaxial midrib; upper surface subsericeous becoming glabrous and granulose; lower surface obscure except for midrib. Conflorescence axillary, simple, erect, 1–4-flowered cluster, opening uncertain; floral rachis 0.5–1 mm long, tomentose. Flowers abaxially oriented. Flower colour: perianth and style red to orange-red. Perianth villous to subsericeous outside, bearded inside; tepals with a median ridge. Pistil 7.5–8.5 mm long; ovary stipitate, villous; style openly pubescent to almost glabrous, scarcely exceeding perianth after release of style-end; pollen-presenter lateral. Follicle ovoid, c. 10 mm long, sparsely villous.

Occurs in south-western W.A., where restricted to an area between Eneabba and Bindi Bindi N of Perth. Grows in shrubland on yellow sand and laterite. Regenerates from seed. Flowers July–Oct. Map 307.

W.A.: Wubin, *W.E.Blackall 3 & C.A.Gardner* (PERTH); Coorow, *W.E.Blackall 2591* (PERTH); Miling to Pithara, *C.A.Gardner 2201* (PERTH); 17.4 km E of Bindi Bindi, *S.Paust 1118* (PERTH); Eneabba area, *E.Wittwer S1612* (KPBG).

Grevillea pinifolia has crowded, strongly ascending leaves. It is easily distinguished from narrow-leaved forms of *G. fasciculata* which have clearly dorsiventral leaves (the submarginal groove or grooves clearly situated on the lower surface, the abaxial midrib not prominent beyond the marginal rolls), and from *G. yorkrakinensis* and *G. disjuncta* which have the style much exceeding the perianth and the pollen-presenter convex (concave in *G. pinifolia*).

This species is recognised as 'Poorly Known' in J.D.Briggs & J.H.Leigh, *Rare or Threatened Australian Plants* (1995).

240. **Grevillea brachystylis** Meisn., *in* J.G.C.Lehmann, *Pl. Preiss.* 1: 538 (1845)

T: ... Mocloy's Plain (Sussex), W.A., 20 Dec. 1839, *L.Preiss 714*; lecto: LD (photo seen), *fide* D.J.McGillivray & R.O.Makinson, *Grevillea* 409 (1993); isolecto: CGE *n.v.* [as *Preiss 911*], G, NY *n.v.*, TCD *n.v.* [as *Preiss 911*].

Low spreading to erect shrub 0.3–1.0 m tall. Leaves linear to narrowly obovate, 1–14 cm long, 2–10 mm wide; margins recurved to revolute; upper surface finely granulose; lower surface usually partly exposed, densely subsericeous to villous. Conflorescence terminal or axillary, usually simple, deflexed, subsecund-umbelloid to wheel-like, 4–8-flowered, opening uncertain; floral rachis 2–7 mm long, villous. Flowers acroscopic. Perianth pubescent or sparsely to densely tomentose outside (densely so on limb), pubescent inside; tepals with a median ridge. Pistil 7–11 (–15) mm long; ovary stipitate, villous; style pubescent, scarcely

Eleven species, south-western Australia and eastern N.S.W. Insect pollinated. Most taxa in this group are characterised by distinctive features of the style-end, which may have a dorsal/apical retrorse to antrorse or ventrally incurved appendage, and which may be either densely hairy like the style, or papillose or glabrous and then (in W.A.) often brightly coloured. This group corresponds to Section *Eriostylis* of Bentham (1870). Affinities are to the *Floribunda* group.

1 Style-end with an apical-dorsal appendage or apiculation

2 Style-end appendage incurled towards or over pollen-presenter (SW of W.A.)

3 Stylar appendage dorsally papillose at apex; upper leaf surface granular-scabrid; base of pollen-presenter projecting from line of style; style strongly incurved **252. G. scabra**

3: Stylar appendage papillose at base, glabrous in upper half; upper leaf surface usually smooth; base of pollen-presenter concurrent with style; style ±straight **251. G. candolleana**

2: Style-end appendage retrorse to erect or antrorsely directed, but not curled in over pollen-presenter

4 Leaves linear-oblong to narrowly elliptic (rarely slightly narrowly oblanceolate); lower leaf surface with a loose villous indumentum, all hairs long and strongly ascending to wavy-erect; pollen-presenter narrowly elliptic, 1–1.6 mm wide; stylar appendage antrorsely directed to erect (very rarely slightly retrorse); pistil 10–13 mm long **243. G. phylicoides**

4: Leaves ovate or slightly obovate or narrowly oblong to elliptic; lower leaf surface with a loosely appressed to weakly ascending indumentum of short often matted hairs, with occasional longer hairs weakly emergent; pollen-presenter round to broadly elliptic, 1.4–3.3 mm wide; style-end appendage retrorsely directed to erect (rarely slightly antrorse); pistil 11–21 mm long **242. G. buxifolia**

1: Style-end lacking an apical-dorsal appendage

5 Stigma a prominent conical to narrowly conical beak, 0.3–0.4 mm long, protruding from face of pollen-presenter; style-end subclavate to subglobose, lacking the long white dorsal hairs of the style and bright red, orange or yellow

6 Dorsal surface of style-end glabrous, smooth **248. G. pilulifera**

6: Dorsal surface of style-end densely papillose

7 Distal edge of pollen-presenter level with or slightly exceeding apex of papillose style-end, not overtopped by it; dorsal tepals lacking an auricular lobe on posterior edge just below limb-cup; leaves usually in short fasciculate clusters **249. G. uncinulata**

7: Distal edge of pollen-presenter narrowly but distinctly overtopped by a slight apical extension of papillose style-end; dorsal tepals with small auricular lobes on posterior edge just below limb-cup; leaves mostly well-spaced, rarely a few in tight clusters **250. G. florida**

5: Stigma a slight boss $\leqslant$ 0.2 mm high, on face of pollen-presenter, rarely more prominent; style-end subdiscoid to flat-ellipsoidal, with white stylar hairs on dorsal side carried right to the apex

8 Longest floral bracts < 2 mm long; pistils 5–13 mm long (south-western Australia)

9 Leaf upper surface smooth; flowers adaxially oriented (styles arching in towards axis of rachis) **246. G. umbellulata**

9: Leaf upper surface granular-scabrid; flowers abaxially oriented (styles arching out, away from axis of rachis) **247. G. occidentalis**

8: Longest floral bracts ⩾ 2 mm long; pistils 9–21 mm long (south-eastern Australia)

10 Flowers adaxially acroscopic, the styles curved in towards the rachis axis; leaves 1–2 mm wide, with most or all of lower surface enclosed **245. G. acerata**

10: Flowers abaxially oriented to basiscopic, the styles curved outwards from the rachis axis; leaves 1.5–8.5 mm wide, with most or all of lower surface exposed

11 Branchlets and lower leaf surfaces subsericeous (hairs appressed) **244. G. sphacelata**

11: Branchlets tomentose to villous (hairs ascending to spreading); lower leaf surfaces loosely subsericeous to tomentose or villous

12 Leaves linear-oblong to narrowly elliptic (rarely slightly narrowly oblanceolate); lower leaf surface with a loose villous indumentum, all hairs long and strongly ascending to wavy-erect; pollen-presenter narrowly elliptic, 1–1.6 mm wide; stylar appendage, when evident, antrorsely directed to erect (very rarely slightly retrorse); pistil 10–13 mm long **243. G. phylicoides**

12: Leaves ovate or slightly obovate or narrowly oblong to elliptic; lower leaf surface with a loosely appressed to weakly ascending indumentum of short often matted hairs, with occasional longer hairs weakly emergent; pollen-presenter round to broadly elliptic, 1.4–3.3 mm wide; style-end appendage, if present, retrorsely directed to erect (rarely slightly antrorse); pistil 11–21 mm long **242. G. buxifolia**

242. Grevillea buxifolia (Sm.) R.Br., *Trans. Linn. Soc. London* 10: 174 (1810)

Embothrium buxifolium Sm., *Spec. Bot. New Holland* 3: 29, t. 10 (1794); *Stylurus buxifolius* (Sm.) Knight, *Cult. Prot.* 115 (1809). T: New South Wales, [*J.*] *White*; lecto: LINN, *fide* D.J.McGillivray & R.O.Makinson, *Grevillea* 409 (1993); isolecto: G, K, MANCH *n.v.*, P *n.v.*, PH *n.v.*

Shrub 1.5–2.5 m tall. Leaves ovate or slightly obovate or narrowly oblong to elliptic, 1–2 cm long, 3–8.5 mm wide; margins recurved; upper surface granulose or rarely smooth; lower surface exposed, with a loosely appressed to weakly ascending indumentum of short often matted hairs, with occasional longer hairs weakly emergent. Unit conflorescence subumbelloid, basipetal. Flowers abaxially oriented, sometimes almost basiscopic. Perianth tomentose to villous outside, sublanate inside. Pistil 11–21 mm long, subvillous or tomentose; style-end either with a retrorsely directed to erect or rarely antrorse dorsal horn-like hairy appendage 2–4 mm long, or appendage absent or obscure; pollen-presenter lateral (sometimes appearing transverse on reflexed style apex), circular or rarely broadly elliptic in face view, slightly convex. Follicle obliquely ovoid, 18–22 mm long, sparsely villous. *Grey Spider Flower.*

Grevillea buxifolia as here defined occurs in central coastal N.S.W. Two subspecies are recognised.

Grevillea buxifolia is one element of a closely related complex of taxa. Despite several recent studies, classification and phylogeny are not yet fully resolved. McGillivray (in McGillivray & Makinson (*Grevillea* 311–315 (1993)) distinguished five races, distributed into a parallel system of three subspecies within a broad-concept *G. buxifolia*, with three populations remaining explicitly unassigned. Races ‘a’ and ‘c’ were grouped under subsp. *buxifolia*, races ‘b’ and ‘d’ under subsp. *phylicoides*, and race ‘e’ as subsp. *sphacelata*.

Olde & Marriott (*Telopea* 5: 711–734 (1994); *Grevillea Book* 2: 73–75; 3: 178–180 (1995)) reinstated *G. sphacelata* R.Br. at species rank, named McGillivray's ‘race a’ as *G. buxifolia* subsp. *ecorniculata*, and maintained McGillivray's *G. buxifolia* subsp. *phylicoides* encompassing races ‘b’ and ‘d’ and with the addition of one of McGillivray's unassigned elements from the N.S.W. South Coast.

Hart & Henwood (*Telopea* 7: 65–76 (1996)) present a phenetic analysis, propose reinstatement of *G. phylicoides s. str.* ('race d' of McGillivray & Makinson) at species rank, and allocate races 'b' and 'c' to a moderately broad *G. buxifolia* subsp. *buxifolia*, while maintaining *G. buxifolia* subsp. *ecorniculata* and *G. sphacelata*. The present treatment agrees with this classification.

Grevillea phylicoides can be distinguished by having all hairs on the lower leaf surface strongly ascending, a scarcely thickened style-end with a usually antrorse appendage, and a narrowly elliptic pollen-presenter; and *G. sphacelata* by the mainly appressed hairs on the branchlets and styles.

Stylar appendage conspicuous, (1.5–) 2–4.0 mm long; hairs on style apex and appendage not longer than those of rest of style; pistil 17–21 mm long **242a.** subsp. **buxifolia**

Stylar appendage absent, or inconspicuous and rarely to 1.3 mm long; hairs on style apex markedly longer than those on rest of style; pistil 11–13 mm long **242b.** subsp. **ecorniculata**

242a. Grevillea buxifolia (Sm.) R.Br. subsp. buxifolia

Embothrium genianthum Cav., *Icon.* 4: 60, t. 387 (1798). T: 'Novae Hollandiae' [Port Jackson area, N.S.W.], *coll. unknown*; holo: MA? *n.v.*, *fide* D.J.McGillivray & R.O.Makinson, *Grevillea* 409 (1993); ?iso: P-JU.

G. buxifolia subsp. *buxifolia* 'race b' and 'race c', of D.J.McGillivray & R.O.Makinson, *Grevillea* 312–314 (1993).

Illustrations: W.R.Elliot & D.L.Jones, *Encycl. Austral. Pl.* 5: 36 (1990); D.J.McGillivray & R.O.Makinson, *Grevillea* 313 (1993); P.M.Olde & N.R.Marriott, *Grevillea Book* 2: 73 (bottom right & 56) (1995).

Shrub 1.5–2.5 m tall. Leaves ovate to elliptic or slightly obovate, 1–1.5 cm long, 5–8.5 mm wide. Pedicels 8–12 mm long, rusty brown. Flower colour: perianth (hairs) rusty brown to fawn outside, white to pale grey or pinkish grey inside; style (hairs) light grey or off-white, with pinkish tones from ground tissue. Pistil 17–21 mm long; style-end with a conspicuous conical horn-like appendage on dorsal side of style apex; appendage (1.5–) 2–4 mm long, usually retrorsely directed or perpendicular relative to line of apical few millimetres of style, or rarely slightly antrorse; hairs on style apex and appendage not longer than those on rest of style; plane of pollen-presenter usually transverse to style or oblique to it at > 60°, occasionally shallowly oblique or parallel to style axis; pollen-presenter round or rarely broadly elliptic in face view. Plate 47.

Occurs in central coastal N.S.W. from the coast N of Tuggerah S to Port Jackson and inland to Kulnura, Culoul Ra. and Windsor; also southern Sydney Basin from Cooks R. to Menai and Waterfall; with disjunct populations in the Pigeon House Mtn area on the N.S.W. south coast near Milton. Grows in sclerophyll woodland or tall heath, usually in skeletal sandy soils over sandstone. Regenerates from seed. Flowers mainly July–Nov., sporadic in other months. Map 311.

N.S.W.: Culoul Ra., between Colo R. and Windsor to Singleton road, c. 10 km NW of Grassy Hill, *B.G.Briggs NSW128844* (NSW); Belrose, *R.Coveny 11049 & P.Hind* (K, NSW); Calga, *L.A.S.Johnson & E.F.Constable NSW20377* (NSW); Royal Natl Park; NW quadrant, c. 300 m S of Farnell Drive entrance station, *R.O.Makinson 1300* (CANB, K, MEL, NSW); Dog Trap Rd, 5 km from Ourimbah, *C.W.E.Moore 6170* (CANB).

In the S of the range (Woronora Plateau including Royal Natl Park), subsp. *buxifolia* grows in mixed stands with *G. sphacelata* without intergrades; *G. sphacelata* can be distinguished by the appressed hairs on the branchlets, less robust flowers (pistils 9–12 mm long), and the stylar appendage lacking or short (≤ 1 mm long) and antrorse.

Populations in the N and NW of the range (Kulnura, Wisemans Ferry and Culoul Ra. to Windsor) show varying degrees of tendency towards narrower leaves, less robust flowers, and an elliptic rather than round pollen-presenter (the plane of which is often more closely parallel to the stylar axis), than do populations to the SE. The trend is suggestive of states more typical of *G. phylicoides s. str.*, but the variance elsewhere in the range of subsp. *buxifolia* makes formal separation of these populations unsustainable on these characters. They share the lower leaf surface indumentum, thickened style-ends, and (usually) erect to reflexed stylar appendages of the remainder of subsp. *buxifolia*. Most of these populations correspond to McGillivray & Makinson's 'race b', regarded by McGillivray & Makinson (*loc. cit.*) and by Olde & Marriott (*loc. cit.*) as one of two components within their shared concept of *G. buxifolia* subsp. *phylicoides*.

242b. Grevillea buxifolia subsp. ecorniculata Olde & Marriott, *Telopea* 5: 709 (1994)

T: Staircase Hill, Putty Road, 82.8 km N of Windsor, N.S.W., 24 Sept. 1989, *R.O.Makinson 384*; holo: NSW; iso: CANB, COLO, HO, LE, MEL, PE.

G. buxifolia subsp. *buxifolia* 'race a', of D.J.McGillivray & R.O.Makinson, *Grevillea* 312–314 (1993).

Illustrations: P.M.Olde & N.R.Marriott, *Grevillea Book* 2: 74 (top right & 57) (1995).

Shrub 1–2 m tall. Leaves elliptic to narrowly so, 1–2 cm long, 3–6 mm wide. Pedicels 6–9 mm long. Flower colour: perianth (hairs) brown to fawn and grey outside, sometimes tinged pink, white to pale grey inside; style (hairs) pale grey to off-white, with pinkish tones from ground tissue. Pistil 11–13 mm long; style-end usually lacking appendage on dorsal side of style apex, or occasionally with an inconspicuous erect apiculation to 0.8 (–1.3) mm long; hairs on style apex markedly longer than those on rest of style; plane of pollen-presenter ±parallel to style axis; pollen-presenter round in face view.

Occurs in eastern subcoastal N.S.W., NW of Sydney between Putty, Gospers Mtn and Wollombi. Grows in shrubby associations in dry eucalypt forest in skeletal sandy soils over sandstone. Regenerates probably from seed only. Flowers mainly July–Nov. Map 312.

N.S.W.: Howes Mtn, 9.7 km NE of Howes Valley, 26 Aug. 1959, *E.F.Constable NSW48253* (BM, K, NSW); S of Putty at 88.5 km by road from Windsor, 28 Sept. 1959, *L.A.S.Johnson & E.F.Constable NSW48295* (BM, K, NSW); 16 km N of Gospers Mtn army airstrip, on military road 14.5 km NE of Glen Davis, *D.J.McGillivray 1170 & A.N.Rodd* (NSW); between Three Ways and Kekeelbon Mtns, 24 km NW of Putty, *D.J.McGillivray 1558* (NSW); Yango Track near Wollombi, *R.Story 6658* (NSW).

Subsp. *ecorniculata* appears to be geographically separated from the northernmost populations of subsp. *buxifolia*, which extends no further N than Kulnura and the Culoul Ra.; no intergrades are known. Species rank could be warranted, but in the light of the several conflicting recent treatments for the complex, any change to ranking should await phylogenetic and genetic studies.

243. Grevillea phylicoides R.Br., *Trans. Linn. Soc. London* 10: 174 (1810)

G. buxifolia subsp. *phylicoides* (R.Br.) McGill., *New Names Grevillea* 3 (1986), *p.p.* T: '... prope Port Jackson; in montibus saxosis', N.S.W. [protologue]; lecto: R.Brown *Iter Austral.* 3323, '22 Grevillea phylicoides prodr. 379 Hills near the banks of the Grose'; lecto: BM, *fide* R.O.Makinson, *Fl. Australia* 17A: 504 (2000); isolecto: BM, ?E *n.v.*, K, ?LE *n.v.*, NSW.

G. buxifolia subsp. *phylicoides* 'race d', of D.J.McGillivray & R.O.Makinson, *Grevillea* 311–314 (1993).

G. buxifolia subsp. *phylicoides* 'Typical form', 'Blue Mountains form', of P.M.Olde & N.R.Marriott, *Grevillea Book* 2: 75 (1995)]

Illustration: D.J.McGillivray & R.O.Makinson, *Grevillea* 314 (1993).

Shrub 1–2 m tall. Leaves elliptic to narrowly so or oblong to slightly and narrowly oblanceolate, 1–2.5 (–3) cm long, 1.5–4 (–6) mm wide; margins recurved; upper surface granulose; lower surface exposed, open-villous with all hairs strongly ascending to wavy-erect. Unit conflorescence subumbelloid, basipetal. Flowers abaxially oriented. Flower colour: perianth grey (hairs) with brown markings, white hairs on inner surface conspicuously displayed; style with white hairs over pinkish ground tissue; style tip sometimes green. Perianth tomentose to villous outside, sublanate inside. Pistil 10–13 mm long, subvillous or tomentose; style-end with an antrorsely-directed to erect (very rarely slightly retrorse) dorsal horn-like hairy appendage (0.3–) 1.0–2 mm long; pollen-presenter lateral, narrowly elliptic in face view, slightly convex. Follicle obliquely ovoid, c. 15 mm long, sparsely villous. *Grey Spider Flower.*

Occurs in coastal ranges of N.S.W., NW to SW of Sydney, mainly in the Blue Mtns from Bilpin and the Mt Hay area, E to the foot of the escarpment and S to Oakdale, and with a disjunct occurrence on the N.S.W. south coast SE of Nowra (Beecroft Peninsula, possibly also Conjola). Grows in dry sclerophyll woodland or heath, usually in skeletal sandy soils over sandstone. Regenerates from seed. Flowers mainly July–Mar., sporadically in other months. Map 313.

N.S.W.: 1.5 km E of Warragamba Dam, *R.Coveny 839* (K, NSW, PERTH); Grose Wold, *R.Coveny 5488* (K, NSW); 8 km from Agnes Banks towards Springwood, *J.Pulley 833* (CANB, NSW); Glenbrook, *C.T.White 8485* (A, BISH, BRI); Kings Tableland, Blue Mtns, *C.L.Wilson 506* (NSW).

Grevillea phylicoides is characterised by strongly spreading hairs on the branchlets and the lower leaf surfaces, the style-end and pollen-presenter scarcely thicker than the style, the pollen-presenter narrowly elliptic and ±parallel with the axis of the apical 2–3 mm of the style, and the stylar appendage erect to antrorsely directed (rarely slightly retrorse). It is distinguished from *G. sphacelata*, which has appressed hairs on the branchlets and lower leaf surfaces, and from *G. buxifolia*, which has more robust flowers (pistils 11–21 mm long), with the style-end and pollen-presenter much thickened, the pollen-presenter round or occasionally broadly elliptic and 1.4–3.3 mm wide (1–1.6 mm wide in *G. phylicoides*), and the stylar appendage (when present) erect to retrorse or rarely slightly antrorse.

Occasional intermediates with *G. buxifolia* subsp. *buxifolia* may occur in the NE of the range (e.g. Mountain Lagoon area). In the S of the main range (Warragamba area) there is a trend towards very short stylar appendages suggesting a grade towards *G. sphacelata*; the species however remain diagnosable and geographically separate. See also notes on *G. scabrifolia* under Doubtful Names (page 449).

244. Grevillea sphacelata R.Br., *Trans. Linn. Soc. London* 10: 174 (1810)

G. buxifolia subsp. *sphacelata* (R.Br.) McGill., *New Names Grevillea* 3 (1986). T: 'In Novae Hollandiae ora orientalis; prope Port Jackson' [protologue]; lecto: ... prope George's R., [N.S.W.], 2 Oct. 1803 [emend. from '1804'], *R.Brown Iter Austral. 3322*; lecto: BM, *fide* D.J.McGillivray & R.O.Makinson, *Grevillea* 409 (1993); remaining syntypes (some possibly isolecto): BM, E, G, G-DC, K, LE *n.v.*, NSW, NY *n.v.*, P *n.v.*

G. walteri Gand., *Bull. Soc. Bot. France* 66: 231 (1919). T: New South Wales, 1902, *ex Herb. C.Walter*; holo: LY *n.v.*

Illustrations: W.R.Elliot & D.L.Jones, *Encycl. Austral. Pl.* 5: 36 (1990); D.J.McGillivray & R.O.Makinson, *Grevillea* 315 (1993); P.M.Olde & N.R.Marriott, *Grevillea Book* 3: 179 (top left & 145A, B) (1995).

Shrub 0.4–2.5 m high. Leaves narrow-elliptic to -oblong or slightly -obovate or sublinear, 0.7–3.5 cm long, 2–4 mm wide; margins recurved to revolute; upper surface glabrous with age and faintly granulose; lower surface mostly or wholly exposed, subsericeous. Unit conflorescence subumbelloid, basipetal. Flowers abaxially oriented. Flower colour: perianth pale brown (hairs) outside, pink with overlying grey hairs inside; style pinkish grey (hairs). Perianth subsericeous to appressed-subvillous outside, villous or lanate inside. Pistil 9–12 mm long, subsericeous to appressed-tomentose; style-end with or without a short antrorse apiculation to c. 1 mm long extending just beyond distal rim of pollen-presenter; pollen-presenter oblong-elliptic in face view. Follicle ovoid, 18–20 mm long, sparsely pubescent. *Grey Spider Flower.*

Occurs in coastal N.S.W., from Port Jackson and Parramatta R. S to West Dapto and Mittagong, with a disjunct population on the N.S.W. south coast between Nowra, Huskisson and Wandandian. Grows in dry sclerophyll woodland or heath, either in sandy soils derived from sandstone or in loamy soils derived from Wianamatta Shale. Regenerates from seed and (sometimes?) lignotuber. Flowers mainly July–Jan. Map 314.

N.S.W.: Field of Mars, Gladesville, Aug. 1914, *J.L.Boorman NSW20243* (NSW); West Dapto, *R.H.Cambage 348* (CANB, SYD); South Coogee, c. 6.5 km N of Botany Bay, *Hj.Eichler 13104* (AD); Royal Natl Park, NW quadrant, c. 300 m S of Farnell Dr. entrance station, *R.O.Makinson 1299* (CANB, K, MEL, NSW); between Appin and Wollongong at turnoff to Wedderburn, 31 Jan. 1973, *J.W.Wrigley CBG048770* (CANB).

McGillivray (*loc. cit.* (1986); and McGillivray & Makinson, *Grevillea* 312–313, 409–410 (1993)) treated *G. sphacelata* as a subspecies of a broadly defined *G. buxifolia*. *Grevillea sphacelata* has appressed hairs on the branchlets and lower leaf surfaces, distinguishing it from *G. buxifolia* and *G. phylicoides* in which these parts are tomentose to subvillous.

245. Grevillea acerata McGill., *New Names Grevillea* 1 (1986)

T: Gibraltar Range Natl Park, 58 miles [93 km] NW of Grafton on the Gwydir Hwy, N.S.W., 8 Feb. 1973, *M.D.Tindale 2069*; holo: NSW; iso: K, NE.

[*G. sphacelata auct. non* R.Br.: N.C.W.Beadle, *Stud. Fl. NE New South Wales* 2: 244, 245, fig. 114D (1973)]

Illustrations: N.C.W.Beadle, *loc. cit.*, as *G. sphacelata*; W.R.Elliot & D.L.Jones, *Encycl. Austral. Pl.* 5: 17 (1990); P.M.Olde & N.R.Marriott, *Grevillea Book* 2: 11 (centre left & 5A–C) (1995).

Shrub 0.6–1.5 m high. Leaves linear to very narrow-elliptic or slightly -obovate, 1–3 cm long, 1.0–1.5 (–2.0) mm wide; margins revolute; upper surface sparsely hairy, soon glabrous and granulate; lower surface mostly enclosed, subsericeous to subvillous. Unit conflorescence umbelloid-subglobose, basipetal. Flowers adaxially acroscopic. Flower colour: perianth with brownish hairs outside especially on limb; inner surface with white to grey indumentum tinged pink by ground tissue; style with white to grey hairs over purplish pink ground tissue. Perianth subvillous outside and inside, often persistent in fruit. Pistil 9–12 mm long, villous; style-end lacking apical appendage; pollen-presenter very oblique to lateral, obovate, flat to slightly convex. Follicle obliquely ovoid, 12 mm long, loosely villous, rugulose.

Occurs in N.S.W. in the Northern Tablelands area E of Glen Innes, known only from Gibraltar Range Natl Park and from nearby Pheasant Ck, Glen Elgin. Grows in sclerophyll woodland, open forest and heath, in granitic soils. Regenerates from seed. Flowers mainly Sept.–Dec., sporadic in other months. Map 315.

N.S.W.: Pheasant Ck, Glen Elgin, Dec. 1913, *J.L.Boorman NSW93434* (CANB, K, NSW); 1.2 km from Milligans [Mulligans] Hut on road to Gwydir Hwy, Gibraltar Range Natl Park, *W.E.Fisher 245* (NSW, PERTH); Gibraltar Ra., NE of Glen Innes, *R.W.Jessup & M.Gray 3061* (CANB, NSW); Wades Rd, Gibraltar Range Natl Park, *N.S.Lander 523* (NSW); Gibraltar Range Natl Park, Little Dandahra Ck, near Mulligans Hut picnic area, *R.O.Makinson 1446 et al.* (AD, BRI, CANB, K, MEL, NSW).

Among eastern Australian relatives *G. acerata* is most likely to be confused with *G. sphacelata*, which has consistently subsericeous branchlets and lower leaf surfaces, usually narrowly obovate-elliptic to oblong-elliptic leaves with the lower surface usually exposed, abaxially oriented flowers, the style-end often slightly horned or apiculate, and the pollen-presenter distinctly rimmed (unrimmed in *G. acerata*). *Grevillea occidentalis* from W.A. is very similar; see under that species for differences.

This species is recognised as 'Rare' in J.D.Briggs & J.H.Leigh, *Rare or Threatened Australian Plants* (1995).

246. Grevillea umbellulata Meisn., *in* J.G.C.Lehmann, *Pl. Preiss.* 2: 252 (1848)

T: Swan R., W.A., *J.Drummond, 2nd Coll. 324*; lecto: NY *n.v.*, *fide* D.J.McGillivray & R.O.Makinson, *Grevillea* 446 (1993); isolecto: A *n.v.*, BM, CGE *n.v.*, G, K, LD *n.v.*, LE *n.v.*, MEL, NY *n.v.*, P *n.v.*

G. acerosa F.Muell., *Fragm.* 1: 136 (1859); *G. umbellulata* subsp. *acerosa* (F.Muell.) Olde & Marriott, *Grevillea Book* 1: 184 (1994). T: Salt R., [W.A.], [*G.*] *Maxwell 448*; holo: MEL; iso: K (*Maxwell s.n.*).

Illustrations: P.M.Olde & N.R.Marriott, *Grevillea Book* 3: 212 (top centre & 173A, B, all as subsp. *umbellulata*), 213 (centre left & 174A, B, all as subsp. *acerosa*) (1995).

Shrub 0.3–1.5 m tall. Leaves linear-subterete to narrowly elliptic, 0.7–5.0 cm long, 0.6–3.5 mm wide; margins revolute; upper surface smooth, glabrous or nearly so; lower surface often enclosed including midvein and then 1-grooved, silky to sublanate when exposed. Unit conflorescence subumbelloid, acropetal. Flowers adaxially acroscopic. Flower colour: perianth with white to cream hairs, often tinged pale grey or pink, also with brownish hairs outside especially at base and on limb; style similar, lacking brown hairs. Perianth villous outside and inside, persistent in fruit. Pistil 5.5–8 mm long, villous; style-end lacking a terminal appendage; pollen-presenter ±lateral, broadly elliptic to round in face view, flat-conical with a narrow rim. Follicle narrowly obloid-ellipsoidal to ovoid, 8–12 mm long, loosely villous, faintly ridged.

Occurs in south-western W.A., where widely distributed from Mt Lesueur S and SE to New Norcia, Wongan Hills, Merredin, Hyden, Newdegate and Ravensthorpe, rare towards Wagin,

Busselton and Manjimup, and with a northern occurrence at Port Gregory. Grows in various habitats including Wandoo or Marri woodlands, mallee scrub and low heath, in sandy or gravelly soils on granite or laterite, often in low moist sites. Regenerates from seed and often lignotuber. Flowers July–Nov. Map 316.

W.A.: Kukerin, *M.Koch 2194* (MEL, PERTH); Merredin, *M.Koch 2844* (CANB, K, MEL, NSW); 19.3 km E of Jerramungup, *K.Newbey 480* (PERTH); Waraninooka, Port Gregory, *A.Oldfield 381* (MEL); 3.2 km NE of Wongan Hills, 13 Sept. 1968, *M.E.Phillips CBG040462* (AD, CANB, NSW).

There is considerable variation in habit and leaf orientation, size, margin recurvature, rigidity and pungency. Populations in the N of the range (NW from about Wongan Hills) are fairly consistent in having most leaves > 2.5 cm long, ±pliable and often not pungent, the margins loosely recurved to revolute with the lower surface often partly exposed; the inflorescences with peduncles 2–5 mm long; and a habit to about 1.5 m tall. These are here recognised as the 'northern form', although corresponding phenotypes also occur sporadically in the southern part of the range; the Type collection appears to be assignable to this form. Plants from the Mogumber area fall within this group except in having conspicuously long peduncles (5–30 mm long); these are designated, following Olde & Marriott (*op. cit.* 3: 212 (1995)) as the 'Mogumber form'.

Southern populations with narrow rigid strongly revolute leaves have in the past been grouped under *G. acerosa* F.Muell., more recently ranked as *G. umbellulata* subsp. *acerosa* by Olde & Marriott (*op. cit.* 1: 184 (1994)), who circumscribe it as having a habit 30–50 cm tall, most leaves 1.0–1.8 cm long and linear-subterete with strongly revolute margins enclosing the whole lower surface, pungent leaf apices, and peduncles up to 2 mm long. This 'acerose form' phenotype is distinctive, but intermediates and intergrades are so widespread that its biological separation is doubtful; formal recognition is not followed here. Many populations and individual plants in mixed populations S from Hyden and Dumbleyung do correspond with the circumscription, but many do not, with intermediates and several apparently mixed populations in the Kellerberrin to Meckering area. Further research is definitely warranted.

247. Grevillea occidentalis R.Br., *Trans. Linn. Soc. London* 10: 173 (1810)

G. occidentalis var. *lanceolata* Meisn., *in* J.G.C.Lehmann, *Pl. Preiss.* 1: 539 (1845), as α *lanceolata*, *nom. illeg.* T: Lewins Land near King George Sound, [W.A.], Dec. 1801, *R.Brown Iter Austral. 3321*; lecto: BM, *fide* D.J.McGillivray & R.O.Makinson, *Grevillea* 431 (1993); isolecto: BM, E, G *n.v.*, G-DC, K, MEL, NSW, P *n.v.*

G. occidentalis var. *linearis* Meisn., *in* J.G.C.Lehmann, *Pl. Preiss.* 1: 539 (1845), as β *linearis*. T: '... ab oppido Albany et ... ad radices occidentales montis Wuljenup, Plantagenet' [protologue]; lecto: Mt Wuljenup (Plantagenet), W.A., 14 Oct. 1840, *L.Preiss 713*; lecto: NY *n.v.*, *fide* D.J.McGillivray & R.O.Makinson, *Grevillea* 431 (1993); isolecto: G, G-DC, HBG *n.v.*, LD *n.v.* (*Preiss s.n.*), LE *n.v.*, MEL, P *n.v.*

Illustrations: W.R.Elliot & D.L.Jones, *Encycl. Austral. Pl.* 5: 86 (1990); D.J.McGillivray & R.O.Makinson, *Grevillea* 310 (1993); P.M.Olde & N.R.Marriott, *Grevillea Book* 3: 61 (bottom right), 62 (44A, B) (1995).

Shrub 0.4–1.8 m high. Leaves linear or narrowly elliptic to narrowly ovate or slightly obovate, 0.3–5 cm long, 0.8–10 mm wide; margins recurved to revolute; upper surface granular-scabrid; lower surface often mostly or entirely enclosed, sericeous. Unit conflorescence umbelloid-subglobose, basipetal. Flowers abaxially basiscopic. Flower colour: perianth (hairs of both surfaces) grey or off-white, tinged brown on limb, after anthesis with pinkish ground tissue showing; style with grey to white hairs over pinkish ground tissue. Perianth villous outside, bearded inside. Pistil 8–11 (–13) mm long, villous (less so on ventral side); style-end lacking a terminal appendage; pollen-presenter strongly oblique to almost lateral, round in face view, flat. Follicle ovoid to obloid, 8–11 mm long, loosely tomentose to villous, rugulose.

Occurs in south-western W.A. in the area bounded by Kojonup, Walpole and Albany. Grows in heath and shrubby woodland habitats (Jarrah, Marri or Karri), in sand, clay or humic soils, sometimes in moist situations. Regenerates from seed. Flowers mainly Aug.–Feb. Map 317.

W.A.: near Cranbrook, Dec. 1927, *W.E.Blackall* (PERTH); 6.4 km S of Narrikup on Albany road, *R.Melville 4393 & R.D.Royce* (AD, BRI, K, NSW, PERTH); Kojonup, 24 Sept. 1953, *N.Rainbow NSW93377* (NSW); 25 km N of Walpole, *P.G.Wilson 6334* (AD, B *n.v.*, BRI, H *n.v.*, K, MEL, NSW, PERTH); Albany, *J.W.Wrigley CBG030307* (CANB, NSW).

Populations in the western part of the range have a tendency to narrower, more linear leaves.

Grevillea umbellulata, the only western species likely to be confused with *G. occidentalis*, has the upper leaf surface smooth, and the flowers adaxially oriented (the ventral suture facing towards the rachis axis). *Grevillea acerata* of north-eastern N.S.W. is very similar but has adaxially oriented flowers, leaves more consistently and strongly revolute (lower surface usually 1-grooved), floral bracts 2.5–6 mm long (1.4–2.0 mm rarely to 3.5 mm long in *G. occidentalis*), an obovate pollen-presenter, and has the base of the leaf-mucro yellowish (black in *G. occidentalis*).

248. Grevillea pilulifera (Lindl.) Druce, *Bot. Soc. Exch. Club Brit. Isles* 4(5) suppl. 2: 625 (1917)

Hakea pilulifera Lindl., *Sketch Veg. Swan R.* xxxvi, n. 178 (1840). T: Swan R., [W.A.], 1839, [*J.*] *Drummond s.n.*; holo: CGE; probable iso: FI *n.v.*, G *n.v.*, K, NSW, P *n.v.*

G. oxystigma Meisn., *in* J.G.C.Lehmann, *Pl. Preiss.* 1: 540 (1845); *G. oxystigma* var. *heterophylla* Meisn., *in* J.G.C.Lehmann, *Pl. Preiss.* 1: 540 (1845), as α *heterophylla*, *nom. illeg.* T: ... prope Mahogany Creek (Darlings range), [W.A.] d. 13 Sept. 1839, *L.Preiss 715*; lecto: NY *n.v.*, *fide* D.J.McGillivray & R.O.Makinson, *Grevillea* 434 (1993); isolecto: CGE (as *Preiss 508 = Preiss 715*), G-DC, LD; remaining syntypes: in sublimosis ad fluvium Caning [Canning R., W.A.], July 1841, *L.Preiss 714 p.p.*; syn: G, K, LD *n.v.*, NY *n.v.*; parts of the syntypes, numbers not checked: B *n.v.*, G-DC *n.v.*, HBG *n.v.*, LE *n.v.*

G. oxystigma var. *acerosa* Meisn., *in* J.G.C.Lehmann, *Pl. Preiss.* 1: 540 (1845), as β *acerosa*. T: 'In region. interior. Australiae merid.-occid. Herb. Preiss 710 et 716' [protologue]; lecto: [W.A.], Oct. 1840, *L.Preiss 716*; lecto: NY *n.v.*, *fide* D.J.McGillivray & R.O.Makinson, *Grevillea* 434 (1993); isolecto: G-DC, LD; remaining syntype: [W.A.], *L.Preiss 710*; syn: G-DC, LD *n.v.*, NY *n.v.*; parts of the syntypes, numbers not checked: B *n.v.*, HBG *n.v.*, LE *n.v.*

G. oxystigma var. *tenella* Meisn., *in* J.G.C.Lehmann, *Pl. Preiss.* 1: 540 (1845), as γ *tenella*. T: Hb. Preiss, No. 714, ex parte, vide supra sub α. (Drummond n. 629, et coll. I [protologue]; lecto: [W.A.], *[J.] Drummond [1st coll.] 629*; lecto: NY *n.v.*, *fide* D.J.McGillivray & R.O.Makinson, *Grevillea* 434 (1993); isolecto: A *n.v.*, BM, CGE *n.v.*, G-DC, K, LE *n.v.*; remaining syntype: ... ad fluvium 'Canning' [Canning R.], W.A., 22 July 1839, *L.Preiss 714 p.p.*; LD, *n.v.*, NY (as Preiss s.n.?, *n.v.*); parts of the syntypes, numbers not checked: LD *n.v.*, NY *n.v.*

G. lycopodina S.Moore, *J. Linn. Soc., Bot.* 45: 191 (1920). T: Kauring, near Greenhills, W.A., 1916, *G.W.Brown* [*Hb. Stoward 614*]; holo: BM.

Illustrations: N.G.Marchant *et al.*, *Fl. Perth Region* 1: 332, fig. 112 (1987); D.J.McGillivray & R.O.Makinson, *Grevillea* 316, fig. 80, 317, col. pl. (1993); P.M.Olde & N.R.Marriott, *Grevillea Book* 3: 94 (bottom right), 95 (72A–C) (1995).

Shrub 0.3–1 m high. Leaves linear-subterete to narrowly oblong-elliptic or -oblanceolate, often varying on a plant, 0.3–6 cm long, 0.5–11 mm wide; margins recurved to revolute; upper surface smooth to granulate; lower surface enclosed or exposed and pubescent to sublanate. Unit conflorescence loosely subumbelloid or few-flowered, basipetal. Flowers obscurely abaxially oriented. Flower colour: perianth and style (hairs) white to cream; style-end naked, bright yellow becoming orange and then red. Perianth villous inside and outside. Pistil 6.5–8.5 mm long, villous (style less so ventrally); style-end glabrous, subglobose with a flat 'face'; pollen-presenter lateral, elliptic, flat with a prominent conical stigma. Follicle obloid, 8–13 mm long, pubescent with loose villous overlay, granulose. *Woolly-flowered Grevillea.*

Occurs in south-western W.A., widespread from Badgingarra S to Busselton and Albany, common in the central Darling Ra. Grows in sclerophyll heath and eucalypt woodland in granitic and lateritic gravels. Regenerates from seed and (at least some populations) from lignotuber. Flowers Apr.–Dec. Map 318.

W.A.: c. 1 km S of Wansborough between Tambellup and Cranbrook, *Hj.Eichler 15949* (AD); 16 km NE of Wagin, *C.A.Gardner 1490 & 1990* (PERTH); Woorooloo, *M.Koch 1386* (A *n.v.*, AD, BRI, NSW, P *n.v.*,

PERTH); 160 km N of Perth on Geraldton Hwy, *R.Melville 4098* (BRI, MEL, NSW, PERTH); 11.3 km SW of Tammin, *K.Newbey 1960* (PERTH).

Leaves are highly variable in shape, size and posture; there is also variation in the position of inflorescences. Olde & Marriott (*op. cit.* 95 (1995)) distinguish three phenotypic 'forms', although there is variation within each, and mixed populations occur. Their 'typical (type) form' occurs mainly N from Perth, and has leaves usually all of about the same size and form on a plant, ±ascending, ⩽ 1 cm long, oblong with strongly revolute margins enclosing most or all of the lower surface, and the branches subcolumnar with the conflorescences borne on short lateral branchlets towards the apices of the larger branches. The 'broad-leaf form' often has a mixture of leaf sizes and shapes on a plant but has at least some leaves 1–6 cm long, 4–11 mm wide, with most of the lower surface exposed, and conflorescences variably serially columnar (terminal on short lateral branchlets along older axes), or in short terminal aggregations; it occurs along the Darling Ra. both N and S of Perth, and intergrades and mixed stands with the 'typical form' are known. There may be no real biological distinction between these previous two 'forms', and the informal name of the first may not be appropriate—a probable isotype of the species, at K, has both broad and narrow leaves.

The 'southern form' occurs mainly in the area bounded by Cranbrook, Borden and Wagin; it often has erect, antrorsely appressed, oblong to linear-subterete leaves ⩽ c. 1 cm long, and few-flowered conflorescences concentrated at the apices of the main branches only (lacking serial short lateral flowering branchlets). The type of *G. lycopodina* S.Moore from Kauring (SE of York) appears to be an extreme example of this form, with very small strongly appressed leaves with long apical spines, and there is a similar collection from 'Coolgardie district' at PERTH. The 'southern form' may warrant formal ranking after further study.

Grevillea pilulifera is very closely related and similar to *G. uncinulata* and *G. florida*. *Grevillea uncinulata* has the style-end papillose, and the leaves usually clustered on short lateral branchlets (in *G. pilulifera* usually alternate and evenly spaced along the main branchlets). *Grevillea florida* has a papillose style-end with a slight apical 'brow' overtopping the pollen-presenter. *Grevillea scabra* and *G. candolleana*, which are also related, have very pronounced terminal appendages on the style. This whole cluster of species requires further research.

249. Grevillea uncinulata Diels, *Bot. Jahrb. Syst.* 35: 152 (1904)

T: Tammin, W.A., 26 July 1901, *E.Pritzel 475* (*F.L.E.Diels 3563*); lecto: B, *fide* D.J.McGillivray & R.O.Makinson, *Grevillea* 446 (1993); isolecto: A *n.v.*, AD, B *n.v.*, BM, E, G, K, NSW, P *n.v.*

G. oxystigma var. *villosa* Benth., *Fl. Austral.* 5: 466 (1870). T: near Belgarup, W.A., *s.d.*, *A.Oldfield*; lecto: K, *fide* D.J.McGillivray & R.O.Makinson, *Grevillea* 446 (1993); remaining syntype: between Perth and K.G. [King Georges] Sound, W.A., April/July 1854, *W.H.Harvey*; syn: K.

Illustrations: D.J.McGillivray & R.O.Makinson, *Grevillea* 319 (1993); P.M.Olde & N.R.Marriott, *Grevillea Book* 3: 214 (top left & 175A, B) (1995); all as *G. uncinulata* subsp. *uncinulata*.

Shrub 0.1–0.3 (–0.5?) m high. Leaves usually in short fascicular clusters, linear-subterete to elliptic or narrowly so, mostly 0.2–1 (–3.5) cm long, 1–2.5 (–6) mm wide; margins refracted; upper surface smooth or granulose; lower surface usually enclosed and 1-grooved, rarely on older growth exposed and glabrous or cobwebby to sublanate. Unit conflorescence umbelloid or a few-flowered cluster, basipetal. Flowers abaxially oriented. Flower colour: perianth and style (hairs) white; style-end yellow, orange or red. Perianth villous outside and inside. Pistil 7–10 mm long, villous (less so ventrally on style); style-end obliquely clavate, densely papillose on dorsal side (occasionally also on ventral side); pollen-presenter lateral, round to elliptic, flat to slightly convex, with marginal rim and prominent narrowly conical stigma; distal edge of pollen-presenter level with or slightly exceeding apex of papillose style-end, not overtopped by it. Follicle obloid-ellipsoidal to slightly obovoid, c. 8 mm long, loosely villous with underlying pubescence, surface faintly ridged. Plate 51.

Occurs in south-western W.A., widespread from the Hill R. (Badgingarra area) S and SE to the Stirling Ra. and Ravensthorpe. Grows in heath shrubland or shrubby woodland in sandy to gravelly lateritic soils. Regenerates from seed and (at least in some populations) from lignotuber. Flowers May–Nov. Map 319.

W.A.: Burabadji, c. 130 km NE of Perth, *A.M.Ashby 1476a* (AD, CANB); 4.2 km NW of Wongan Hills towards Piawaning, *R.Coveny 7802 & B.R.Maslin* (K, NSW); near road from Wagin to Katanning, *Hj.Eichler 15888* (AD, B *n.v.*, G *n.v.*, L *n.v.*, W *n.v.*); sources of the Hill R., N from Dandaragan, *C.A.Gardner 9003* (PERTH); 26 km by road E of Lake Grace P.O. towards Lake King, *D.J.McGillivray 3545 & A.S.George* (B *n.v.*, K, NSW, PERTH, US *n.v.*).

Populations in the Hill R. area tend to have some or most leaves at the large end of the range (2–3.5 cm long, up to 6 mm wide); these constitute a 'larger-leaved form'. Populations elsewhere usually have leaves 1–1.5 cm long and 1–2 mm wide. Material from the Coomallo area needs further study.

Grevillea uncinulata has floral bracts 0.4–0.6 mm long, and the dorsal tepals lack a small auricular lobe on the posterior edge just below the limb-cup. These features assist in distinguishing it from the very similar *G. florida*; see notes under that species.

250. **Grevillea florida** (McGill.) Makinson, *Fl. Australia* 17A: 504 (2000)

G. uncinulata subsp. *florida* McGill., *New Names Grevillea* 15 (1986). T: 74 mile peg on Perth to New Norcia road [c. 118 km N of Perth], W.A., 11 Sept. 1963, *F.G.Smith 1732*; holo: PERTH.

Illustrations: P.M.Olde & N.R.Marriott, *Grevillea Book* 3: 214 (centre right), 215 (176A, C) (1995), as *G. uncinulata* subsp. *florida*.

Low shrub to 1 m high. Leaves linear-subterete to narrowly elliptic, 0.5–3 cm long, 1–3 (–6) mm wide, rarely a few clustered together; margins smoothly revolute or recurved; upper surface smooth to granulose; lower surface enclosed and 1-grooved, or exposed and glabrous or sublanate. Unit conflorescence subumbelloid or few-flowered, basipetal. Flowers abaxially oriented. Flower colour: perianth and style (hairs) white to cream-yellow; style-end bright yellow or orange, deepening to red after anthesis. Perianth villous outside and inside. Pistil 7–10 mm long, villous (less so ventrally on style); style-end obliquely clavate, densely papillose dorsally and also ventrally below base of pollen-presenter; pollen-presenter lateral, round to elliptic, flat to slightly convex, with a prominent narrowly conical stigma; distal edge of pollen-presenter narrowly but distinctly overtopped by a slight apical extension of papillose style-end. Follicle obloid-ellipsoidal to slightly -obovoid, c. 13 mm long, pubescent with loose villous overlay, faintly ridged.

Occurs in south-western W.A., from just N of Perth NE to the Wannamal to New Norcia area and (not seen) N to Badgingarra and E to Yoting. Grows in heath shrubland or shrubby woodland in sandy to gravelly lateritic soils. Regenerates from seed. Flowers Aug.–Dec. Map 320.

W.A.: 120 km from Perth towards New Norcia along Geraldton Hwy, *E.M.Canning CBG057379* (CANB, NSW, PERTH); 16 km N of Perth on Perth to Moora road, *H.Davis 54* (AD); c. 115.9 km N of Perth, *C.H.Gittins 1731* (BRI, K, NSW, PERTH); Great Northern Hwy, 3.6 km N of Calingiri turnoff, *S.Patrick 309* (K, PERTH).

Grevillea florida is narrowly distinct from *G. uncinulata*, but the differences seem adequate to justify species ranking. *Grevillea florida* has leaf apices with a straight to slightly deflexed point; leaf margins ±smoothly revolute, or on lower leaves smoothly recurved, often with some or most of the lower surface exposed; pedicels 8–14 mm long; floral bracts 0.5–2.0 mm long; the dorsal tepals each with a small auricular lobe on the posterior edge just below the limb-cup; style-end with a dense covering of minute simple papilloid hairs on both dorsal side and (below base of pollen-presenter) on ventral side; and the distal edge of pollen-presenter narrowly (by about 0.3–0.4 mm) but distinctly overtopped by the extended apex of the papillose style-end.

By contrast, *G. uncinulata* has most leaves tightly clustered on short lateral branchlets; leaf apices always with a strongly deflexed point; leaf margins angularly and tightly revolute, only rarely some undersurface exposed on lower leaves; floral bracts 0.4–0.6 mm long; pedicels 4–6(–9) mm long; dorsal tepals lacking posterior auricular lobes below the limb-cups; style-end with papillae usually restricted to the dorsal side; and the distal edge of pollen-presenter level with or slightly exceeding (not overtopped by) the apex of the style-end.

Grevillea uncinulata has ?mature fruits c. 8 mm long; *G. florida* has fruits up to 13 mm long. It is not clear from the small sample available whether there is a consistent difference. The general ranges of the two taxa overlap. No mixed populations or intermediates are known.

Grevillea florida is recognised, under *G. uncinulata* subsp. *florida*, as 'Poorly Known' in J.D.Briggs & J.H.Leigh, *Rare or Threatened Australian Plants* (1995).

251. **Grevillea candolleana** Meisn., *in* J.G.C.Lehmann, *Pl. Preiss.* 1: 541 (1845)

T: Swan R., W.A., *J.Drummond [prob. 1st coll.] 628*; lecto: NY *n.v.*, *fide* D.J.McGillivray & R.O.Makinson, *Grevillea* 410 (1993); isolecto: A *n.v.*, CGE *n.v.*, E *n.v.*, G *n.v.*, ?G-DC, K, MEL, P; remaining syntypes: ad flumen Cygnorum [Swan R. colony, W.A.], *L.Preiss 2625*; syn: not found.

Illustrations: P.M.Olde & N.R.Marriott, *Grevillea Book* 2: 84 (centre right & 66A), 85 (66B, C) (1995).

Shrub 0.2–0.8 m high. Leaves narrowly ovate to narrowly elliptic or linear, 1–3.5 cm long, 1–9 mm wide; margins shortly recurved to revolute; upper surface villous when young, later glabrous and smooth or very finely granulose; lower surface exposed, felty-pubescent. Unit conflorescence subumbelloid or few-flowered, basipetal. Flowers abaxially oriented. Flower colour: perianth and style (hairs) white to cream; glabrous stylar appendage yellow, deepening to orange and red with age. Perianth villous outside and inside; dorsal tepals overtopping limb. Pistil 9.5–11.5 mm long, villous; style-end with a tongue-shaped, incurved, apically emarginate appendage 2–3 mm long, overtopping pollen-presenter, the appendage basally papillose becoming glabrous in upper half; pollen-presenter lateral, broadly elliptic, slightly concave. Follicle ellipsoidal or ovoid, 8.5–11 mm long, pubescent, ridged along flanks. *Toodyay Grevillea.*

Occurs in south-western W.A. where restricted to the Toodyay area NE of Perth, with one uncorroborated record from Kalamunda near Perth. Grows in *Eucalyptus wandoo* woodland on lateritic loam. Regeneration mode unknown. Flowers Aug.–Nov.? Map 321.

W.A.: 67.5 km from Perth towards Toodyay, *E.M.Canning 2798* (CANB, NSW); 9 km from Toodyay P.O. on Red Hill Rd, *D.J.McGillivray 3457 & A.S.George* (NSW); District Avon, Aug. 1901, *E.Pritzel 537* (K); Toodyay, *R.D.Royce 4313* (PERTH); Kalamunda, received Feb. 1916, *F.Stoward 650* (K).

Grevillea candolleana is similar to *G. scabra*, which has a scabrid upper leaf surface, a strongly incurved style (±straight in *G. candolleana*) with the terminal appendage also more strongly recurved, and the base of the pollen-presenter jutting outward from the style (concurrent with style in *G. candolleana*). *Grevillea uncinulata* and *G. florida* are superficially similar but lack the conspicuous stylar appendage.

This species is recognised as 'Poorly Known' in J.D.Briggs & J.H.Leigh, *Rare or Threatened Australian Plants* (1995).

252. **Grevillea scabra** Meisn., *in* J.G.C.Lehmann, *Pl. Preiss.* 1: 541 (1845)

T: Swan R., W.A., *J.Drummond 1st coll. 627*; lecto: NY *n.v.*, *fide* D.J.McGillivray & R.O.Makinson, *Grevillea* 440 (1993); isolecto: A *n.v.*, B *n.v.*, BM, G, K, LE *n.v.*, MEL, P *n.v.*, PERTH.

Illustrations: D.J.McGillivray & R.O.Makinson, *Grevillea* 320, fig. 81 (1993); P.M.Olde & N.R.Marriott, *Grevillea Book* 3: 160 (top centre & 129A–C) (1995).

Shrub to c. 1 m tall with ascending branches. Leaves clustered on short lateral branchlets, narrowly elliptic to linear-subterete, 0.5–3.5 cm long, 1–6 mm wide; margins revolute; upper surface soon glabrous and granular-scabrid; lower surface often enclosed and 1-grooved, when exposed then loosely villous to sublanate. Unit conflorescence subumbelloid, opening order indeterminate, 2–6-flowered, terminal on short lateral branchlets. Flowers abaxially oriented. Flower colour: perianth and style (hairs) white to cream; style-end yellow, aging to reddish. Perianth villous outside and inside. Pistil 8–13.5 mm long, villous; style-end with a conspicuous subspathulate strongly curved terminal appendage (C-shaped in side view) 1.5–3 mm long, arching over pollen-presenter; appendage dorsally papillose to apex; pollen-presenter projecting from line of style, broadly oblong. Follicle obloid-ellipsoidal, 10–13 mm long, pubescent to hispid. *Rough-Leaved Grevillea.* Plate 49; Fig. 30D–F.

Figure 30. *Grevillea*. **A–C**, *G. albiflora*. **A**, flowering branch; **B**, flower; **C**, pistil (**A–C**, G.H.Allen 74, CANB). **D–F**, *G. scabra*. **D**, flowering branch; **E**, flower bud; **F**, pistil and half perianth (**D–F**, S.D.Hopper 5992, PERTH). **G–J**, *G. georgeana*. **G**, flowering branch; **H**, leaf; **I**, flower; **J**, pistil (**G–J**, K.Newbey 8875, PERTH). Scale bars: **A**, **D**, **G–J** = 1 cm; **B–C** = 5 mm; **E–F** = 3 mm. Drawn by: **A–C**, D.Boyer; **D–F**, D.Mackay; **G–J**, D.Fortescue.

Occurs in south-western W.A., NE of Perth, rare in the area bounded by York, Toodyay, Bolgart and near Goomalling; also one old record from Lake Wagin. Grows in open eucalypt forest or woodland on rises in lateritic loam. Regeneration mode uncertain, regenerates at least from seed. Flowers c. Oct. Map 322.

W.A.: near L. Wagin, 1895, *M.Cronin* (MEL); near Bolgart, *C.A.Gardner 8693* (PERTH).

See under *G. candolleana* for differences.

This species is recognised as 'Poorly Known' in J.D.Briggs & J.H.Leigh, *Rare or Threatened Australian Plants* (1995).

Longistyla Group

Shrubs. Leaves divided or toothed or rarely entire, dorsiventral; surfaces dissimilar; margins recurved to revolute. Conflorescence erect or rarely spreading or decurved, terminal or axillary, simple or sparingly branched; unit conflorescence subglobose to cylindrical or occasionally semi-secund or 1–3-flowered, basipetal or rarely acropetal. Flowers basiscopic or acroscopic or irregular. Torus oblique. Perianth zygomorphic, glabrous or hairy outside, hairy inside; tepals remaining coherent and held ventrally. Pistil 11–52 mm long; ovary hairy, stipitate; style glabrous or hairy, exserted (sometimes weakly) from late bud; pollen-presenter very oblique to lateral, convex to slightly concave. Follicle usually loosely villous with hairs falling easily; pericarp c. 1 mm thick, bony textured. Seed biconvex, narrowly winged all around; inner face with central ridge and intramarginal channel.

A group of 11 species, occurring mostly in the south-west of W.A., with two species in eastern Australia. All bird-pollinated. Closest affinities appear to be to the *Floribunda* Group.

In this group, except in *G. fulgens*, the stipe base is adnate to the torus, then sharply inflexed at dorsal margin of torus.

1 All unit conflorescences 1–3-flowered

2 Perianth limb glabrous outside; adult leaves deeply pinnatipartite (almost pinnatisect) with short regular lobes **263. G. involucrata**

2: Perianth limb hairy outside; adult leaves entire or with few teeth **262. G. fulgens**

1: Most or all unit conflorescences ≥ 4-flowered

3 Leaves deeply and divaricately divided

4 Outer perianth surface hairy **258. G. dissecta**

4: Outer perianth surface glabrous

5 Pistil 20–27 mm long; floral rachis 15–70 mm long; style usually sparsely villous near base, or glabrous; pedicels 9–11 (–14) mm long **257. G. georgeana**

5: Pistil 30–36 mm long; floral rachis 4–16 (–30) mm long; style with at least some hairs persisting to near apex; pedicels 12–26 mm long **256. G. wilsonii**

3: Leaves (when divided) with lobes not divaricate

6 Simple leaves, or lobes of divided leaves, linear

7 Leaves ≤ 12 cm long, all divided; longest leaf lobes < 8 cm long; style loosely villous for more than half its length **255. G. erectiloba**

7: Leaves usually > 12 cm long, all divided or some or all simple and entire; longest ultimate lobes (or simple leaves) usually > 8 cm long; style subsericeous in basal few millimetres, glabrous above

8 Lower surface of leaves or leaf lobes with lamina on either side of midvein usually enclosed by revolute margins; outer surface of perianth with reddish biramous hairs only (N.S.W.) **254. G. johnsonii**

8: Lower surface of leaves or leaf lobes with lamina on either side of midvein usually narrowly but clearly exposed; outer surface of perianth usually with both pale biramous and simple erect glandular hairs (Qld) **253. G. longistyla**

6: Leaves or leaf lobes not linear

9 Upper surface of leaves glabrous **261. G. insignis**

9: Upper surface of leaves pubescent with mainly short, erect, simple, glandular hairs

10 Leaves nearly sessile, star-like with palmate division; conflorescences mostly 4–6-flowered **260. G. asteriscosa**

10: Leaves clearly petiolate, dentate to pinnatifid division; conflorescences mostly ⩾ 8-flowered **259. G. pilosa**

253. Grevillea longistyla Hook., *in* T.L.Mitchell, *J. Exped. Trop. Australia* 343 (1848)

T: sub-tropical New Holland [Qld, Mt Pluto area], 1846, *T.L.Mitchell 316*; lecto: K, *fide* D.J.McGillivray & R.O.Makinson, *Grevillea* 427 (1993); isolecto: CGE *n.v.* Remaining syntypes: 'Mitchell 1846. 175'; syn: MEL; 'Mitchell 1846. 400'; syn: K; (with unnumbered Mitchell duplicates at B, CGE, NSW, TCD).

G. neglecta R.Br., *in* C.Sturt, *Narr. Exped. C. Australia* 2: app. 87 (1849). T: none cited; BM? *n.v.*

Illustrations: K.A.W.Williams, *Native Pl. Queensland* 1: 137 (1982); S.Pearson & A.Pearson, *Pl. Central Queensland* 209 (1990); P.M.Olde & N.Marriott, *Grevillea Book* 2: 246 (top left, 205A, B), 247 (205C) (1995).

Shrub 1.5–5 m tall, sometimes several stems from a lignotuber. Leaves 12–30 cm long, entire or 2–6-partite with ascending lobes to 17 cm long; simple leaves and lobes linear, (1–) 1.5–4 mm wide, not pungent; margins recurved to revolute; lower surface subvillous, usually with a narrow strip of lamina exposed on either side of midvein. Conflorescence erect, usually terminal, simple to few-branched; unit conflorescence basipetal, loosely cylindrical (sometimes shortly so); ultimate rachis (5–) 25–85 (–125) mm long. Flowers basiscopic. Flower colour: perianth red to orange-red or bright pink, occasionally orange-pink or creamy pink; style red. Perianth with simple, erect, glandular hairs outside, usually mixed with pale appressed to ascending biramous hairs, sometimes the latter only, subvillous inside above ovary. Pistil (28–) 40–52 mm long; ovary villous; style glabrous except for appressed hairs in basal 1–5 mm. Follicle obloid-ellipsoidal with a short apiculum, 10–14 mm long, rugulose; indumentum open to sparse, persistent. Plate 50.

Occurs in inland south-eastern and central Qld, from Gurulmundi and Chinchilla area north to Blackdown Tableland near Emerald. Grows usually in eucalypt woodland or forest in poor sandy soils, often on sandstone. Regenerates from seed and sometimes a lignotuber. Flowers year round, mainly Aug.–Nov. Map 323.

Qld: Blackdown Tableland, c. 35 km SE of Blackwater, *R.J.Henderson 1057* (BRI, CANB, K, NSW, PERTH); Eisvold–Mundubberra road, 14.5 km SE of Eidsvold, *R.W.Johnson 2834* (BRI, CANB); 25 km N of Jandowae on road to Durong, *R.O.Makinson 261* (K, NSW); 33.8 km SE of Bedourie HS, *N.H.Speck 1844* (CANB, K, NSW); Planet Ck, c. 48 km NE of Rolleston Township, *R.Story & Yapp 296* (CANB, K).

Grevillea longistyla is very similar to *G. johnsonii*, which is non-lignotuberous, has a shorter floral rachis (0.4–1.2 cm long), lacks glandular hairs on the perianth, and has the leaf lower surface on either side of the midveins always completely enclosed by the margins.

A 'Blackdown Tableland form' lacks glandular hairs on the conflorescences, and has a higher frequency of pale-coloured flowers. A collection with pistils c. 28 mm long from near Jandowae may or may not represent a small-flowered population.

Ra.). Grows in open shrub associations in well-drained situations in shallow stony soils on ironstone. Regenerates from seed. Flowers (Mar.–) July–Oct. (–Dec.?). Map 327.

W.A.: Mt Manning Ra., *A.S.George 15117* (PERTH); northern hills of Helena and Aurora Ra., NE of Koolyanobbing, *G.J.Keighery 4434* (NSW); on N side of Die Hardy Ra., c. 140 km directly N of Southern Cross, *D.J.McGillivray 3672 & A.S.George* (K, NSW); Finnerty, c. 60 km ENE of Koolyanobbing, *K.Newbey 8875* (PERTH); Die Hardy Ra., 28 Sept. 1975, *B. & M.Smith* (PERTH).

Collections from Mt Manning Ra. have pistils shorter (20.5–23 mm long) than other material, a shorter and more open indumentum on the inner surface of the perianth, longer pedicels (11–15 mm long) and may lack hairs on the style. *Grevillea georgeana* is similar to *G. wilsonii*, which has shorter subumbelloid conflorescences (rachises 4–30 mm long) with pedicels 12–26 mm long (9.5–14 mm long, and usually not more than 11 mm, in *G. georgeana*), the perianth (including limb) all red, the pistil 30–36 mm long, and the style more hairy.

This species is recognised as 'Poorly Known' in J.D.Briggs & J.H.Leigh, *Rare or Threatened Australian Plants* (1995).

258. Grevillea dissecta (McGill.) Olde & Marriott, *Nuytsia* 9: 282 (1993)

G. pilosa subsp. *dissecta* McGill., *New Names Grevillea* 12 (1986). T: Lake Barker area, W.A., 13 Feb. 1973, *W.H.Butler*; holo: PERTH (specimen at lower right of sheet only).

Illustrations: D.J.McGillivray & R.O.Makinson, *Grevillea* 370 (1993), as *G. pilosa* subsp. *dissecta*; P.M.Olde & N.Marriott, *Nuytsia* 9: 284, fig. 15A–I (1993); P.M.Olde & N.Marriott, *Grevillea Book* 2: 130 (top left & 104A–C) (1995).

Low dense shrub to 1 m tall. Leaves 1–2 cm long, 20–40 mm wide, divaricate, deeply 3–9-partite with primary lobes usually again 3-partite, sometimes with some similar third-order division; ultimate lobes linear to subulate, 7–15 mm long, 1–1.5 mm wide, pungent; margins revolute, enclosing lower surface except for midveins of leaf and lobes; lower surface bisulcate, silky in grooves. Conflorescence erect; terminal, usually simple, ?basipetal, a loose, often semisecund cluster; rachis 2–7 mm long. Flowers acroscopic. Flower colour: perianth rose pink with a white to cream limb; style pinkish red. Perianth tomentose or sparsely villous outside with mixed biramous and simple, erect, glandular hairs, bearded inside. Pistil 18–20 mm long; ovary villous; style loosely villous. Follicle subglobose to ellipsoidal, 10–12 mm long, sparsely villous with hairs easily caducous.

Occurs in the inland south-west of W.A., between Moorine Rock and Lake Barker and south to Mt Holland. Grows in mallee shrubland and heath in sandy or gravelly loam soils. Regenerates from seed. Flowers ?Sept.–Feb. Map 328.

W.A.: Southern Cross [area], SE of Edwards Find L3, *E.A.McKinnon S1453* (KPBG); Site M, Mt Holland area, *W.Martinick & K.Tinley 3* (PERTH); 54 km N of L. Cronin, *P.Olde 86/876* (NSW *n.v.*).

Grevillea dissecta is florally very similar to *G. pilosa* and was formerly included as a subspecies of the latter by McGillivray (*loc. cit.* (1986)) and McGillivray & Makinson (*loc. cit.* (1993)), before being raised to species rank by Olde & Marriott (*loc. cit.* (1993)). *Grevillea dissecta* has branchlets subsericeous with biramous hairs only, deeply dissected leaves with divaricate lobes, upper leaf surface glabrous, and floral rachises and pedicels subsericeous to almost glabrous. *Grevillea pilosa* differs in having the branchlets tomentose with mixed biramous and simple glandular hairs, relatively shallowly divided leaves with the teeth or lobes not divaricate; the upper surface of the leaf with an open indumentum of erect simple glandular hairs; and floral rachises and pedicels subvillous. *Grevillea wilsonii* and *G. georgeana* are superficially similar but have the outer surface of the perianth glabrous.

This species is recognised as 'Rare' in J.D.Briggs & J.H.Leigh, *Rare or Threatened Australian Plants* (1995).

259. Grevillea pilosa A.S.George, *W. Austral. Naturalist* 10: 32 (1966)

G. rufa C.A.Gardner, *J. Roy. Soc. W. Australia* 10: 32 (1966), *nom. illeg. non* (Warb.) Sleumer (1939). T: gravelly rises, Pallarup Rocks, W.A., Sept. 1930, *C.A.Gardner*; lecto: PERTH, *fide* D.J.McGillivray & R.O.Makinson, *Grevillea* 434 (1993); isolecto: PERTH (2 sheets).

Dense shrub 0.5–2 m tall. Leaves cuneate-obovate to oblong-ovate, 1–5.5 cm long, (5–) 10–60 mm wide, dentate to pinnatifid with 3–19 marginal teeth or lobes, sometimes these with shallow secondary division, rarely a few leaves entire and narrowly obovate; teeth or lobes pungent, subtriangular, 2–7 mm long, 2–4 mm wide; margins recurved to shortly revolute; lower surface mostly exposed, loosely to densely subsericeous. Conflorescence erect to spreading or decurved, terminal or axillary, usually simple, weakly basipetal, loosely subglobose or semi-secund; rachis 2–7 mm long. Flowers acroscopic. Perianth outer surface tomentose below limb with mixed biramous and simple glandular hairs, villous on limb with spreading rusty biramous hairs; inner surface pubescent to pilose. Pistil 20–27 mm long; ovary villous; style villous. Follicle subglobose to obloid-ovoid, c. 12 mm long, loosely villous with caducous biramous hairs and more persistent simple glandular hairs.

The upper surface of the leaf has a distinctive open indumentum mainly of minute, erect, simple, glandular hairs, a feature shared among related species only with *G. asteriscosa*. *Grevillea pilosa* differs from *G. astericosa* in its clearly petiolate leaves (nearly sessile in *G. astericosa*). For differences from *G. dissecta*, see notes under that species.

McGillivray (*loc. cit.* (1986)) and McGillivray & Makinson (*loc. cit.* (1993)) included in *G. pilosa* his subsp. *dissecta* McGill., here recognised at species rank as *G. dissecta*. Following Olde & Marriott (*loc. cit.* (1993); *loc. cit.* (1995)), two narrowly distinct subspecies are recognised within the remaining circumscription of *G. pilosa*.

Perianth 3.5–6.5 mm wide; limb of bud 3–4 mm wide; outer surface of tepals densely villous; most leaves > 20 mm wide, with base shortly cuneate to truncate, and lobing shallow **259a.** subsp. **pilosa**

Perianth 2.5–3 mm wide; limb of bud 2.5–3 mm wide; outer surface of tepals often loosely villous; most leaves < 15 mm wide with a long-cuneate base and shallow division, some up to 25 mm wide with deeper division **259b.** subsp. **redacta**

259a. Grevillea pilosa A.S.George subsp. pilosa

Illustrations: D.J.McGillivray & R.O.Makinson, *Grevillea* 369, fig. 91 & col. pl. (1993); P.M.Olde & N.Marriott, *Nuytsia* 9: 13, fig. 13 (1993); P.M.Olde & N.Marriott, *Grevillea Book* 3: 93 (top centre & 70A–C) (1995).

Shrub 0.5–2 m tall. Branchlets tomentose to subvillous with mainly biramous hairs intermixed with erect glandular hairs. Leaves 15–60 mm wide, broadly cuneate-obovate to oblong-ovate, shortly cuneate to truncate at base, with 3–11 (–19) teeth or shallow triangular lobes. Flower colour: perianth pale pink to red; style red; both with red or rusty brown hairs. Perianth 3.5–6.5 mm wide; limb of bud 3–4 mm wide; outer surface of tepals densely tomentose, becoming villous on limb.

Occurs in inland south-western W.A., in the area Newdegate–Mt Holland–Ravensthorpe of the southern wheatbelt. Grows in mallee shrubland or heath associations in various soils including granitic loam and sand over laterite. Regenerates from seed. Flowers June–Dec. Map 329.

W.A.: Lake King to Newdegate road, 38.6 km E of Newdegate, *R.Filson 9375* (MEL); towards Diggers Rocks, 9 Dec. 1964, *C.A.Gardner s.n.* (PERTH); near Pallarup, next to Mt Short, *C.A.Gardner 14812* (PERTH); 130 km W of Daniell, *R.H.Kuchel 1810* (AD, L *n.v.*, PERTH); 18.5 km by road E of Newdegate P.O., *D.J.McGillivray 3561 & A.S.George* (K, NSW, PERTH, US *n.v.*).

259b. Grevillea pilosa subsp. redacta Olde & Marriott, *Nuytsia* 9: 280 (1993)

T: 13 km N of Lake Cronin, W.A., 10 Dec. 1964, *C.A.Gardner 15915*; holo: PERTH.

Illustrations: P.M.Olde & N.Marriott, *Nuytsia* 9: 281, fig. 14 (1993); P.M.Olde & N.Marriott, *Grevillea Book* 3: 94 (top centre & 71) (1995).

Shrub c. 2 m tall. Branchlets tomentose with mainly biramous hairs intermixed with a few erect glandular hairs. Leaves (5–) 10–15 (–25) mm wide, usually with most leaves narrowly obovate with a long-cuneate base and 3–5 shallow teeth (sometimes a few leaves entire and narrowly obovate), and some broader leaves up to 25 mm wide with 5–12 deeper linear to narrowly triangular lobes. Flower colour: perianth rose-pink; style pinkish red; both with pale hairs. Perianth 2.5–3 mm wide; limb of bud 2.5–3 mm wide; outer surface of tepals sparsely to moderately densely tomentose, becoming villous on limb.

Occurs in the inland south-west of W.A., where confined to a small area from just north of Lake Cronin to a little N of Mt Holland. Grows in mallee shrubland or heath, and is reported from gravelly rises in brown loam, and in sand over laterite. Regenerates from seed. Flowers Aug.–Dec. Map 330.

W.A.: Mt Holland, *J.S.Beard 3870* (PERTH); Holts Rock, *C.Davies 110* (K, PERTH); 8 km N of Mt Holland, Dec. 1964, *A.R.Main s.n.* (PERTH); 11.4 km N of Crossroads near L. Cronin on road to Mt Holland, *P.Olde 86/782* (NSW *n.v.*).

The distinctions from the type subspecies are very narrow, and some overlap is to be expected on most features.

This subspecies is recognised as 'Poorly Known' in J.D.Briggs & J.H.Leigh, *Rare or Threatened Australian Plants* (1995).

260. Grevillea asteriscosa Diels, *Bot. Jahrb. Syst.* 35: 151 (1904)

T: 100 miles nördlich vorn Stirling Range, W.A., [1879], leg. [T.] Muir (*L.Diels 7553*); holo: B *n.v.*; iso: K, MEL.

Illustrations: R.Erickson *et al.*, *Fl. & Pl. W. Australia* 120, pl. 366 (1973, 1979); D.J.McGillivray & R.O.Makinson, *Grevillea* 371, col. pl. & fig. 92 (1993); P.M.Olde & N.Marriott, *Grevillea Book* 2: 43 (bottom left & 30A–C) (1995).

Shrub 0.3–2 m tall. Leaves subsessile, star-shaped, 5–10 mm long, 8–18 mm wide, palmately divided with 3–9 spreading, rigid, pungent, ±triangular lobes (3–) 5–8 mm long, 1.5–3 mm wide; margins recurved; lower surface mostly exposed, glabrous to pilose, with simple erect glandular hairs along veins. Conflorescence erect, terminal, simple, weakly basipetal, a loose (2–) 4–10-flowered cluster; rachis 1–3 mm long. Flowers ±acroscopic. Flower colour: perianth and style bright red. Perianth openly pubescent outside with glandular hairs or sometimes glabrous, bearded inside opposite ovary. Pistil 15.5–19.5 mm long; ovary appressed-villous; style pilose to subsericeous on dorsal side. Follicle obloid-ellipsoidal, 9–12 mm long, slightly ridged dorsally, loosely tomentose. *Star-leaf Grevillea.*

Occurs in the south-west of W.A. in the central wheatbelt, from c. 15 km north of Muntadgin to near Pingaring and to c. 15 km east of Bullaring. Grows in heath or tall shrub associations in gravelly (often granitic) soils. Regenerates from seed. Flowers May–Nov. Map 331.

W.A.: Bruce Rock district, July 1932, *E.T.Bailey* (PERTH); 15 km N of Muntadgin, *R.H.Kuchel 2051* (AD, L *n.v.*, PERTH); 16 km E of Bullaring, *F.Lullfitz 1780* (KPBG, PERTH); Int[erior] SW Australia, *s.d.*, *J.S.Roe* (K); N of Narembeen, *R.D.Royce 6671* (CANB, K, NBG *n.v.*, PERTH).

Similar to *G. pilosa*, see under that species for details.

This species is recognised as 'Rare' in J.D.Briggs & J.H.Leigh, *Rare or Threatened Australian Plants* (1995).

261. Grevillea insignis Kippist ex Meisn., *Hooker's J. Bot. Kew Gard. Misc.* 7: 76 (1855)

T: Swan River Colony, W.A., *J.Drummond coll. 5, Suppl. n. 12*; holo: K; iso: BM, K, MEL, NSW, NY *n.v.*, P *n.v.*, TCD *n.v.*

Shrub 1–4 m high. Leaves ±oblong in general outline, 2.5–9 cm long, 2–4 cm wide, serrato-dentate to shallowly pinnatifid with 7–17 pungent subtriangular teeth or lobes 3–8 mm long, 3–8 mm wide; margins ±flat; lower surface fully exposed, glabrous, glaucous. Conflorescence erect to decurved, terminal, simple or few-branched; unit conflorescence weakly basipetal, subglobose to loosely and shortly subcylindrical; ultimate rachis 10–25 mm long. Flowers acroscopic. Perianth glabrous outside, pilose to subvillous inside. Pistil 11–20 mm long; ovary densely villous; style loosely villous in lower half, becoming glabrous above. Follicle obloid-ellipsoidal, 10–14 mm long, ±smooth, loosely villous soon becoming glabrous and glaucous.

The leaves (upper and lower surfaces glabrous and only slightly dissimilar) are distinctive within the group. Occurs in the south-west of W.A. Two subspecies are recognised.

Branchlets glabrous and glaucous; leaves usually ±truncate at base **261a.** subsp. **insignis**

Branchlets glabrous but not glaucous; leaves cuneate at base **261b.** subsp. **elliotii**

261a. Grevillea insignis Kippist ex Meisn. subsp. insignis

Illustrations: D.J.McGillivray & R.O.Makinson, *Grevillea* 231 (1993), as *G. insignis*; P.M.Olde & N.Marriott, *Nuytsia* 9: 286, fig. 16B–L (1993); P.M.Olde & N.Marriott, *Grevillea Book* 2: 212 (bottom centre & 175) (1995).

Branchlets glabrous and glaucous. Leaves 2.5–9 cm long, dentate with shallowly arcuate sinuses, bluish grey (usually glaucous both sides); base ±truncate or occasionally cuneate. Flower colour: perianth cream becoming pink; style deep pink to red. Plate 52.

Occurs in the inland south-west of W.A. between Tammin, Nyabing and Tarin Rock. Grows in mallee and heath-shrubland associations in gravelly lateritic or ironstone soils in scattered populations. Regenerates from seed. Flowers Aug.–Dec. Map 332.

W.A.: 6 km N of Balkuling, *J.S.Beard 8100* (PERTH); Youndigin Hill, 20 km SE of Cunderdin, *M.D.Crisp 6198* (CANB, NSW, PERTH); Narrogin, 12 Sept. 1962, *C.J.Cromini* (PERTH); 4.5 km N of Nyabing, *D.J.McGillivray 3528 & A.S.George* (NSW, PERTH); 25 km W of Lake Grace near Tarin Rock, *D.J.E.Whibley 5325* (AD, PERTH).

See notes under subsp. *elliotii*.

This subspecies is recognised as 'Poorly Known' in J.D.Briggs & J.H.Leigh, *Rare or Threatened Australian Plants* (1995).

261b. Grevillea insignis subsp. elliotii Olde & Marriott, *Nuytsia* 9: 285 (1993)

T: Digger Rocks, E of Varley, W.A., 4 Oct. 1986, *P.M.Olde 86/757*; holo: NSW; iso: CANB, PERTH.

Illustrations: P.M.Olde & N.Marriott, *Nuytsia* 9: 286, fig. 16A (1993); P.M.Olde & N.Marriott, *Grevillea Book* 2: 213 (top centre & 176) (1995).

Branchlets glabrous but not glaucous. Leaves 2.5–6 cm long, dentate to shallowly pinnatifid with deeply arcuate sinuses, green (not glaucous or only faintly so); base cuneate. Flower colour: perianth cream becoming pink; style deep pink.

Occurs in the inland south-west of W.A., restricted to an area east of Varley between Hatters Hill and Middle Ironcap (S of Lake Cronin). Grows on laterite outcrops in well-drained, dry sites in deeply laterised loam in eucalypt woodland or mixed shrub associations. Regenerates from seed. Flowers Aug.–Dec. Map 333.

W.A.: Middle Ironcap, *J.W.Green 5558* (CANB, PERTH); Hatters Hill, c. 41 km NE of Lake King, *K.Newbey 6551* (PERTH); South Ironcap, *P.Olde 92/252* (NSW).

Narrowly distinct from subsp. *insignis* in having its branchlets and (usually) leaves non-glaucous, and leaves greener in colour and generally smaller, and usually a little more deeply divided, with a consistently cuneate to broadly cuneate base. The ranges of the two subspecies are separated by approximately 200 km.

This subspecies is recognised as 'Poorly Known' in J.D.Briggs & J.H.Leigh, *Rare or Threatened Australian Plants* (1995).

262. Grevillea fulgens C.A.Gardner, *J. Roy. Soc. W. Australia* 47: 55 (1964)

T: montem Desmond ... [Mt Desmond, W.A.], 30 Aug. 1962, *C.A.Gardner 14070*; holo: PERTH.

Illustrations: P.M.Olde, *Austral. Pl.* 13 (108): 377 (1986); D.J.McGillivray & R.O.Makinson, *Grevillea* 234, col. pl. & fig. 63 (1993); P.M.Olde & N.Marriott, *Grevillea Book* 2: 171 (top right & 140A, B) (1995).

Shrub 0.6–3.0 m tall. Leaves (2–) 4–11 cm long, entire and linear and 1–4 mm wide, and/or sparingly toothed or (usually juvenile leaves only) pinnatifid with 2–11 subtriangular often slightly twisted lobes 1–5 mm long, 1–1.5 mm wide; apices not pungent; margins revolute, usually enclosing lower surface except midvein; lower surface bisulcate, silky in grooves or where lamina exposed. Conflorescence erect, axillary or terminal, usually simple, 1-flowered; rachis 0.5–2 mm long. Flower orientation irregular. Flower colour: perianth mid- to deep pink or reddish, with conspicuous white hairs on limb; style red. Perianth glabrous outside below limb, with a villous white beard of biramous hairs at limb apex, bearded inside in lower half. Pistil 23–26 mm long; ovary villous; style villous, sparsely so towards apex. Follicle ovoid and apically attenuate, 12–15 mm long, villous when young, with hairs soon falling.

Occurs in the south-west of W.A., restricted to the Ravensthorpe area from Mt Short to Mt Desmond, with one unconfirmed record from the Parker Ra. Grows in shrubland or mallee heath in shallow gravelly soil on laterite. Regenerates from seed. Flowers June–Oct. Map 334.

W.A.: 1 km E of Mt Desmond summit, *B.Barnsley 447* (CANB, NSW); E side of hill behind Elverdton Copper Mine, S of Ravensthorpe, *A.S.George 1637* (PERTH); near base of N slope of Mt Short, *D.J.McGillivray 3577 & A.S.George* (B *n.v.*, CANB, K, NSW, NY *n.v.*, PERTH, PRE *n.v.*, US *n.v.*).

The tufted beard on the limb of the bud is a distinctive character, as are the very large (3–9 mm long, 3–8.5 mm wide) obovate to obcordate, subsericeous to villous floral bracts; these enclose the very young conflorescences, and fall when the buds are c. 5 mm long. There is also a characteristic contrast between the white hairs of the ovary and the rusty brown hairs of the style. *Grevillea involucrata* is closely related and somewhat similar; it has regularly shortly pinnate leaves, glabrous outer surface of perianth (including limb), glabrous pedicels (villous in *G. fulgens*) and persistent reddish floral bracts condensed in a pseudo-whorl.

This species is recognised as 'Poorly Known' in J.D.Briggs & J.H.Leigh, *Rare or Threatened Australian Plants* (1995).

263. Grevillea involucrata A.S.George, *Nuytsia* 1: 372 (1974)

T: between Hyden and Lake Varley, W.A., 30 June 1970, *A.S.George 9890*; holo: PERTH; iso: AD, CANB, L *n.v.*, NSW, PERTH.

Illustrations: J.Leigh *et al.*, *Extinct & Endangered Pl. Australia* col. pl. facing p. 189 (1984); D.J.McGillivray & R.O.Makinson, *Grevillea* 234, fig. 64, 235, col. pl. (1993); P.M.Olde & N.Marriott, *Grevillea Book* 2: 216 (top left & 179A, B) (1995).

Open prostrate to decumbent shrub 0.15–0.5 m tall, to 2 m across. Leaves 1.5–4 cm long, 6–10 mm wide, secund, pinnatipartite (almost pinnatisect) with 5–15 short mutually aligned oblong to linear lobes 1–8 mm long, 0.8–1.2 mm wide, not pungent; margins revolute; lower surface enclosed except for midveins, bisulcate, sometimes with a few scattered appressed hairs. Conflorescence erect, terminal or axillary, simple, 1–3-flowered, rachis 0.5–1 mm long. Flowers erect. Flower colour: perianth pale pearly pink to dark pink; style deep pinkish red. Perianth glabrous outside, pilose to villous inside. Pistil 23–25 mm long; ovary villous; style loosely villous, becoming sparsely so towards apex. Follicle obloid to ellipsoidal and apiculate, 13–15 mm long, loosely villous.

Occurs in the south-west of W.A. where scattered in a small area between Lake Grace, Hyden, Holt Rock and Lake Magenta. Grows in low open shrubland on well-drained gravelly sand over laterite. Regenerates from seed. Flowers June–Oct. Map 335.

W.A.: c. 10 km NW of L. Magenta, *K.Newbey 3388* (PERTH); Lake Grace, *E.Wittwer 1835* (CANB, NFLD *n.v.*, PERTH).

The persistent large floral bracts (5–9 mm long, 2.5–4 mm wide) aggregated in short pseudo-whorls are distinctive. Closely related and with some superficial similarity to *G. fulgens*; see under that species for differences.

This species is recognised as 'Vulnerable' in J.D.Briggs & J.H.Leigh, *Rare or Threatened Australian Plants* (1995).

Marriottii Group

Shrubs. Leaves entire or coarsely toothed or divided, dorsiventral; surfaces dissimilar; margins recurved to revolute. Conflorescence erect, terminal or axillary, simple or few-branched; unit conflorescence regular-umbelloid, acropetal or basipetal. Flowers adaxially oriented. Torus slightly oblique. Perianth zygomorphic, hairy on both surfaces; tepals (dorsal) weakly everted, then all separate. Pistil 7–12 mm long; ovary hairy (biramous hairs), stipitate; style with minute hairs or papillae at least in apical 1–2 mm, sometimes over whole style, exserted from late bud; pollen-presenter oblique, flat to convex. Follicle loosely to sparsely hairy (hairs soon falling), ±smooth; pericarp thin, crustaceous. Seed narrowly ellipsoidal, with a membranous border along one side and an apical elaiosome.

A group of three species, all occurring in south-western W.A. Insect pollinated. Affinities uncertain, but probably closely related to the *Costata* and perhaps *Speciosa* subgroups of the *Linearifolia* group, with which it shares the presence of minute hairs or papillae at the style apex. There may be a more distant connection with the *Hakeoides* group.

1 Leaf upper surface smooth **265. G. lissopleura**

1: Leaf upper surface scabrous

  2 All leaves simple and entire, usually ±strongly curved; stipe of follicle ±straight (follicle erect) **266. G. scabrida**

  2: At least some leaves usually divided or toothed, ±straight, rigid; stipe of follicle strongly incurved **264. G. marriottii**

264. Grevillea marriottii Olde, *W. Australian Naturalist* 18: 75 (1990)

T: ... S of Mt Holland, W.A., 4 Oct. 1988, *P.M.Olde & N.R.Marriott s.n.*; holo: NSW.

Illustrations: P.M.Olde & N.R.Marriott, *Grevillea Book* 3: 18 (bottom right), 19 (10A–C) (1995).

Shrub 0.8–1.2 m high. Leaves erect to spreading, mixed entire and divided; entire leaves linear, (0.5–) 1–5 cm long, 1.5–3.5 (–12 in juveniles) mm wide; divided leaves apically 2- or 3-fid or -partite, with weakly divaricate lobes to 10 mm long, to 2.5 mm wide; upper surface conspicuously granulose, with midvein weakly recessed and edge veins faintly ridged; margins smoothly tightly revolute; lower surface usually partly exposed either side of midvein, villous. Conflorescence erect, terminal, simple, ±sessile, umbelloid, ±acropetal; rachis 3–5 mm long, villous. Flower colour: perianth whitish; style white, rarely pink-tinged with age. Perianth appressed-villous outside, minutely pubescent or papillose inside; tepals independently recurved after anthesis. Pistil 8–12 mm long; ovary villous; style villous at base, ±glabrous or papillose above, exserted from late bud. Follicles oblique on strongly incurved stipe, obloid to obloid-ellipsoidal with a rounded apiculum, 10–14 mm long, with 3 longitudinal ridges, pubescent when young, soon glabrous.

Occurs in south-western W.A. in a small area near Mt Holland, S of Southern Cross, growing in sand over laterite. Regenerates from seed and basal suckers. Flowers July–Oct. Map 336.

No specimens cited because of the rarity of the species.

Placement of the divided-leaved *G. marriottii* in this group is somewhat tentative. The presence of leaf division, together with the umbelloid conflorescences, suggests a possible connection with the *Hakeoides* group. The other members of the group have all leaves entire, a feature shared with the apparently most-closely related taxa (*Linearifolia* group).

265. Grevillea lissopleura McGill., *New Names Grevillea* 9 (1986)

T: about 25 km NNW of Mt Holland, c. 86 km NE of Hyden, W.A., 23 Aug. 1979, *K.Newbey 5808*; holo: PERTH.

Illustrations: P.M.Olde & N.R.Marriott, *Grevillea Book* 2: 242 (bottom right), 243 (202A, B) (1995).

Erect shrub to 1.2 m tall. Leaves entire, linear, straight to slightly curved, 1–3.5 (–7) cm long, 1.0–1.2 mm wide; upper surface with 3–7 smooth prominent longitudinal ridges; margins angularly revolute to midvein; lower surface 2-grooved, glabrous. Conflorescence erect, subsessile, terminal or in upper axils, simple, umbelloid, acropetal; rachis 1–2 mm long, tomentose. Flower colour: perianth and style white to cream. Perianth loosely subsericeous outside, bearded inside opposite ovary; tepals independently recoiled after anthesis. Pistil 7–8 mm long; ovary subsericeous to tomentose; style hairy for c. 1 mm above ovary, glabrous above except for short erect hairs or papillae immediately below style-end, weakly exserted in late bud. Follicles erect on erect stipe, ovoid, 6–9 mm long, minutely subsericeous, becoming glabrous.

A rare species found only between Southern Cross and Mt Holland, W.A., growing in open scrub in rocky loam. Regenerates probably from seed only. Flowers recorded for Aug. Map 337.

No specimens cited because of the rarity of the species.

Grevillea lissopleura is similar to *G. scabrida*, which differs in having scabrid leaf veins and often ±strongly recurved leaves, and in its angular ridged branchlets (terete in *G. lissopleura*).

This species is recognised as 'Poorly Known' in J.D.Briggs & J.H.Leigh, *Rare or Threatened Australian Plants* (1995).

266. Grevillea scabrida C.A.Gardner, *J. Roy. Soc. W. Australia* 22: 120 (1936)

T: Mt Singleton, W.A., 9 July 1931, *C.A.Gardner 2210*; lecto: PERTH, *fide* D.J.McGillivray & R.O.Makinson, *Grevillea* 440 (1993); isolecto: CANB, K, MEL, PERTH; remaining syntype: Mt Singleton, W.A., *W.E.Blackall 17 & C.A.Gardner*; syn: PERTH.

Illustrations: P.M.Olde & N.R.Marriott, *Grevillea Book* 2: 161 (centre right & 130A, B) (1995).

Shrub to c. 1 m high. Leaves entire, linear, often variously recurved or recurled, 1–5 (–8) cm long, 0.5–1.3 (–1.5) mm wide; upper surface with 3–5 (–7) conspicuous prominent longitudinal veins, scabrous along veins; margins angularly revolute; lower surface usually narrowly exposed beside midvein, subsericeous. Conflorescence erect, usually pedunculate, usually terminal on short lateral branches, occasionally axillary, simple or few-branched; unit conflorescence usually regular and umbelloid, sometimes weakly secund, acropetal; rachis 0.8–2 mm long, subsericeous. Flower colour: perianth greenish yellow in bud, becoming white; style whitish, pink-tinged (?with age). Perianth subsericeous outside, shortly bearded inside opposite ovary; tepals independently and widely recoiled after anthesis. Pistil 7–8.5 mm long; ovary subsericeous to subvillous at least near apex or on dorsal side; style subsericeous in basal c. 2 mm, otherwise glabrous except for papillae immediately below style-end, exserted from late bud. Follicles erect on erect stipe, 9–11.5 mm long, ±ovoid to ellipsoidal, faintly longitudinally ridged, sparsely subsericeous.

Found only in the Mt Singleton–Mt Gibson area, W of Lake Moore, W.A., growing in red loam and laterite. Regenerates probably from seed only. Flowers recorded in July. Map 338.

W.A.: 28.8 km ESE along Mt Gibson HS road off Wubin to Paynes Find road, *R.Coveny 7896 & B.R.Maslin* (K, NSW, PERTH); top of Mt Singleton, *D.J.McGillivray 3395 & A.S.George* (CANB, K, NSW, PERTH, US); Mt Singleton, *B.H.Smith 430* (MEL); 106.2 km N [NE?] of Wubin [probably on Paynes Find road], *E.Wittwer 1248* (PERTH).

Grevillea scabrida has distinctive reddish angular branchlets. It is similar to *G. lissopleura*, which differs in its straight to gently incurled leaves with the upper surface smooth (not scabrid) and its terete branchlets.

This species is recognised as 'Poorly Known' in J.D.Briggs & J.H.Leigh, *Rare or Threatened Australian Plants* (1995).

Cirsiifolia Group

Prostrate shrub. Leaves entire, toothed or divided, dorsiventral; surfaces dissimilar; margins recurved. Conflorescence erect, terminal or axillary, simple or 2-branched; unit conflorescence conico-cylindrical or semi-secund, acropetal. Flowers acroscopic. Torus transverse. Perianth zygomorphic, hairy outside, glabrous inside; tepals everting along dorsal suture below limb, otherwise remaining ±coherent and held ventrally. Pistil 7–8 mm long; ovary obscurely stipitate, densely hairy; style subsericeous at base, glabrous above, shortly exserted from late bud; pollen-presenter lateral, flat to convex. Follicle rugulose, subsericeous; pericarp moderately thick, crustaceous. Seed not seen.

One species, endemic to south-western W.A. Insect pollinated. Affinities uncertain. McGillivray (*in* McGillivray & Makinson, *Grevillea* 102 (1993)) placed *G. cirsiifolia* as a monotypic subgroup at the end of his Group 1 (the 'toothbrush alliance', equivalent in general terms to the *Pteridifolia* group of the present work). His placement of it there was largely on the basis of the hairy ovary combined with the glabrous inner surface of the perianth; its insect pollination syndrome in a primarily bird-pollinated group being seen as derived (a western analogue to the eastern *Ramosissima* subgroup). In the present work the indumental states are not seen as sufficient to establish a connection with those species. *Grevillea prostrata*, itself an isolated taxon, may be a moderately close relative.

267. **Grevillea cirsiifolia** Meisn., *in* J.G.C.Lehmann, *Pl. Preiss.* 2: 253 (1848)

T: Swan River, W.A., *J.Drummond III: 267*; holo: NY? *n.v.*; iso: BM, CGE *n.v.*, G, G-DC, K, MEL, ?NY *n.v.*, P, PERTH, TCD *n.v.*

Illustrations: B.L.Rye & S.D.Hopper, *Guide Gazetted & Rare Fl. W. Australia* 111–113 (1981); P.M.Olde & N.R.Marriott, *Grevillea Book* 3: 92 (bottom left & 73A, B) (1995).

Prostrate shrub to 2 m across. Leaves spreading to ascending, (5–) 7–16 cm long, (2.5–) 5–25 (–30) mm wide, commonly pinnatifid to pinnatipartite with 8–30 lobes, or occasionally toothed, or entire and then linear or narrowly obovate; lobes linear to subtriangular, 1.2–6 cm long, 1.5–5 mm wide, entire or coarsely serrate; margins shortly recurved; lower surface mostly exposed, densely subsericeous to villous. Conflorescence erect, pedunculate, terminal or axillary, simple or few-branched; unit conflorescence lax, conico-cylindrical, acropetal; rachis (10–) 25–70 mm long, subsericeous to villous. Flowers acroscopic. Flower colour: perianth creamy white outside, brighter yellow inside; style white to pale yellow with green pollen-presenter. Perianth subsericeous to subvillous outside, glabrous inside; tepals reflexed in 2 lateral pairs and displaying inner surface in late bud, remaining thus after release of style-end. Pistil 7–8 mm long; ovary stipitate, sericeous; style sericeous to c. 1.5 mm above base, glabrous above, weakly exserted in late bud; pollen-presenter lateral. Follicles obovoid to ellipsoidal, 10–11.5 mm long, rugulose, subsericeous.

Occurs in south-western W.A., from about Darkan to Mt Lindesay near Denmark, in Jarrah or Wandoo woodland, often in winter-wet areas in sandy to clayey soils. Regeneration strategy unknown. Flowers Oct.–Dec. Map 339.

W.A.: Albany Hwy, S of 177 mile peg [c. 284.9 km], *A.S.George 6455* (PERTH); 27 km SSE of Kojonup on Albany Hwy, *McGillivray 3480 & A.S.George* (NSW); 23.3 km from Arthur R. towards Darkan, *M.E.Phillips WA68/1991* (CANB, L, NSW, PERTH); Mt Lindesay, *W.Webb 14* (MEL); corner Corbalup Rd & Smith Rd, 13 Oct. 1976, *M.Wittwer* (PERTH).

Olde & Marriott (*loc. cit.*) recognise two horticultural forms ('small-leaved' and 'large-leaved'), differing in leaf size, colour, lobe shape and position and inflorescence posture. It is unclear whether these reflect populational differences in the wild. *Grevillea cirsiifolia* is sometimes misidentified as *G. leptobotrys*, which has a similar tepal eversion but differs in having pink flowers and a glabrous ovary. It is also occasionally confused with *G. quercifolia*, which also has pink flowers, a glabrous pistil, with the style swollen over the apical 1/2 to 2/3, and broader leaves and lobes.

This species is recognised as 'Rare' in J.D.Briggs & J.H.Leigh, *Rare or Threatened Australian Plants* (1995).

Pythara Group

Shrub. Leaves entire, dorsiventral; surfaces dissimilar; margins recurved to revolute. Conflorescence erect, terminal, simple, a loose, 2–8-flowered cluster, opening order uncertain. Flowers obscurely acroscopic. Torus very oblique to lateral. Perianth zygomorphic, hairy outside and inside; tepals remaining loosely coherent and held vertically. Pistil 20–22 mm long; ovary long-stipitate, hairy; style hairy, exserted from late bud; pollen-presenter lateral, flat. Follicle and seed not seen.

One species, endemic to south-western W.A. Pollinator unknown. Affinities uncertain, although some characters of the pistil and the perianth suggest a possible relationship to the *G. brachystylis* alliance within the *Floribunda* group. There may also be some connection to the *Longistyla* group.

268. Grevillea pythara Olde & Marriott, *Nuytsia* 9: 293 (1993)

T: near Pithara, W.A., 29 Sept. 1992, *P.M.Olde 92/173*; holo: PERTH; iso: NSW.

Illustrations: P.M.Olde & N.R.Marriott, *Nuytsia* 9: 295, fig. 19A–G (1993); P.M.Olde & N.R.Marriott, *Grevillea Book* 3: 124 (top right & 96A, B) (1995); A.Brown *et al.* (eds), *W. Australia's Threatened Flora* 98 (1998).

Suckering shrub 6–30 cm high, with columnar foliage. Branchlets terete, villous. Leaves 7–16 mm long, 1.5–4 mm wide, entire, villous, broadly linear to narrowly elliptic or sometimes obovate, crowded, slightly discolorous; venation obscure, the midvein sometimes evident on lower surface; margins shortly recurved to strongly revolute, sometimes enclosing lower surface; apex obtuse, mucronate; upper and lower leaf surfaces appressed subvillous. Conflorescence erect, terminal, simple, ±sessile, 4–8-flowered, subsynchronous, loosely subsecund; rachis c. 3 mm long, villous. Flowers acroscopic. Flower colour: perianth red, blue around tepal margins near limb; style red. Perianth loosely tomentose outside, glabrous inside in basal pouch, densely bearded above; tepals remaining coherent after release of style-end except along dorsal suture. Pistil 20–22 mm long; stipe 7–8 mm long, sparsely villous, adnate to torus at base; ovary villous; style open-tomentose, with mainly biramous hairs, near apex also with erect multicellular glandular hairs, exserted in late bud; pollen-presenter lateral. Follicles not seen. Plate 53.

W.A., confined to the type locality near Pithara, just south of Dalwallinu, where it occurs in two small populations. Possibly monoclonal, growing in remnant vegetation along roadside in gravelly sand. Probably reproducing by rhizomes only. Flowers May–Oct., possibly sporadically throughout the year. Map 340.

No specimens cited because of rarity.

This species is recognised as 'Endangered' in J.D.Briggs & J.H.Leigh, *Rare or Threatened Australian Plants* (1995).

Pterosperma Group

Robust shrubs or small trees. Leaves entire or divided, dorsiventral; surfaces dissimilar; margins revolute. Conflorescence erect, usually terminal, simple or 2–several-branched; unit conflorescence usually cylindrical (rarely umbelloid-globose), dense, basipetal or synchronous, rarely weakly acropetal. Flowers oriented transversely to rachis or (*G. globosa*) irregularly and adaxially acroscopic. Torus transverse to oblique (rarely slightly reversely oblique in *G. albiflora*). Perianth zygomorphic, hairy outside, glabrous or hairy inside; tepals separating and partly everting along dorsal suture while limb segments remaining coherent, later all free. Pistil 10–22 mm long; ovary shortly stipitate, densely hairy; style incurved, exserted from late bud; pollen-presenter oblique, flat to broadly conical. Follicle thickly lenticular to subglobose, hairy with biramous and sometimes also simple glandular hairs, lacking indumental markings; pericarp moderately thick, bony. Seed usually strongly laterally compressed, oblong to ±obovate or (*G. eriobotrya*) almost hemispherical, peripterous.

A group of four species, found in inland areas of southern Australia. Usually insect-pollinated, also bird-pollinated, at least in *G. eriobotrya*. Affinities uncertain, possibly a distant relationship to the *Hilliana* and/or *Longistyla* groups.

1 Unit conflorescences subglobose-umbelloid **272. G. globosa**

1: Unit conflorescences cylindrical

2 Style villous over all its length, at least on dorsal side; perianth glabrous inside; seeds ±hemispherical **271. G. eriobotrya**

2: Style glabrous or with hairs restricted to lower third; perianth hairy or glabrous inside; seeds strongly compressed

3 Style glabrous or with hairs extending only c. 2 mm above ovary; floral bracts 4–7 (–9?) mm long; most or all leaves simple **269. G. pterosperma**

3: Style villous in basal third; floral bracts < 2.5 mm long; most or all leaves divided **270. G. albiflora**

269. Grevillea pterosperma F.Muell., *Trans. Philos. Soc. Victoria* 1: 22 (1855)

T: Murray River towards junction with Murrumbidgee River, Vic., Dec. 1854 (error for 1853), *F.Mueller*; lecto: MEL, *fide* D.J.McGillivray & R.O.Makinson, *Grevillea* 436 (1993); isolecto: MEL; remaining syntypes: Murray, *F.Mueller*; syn: K.

G. sericostachya Meisn., in A.L.P.P. de Candolle, *Prodr.* 14: 384 (1856). T: S.W. Australia, *J.Drummond Suppl. 5th Coll. 10*; holo: NY *n.v.*; iso: BM, CGE *n.v.*, MEL, P *n.v.*, TCD *n.v.*

G. simulans A.Morrison, *J. Bot.* 50: 277 (1912). T: Uaroo and Minderoo, Asburton R. W.A.; holo: not found.

Illustrations: D.J.McGillivray & R.O.Makinson, *Grevillea* 225, col. pl. (1993); P.M.Olde & N.R.Marriott, *Grevillea Book* 3: 117 (top left & 89A–C) (1995).

Erect shrub 2–4 m high. Leaves ascending, (3–) 6–18 cm long, entire and ±linear and 0.8–2 mm wide, or occasionally 2–6-partite with entire linear lobes 1–1.5 mm wide; upper surface with 3–5 longitudinal ridges; margins ±angularly revolute; lower surface enclosed except for midvein, 2-grooved. Conflorescence erect, usually simple, sometimes few-branched; unit conflorescence dense, cylindrical, 6–8 cm long, basipetal or sometimes subsynchronous. Flowers oriented transversely to rachis. Flower colour: perianth greyish white outside, creamy white inside; style cream to pale yellow. Perianth subsericeous-villous outside, glabrous or with a few hairs at base inside; tepals lax and recurved after anthesis. Pistil 12–21 mm long; ovary stiffly villous; style glabrous or with hairs in basal c. 2 mm only; pollen-presenter oblique. Follicles 15–20 mm long, transverse to oblique on pedicel, thickly lenticular, subvelutinous to tomentose with short biramous hairs often also with simple erect glandular viscid hairs.

A mostly inland species occurring in all mainland states except Qld; in W.A. south from c. 23°S from Giles west to the lower Murchison and south to Mt Holland; in the far south-west of N.T.; in S.A. from north of the Nullarbor Plain through the northern Eyre Peninsula and in the south-east from the Murray R. to Bordertown; in N.S.W. in the south-west, extending to north-western Vic. Grows in a variety of habitats in mallee, heath, open woodland and spinifex, in rocky loam to deep sand. Regenerates from seed only. Flowers June–Jan. Map 341.

W.A.: 19.3 km N of Wiluna, *N.H.Speck 1330* (CANB, MEL, NSW, PERTH). N.T.: George Gill Ra., Wallaby Gorge area, ±1.5 km NE of Reedy Rock Hole, *A.C.Beauglehole 26551* (AD, CANB, NSW). S.A.: lake complex S of L. Bring, ±160 km W of Commonwealth Hill Stn HS, *D.E.Symon 3412* (AD, CANB, NSW). N.S.W.: 28 km W of Pan Ban HS on Pooncarie Rd, *M.Fox 8310389 & H.Fallding* (NSW). Vic.: Wyperfeld Natl Park, c. 1.6 km S of Moorong Rise, *A.C.Beauglehole 29355* (MEL, NSW).

Similar to *G. albiflora* which differs in having all or most leaves deeply divided, smaller floral bracts ⩽ 2.2 mm long (4–7 (–9) mm in *G. pterosperma*), a ±transverse torus (oblique in *G. pterosperma*) and the style villous over the basal third. *Grevillea eriobotrya* has more conspicuously hairy flowers, pistils 10–13 mm long and a less compressed follicle with a subhemispherical seed (seeds highly compressed in *G. pterosperma* and *G. albiflora*).

270. Grevillea albiflora C.T.White, *Proc. Roy. Soc. Queensland* 55: 79 (1944)

T: Gilruth Plains, E of Cunnamulla, Qld, 20 May 1939, *S.T.Blake 14065*; holo: BRI; iso: BRI, K, NSW.

Illustrations: P.M.Olde & N.R.Marriott, *Grevillea Book* 2: 18 (bottom right), 19 (11A, B) (1995).

Spreading to erect shrub or small tree 3–6 (–8) m high. Leaves 8–30 cm long, ascending, usually (3–) 5–9-partite or occasionally entire; simple leaves and lobes linear, 1–2 mm wide; lower primary lobes often 2- or 3-partite; upper surface not ridged; margins tightly revolute; lower surface enclosed except for midvein, 2-grooved. Conflorescence erect, terminal or occasionally axillary, usually 2–5-branched; unit conflorescence dense, cylindrical, 5–10 cm long, basipetal to synchronous; floral bracts 1.5–2.2 mm long. Flowers transversely oriented to rachis. Flower colour: perianth white to greenish cream; style white to creamy yellow. Perianth ±villous outside, open pilose towards base inside; tepals loosely connate after anthesis except along dorsal suture. Pistil 15–21.5 mm long; ovary villous; style villous in basal third, glabrous above; pollen-presenter oblique. Follicles 20–25 mm long, thickly ±lenticular, velutinous. Fig. 30A–C.

Occurs in N.T., Qld and N.S.W., in two widely disjunct areas, one in southern N.T. from near Uluru to Rainbow Valley and the other from Cunnamulla to W of St George in Qld, and south to Bourke in N.S.W. Grows on sand ridges or sand plain, in deep red sand. Probably regenerates from seed only. Flowers fragrant. Flowers mainly Nov.–Jan., occasionally in other months. Map 342.

N.T.: Rainbow Valley *J.R.Maconochie 2275* (DNA, NSW, PERTH). Qld: 16 km E of Gilruth, *G.H.Allen 74* (CANB); 36.4 km E of Cunnamulla, *D.Hangar 1970* (BRI); 25.7 km W of St George, *D.M.Gordon 360* (BRI). N.S.W.: Nulty Springs, Enngonia, *G.M.Cunningham & P.L.Milthorpe 4174* (NSW).

Differs from *G. eriobotrya*, in which the style is hairy throughout and the perianth entirely glabrous inside, and from *G. pterosperma*, which has usually simple linear leaves, style glabrous and longer floral bracts. Plants from N.T. are usually more robust than those from Qld and N.S.W., with broader leaf segments and larger flowers.

271. Grevillea eriobotrya F.Muell., *Fragm.* 10: 44 (1876)

T: near Mt Churchman, W.A., [spring 1875], *J.Young s.n.*; holo: MEL; iso: B (as *Diels 7450*).

G. victori A.Morrison, *J. Bot.* 50: 276 (1912), as *G. Victori*. T: 14 miles [22.5 km] north of Kununoppin, W.A., Oct. 1911, *F.E.Victor*; lecto: E *fide* D.J.McGillivray & R.O.Makinson, *Grevillea* 416 (1993); isolecto: E *n.v.*

Illustrations: D.J.McGillivray & R.O.Makinson, *Grevillea* 227, col. pl. (1993); P.M.Olde & N.R.Marriott, *Grevillea Book* 2: 148 (bottom right), 149 (120A–C) (1995).

Dense spreading erect shrub to c. 3 m high. Leaves (5–) 8–18 cm long, 0.9–1.5 (–2) mm wide, ascending, usually entire and linear, sometimes 2- or 3-partite with linear lobes; upper surface with 3–5 longitudinal ridges; margins revolute; lower surface enclosed except for midvein, 2-grooved. Conflorescence erect, terminal or sometimes axillary, simple or few-branched; unit conflorescence dense, cylindrical, 6–8 cm long, subsynchronous to weakly and 1-sidedly acropetal. Flowers oriented transversely to rachis. Flower colour: perianth and style white to creamy white. Perianth densely villous-woolly outside, glabrous inside; tepals remaining loosely connate after release of style-end. Pistil 10–13 mm long; ovary villous; style villous throughout, at least on dorsal side; pollen-presenter oblique. Follicles 20–23 mm long, transverse on pedicel, thickly lenticular to subglobose, densely pubescent.

Occurs in the northern wheatbelt of W.A., from Lake Moore and Mt Churchman S to Koorda and Campion; grows on low hills in tall shrubland, mostly in yellow sand. Regenerates from seed only. Flowers mainly Sept.–Dec. Map 343.

W.A.: N of Bencubbin, *W.E.Blackall 3325* (PERTH); near Mt Churchman, *C.A.Gardner 13549* (PERTH); c. 2 km S of Waddouring Hill, *D.J.McGillivray 3433 & A.S.George* (NSW, PERTH); near S border of Reserve 23029, c. 30 km due W of Beacon Siding, just N of Mollerin L., *B.G.Muir 547* (CANB, PERTH); N of Boorabin, Sept. 1931, *E.S.Simpson s.n.* (PERTH).

Grevillea eriobotrya is very distinctive in having a very lightweight subhemispherical seed with a spongy testa, perhaps adapted for wind-dispersal. It differs from *G. albiflora*, which has its pistil glabrous in upper part, perianth less densely villous outside and leaves without longitudinal ridges on upper surface. Also similar to *G. pterosperma*, which differs in its longer, mostly glabrous pistil and less densely villous outer perianth surface.

This species is recognised as 'Poorly Known' in J.D.Briggs & J.H.Leigh, *Rare or Threatened Australian Plants* (1995).

272. Grevillea globosa C.A.Gardner, *J. Roy. Soc. W. Australia* 47: 55 (1964)

T: 32 km north of Pindar, W.A., Jan. 1963, *F.Lullfitz 2241*; holo: PERTH.

Illustrations: D.J.McGillivray & R.O.Makinson, *Grevillea* 373, fig. 93 & col. pl. (1993); P.M.Olde & N.R.Marriott, *Grevillea Book* 2: 178 (bottom centre), 179 (146A–C) (1995).

Spreading shrub to c. 2 m high. Leaves ascending, (4–) 9–18 cm long, pinnatipartite, almost pinnatisect; lobes 3–9, ascending, linear or occasionally subterete to trigonous, 0.5–1 mm wide; margins tightly revolute against prominent midvein; lower surface 2-grooved. Conflorescence erect, on long peduncles, terminal or axillary, simple or 2-branched; unit conflorescence a dense globose head, weakly basipetal; rachis 5–8 mm long, villous. Flower colour: perianth (hairs) pale green to whitish below curve, red-brown on limb outside, inside exposed, creamy green, turning glossy black after anthesis; style greenish with white hairs. Perianth loosely tomentose to subvillous outside, pilose inside for c. 3 mm from base, glabrous above; tepals everting before release of style-end and displaying inner surface. Pistil 13–22 mm long; ovary densely villous; style villous at base, with long, scattered hairs above to near apex, exserted in late bud. Follicles erect on pedicels, obloid-ellipsoidal, 9–12.5 (–14) mm long, tomentose, subdorsally ridged, rugulose elsewhere. Fig. 39D–F.

Occurs in the northern wheatbelt of W.A., from around Lake Moore to N of Pindar. Rare, only three populations known. Grows in Mulga shrubland or mallee woodland in loam or sand. Regenerates from seed only. Flowers sporadically through the year, perhaps concentrated in late spring to summer. Flowers honey-scented. Map 344.

W.A.: 48 km SW of Yalgoo, *J.S.Beard 3134* (KPBG, PERTH); 22 km WSW of Paynes Find on Great Northern Hwy, *D.J.McGillivray & A.S.George* (NSW, PERTH).

Grevillea globosa was regarded by McGillivray & Makinson (*loc. cit.*) and Olde & Marriott (*loc. cit.*) as a species of uncertain affinity. There is however concordance in most features with other members of the *Pterosperma* group, the main difference being in the much shorter subglobose conflorescences, perhaps a result of simple abbreviated rachis meristem development rather than a ancestral condition. The unit conflorescence and floral orientation are not fully regular, and there appears to be a residual tendency (at least at the base of the conflorescence) towards the transverse floral orientation that is typical of other members of the group. The seed is somewhat intermediate between the compressed type found in *G. pterosperma* and *G. albiflora*, and the spongy subhemispherical type of *G. eriobotrya*.

Grevillea globosa is rarely confused with *G. bracteosa*, which has simple-linear or rarely 2-or 3-lobed leaves, a subapically swollen style and large conspicuous persistent floral bracts 7–14 × 5–13 mm (c. 2 × 1.5–1.8 mm in *G. globosa*).

This species is recognised as 'Poorly Known' in J.D.Briggs & J.H.Leigh, *Rare or Threatened Australian Plants* (1995).

Petrophiloides Group

Shrubs. Leaves divided or sometimes some entire, terete-unifacial or dorsiventral; surfaces often ±similar; margins flat to incurved or obscure. Conflorescence usually terminal, erect or rarely decurved, sometimes emergent on long canes, simple or branched; unit conflorescence cylindrical, basipetal. Flowers obscurely acroscopic. Torus straight or slightly oblique. Perianth glabrous, actinomorphic until shortly before release of style-end, then limb usually becoming deflexed; tepals all separating below the coherent limb in late bud, then all free and independently lax. Pistil 14–22 mm long; ovary shortly stipitate, glabrous or minutely glandular-pubescent, the hairs sometimes not apparent until after anthesis; style exserted from late bud, later erect, often sparsely papillose in basal half; pollen-presenter erect, narrowly conical-subulate, often with a basal bulb. Follicle tuberculate or colliculate, usually erupting into pustular often viscid craters, glabrous or with simple glandular and sometimes also biramous hairs, lacking indumental markings, pericarp firmly crustaceous to bony. Seed ellipsoidal, peripterous.

Four species, endemic to south-western W.A., extending N to Shark Bay. All bird-pollinated. Affinities uncertain but possibly to the *Hilliana* group.

1 Leaves divaricately lobed; leaf rachises deflexed at each node; floral rachises densely white-villous **276. G. paradoxa**

1: Leaves not divaricately lobed; leaf rachises not deflexed at each node; floral rachises glabrous or with an inconspicuous usually open indumentum

2 Leaves subspathulate to narrowly so, with a long cuneate base, apically toothed or shallowly few-lobed, and/or entire and oblanceolate and then < 5 mm wide; ovarian stipe > 2 mm long; pedicels ⩽ 1 mm long **275. G. rogersoniana**

2: Leaves deeply divided, 2–9-sect, or entire and then < 3.5 mm wide and linear-subterete to oblanceolate; ovarian stipe ⩽ 2 mm long; pedicels > 1 mm long

3 Leaves mostly entire or 2- or 3- (–5)-partite, rarely with a secondary lobe or two; ultimate lobes < 10 per leaf; entire leaves and lobes linear to linear-subterete, often glaucous; branchlets usually secund; inflorescences erect to decurved, not emergent on long canes, held within or below foliage; floral rachises 30–75 mm long **274. G. oligomera**

3: Leaves mostly divided, 3–9-partite, often with secondary or sometimes tertiary lobing; ultimate lobes usually > 10 per leaf (rarely fewer); entire leaves and lobes linear-subterete or trigonous-linear or flat and then narrowly oblong to oblanceolate, not glaucous; branchlets not secund; inflorescences usually exserted above foliage on long canes; floral rachises 60–200 mm long **273. G. petrophiloides**

273. Grevillea petrophiloides Meisn., *in* J.G.C.Lehmann, *Pl. Preiss.* 2: 257 (1848)

T: Swan R., W.A., *J.Drummond 3rd coll., 300*; lecto: NY *n.v.*, *fide* D.J.McGillivray & R.O.Makinson, *Grevillea* 433 (1993); isolecto: BM, CGE *n.v.*, K, MEL, NY *n.v.*, P *n.v.*

Shrub 1–4 m tall; branchlets occasionally secund, glaucous or not. Leaves (3–) 6–25 cm long, mostly or all 3–9-sect, the primary lobes often again 3–7-partite, rarely with tertiary division, sometimes a few leaves entire; ultimate lobes ascending, usually > 10 per leaf, subterete-linear or trigonous-linear to -oblanceolate, 1–13 cm long, 0.5–3.5 mm wide; margins incurved, flat, or (subterete lobes) obscure; surfaces silky when young, soon glabrous, granulose. Conflorescence erect, terminal on emergent canes, 0–5-branched; unit conflorescence cylindrical, basipetal; floral rachis 60–200 mm long, glabrous or occasionally with a few glandular hairs near base. Pistil 14–21.5 mm long; style-end with a subapical globose dilation then constricted below erect narrowly conical pollen-presenter. Follicle ovoid to subglobose, 7–18 mm long, glabrous or with glandular hairs, developing pustule-like usually viscid craters. *Rock Grevillea, Poker Grevillea.*

Occurs in south-western W.A. Three subspecies are recognised.

Very variable in leaf form and some fine details of flowers and fruits. McGillivray's (*New Names Grevillea*, 11 (1986)) and McGillivray & Makinson's (*Grevillea*, 387 (1993)) three subspecies were all regarded as full species by Olde and Marriott (The *Grevillea Book*, vols 1 (1994) & 3 (1995)), but with the exception of the taxon accepted here at species rank as *G. oligomera*, species ranking is difficult to sustain. Subsp. *petrophiloides* has much variation, and further research into the complex is desirable.

1 Leaf lobes oblanceolate or oblong **273a.** subsp. **petrophiloides**

1: Leaf lobes subterete to linear or linear-trigonous

2 Peduncles (and usually branchlets) strongly glaucous; fruit surface usually with many erect simple glandular hairs 0.9–1.2 mm long **273b.** subsp. **magnifica**

2: Peduncles and branchlets not glaucous; fruit surface glabrous or with erect simple glandular hairs c. 0.3–0.6 mm long

3 Leaves with 3–9 primary lobes, usually > 10 ultimate lobes; style usually pink to reddish, drying to red-brown, or rarely perianth and style white; fruit surface glabrous or glandular-pubescent, pustules viscid **273a.** subsp. **petrophiloides**

3: Leaves usually with 3 primary lobes, ⩽ 10 ultimate lobes; style white to pale pink, drying to straw-coloured; fruit surface glabrous, pustular craters ?not viscid **273c.** subsp. **remota**

273a. Grevillea petrophiloides Meisn. subsp. petrophiloides

Illustrations: A.S.George, *Introd. Proteaceae W. Australia* 57, t. 80 (1984), as *G. petrophiloides*; N.R.Marriott, *Austral. Pl.* 13 (108), cover pl. (1986), as *G. petrophiloides*; D.J.McGillivray & R.O.Makinson, *Grevillea* 387, 388, col. pl. & fig. 103 (1993).

Shrub, with emergent flowering branches to 3 m tall. Branchlets regular to subsecund, rarely the youngest branchlets glaucous. Leaves 5–13 cm long, not glaucous, mostly with 3–9 primary lobes, these again 3–5 (–9)-partite, occasionally some tertiary division; ultimate lobes usually > 10 per leaf, (10–) 20–60 (–90) mm long, 0.5–2 (–3.5) mm wide, linear-subterete, linear-trigonous or flat and then narrowly oblong to oblanceolate; margins incurved to flat or (on subterete lobes) obscure; upper surface convex to flat or with a broad longitudinal channel. Peduncles not glaucous. Unit conflorescence 6–10 cm long. Flower colour: perianth pale pink to reddish pink with a blue-grey limb, or rarely creamy white with a greenish grey limb; style pink or red or with a pale tip, drying to red brown, rarely (with perianth) white. Pistils 14–18 mm long. Fruit surface with viscid pustules, glabrous or with erect glandular hairs 0.3–0.6 mm long. Plate 56.

Occurs in south-western W.A., from the lower Murchison R. S to Quairading and about Hyden. Grows mainly in sand-plain habitats in sclerophyll shrubland or heath in sandy or gravelly soils. Regenerates from seed. Flowers year round, peaking in Aug.–Sept. Map 345.

W.A.: Hyden, *M.Barrow 111* (PERTH); 54 km E of Geraldton on Mullewa road, *B.G.Briggs 6435* (NSW); S of Ajana sandplain, *N.T.Burbidge 2183* (CANB); Northwest Coastal Hwy, 16 km S of Northampton, *R.Filson 8671* (MEL); 3.1 km E of Wyalkatchem, *M.D.Tindale 108* (NSW, PERTH).

Juvenile leaves on the first few nodes of a seedling are (?sometimes) cuneate and apically toothed. Occurrence of leaves with lobes > 2 mm wide in adults may be a neotenic character state.

273b. Grevillea petrophiloides subsp. magnifica McGill., *New Names Grevillea* 11 (1986)

G. magnifica (McGill.) Olde & Marriott, *Grevillea Book* 1: 179 (1994); *G. magnifica* subsp. *magnifica* Olde & Marriott *loc. cit.* T: Mt Caroline S of Kellerberrin, W.A., 12 Sept. 1978, *D.F.Blaxell 1722*; holo: NSW.

Illustrations: P.M.Olde & N.R.Marriott, *Grevillea Book* 3: 11 (bottom right), 12 (3A, 3B) (1995), as *G. magnifica* subsp. *magnifica*.

Shrub, with emergent flowering branches to 4 m tall. Branchlets not secund, usually glaucous. Leaves (13–) 18–25 cm long, usually with 3–5 primary lobes, these usually again 3–7-partite, sometimes with tertiary division; ultimate lobes > 10 per leaf, 3–12 cm long, 0.5–1.0 mm wide, linear-subterete, not glaucous; margins incurved to flat or obscure; upper surface flat or with a broad longitudinal channel. Peduncles strongly glaucous. Unit conflorescence usually 10–20 cm long. Flower colour: perianth pale to mid-pink with blue-grey limb; style straw-white to very pale pink. Pistils 16–21.5 mm long. Follicle surface with viscid pustules and with many erect simple glandular hairs 0.9–1.2 mm long (rarely surface glabrous or almost so).

Occurs in south-western W.A., restricted to the area between Tammin and about Pantapin. Grows at bases of granite outcrops or in crevices in same, in deep to skeletal granitic loams. Regenerates from seed. Flowers mainly June–Aug. Map 346.

W.A.: *s. loc.*, *s.d.*, *J.Drummond 8* (BM); Mt Stirling, July 1941, *C.A.Gardner s.n.* (PERTH); Mt Caroline, July 1963, *C.A.Gardner s.n.* (PERTH); Kondacutten Rock, *D.J.McGillivray 3450 & A.S.George* (K, NSW, PERTH).

This subspecies differs from subsp. *petrophiloides* in its distinctive habitat, and in having a somewhat more robust habit with glaucous branchlets and peduncles, more heavily divided leaves with rather longer ultimate lobes, more strongly emergent inflorescences, generally longer unit conflorescences, and somewhat longer and paler pistils. There is however, overlap on most features, and species rank is not appropriate based on the characters used to date.

273c. Grevillea petrophiloides subsp. remota (Olde & Marriott) Makinson, *Fl. Australia* 17A: 504 (2000)

G. magnifica subsp. *remota* Olde & Marriott, *Grevillea Book* 1: 179 (1994). T: Cave Hill Nature Reserve, W.A., 30 Sept. 1988, *P.M.Olde 88/18 & N.R.Marriott*; holo: PERTH; iso: NSW.

Shrub, with emergent flowering branches to c. 2.5–3 m tall. Branchlets not usually secund, not glaucous. Leaves 12–20 cm long, usually with 3 primary lobes, these often again 3-partite; ultimate lobes < 10 per leaf, 3–12 cm long, 0.5–1.0 mm wide, linear-subterete, non-glaucous; upper surface flat or with a broad longitudinal channel; margins incurved or flat or obscure. Peduncles not glaucous. Unit conflorescence 8–15? cm long. Flower colour: perianth pale to mid-pink with a blue-grey limb; style straw-white to very pale pink, drying to straw-coloured. Pistils 16–18 mm long. Follicles glabrous.

Occurs in inland south-western W.A., known only from a few sites from near Varley to Norseman. Grows at bases of granite outcrops or in crevices in same, in deep to skeletal granitic loams. Regenerates from seed. Flowers mainly ? June–Oct. Map 347.

W.A.: Cave Hill Nature Reserve, *P.M.Olde 88/18* (NSW *n.v.*); Lily McCarthy Rock, E of Varley, *P.M.Olde 92/247* (NSW *n.v.*).

This taxon is poorly known. While in the sparing leaf division it shows some similarity to *G. oligomera* (here retained at species rank on the basis of its secund foliage and scarcely emergent, usually decurved to pendulous inflorescences), *G. petrophiloides* subsp. *remota* appears to be better placed in *G. petrophiloides s. lat.* as circumscribed here, with which it shares erect strongly emergent inflorescences, and with a very close relationship to subsp. *magnifica*, with which it shares habitat, style colour, and a general tendency to robustness by comparison with subsp. *petrophiloides*. It differs from subsp. *magnifica* mainly in its less divided leaves and non-glaucous branchlets and peduncles. From the limited fruiting material available, the pustules are at least sometimes not viscid. With reduction in the present treatment of *G. magnifica* to subspecies rank, transfer of the *remota* entity at the same rank is adopted pending future analysis of the whole superspecies.

274. Grevillea oligomera (McGill.) Olde & Marriott, *Grevillea Book* 1: 179 (1994)

G. petrophiloides subsp. *oligomera* McGill., *New Names Grevillea* 12 (1986). T: 59 mile post [95 km] N of Kalgoorlie, W.A., 4 Oct. 1973, *H.Demarz 4564*; holo: PERTH.

Illustrations: P.M.Olde & N.R.Marriott, *Grevillea Book* 3: 66 (bottom right), 67 (48A–C) (1995).

Shrub 1–2 m tall. Branchlets usually secund, sometimes glaucous. Leaves 3–16 cm long, entire and linear or linear-subterete, or 2 or 3 (–5)-sect with primary lobes entire or rarely again bipartite; leaves and lobes ascending, 1.0–1.3 mm wide, sometimes glaucous; divided leaves with ultimate lobes < 10 per leaf, 2.5–9 cm long; margins incurved to flat or obscure; surfaces silky when young, soon glabrous, granulose. Conflorescence erect to decurved or pendulous, terminal, not or scarcely emergent from foliage, usually simple; unit conflorescence cylindrical, basipetal; floral rachis 30–75 mm long, glabrous, glaucous. Flower colour: perianth reddish pink with blue-grey limb; style pinkish red with a pale tip.

Figure 31. *Grevillea*. **A–C**, *G. paradoxa*. **A**, flowering branch; **B**, flower; **C**, pistil (**A–C**, J.H.Ross 2755, CANB). **D–F**, *G. coriacea*. **D**, flowering branch; **E**, flower; **F**, pistil (**D–F**, B.J.Conn & J.de Campos 1368, CANB; leaves on **D**, B.Hyland 5941, CANB). Scale bars: **A**, **E–F** = 1 cm; **B–C** = 5 mm; **D** = 2 cm. Drawn by D.Boyer.

Pistil 14–15 mm long; style-end with a small subapical globose dilation then constricted below the erect narrowly conical pollen-presenter. Follicle ovoid to subglobose, c. 7 mm long, glabrous or with simple glandular hairs, and developing pustular usually viscid craters.

Occurs in inland south-western W.A., from Merredin to Coolgardie and N to Menzies. Grows in sclerophyll shrubland, usually on lateritic rises. Regenerates from seed. Flowers mainly July–Dec. Map 348.

W.A.: 20.9 km N of Broad Arrow on Menzies road, *J.Bale 217* (PERTH); N of Bullabulling, *J.S.Beard 3322* (KPBG, PERTH); Yorkrakine, Sept. 1920, *C.A.Gardner s.n.* (PERTH); Comet Vale, 11 Sept. 1917, *F.Schock* (A *n.v.*); Ora Banda, *E.Wittwer 1314* (KPBG, PERTH).

Narrowly distinct from *G. petrophiloides*, differing mainly in the sparing or absent leaf division and the scarcely emergent, often decurved to pendulous, usually unbranched inflorescences. The branchlets are usually secund (rarely so in the allied species).

275. Grevillea rogersoniana C.A.Gardner, *J. Roy. Soc. W. Australia* 47: 56 (1964)

T: near Shark Bay, W.A., 24 Aug. 1961, *C.A.Gardner 13517*; lecto: PERTH, *fide* D.J.McGillivray & R.O.Makinson, *Grevillea* 438 (1993); isolecto: PERTH.

Illustrations: D.J.McGillivray & R.O.Makinson, *Grevillea* 386, fig. 102, fruit (1993); P.M.Olde & N.R.Marriott, *Grevillea Book* 3: 145 (top left & 115A–C) (1995).

Erect robust shrub 1–4 (–8) m tall. Leaves 4–10 cm long, 0.4–3.5 cm wide, entire and spathulate or narrowly so, or spathulate-cuneate and with 3–5 rounded apical teeth or shallow lobes, always with a long narrowly cuneate base; margins flat or incurved; surfaces sparsely silky, soon glabrous. Conflorescence erect, terminal or rarely cauline, scarcely emergent from foliage, usually 2- or 3-branched; unit conflorescence cylindrical, subsynchronous; floral rachis 40–60 mm long, glabrous or with scattered to dense but inconspicuous biramous and/or simple glandular hairs. Flower colour: perianth bronze in bud, reddish pink at maturity; style pink with a cream tip. Pistil 14–18 mm long; pollen-presenter a slightly oblique to erect narrow cone, slightly dorsally bulbous at base. Follicle broadly ellipsoidal or subglobose, 13.5–20 mm long, glabrous, tuberculate with viscid pustular craters developing. *Rogerson's Grevillea.*

Occurs on the central west coast of W.A., restricted to an area S of Shark Bay between Nanga, Hamelin Pool and Johnson Bore (Tamala Stn). Grows in tall woodland or *Banksia* scrub, on dunes in red calcareous sand. Regenerates from seed. Flowers Aug.–Oct. Map 349.

W.A.: between Hamelin and Nanga, *J.S.Beard 6764* (NSW, PERTH); 40.2 km SW of Hamelin Pool, *A.S.George 3245* (PERTH); 57.9 km S of Denham, *A.S.George 9558* (PERTH); 1.6 km SE of Johnson Bore, *A.S.George 9578* (PERTH); 516 mile peg [830.4 km] on Hamelin Pool turnoff, *F.Lullfitz 2840* (KPBG, PERTH).

The leaf shape is distinct from other species in the group. The ovary stipe (2–3.2 mm long) is slightly longer than in other taxa of the group (all not more than 2 mm long), and the pedicels (0.5–1 mm long) are shorter.

This species is recognised as 'Poorly Known' in J.D.Briggs & J.H.Leigh, *Rare or Threatened Australian Plants* (1995).

276. Grevillea paradoxa F.Muell., *Fragm.* 6: 246 (1868)

T: Western Australia, *J.Drummond [coll. 5] 11*; lecto: MEL, *fide* D.J.McGillivray & R.O.Makinson, *Grevillea* 433 (1993); isolecto: BM, FI, K, MEL, P.

Illustrations: J.W.Wrigley & M.Fagg, *Banksias, Waratahs & Grevilleas* 279 (1989); D.J.McGillivray & R.O.Makinson, *Grevillea* 389 (1993); P.M.Olde & N.R.Marriott, *Grevillea Book* 3: 75 (bottom right), 76 (55A–C) (1995).

Erect open shrub 1–2 m tall. Leaves 15–55 mm long, divaricately pinnatipartite with 3–9 primary lobes, these mostly again 3-partite, sometimes with tertiary division; ultimate lobes linear-subterete, pungent, 5–20 (–27) mm long, 0.5–1.5 mm wide; margins not evident; surfaces pubescent when young, soon glabrous and granulate. Conflorescence erect, terminal or occasionally also axillary, simple, somewhat emergent above foliage; unit conflorescence cylindrical, basipetal to subsynchronous; floral rachis stout, 4–8 cm long, densely and conspicuously white-villous. Flower colour: perianth pale to dark pink or cream; styles pinkish red with a pale tip, or occasionally pale pink to cream. Pistil (12–) 15–18 mm long; pollen-presenter erect, narrowly conical to subterete. Follicle ovoid, slightly compressed, 8–13 mm long, tuberculate, loosely subvillous with biramous hairs and underlying simple glandular hairs. *Bottlebrush Grevillea.* Plate 55; Fig. 31A–C.

Occurs in south-western W.A., widespread in drier parts in the area bounded by Mullewa, Wubin, Kondinin, Kalgoorlie and Die Hardy Ra. Grows in mallee scrub and shrubland, in sandy to clayey or gravelly soils over laterite and sandstone. Regenerates from seed. Flowers year round, mainly June–Oct. Map 350.

W.A.: 48 km W of Hyden, *M.L.Clark 222* (PERTH); Kurrawang, Sept. 1915, *J.B.Cleland s.n.* (AD); Cowcowing, *M.Koch 1103* (AD, NSW, P *n.v.*); N side of Die Hardy Ra., c. 140 km N of Southern Cross, *D.J.McGillivray 3674 & A.S.George* (NSW, PERTH); 3.1 km E of Wyalkatchem, *M.D.Tindale 104* (K, NSW, PERTH, RSA).

Hilliana Group

Robust shrubs or small trees. Leaves entire or divided, dorsiventral with surfaces similar or dissimilar and margins flat to shortly revolute, or unifacial with surfaces ±similar and edges flat or obscure. Conflorescence usually terminal, sometimes also axillary, erect or occasionally decurved, usually paniculately branched; unit conflorescence cylindrical to shortly so, acropetal to synchronous or occasionally basipetal. Flowers acroscopic or transverse on rachis, rarely basiscopic. Torus moderately oblique to transverse. Perianth zygomorphic, glabrous or occasionally hairy outside and/or inside; tepals either remaining coherent and held ventrally, or partially everting then independently lax or recoiled, or rarely everted with limb remaining coherent. Pistil 3–42 mm long; ovary stipitate, glabrous or rarely sparingly hairy; style glabrous, exserted from late bud; pollen-presenter usually erect-conical, sometimes oblique and broadly conical to flat. Fruit usually slightly (rarely strongly) compressed, glabrous or with hairs soon falling, often viscid when young and/or decorticating when older; pericarp usually thick-walled, > c. 1 mm thick and horny, sometimes thinner and crustaceous. Seed flat-ellipsoidal, usually winged all around, sometimes unwinged and then ellipsoidal or rarely hemispherical with a spongy testa.

A group of 25 species found in subtropical rainforests, monsoon tropics, south-west temperate, and eremaean zones; absent from south-eastern temperate areas. *Grevillea glauca* also occurs in Papua New Guinea. Mostly insect-pollinated, some larger-flowered species (including *G. parallela*, *G. coriacea*, *G. calcicola*) bird-pollinated. *Grevillea meisneri* and *G. gillivrayi* (both New Caledonia) have their closest affinities with this group. The fruit of *G. candicans* is unique in the genus in being a nut (indehiscent); the fruits of the other species in the group are follicles.

Within the group, species 285–289 (*G. pyramidalis* to *G. calcicola*) and species 290–292 (*G. glauca* to *G. donaldiana*) form distinctive assemblages of closely related taxa. The group as a whole is closely related to the *Trifida* and *Integrifolia* groups.

1 Leaves unifacial (adaxial surface reduced to a narrow usually pale dorsal stripe or ridge along upper edge; most of leaf made up of the expanded abaxial midvein) and vertically inserted on stem, or leaves dorsiventral with both faces similar, or terete with multiple grooves

2 Perianth with hairs (sometimes few) on outer surface

3 Conflorescences simple, decurved to pendulous; pollen-presenter an erect narrow cone **290. G. glauca**

3: Conflorescences erect, if branched then branches spreading or erect; pollen-presenter an erect to oblique broad cone

29 Perianth outer surface glabrous

30 Lower surface of leaves on either side of midvein at least partly exposed; single-stemmed shrub or small tree 2.5–15 m tall **281. G. parallela**

30: Lower surface of leaves with margins tightly revolute to midvein, concealing lamina; shrub, much branched towards base, 2–4 m tall

31 Pollen-presenter a prominent cone or dome; nectary usually annular, sometimes U-shaped; leaf lobes ⩽ 1 mm wide; stipe of ovary > 2 mm long **293. G. candelabroides**

31: Pollen-presenter flat to concave; nectary U-shaped; leaf lobes 1.3–2.5 mm wide; stipe of ovary ⩽ 2 mm long **289. G. calcicola**

29: Perianth outer surface with hairs (sometimes a few only on limb)

32 Lower surface of leaves on either side of midvein at least partly exposed and with 1–7 parallel longitudinal veins (monsoon tropics) **281. G. parallela**

32: Lower surface of leaves with margins usually tightly revolute to midvein, concealing lamina, only midvein evident (arid zone)

33 Ovary glabrous; conflorescence usually 5–12 (–20)-branched; longest ultimate floral rachis > 6 cm long; follicle crustaceous with exocarp flaking away soon after maturity to reveal pale mesocarp **294. G. stenobotrya**

33: Ovary usually with a few biramous hairs (soon falling); conflorescence 2–6-branched; longest ultimate floral rachis ⩽ 6 cm long; follicle bony-textured, with exocarp not flaking away **283. G. berryana**

277. Grevillea helmsiae F.M.Bailey, *Queensland Agric. J.* 4: 195, t. LXXXIX, XC (1899)

T: Childers, Qld, 14 Apr. 1908, *S.Helms s.n.*; holo: BRI; iso: K, MEL.

Illustrations: F.M.Bailey, *loc. cit.*; J.W.Wrigley & M.Fagg, *Banksias, Waratahs & Grevilleas* 257 (1989); P.M.Olde & N.R.Marriott, *Grevillea Book* 2: 193 (lower right), 194 (159A, B) (1995).

Shrub or tree, 3–10 m tall. Leaves dorsiventral, entire, elliptic to narrowly so or narrowly obovate, 5–20 cm long, 7–40 mm wide; margins shortly recurved; surfaces dissimilar, with lower surface densely to sparsely subsericeous or rarely glabrous. Conflorescence usually terminal, erect, usually simple, shortly cylindrical to subglobose, 2–4.5 cm long, basipetal; floral rachis 4–20 mm long. Flowers basiscopic. Flower colour: perianth white to cream; style green. Perianth subsericeous outside, bearded inside. Nectary obconical. Pistil 12.5–20 mm long, glabrous; pollen-presenter very oblique, convex. Follicle compressed-ellipsoidal, 20–31 mm long, glabrous, not viscid. Fig. 32D–F.

Occurs in south-eastern Qld near the coast and coastal ranges from the Brisbane area to N of Rockhampton. Grows on rainforest margins and in dry rainforest scrubs and brigalow communities on various substrates. Regenerates from seed, and suckers from base. Flowers Oct.–Apr. Map 351.

Qld: 11.3 km NE of 'Wollombi' Stn, *L.G.Adams 1098* (BRI, CANB); Isis, *S.Helms 130* (C *n.v.*, K); 5 km E of Ban Ban Springs on road to Biggenden, *A.N.Rodd 4373 & M.Hardie* (BRI, CANB, NSW, QRS); Goodnight Scrub, c. 65 km SW of Bundaberg, *L.S.Smith 09853* (BRI); Grevillea Plateau, Callide Valley, *C.T.White 10767* (BRI, K).

Grevillea helmsiae has close affinities to *G. elbertii* Sleumer (from Sulawesi, Indonesia), and perhaps to *G. papuana* Diels (from Papua New Guinea).

278. Grevillea hilliana F.Muell., *Trans. & Proc. Philos. Inst. Victoria* 2: 72 (1858)

T: Pine River, Moreton Bay, [Qld], *s.d.*, *W.Hill & F.Mueller*; holo: MEL; iso: K.

Illustrations: D.J.McGillivray & R.O.Makinson, *Grevillea* 125 (1993); P.M.Olde & N.R.Marriott, *Grevillea Book* 2: 194 (bottom centre), 195 (160A, B) (1995).

Tree 5–30 m tall. Leaves dorsiventral. Juvenile and intermediate leaves 28–40 cm long, 150–300 mm wide across lobes, 3–5-fid or 3–5 (–10)-partite with narrowly oblong to narrowly ovate ascending lobes 10–25 cm long, 10–50 mm wide, rarely some secondary division. Adult leaves usually entire, narrowly oblong to slightly narrowly obovate, 9–24 cm long, 15–60 mm wide; margins shortly recurved; surfaces dissimilar, with lower surface sericeous, silvery grey. Conflorescence terminal or axillary, erect to decurved, simple or rarely few-branched; unit conflorescence cylindrical, 9–22 cm long, weakly acropetal to subsynchronous. Flowers transverse on rachis. Flower colour: perianth green in late bud becoming white to cream; style cream to white. Perianth sericeous outside, glabrous inside or papillose to minutely pubescent towards base. Nectary arcuate. Pistil 13.5–16 mm long, glabrous; pollen-presenter very oblique, convex. Follicle compressed-ovoid to -broadly ellipsoidal, 17–26 mm long, glabrous, rugulose, not viscid. *White Silky Oak, White Yiel Yiel.*

Occurs in near coastal eastern Australia from Cooktown, Qld, S to the Brunswick R. in northern N.S.W. Grows in and on margins of rainforest communities, often in deep fertile soils. Regenerates from seed. Flowers erratically, mainly July–Oct.? Map 352.

Qld: Dalrymple Heights and vicinity, Nov. 1947, *M.S.Clemens s.n.* (BRI, K); Strathdickie North near Proserpine, *K.Macpherson 125* (BRI, QRS); Eungella Ra., *C.T.White 12860* (BRI, CANB). N.S.W.: Brunswick R., 1.6 km NW of Brunswick Heads, 12 June 1957, *L.A.S.Johnson & E.F.Constable* (K, NSW); Brunswick R. next to Pacific Hwy, c. 3.2 km N of Brunswick [Heads], *R.F.Thorne 21932* (N *n.v.*, NSW, RSA *n.v.*).

Grevillea hilliana can be confused with *G. baileyana* (both have leaves penninerved below), but *G. baileyana* has rusty brown hairs on the lower leaf surface, usually much-branched conflorescences, a shorter pistil 9–13.5 mm long, and pericarp < 1 mm across at the suture (> 1.5 mm across in *G. hilliana*). The timber of *G. hilliana* has been used in the past for cabinet work and veneer.

279. Grevillea baileyana McGill., *New Names Grevillea* 2 (1986)

Kermadecia pinnatifida F.M.Bailey, *Queensland Woods* 69 n. 332a (1886); *G. pinnatifida* (F.M.Bailey) F.M.Bailey, *Occas. Pap. Queensland Fl.* 6 (1886), *nom. illeg. non* Jacques (1843). T: [cultivated in Botanic Gardens, Brisbane, Qld]; neo: BRI 145218, *fide* D.J.McGillivray & R.O.Makinson, *Grevillea* 407 (1993).

Illustrations: F.M.Bailey, *Queensland Pl.* 4, t. LVIII (1901), as *G. pinnatifida*; W.R.Elliot & D.L.Jones, *Encycl. Austral. Pl.* 5: 29 top right (1990); P.M.Olde & N.R.Marriott, *Grevillea Book* 2: 47 (top left & 33A, B) (1995).

Tree 6–30 m tall. Leaves dorsiventral, 6–30 cm long. Juvenile leaves pinnatipartite with 5–9 ascending oblong-ovate to lanceolate simple lobes 3–11 cm long, 15–30 mm wide. Adult leaves usually entire, narrowly elliptic or narrowly oblong-obovate, 10–60 (–100) mm wide, or with a few 2- or 3-fid leaves persisting; margins flat or very slightly recurved; surfaces dissimilar, with lower surface rusty brown-sericeous (rarely silver-grey). Conflorescence terminal, erect, usually paniculately 5–16-branched, or sometimes axillary and simple; unit conflorescence cylindrical, 6–14 cm long, acropetal. Flowers acroscopic. Flower colour: perianth pale greenish in bud becoming cream to very pale yellow; style cream. Perianth sericeous to loosely so outside, pilose inside below ovary. Nectary arcuate. Pistil 9–13.5 mm long, glabrous; pollen-presenter strongly oblique, convex to broadly conical. Follicle compressed-obloid to -obovoid, 13–19 mm long, ±smooth, glabrous, not viscid. *Scrub Beefwood, White Oak, Brown Silky Oak.* Plate 54.

Occurs in north-eastern Qld within c. 50 km of the coast, mainly from about Ingham N to the Cooktown area and in the Coen–McIlwraith Ra. area; also in Papua New Guinea. Grows in rainforests and margins, sometimes common in regrowth areas, various sites including ridges and creek banks, often in granitic soils. Regenerates from seed. Flowers Aug.–Dec. Map 353.

Qld: Mowbray R., *L.J.Brass 1989* (A *n.v.*, B *n.v.*, BISH *n.v.*, BRI, K); State Forest Reserve 933, Little Pine Logging Area, *B.Hyland 9210* (K, QRS); Daintree R., *S.F.Kajewski 1453* (A *n.v.*, B *n.v.*, E, K, NY *n.v.*, P); 25.8 km SE of Coen, *L.S.Smith 11935* (BRI, K); Etty Bay, *C.T.White 11741* (A *n.v.*, BRI).

Wood from this species has been used in cabinet making. The flowers are fragrant or sweet-scented. There is some variation, plants from the Iron Ra. area having leaves with broader juvenile leaf lobes; populations in Papua New Guinea are more prone to retaining divided leaves on adult plants.

Grevillea baileyana can be confused with *G. hilliana* (both species have leaves penninerved below) but *G. hilliana* always has silvery grey hairs on the lower leaf surface, usually simple conflorescences, pistil 13.5–16 mm long, and pericarp > 1.5 mm across at suture (< 1 mm across in *G. baileyana*).

280. Grevillea striata R.Br., *Trans. Linn. Soc. London* 10: 177 (1810)

G. striata var. *typica* Domin, *Biblioth. Bot.* 89: 36 (1921), *nom. illeg.* T: 'In Novae Hollandiae ora septentrionale; Carpentaria [Qld or N.T.]: prope littora' [protologue]; type not found (possibly BM?).

G. lineata R.Br., *in* C.Sturt, *Exped. Central Australia* 2: App. 87 (1849); *G. striata* var. *lineata* (R.Br.) Domin, *Biblioth. Bot.* 89: 36 (1921). T: 'it takes the place of the Gum tree in the creeks in about Lat 29°30' S', [S.A.], *s.d.*, [*C.*]*Sturt 31*; holo: BM.

Illustrations: D.J.McGillivray & R.O.Makinson, *Grevillea* 127 (1993); A.A.Mitchell & D.G.Wilcox, *Arid Shrubland Pl. Western Australia* 442, 443 (1994); P.M.Olde & N.R.Marriott, *Grevillea Book* 3: 187 (top right & 152A, B) (1995).

Erect spindly shrub or robust tree 3–15 m tall. Leaves dorsiventral, entire, linear to broadly so or strap-like, often wavy, 10–45 cm long, 2–15 mm wide; margins flat or shortly recurved; surfaces rather similar; lower surface longitudinally 5–13 striate, sericeous especially in grooves. Conflorescence usually terminal, erect, paniculately 4–12-branched or rarely simple; unit conflorescence cylindrical, 5–8 cm long, subsynchronous or weakly basipetal or acropetal. Flowers ±acroscopic. Flower colour: perianth and style white to cream or pale yellow. Perianth subsericeous to tomentose outside, pilose inside near ovary. Nectary minute, arcuate. Pistil 6–10 mm long, glabrous; pollen-presenter basally transverse to oblique, erect-conical. Follicle compressed obloid to ovoid, shortly subapically beaked, 13–21 mm long, glabrous, almost smooth, not viscid. *Beefwood, Silver Honeysuckle.*

Occurs in all mainland States except Vic.; widespread in dry inland areas, occasionally extending to near the coast in drier northern areas. Grows in open mixed *Acacia* and *Eucalyptus* woodland, *Triodia*-shrubland communities, in a range of soil types including red sands, gravelly loams, and occasionally black soils or calcareous clays. Regenerates from seed, some epicormic shooting capability. Flowers mainly Aug.–Dec., also opportunistically. Map 354.

W.A.: near Mt Marmion, *W.V.Fitzgerald 1151* (PERTH). N.T.: 16 km W of Borroloola, *J.R.Maconochie 857* (AD, DNA, NSW, PERTH). S.A.: 66 km E of Everard Park Stn and on Granite Downs Stn, *D.E.Symon 3352* (AD, CANB, HUJ *n.v.*, K, NSW). Qld: 1.6–2.4 km N of Yabulu, *L.S.Smith T. 170* (A *n.v.*, BRI, CANB). N.S.W.: Calindry Stn, *P.Martensz 252* (CANB, NSW).

Fairly uniform throughout most of the range. A 'narrow-leaved form' occurs in W.A. from Wiluna to the Gascoyne R. area; this has leaves usually < 5 mm wide, with fewer (usually c. 5) nerves; flowering material of this form has not been seen.

Grevillea striata may occasionally be confused with *G. parallela* or *G. coriacea*, which have longer pistils (> 13 mm long), thicker-walled fruits (pericarp 1–2 mm thick), and more strongly dissimilar upper and lower leaf surfaces.

A sheet at BM with a label 'No 6 spie Carpentaria Island c Nov 19 - 1802', while not a mixed collection, also bears a label with conflicting [Qld] localities, and bears only budding material; in describing *G. striata* Brown clearly saw mature flowers. The sheet may nevertheless be part of the type material.

281. Grevillea parallela Knight, *Cult. Prot.* 121 (1809)

T: Endeavour River, Qld, 1770, *J.Banks*; lecto: BM, *fide* D.J.McGillivray & R.O.Makinson, *Grevillea* 433 (1993); possible isolecto: B *n.v.*, BM, E, MEL, NSW, P *n.v.*, US *n.v.*

G. ceratophylla R.Br., *Trans. Linn. Soc. London* 10: 177 (1810); *G. refracta* var. *ceratophylla* (R.Br.) Benth., *Fl. Austral.* 5: 458 (1870). T: Carpentaria Islands, [Morgan Island, c. 13°S, 136°E, N.T.], 20 Jan. 1803, *R.Brown Iter Austral. 3318*; lecto: BM, *fide* D.J.McGillivray, *Telopea* 1: 29 (1975).

G. polystachya R.Br., *Trans. Linn. Soc. London* 10: 177 (1810). T: Banks Island, Shoal Bay Passage, Qld., 28 Aug. 1802, *R.Brown Iter Austral. 3316*; lecto: BM, *fide* D.J.McGillivray & R.O.Makinson, *Grevillea* 433 (1993); ?isolecto: BRI, FI *n.v.*, NSW, P *n.v.*, S *n.v.*

G. angustata R.Br., *Suppl. Prodr. Fl. Nov. Holl.* 24 (1830). T: Cape Cleveland, Qld, 1819, *A.Cunningham, 361* (2nd voyage of *Mermaid*); holo: BM.

G. polystachya var. *hebestachya* Benth., *Fl. Austral.* 5: 459 (1870). T: Cape York, Qld, *s.d.*, *E.Daemel*; syn: BM, K; Dayman's Island, Qld, *s.d.*, *W.Hill*; syn: K.

G. polybotrya F.Muell., *Hooker's J. Bot. Kew Gard. Misc.* 9: 23 (1857), *nom. illeg. non* Meisn. (1856); *G. muelleriana* H.Buek, *Gen. Sp. Syn. Candolleana* 4: iv (1874). T: McAdam's Range, N.T., *F.Mueller?*; holo: ?K or ?MEL.

G. heteroneura W.Fitzg., *Western Mail (Perth)* 21 (1066): 10, 28 incl. t (1906), as *G. Hetroneura*. T: Tabletop Mountain, W.A., Sept. 1905, *W.V.Fitzgerald*; lecto: NSW, *fide* D.J.McGillivray, *Telopea* 1: 29 (1975); ?isolecto: E, NSW, PERTH.

Illustrations: K.A.W.Williams, *Native Pl. Queensland* 2nd edn, 1: 137 (1980); J.Brock, *Top End Native Pl.* 205 (1988); P.M.Olde & N.R.Marriott, *Grevillea Book* 3: 77 (56A, B), 78 (top left) (1995).

Single-stemmed shrub or tree, 2.5–15 m tall. Leaves dorsiventral, 10–40 cm long, simple and linear to strap-like and 1.5–10 mm wide, or pinnatipartite to pinnatisect with 3–9 ascending linear (rarely subtriangular) lobes 2–20 cm long and 0.9–10 mm wide; margins recurved to shortly revolute; surfaces dissimilar with lower surface sericeous, with 1–7 parallel longitudinal veins. Conflorescence usually terminal or occasionally axillary, erect, usually paniculately 3–7-branched; unit conflorescence cylindrical, 6–10 cm long, acropetal to subsynchronous. Flowers transverse on rachis. Flower colour: perianth and style white or cream to pale yellow-green. Perianth glabrous outside or occasionally subsericeous to tomentose, partly pilose inside or papillose or rarely tuberculate to smooth. Nectary arcuate. Pistil 13–26 mm long, glabrous; pollen-presenter oblique, conical to convex. Follicle compressed-ellipsoidal to lenticular, 14–29 mm long, glabrous, granulose, not viscid. *Silver Oak.* Plate 57.

Widely distributed in the monsoon tropics of northern Australia; in W.A. in the Kimberley and S to Derby; in N.T. N of 19°S; in Qld N from Tambo area. Grows in open sclerophyll forest or woodland, often with grass understorey on level plains or river levees, in sand or granitic or lateritic loam or cracking clays. Regenerates mainly from seed, some epicormic sprouting. Probably bird-pollinated. Flowers predominantly June–Oct. Map 355.

W.A.: 115.9 km NNW of Gibb River Stn, Kimberleys, *N.H.Speck 4948* (BRI, CANB, PERTH). N.T.: 97.2 km N of Top Springs Store, *G.Chippendale 3721* (AD, CANB, DNA, NSW, PERTH); c. 8 km SE Fish River HS turnoff, *J.R.Maconochie 2010* (CANB, DNA, NSW). Qld: 9.7 km W of Westmoreland Stn, *R.A.Perry 1325* (AD, BRI, CANB, NSW, PERTH); 34.6 km NW of Marlborough, *N.H.Speck 1742* (BRI, CANB, PERTH).

Similar to and easily confused with *G. coriacea*, which has pistils 33–42 mm long, and always simple leaves with fewer (usually 3) longitudinal veins. For differences from *G. striata* see under that species.

282. Grevillea coriacea McGill., *Telopea* 1: 19 (1975)

T: 58 miles [93 km] from Coen towards Moreton Telegraph Station, Qld, July 1968, *C.H.Gittins 1820*; holo: NSW.

Illustrations: P.M.Olde & N.R.Marriott, *Grevillea Book* 2: 99 (bottom right), 100 (79A, B) (1995).

Slender tree 4–15 m tall. Leaves dorsiventral, entire, linear to strap-like, 8–30 cm long, 4–10 mm wide; margins recurved to loosely revolute; surfaces dissimilar, with lower surface sericeous with prominent midvein and intramarginal veins. Conflorescence terminal, erect, simple or 2- or 3-branched; unit conflorescence loosely cylindrical, 8–15 cm long, acropetal. Flowers basiscopic. Flower colour: perianth and style white to cream; style sometimes pink with age. Perianth glabrous outside, pilose inside near base. Nectary arcuate. Pistil 33–42 mm long, glabrous; pollen-presenter very oblique, broadly conical. Follicle lenticular-ellipsoidal, 21–28 mm long, smooth and glaucous, glabrous. Fig. 31D–F.

Occurs in north-eastern Qld, from Petford to Mt Molloy and Mt Mulligan, and NW from Cooktown. Grows in open forest or savanna woodland often on dry slopes, in loam to clay soils over often slatey substrates. Regenerates from seed. Probably bird-pollinated. Flowers July–Sept. Map 356.

Qld: Desailly Ra. on Lakeland Downs to Cooktown road, *D.F.Blaxell 1182* (NSW); Mareeba–Mt Molloy road, *L.J.Brass 33859* (QRS); inland from Bathurst Bay, *B.P.M.Hyland 4803* (NSW, QRS); 5 km W of Mt Carbine, *B.P.M.Hyland 5941* (NSW, QRS); 4 km E of Mareeba on Tinaroo Ck road, *L.J.Webb & J.G.Tracey 6179* (BRI, CANB).

Similar and very closely related to *G. parallela*, which has pistils 13–26 mm long, leaves often divided, and usually more numerous longitudinal veins on the lower leaf surface. For differences from *G. striata*, see under that species.

283. **Grevillea berryana** Ewart & Jean White, *Proc. Roy. Soc. Victoria* n. ser. 22(1): 14, t. 8 (1909)

T: Malcolm, W.A., Dec. 1907, *F.A.Rodway 321*; holo: MEL; iso: MEL, PERTH.

Illustrations: D.J.McGillivray & R.O.Makinson, *Grevillea* 131, fig. 24 (1993); A.A.Mitchell & D.G.Wilcox, *Arid Shrubland Pl. W. Australia* 438, 439 (1994); P.M.Olde & N.R.Marriott, *Grevillea Book* 2: 61 (bottom centre & 45) (1995).

Shrub or tree, 1.5–7 m tall. Leaves dorsiventral, 6–20 (–30) cm long, usually irregularly pinnatipartite with 2–7 lobes, sometimes (especially on sucker growth) entire; simple leaves and lobes linear, 4–15 cm long, 0.7–2 mm wide; margins revolute to midvein; surfaces dissimilar with lower surface 2-grooved and lamina enclosed except at sinuses, subsericeous. Conflorescence terminal, erect, paniculately 2–6-branched; unit conflorescence cylindrical, 2–6 cm long, ?subsynchronous. Flowers transverse on rachis. Flower colour: perianth and style pale cream to yellow. Perianth sparsely villous or subsericeous to almost glabrous outside, glabrous or with scattered erect hairs inside. Nectary arcuate. Pistil 10.5–13 mm long, glabrous or with loose hairs about ovary and base of style, soon falling; pollen-presenter strongly oblique, conical. Follicle compressed-ovoid to subdiscoid, apiculate, 10–17 mm long, rugulose, glabrous, not viscid. Fig. 32A–C.

Occurs in W.A., where widely distributed in the eremaean zone from Menzies NE to the eastern Gibson Desert, and from Mount Magnet NW to the lower Fortescue R. Grows in various habitats, mulga and mixed shrublands, and *Triodia* associations, flat to rocky situations, usually in red lateritic loam soils. Regenerates from remote rhizomes and seed. Flowers sweet-scented. Flowers Dec.–Feb. Map 357.

W.A.: Bee Hill, *G.Davis 184* (PERTH); near upper Rudall R., *A.S.George 10749* (PERTH); c. 190 km N of Neale Junction, Gibson Desert, E of Miss Gibson Hill, *A.S.George 12003* (PERTH); 5 km from Mount Magnet towards Wubin on Great Northern Hwy, *D.J.McGillivray 3375 & A.S.George* (CANB, K, MEL, NSW, PERTH); Ashburton R., 10 km below highway, *A.S.Weston 10852* (PERTH).

Grevillea berryana can be confused with *G. nematophylla* which has smooth silvery adult bark (rough in *G. berryana*), the leaves without an obvious margin, and the abaxial midvein making up the bulk of the leaf.

284. **Grevillea nematophylla** F.Muell., *Fragm.* 1: 136 (1859)

T: [Mt Murchison near] Darling River, [N.S.W.], [*J.*] *D*[*allachy*]; lecto: NSW 117255, *fide* D.J.McGillivray & R.O.Makinson, *Grevillea* 431 (1993); isolecto: CGE *n.v.*, E, K, LD *n.v.*, MEL, NA *n.v.*, NY *n.v.*, P *n.v.*

Illustrations: J.P.Jessop & H.R.Toelken (eds), *Fl. S. Australia*, 4th edn, 1: 131, fig. 70C (1986), not assignable to subspecies; P.M.Olde & N.R.Marriott, *Grevillea Book* 3: 48 (bottom right), 49 (34A, B) (1995), probably subsp. *nematophylla*.

Figure 32. *Grevillea*. **A–C**, *G. berryana*. **A**, flowering branch; **B**, flower; **C**, pistil (**A–C**, K.Allen 211, PERTH). **D–F**, *G. helmsiae*. **D**, flowering branch; **E**, flower; **F**, pistil (**D–F**, R.O.Makinson 1397 *et al.*, CANB). Scale bars: **A** = 2 cm; **B–C** = 2 mm; **D** = 1 cm; **E–F** = 4 mm. Drawn by: **A–C**, D.Fortescue; **D–F**, L.Spindler.

Shrub or tree, 1.5–6 (–10) m tall. Leaves unifacial or almost so (with margins often obscure and adaxial surface reduced to a 2-furrowed strip or narrow linear crest atop a rotund or vertically flattened abaxial midvein), 4–30 cm long, simple and subterete or linear to very narrowly obovate, or pinnatisect with 3–10 ascending linear lobes; simple leaves and lobes 0.5–1.5 mm wide; surfaces openly subsericeous. Conflorescence terminal, erect, paniculate with 5–10 widely diverging branches or occasionally simple; unit conflorescence cylindrical, 6–14 cm long, acropetal to subsynchronous. Flowers transverse on rachis. Flower colour: perianth and style cream (?occasionally tinged yellow or pink). Perianth glabrous or subsericeous outside, glabrous or sparsely bearded or minutely papillose inside. Nectary annular or nearly so. Pistil 5–15 mm long, glabrous; pollen-presenter erect, oblique-conical. Follicle compressed ellipsoidal to sublenticular, 13–24 mm long, glabrous, rugulose, glossy, ?not viscid. *Water Bush, Silver (-Leaved) Water Bush.*

Occurs in southern eremaean Australia, widespread in W.A. south from about 26°S and in inland S.A.; rare in southern parts of N.T.; rare (possibly extinct) in far western N.S.W.; not known to occur in Qld or Vic. Grows in varied habitats including *Acacia* (Mulga) or *Eucalyptus* woodland, *Triodia* associations, tall shrubland or bluebush and saltbush plains, often in low-lying or somewhat swampy situations or along watercourses in clay to sand-loam soils, sometimes over limestone, laterite or ironstone. Regenerates from seed. Three subspecies are recognised.

Grevillea nematophylla has distinctively silver-grey narrow leaves or leaf lobes and smooth silvery adult bark. It lacks obvious grooves on the lower leaf surface, in contrast to *G. berryana* which is 2-grooved. *Grevillea nematophylla* also has the stipe of the ovary clearly less than half the length of the style, slender branchlets and leaf bases and leaf scars < 2 mm across, features which distinguish it from *G. pyramidalis* and *G. erythroclada* which have the ovary stipe at least half the length of the style, stout branchlets and leaf bases and leaf scars > 2 mm across.

The differences in leaf form appear to represent variations of intermediacy between a dorsiventral pattern and the unifacial pattern. Further studies are needed on all the variants, including the anatomy and morphology of their juvenile leaves, and ecological preferences. The specimen base for this species is unusually poor, but the foliar differences appear consistent and geographically based, and on that basis three subspecies are recognised. Two collections of *G. nematophylla* from the southern margin of the range of subsp. *nematophylla* have the leaves with the cross-sectional shape typical of subsp. *supraplana* (see McGillivray & Makinson, *Grevillea* 134 (1993) for details) and are here treated as unassigned to subspecies.

1 Leaves simple, subterete; upper surface with a narrow rounded longitudinal ridge (much narrower than leaf), the base of the ridge usually with a shallow inconspicuous groove on either side (pistil (6–) 8–15 mm long)	**284a.** subsp. **nematophylla**
1: Some or all leaves divided, or if all simple then either vertically flattened (narrowly elliptic to oblong in cross-section), or three-quarters round in cross-section with upper surface a broad shallowly concave strip almost as wide as leaf, with grooves under outer edges of strip	
2 Simple leaves or lobes three-quarters round in cross-section with upper surface a broad shallowly concave strip as wide as leaf, with grooves under outer edges of strip; pistil 5–6 mm long	**284b.** subsp. **supraplana**
2: Simple leaves or lobes laterally flattened (narrowly elliptic to oblong in cross-section), pistil c. 7–9 mm long	**284c.** subsp. **planicosta**

284a. Grevillea nematophylla F.Muell. subsp. nematophylla

G. purdieana Diels, *Bot. Jahrb. Syst.* 35: 154, fig. 15 (1904). T: Yilgarn, 2.5 km westlich von Southern Cross, W.A., 29 Nov. 1900, *L.Diels 1717*; lecto: B *n.v.*, *fide* D.J.McGillivray & R.O.Makinson, *Grevillea* 431 (1993); remaining syntypes: Huegel bei Southern Cross (Yilgarn Goldfields), W.A., Nov. 1900, *E.Pritzel 78*; syn: B *n.v.*, E; *L.Diels 1717a*; syn: B? *n.v.*; *L.Diels & E.Pritzel 39*; syn: PERTH; *E.Pritzel 78*; syn: A *n.v.*, AD, B *n.v.*, G, K, L *n.v.*, MO *n.v.*, NSW, P *n.v.*, US *n.v.*

G. nematophylla 'Simple-leaved race', of D.J.McGillivray & R.O.Makinson, *Grevillea* 132–135 (1993).

G. nematophylla 'Simple-leaf form, Typical form', of P.M.Olde & N.R.Marriott, *Grevillea Book* 3: 48–50 (1995).

Illustration: D.J.McGillivray & R.O.Makinson, *Grevillea* 133, fig. 25 c1, c2 (1993).

Densely foliaged shrub or small tree 2–6 (very rarely to 10?) m tall. Leaves simple, subterete, 0.5–1.5 mm diam.; upper surface with a narrow rounded longitudinal ridge (much narrower than leaf), usually with a recessed groove along each side at the base of the ridge. Pedicels (0.5–) 0.7–3.0 mm long. Flowers cream. Pistil (6–) 8–15 mm long.

Almost transcontinental, occurring in W.A. in the area bounded by Southern Cross, Lake Barlee, Menzies and Norseman; in N.T., Finke R. area; in S.A. widespread in the dry inland including the Nullarbor Plain; in N.S.W. now possibly extinct, known only from one collection (the type of the name *G. nematophylla*) from Mt Murchison on the Darling R. near Wilcannia. Flowers recorded in Nov. and Dec. Map 358.

W.A.: Die Hardy Ra., *C.A.Gardner 19127* (PERTH); 51 km N of Bullfinch on road to Diemals, *B.H.Smith 1450* (AD, CANB, HO, MEL, NSW, PERTH). N.T.: Hermannsburg, Finke River, 1906–1908, *Strehlow* (B). S.A.: 40 km NW of Mulgaria HS, *F.J.Badman 1025* (AD, NSW); c. 117 km W of Mabel Creek HS on road to Emu, *M.Lazarides 8356* (CANB, NSW).

Populations towards the east of the range, in S.A., tend to have smaller flowers (pistils 6–10 mm long) than those in W.A. (9–15 mm long). The common name 'Water Bush' is supposed to derive from use by Aborigines as a water source; the lateral near-surface roots yielding water when sectioned and drained. The tendency of the plant to occur along drainage lines or near soaks may contribute to currency of the name.

284b. **Grevillea nematophylla** subsp. **supraplana** Makinson, *Fl. Australia* 17A: 504 (2000)

T: 30 miles [50 km] N from Mt Churchman, W.A., 10 Dec. 1891, *R.Helms* ['14' on some sheets]; holo: K, *Helms 14*, excluding detached inflorescence at right of sheet; iso: AD, MEL, NSW, PERTH.

G. nematophylla 'Divided-leaved race', of D.J.McGillivray & R.O.Makinson, *Grevillea* 132–135 (1993).

G. nematophylla 'Divided-leaf form', of P.M.Olde & N.R.Marriott, *Grevillea Book* 3: 48–50 (1995).

Illustration: D.J.McGillivray & R.O.Makinson, *Grevillea* 133, fig. 25 b1, b2 (1993).

Shrub or small tree 2–5? m tall. Leaves simple or divided; leaves or lobes c. 0.8–1.5 mm wide, three-quarters round in cross-section, with upper surface a broad shallowly concave strip as wide as leaf, with grooves under outer edges of strip. Pedicels 0.5–0.6 mm long. Flowers cream. Pistil 5–6 mm long.

Occurs in W.A. from Meekatharra and Yalgoo to E of Rawlinna, and in S.A. on the Nullarbor Plain. Flowers recorded Dec. (a single record). Map 359.

W.A.: Wubin to Paynes Find, *F.Lullfitz 3166* (KPBG); 19 km W of Yalgoo on road to Mullewa, *D.J.McGillivray 3369 & A.S.George* (K, NSW, US); Boolardy Stn, *R.Melville 6* (K); Mt Narryer, upper Murchison R., 1899, [*J.?*]*Tyson s.n.* (K). S.A.: Duner Camp, Hughes to Oil Bore Rd, 11 July 1967, *S.Reid* (AD, CANB).

284c. **Grevillea nematophylla** subsp. **planicosta** Makinson, *Fl. Australia* 17A: 505 (2000)

T: Comet Vale, W.A., 14 Dec. 1961, *C.A.Gardner 13862*; holo: K; iso: PERTH.

G. nematophylla 'Flat-leaved race', of D.J.McGillivray & R.O.Makinson, *Grevillea* 132–135 (1993).

G. nematophylla 'Flat-leaf form', of P.M.Olde & N.R.Marriott, *Grevillea Book* 3: 48–50 (1995).

Illustrations: D.J.McGillivray & R.O.Makinson, *Grevillea* 133, fig. 25d (1993); P.M.Olde & N.R.Marriott, *Grevillea Book* 3: 49 (34C) (1995).

Shrub 1–3? m tall. Leaves simple or divided; leaves or lobes (1–) 2–4 mm broad, laterally flattened with upper edge (adaxial surface) a narrow, pale, finely 2-grooved, linear crest sitting atop the flattened abaxial midrib (leaves and lobes like a short-armed T in cross-section). Pedicels < 0.5 mm long. Flowers cream. Pistil c. 7–9 mm long.

Occurs in southern inland W.A., from near Balladonia to the Die Hardy Ra. N of Kalgoorlie, and NE of Laverton. Flowers recorded Dec. (a single record). Map 360.

W.A.: Laverton–Warburton road, 100 km E of Cosmo Newberry, *A.S.George 12192* (PERTH); 6.4 km W of Zanthus, *R.D.Royce 5267* (PERTH).

Leaves or lobes appear to be broadest in the Zanthus area, where plants are reportedly low shrubs less than 2 m tall. This subspecies is very poorly collected, and the variation indicated above is based on a small sample only.

285. **Grevillea pyramidalis** A.Cunn. ex R.Br., *Suppl. Prodr. Fl. Nov. Holl.* 25 (1830)

T: Regents River, [W.A.], Aug. 1821, *A.Cunningham, Kings 4th Voyage, 73*; lecto: BM, *fide* D.J.McGillivray & R.O.Makinson, *Grevillea* 436 (1993); isolecto: K.

Illustrations: D.J.McGillivray & R.O.Makinson, *Grevillea* 146, 147, 148, fig. 31, 32 (1993), indet. to subspecies.

Erect spindly shrub or small tree 2–6 m tall. Leaves unifacial, 15–42 cm long, simple and entire or pinnatisect with 2–20 ascending primary lobes, the lower ones often again 2–8-sect, rarely some tertiary lobing; ultimate lobes laterally flattened, linear to narrowly obovate, 5–29 cm long, (0.6–) 1–21 mm broad; edges flat (true margins obscure); surfaces almost smooth but with very fine ±continuous longitudinal wrinkles, glabrous or with a sparse to dense appressed indumentum. Conflorescence usually terminal, erect, paniculate with 3–11 spreading branches, sometimes some secondary (rarely tertiary) branching, and then secondary and tertiary peduncles with black triangular bases; unit conflorescence cylindrical, 6–20 cm long, weakly acropetal. Flowers acroscopic. Perianth glabrous outside, papillose inside. Nectary U-shaped. Pistil 5–8 (–9.5) mm long, glabrous; pollen-presenter moderately oblique to almost transverse, conical. Follicle compressed-ellipsoidal to -obovoid or suborbicular, 18–24 mm long, glabrous, viscid. *Caustic Bush.*

Occurs in north-western W.A. and northern N.T. There are three subspecies.

Very similar and closely related to *G. erythroclada*, which has reddish to orange branchlets (brownish in *G. pyramidalis*), the leaves or lobes sometimes terete rather than flattened, < 2 mm broad, consistently glabrous or nearly so, and with the longitudinal wrinkles deep, wavy, relatively short and discontinuous. See also notes under *G. nematophylla* for differences.

McGillivray & Makinson (*Grevillea* 436 (1993)) synonymised *G. leucadendron* A.Cunn. ex R.Br. and *G. longiloba* F.Muell. as races under *G. pyramidalis*. Olde & Marriott (*Grevillea Book* 1: 178 (1994); 2: 235–236 (1995); 3: 122–123 (1995)), raised *G. longiloba* to subspecies rank and reinstated *G. leucadendron* at specific rank, distinguished on foliar features only, but there are numerous intergrades between the latter and *G. pyramidalis s. str.*

1 Leaves mostly < 30 cm long, usually 3–20-sect and often with some secondary division; surfaces of adult leaves glabrous or sparsely to densely sericeous

2 Adult leaves glabrous or nearly so; lobes 3–21 mm broad, linear to very narrowly obovate; leaf rachis usually flattened above first node **285a.** subsp. **pyramidalis**

2: Adult leaves usually densely to openly sericeous with lobes 1–5 (–12) mm broad, rarely glabrous or nearly so and then lobes 0.6–1.0 mm broad; leaf rachis ±terete above first node **285b.** subsp. **leucadendron**

1: Leaves mostly 30–42 mm long, simple and linear or 2- or 3-sect with linear to strap-like lobes; surfaces of adult leaves glabrous or almost so **285c.** subsp. **longiloba**

285a. **Grevillea pyramidalis** A.Cunn. ex R.Br. subsp. **pyramidalis**

G. pyramidalis Broad-lobed form, form ‘a’, of D.J.McGillivray & R.O.Makinson, *Grevillea* 146–147 (1993).

Illustrations: J.R.Wheeler (ed.) *et al.*, *Fl. Kimberley Region* 472, fig. 145F (1992), as *G. pyramidalis*; P.M.Olde & N.R.Marriott, *Grevillea Book* 3: 122 (bottom right), 123 (94A–C) (1995).

Leaves mostly 15–32 cm long, most or all pinnatisect with 2–12 ascending primary lobes, these often again 2–4 (–7)-sect, rarely a few leaves entire; simple leaves and ultimate lobes

(3–) 7–21 mm broad, very narrowly obovate to linear or strap-like (laterally compressed), often curved; surfaces of adult leaves glabrous or with scattered appressed biramous hairs; leaf rachis above first lobe usually flattened or occasionally terete. Flower colour: perianth and style white to cream or pale yellow.

Occurs in north-western W.A., from about Broome N to the Prince Regent R. and often within 50 km of the coast, with an isolated occurrence in the McLarty Hills. Grows in low woodland or open *Triodia* associations in sandy or gravelly loam soils. Regenerates from seed and basal epicormic shoots. Flowers c. May–July. Map 361.

W.A.: junction of Stuart & Robinson R., c. 72 km NNE of Derby, *A.C.Beauglehole 52777* (NSW); Glenelg R., *H.Basedow 145* (NSW); McLarty Hills, Great Sandy Desert, *A.S.George 14668* (PERTH); 11.3 km E of Oobagooma Stn, *M.Lazarides 6599* (CANB, PERTH); Kimbolton, *E.Wittwer 23336* (PERTH).

Intermediates with subsp. *leucadendron* occur, usually with leaf lobes 6–12 mm wide with a dense to open indumentum, and leaf rachises slightly flattened.

285b. Grevillea pyramidalis subsp. leucadendron (A.Cunn. ex R.Br.) Makinson, *Fl. Australia* 17A: 505 (2000)

G. leucadendron A.Cunn. ex R.Br., *Suppl. Prodr. Fl. Nov. Holl.* 25 (1830); *G. pyramidalis* var. *leucadendron* (A.Cunn. ex R.Br.) C.A.Gardner ex E.M.Watson, *J. Roy. Soc. W. Australia* 30: 89 (1946). T: Cambridge Gulf, [W.A.], Sept. 1819, *A.Cunningham*, (2nd Voyage of *Mermaid*), *412*; lecto: BM, *fide* D.J.McGillivray & R.O.Makinson, *Grevillea* 436 (1993); isolecto: BM, K.

G. obliqua R.Br., *Suppl. Prodr. Fl. Nov. Holl.* 25 (1830). T: 'Ora septent.-occid.? 1801. in it. *Baudin* ex Herb. Mus. Paris' [protologue]; holo: BM (Nov. Holl. or. occident Sharks Bay [W.A., 1801, *N.Baudin*] Herb. Mus. Paris.). [There is no confirmation of occurrence of the species at Shark Bay].

G. viscidula C.A.Gardner, *Forests Dept W. Australia Bull.* 32: 44 (1923). T: Carson River near Hunter River and Napier Broome Bay, W.A., 31 Aug. 1921, *C.A.Gardner 1588* [or *1558*?]; lecto: PERTH, *fide* D.J.McGillivray & R.O.Makinson, *Grevillea* 437 (1993); isolecto: NSW, PERTH.

Illustrations: J.Brock, *Top End Native Pl.* 207 (1988), as *G. pyramidalis*; C.R.Dunlop *et al.*, *Fl. Darwin Region* 2: 124, fig. 36 *p.p.* (1995), as *G. pyramidalis*; P.M.Olde & N.R.Marriott, *Grevillea Book* 2: 235 (bottom right), 236 (196A, B) (1995).

Leaves 15–30 cm long, pinnatisect, usually with 7–20 primary lobes, lower ones often again 2–8-partite, sometimes some tertiary division, rarely a few leaves entire and linear; ultimate lobes ascending; simple leaves and lobes (0.6–) 1–5 (–12) mm broad, linear to broadly so, rarely broader towards apex, laterally compressed, straight to curved, openly to densely sericeous or rarely (when lobes < 1 mm broad) sparsely silky to glabrous; leaf rachis terete. Flower colour: perianth and style white to cream or pale yellow.

Occurs in north-western W.A. (Pilbara and Hamersley Ra. to Halls Creek and northern and eastern Kimberley), and western N.T. (scattered in the Victoria R. district and almost to Katherine). Not known to occur at Shark Bay as given on the holotype of *G. obliqua*. Grows in open grassy eucalypt woodland in loamy soils. Regenerates from seed and basal epicormic shoots. Flowers c. May–July. Map 362.

W.A.: 16 km E of Nullagine, *J.V.Blockley 393* (CANB, KPBG, PERTH); near Lake Argyle Spillway, near Kununurra, *K.F.Kenneally* 1918 (CANB, NSW, PERTH); near Mabel Downs Stn, *R.A.Perry 2504* (BRI, CANB, MEL, NSW); Kirkimbie Stn, *C.S.Robinson 86* (DNA). N.T.: 11 km N of the Nicholson turnoff on road from Katherine to Kununurra, *D.J.McGillivray 3782 & A.S.George* (DNA, K, NSW, US).

Subsp. *leucadendron* usually has leaf lobes 1–6 mm broad and moderately to densely hairy. Some populations in the northern Kimberley area of W.A. (near Kalumburu) have lobes mostly 0.6–1.0 mm wide and sparsely sericeous to almost glabrous. This 'Kalumburu form' nevertheless seems best placed with subsp. *leucadendron*.

285c. Grevillea pyramidalis subsp. **longiloba** (F.Muell.) Olde & Marriott, *Grevillea Book* 1: 178 (1994)

G. longiloba F.Muell., *Fragm.* 1: 136 (1859). T: '... prope originem fluviorum Roper, Wickham et MacArthur River.' [protologue]; lecto: Gulf of Carpentaria, [N.T.], *s.d.*, *Dr M[ueller]*; lecto: K, *fide* D.J.McGillivray & R.O.Makinson, *Grevillea* 436 (1993); isolecto: A *n.v.*, MEL, TCD *n.v.*

Leaves mostly 30–42 cm long, entire or 2- or 3-sect with very strongly ascending simple lobes; lobes of divided leaves 4–20 cm long; simple leaves and ultimate lobes 2.5–5 mm wide, linear to strap-like (laterally compressed), usually curved, usually glabrous or almost so; leaf rachis below first lobe ±flattened. Flower colour: perianth and style white to cream or pale yellow.

Occurs in N.T., mainly in the Katherine area (E of Scott Ck to Katherine Gorge Natl Park) on the S end of the Pinkerton Ra., and one record from c. 33 km SW of Borroloola. Grows in open grassy woodland in gravelly loam soils. Regenerates from seed and basal epicormic shoots. Flowers c. May–July. Map 363.

N.T.: c. 33.8 km SW of Borroloola, *J.R.Maconochie & S.Parker 134* (DNA); 46.7 km E of Scott Ck on the Kununurra to Katherine road, *D.J.McGillivray 3896* (NSW); SW end of Pinkerton Ra., *D.J.McGillivray 3789 & A.S.George* (DNA, NSW); Katherine Gorge Natl Park, 27 July 1976, *A.Wood* (DNA, NSW).

Differs from subsp. *pyramidalis* in having the leaves often all or mostly simple, if divided then with 2 or 3 lobes only, and leaves and lobes 2.5–5 mm wide. Simple-leaved specimens may be superficially similar to *G. striata*, which has biramous hairs on the outer surface of the perianth and very regular continuous striation on the lower surface of the dorsiventral leaves. *Grevillea mimosoides*, which occurs in the same general area, has always simple leaves which are (6–) 15–50 mm broad.

Collections from the Pinkerton Ra. have inconspicuously open-sericeous leaf surfaces. A fruiting collection of uncertain locality (inter Adelaide River et South Alligator River, 11 Aug. 1981, *R.Letouzey AUS 2* (K, P)), assignable to *G. pyramidalis* and close to the range of subsp. *longiloba*, has 5-partite densely subsericeous leaves very similar to those of subsp. *leucadendron*, secondary conflorescence axes ascending rather than spreading, and the primary and secondary peduncles densely clad with short erect glandular hairs: this specimen may represent a distinct taxon.

286. Grevillea erythroclada W.Fitzg., *Western Mail (Perth)* 21 (1066): 10, 28 incl. t. (1906)

T: Isdell River near Mt Barnett Homestead, W.A., Sept. 1905, *W.V.Fitzgerald 1487*; lecto: PERTH, *fide* D.J.McGillivray & R.O.Makinson, *Grevillea* 416 (1993); isolecto: E, K, NSW.

Illustrations: J.Brock, *Top End Native Pl.* 201 (1990); D.J.McGillivray & R.O.Makinson, *Grevillea* 145 (1993); P.M.Olde & N.R.Marriott, *Grevillea Book* 2: 152 (123A, B), 153 (top left) (1995).

Shrub or tree, 2.5–8 m tall. Leaves unifacial, (14–) 25–52 cm long, usually pinnatisect to bipinnatisect with 5–13 ascending primary lobes, the lower ones sometimes again up to 4-partite, rarely a few leaves entire; ultimate lobes 10–36 cm long, 0.5–1.8 mm wide, subterete to flat-linear; margins obscure; surface with deep longitudinal discontinuous wrinkles, glabrous or with scattered appressed hairs. Conflorescence usually terminal, erect, exceeding foliage, paniculate with up to 8 spreading branches or occasionally simple; secondary peduncles often with blackish triangular bases; unit conflorescence cylindrical, 6–18 cm long, weakly acropetal to subsynchronous. Flowers acroscopic. Flower colour: perianth and style cream to pale yellow. Perianth glabrous outside, papillose inside. Nectary U-shaped. Pistil 6.5–9 mm long, glabrous; pollen-presenter transverse to oblique, conical. Follicle compressed-ellipsoidal, 19–29 mm long, glabrous, tuberculate to almost smooth, caustic-viscid when young. *Needle-leaf Grevillea.*

Occurs in W.A. (scattered locations in E and central Kimberley), and N.T. (Katherine area, and possibly SW of Borroloola). Grows in open eucalypt woodland, often on river levees or near creeks, in sandy alluvial or red volcanic soils. Regenerates from seed and epicormic buds. Flowers Sept.–Oct. Map 364.

W.A.: new Cockatoo Sand site, CSIRO Kununurra, *M.H.Andrew 51* (NSW); 3 km E of Gibb River Stn gate on road to Wyndham; *D.J.McGillivray 3844 & A.S.George* (NSW); 30 km W of Dunham R. and 1.4 km E of Middle Ck on road from Kununurra to Wyndham, *D.J.McGillivray 3890* (NSW). N.T.: near Katherine, *S.T.Blake 17203* (CANB).

A collection (*G.F.Hill, 572* (MEL, NSW)) from 32 km SW of Borroloola is possibly assignable to this species, but its occurrence at that locality has not been confirmed; narrow-leaved plants of *G. mimosoides* also occur in the area.

Grevillea erythroclada is very similar to narrow-lobed forms of *G. pyramidalis* subsp. *pyramidalis* and subsp. *leucadendron*. The latter two taxa have the branchlets usually brown or grey (orange to red in *G. erythroclada*) and the leaves consistently laterally flattened, with fine ±continuous wrinkles only (wrinkles deep, short, and discontinuous in *G. erythroclada*). See notes under *G. nematophylla* for differences from that species.

287. Grevillea mimosoides R.Br., *Trans. Linn. Soc. London* 10: 177 (1810)

T: Gulf of Carpentaria, Island a [Sweers Island, Qld], 17 Nov. 1802, *R.Brown 3317*; lecto: BM, *fide* D.J.McGillivray & R.O.Makinson, *Grevillea* 429 (1993).

G. pachypoda O.Schwarz, *Feddes Repert.* 24: 83 (1927). T: Darwin, near 2½ miles railway crossing, N.T., 15 July 1925, *F.A.K.Bleeser 356*; lecto: B *n.v.*, *fide* D.J.McGillivray & R.O.Makinson, *Grevillea* 429 (1993); remaining syntype: same locality, 9 Sept. 1926, *F.A.K.Bleeser 720*; syn: B *n.v.*

Illustrations: A.S.George, *Introd. Proteaceae W. Australia* 61, t. 86 (1984); D.J.McGillivray & R.O.Makinson, *Grevillea* 149, fig. 33 & col. pl. (1993); P.M.Olde & N.R.Marriott, *Grevillea Book* 3: 28 (top centre & 17A, B) (1995).

Shrub to tree, 2–10 m tall. Leaves unifacial, detaching readily, entire, 6.5–40 cm long, 6–50 mm wide, obliquely narrowly elliptic to obovate, usually falcate; edges flat (true margins obscure); surface longitudinally veined and glabrous to inconspicuously open-sericeous. Conflorescence terminal or rarely axillary, erect, paniculate with 3–11 spreading branches or occasionally unbranched; secondary peduncles with black triangular bases; unit conflorescence cylindrical, 12–15 cm long, acropetal. Flowers ±acroscopic. Flower colour: perianth and style greenish white to cream or pale yellow. Perianth glabrous outside, papillose to pubescent inside. Nectary U-shaped. Pistil 5–11 mm long, glabrous; pollen-presenter erect to moderately oblique, conical. Follicle compressed-ellipsoidal to -obovoid, 14–25 mm long, glabrous, ±smooth, caustic-viscid when young. *Caustic Bush.*

Occurs in the monsoon tropics of W.A., N.T. and Qld. In W.A. in the northern and western Kimberley; in N.T. N of about 18°S; in Qld, N from a line between Cloncurry, Chillagoe and Mareeba. Grows in tall open shrubland or eucalypt woodland, often in seasonally inundated areas, in heavy textured clay- to gravel-loams or sandy clays over laterite, basalt, argillaceous limestone or occasionally sandstone. Regenerates from seed and basal suckers, also epicormic shoots and possibly short rhizomes. Flowers year round, predominantly July–Sept. Map 365.

W.A.: 96.6 km S of Kalumburu Mission, *N.H.Speck 4843* (BRI, CANB, PERTH). N.T.: 141.5 km N of Top Springs, *G.Chippendale 3761* (AD, BRI, CANB, DNA, MEL, NSW); Kakadu Natl Park, 3.5 km NW of Koongarra Saddle, *I.R.Telford 8414* (CANB, NSW). Qld: Dimbulah–Chillagoe road, c. 12.3 km from Dimbulah, *V.K.Moriarty 1121* (CANB, NSW); near Conjuboy Stn, *R.A.Perry 3743* (BRI, CANB, MEL, NSW).

Populations in the Kimberley of W.A. generally tend to have leaves at the broad end of the range (broadest leaves 15–50 mm broad), and smaller flowers (pistils ⩽ 7 mm long); there is, however, overlap with populations elsewhere and formal status is not warranted on the information available.

Grevillea mimosoides is similar and very closely related to *G. dimidiata*, which has usually broader leaves (25–75 mm broad) of a more blue-green hue, and longer pistils (16–21 mm long). *Grevillea pyramidalis* is florally similar to *G. mimosoides* but has leaves either divided or if simple then < 1 cm broad.

288. Grevillea dimidiata F.Muell., *Fragm.* 3: 146 (1863)

T: Roper River, [N.T.], 1862, J. Macd. Stuart's Exped. *?Sturt*; holo: MEL.

Illustrations: J.Brock, *Top End Native Pl.* 200 (1988); D.J.McGillivray & R.O.Makinson, *Grevillea* 150 (1993); P.M.Olde & N.R.Marriott, *Grevillea Book* 2: 125 (top centre & 100A, B), 126 (100C) (1995).

Tree or shrub, 2–6 (–12?) m tall. Leaves unifacial, detaching readily, entire, 9–33 cm long, 8–75 mm wide, sickle-shaped to semi-obovate, usually falcate; edges flat and usually undulate (true margins obscure); surface longitudinally wrinkled, glabrous, glaucous. Conflorescence terminal, erect, paniculate with 3–8 primary branches, sometimes with up to third-order branching; secondary peduncles usually with blackish triangular bases; unit conflorescence cylindrical, 4–11 cm long, acropetal. Flowers acroscopic. Flower colour: perianth and style greenish white to cream, sometimes with a pink tinge. Perianth glabrous outside and inside (sometimes papillose inside in lower half). Nectary U-shaped. Pistil 16.5–21 mm long, glabrous; pollen-presenter oblique on style, obliquely conical. Follicle spherical to fat-lenticular, 16–23 mm long, glabrous, ±smooth, viscid when young. *Caustic Bush.* Fig. 33A–D.

Occurs in W.A. between Fitzroy Crossing, Mt Wittenoom and Joseph Bonaparte Gulf, and in N.T. north of a line from Wave Hill to Gove. Grows in open eucalypt grassy woodland in light sandy soils on laterite, basalt or sandstone. Regenerates from seed and epicormic shoots. Flowers predominantly May–Oct. Map 366.

W.A.: 48 km E of Fitzroy Crossing township, *M.Lazarides 6476* (BRI, CANB, NSW); 11.5 km SE of El Questro turn-off on Wyndham–Kununurra road, *D.J.McGillivray 3878 & A.S.George* (DNA, K, NSW, PERTH, US). N.T.: Gibbie Ck, 35 km from new Humbert River Stn HS, *D.J.McGillivray 3900* (CANB, DNA, K, LE, NSW, PERTH, US); 11.3 km SW of Wilton R., Bulman Crossing, *J.R.Maconochie 1434* (DNA, MO *n.v.*, NSW, PERTH); 74 km SW Birrimbah Out-station, *R.A.Perry 2073* (AD, BRI, CANB, NSW).

Similar to *G. mimosoides*; see under that species for differences.

289. Grevillea calcicola A.S.George, *J. Roy. Soc. W. Australia* 50: 97, 100, fig. 2A–F (1967)

T: Charles Knife Road, Cape Range, N of Learmonth, W.A., 30 Aug. 1961, *A.S.George 1331*; holo: PERTH; iso: PERTH.

Illustrations: A.S.George, *loc. cit.*; D.J.McGillivray & R.O.Makinson, *Grevillea* 144 (1993); P.M.Olde & N.R.Marriott, *Grevillea Book* 2: 78 (top right & 61A), 79 (61B, C) (1995).

Much-branched shrub 2–4 m tall. Leaves dorsiventral, 7–20 cm long, usually 2–7-sect with ascending lobes, sometimes entire; simple leaves and lobes linear, 1.3–2.5 mm wide; margins revolute to midvein; surfaces dissimilar, with lower surface enclosed except for midvein, villous in grooves. Conflorescence terminal, erect, paniculately 2–6-branched; bases of secondary peduncles sometimes blackish and triangular; unit conflorescence cylindrical, 6–7 cm long, acropetal. Flowers acroscopic. Flower colour: perianth and style off-white to cream. Perianth glabrous outside, sparsely villous inside. Nectary U-shaped. Pistil 11.5–12 mm long, glabrous; pollen-presenter strongly oblique, flat to concave. Follicle ellipsoidal to obloid, not or scarcely compressed, 21–27 mm long, glabrous, almost smooth, viscid when young.

Occurs in W.A., where restricted to the Cape Ra. W of Exmouth Gulf. Grows in low mallee-shrubland and *Triodia* associations on limestone, often on hilltops. Regeneration mode unknown, possibly from seed only. Flowers May–Aug. Map 367.

W.A.: Cape Ra., Charles Knife Rd, *J.S.Beard 3582* (KPBG); Cape Ra. 4.8 km S of Yardie Creek HS, *A.S.George 6613* (PERTH); 16 km W of Learmonth, *D.W.Goodall 1168* (PERTH).

Finer-lobed variants of *G. pyramidalis* are similar but have unifacial leaves without a distinct margin, and an erect-conical pollen-presenter. There is a superficial similarity to *G. candelabroides*, which has the ovary stipe > 2 mm long (< 2 mm long in *G. calcicola*).

This species is recognised as 'Poorly Known' in J.D.Briggs & J.H.Leigh, *Rare or Threatened Australian Plants* (1995).

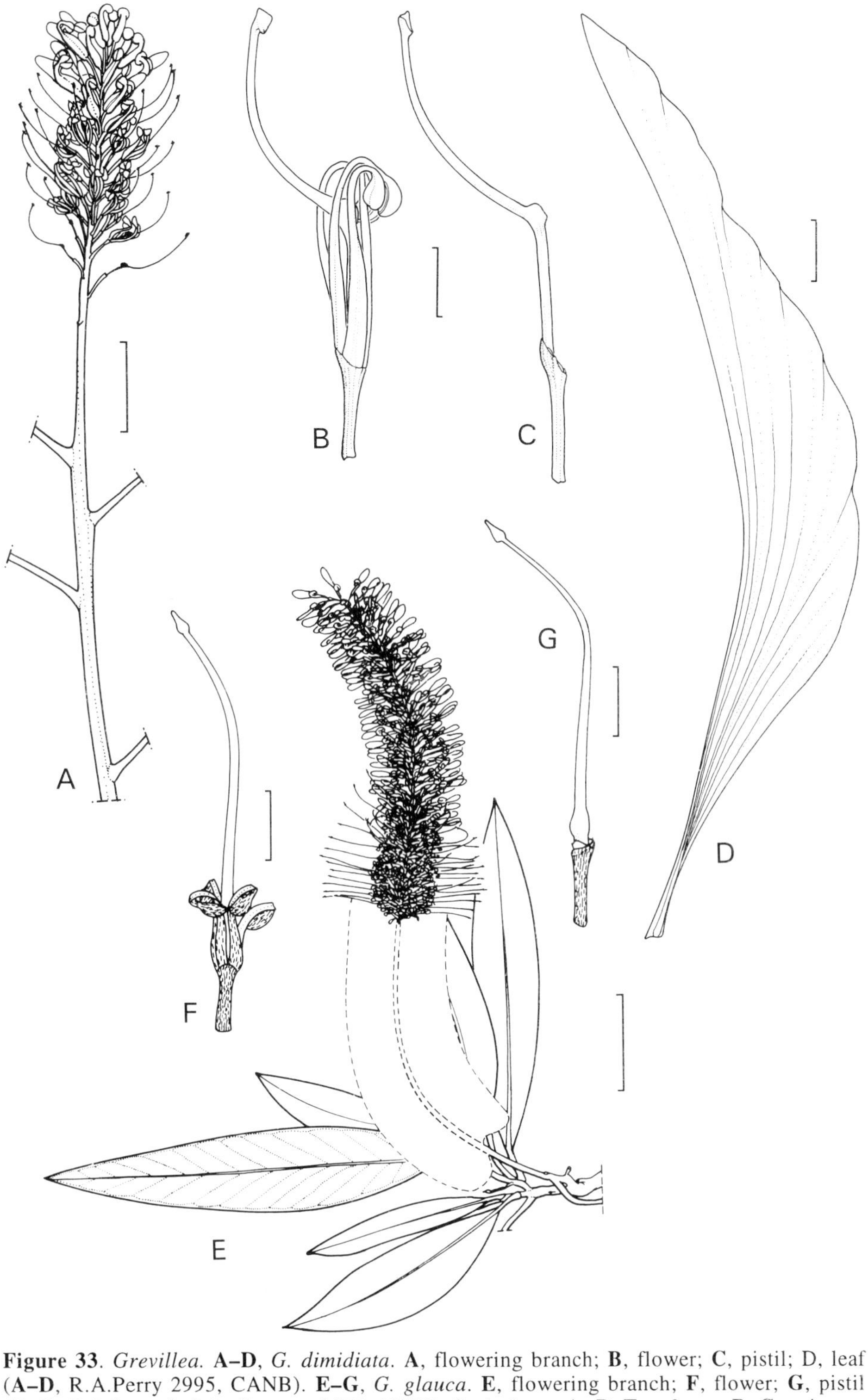

Figure 33. *Grevillea*. **A–D**, *G. dimidiata*. **A**, flowering branch; **B**, flower; **C**, pistil; D, leaf (**A–D**, R.A.Perry 2995, CANB). **E–G**, *G. glauca*. **E**, flowering branch; **F**, flower; **G**, pistil (**E–G**, B.J.Conn & J.de Campos 1254, NSW). Scale bars: **A**, **D–E** = 2 cm; **B–C** = 4 mm; **F–G** = 3 mm. Drawn by: **A–D**, D.Boyer; **E–G**, D.Fortescue.

290. Grevillea glauca Banks & Solander ex Knight, *Cult. Prot.* 121 (1809)

G. gibbosa R.Br., *Trans. Linn. Soc. London* 10: 177 (1810), *nom. illeg.* T: [near Endeavour River, Qld], 1770, *J.Banks & D.Solander*; lecto: BM, *fide* D.J.McGillivray & R.O.Makinson, *Grevillea* 418 (1993); probable isolecto: B *n.v.*, E, MEL, NSW, P, PERTH.

Illustrations: J.W.Wrigley & M.Fagg, *Banksias, Waratahs & Grevilleas* 226 (1989); D.J.McGillivray & R.O.Makinson, *Grevillea* 122, fig. 22 (1993); P.M.Olde & N.R.Marriott, *Grevillea Book* 2: 177 (bottom centre & 145A), 178 (145B, C) (1995).

Spindly erect shrub or small tree 2–10 (–15?) m tall. Leaves dorsiventral, entire, narrowly ovate to elliptic, 6–20 cm long, 10–65 mm wide; margins flat; surfaces similar, pubescent tending more appressed below. Conflorescence terminal, usually 2–8-branched from near base; primary peduncle ±erect but branches usually decurved to pendulous; unit conflorescence decurved to pendulous, cylindrical, 6–18 cm long, subsynchronous to weakly basipetal. Flowers transversely oriented with sutures extrorse. Flower colour: perianth and style creamish to greenish white. Perianth subsericeous to tomentose outside, glabrous inside. Nectary arcuate. Pistil (10–) 14–16.5 mm long, glabrous; pollen-presenter an erect narrow cone. Follicle subglobose to subovoid, 24–40 mm long, rugose, glabrous. *Bushman's Clothes-Peg.* Plate 58; Fig. 33E–G.

Occurs in north-eastern Qld from Jericho and Torrens Ck to Normanton and Cape York; also in Papua New Guinea. Grows in a wide range of habitats including seasonally dry open forest, woodland, savanna, sometimes in coastal dune heath, in light sandy or gravelly often granitic soils, sometimes in low waterlogged areas. Regenerates from seed and epicormic buds. Flowers mainly Apr.–Aug. Map 368.

Qld: Mt Molloy, *L.J.Brass 2506* (A *n.v.*, B *n.v.*, BISH *n.v.*, BRI, MEL); 43.5 km NNE of Normanton Rd towards Lotus Vale Stn, *P.Ollerenshaw 1459* (BRI, CANB, NSW); 9.7 km E of Torrens Creek Township, *N.H.Speck 4557* (BRI, CANB).

Grevillea glauca shows little variation over most of its range. Occasional specimens from the Stannary Hills near Herberton have unusually small flowers (pistils c. 10 mm long); it is unclear whether these represent a consistent population; plants with normal-sized flowers (pistils ⩾ 14 mm long) are common in the same general area.

The timber has been used in boomerang manufacture, and the open gaping fruits were once used as improvised clothes pegs.

291. Grevillea myosodes McGill., *New Names Grevillea* 10 (1986)

T: 33 miles [c. 53 km] W of Tableland Station, W.A., 24 July 1959, *M.Lazarides 6420*; holo: CANB; iso: BRI, CANB, K, NSW, PERTH.

Illustrations: J.R.Wheeler (ed.) *et al.*, *Fl. Kimberley Region* 467, fig. 143C (1992); D.J.McGillivray & R.O.Makinson, *Grevillea* 123, fig. 23 & col. pl. (1993); P.M.Olde & N.R.Marriott, *Grevillea Book* 3: 45 (top centre & 31A–C) (1995).

Shrub 1–2 m tall, to 2 m across. Leaves dorsiventral, entire, obliquely elliptic, sometimes falcate, 8–13 cm long, 20–30 mm wide; margins flat; surfaces similar, finely and densely sericeous with sparkling hairs. Conflorescence erect, terminal, paniculately 2–5-branched; unit conflorescence cylindrical, 6–12 cm long, weakly acropetal. Flowers transversely oriented with sutures extrorse. Flower colour: perianth and style cream after anthesis, sometimes tending very pale yellow. Perianth glabrous inside and out. Nectary arcuate to semi-annular. Pistil 10–13 mm long, glabrous; pollen-presenter an erect cone. Follicle 19–22 (–35?) mm long, compressed-ellipsoidal, shortly apiculate, rugulose, glabrous. Fig. 34 I–K.

Occurs in W.A. (eastern Kimberley, from Gibb R. to the Durack Ra.), and N.T. (south-eastern Arnhem Land, Bloomfield Spring area, and also Bradshaw Stn in the Victoria River District). Grows in woodland or mixed shrubland in sandy loam soils. Regenerates from seed and lignotuber. Flowers May–July. Map 369.

W.A.: 53.1 km W of Tableland Stn, *M.Lazarides 6420* (BRI, CANB, K, NSW, PERTH); c. 2 km W from Traine R. on road W from Tableland Stn, *D.J.McGillivray 3838 & A.S.George* (B *n.v.*, CANB, K, MEL, NSW, PERTH, PRE *n.v.*); 33.5 km E from Gibb River Stn to Drysdale River Stn road, on the road to Wyndham, *D.J.McGillivray 3869 & A.S.George* (BRI, DNA, K, LE *n.v.*, MEL, NSW, NY *n.v.*).

N.T.: Kakadu Natl Park, *T.M.Orr 334* (AD, BRI, CANB, DNA, K, L, MEL, NSW, PERTH); Bradshaw Stn, *T.M.Orr 440* (BRI, CANB, DNA, MEL).

Flowers of the W.A. populations have a very strong mousy (?amine) odour, sometimes persistent after drying. This has not been confirmed for the N.T. populations, but these appear morphologically congruent with *G. myosodes.*

This species is recognised as 'Poorly Known' in J.D.Briggs & J.H.Leigh, *Rare or Threatened Australian Plants* (1995).

292. Grevillea donaldiana Kenneally, *W. Australian Naturalist* 17: 115 (1988)

T: banks of the Sale River, Kimberley, W.A., 15°59'S, 124°39'E, 16 May 1986, *K.F.Kenneally 9676*; holo: PERTH; iso: BRI, CANB, K, NSW.

Illustrations: K.F.Kenneally, *op. cit.* 17: 112–113, 116 (1988); J.R.Wheeler (ed.) *et al., Fl. Kimberley Region* 468, fig. 144A1–A5 (1992); P.M.Olde & N.R.Marriott, *Grevillea Book* 2: 134 (top right, except for hairs shown on style in error, & 108A, B) (1995).

Erect shrub or tree to 10 m tall. Leaves dorsiventral, entire, elliptic, straight to slightly falcate, (5–) 9–15 cm long, 4–10 mm wide; margins flat; surfaces similar, finely and densely sericeous. Conflorescence terminal, erect, with several paniculate branches; unit conflorescence cylindrical, c. 7 cm long, subsynchronous to weakly acropetal. Flowers acroscopic. Flower colour: perianth and style creamy white. Perianth 3-merous, glabrous inside and out. Nectary ±arcuate. Pistil 9–12 mm long, glabrous; pollen-presenter erect-conical. Follicle 20–23 mm long, fat-lenticular, shortly apiculate, rugulose, glabrous.

Occurs in W.A. in the western Kimberley, where known from one locality on the Sale R. Grows on estuary cliff lines and steep rocky gullies, on sandstone. Regeneration mode unknown. Flowers May–?July. Map 370.

W.A.: Sale R., 12 May 1987, *K.Coate s.n.* (BRI, CANB, DNA, K, NSW, NY *n.v.*, PERTH); Sale R., *K.F.Kenneally 9575* (PERTH).

This species is highly unusual in the reduction of the number of tepals to three, and in the irregularly laciniate (usually 4–9-lobed) appendages spreading from the toral rim.

This species is recognised as 'Poorly Known' in J.D.Briggs & J.H.Leigh, *Rare or Threatened Australian Plants* (1995).

293. Grevillea candelabroides C.A.Gardner, *J. Roy. Soc. W. Australia* 47: 56 (1964)

T: 'in arenosis prop. Ajana, etiam Binnu, et Yuna', W.A., 4 Jan. 1959, *C.A.Gardner 12062*; holo: PERTH.

Illustrations: A.Fairall, *W. Australian Native Pl. Cult.* 135 (1970); D.J.McGillivray & R.O.Makinson, *Grevillea* 188, fig. 46 & col. pl. (1993); P.M.Olde & N.R.Marriott, *Grevillea Book* 2: 82 (top centre & 64A, B) (1995).

Shrub 2–4 m tall. Leaves dorsiventral, 13–26 cm long, pinnatisect with 7–14 ascending linear lobes 5–19 cm long, 0.7–1.0 mm wide; margins revolute to midveins; surfaces dissimilar, with lower surface 2-grooved, hairy in grooves. Conflorescence terminal, erect, paniculately 2–14-branched with branches strongly ascending; unit conflorescence cylindrical, 10–25 cm long, acropetal. Flowers acroscopic. Flower colour: perianth and style deep cream to white. Perianth glabrous outside, shortly pubescent inside near base. Nectary usually annular and shortly cylindrical or sometimes U-shaped. Pistil 11–15 mm long, glabrous; pollen-presenter oblique, ±conical. Follicle compressed-ovoid, 13–15 mm long, glabrous, viscid when young, decorticating later. Fig. 34E–H.

Occurs in south-western W.A., in the northern wheatbelt from near Wannoo S to near Coorow. Grows in open heath or tall shrubland in deep yellow or white sand, often over laterite. Regenerates from seed. Flowers usually Aug.–Jan., sweet honey-like scent. Map 371.

W.A.: 3.2 km W of Wicherina, *J.W.Green 476* (PERTH); 15 km SE of Mingenew on Geraldton highway, *T.A.Halliday 132* (PERTH); N of Binnu, Dec. 1960, *L.A.S.Johnson W39* (NSW); near Shark Bay, Oct. 1877, *F.Mueller* (MEL); 4.8 km W of Indarra, *K.Newbey 2150* (PERTH).

Grevillea candelabroides is similar and related to *G. stenobotrya*, which has the outer surface of the perianth always with at least a few hairs, leaves usually simple, and the stipe of the ovary < 2 mm long (2–3.5 mm long in *G. candelabroides*).

This species is recognised as 'Poorly Known' in J.D.Briggs & J.H.Leigh, *Rare or Threatened Australian Plants* (1995).

294. Grevillea stenobotrya F.Muell., *Fragm.* 9: 3 (1875)

T: M Donnells [Macdonnell] Range, [N.T.], *s.d.*, *E.Giles*; holo: MEL; iso: MEL.

G. simulans Morrison, *J. Bot.* 50: 277 (1912). T: Uaroo near Ashburton River, W.A., 2 Oct. 1905, *A.Morrison*; lecto: E, *fide* D.J.McGillivray & R.O.Makinson, *Grevillea* 442 (1993); remaining syntype: Minderoo, Ashb. [Ashburton] R., 11 Oct. 1905, *A.Morrison*; syn: E.

G. livea Ewart & M.E.L.Archer, *in* A.J.Ewart & O.B.Davies, *Fl. N. Terr.* 84, t. VIII (1917). T: 70 miles [112 km] N of Camp IV, N.T., 28 June 1911, *G.F.Hill 231a*; lecto: MEL, *fide* D.J.McGillivray & R.O.Makinson, *Grevillea* 443 (1993).

Illustrations: G.M.Cunningham *et al.*, *Pl. W New South Wales* 212 (1981); D.J.McGillivray & R.O.Makinson, *Grevillea* 189 (1993); P.M.Olde & N.R.Marriott, *Grevillea Book* 3: 183 (top right & 149A), 184 (149B, C) (1995).

Shrub 1.5–6 m tall. Leaves dorsiventral, 6–28 cm long, usually entire and linear, 0.7–2.5 mm wide, rarely some or all leaves 2–7-sect with ascending linear lobes 1–2 mm wide, rarely lower lobes again divided; margins tightly and smoothly to angularly revolute; surfaces dissimilar, with lower surface sericeous or often enclosed except for midvein and then 2-grooved. Conflorescence terminal, erect, paniculate with 5–12 (–20) ascending to spreading branches, occasionally the lower ones again 2–7-branched; unit conflorescence cylindrical, 7.5–10 cm long, acropetal to subsynchronous. Flowers irregularly acroscopic. Flower colour: perianth and style creamy white to pale yellow, rarely creamy pink. Perianth subsericeous to subvillous or sparsely so to almost glabrous outside, bearded near base inside or rarely glabrous. Nectary U-shaped or annular. Pistil 5–11 mm long, glabrous; pollen-presenter very oblique, broadly conical. Follicle compressed oblong-ellipsoidal with a decurved apiculum, (8–) 11–16.5 mm long, glabrous, usually faintly viscid when young, smooth at dehiscence, crustaceous with exocarp flaking away soon after maturity to reveal pale mesocarp. *Sandhill Grevillea, Rattle-Pod Grevillea.*

Occurs in all mainland States except Vic., widespread in arid areas: in W.A. in the Gibson and Great Sandy Deserts and near the coast from Shark Bay to Onslow; in N.T. southwards from Tennant Creek; in S.A. in the NW third of the State; in Qld W from Charleville and Windorah; in N.S.W. in the far NW corner only. Grows in red sandhill country with shrub and *Triodia* communities. Regenerates from seed. Flowers May–Dec. Map 372.

W.A.: Gascoyne Junction, *N.H.Speck 1507* (AD, CANB, PERTH). N.T.: 43.5 km N of Barrow Ck, *N.Forde 216* (AD, CANB, DNA, MEL, PERTH). S.A.: 11.3 km N of Emu, *N.Forde 499* (CANB, MEL, NSW). Qld: 48 km W of Charleville, 8 Sept. 1968, *G.Bates & Rose* (QRS). N.S.W.: Binerah Downs, between Fort Grey and Waka, *P.L.Milthorpe 552* (AD, NSW).

There is considerable variation in several features, and further research is warranted. Some populations between North West Cape and about Coral Bay in W.A. have divided leaves, which are very rare elsewhere. Most populations between Binnu and Shark Bay, W.A., have very small fruits (8–9 mm long) and small flowers (pistils 5–6 mm long); these features are only partially correlated with divided leaves. Nectary structure is annular in some W.A. plants and perhaps populations. Longer ovarian stipes occur in the Gascoyne R. area of W.A. Indumentum on the outer surface of the perianth is denser in the west of the range; flowers are largest in the Tanami area of N.T., and smallest at the southern end of the range in W.A. (Hamelin–Tamala area) where the inner surface of the perianth is also glabrous (elsewhere always with some hairs).

Figure 34. *Grevillea*. **A–D**, *G. annulifera*. **A**, flowering branch; **B**, flower; **C**, pistil; **D**, leaf (**A–?D**, N.T.Burbidge 6430, CANB). **E–H**, *G. candelabroides*. **E**, flowering branch; **F**, flower; **G**, pistil; **H**, leaf (**E–H**, T.H.Halliday 132, CANB). **I–K**, *G. myosodes*. **I**, flowering branch; **J**, flower; **K**, pistil (**I–K**, D.J.McGillivray 3869 & A.S.George, NSW). Scale bars: **A**, **E**, **H–I** = 2 cm; **B–C**, **F–G** = 5 mm; **D** = 1 cm; **J–K** = 2 mm. Drawn by: **A–H**, D.Boyer; **I–K**, D.Fortescue.

Illustrations: F.Mueller, *Fragm.* 4: t. xxxii (1864), as *G. Martini*; A.S.George, *Introd. Proteaceae W. Australia* 56, t. 78 (1984); P.M.Olde & N.R.Marriott, *Grevillea Book* 3: 105 (top right & 80A, B), 106 (80C) (1995).

Shrub, 1.2–4 m tall, with sparsely leafy emergent floral branches. Leaves dorsiventral, entire, obovate or oblanceolate to oblong, rarely subrotund, 1.2–5 cm long, 4–9 mm wide; margins thickened, flat to slightly recurved; surfaces similar, faintly wrinkled with scattered appressed hairs especially near margins, sometimes glaucous and almost glabrous. Conflorescence terminal, erect, paniculate with 10–20 ascending branches; unit conflorescence cylindrical, 6–8 cm long, irregular to subsynchronous. Flowers acroscopic. Flower colour: perianth and style pink cream in bud, becoming white to cream, rarely remaining pink or becoming pale yellow. Perianth subsericeous or glabrous outside, glabrous inside. Nectary U-shaped. Pistil 7.5–10 mm long, glabrous; pollen-presenter erect, conical with truncate apex. Follicle slightly compressed obovoid, 11–15 mm long, apiculate, with a flange-like lip along each valve, glabrous, granulate and sometimes viscid.

Occurs in south-western W.A. from the Geraldton area S to Moore R. and inland to Manmanning and Dalwallinu. A putative record from Glenelg R. in north-western W.A. (type of *G. martinii* F.Muell.) almost certainly results from a labelling error. Grows in sand heath in sandy soils over clay. Regenerates from seed. Flowers mainly Sept.–Dec.; caramel-like floral scent. Map 376.

W.A.: Gunyidi, *J.S.Beard 2494* (KPBG, PERTH); near Yorkrakine, Feb. 1940, *C.A.Gardner s.n.* (PERTH); Dingo Rocks, 11.3 km W of Manmanning, *A.S.George 11636* (PERTH); 33 km from Strawberry on Burma Rd, 7 Oct. 1963, *P.Luff & P.Birrel* (AD).

There is variation in flower colour, with pink-flowered plants known from the Marchagee Track, and yellow-flowered plants from S of Eneabba.

Not easily confused with other species; *G. makinsonii* is similar, but has the leaves ±densely clad with sparkling appressed hairs, and pedicels > 2 mm long (leaves duller, subsericeous and pedicels c. 0.5 mm long in *G. polybotrya*). *Grevillea cheilocarpa* appears to be closely related and is similar in leaf shape and flower size (see under that species for differences).

This species is recognised as 'Poorly Known' in J.D.Briggs & J.H.Leigh, *Rare or Threatened Australian Plants* (1995).

299. Grevillea cheilocarpa Makinson, *Fl. Australia* 17A: 505 (2000)

T: Dragon Rock Nature Reserve, 36128, 75 km E of Kulin on roadside, W.A., 32°38'30"S, 119°01'30"E, 7 Sept. 1984, *J.M.Brown 089*; holo: PERTH.

Shrub 1?–3 m tall. Leaves dorsiventral, entire, obovate to subspathulate, (1.0–) 1.2–3.2 cm long, 7–13 mm wide; margins flat; surfaces similar, densely sericeous with shiny almost transparent biramous hairs. Conflorescence terminal, scarcely exceeding foliage, erect, simple or with 2 or 3 ascending basal branches; unit conflorescence cylindrical to shortly so, 2–4 cm long, subsynchronous-irregular. Flowers acroscopic. Flower colour not known, probably cream. Perianth loosely open-villous outside, glabrous inside. Nectary minutely linguiform. Pistil 6.5–8 mm long; stipe and ovary loosely villous on dorsal side (ground tissue usually visible); style glabrous; pollen-presenter erect, conical, with a broad basal rim. Follicle compressed ellipsoidal, 10–12 mm long, with a lip-like flange along ventral edge of each valve, loosely hairy when young, viscid.

Occurs in south-western W.A., where known from only two localities: between Hyden and Lake Varley, and Dragon Rock Nature Reserve E of Kulin. Recorded from low heath in sandy soil over laterite. Regeneration mode unknown. Flowering recorded for Sept. Map 377.

W.A.: Dragon Rocks Nature Reserve No 36128, *A.M.Coates 2878* (PERTH); between Hyden & L. Varley, *A.S.George 9891* (NSW, PERTH).

Grevillea cheilocarpa has a dense even silky indumentum of closely appressed almost transparent biramous hairs on both surfaces of the leaves, tomentose branchlets, and the floral rachis, pedicels, outer surface of perianth and the stipe and ovary loosely villous. It is similar to *G. polybotrya* which has the leaves with hairs few and mostly near the margins, the branchlets, floral rachis, pedicels and outer surface of the perianth glabrous or with a sparse to dense appressed indumentum and the pistil glabrous. *Grevillea makinsonii* has glabrous pedicels, perianth and pistil.

300. Grevillea makinsonii McGill., *New Names Grevillea* 9 (1986)

T: Arrino, S of Mullewa, W.A., 1969, from *N.H.Speck* collection; holo: CANB.

Illustrations: D.J.McGillivray & R.O.Makinson, *Grevillea* 136, fig. 27 & col. pl. (1993); P.M.Olde & N.R.Marriott, *Grevillea Book* 3: 13 (bottom left & 5A), 14 (5B, C) (1995).

Shrub 0.6–1.6 m tall. Leaves dorsiventral, entire, obovate or subspathulate, 1–3 cm long, 3–8 mm wide; margins flat; surfaces similar, sericeous with small sparkling hairs; upper surface concave. Conflorescence usually terminal, erect, simple or 2-branched; unit conflorescence conico-cylindrical, 3–6 cm long, acropetal. Flowers acroscopic. Flower colour: perianth and style cream to very light yellow. Perianth glabrous inside and out. Nectary obscure, ±linguiform. Pistil 5.5–6 mm long, glabrous; pollen-presenter ±erect, conical. Follicle angularly obovoid, c. 6 mm long, markedly rugose to muricate, glabrous, not viscid.

Occurs in south-western W.A., restricted to the Arrino–Three Springs–Eneabba area and W of Watheroo. Grows in sandplain heath in white sands over laterite. Regeneration mode unknown, probably regenerates from seed only. Flowers c. Sept. Map 378.

W.A.: Three Springs to Eneabba, 23 Sept. 1969, *D.Clyne NSW125904* (NSW); Arrino, Sept. 1903, *W.V.Fitzgerald* (PERTH); W of Watheroo, *C.A.Gardner 1950* (PERTH); Watheroo, 9 Nov. 1906, *A.Morrison* (E).

Superficially similar to *G. polybotrya*, which has heavily branched conflorescences, pedicels < 1 mm long (2.7–4.3 mm long in *G. makinsonii*), and longer pistils 7.5–10 mm long; and to *G. didymobotrya* subsp. *involuta*, which has hairs on the outer surface of the perianth.

This species is recognised as 'Poorly Known' in J.D.Briggs & J.H.Leigh, *Rare or Threatened Australian Plants* (1995).

301. Grevillea didymobotrya Meisn., *in* A.L.P.P. de Candolle, *Prodr.* 14: 386 (1856)

T: 'In colonia Swan River, [W.A.] (Drumm. [*J. Drummond*] *coll. 4, n. 280!*)' [protologue]; holo: NY *n.v.*; iso: G, MEL, P, PERTH.

G. didymobotrya Meisn., *Hooker's J. Bot. Kew Gard. Misc.* 4: 186 (1852), *nom. nud.*

Shrub 1–3 (–5?) m tall. Leaves entire, either unifacial (subsp. *didymobotrya*) or dorsiventral (subsp. *involuta*). Unifacial leaves subterete to linear, 2.5–17 cm long, 0.5–1.3 mm wide; margins obscure; surface with 7–17 longitudinal ridges and appressed hairs in grooves. Dorsiventral leaves narrowly elliptic to obovate (rarely a few sublinear), 1–7 cm long, (1–) 3–7 mm wide; margins incurved to involute, occasionally enclosing upper surface; surfaces somewhat similar; lower surface with 7–11 longitudinal ridges and an open appressed indumentum. Conflorescence terminal and occasionally also axillary, erect, simple or paniculately 2–5-branched, sometimes with a little secondary branching; unit conflorescence cylindrical, 3.5–6 cm long, ±synchronous. Flowers acroscopic. Perianth sericeous outside, glabrous inside. Nectary inconspicuous and arcuate or sometimes absent. Pistil 3–6 mm long, glabrous; pollen-presenter erect to oblique, conical. Follicle mostly semiobcordate (cuneate in side view), often attenuate to style base, 5–8.5 mm long, glabrous, smooth, not viscid.

Occurs in south-western Australia from Shark Bay to Esperance. There are two subspecies.

Leaves unifacial, subterete to linear ..... **301a.** subsp. **didymobotrya**

Leaves dorsiventral, narrowly elliptic to obovate (rarely involute-sublinear) ..... **301b.** subsp. **involuta**

301a. Grevillea didymobotrya Meisn. subsp. **didymobotrya**

Illustrations: J.W.Wrigley & M.Fagg, *Banksias, Waratahs & Grevilleas* 225 (1989); D.J.McGillivray & R.O.Makinson, *Grevillea* 142, fig. 30 fruit only (1993); P.M.Olde & N.R.Marriott, *Grevillea Book* 2: 119 (top centre & 95A), 120 (95B, C) (1995).

Shrub 1–3 (–5?) m tall. Leaves unifacial (adaxial surface reduced to a narrow dorsal strip) and subterete to linear, 2.5–17 cm long, 0.5–1.3 mm wide; margins obscure; surface with 7–17 longitudinal ridges and appressed hairs in grooves. Flower colour: perianth and style usually bright yellow or occasionally cream; style sometimes reddening after anthesis.

Occurs in south-western W.A., widespread from the Shark Bay area inland to Cundeelee and S to Balladonia and Ravensthorpe. Grows in heath or sclerophyll shrubland or open mallee on sandplain. Regenerates from seed. Flowers mainly Aug.–Jan., sporadically in other months. Map 379.

W.A.: Kalbarri Natl Park, 30 Aug. 1967, *A.M.Ashby* (AD, PERTH); between Ravensthorpe & Lake King, *E.M.Canning CBG034252* (CANB, NSW); W of Moorine Rock towards Nulla Nulla, *M.E.Phillips CBG040396* (CANB, NSW); Baanga Hill E of Lake King, *R.A.Saffrey 447* (PERTH); 11.3 km S of Coolgardie, *C.T.White 5472* (A, BRI).

Within subsp. *didymobotrya* there is significant subtle variation which needs further research. In the area from Northampton to Wannoo (the north-western part of the species range), plants have linear rather than subterete leaves (i.e. ±oblong in cross-section), and the style inserted towards the dorsal edge of the style-end. In the area stretching from Cowcowing to Coolgardie and Comet Vale, subterete leaves (rounded in cross-section) predominate, and many populations have the style-end centrally positioned on the style. Further south, subterete leaves and centrally positioned style-ends are the rule. Plants with particularly narrow leaves (0.5–0.7 mm wide) occur near Mt Holland. Olde & Marriott (*op. cit.* 119) report a 'silver-leaved form', a low shrub to 1.5 m tall, with densely silky leaves that are usually short and terete but occasionally (not seen) elliptic; the status of this population is uncertain.

301b. Grevillea didymobotrya subsp. **involuta** McGill., *New Names Grevillea* 4 (1986)

T: in fruticetis arenosis inter flumina Moore et Murchison, W.A., Sept. 1901, *E.Pritzel 608*; holo: PERTH; iso: A *n.v.*, AD *n.v.*, E, G, NSW, P *n.v.*

Illustrations: D.J.McGillivray & R.O.Makinson, *Grevillea* 142, col. pl. (1993); P.M.Olde & N.R.Marriott, *Grevillea Book* 2: 120 (bottom right & 96) (1995).

Shrub to 1.5 m tall. Leaves dorsiventral, narrowly elliptic to obovate (rarely a few sublinear) occasionally sigmoid in lateral view, 1–7 cm long, usually 3–7 mm wide; margins evident, incurved to involute, occasionally enclosing upper surface; surfaces subsericeous with indumentum less dense on lower surface; lower surface with 7–11 prominent longitudinal ridges. Flower colour: perianth and style yellow; styles (sometimes?) reddening after anthesis.

Occurs in south-western W.A., between Geraldton and Mullewa. Grows in heath or sclerophyll shrubland or open mallee on sandplain. Regeneration mode not known. Flowers Aug.–Sept. Map 380.

W.A.: W of Mullewa, *J.Galbraith 443* (MEL); 8 km W of Tenindewa on Geraldton to Mullewa road, *N.Marriott 69* (NSW); 5 km W of Indarra, *K.Newbey 2151* (PERTH).

Trifida Group

Shrubs. Leaves entire or divided, dorsiventral to occasionally dipleural; surfaces dissimilar or similar; margins flat to revolute. Conflorescence terminal or axillary, erect or rarely decurved, simple or few-branched; unit conflorescence regular and umbelloid to cylindrical or conico-cylindrical, or secund, basipetal to acropetal. Flowers acroscopic or rarely transverse. Torus usually transverse, rarely oblique. Perianth zygomorphic, glabrous outside (except occasionally in *G. leptobotrys*), glabrous or hairy inside; tepals either loosely cohering except along dorsal suture and held ventral to style, or separating and partly everting along dorsal suture while limb segments remain coherent (later all separating completely). Pistil 4–10 (–36) mm long, glabrous; ovary stipitate; style usually exserted from late bud, sometimes exposed but not exserted, often retrorse after anthesis or sometimes incurved; pollen-presenter erect to oblique, usually conical. Follicle glabrous, usually smooth or faintly rugulose, sometimes tuberculate, rarely muricate or echinate; pericarp crustaceous to weakly bony. Seed ellipsoidal; inner face usually with a submarginal channel and a waxy border.

Nineteen species, all occurring in the south-west of W.A. Insect pollinated, except probably *G. dielsiana* and perhaps *G. teretifolia*. The nectary is obscure in several species. Placement of *G. crithmifolia* and *G. trachytheca* in this group is tentative: their muricate fruits differ strongly from the other member species. General affinities are with the *Hilliana* group, and probably also with the *Rudis* group. *Grevillea obliquistigma* and its close relative *G. zygoloba* are here, as in other recent works, classed within the *Trifida* group. While there is certainly some relationship, and strong similarities in flower form, late consideration of some fruit and leaf characters suggests that closest affinities of these two species may rather be to the *Hilliana* group.

1 Unit conflorescences regular and umbel-like or subglobose, no longer than wide

2 Pistil 26–36 mm long **317. G. dielsiana**

2: Pistil ≤ 8 mm long

3 Inner surface of perianth sparsely villous; simple leaves or lobes of divided leaves < 2 mm wide, linear to narrowly oblong; leaf lower surface completely enclosed by revolute margins except for midvein; fruit surface muricate-echinate **319. G. crithmifolia**

3: Inner surface of perianth glabrous, smooth to papillose; simple leaves, or teeth or lobes of divided leaves, 1–4 mm wide, triangular to ovate or occasionally linear; leaf lower surface usually at least narrowly exposed on either side of midveins; fruit surface almost smooth

4 Unit conflorescences umbelloid, axillary or terminal on short lateral branches, often aggregated along branches, enclosed by foliage; leaves of flowering branches usually similar in form to those on vegetative branches **302. G. trifida**

4: Unit conflorescences subglobose to umbelloid, terminal or near-terminal axillary at upper nodes of branchlets only, often exceeding foliage; leaves of flowering branches smaller and with fewer (or no) teeth relative to those of vegetative branches **303. G. muelleri**

1: Unit conflorescences longer than wide, either secund, or regular and ovoid to conical or cylindrical (sometimes shortly so) or obconical or obovoid

5 Either all leaves simple and entire, or some or all leaves toothed or divided with lobes not divaricate

6 Unit conflorescences regular, strongly acropetal (basal flowers opening first); simple leaves, and lobes of divided leaves, linear to very narrowly oblong

7 Floral rachis 6–12 (–24) mm long, villous; unit conflorescences umbelloid to corymbose or rarely conical **319. G. crithmifolia**

7: Floral rachis 20–100 mm long, glabrous or sparsely subsericeous or villous; unit conflorescences narrowly conical to cylindrical

8 Longest leaves < 4 cm long; floral rachis villous **320. G. trachytheca**

8: Longest leaves > 4 cm long; floral rachis glabrous or subsericeous

9 Leaves all simple and entire, or with 2 or 3 simple lobes **308. G. obliquistigma**

9: Leaves all pinnatipartite with 3–12 lobes, sometimes with secondary division **309. G. zygoloba**

6: Unit conflorescences either secund, or regular and then basipetal (apical flowers opening first) to subsynchronous, ovoid or cylindrical or obovoid in overall shape; simple leaves and leaf lobes variously shaped

10 Unit conflorescences secund or secund-obconical, loose; outer surface of perianth hairy or glabrous; flowers pink **318. G. leptobotrys**

10: Unit conflorescences either regular and cylindrical to shortly so and dense, or secund and dense; outer surface of perianth always glabrous; flowers white to cream or pale yellow

11 Leaves ovate to oblong-elliptic in general outline, reticulum evident below, coarsely dentate to pinnatifid with teeth distributed around margins; unit conflorescences ± secund **307. G. monticola**

11: Leaves variously shaped, reticulum obscure, primary division deep or if shallow then teeth or lobes mainly towards apex; unit conflorescences secund, or regular and cylindrical

12 Leaf rachis strongly flexuose; primary leaf lobes usually > 7; fruit c. 20 mm long **306. G. flexuosa**

12: Leaf rachis straight or very weakly flexuose; primary lobes ⩽ 7; fruit 8–13 mm long

13 Unit conflorescences cylindrical; widest ultimate leaf lobes 2–12 mm wide; pollen-presenter conical, usually lacking a projecting basal flange or rim, if flange present then straight (not cupped), not toothed, and not marginally papillose **305. G. synapheae**

13: Unit conflorescences usually ±strongly secund; widest ultimate leaf lobes ⩽ 2 mm wide; pollen-presenter narrowly conical, with a narrow projecting cupped basal flange or rim, its edge slightly toothed and minutely papillose **304. G. prominens**

5: Some or all leaves divided; leaf lobes divaricate

14 Pistils ⩾ 10 mm long

15 Pistils 10–17 mm long **316. G. teretifolia**

15: Pistils 26–36 mm long **317. G. dielsiana**

14: Pistils ⩽ 9 mm long

16 Floral rachises densely villous; fruit surface muricate-echinate; unit conflorescences ovoid **319. G. crithmifolia**

16: Floral rachises glabrous or subsericeous or with an open indumentum of appressed to ascending hairs; fruit surface smooth to rugose or rarely rugose-tuberculate to muricate; unit conflorescences secund or cylindrical to conical or rarely ovoid

17 Unit conflorescences loose, secund-obconical, basipetal (apical flowers opening first); dorsal tepals widely everted in late bud to form a flat platform; flowers pink at anthesis **318. G. leptobotrys**

17: Unit conflorescences dense, either secund or regular and then conical to cylindrical or obconical; dorsal tepals widely everted in late bud, or not; flowers white to cream or pale yellow at anthesis

18 Branchlets densely hairy

19 Branchlets and floral rachises with spreading hairs; pedicels 2–4 mm long **320. G. trachytheca**

19: Branchlets and floral rachises with appressed hairs; pedicels 3.5–7 mm long

20 Floral rachises (30–) 40–80 mm long; unit conflorescences regular, cylindrical **311. G. subtiliflora**

20: Floral rachises 15–30 mm long; unit conflorescences often ±secund **315. G. kenneallyi**

18: Branchlets glabrous or with an open to sparse indumentum (ground tissue visible between hairs)

21 Leaf margins recurved to shortly revolute, clearly evident; abaxial surface of lamina on either side of midveins exposed

22 Pollen-presenter cone with a narrow, cupped basal flange or rim, the margin of the flange slightly toothed and minutely papillose; leaf deeply ternate **304. G. prominens**

22: Pollen-presenter cone lacking a basal flange or rim, or flange occasionally (subsp. *minyulo*) present but then not cupped, not toothed, and not marginally papillose; leaf deeply 3–7-fid to 3–7-partite **305. G. synapheae**

21: Leaf margins tightly revolute, or obscure; lower surface on either side of midveins not exposed

23 Inner surface of perianth glabrous

24 Longest ultimate leaf lobes > 3 cm long; unit conflorescences ±conical, strongly acropetal (basal flowers opening first) **312. G. intricata**

24: Longest ultimate leaf lobes < 3 cm long; unit conflorescences ±cylindrical, weakly basipetal (apical flowers opening first) to weakly acropetal **313. G. minutiflora**

23: Inner surface of perianth with hairs (sometimes inconspicuous and only near base)

25 Ultimate leaf lobes ±rectangular in cross-section, with 2 grooves on lower surface (one either side of the scarcely prominent abaxial midvein); young branchlets usually glaucous **310. G. leucoclada**

25: Ultimate leaf lobes ±rounded in cross-section, with a lateral groove along each side; abaxial midvein very prominent, making up half or more of leaf thickness; branchlets not usually glaucous

26 Pollen-presenter erect on style, conical, 0.6–0.7 mm high; pedicels usually 3.5–5.5 mm long; unit conflorescences regular, conico-cylindrical or rarely ovoid **312. G. intricata**

26: Pollen-presenter oblique on style, convex to broadly conical, < 0.5 mm high; pedicels 6–8.5 mm long; unit conflorescences usually secund, occasionally tending conico-cylindrical **314. G. leptopoda**

302. Grevillea trifida (R.Br.) Meisn., *in* J.G.C.Lehmann, *Pl. Preiss.* 1: 553 (1845)

Anadenia trifida R.Br., *Trans. Linn. Soc. London* 10: 167 (1810). T: 'In Novae Hollandiae ora australi; Leewins Land ...' [protologue]; lecto: ... propr Portum Georgii IIIm Decr 1801 [King Georges Sound, W.A.], *R.Brown Iter Austral. 3312*; lecto: BM, *fide* D.J.McGillivray & R.O.Makinson, *Grevillea* 445 (1993); isolecto?: BM, E, K.

G. brevicuspis Meisn., *in* J.G.C.Lehmann, *Pl. Preiss.* 2: 256 (1848). T: Swan River, [W.A.], *J.Drummond [2nd coll.] 321*; lecto: NY *n.v.*, *fide* D.J.McGillivray & R.O.Makinson, *Grevillea* 445 (1993); isolecto: BM, CGE *n.v.*, G, LD *n.v.*, LE *n.v.*, P.

Illustrations: D.J.McGillivray & R.O.Makinson, *Grevillea* 165 (1993); P.M.Olde & N.R.Marriott, *Grevillea Book* 3: 207 (top right & 169A, B), 208 (169C) (1995).

differs in its shorter (c. 1 cm long) umbelloid unit conflorescences, which are mostly held within the foliage in the upper leaf axils or on short lateral subapical branchlets, which are in turn often aggregated in columnar fashion.

This species is recognised as 'Poorly Known' in J.D.Briggs & J.H.Leigh, *Rare or Threatened Australian Plants* (1995).

305. Grevillea synapheae R.Br., *Suppl. Prodr. Fl. Nov. Holl.* 23 (1830)

T: 'Ora merid.-occid., Swan River, 1827. *D.Fraser.*' [protologue]; Swan River, [W.A.], *s.d.*, *[C.]Fraser s.n.*; holo: BM.

Prostrate to erect shrub 0.2–1.5 m tall. Branchlets glabrous to sparsely tomentose. Leaves dorsiventral, (2–) 4–18 cm long, 1–6 cm wide, very variable in form from 3–7-fid to deeply 3–7-partite, frequently with shallow to deep secondary division, occasionally a few leaves entire and narrowly obovate; ultimate lobes ascending or rarely spreading, broadly to narrowly triangular or sublinear, 10–25 mm long, 2–12 mm wide; margins flat to recurved or rarely shortly revolute; lower surface mostly or wholly exposed, glabrous or sparsely sericeous, sometimes glaucous, minutely pitted. Conflorescence axillary or terminal, erect, simple or to 6-branched; unit conflorescence cylindrical or shortly so, basipetal to subsynchronous; ultimate floral rachis 10–60 mm long, glabrous to sericeous. Flowers ±acroscopic. Perianth mealy to papillose inside. Pistil 4–5 mm long; style exserted from late bud; pollen-presenter erect to slightly oblique, broadly conical, usually lacking projecting basal flange, rarely (subsp. *minyulo*) flange present. Follicle obliquely ellipsoidal to obliquely obovoid, 8–13 mm long, rugulose. *Catkin Grevillea.*

Occurs in south-western W.A. Four formal subspecies, and one informal subspecies, are recognised here.

Closely related to *G. trifida*, *G. muelleri* and *G. prominens*, all of which have the pollen-presenter with a projecting basal flange or 'brim'; the first two also have an umbelloid unit conflorescence. In *G. synapheae* the pollen-presenter lacks the basal flange (except in subsp. *minyulo*), and the unit conflorescence is subcylindrical. *Grevillea synapheae* is also close to *G. flexuosa* which has larger basal leaves with 7–18 lobes, strongly flexuose leaf rachises (straight or very weakly flexuose in *G. synapheae*), and larger fruits c. 20 mm long.

There is a general south to north trend for longer peduncles and rachises, complicated by other characters. Further research is required.

1 Longest floral rachises 1–2 cm long; floral bracts falling in early to mid bud stage

2 Leaf lower surface with minute ascending or rarely appressed hairs openly to sparsely but evenly distributed on the lamina away from the main veins, the hairs persisting (sometimes sparsely) on adult leaves **305a.** subsp. **synapheae**

2: Leaf lower surface glabrous or with very scattered hairs mainly on midveins when young

3 Leaf rachis straight or gently and evenly downcurved, not laterally flexuose; leaf lobes coplanar with rachis or only weakly divaricate; plant mounded to spreading or erect

4 Primary leaf lobes 5–15 mm wide; ultimate lobes usually shallow and broadly triangular **305b.** subsp. **latiloba**

4: Primary leaf lobes 2.5–6 mm wide; ultimate lobes narrowly triangular or oblong or ovate, rarely broadly triangular **305a.** subsp. **synapheae**

3: Leaf rachis angularly deflexed at each node, laterally flexuose (zigzag); leaf lobes strongly divaricate; plant sprawling or semi-prostrate **305e.** subsp. **A**

1: Longest floral rachises 2–6 cm long; floral bracts often persistent until late bud stage or until after anthesis, occasionally falling in early bud

5 Ultimate leaf lobes straight, 3–11 mm wide; lower surface of adult leaves with lamina on either side of midvein usually glabrous, rarely with a few scattered hairs; pollen-presenter conical, lacking a projecting basal flange or rim **305c.** subsp. **pachyphylla**

5: Ultimate leaf lobes gently curved, 2–3 (–5?) mm wide; lower surface of adult leaves usually with an open indumentum of appressed hairs; pollen-presenter conical with a distinct basal flange or rim **305d.** subsp. **minyulo**

305a. Grevillea synapheae R.Br. subsp. synapheae

Anadenia gracilis Lindl., *Sketch Veg. Swan R.* xxxi, n. 144 (1839). T: Swan River, [W.A.], *s.d.*, *[J.] Drummond s.n.*; holo: CGE (photo seen).

G. flexuosa var. *pauciloba* Benth., *Fl. Austral.* 5: 480 (1870). T: 'Darling range, Oldfield' [protologue]; lecto: Darling Range, [W.A.], *s.d.*, *A.Oldfield*; lecto: K, *fide* D.J.McGillivray & R.O.Makinson, *Grevillea* 443 (1993); remaining syntypes: K (*A.Oldfield*, W. Australia); MEL (*A.Oldfield* 180b, 347A, B?).

G. synapheae subsp. *synapheae* 'Typical form', of P.M.Olde & N.R.Marriott, *Nuytsia* 9: 257 (1993); *Grevillea Book* 3: 193 (1995).

Illustrations: D.J.McGillivray & R.O.Makinson, *Grevillea* 163 (1993), as *G. synapheae*; P.M.Olde & N.R.Marriott, *Nuytsia* 9: 256, fig. 5A–N (1993); P.M.Olde & N.R.Marriott, *Grevillea Book* 3: 193 (top right & 157B, C) (1995).

Prostrate to spreading low shrub. Branchlets terete to angular. Leaves 5–12 (–18) cm long; most leaves deeply 3–7-partite or occasionally 3–7-fid (sometimes a few simple and narrowly obovate); primary lobes straight to slightly curved, (2–) 3–5 mm wide, often again apically 3 (–5)-toothed to -fid; ultimate teeth or lobes straight, usually flat or rarely divaricate, narrowly oblong-ovate to -triangular, 2–17 (–32) mm long, 2–5 mm wide; leaf rachis straight or gently and evenly downcurved; lower surface with minute ascending or rarely appressed hairs openly to sparsely but evenly distributed on the lamina away from the main veins, persisting (sometimes sparsely) on adult leaves. Unit conflorescences not exceeding the nearby leaves; ultimate floral rachises 10 (–20) mm long; floral bracts falling in early bud stage. Flower colour: perianth and style white to creamy yellow.

Occurs in the south-west of W.A., where fairly common in and east of the Darling Ra., from Perth suburbs south-east to Narrogin and Williams, north almost to Gingin, and north-east to near Toodyay. Grows in eucalypt (Jarrah, Marri or Wandoo) woodland or in shrub associations, in sandy or gravelly soils over laterite or ?granite. Regenerates from seed and lignotuber (or rhizomes?). Flowers July–Oct. Map 384.

W.A.: Parkerville, *R.Coveny 8040* (K, NSW, PERTH); Karragullen, intersection of Irymple Rd and Brookton Hwy, *J.D'Alonzo 26* (K, PERTH); Wooroloo, *M.Koch 1394* (A *n.v.*, AD, BM, BRI, E *n.v.*, K, MEL, NSW, P *n.v.*, PERTH); 53 km from Perth towards Narrogin, *M.E.Phillips CBG022086* (CANB, NSW); Darling Ra., *E.Pritzel 500* (A *n.v.*, AD, B *n.v.*, E, G *n.v.*, NSW, P *n.v.*, PERTH).

Subspecies *synapheae* occurs at the southern end of the species range, and is characterised by thin-textured leaves, relatively narrow primary leaf lobes, and (in some populations, mostly from Mundaring southwards) by an open even indumentum on the leaf lamina lower surface. Leaves are usually relatively flat with the ultimate lobes not or rather weakly divaricate. The name *G. flexuosa* var. *pauciloba* Benth. is typified on plants of this subspecies with particularly long and narrow, mostly simple, primary lobes (10–32 mm long, 2–3 mm wide); the leaf rachis is angularly deflexed at the nodes and the lobes are divaricate with respect to the rachis.

305b. Grevillea synapheae subsp. latiloba (Meisn.) Makinson, *Fl. Australia* 17A: 505 (2000)

G. synapheae var. *latiloba* Meisn., *in* J.G.C.Lehmann, *Pl. Preiss.* 2: 259 (1848). T: Swan River, W.A., *[J.] Drummond [2nd coll.] 313*; neo: NY *n.v.*, *fide* D.J.McGillivray & R.O.Makinson, *Grevillea* 443 (1993); isoneo: A *n.v.*, E, G, K, LD *n.v.*, LE *n.v.*, MEL, NSW.

G. synapheae subsp. *synapheae*, 'broad-leaved form', of P.M.Olde & N.R.Marriott, *Nuytsia* 9: 257 (1993).

G. synapheae subsp. *synapheae*, 'Mogumber form', of P.M.Olde & N.R.Marriott, *Grevillea Book* 3: 193 (1995).

Mounded to spreading or erect shrub to 1.5 m tall. Branchlets sharply angular. Leaves 5–12 cm long; most leaves bipinnatifid or with 3–5 deeply divided primary lobes 5–15 mm wide, these usually again apically 3 (–5)-toothed to -fid, occasionally a few leaves obovate with 1 or 2 marginal teeth; ultimate teeth or lobes straight, flat to divaricate, usually triangular to oblong-ovate, 2–17 (–32) mm long, 2–7 (–12) mm wide; leaf rachis ±straight; lower surface glabrous or with a very few hairs on midveins. Unit conflorescences scarcely exceeding the nearby leaves; ultimate floral rachises 10–20 mm long; floral bracts falling in early- to mid-bud stage. Flower colour: perianth and style white to creamy yellow.

Occurs in south-western W.A., N of Perth between about Bindoon and Mogumber. Grows in woodland or shrub associations, in sandy or gravelly soils. Regeneration mode unknown. Flowers ?Aug.–?Oct. Map 385.

W.A.: S.W. Australia, *[J.] Drummond 32/1843* (K); Wannamal, 109 km N of Perth, *R.Melville 69 et al.* (K); Mogumber, Nov. 1905, *J.Staer* (E).

The distribution of subsp. *latiloba* appears to directly abut that of subsp. *pachyphylla* in the Mogumber area, and possible intermediates are recorded from between Mogumber and about New Norcia. There are few recent collections.

305c. **Grevillea synapheae** subsp. **pachyphylla** Olde & Marriott, *Nuytsia* 9: 257 (1993)

T: Brand Hwy, 8 km S of Half Way Mill Roadhouse, W.A., *P.Olde 91/87*, 14 Sept. 1991; holo: PERTH; iso: NSW.

G. synapheae subsp. *pachyphylla* 'Typical form', of P.M.Olde & N.R.Marriott, *Grevillea Book* 3: 193, 194 (1995).

Illustrations: P.M.Olde & N.R.Marriott, *Nuytsia* 9: 258, fig. 6 (1993); P.M.Olde & N.R.Marriott, *Grevillea Book* 3: 194 (top left & 158A, B), 195 (158C) (1995).

Shrub 0.3–1 m tall. Branchlets sharply angular. Leaves 3.5–7 cm long; most leaves divided and ±flat with straight non-divaricate lobes, apically 3–5-fid or -partite, sometimes a few leaves simple and narrowly obovate; primary lobes 3–11 mm wide, simple or shallowly and apically again 2–3-toothed or -fid, 2.5–6 mm wide; ultimate lobes ovate to oblong or short-triangular (2–) 10–22 mm long, 2–6 mm wide; leaf rachis straight or gently and evenly downcurved; lower surface completely glabrous or on young leaves with scattered appressed to ascending hairs almost exclusively on midveins, soon falling. Unit conflorescences exceeding foliage; ultimate floral rachises 2–4 cm long; floral bracts usually persistent at or after anthesis. Flower colour: perianth and style white to creamy yellow.

Occurs in the N of the species range, from the Badgingarra area N to Eneabba. Grows in open heath associations, usually on rises in lateritic loam or sand over laterite. Regenerates from seed. Flowers July–Sept. Map 386.

W.A.: Mt Lesueur Reserve, *J.D'Alonzo 74* (K, PERTH); hills S from Three Springs, *C.A.Gardner 9052* (PERTH); 0.5 km E of Badgingarra, *D.J.McGillivray 3283 & A.S.George* (NSW, PERTH); 24 km NNW of Dinner Hill, *K.Newbey 2297* (PERTH); Bibby Rd, Cervantes, 5 km from highway, *D.Woolcock G33* (NSW).

Subspecies *pachyphylla* as erected by Olde & Marriott (*op. cit.* 259 (1993); 194 (1995)) included (as the 'Minyolo form') the taxon here referred to subsp. *minyulo*. The latter differs from subsp. *pachyphylla* in the more restricted sense used here in having a lignotuberous habit, leaves secund and almost pinnatisect with 5–7 strongly divaricate leaf lobes ⩽ 3 mm wide, a linear to narrowly cuneate leaf base, lower surface of the leaf with a usually persistent open indumentum of large appressed biramous hairs on the veins and the lamina, and the pollen-presenter conical with a distinct projecting basal flange or 'rim'.

Occasional collections from near Jurien, assignable overall to subsp. *pachyphylla*, are atypical in having scattered ascending hairs on the lower surface of apparently adult leaves.

305d. Grevillea synapheae subsp. **minyulo** Makinson, *Fl. Australia* 17A: 505 (2000)

T: Walyering Road, 1 km S of Minyulo Brook Reserve, W.A., 20 Sept. 1990, *M.Pieroni 90/4*; holo: PERTH; iso: NSW [plus 2 at PERTH yet to be allocated].

G. synapheae subsp. *pachyphylla* 'Minyolo form', of P.M.Olde & N.R.Marriott, *Grevillea Book* 3: 194 (1995).

Illustrations: P.M.Olde & N.R.Marriott, *Nuytsia* 9: 258, fig. 6 (B only) (1993); P.M.Olde & N.R.Marriott, *Grevillea Book* 3: 195 (158C) (1995).

Spreading to sprawling low shrub 20–30 (–50?) cm tall. Branchlets strongly angular. Leaves 2.5–6 cm long, almost pinnatisect, usually with 7 primary lobes 1.5–3 mm wide, these usually again 2–3-lobed; ultimate lobes usually divaricate, linear, gently curved, 7–30 mm long, 2–3 (–5?) mm wide; leaf rachis often slightly deflexed at each node, sometimes slightly flexuose; lower surface of leaf with a usually persistent indumentum of large appressed biramous hairs on veins and lamina. Unit conflorescence on strongly emergent peduncle; ultimate floral rachis 3–6 cm long; floral bracts persistent to late bud stage or sometimes to anthesis. Flower colour: perianth and style white to creamy yellow. Pollen-presenter with narrow but distinct basal projecting flange or rim.

Occurs in SW of W.A., in the Dandaragan–Cataby area. Grows in heath associations in gravelly lateritic soil. Regeneration at least sometimes from lignotubers (or rhizomes?). Flowers at least Aug.–Sept. Map 387.

W.A.: Minyolo Rd, W of Dandaragan, *E.A.Griffin 5060* (PERTH); Dandaragan West, *A.Kanis 1576* (CANB); Cataby, *P.M.Olde 88/95* (NSW).

The presence of a basal 'flange' on the pollen-presenter is unique within *G. synapheae* and is more characteristic of species such as *G. trifida*. The pinnatisect and usually secund leaves are also unusual within the species *s. lat.* Nevertheless, subsp. *minyulo* appears closer to *G. synapheae* than to any other species and is grouped with it for the present. A collection at PERTH (*Griffin 4912*, only photo seen) has more triangular lobes and apparently flatter leaves, but has the long unit conflorescences on emergent peduncles that are characteristic of subsp. *minyulo*: it may represent an intergrade to subsp. *pachyphylla*.

305e. Grevillea synapheae subsp. **A**

based on: N edge of Mt Misery, W of Dandaragan, *S.Hopper 6333* (PERTH).

G. synapheae subsp. *synapheae* 'Mt Misery form', of P.M.Olde & N.R.Marriott, *Grevillea Book* 3: 194 (1995).

Illustrations: P.M.Olde & N.R.Marriott, *Grevillea Book* 3: 193 (157A, C) (1995).

Semi-prostrate sprawling shrub 20–30 cm tall. Branchlets strongly angular. Leaves deeply 3–7-partite with very narrow segments; primary lobes 2.5–6 mm wide, simple or again divaricately 2- or 3-lobed; ultimate lobes straight, rigid, narrowly triangular to subulate, 5–20 mm long, 2–3 mm wide; leaf rachis distinctly deflexed at each node, flexuose; lower surface of leaf glabrous. Unit conflorescences scarcely emergent from foliage; ultimate floral rachises c. 10 mm long; floral bracts falling in early bud stage or sometimes persistent almost to anthesis. Flower colour: perianth and style white to creamy yellow.

Occurs in south-western W.A., known only from Mt Misery near Dandaragan. Grows in heathy associations in gravelly loam soil. Regeneration mode unknown. Flowering known for Sept. Map 388.

W.A.: Mt Misery, *P.Olde 86/843* (NSW *n.v.*); Mt Misery, *P.Olde 91/82* (NSW).

Grevillea synapheae subsp. *A*, the 'Mt Misery form', resembles subsp. *synapheae* in length of unit conflorescences, and approaches divaricately lobed variants of that subspecies in leaf form. However, it lacks the open minute ascending indumentum on the leaf lower surface that is a strong diagnostic character for many populations of subsp. *synapheae*, and has narrower leaf lobes that are more strongly divaricate, rigid and pungent than subsp. *synapheae*. The timing of floral bract fall appears to be variable. The Mt Misery locality is close to the area of occurrence of subsp. *minyulo*. Its affinities are unclear and it is left as informal pending further investigation.

306. Grevillea flexuosa (Lindl.) Meisn., *in* J.G.C.Lehmann, *Pl. Preiss.* 1: 553 (1845)

Anadenia flexuosa Lindl., *Sketch Veg. Swan R.* xxxi, n. 142 (1839). T: Swan River, W.A., 1839, *[J.] Drummond s.n.*; holo: CGE (photo seen); iso: [*Drummond 1st coll. 618*] BM, E? (*s.n.*), G, G-DC, K, LE *n.v.*, MEL, NSW, P, PERTH, NY (photo seen at NSW).

A.Brown *et al.* (eds), *W. Australia's Threatened Flora* 162 (1998).

Erect shrub to 1.5 m tall. Branchlets glabrous. Leaves dorsiventral, those of flowering branches 5–10 cm long, 50–70 mm wide, on vegetative branches 15–26 cm long, 100–160 mm wide, subpinnatisect with 7–18 primary lobes, these 3–5-fid, not divaricate; ultimate lobes broadly to narrowly triangular, 0.5–3 cm long, 5–10 mm wide; leaf rachis strongly flexuose, less so on primary lobes; margins shortly recurved; lower surface glabrous, glaucous. Conflorescence erect, terminal or in upper axils, simple or to 3-branched; unit conflorescence cylindrical to obovoid, basipetal; ultimate floral rachis 35–65 mm long, glabrous. Flowers acroscopic. Flower colour: perianth and style cream to pale yellow. Perianth papillose inside. Pistil 5–8.5 mm long; style exposed but scarcely exserted from late bud; pollen-presenter erect, convex to conical with projecting basal flange. Follicle ovoid, apically attenuate, c. 20 mm long, rugulose. *Zig Zag Grevillea.*

Occurs in south-western W.A. where known only from populations E and NE of Perth between Stoneville and Toodyay, with a single recent southern collection (not seen) from Mt Vincent E of Jarrahdale. Map 389.

W.A.: Swan R., New Holland, *s.d.*, *[J.] Drummond s.n.* (K — two sheets, possibly part of the Type); Berry Reserve, Stoneville, *P.Olde 91/75* (NSW); Swan R., *s.d.*, *Veitch s.n.* (K).

McGillivray (*in* McGillivray & Makinson, *Grevillea* 162–163, 443 (1993)) regarded *G. flexuosa*, known to him only from the Type material, as a synonym of *G. synapheae*, but further material seen since its rediscovery in the wild confirms its separate status. The two species grow together near Stoneville. *Grevillea synapheae* has smaller fruits 8–13 mm long, tending to obovoid in shape, with a thinner pericarp < 1 mm thick at the suture (cf. 1–2 mm thick in *G. flexuosa*); it also has smaller basal leaves, mostly < 10 cm long, with 3–7 primary lobes and a straight to scarcely flexuose leaf rachis. *Grevillea flexuosa* is distinctive in its large, strongly flexuose leaves which vary in size, those of the flowering branches 5–10 cm long, and 50–70 mm wide, on vegetative branches 15–26 cm long, 100–160 mm wide.

This species is recognised as 'Endangered' in J.D.Briggs & J.H.Leigh, *Rare or Threatened Australian Plants* (1995).

307. Grevillea monticola Meisn., *in* J.G.C.Lehmann, *Pl. Preiss.* 2: 259 (1848)

Replaced syn.: *Anadenia aquifolium* Lindl., *Sketch Veg. Swan R.* xxxi, n. 145 (1839); *G. aquifolium* (Lindl.) Meisn., *in* J.G.C.Lehmann, *Pl. Preiss.* 1: 551 (1845), *nom. illeg. non* Lindl. (1838). T: Swan River, [W.A.], 1839, *[J.] Drummond s.n.*; holo: CGE *n.v.*; iso: E (*J.Drummond 62*), FI *n.v.*, G *n.v.*, G-DC *n.v.*, K.

Illustrations: D.J.McGillivray & R.O.Makinson, *Grevillea* 161, fig. 40, 162, col. pl. (1993); P.M.Olde & N.R.Marriott, *Grevillea Book* 3: 36 (bottom right), 37 (24A, B) (1995).

Shrub 0.6–2 m high. Branchlets glabrous or subsericeous. Leaves dorsiventral, ovate, to oblong-elliptic in outline, 2.5–6.5 cm long, 15–40 mm wide, coarsely dentate to pinnatifid with 5–13 marginal teeth or shallow lobes; margins flat to slightly recurved; lower surface sericeous or glabrous and glaucous. Conflorescence terminal and axillary, simple or to 3-branched; unit conflorescence decurved, strongly to loosely secund, basipetal; ultimate floral rachis 15–40 mm long, glabrous. Flowers acroscopic. Flower colour: perianth and style pale cream to yellowish cream. Perianth glabrous inside. Pistil 6.5–8.5 mm long; style dorsally exposed but scarcely exserted from late bud; pollen-presenter erect, conical with crenate basal flange. Follicle ovoid to ellipsoidal, 8–12 mm long, rugulose.

Occurs in the south-west of W.A., mainly in the Darling Ra. E of Perth, in the area between Kelmscott, Beverley, Pingelly and Wandering. Grows usually in open eucalypt (Jarrah, Wandoo) forest or woodland, in gravelly, sandy or loamy soils over laterite, granite or ironstone. Regenerates from seed. Flowers June–Oct. Map 390.

W.A.: c. 8 km from Wandering, *N.T.Burbidge 2276* (PERTH); E of Kulyalling, N of Pingelly, *A.S.George 14612* (PERTH); Boyagin Reserve, *D.J.McGillivray 3714 & A.S.George* (NSW, PERTH); on Brookton Hwy near track to Warradale, *D.J.McGillivray 3722 & A.S.George* (NSW, PERTH); Talbot Brook near York, *O.H.Sargent 1322* (MEL).

Grevillea monticola is rarely confused with other taxa; the dentate or shallowly lobed leaves are distinctive among related species, as are the greenish orange (when fresh) developing fruits.

308. Grevillea obliquistigma C.A.Gardner, *J. Roy. Soc. W. Australia* 27: 169 (1942)

T: between Pindar and Wuraga, W.A., 19 Sept. 1931, *C.A.Gardner 2620*; holo: PERTH; iso: PERTH.

Shrub 0.5–6 m tall. Branchlets sericeous or glabrous and glaucous. Leaves dorsiventral or dipleural, 4–22 cm long, entire and linear, or rarely 2- or 3-sect with ascending linear lobes; simple leaves 0.8–2.0 mm wide; ultimate lobes of divided leaves 0.7–1.2 mm wide; margins revolute; lower surface enclosed except for midveins, 2-grooved (grooves lateral in subsp. *cullenii*). Conflorescence terminal and axillary, erect, simple or to 6-branched; unit conflorescence narrowly conical to cylindrical, acropetal; ultimate floral rachis 50–100 mm long, glabrous or sparsely silky. Flowers acroscopic. Perianth pilose inside near base. Pistil 5–9.5 mm long; style exserted from late bud; pollen-presenter oblique at c. 45°, truncate-conical with a basal flange. Follicle obloid-ellipsoidal to slightly obovoid, (5–) 7–15 mm long, tuberculate to almost smooth; pericarp usually weakly exfoliating and sometimes faintly viscid.

Occurs in the inland south-west of W.A. Three subspecies are currently recognised.

There is considerable variation in leaves and fruits, and further investigation is needed, encompassing the related *G. zygoloba*. See note on group placement under the group description above.

1 Leaves dorsiventral, with the grooves between the revolute margins and the abaxial midvein clearly on the topological lower side of leaf or lobes

  2 Leaves or lobes 0.7–1.0 mm wide with angularly revolute margins; floral bracts 2–3.5 mm long, all falling in early bud stage ..... **308a.** subsp. **obliquistigma**

  2: Leaves or lobes 1.0–2.0 mm wide with smoothly (rarely angularly) revolute margins; floral bracts 5–6.5 mm long, often some persisting at anthesis ..... **308b.** subsp. **funicularis**

1: Leaves dipleural, with the grooves between revolute margins and abaxial midvein situated laterally on leaf or lobes ..... **308c.** subsp. **cullenii**

308a. Grevillea obliquistigma C.A.Gardner subsp. obliquistigma

Illustrations: D.J.McGillivray & R.O.Makinson, *Grevillea* 153 (1993), as *G. obliquistigma*; P.M.Olde & N.R.Marriott, *Grevillea Book* 3: 55 (bottom right), 56 (39A, B) (1995).

Dense spreading shrub or spindly small tree (1–) 2–6 m tall. Branchlets glabrous or sericeous. Leaves entire or 3-sect, dorsiventral (2-grooved below); simple leaves and lobes 0.7–1.0 mm wide, ±rectangular in cross-section; margins angularly revolute. Unit conflorescence 5–10 cm long; floral bracts 2–3.5 mm long, 0.6–1.2 mm wide, glabrous with ciliate margins, all falling in early bud stage as soon as rachis begins expansion. Flower colour: perianth and style cream-white to yellow-cream, ?rarely with pink tinges. Torus slightly oblique to pedicel axis. Nectary obscure. Pistil 5–7.5 mm long; pollen-presenter with conspicuous basal flange. Follicle 7–15 mm long. Plate 61.

Occurs in south-western W.A., where widespread from the lower Murchison R. inland to Mount Magnet and S towards Bullfinch. Grows in open sclerophyll shrubland in sandy or clay-loam soils over granite or ironstone. Regenerates from seed. Flowers July–Dec. Map 391.

W.A.: 16 km WSW of Paynes Find, *A.C.Beauglehole 49168* (NSW, PERTH); 8 km NE of Mt Gibson HS turnoff on Wubin–Paynes Find road, *R.Coveny 7900* (K, NSW); 189.4 km WSW of Yalgoo Rly Siding towards Mullewa, *R.Coveny 7948 & B.R.Maslin* (K, NSW); Bullfinch, *F.Lullfitz 3102* (KPBG, PERTH).

The normal ('type') form of subsp. *obliquistigma* occurs in the northern part of the range, and is often 3 m or more tall; it usually has fruits at the large end of the size range (usually 9–15 mm long), the surface very dark but not viscid, and weakly tuberculate to almost smooth, and with the pericarp exfoliating only partially and very late, or not at all. The 'Parker Range form' is of shrubby habit usually 1–2 m tall, with glabrous and often glaucous branchlets, stiffer glaucous leaves, and smaller fruits 5–7 mm long that are more strongly tuberculate, sometimes faintly viscid, with the pericarp exfoliating at maturity over much of the surface. In fruit features, the 'Parker Range form' shows a strong approach to subsp. *funicularis*. Specimens apparently assignable to the 'type form' but with fruits < 8 mm long occur in the ranges near Mt Gibson, distant from Parker Range.

308b. **Grevillea obliquistigma** subsp. **funicularis** Olde & Marriott, *Grevillea Book* 1: 180 (1994)

T: 1.3 km S of Perenjori, W.A., 20 Sept. 1991, *P.M.Olde 91/165*; holo: NSW; iso: PERTH.

Illustrations: P.M.Olde & N.R.Marriott, *Grevillea Book* 3: 57 (top right & 40A, B) (1995).

Dense spreading shrub 0.5–2 m tall. Branchlets glabrous or sericeous. Leaves entire or 2- or 3-sect, dorsiventral (2-grooved below), simple leaves and lobes 1–2 mm wide, ±rounded-elliptic in cross-section; margins smoothly (rarely angularly) revolute. Unit conflorescence 4.5–10 cm long; floral bracts 5–6.5 mm long, 1.8–2.4 mm wide, glabrous along midline and villous towards ends, often persisting to anthesis. Flower colour: perianth and style cream-white to yellow-cream, sometimes flushed pale pink; ovary (?sometimes) becoming pink with age. Torus straight to slightly oblique to pedicel axis. Nectary obscure. Pistil 5–8 mm long; pollen-presenter with conspicuous basal flange. Follicles 7–8 mm long, strongly tuberculate, pericarp (?sometimes) exfoliating.

Occurs in the inland south-west of W.A., in the area bounded by Morawa, Mingenew and Ballidu. Grows in open shrubland in sandy soils, sometimes with laterite. Regenerates from seed. Flowers c. Sept.–Dec. Map 392.

W.A.: between Perenjori and Dalwallinu, *W.E.Blackall 2823* (PERTH); 4 km E of Dalwallinu on road to Kalanie, *M.G.Corrick 7787* (MEL, PERTH); Caron, *W.Forbes 6* (PERTH); c. 13 km ENE of Wubin towards Paynes Find on Great Northern Hwy, *D.J.McGillivray 3407 & A.S.George* (K, NSW, PERTH); Spencer Rd, Latham, *P.Olde 91/265 & J.Cullen* (PERTH).

Narrowly distinct from subsp. *obliquistigma*. Some material from near Morawa has smoothly revolute leaf margins but the bracts falling early, and may be intermediate.

308c. **Grevillea obliquistigma** subsp. **cullenii** Olde & Marriott, *Grevillea Book* 1: 180 (1994)

T: 41 km NE of Laverton, W.A., 29 Sept. 1992, *J.Cullen*; holo: PERTH.

Illustration: P.M.Olde & N.R.Marriott, *Grevillea Book* 3: 56 (centre right only) (1995).

Shrub c. 0.5 m tall. Branchlets sericeous. Leaves entire, dipleural, c. 0.8 mm wide; margins smoothly revolute. Unit conflorescence c. 5–6 cm long; floral bracts c. 1.5 mm long, 0.8 mm wide, glabrous except for ciliate margins, falling in early bud stage. Flower colour: perianth and style cream-white to yellow-cream. Torus oblique at c. 45° to pedicel axis. Nectary evident. Pistil 7.5–9.5 mm long; pollen-presenter with an obscure basal flange. Follicles not seen.

Occurs in the inland south-west of W.A.; known from near Laverton and possibly also Cosmo Newberry (cited in D.J.McGillivray & Makinson, *Grevillea* 153 (1993), specimen not seen). Grows in mallee and *Triodia*-shrub associations in red sandy soils. Regeneration mode not known. Flowers Sept.–Oct. Map 393.

No other specimens seen.

309. **Grevillea zygoloba** Olde & Marriott, *Grevillea Book* 1: 180 (1994)

T: Bungalbin Hill, N of Southern Cross, W.A., 10 Sept. 1991, *P.M.Olde 91/48*; holo: NSW; iso: PERTH.

Illustrations: P.M.Olde & N.R.Marriott, *Grevillea Book* 3: 238 (top right & 195A), 239 (195B, C) (1995).

Shrub 2 m tall; branches erect. Branchlets sericeous or glabrous and glaucous. Leaves dorsiventral, 5–8 cm long, 3–7 (–12)-partite with ascending parallel (non-divaricate) primary lobes; basal lobes rarely again 2–4-partite; ultimate lobes linear, 3–5 cm long, 0.7–1.0 mm wide; margins angularly revolute to midveins; lower surface enclosed except for midveins, 2-grooved, sometimes silky in grooves. Conflorescence terminal or axillary, erect, simple or to 3-branched; unit conflorescence conico-cylindrical, acropetal; ultimate floral rachis 20–40 mm long, glabrous or with sparse appressed hairs. Flowers acroscopic; pedicels 4–6 mm long. Flower colour: perianth and style creamy white. Perianth bearded inside near base. Pistil 6.5–7.5 mm long; style exserted from late bud; pollen-presenter erect with a strongly oblique base, conical. Follicle ellipsoidal, 5–7 mm long, strongly tuberculate, with projections glandular and viscid; pericarp exfoliating.

Occurs in south-western W.A., in an area N of Southern Cross between Koolyanobbing and Die Hardy Ra. Grows in eucalypt woodland or shrubland in lateritic soils. Regenerates from seed (only?). Flowers Sept.–Nov. Map 394.

W.A.: Mt Dimer, Jaudi Stn, *A.C.Chapman 27 et al.* (PERTH); Koolyanobbing, June 1972, *R.B.Hacker KY20* (PERTH); Die Hardy Ra., *D.J.McGillivray 3671 & A.S.George* (NSW, PERTH); Bungalbin Hill, N of Southern Cross, *P.Olde 86/194* (NSW); near Koolyanobbing, *P.Olde 91/53* (NSW).

The pedicels of the developing fruits (at least sometimes?) twist to orient the fruits below the rachis; this may prove to be a distinguishing character. McGillivray & Makinson's (*Grevillea* 153 (1993)) reference (under *G. obliquistigma*) to specimens from Koolyanobbing relates to this species. See also note under group description above.

310. **Grevillea leucoclada** McGill., *New Names Grevillea* 8 (1986)

T: c. 23.5 km from North West Coastal Highway on road to Kalbarri, W.A., 13 June 1976, *D.J.McGillivray 3348 & A.S.George*; holo: NSW; iso: CANB, K, LE, PERTH, US (all distributed as *G. intricata* Meisn.).

[*G. intricata auct. non* Meisn.: G.Bentham, *Fl. Austral.* 5: 481 (1870), *p.p.*]

Illustrations: D.J.McGillivray & R.O.Makinson, *Grevillea* 154, fig. 34 (1993); P.M.Olde & N.R.Marriott, *Grevillea Book* 2: 237 (top left & 197A, B) (1995).

Shrub 0.8–3 m high. Branchlets glaucous, glabrous or with scattered appressed hairs. Leaves dorsiventral, 6–14 cm long, 3–7-sect, with lateral primary lobes often again 3-sect; ultimate lobes divaricate, linear or linear-subulate, 1–6 cm long, 0.8–1.5 (–2.0) mm wide; leaf rachis angularly deflexed at each node; margins angularly revolute, enclosing lower surface except for midveins; lower surface 2-grooved, silky in grooves. Conflorescence terminal and axillary, erect, simple or to 6-branched; unit conflorescence conico-cylindrical, acropetal; ultimate floral rachis 40–60 mm long, glabrous or openly subsericeous. Flowers mostly acroscopic, or sometimes oblique to rachis. Flower colour: perianth and style white to cream. Perianth bearded inside near base. Pistil 5.5–6 mm long; style exserted from late bud; pollen-presenter oblique to style, obliquely conical. Follicle obloid-ellipsoidal, 8.5–13.5 mm long, rugulose.

Occurs in the south-west of W.A., restricted to the lower Murchison R. area. Grows in open shrub associations, often in rocky sites, sometimes on level ground, in gravelly or sandy soil. Regenerates from seed. Flowers Aug.–Dec. Map 395.

W.A.: Ross Graham Lookout, Murchison R., E of Kalbarri, *H.Demarz 924* (PERTH); 44 km SE of Kalbarri, *R.Filson 8662* (MEL, PERTH); lower Murchison R., *C.A.Gardner 12256* (PERTH); Murchison R., June 1966, *F.Lullfitz NSW100595* (NSW); Murchison R., *s.d. A.Oldfield* (MEL).

Can be confused with *G. intricata*, which has non-glaucous branchlets, an erect pollen-presenter cone, perianth glabrous inside, and relatively pliable leaf lobes.

This species is recognised as 'Poorly Known' in J.D.Briggs & J.H.Leigh, *Rare or Threatened Australian Plants* (1995).

311. **Grevillea subtiliflora** McGill., *New Names Grevillea* 14 (1986)

T: Mt Singleton, W.A., 7 Sept. 1973, *J.S.Beard 6456*; holo: PERTH; iso: NSW.

Illustrations: P.M.Olde & N.R.Marriott, *Grevillea Book* 3: 189 (centre right & 154A), 190 (154B, C) (1995).

Open erect shrub 1–2.5 m tall. Branchlets densely sericeous. Leaves dorsiventral, 2.5–4.5 cm long, pinnatisect with 3–11 primary lobes, these usually again 2- or 3-sect; ultimate lobes divaricate, linear, 0.5–2 cm long, 0.5–1.0 mm wide; leaf rachis deflexed at nodes; margins angularly revolute; lower surface enclosed except for midveins, 2-grooved, silky in grooves. Conflorescence terminal and axillary, erect to decurved, simple or to 6-branched; unit conflorescence cylindrical, subsynchronous to weakly basipetal; ultimate floral rachis (30–) 40–80 mm long, loosely to sparsely sericeous. Flowers acroscopic to irregularly oriented. Flower colour: perianth and style white; limb of bud yellow-green. Perianth loosely bearded inside near base. Pistil 4–5 mm long; style exserted from late bud; pollen-presenter oblique at c. 30° to style, obliquely conical. Follicle obloid-obovoid or slightly obovoid, 8–10 mm long, smooth to faintly rugulose.

Occurs in the inland south-west of W.A., restricted to the Paynes Find area. Grows in mixed shrubland in red clay-loam soils. Regenerates probably from seed only. Flowers July–Oct. Map 396.

W.A.: 80 km S of Paynes Find on Great Northern Hwy, Sept. 1938, *W.E.Blackall* (PERTH); near Mt Singleton, *C.A.Gardner 14360* (PERTH); 38 km from Paynes Find towards Wubin on Great Northern Hwy, *D.J.McGillivray 3393 & A.S.George* (NSW); Mt Singleton, *B.H.Smith 431* (CANB, MEL, PERTH).

Can be confused with *G. intricata*, which has perianth inner surface glabrous, a conical (strongly acropetal) unit conflorescence, more tangled foliage with longest ultimate leaf lobes (2–) 3–8 cm long, and a less conspicuous nectary.

This species is recognised as 'Poorly Known' in J.D.Briggs & J.H.Leigh, *Rare or Threatened Australian Plants* (1995).

312. **Grevillea intricata** Meisn., *Hooker's J. Bot. Kew Gard. Misc.* 7: 74 (1855)

T: interior north of Swan River, a. [W.A., ?Bowes River district, c. 1850–51], *[J.] Drummond coll. IV: 189*; lecto: NY (*n.v.*), *fide* D.J.McGillivray & R.O.Makinson, *Grevillea* 423 (1993); isolecto: BM; possible isolecto: CGE *n.v.*, FI *n.v.*, G-DC, K, LD *n.v.*, MEL, P *n.v.*

Illustrations: J.W.Wrigley & M.Fagg, *Austral. Native Pl.* opp. 256 (1979); D.J.McGillivray & R.O.Makinson, *Grevillea* 155, fig. 35 (1993); P.M.Olde & N.R.Marriott, *Grevillea Book* 2: 214 (bottom right), 215 (178A–C) (1995).

Dense shrub 0.5–3 m tall with tangled foliage. Branchlets with an open appressed indumentum. Leaves 8–16 cm long (when straightened), pinnatisect to partly quadripinnatisect, with all lobes widely divaricate; ultimate lobes linear-subterete, dipleural, 2–8 cm long, 0.4–1.1 mm across; margins very tightly revolute to sides of midvein; segments with a lateral silky groove each side; lower surface obscure except for prominent midveins. Conflorescence usually terminal, simple or to 4-branched, erect; unit conflorescence ovoid to narrowly conico-cylindrical, strongly acropetal; rachis 50–80 mm long, glabrous or with scattered appressed hairs. Flowers acroscopic to transverse. Flower colour: buds green becoming yellow; perianth after anthesis pale greenish white to light cream; style white. Perianth usually glabrous inside (rarely papillose or sparsely pilose near base). Pistil 4.5–5.5 mm long; style exserted from late bud; pollen-presenter erect with an oblique base, obliquely truncate-conical, with basal flange. Follicle obloid-ellipsoidal, 11–15 mm long, rugose-tuberculate.

Occurs in inland south-western W.A., from Northampton to Ajana and inland to the Chapman East R. Grows in heath, mallee woodland or tall shrubland, in sandy often granitic soils. Regenerates from seed (and lignotuber?). Flowers May–Oct. Map 397.

W.A.: 10.5 km S of Northampton, *E.M.Bennett 2221* (PERTH); Northampton, *C.A.Gardner 2604* (PERTH); 8 km SE of Northampton, *J.W.Green 465* (PERTH); 6 km N of Buller R. on North West Coastal Hwy, *D.J.McGillivray 3317 & A.S.George* (K, NSW, PERTH, US); Northampton–Ajana region, *F.G.Smith 1700* (PERTH).

This species is recognised as 'Poorly Known' in J.D.Briggs & J.H.Leigh, *Rare or Threatened Australian Plants* (1995).

313. **Grevillea minutiflora** McGill., *New Names Grevillea* 10 (1986)

T: 18 km directly W of Mukinbudin on road E from Waddouring, W.A., 17 June 1976, *D.J.McGillivray 3434 & A.S.George*; holo: NSW; iso: CANB, K, NY, PERTH.

Illustrations: D.J.McGillivray & R.O.Makinson, *Grevillea* 156, fig. 36 (1993); P.M.Olde & N.R.Marriott, *Grevillea Book* 3: 30 (bottom centre), 31 (19A–C) (1995).

Shrub 1–2 m tall. Branchlets glabrous or with open appressed indumentum. Leaves entangled, 3–6 cm long, partly divaricately bipinnatipartite with 5 primary lobes, these again 3-lobed; ultimate lobes sublinear to subterete-dipleural, 1–2.5 cm long, 0.4–0.8 mm wide; leaf rachis deflexed at nodes; margins obscure, flat to slightly recurved against midrib; lower surface obscure except for prominent midveins and a lateral silky groove on each side of segments. Conflorescence terminal, simple or to 4-branched, erect to decurved; unit conflorescence ±cylindrical, weakly basipetal to weakly acropetal; ultimate floral rachis 20–30 mm long, glabrous or sparsely silky. Flowers ±acroscopic. Flower colour: perianth and style creamy white; pedicels and style reddening ?after fertilisation. Perianth glabrous inside. Pistil 4.5 mm long; style exserted from late bud; pollen-presenter oblique at 20°–30° to style, obliquely truncate-conical with basal flange. Follicle obloid-ellipsoidal, 7.5–8.5 mm long, faintly rugose.

Occurs in inland south-western W.A., restricted to the Mukinbudin area. Grows in *Acacia* shrubland in clay-loam or sand-loam soils over granite. Regeneration probably from seed only. Flowers Apr.–Sept. Map 398.

W.A.: 15 km W of Mukinbudin, *S.Donaldson 1381 & G.Flowers* (CANB, PERTH); 17 km E of Trayning, *K.Newbey 1548* (PERTH).

This species is recognised as 'Poorly Known' in J.D.Briggs & J.H.Leigh, *Rare or Threatened Australian Plants* (1995).

314. **Grevillea leptopoda** McGill., *New Names Grevillea* 8 (1986)

T: 7 miles [c. 11 km] NW of Carnamah on Geraldton Highway, W.A., 18 July 1953, *R.Melville 4126 & J.Calaby*; holo: NSW.

Illustrations: D.J.McGillivray & R.O.Makinson, *Grevillea* 157 (1993); P.M.Olde & N.R.Marriott, *Grevillea Book* 2: 234 (bottom right), 235 (195A–C) (1995).

Spreading shrub 0.6–1.5 m tall. Branchlets openly sericeous. Leaves 4–8 cm long, divaricately pinnatisect with 3–5 primary lobes, the lateral ones usually again 3-sect; ultimate lobes linear to subulate, 0.5–4.5 cm long, 0.6–1.2 mm wide; margins obscure, shortly revolute against sides of abaxially prominent midveins; lower surface obscure except midveins; leaf segments almost dipleural, with a silky groove along each side. Conflorescence terminal and axillary, simple or to 5-branched, erect; unit conflorescence usually secund, occasionally tending conical-cylindrical, basipetal; ultimate floral rachis 40–60 mm long, sericeous to almost glabrous. Flowers acroscopic. Flower colour: perianth and style white to cream; buds pinkish. Perianth bearded inside near base. Pistil 7.5–9 mm long; style exserted from late bud; pollen-presenter strongly oblique to style, convex to broadly conical. Follicle obloid-ellipsoidal, 9–10.5 mm long, rugulose.

Occurs in the south-west of W.A. from Kalbarri south to Moora. Grows in mallee, heath or shrubland, in sandy or gravelly lateritic soils. Regenerates from seed. Flowers Aug.–Nov. Map 399.

W.A.: Three Springs, *W.E.Blackall 4868* (PERTH); 8.7 km SW of Three Springs towards Eneabba, *R.Coveny 7972 & B.R.Maslin* (K, NSW, PERTH); Mingenew, Sept. 1903, *W.V.Fitzgerald NSW129151* (NSW); Horrocks Beach Rd, Geraldton district, Aug. 1964, *J.Galbraith* (MEL); 80 km N of Moora, *F.W.Went 221* (PERTH).

Can be confused with *G. teretifolia*, which has a shorter floral rachis 10–30 mm long, longer pistils 10–17 mm long, and the inner surface of the perianth with the beard densest in the upper half.

This species is recognised as 'Poorly Known' in J.D.Briggs & J.H.Leigh, *Rare or Threatened Australian Plants* (1995).

317. Grevillea dielsiana C.A.Gardner, *J. Roy. Soc. W. Australia* 27: 169 (1942)

T: Murchison River, W.A., 20 Aug. 1931, *C.A.Gardner 2590*; holo: PERTH.

Illustrations: A.S.George, *Introd. Proteaceae W. Australia* 51, t. 70 (1984); D.J.McGillivray & R.O.Makinson, *Grevillea* 160, col. pl., 161, fig. 39 (1993); P.M.Olde & N.R.Marriott, *Grevillea Book* 2: 21 (top centre & 97A), 122 (97B, C) (1995).

Shrub 0.3–2.5 m tall. Branchlets glabrous and sometimes glaucous, or sparsely subsericeous. Leaves 3–8 cm long, divaricately 3-sect, with primary lobes usually again 3-sect; ultimate lobes linear-subulate, 0.2–2.5 cm long, 0.8–1.3 mm wide; margins obscure; lower surface obscure except for prominent abaxial midveins; leaf segments with silky lateral groove on either side. Conflorescence terminal, simple or to 3-branched; unit conflorescence spreading to decurved, semi-secund or a loose hemispherical cluster, weakly acropetal; ultimate floral rachis 20–40 mm long, sparsely sericeous to glabrous. Flowers acroscopic. Flower colour: perianth usually red to red-orange or yellow-orange often with pink and cream streaks; style red or pink at base, becoming pale pink or white apically. Perianth pubescent to pilose inside. Pistil 26.5–36 mm long; style strongly exserted from late bud; pollen-presenter strongly oblique, convex. Follicle obloid to obloid-ellipsoidal, 10–15 mm long, rugulose.

Occurs in the south-west of W.A., from Geraldton and Mullewa N almost to Shark Bay. Grows in heath or shrubland, in sandy soils. Regenerates from seed. Flowers mainly Aug.–Sept. Map 402.

W.A.: near Tenindewa, *A.M.Ashby 1485* (AD); c. 16 km E of Nabawa, *A.S.George 2392* (PERTH); 71 km S of Wannoo Roadhouse, *C.H.Gittins 1565* (BRI, NSW, PERTH); Kalbarri Natl Park, Great Northern Hwy, 391 mile peg [629.6 km], *R.Hnatiuk 760419* (PERTH); 7 km N of Whelarra on road to Balla, *D.J.McGillivray 3355 & A.S.George* (AD, MEL, MO *n.v.*, NSW, PERTH).

Variable in glaucosity of leaves, habit and flower colour. Rarely confused with other species; its closest relative is *G. teretifolia* which has shorter pistils 10–17 mm long.

318. Grevillea leptobotrys Meisn., *in* J.G.C.Lehmann, *Pl. Preiss.* 2: 256 (1848)

T: [W.A.], *[J.] Drummond (3rd coll.) 268*; lecto: NY *n.v.*, *fide* D.J.McGillivray & R.O.Makinson, *Grevillea* 425 (1993); isolecto: B, BM, CGE, E, G-DC, K, LE, MEL, P, TCD, all *n.v.* except BM, E, K, MEL.

G. leptobotrys var. *simplicior* F.Muell., *Fragm.* 6: 209 (1868). T: '... varietatum simpliciorum' [protologue]; Gordon River, [W.A.], *[G.]Maxwell*; holo: MEL; ?iso: PERTH.

[*G. flexuosa auct. non* (Lindl.) Meisn.: J.S.Beard, *Descr. Cat. W. Austral. Pl.* 21 (1965); R.Erickson *et al.*, *Fl. Pl. W. Australia* 128, incl. plate 399 (1973, 1979).]

Illustrations: A.S.George, *Introd. Proteaceae W. Australia* 55, t. 76 (1984); D.J.McGillivray & R.O.Makinson, *Grevillea* 166, fig. 42 & col. pl. (1993); P.M.Olde & N.R.Marriott, *Grevillea Book* 2: 233 (top right & 194A–D) (1995).

Prostrate to low shrub 0.1–0.4 m high. Branchlets glabrous or sparsely to densely subsericeous to tomentose. Leaves dorsiventral, 3–29 cm long, flexuose or not, very variable, simple and toothed, rarely entire, to compound pinnatifid to partly tripinnatisect with up to 3 orders of division; primary lobes 7–15; ultimate lobes occasionally divaricate, triangular to ovate, oblong, or sublinear, 0.2–3.5 (–11) cm long, 1–5 mm wide; margins flat to revolute; lower surface glabrous or sparsely to densely silky to tomentose. Conflorescence terminal, erect, simple or to 3-branched; unit conflorescence secund to secund-obconical, loose, basipetal; ultimate floral rachis 30–50 mm long, glabrous or sericeous. Flowers acroscopic to transverse. Flower colour: perianth and style pale pink to deep lilac-pink. Perianth glabrous outside or with appressed to ascending hairs, minutely papillose inside; dorsal tepals flaring widely and partially everted before anthesis, forming a small platform. Pistil 4–6 mm long; style exposed but scarcely exserted before anthesis; pollen-presenter oblique to style, conical, basally flanged. Follicle obovoid to subclavate, 9–14 mm long, smooth to rugulose, red. *Tangled Grevillea.* Fig. 35G–K.

Occurs in south-western W.A., scattered populations in the area bounded by Cranbrook, Brookton, North Bannister and Shannon. Grows in eucalyptus woodland or open forest with shrubby understorey, in well-drained often lateritic soils. Regenerates from lignotuber, rhizomes, and ?seed; degree of exclusively vegetative propagation uncertain. Flowers year round, mainly Oct.–Dec. Map 403.

W.A.: watershed off Brookton Hwy, 16 Dec. 1969, *R.J.Edmiston* (PERTH); 49 km SW of Kojonup on Frankland Rd, *A.S.George 15242* (PERTH); 21 km W of Kojonup, *K.Newbey 1784* (PERTH); 57.5 km SE of Mayanup towards Cranbrook, *B.R.Maslin 3991* (PERTH); 6 km S of Wickepin, *E.Wittwer 886* (KPBG, PERTH).

This species is exceedingly variable in leaf form. The very strong eversion of the dorsal tepals, forming a flat platform, together with the pink flower colour and short glabrous pistil, are distinctive. *Grevillea prostrata* and *G. cirsiifolia* have flowers of a similar configuration. *Grevillea prostrata* has pinkish white flowers but has leaves 3–7 cm long, secund with very regular oblong lobes; *G. cirsiifolia* has a hairy ovary, a flat pollen-presenter and yellow flowers.

Certain of the commoner forms have a degree of geographical consistency. The 'Tutanning form' is prostrate and mat-forming and has broad flexuose grey-pubescent leaves, hairy branchlets, floral rachises and perianths, and broadly linear leaf lobes. The 'Brookton form' has a clumping habit, glabrous leaves and branchlets (also often glabrous rachises and perianths), and linear leaf lobes with strongly revolute margins. The 'Dryandra form' is similar to the 'Brookton form' but has shorter, ±triangular bluish grey leaf lobes. The 'Williams form' is lignotuberous with prostrate branches and once- to twice-pinnatifid leaves with oblong lobes. These forms do not cover the whole variance and further research and delimitation of genetic isolates is desirable.

319. Grevillea crithmifolia R.Br., *Suppl. Prodr. Fl. Nov. Holl.* 23 (1830)

T: Swan River, [W.A., 1827], *[C.]Fraser*; lecto: BM, *fide* D.J.McGillivray & R.O.Makinson, *Grevillea* 412 (1993); isolecto: BM.

Illustrations: P.M.Olde & N.R.Marriott, *Grevillea Book* 2: 103 (top right), 104 (83A–C) (1995).

Dense shrub 0.6–2.5 (–5) m tall, or rarely prostrate. Branchlets villous. Leaves dorsiventral, 1–2 (–3.5) cm long, crowded, entire and c. 0.7 mm wide, or 2–5-partite and 5–10 mm wide, often simple and divided leaves mixed; entire leaves and lobes linear to narrowly oblong, lacking secondary division; lobes 0.7–1.7 mm wide, often gently curved, sometimes weakly divaricate; margins revolute; lower surface enclosed, 2-grooved, with grooves and midveins pilose to villous. Conflorescence usually terminal, simple, erect, umbelloid or subcorymbose or rarely conical, acropetal; rachis 6–12 (–24) mm long, densely villous. Flowers acroscopic to irregular. Flower colour: perianth sometimes pink in bud becoming paler pink to creamy white after anthesis; style cream-white. Perianth sparsely villous inside. Pistil 4.8–6 mm long; style exposed but scarcely exserted from late bud; pollen-presenter oblique, conical with basal rim. Follicle ellipsoidal to ovoid, 12–15 mm long, muricate-echinate.

Occurs in the south-west of W.A., from Wanneroo to Yalgorup Natl Park between Lake Preston and Lake Clifton, with an isolated occurrence near Dongara. Grows in subcoastal woodland or coastal scrub in calcareous sand and limestone soils on dunes, slopes and sandplains, often near to the coast. Regenerates from seed and basal suckers or lignotuber. Flowers June–Sept. (–Dec.). Map 404.

W.A.: near Dongara, Nov. 1963, *F.W.Humphreys* (PERTH); Mt Eliza, Perth, *s.d.*, *A.Oldfield* (MEL); Fremantle, *L.Preiss 690* (B, G-DC, HBG, LE, MEL, P (all *n.v.* except MEL)), also same collection as *L.Preiss 51* (CGE, TCD); between L. Preston & L. Clifton, *S.Paust 1381* (PERTH); 1.6 km N of Mandurah, *H.Salasoo 4033* (NSW); Nedlands near Perth, *C.T.White 5182* (BRI).

Olde & Marriott (*op. cit.* 103 (1995)) list horticultural forms. *Grevillea crithmifolia* can be confused with *G. murex*, which has branchlets angular in cross-section (terete in *G. crithmifolia*), smaller leaves 0.4–1 cm long, and a flat pollen-presenter. The strongly muricate-echinulate fruits are unusual in this group, which otherwise (with the possible exception also of *G. trachytheca*, *G. zygoloba* and *G. obliquistigma*) forms a natural group; the inclusion of *G. crithmifolia* in it is provisional.

320. Grevillea trachytheca F.Muell., *Fragm.* 6: 207 (1868)

T: 'Ad flumen Murchison ... Oldf.'; lecto: ... Sandy places Murchison [Murchison R., W.A.], *s.d.*, [*A.F.Oldfield*]; MEL, *fide* D.J.McGillivray & R.O.Makinson, *Grevillea* 445 (1993); possible isolecto: MEL, K.

Illustrations: D.J.McGillivray & R.O.Makinson, *Grevillea* 380 (1993); P.M.Olde & N.R.Marriott, *Grevillea Book* 3: 204 (bottom right), 205 (167A–C) (1995).

Shrub 0.6–2 m high. Branchlets tomentose to velvety. Leaves dorsiventral, 1.5–3.5 cm long, usually simple and entire, sometimes a few 2–5-partite; simple leaves and lobes broadly linear, 1–2 mm wide; margins refracted, enclosing most or all of lower surface except for midveins; lower surface tomentose. Conflorescence terminal, erect, usually simple, narrowly conical to subcylindrical, acropetal; ultimate floral rachis 20–60 mm long, villous. Flowers ±acroscopic. Flower colour: perianth white to cream with a yellow limb; style white. Perianth glabrous inside. Pistil 5.5–7 mm long; style slightly exserted from late bud; pollen-presenter erect on a slightly oblique base, conical, with a basal flange. Follicle ovoid to ellipsoidal, 11–14 mm long, muricate.

Occurs in south-western W.A., restricted to the Kalbarri area on the lower Murchison R. Grows in shrub associations in calcareous sandy soils. Regenerates from seed. Flowers May–Oct. Map 405.

W.A.: 55 km from North West Coastal Hwy on road to Kalbarri, *D.J.McGillivray 3334, 3335 & A.S.George* (NSW); Red Bluff, Kalbarri Natl Park, *M.E.Phillips CBG036031* (CANB, PERTH).

Grevillea trachytheca appears to be closely related to *G. crithmifolia*, which differs in its shorter unit conflorescence and mostly divided leaves. See comments on wider affinities under *G. crithmifolia*.

This species is recognised as 'Poorly Known' in J.D.Briggs & J.H.Leigh, *Rare or Threatened Australian Plants* (1995).

Integrifolia Group

Shrubs. Leaves entire, unifacial or dorsiventral; surfaces similar; margins flat or involute. Conflorescence usually terminal, erect, usually branched; unit conflorescence cylindrical or umbel-like, synchronous or acropetal to basipetal. Flowers adaxially oriented or acroscopic. Torus straight to slightly oblique. Perianth actinomorphic with limb erect in bud, glabrous outside and inside; tepals separating below limb in late bud and arched outwards, later all free. Pistil 5–10 mm long, glabrous; ovary stipitate; style undilated below style-end, ±exserted near base from late bud; pollen-presenter erect, conical to cylindrical. Follicle often persistent, variably obovoid with an apical beak, smooth, glabrous; pericarp thin, crustaceous. Seed obovoid with a membranous band at apex; outer face convex, inner flat.

A group of eight species in W.A., seven endemic in the SW, and one in the E Kimberley. Insect pollinated. The group is characterised by the erect, regular perianth, the glabrous pistil without stylar swelling, the erect pollen-presenter, and (except *G. psilantha*) the obliquely obovoid to narrowly obovoid or subcylindrical follicle (which is often also weakly sigmoid). The nectary (except in *G. psilantha*) is absent or minute. Affinities are to the *Trifida* group and the *Hilliana* group (especially *G. didymobotrya*) and to the *Rudis* group.

1 Ovarian stipe > 1.5 mm long; nectary clearly evident, V-shaped; follicle obliquely compressed-ellipsoidal or -obovoid **328. G. psilantha**

1: Ovarian stipe < 1.5 mm long; nectary absent to obscure, not V-shaped; follicle narrowly obovoid-clavate

2 Adult leaves subterete or plumply linear, < 3.0 cm long, ±incurved

3 Unit conflorescences umbel-like or shortly cylindrical; ultimate floral rachises 10–20 mm long; flowers golden yellow or rarely creamy yellow; leaves densely crowded; follicle 10–16 mm long **325. G. incrassata**

3: Unit conflorescences cylindrical; ultimate floral rachises (25–) 40–50 mm long; flowers cream-white; leaves not densely crowded; follicle 9–10 mm long **326. G. incurva**

2: Adult leaves ±flat, linear to obovate, elliptic or ovate, 1–16 cm long, straight or incurved

4 Leaves and branchlets with indumentum of ascending to spreading hairs; floral bracts 2–5 mm long, loosely villous; floral rachises glabrous, yellow when fresh, drying yellow or rarely to dark brown **323. G. ceratocarpa**

4: Leaves and branchlets glabrous or with dense to sparse indumentum of ±appressed hairs; floral bracts ≤ 2 mm long, densely sericeous to tomentose; floral rachises sericeous or rarely glabrous, yellow or not

5 Longest adult leaves linear to narrowly linear-obovate or narrowly elliptical

6 Floral rachises yellow, glabrous or nearly so; follicles 12–18 mm long, 1.5–2.5 mm wide **324. G. eremophila**

6: Floral rachises brown and subsericeous, rarely glabrous and then yellow or brown; follicles usually 7–13 mm long, 2.5–4 mm wide **322. G. biformis**

5: Longest adult leaves orbicular, obovate or elliptic

7 Unit conflorescences acropetal; leaf margins flat; tepal limb segment not ribbed along midline at base; follicle usually 4–6 mm wide; leaf lower surface usually glabrous, rarely sericeous **327. G. shuttleworthiana**

7: Unit conflorescences subsynchronous or occasionally basipetal; leaf margins flat or slightly to strongly involute; tepal limb obscurely ribbed along midline at base; follicle usually 2.5–4 mm wide; leaf lower surface sericeous

8 Adult leaves with sericeous upper surface **321. G. integrifolia**

8: Adult leaves with glabrous upper surface **322. G. biformis**

321. Grevillea integrifolia (Endl.) Meisn., *in* A.L.P.P. de Candolle, *Prodr.* 14: 385 (1856)

Anadenia integrifolia Endl., *Stirp. Herb. Hügel.* 21 (1838). T: Tuttaning Reserve, Possum Rd, E of Pingelly, W.A., 1 Oct. 1967, *G.Heinsohn 14*; neo: PERTH.

G. integrifolia subsp. *integrifolia*, of D.J.McGillivray & R.O.Makinson, *Grevillea* 138 (1993).

Illustrations: D.J.McGillivray & R.O.Makinson, *Grevillea* 138, col. pl. (1993); P.M.Olde & N.R.Marriott, *Grevillea Book* 2: 213 (bottom right), 214 (177A–C) (1995).

Erect shrub 1–3 m high. Branchlets subsericeous. Leaves bifacial, entire, 1–5 cm long, 2–10 mm wide, ascending, sessile, obovate, sometimes recurved; upper and lower surfaces similar, shiny-sericeous; margins shortly and tightly involute, often also undulate; venation obscure on upper surface, the midvein and usually 2 intramarginal veins prominent below, sometimes also acutely angled lateral veins evident. Conflorescence erect, terminal or usually upper-axillary, simple or branched; unit conflorescence 3–4.5 (–7.5) cm long, cylindrical, ±synchronous; floral rachis subsericeous, occasionally sparsely so, brownish. Flowers adaxially oriented. Flower colour: perianth white to creamy white or occasionally pale yellow, sometimes with pinkish tinges; style white to creamy white. Perianth glabrous; tepals free and ascending after anthesis. Pistil 5.5–6.5 mm long, glabrous; pollen-presenter conico-cylindrical. Follicles narrowly obliquely obovoid, 9–11 mm long, smooth.

Occurs in south-western W.A. around Burracoppin, Quairading, Kukerin and Corrigin, usually in open heath in sand over laterite. Regenerates from seed. Flowers mainly Oct.–Dec. Map 406.

W.A.: near Tammin Reserve, *T.E.H.Aplin 3032* (CANB, PERTH); between Booraan and Burracopping, *H.F. & M.Broadbent NSW125488* (NSW); 0.5 km E of Harrismith, *M.D.Crisp 6158 et al.* (CANB, NSW,

PERTH); 16.5 km S of Cunderdin, *B.R.Maslin 4415* (K, NSW, PERTH); 17 km S of Tammin, (Charles Gardner Flora Reserve area), *R.W.Purdie 5303* (CANB).

Grevillea integrifolia can be confused with *G. shuttleworthiana*, but the latter species has leaves with flat (not involute) margins and wider fruit. *Grevillea didymobotrya* is also similar, but can be distinguished by its zygomorphic perianth with a hairy outer surface.

322. **Grevillea biformis** Meisn., *in* J.G.C.Lehmann, *Pl. Preiss.* 2: 258 (1848)

G. integrifolia subsp. *biformis* (Meisn.) McGill., *New Names Grevillea* 7 (1986). T: interior N of Swan River, W.A., *J.Drummond coll. 3, 265*; lecto: NY (photo at NSW), *fide* D.J.McGillivray & R.O.Makinson, *Grevillea* 422 (1993); isolecto: A *n.v.*, CGE *n.v.*, G, G-DC *p.p.*, K, LE *n.v.*, MEL, TCD *n.v.*

Erect shrub 1.5–2.5 m high. Branchlets subsericeous. Leaves entire, linear (adult) or obovate (juvenile), occasionally both present, 3–16 cm long, 1–12 mm wide, sericeous on one or both surfaces; linear leaves with longitudinal ridges and obscure margins; obovate leaves with a few subparallel ridges on lower surface only and flat to slightly involute margins. Conflorescence erect, terminal or upper axillary, simple or with few ascending branches; unit conflorescence (2.5–) 4–13 (–18.5) cm long, cylindrical, ±synchronous or occasionally weakly basipetal; floral rachis subsericeous to sparsely so, usually brown (sometimes yellow) beneath indumentum, occasionally sparsely subsericeous or glabrous. Flowers adaxially acroscopic. Perianth glabrous; tepals free and ascending after anthesis. Pistil 5.5–8.5 mm long, glabrous; style ±smooth; pollen-presenter narrowly conico-cylindrical with a truncate apex. Follicles 7–13 mm long, narrowly and obliquely obovoid, ±smooth to faintly colliculate.

Widespread in W.A. between Shark Bay and Hyden. There are two subspecies.

Grevillea ceratocarpa differs in its longer, narrower fruit, spreading foliar and branchlet indumentum and glabrous (or nearly so) floral rachises. *Grevillea eremophila* has longer fruits, a glabrous floral rachis and usually slightly longer pistils. *Grevillea incrassata* and *G. incurva* differ in their shorter (to 4.5 cm long) often subterete leaves and shorter unit conflorescences.

Adult leaves ±linear; upper and lower leaf surfaces similar with longitudinal ridges and appressed hairs (at least in grooves); follicle 2.5–3 (–4) mm wide (obovate juvenile leaves, if present, sericeous on both surfaces) **322a.** subsp. **biformis**

Adult leaves obovate to narrowly so; upper leaf surface glabrous and not or scarcely ridged; lower surface few-ridged and sericeous; follicle 3–4 mm wide **322b.** subsp. **cymbiformis**

322a. **Grevillea biformis** Meisn. subsp. **biformis**

G. stenocarpa Benth., *Fl. Austral.* 5: 485 (1870), *nom. illeg.* T: as for *G. biformis* Meisn.

G. integrifolia var. *stenocarpa* (Benth.) Grieve, *How to Know W. Austral. Wildfl.* 1: 112 (1954), *nom. nud.*

Illustrations: P.M.Olde & N.R.Marriott, *Grevillea Book* 2: 62 (46A, B), 63 (top left) (1995).

Juvenile leaves broadly to narrowly obovate or elliptical, 2.5–3.5 cm long, 6–22 mm wide; upper surface sericeous to openly so; lower surface sericeous, with prominent midvein and reticulum. Adult leaves 5–16 cm long, 1–2.5 mm wide, ±linear, straight or slightly incurved; margins obscure, flat; surfaces ±similar, with several longitudinal parallel ridges and appressed hairs at least in grooves. Unit conflorescence (2.5–) 4–13 (–18.5) cm long; floral rachis silver-sericeous to sparsely so or rarely glabrous, usually brown or sometimes yellow beneath indumentum. Flower colour: perianth and style creamy white, rarely the perianth pale pink. Pistil 5.5–8.5 mm long. Follicles 2.5–3 (–4) mm wide, sometimes faintly colliculate.

Widespread in south-western W.A., in the northern, central and southern wheatbelt regions from Shark Bay and Murchison R. to Lake Grace and Hyden. Found in a variety of habitats from sandplain to shrubland and mallee, occasionally poorly drained. Probably regenerates by seed only. Flowers mainly June–Oct. Map 407.

W.A.: N of Wongan Hills & c. 185–187 km N of Perth, *E.M.Canning CBG028185* (CANB, NSW); 5.1 km E along road that is 1 km N of Binnu on North West Coastal Hwy, *J.M.Fox 88/111* (CANB, NSW, PERTH);

10 km along road from Lake Grace towards Dumbleyung at Lake Grace Lookout, *A.Lyne 1008 et al.* (AD, BRI, CANB, MEL, NSW, PERTH); 10 km by road NE of Kukerin towards Tarin Rock, *D.J.McGillivray 3536 & A.S.George* (NSW).

Two forms are recognised, although the division is imperfect and there is a broad intergrade zone from Merredin to Wongan Hills. The 'northern form' occurs from Shark Bay and the Murchison R. S to Wongan Hills; it has leaves usually 7–13 cm long and 1.5–2.5 mm wide, with very prominent and usually glabrous ridges, a usually long (7–18.5 cm long) unit conflorescence, pedicels 2.5–4 mm long and the style often obscurely subpapillose; the leaves often dry to a distinctive yellow-green colour. The 'southern form' occurs from Bendering, Hyden and Kulin S and E to Lake Grace area; it has the leaves generally shorter (5–9 cm long) and often narrower (1–2 mm wide), with more subdued ridges which are covered with a subsericeous indumentum; the unit conflorescence is usually shorter (3–6 cm long), as are the pedicels (1.2–2 mm long), and the style is smooth; the leaves usually dry to a pale, slightly grey-green colour.

322b. **Grevillea biformis** subsp. **cymbiformis** Olde & Marriott, *Grevillea Book* 1: 176 (1994)

T: Erindoon Rd, 11.4 km S of Eneabba–Coolimba road, S of Eneabba, W.A., 15 Sept. 1991, *P.M.Olde 91/103*; holo: NSW; iso: PERTH.

G. integrifolia 'unassigned 2' of D.J.McGillivray & R.O.Makinson, *Grevillea* 140 (1993).

Illustrations: P.M.Olde & N.R.Marriott, *Grevillea Book* 2: 63 (centre right & 47) (1995).

Juvenile leaves not seen. Adult leaves 3–9 cm long, 1–12 mm wide, usually obovate and upwardly concave; upper surface glabrous, smooth, sericeous; lower surface with prominent midvein; upper leaves sometimes narrowly linear with a distinct narrow glabrous groove along adaxial midline, ridged to either side; margins flat or slightly involute. Unit conflorescence 4–13 cm long; floral rachis silver-sericeous to sparsely so, usually brown beneath indumentum. Flower colour: perianth and style creamy white. Pistil 7.5–8 mm long. Follicles 3–4 mm wide, obovoid, rugulose.

Found in a small area S and SSW of Eneabba, W.A. A specimen at CANB (*Speck s.n.*, CANB195211) is allegedly from Hill R.: the locality may be in error. Grows in low heath, in sand over laterite. Regenerates from seed. Flowers mainly Oct. Map 408.

W.A.: 3 km W of L. Indoon, W of Eneabba, *E.A.Griffin & M.Blackwell 3030* (PERTH); 100 m from Brand Hwy after corner of road to Coolimba, *P.M.Olde 86/605* (NSW, PERTH); 2.4 km W of Campbell-White Rd off Erindoon Rd, S of Eneabba, *P.M.Olde 92/138* (NSW).

This subspecies is recognised as 'Poorly Known' in J.D.Briggs & J.H.Leigh, *Rare or Threatened Australian Plants* (1995).

323. **Grevillea ceratocarpa** Diels, *Bot. Jahrb. Syst.* 35: 157 (1904)

G. integrifolia subsp. *ceratocarpa* (Diels) McGill., *New Names Grevillea* 7 (1986). T: Bronti, near Southern Cross, W.A., 20 Nov. 1901, *L.Diels 5980*; lecto: B, *fide* D.J.McGillivray & R.O.Makinson, *Grevillea* 422 (1993); remaining syntype: Boodalin, *s.d.*, *A.Forrest*; syn: MEL.

G. integrifolia subsp. *ceratocarpa* 'tomentose form', of D.J.McGillivray & R.O.Makinson, *Grevillea* 139 (1993).

Illustrations: P.M.Olde & N.R.Marriott, *Grevillea Book* 2: 88 (top left & 69) (1995).

Erect shrub 0.5–1.5 m high. Branchlets pubescent to tomentose (hairs ±spreading). Leaves usually entire, ascending to erect, leathery, usually narrowly elliptic or narrowly obovate, rarely linear, 5.5–8.5 cm long, 2–10 mm wide (rarely broadly elliptic and to 22 mm wide in juveniles); margins flat to shortly incurved; upper surface densely pubescent to tomentose with venation obscured; lower surface more sparsely tomentose and faintly ribbed. Conflorescence erect, terminal or upper axillary, simple or branched; unit conflorescence 7–9 (–15) cm long, cylindrical, ±synchronous or sometimes basipetal; floral rachis initially yellow, fading to brown, glabrous or with a few appressed hairs. Flowers adaxially oriented. Flower colour: perianth and style creamy white. Perianth glabrous; tepals free and ±spreading after anthesis. Pistil 7.5–10 mm long, glabrous; pollen-presenter sub-cylindrical. Follicles narrowly and obliquely obovoid-cylindrical, 12–18 mm long, smooth.

venation evident to obscure on upper surface, prominently raised or occasionally obscure below. Conflorescence erect, terminal or upper axillary, simple or few-branched; unit conflorescence 3–4.5 (–7.5) cm long, cylindrical, acropetal; floral rachis usually loosely to densely ±white-sericeous, occasionally yellow and glabrous or sparsely sericeous. Flowers adaxially oriented. Perianth glabrous; tepals independently reflexed after anthesis. Pistil 4.5–6.5 mm long, glabrous; pollen-presenter conico-cylindrical, with or without flanged base. Follicles obovoid to obpyramidal, 7–17 mm long, smooth to rugulose.

Occurs in south-western W.A., from near Kalbarri on the sandplains N of Perth, E to Mt Jackson (N of Southern Cross) and S to Ravensthorpe Ra. Three subspecies are recognised.

The branchlets of *G. shuttleworthiana* are often angular in cross-section; all other species in the group have subterete branchlets. *Grevillea integrifolia* and *G. biformis* subsp. *cymbiformis* are similar: both differ from *G. shuttleworthiana* in their synchronous to weakly basipetal flower-opening and have a ridge on the tepal midline on the limb segment (not ridged in *G. shuttleworthiana*). *Grevillea integrifolia* also differs in having leaves sericeous on both surfaces and with tightly and strongly involute margins. *Grevillea biformis* subsp. *cymbiformis* has upwardly concave leaves which are glabrous on the upper surface and sericeous below.

1 Floral rachises glabrous or nearly so and bright yellow to yellow-brown (best seen on fresh material)

2 Leaves yellow-green in dried state and glabrous on both surfaces; longest leaves usually 3–6 cm long; lateral veins (and often reticulum) moderately prominent on lower leaf surface; unit conflorescences 3.5–7.5 cm long; pistil 5.5–6.5 mm long; fruit 13–17 mm long **327b.** subsp. **canarina**

2: Leaves grey-green (rarely to yellow-green) in dried state and usually with an open even sprinkling of appressed hairs on both surfaces; longest leaves 1–4 cm long; venation on lower leaf surface obscure (wrinkles present); unit conflorescences 1–3.5 cm long; pistil 4.8–5.5 mm long; fruit 7–10 mm long **327c.** subsp. **obovata**

1: Floral rachises densely to loosely subsericeous and brown to greenish (rarely yellowish) beneath the indumentum

3 Leaves 11–27 mm wide; leaf lower surface with lateral veins (and often reticulum) moderately prominent, nonvenous wrinkling often also present; follicle 4–6 mm wide **327a.** subsp. **shuttleworthiana**

3: Leaves 4–17 mm wide; leaf lower surface with venation obscure, nonvenous wrinkling present; follicle 3–5 mm wide **327c.** subsp. **obovata**

327a. Grevillea shuttleworthiana Meisn. subsp. shuttleworthiana

G. cordata C.A.Gardner, *J. & Proc. Roy. Soc. W. Australia* 9: 35 (1923); *G. shuttleworthiana* subsp. *cordata* (C.A.Gardner) Grieve, *How to Know W. Austral. Wildfl.* 1: 112 (1954), *nom. inval.* T: Yorkrakine, W.A., 4 Sept. 1922, *C.A.Gardner 1727*; lecto: PERTH, *fide* D.J.McGillivray & R.O.Makinson, *Grevillea* 422 (1993); isolecto: CANB, MEL, PERTH.

Illustrations: P.M.Olde & N.R.Marriott, *Grevillea Book* 3: 172 (139A, B), 173 (top left & 139C) (1995).

Leaves bluish green, soon glabrous, usually glaucous, 1–4.5 cm long, 11–27 mm wide, elliptic to almost orbicular or occasionally ovate or obovate, sometimes cordate to subamplexicaul; venation evident on lower surface, with spreading reticulum, wrinkled or not. Unit conflorescence 2–4.5 cm long; floral rachis white-sericeous, pale brown below indumentum. Flower colour: perianth and style pale yellow. Pistil 5–6 mm long; pollen-presenter 0.8–1 mm long, conical, with basal flange lacking or obscure. Fruit 8–13 mm long, 4–6 mm wide.

Found in south-western W.A. from Perenjori and Bonnie Rock S to Yorkrakine; grows in low heath on white sand with gravel. Regenerates from seed and ?lignotuber. Flowers July–Sept. Map 413.

W.A.: Perenjori, *A.Carn s.n.* (PERTH); c. 3.2 km N of Wongan Hills, Research Stn vicinity, *M.E.Phillips 872* (CANB, NSW); c. 4 km S of Manmanning, Telephone post 31-08, *B.H.Smith 412* (CANB, HO, NSW, PERTH); Wongan Hills rly [line], c. 3 km SE of Elphin, *B.H.Smith 854* (CANB, LEN[?], NSW, PERTH).

A specimen from Maya railway siding (*N.Hoyle 309*, CANB, PERTH) seems to represent an intergrade with subsp. *canarina*, having the leaves and floral rachis of subsp. *shuttleworthiana* and the flanged pollen-presenter of subsp. *canarina*.

327b. Grevillea shuttleworthiana subsp. canarina Olde & Marriott, *Grevillea Book* 1: 177 (1994)

T: Erindoon Rd, 11.4 km S of Eneabba–Coolimba road, S of Eneabba, W.A., 15 Sept. 1991, *P.M.Olde 91/104*; holo: NSW; iso: PERTH.

G. shuttleworthiana subsp. *shuttleworthiana* 'long-leaved form', of D.J.McGillivray & R.O.Makinson, *Grevillea* 139 (1993).

Illustrations: P.M.Olde & N.R.Marriott, *Grevillea Book* 3: 173 (centre & 140A), 174 (140B) (1995).

Leaves yellowish green, (1.5–) 2.5–6 cm long, obovate-cuneate, glabrous or with a very few appressed hairs when young; venation moderately conspicuous with prominent midvein, a very acute, ascending reticulum and conspicuous wrinkling on lower surface. Unit conflorescence (3.5–) 4.5–7.5 cm long; floral rachis bright yellow, glabrous or sparsely sericeous. Flower colour: perianth and style pale yellow. Pistil 5.5–6.5 mm long; pollen-presenter 1–1.5 mm long, conical with basal flange narrow to obscure. Fruit 13–17 mm long, 4–4.5 mm wide.

Found in south-western W.A. from Binnu (SE of Kalbarri) S to c. Mullering Brook and E to Piawaning; growing in white sand with gravel, usually dominant in low heath. Regenerates from seed and ?lignotuber. Flowers Aug.–Oct. Map 414.

W.A.: Watheroo Natl Park, intersection of Marchagee Track Rd and Coatara Rd, *N.Hoyle 838* (CANB, PERTH); c. 15 km from Brand Hwy E along Moore River Rd towards Mogumber, *R.W.Purdie 5088* (CANB); Mullering Brook, *D.Woolcock G32b* (NSW).

327c. Grevillea shuttleworthiana subsp. obovata (Benth.) Olde & Marriott, *Grevillea Book* 1: 177 (1994)

G. integrifolia var. *obovata* Benth., *Fl. Austral.* 5: 485 (1870). T: interior, south-western W.A., 183-, *J.S.Roe s.n.*; lecto: K, *fide* D.J.McGillivray & R.O.Makinson, *Grevillea* 422 (1994); remaining syntype: Swan River, W.A., *J.Drummond coll. 3, 265*; syn: A, CGE, G, G-DC, K, LE, MEL, TCD.

Illustrations: P.M.Olde & N.R.Marriott, *Grevillea Book* 3: 174 (top right & 141A, B) (1995).

Leaves bluish green to grey (occasionally yellow-green), 0.5–3 (–4) cm long, 4–17 mm wide, orbicular to narrowly obovate or spathulate, with a sparse to open (rarely dense) indumentum of appressed hairs on both surfaces; venation usually inconspicuous, sometimes prominent on upper surface, on lower surface obscure with conspicuous wrinkling. Unit conflorescence (1–) 2–3.5 cm long; floral rachis greenish to brown or yellow-brown, white-sericeous or sparsely so. Flower colour: perianth and style pale yellow. Pistil 4.5–5.5 mm long; pollen-presenter cylindrical, 0.8–1 mm long, lacking a basal flange but sometimes offset from line of style. Fruit 7–10 mm long, 3–5 mm wide.

Occurs in south-western W.A. from Mt Jackson and Bronti S to the Ravensthorpe Ra., and W to Lake Grace; grows in yellow sand or occasionally laterite, in low heath. Regenerates from seed and ?lignotuber. Flowers late July–Dec. (–Jan.). Map 415.

W.A.: L. Varley, SE of Hyden, *M.Barrow 77* (CANB, PERTH); c. 14.5 km E of Newdegate, 4 Nov. 1947, *N.T.Burbidge s.n.* (CANB); Yellowdine, *R.J.Cranfield 704* (CANB, PERTH); Mt Short, c. 16 km N of Ravensthorpe, *J.W.Wrigley CBG036593* (CANB, DNA, NSW).

There is considerable variation in leaf form and indumentum, and in pollen-presenter shape within this subspecies. The distinctive 'Ravensthorpe Range form' has long (3–4 cm) obovate to spathulate leaves which are nonglaucous and dry to a mid- to yellow-green colour, with a sprinkling of very fine short sparkling hairs on both surfaces. This form resembles subsp. *canarina* but apparently has consistently hairy floral rachises that are brownish beneath the indumentum, and an open (rarely dense) inconspicuous indumentum of minute appressed hairs evenly distributed over both leaf surfaces. Olde & Marriott (*Grevillea Book* 3: 174–175 (1995)) identify two further forms. Their 'typical form' occurs in the Southern Cross area and S towards Hyden, and has 'small, blue-green, orbicular to obovate, glabrous, usually glaucous leaves'; in practice, leaves of this form often have a sprinkling of appressed hairs, especially about the margins, and rarely have dense hairs across the lamina. Their 'silky-leaved form' occurs from NE of Ravensthorpe to Newdegate, Hyden and the Mt Holland area; it has an even, relatively dense, sericeous indumentum on the leaf lamina. Delimitation of these two forms remains somewhat problematic, and further study is needed.

A collection from Wyalkatchem (*J.M.Fox, 86/356*, CANB) seems, on foliar features, to be an intergrade with subsp. *canarina*.

328. Grevillea psilantha McGill., *New Names Grevillea* 12 (1986)

Piccaninny Creek Gorge, 15 km SE of Bungle Bungle outcamp, E Kimberley, W.A., 12 July 1984, *K.F.Kenneally 9282*; holo: PERTH; iso: NSW.

Illustrations: D.J.McGillivray & R.O.Makinson, *Grevillea* 137, fig. 28 (1993); P.M.Olde & N.R.Marriott, *Grevillea Book* 3: 113 (bottom right), 114 (87A, B) (1995).

Erect, open to mid-dense shrub to 1 m high. Branchlets subsericeous. Leaves entire, 7–15 cm long, 2.5–7.2 mm wide, dorsiventral, ascending, petiolate, concolorous, linear to narrowly obovate; upper and lower surfaces similar; margins flat; venation of both surfaces consisting of several faint parallel veins. Conflorescence erect, terminal, branched; unit conflorescence cylindrical, open, 9–11 cm long, acropetal; floral rachis glabrous or with scattered appressed hairs or sparsely pubescent. Flowers adaxially oriented. Flower colour: perianth and style white to cream. Perianth glabrous; tepals independently revolute after anthesis. Pistil 7–8 mm long, glabrous; pollen-presenter shortly conical to ellipsoidal. Follicles laterally compressed-obovoid, 9–14 mm long, ridged submarginally all around, otherwise rugulose. Fig. 35A–C.

Occurs in northern W.A. in the Bungle Bungle Ra. (E Kimberley area); grows in crevices in walls of sandstone gorges. Regenerates from seed. Flowers Apr.–July. Map 416.

W.A.: Piccaninny Ck gorge, *S.J.Forbes 2643* (CANB, DNA, MEL, NSW); Piccaninny Gorge, *K.Hill 3493* (CANB, NSW, PERTH); Bungle Bungle Ra., side gorge off Cathedral Gorge, *R.W.Purdie 3336* (CANB, PERTH).

Grevillea psilantha differs from all other species in this group in several features. Its follicles are retrorse on incurved stipes, are laterally compressed-obovoid, and have a clear lacquer-like subviscid coating on the surface (follicles erect on ±straight stipe, obliquely obovoid to narrowly so to subcylindrical and lacking a subviscid coating in other species of the group). *Grevillea psilantha* also has pedicels 4.5–5 mm long (cf. up to 4 mm long) and ovarian stipes c. 1.6–1.7 mm long (cf. < 1 mm long). While clearly closely related to other members of this group, there may also be an affinity with *G. stenobotrya*.

This species is recognised as 'Poorly Known' in J.D.Briggs & J.H.Leigh, *Rare or Threatened Australian Plants* (1995).

Triloba Group

Shrubs. Leaves divided or rarely entire, dorsiventral; surfaces dissimilar; margins flat to revolute. Conflorescence terminal and/or axillary, often aggregated towards end of branches, erect to spreading, simple or few-branched; unit conflorescence erect, subglobose to subumbelloid, rarely shortly cylindrical or subconical, usually acropetal. Torus transverse to slightly oblique. Flowers obscurely adaxially oriented. Perianth actinomorphic, glabrous or rarely subsericeous outside, glabrous or sometimes sparsely pilose inside; tepals separating below limb in late bud and arched outwards, later all separating and independently recoiled. Pistil 3–6 mm long (not or scarcely longer than tepals), glabrous; ovary stipitate, with stipe often sigmoid; style not exserted from late bud, usually constricted at base, dilated above into a usually ovoid swelling, then narrowing below style-end; pollen-presenter ±erect, conical or shortly cylindrical. Follicle obloid to ellipsoidal, glabrous; pericarp thin- or rarely thick-walled. Seed ellipsoidal, inner face with a membranous coating usually forming a narrow marginal rim.

A group of 21 species, occurring in south-western W.A. with one (*G. anethifolia*) also in S.A. (Eyre Peninsula) and N.S.W. (Griffith to near Mt Hope). All insect-pollinated. A very distinctive group, corresponding to Section *Manglesia* of Bentham (1870). This group is easily diagnosed by the regular perianth, erect pollen-presenter, swollen style, and small flowers with the filamentous pedicels usually much longer than the perianth. In several species, the limb of the bud is yellow, pink or brown, contrasting with the white of the rest of the perianth, a colour contrast also seen in some taxa in the *Rudis* group. Species differences are often very fine, and delimitation and ranking remain problematic in the complexes around *G. amplexans* and *G. paniculata*. A number of problematic specimens may represent distinct undescribed taxa. The group has several highly derived character states. It is probably descended from a stock similar to the *Rudis* group, which seems likely to represent the most closely related group.

1 Pedicels and outer surface of perianth with appressed biramous hairs

2 Leaves all deeply pinnatipartite (almost pinnatisect); basal section (leaf base to first node) linear; ultimate lobes linear, ⩽ 1 mm wide; lower surface (except for main veins) completely concealed by tightly revolute margins **348. G. erinacea**

2: Leaves pinnatifid on vegetative branches, becoming coarsely pinnatipartite on flowering branches; basal section cuneate; ultimate lobes triangular to subulate, c. 1–3 mm wide; lower surface on either side of main veins mostly exposed **336. G. metamorpha**

1: Pedicels and outer surface of perianth glabrous

3 Either some or all leaves dentate to shallowly divided and ±flat, or leaves with deep primary division but then lobes not (or scarcely) divaricate and with lower surface of leaf lamina on either side of lobe midveins mostly or wholly exposed; ultimate lobes or teeth triangular to ovate or oblong or rarely broadly sublinear

4 All leaves basally cordate-auriculate and stem-clasping **333. G. amplexans**

4: Most or all leaves not basally stem-clasping

5 Leaves dimorphic, those of basal vegetative branches obovate to obtrullate with shallow apical teeth, those of flowering branches smaller and deeply tripartite with simple rigid narrowly triangular lobes; pollen-presenter narrowly truncate-conical to fusiform, with base scarcely broader than apex of style **335. G. acrobotrya**

5: Leaves of vegetative and flowering branches all ±similar in form; pollen-presenter conical to subcylindrical, with base much broader to scarcely broader than apex of style

6 Leaf base truncate or very broadly cuneate (rarely a few slightly amplexicaul); leaves ±ovate, with 7–11 teeth or shallow triangular lobes (to 5 mm long) spaced regularly around margins **334. G. uniformis**

6: Leaf base cuneate to narrowly linear (not truncate); leaves with teeth or lobes towards apex only, or with deep primary division and sometimes shallow to deep secondary division

7 Branchlets either glabrous or with appressed hairs

8 Upper surface of leaf somewhat concave (lateral and terminal lobes often gently up-curved), with venation obscure, sometimes midvein apparent only as an impressed groove; primary leaf lobes usually simple, oblong to broadly linear with apex rounded below point; fruit with pericarp c. 1 mm across at suture, bony-textured **339. G. curviloba**

8: Upper surface of leaf ±flat (lobes not up-curved), with venation evident and midveins sometimes prominent; primary leaf lobes usually with some secondary division, triangular to narrowly so with apex usually not rounded below point; fruit with pericarp c. 0.5 mm across at suture, crustaceous

9 Branchlets, leaves and floral rachises all glabrous; apices of leaf lobes scarcely to moderately pungent **331. G. manglesii**

9: Branchlets, leaves and floral rachises with hairs on some or all of these parts; apices of leaf lobes very pungent **332. G. phanerophlebia**

7: Branchlets with ascending to spreading hairs

10 Upper surface of leaf shallowly concave (lateral and terminal lobes often gently up-curved), with venation obscure, sometimes midvein apparent only as an impressed groove; primary leaf lobes usually simple, oblong to broadly linear with apex rounded below point; fruit with pericarp c. 1 mm across at suture, bony-textured **339. G. curviloba**

10: Upper surface of leaf shallowly concave or flat to gently convex (lobes not up-curved), with venation usually evident and midveins sometimes prominent; primary leaf lobes with or without some secondary division; ultimate teeth or lobes triangular to narrowly so with apex rounded or not rounded below point; fruit with pericarp c. 0.5 mm across at suture, crustaceous

11 Longest ultimate lobes > 20 mm long; leaves deeply divided into 2 or 3 simple narrowly triangular lobes each 1–8 mm wide; upper surface of leaf rachis and lobes with many acutely angled lateral veins at 20°–30° to midveins **329. G. triloba**

11: Longest ultimate lobes ⩽ 20 mm long; leaves variously divided, shallowly to deeply 3–7-toothed or -lobed and sometimes with secondary division; upper surface of leaf rachis and lobes with few or no lateral veins apart from lobe midveins

12 Stylar swelling ovoid; pollen-presenter conical with base distinctly broader than apex of style; floral rachises with a sparse to dense indumentum of ±spreading hairs **330. G. vestita**

12: Stylar swelling narrowly ovoid to narrowly subcylindrical; pollen-presenter subcylindrical to narrowly truncate-conical with base scarcely broader than apex of style; floral rachises glabrous **338. G. stenogyne**

3: Either all leaves entire, or some or all leaves with very deep primary or secondary division (pinnatipartite, often almost pinnatisect) and then with lobes strongly divaricate from leaf midrib, and lower surface of leaf 2-grooved; entire leaves linear to narrowly oblanceolate; ultimate lobes of divided leaves subulate to linear or narrowly triangular

13 All leaves divided, with distance from branchlet to base of first lobe ⩽ 5 mm

14 Pedicels 2–5 mm long; leaves mostly with 5 primary lobes; ultimate lobes not or weakly pungent, often curved **340. G. rara**

14: Pedicels 5–10 mm long; leaves with 3 or 5–7 primary lobes; ultimate lobes strongly pungent or not, straight

15 Pollen-presenter narrowly truncate-conical to fusiform, with its base much narrower than its height and scarcely broader than apex of style

16 Basal section of leaf with a narrow wing-like strip of flat lamina extending all the way to the branchlet, its margins flat and projecting sideways; ultimate leaf lobes 8–35 mm long; fruit rugose **341. G. corrugata**

16: Basal section of leaf lacking wing-like lamina, if laminal tissue present then its margins strongly revolute; ultimate leaf lobes ⩽ 7 mm long; fruit smooth or obscurely colliculate **335. G. acrobotrya**

15: Pollen-presenter broadly conical, its base broader than its height

17 Branchlets pubescent to tomentose; fruits rugulose **349. G. spinosissima**

17: Branchlets glabrous; fruits ±smooth **337. G. roycei**

13: Either all leaves entire, or some or all leaves divided and then with distance from branchlet to base of first lobe > 5 mm

18 Leaves divided; basal section of leaf (below first lobes) with a narrow wing-like strip of flat lamina extending all the way to the branchlet, its margins flat and projecting sideways; fruits rugose with short irregular peaks; leaf lobes 0.7–0.8 (–1.2) mm wide **341. G. corrugata**

18: Leaves entire or divided; basal section of divided leaves (below first lobes) lacking wing-like lamina, if laminal tissue present then its margins strongly revolute; fruits smooth to rugose; lobes of divided leaves mostly 0.3–2.8 (–3.4) mm wide

19 Branchlets glabrous or with sparse to dense straight appressed hairs

20 Upper surface of leaf with venation obscure, sometimes midvein apparent only as an impressed groove; ultimate leaf lobes often curved, ⩽ 20 mm long; fruit rugose with thick bony pericarp c. 1 mm thick at suture **339. G. curviloba**

20: Upper surface of leaves with venation (at least midvein) usually evident, often slightly prominent, rarely obscure; ultimate leaf lobes usually straight, 5–45 (–70) mm long; fruit smooth to rugose with a relatively thin crustaceous pericarp c. 0.5 mm thick at suture

21 Longest floral rachises 20–55 mm long; unit conflorescences subcylindrical or narrowly conico-cylindrical; pedicels often spreading

22 Floral rachises and outer surface of floral bracts both densely tomentose; fruits faintly rugose with rounded bumps **343. G. biternata**

22: Floral rachises glabrous or sparsely silky at base; outer surface of floral bracts glabrous, with margins ciliate; fruits smooth or rugulose

23 Floral bracts conspicuous, 2.8–3.4 mm long, 3–4 mm wide, persistent almost to anthesis; basal pedicels 2.2–3.2 mm long; leaf lobes c. 0.8 mm wide; lower surface of leaf with margins tightly revolute to prominent midveins, completely obscuring lamina **346. G. elongata**

23: Floral bracts inconspicuous, 0.8–1.0 mm long, 0.3–0.8 mm wide, usually falling in early bud stage; basal pedicels 3.5–6.5 mm long; leaf lobes 0.8–1.4 mm wide; leaf lower surface often with a little lamina exposed between margins and midveins at least at sinuses **345. G. xiphoidea**

21: Longest floral rachises ⩽ 20 mm long; unit conflorescences usually subglobose or subumbelloid, rarely shortly cylindrical; pedicels ascending

24 Branchlets densely hairy; fruit rugulose-colliculate or very strongly rugose, not smooth

25 Floral bracts glabrous on outer surface, with ciliate margins; fruit very strongly rugose with ridges drawn into short irregular peaks, or muricate **342. G. paniculata**

25: Floral bracts usually densely hairy on outer surface, occasionally sparsely so or glabrous; fruit rugulose-colliculate

330. Grevillea vestita (Endl.) Meisn., *in* J.G.C.Lehmann, *Pl. Preiss.* 1: 548 (1845)

Manglesia vestita Endl., *in* S.L.Endlicher & E.Fenzl, *Nov. Stirp. Dec.* 4: 26 (1839); *G. vestita* var. *dilatata* Meisn., *in* J.G.C.Lehmann, *Pl. Preiss.* 1: 548 (1845), *nom. illeg.* T: 'Habitat in Novae Hollandiae austro-occidentalis colonia King-Georges-Sound [W.A.] (Hügel)' [protologue]; type not found.

Spreading shrub 0.6–4.5 m high, sometimes rhizomatous and gregarious. Branchlets tomentose to subvillous. Leaves 1–4 (–5.5) cm long, 6–35 mm wide (across lobes, flattened); either (subsp. *vestita*) 3 (rarely –6)-fid or -toothed (rarely bifid or entire and then linear to oblanceolate), with primary lobes ascending, divaricate or not, sometimes again 2- or 3-fid, and ultimate lobes ovate to triangular; or (subsp. *isopogoides*) 3–5-partite (rarely a few leaves entire and linear), with primary lateral lobes often again 2- or 3-partite, and ultimate lobes divaricate, usually narrowly triangular; apices often pungent; margins shortly recurved to refracted; lower surface variably exposed or enclosed, subvillous. Unit conflorescence acropetal, subglobose to domed-subumbelloid; floral rachis tomentose or rarely almost glabrous. Perianth glabrous outside. Pistil 4–6.1 mm long; stylar swelling ovoid; pollen-presenter conical, with base much broader than apex of style. Follicle oblong-ovoid to oblong-obovoid, 8–10.5 mm long, nearly smooth. n = 10 [probably subsp. *vestita*], H.P.Ramsay, *Austral. J. Bot.* 11: 5 (1963).

This species occurs in south-western W.A. and has two subspecies.

A few collections of *G. vestita* from the S of the range (Lancelin, Spearwood, Beverley) are not assignable to subspecies, having ±intermediate foliar features; it is unclear whether these represent morphologically consistent populations. Digynous flowers are also moderately common, and at least three instances are known of flowers with the anthers borne on free filaments.

There are some inconsistencies between the description in the protologue and features of the taxon long recognised as *G. vestita*; in addition the species is not known to occur at the type location of King Georges Sound, although that place name was often used loosely. In the absence of the type, past usage is maintained. See protologue and McGillivray & Makinson (*Grevillea* 447 (1993)) for details.

Lower surface of leaves mostly exposed; leaves 3-fid to 3-dentate (rarely to 6-fid or -dentate) and ±flat **330a.** subsp. **vestita**

Lower surface of leaves mostly or wholly obscured (except for midvein of leaf and lobes); leaves deeply divided, 3–5-partite with divaricate lobes **330b.** subsp. **isopogoides**

330a. Grevillea vestita (Endl.) Meisn. subsp. **vestita**

Illustrations: N.G.Marchant *et al.*, *Fl. Perth Region* 1: 334, fig 114A–D (1987), as *G. vestita*; D.J.McGillivray & R.O.Makinson, *Grevillea* 169 (1993); P.M.Olde & N.R.Marriott, *Grevillea Book* 3: 221 (bottom right), 222 (182A–C) (1995).

Leaves 1–2.5 (–5.5) cm long, 6–35 mm wide across lobes, cuneate and almost flat, usually 3 (–6)-fid or -toothed, or rarely bifid or simple and then obovate; basal section cuneate; primary lobes ascending, not divaricate, simple or up to 3-fid; ultimate lobes ovate to triangular or oblong, 2–19 mm long, 1.5–8 mm wide; apices often pungent; margins shortly recurved to refracted; lower surface mostly exposed, open-tomentose to loosely subvillous. Floral bracts 0.3–0.9 mm long. Flower colour: perianth white to cream, sometimes tinged pink, with limb greenish to yellow or orange; style white. Figs 3C, 36A–C.

Occurs in south-western W.A., inland from Badgingarra to Pingelly and Williams, and along the coast from Yanchep to Cape Naturaliste. Grows in mixed heath or Jarrah–Marri woodland in sandy to loamy acid to calcareous soils over laterite, granite or limestone. Regenerates from seed and rhizomes. Flowers year round, mainly June–Oct. Map 418.

W.A.: Cottesloe, *C.Andrews [1st coll.] 789* (K); Narrogin, 30 Aug. 1926, *J.B.Cleland* (AD); S of Mandogalup [?Mandalup Hill], *R.Filson 85* (MEL); Meenaar, *C.A.Gardner* 7602 (PERTH); 13.8 km along Marchagee Track from Brand Hwy, *N.Marriott 75* (NSW); c. 10 km from Toodyay towards Goomalling, 18 Sept. 1968, *M.E.Phillips* (AD, CANB, NSW).

There is minor variation within this subspecies, some populations being highly gregarious and rhizomatous and others less so. Occasional specimens have hairs on the pedicels or the inner surface of the perianth. Olde & Marriott (*op. cit.* 221) report a 'Gillingarra form' (not seen) which has the leaf base almost truncate and the lobes spreading and often with secondary division. In the S of the range (S from Pinjarra and Pingelly) the leaves are sometimes longer, 2.5–3 (–5.5) cm long, and more narrowly cuneate with more narrowly triangular lobes; the leaf remains flat and the margins shortly recurved.

330b. Grevillea vestita subsp. isopogoides F.Muell. ex McGill., *New Names Grevillea* 16 (1986)

T: 3.5 km N of Binnu on North West Coastal Hwy, W.A., 12 June 1976, *D.J.McGillivray 3326 & A.S.George*; holo: PERTH; iso: K, NSW, US *n.v.*

Illustrations: P.M.Olde & N.R.Marriott, *Grevillea Book* 3: 222 (bottom right), 223 (183A–C) (1995).

Leaves 1–2 (–2.7) cm long, 10–30 mm wide when flattened, deeply 3–5-partite or rarely a few leaves simple and linear; basal section narrowly cuneate; primary lobes ascending, divaricate, simple or up to 3-partite; ultimate lobes narrowly triangular, 0.4–1.5 cm long, 1.3–2.3 (–3.4) mm wide; apices pungent; margins angularly refracted; lower surface mostly or wholly enclosed on either side of midveins, loosely subvillous. Floral bracts 1.0–1.2 mm long. Flower colour: perianth and style white to cream, sometimes tinged pink. Fig. 36D.

Occurs in south-western W.A. in the area bounded by Kalbarri, Three Springs, Mullewa and Mingenew, and possibly also to the S near the Irwin R. Grows in heath and mixed sclerophyll scrub in sandy or gravelly soils. Regenerates from seed or rhizomes. Flowers (Mar.–) June–Sept. Map 419.

W.A.: Yuna, 78 km NE of Geraldton, *A.C.Burns [1967] 21 bis* (MEL, NSW, PERTH); 246 km from Mt Magnet on Geraldton road, *D.W.Goodall 1780* (PERTH); c. 10 km SW of Mullewa, *R.Melville 4216* (AD, BRI, K, MEL, NSW, PERTH); c. 8 km S of Murchison R. Bridge, North West Coastal Hwy, *M.Tindale 1320* (K, NSW, PERTH, US).

Two forms of subsp. *isopogoides* are distinguished. The 'broader-lobed form', which includes the type of the subspecies, has leaf lobes > 2 mm wide, the margins not fully revolute to the midveins (up to nearly half of the leaf lower surface exposed), and a slightly hairier leaf upper surface. This form occurs in the NW of the subspecies range, from c. 20 km S of Kalbarri E to Ajana and N to the Mary Springs area, with an outlying occurrence NE of Yuna, and one collection far to the S from 'near Irwin R.'. The 'narrower-lobed form' has most leaf lobes ≤ 2 mm wide, the margins tending to be fully revolute to the midvein, and a less dense indumentum on the leaf upper surface. It occurs in the area between Yuna, Mullewa, Wicherina and Mingenew.

331. Grevillea manglesii (Graham) Planch., *Hort. Donat.* 96 (1858)

Anadenia manglesii Graham, *Edinburgh New Philos. J.* 27(53): 189 (1839); *G. manglesii* (Graham) McGill., *in* D.J.McGillivray & R.O.Makinson, *Grevillea* 427 (1993), *comb. superfl.* T: 'Anadenia manglesii 25 July 1839. Under this name I have somewhere published it ...' [cultivated from Clapton Nursery; R.Graham script; protologue]; neo: E, *fide* D.J.McGillivray & R.O.Makinson, *Grevillea* 427 (1993).

Shrub 1.5–3 (–5) m high. Branchlets glabrous, sometimes glaucous. Leaves 2–8 cm long, 10–50 mm wide, trifid to tripartite with ascending to spreading primary lobes and often with shallow secondary division (subsp. *manglesii*), or basally long-cuneate and apically 2- or 3-lobed (subsp. *ornithopoda*), or biternate with spreading primary lobes and deep secondary division (subsp. *dissectifolia*); ultimate lobes triangular or sometimes rounded; apices scarcely to moderately pungent; margins flat to recurved; lower surface exposed, glabrous. Unit conflorescence acropetal, shortly subconical to subglobose or shortly cylindrical; floral rachis glabrous. Perianth glabrous outside. Pistil 3.3–5.5 mm long; stylar swelling ovoid; pollen-presenter conical with base broader than apex of style. Follicle obloid-ovoid to slightly obovoid, 7–10.5 mm long, rugose-tuberculate.

Occurs in south-western W.A. There are three subspecies.

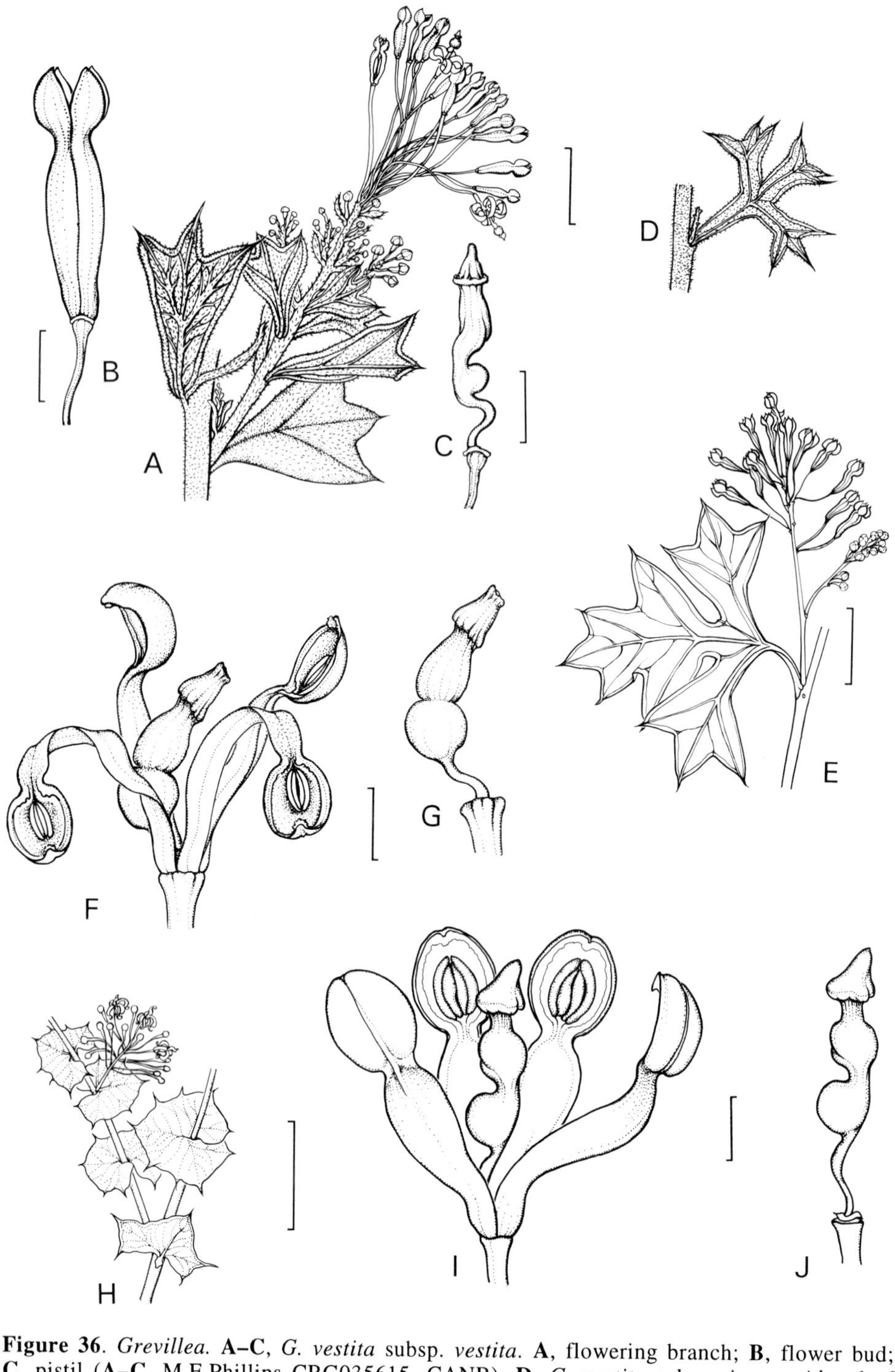

Figure 36. *Grevillea*. **A–C**, *G. vestita* subsp. *vestita*. **A**, flowering branch; **B**, flower bud; **C**, pistil (**A–C**, M.E.Phillips CBG035615, CANB). **D**, *G. vestita* subsp. *isopogoides*, leaf (M.E.Phillips CBG005406, CANB). **E–G**, *G. manglesii* subsp. *manglesii*. **E**, flowering branch; **F**, flower; **G**, pistil (**E–G**, M.E.Phillips CBG022023, CANB). **H–J**, *G. amplexans* subsp. *amplexans*. **H**, flowering branch; **I**, flower; **J**, pistil (**H–J**, S.J.Forbes 1707, CANB). Scale bars: **A**, **D–E**, **H** = 1 cm; **B–C** = 3 mm; **F–G**, **I–J** = 1 mm. Drawn by C.Payne.

Note: many loan specimens were determined by McGillivray between c. 1976 and 1986 as subspecies within *G. glabrata* (Lindl.) Meisn., a name since superseded by *G. manglesii* (Graham) Planch. The correspondence of these subspecific taxa with the subspecies here recognised is exact.

1 Most leaves 4–8 cm long, basally narrowly long-cuneate and apically trifid or bifid, with primary division only ... **331b.** subsp. **ornithopoda**

1: Most leaves 2–5 cm long, basally cuneate to broadly so to almost truncate, 3–7-fid or -partite; primary leaf lobes often divided

2 Leaves with primary lobes coarsely dentate or shallowly 3-fid; ultimate teeth or lobes triangular to broadly triangular, 2–8 mm wide ... **331a.** subsp. **manglesii**

2: Leaves with primary lobes deeply 3-partite (sinuses extending almost to midvein of primary lobe); ultimate leaf lobes sublinear to narrowly triangular, c. 1.5–2.5 mm wide ... **331c.** subsp. **dissectifolia**

331a. Grevillea manglesii (Graham) Planch. subsp. manglesii

Manglesia glabrata Lindl., *Sketch Veg. Swan R.* xxxvii (1840); *G. glabrata* (Lindl.) Meisn., *in* J.G.C.Lehmann, *Pl. Preiss.* 1: 549 (1845). T: Swan River, [W.A.], 1839, *J.Drummond*; holo: CGE *n.v.*, *fide* D.J.McGillivray & R.O.Makinson, *Grevillea* 427 (1993); ?iso: Swan River, W.A., [*J.Drummond 1st coll. 621*]; CGE *n.v.*, G, G-DC, K, LE *n.v.*, MEL.

Manglesia cuneata Endl., *Nov. Stirp. Dec.* 4: 25 (1839), *nom. nud.*; *G. cuneata* (Endl.) Druce, *Bot. Soc. Exch. Club Brit. Isles* 4(5) Suppl. 2: 625 (1917), *nom. nud.*

Illustrations: N.G.Marchant *et al.*, *Fl. Perth Region* 1: 332, fig. 111A–D (1987), as *G. glabrata*; P.M.Olde & N.R.Marriott, *Grevillea Book* 3: 15 (bottom left & 6A, B) (1995).

Leaves 2–5 cm long, 1–4 cm wide, mostly with 3 (–7) deeply cleft primary lobes, these simple to apically 3-fid or 3-partite, or sometimes leaves cuneate and apically 3–5-fid with ascending lobes, rarely a few leaves entire and obovate; ultimate lobes 0.3–1.0 (–3.0) mm long, 3–8 mm wide. Flower colour: perianth and style white to cream; limb of bud sometimes pale pink or yellowish. Fig. 36E–G.

Occurs in south-western W.A., in the northern part of the species range, in the Darling Ra. between Wooroloo and Mundaring Weir–Helena Valley, and S to Mt Dale. Grows in eucalypt woodland or open forest, often in moist situations along creeks. Regenerates from seed. Flowers year round, mainly June–Nov. Map 420.

W.A.: Parkerville, *R.Coveny 8048* (NSW); Wooroloo, *M.Koch 1518* (E, K, MEL, NSW, P, PERTH); c. 3.2 km downstream from Mundaring Weir in valley of Helena R., *B.R.Maslin 673* (MEL, NSW, PERTH); banks of Helena R., Midland Junction, 10 Apr. 1900, *A.Morrison* [*10387*] (K).

331b. Grevillea manglesii subsp. ornithopoda (Meisn.) McGill., *in* D.J.McGillivray & R.O.Makinson, *Grevillea* 427 (1993)

G. ornithopoda Meisn., *in* J.G.C.Lehmann, *Pl. Preiss.* 2: 256 (1848); *G. glabrata* subsp. *ornithopoda* (Meisn.) McGill., *New Names Grevillea* 6 (1986). T: Swan River, [W.A.], [*J.*] *Drummond 2nd coll., 314*; lecto: NY *n.v.*, *fide* D.J.McGillivray & R.O.Makinson, *Grevillea* 427 (1993); isolecto: AD *n.v.*, B *n.v.*, E *n.v.*, G, K, LD *n.v.*, LE *n.v.*, MEL.

Illustrations: P.M.Olde & N.R.Marriott, *Grevillea Book* 3: 16(8), 17 (top left) (1995).

Leaves 5–8 cm long, 1.5–3.5 cm wide, narrowly cuneate, apically trifid or occasionally bifid with ascending simple narrowly triangular lobes 6–35 mm long, 2–4 mm wide. Flower colour: perianth and style white; limb of early bud greenish brown becoming yellow. *Birds-foot Grevillea.*

Occurs in south-western W.A., in the SW part of the species range, in the Darling Ra. from Jarrahdale and Pinjarra to Mt William E of Cookerup. Grows in Jarrah woodland and forest in moist situations, often along creeks. Regenerates from seed. Flowers May–Nov. Map 421.

W.A.: vicinity of Mt William, Nov. 1842, *Gilbert 323* (FI); Pinjarra, Murray R., 23 Sept. 1897, *R.Helms* (NSW, PERTH).

This subspecies is recognised as ‘Poorly Known’ in J.D.Briggs & J.H.Leigh, *Rare or Threatened Australian Plants* (1995).

331c. Grevillea manglesii subsp. **dissectifolia** (McGill.) McGill., *in* D.J.McGillivray & R.O.Makinson, *Grevillea* 427 (1993)

G. glabrata subsp. *dissectifolia* McGill., *New Names Grevillea* 6 (1986). T: 1 km N of North Bannister on the Albany Hwy, W.A., 23 June 1976, *D.J.McGillivray 3468 & A.S.George*; holo: NSW; iso: K, MEL, NY *n.v.*, PERTH.

Illustrations: D.J.McGillivray & R.O.Makinson, *op. cit.*, 173; P.M.Olde & N.R.Marriott, *Grevillea Book* 3: 15 (bottom right), 16 (7A, B) (1995).

Leaves 2–4.5 cm long, 2–5 cm wide, deeply 3 (5)-partite (divided almost to midvein), with primary lobes ±spreading and usually again 2- or 3-partite; ultimate lobes linear or very narrowly triangular, 7–20 mm long, 1.3–2.8 mm wide. Flower colour: perianth and style white, sometimes tinged pink; limb of bud green becoming yellow.

Occurs in south-western W.A., in the SE part of the species range, in the Darling Ra. from Dwellingup and North Bannister N to Gleneagle. Grows in eucalypt forest or woodland, often in moist situations. Regenerates from seed. Flowers June–Nov. Map 422.

W.A.: 31 Mile Rd, Gleneagle, *R.J.Edmiston H738* (PERTH); Big Brook crossing on NE road in Forest Dept Reserve, Dwellingup, *P.C.Kimber 227* (PERTH); 11 km N of North Bannister on Albany Hwy, *D.J.McGillivray 3468 & A.S.George* (K, MEL, NSW, NY, PERTH).

This subspecies is recognised as 'Poorly Known' in J.D.Briggs & J.H.Leigh, *Rare or Threatened Australian Plants* (1995).

332. Grevillea phanerophlebia Diels, *Bot. Jahrb. Syst.* 35: 157 (1904)

T: N of Mingenew, W.A., 9 June 1901, *L.Diels 3045*; lecto: B *n.v.*, *fide* D.J.McGillivray & R.O.Makinson, *Grevillea* 434 (1993); isolecto: B *n.v.*

Illustrations: P.M.Olde & N.R.Marriott, *Grevillea Book* 3: 90 (bottom right), 91 (68A, B) (1995).

Spreading to sprawling shrub to 1.5 m high. Branchlets with sparse to moderately dense indumentum (rarely almost glabrous) of ascending to appressed hairs. Leaves 2.5–4.5 cm long, 3–5 cm wide across lobes, ±flat, fan-shaped in outline, tripartite with primary (especially lateral) lobes again 2- or 3-partite; ultimate lobes narrowly triangular to sublinear, 1–2.5 (–3.5) cm long, 1.5–2.5 (–3.5) mm wide; apices pungent; margins loosely recurved or revolute, with lamina mostly exposed on broader parts of leaf and lobes, sometimes enclosed on narrower lobes; lower surface with an open to sparse appressed to spreading indumentum. Unit conflorescence acropetal, loosely subglobose to dome-shaped; floral rachis subsericeous to tomentose or ?glabrous. Flower colour: perianth white with cream to yellow limb; style white. Perianth glabrous outside. Pistil 3.5–3.9 mm long; stylar swelling ovoid to ellipsoidal; pollen-presenter conical to truncate-conical, with base slightly wider than apex of style. Follicle obloid-ellipsoidal, 8–12 mm long, smooth to rugose.

Occurs in south-western W.A. between Mullewa and Mingenew and at Eradu. Grows in open scrub or heath on sandplain. Probably regenerates from seed only. Flowers mainly Aug.–Sept.? Map 423.

W.A.: Mullewa plains, *W.E.Blackall 699* (PERTH); Mingenew, Sept. 1903, *W.V.Fitzgerald* (NSW); Railway Reserve Eradu, *S.Patrick 2134, 2135* (PERTH).

Grevillea phanerophlebia is very poorly collected and believed to be extremely rare and endangered; it is recognised as 'Poorly Known' in J.D.Briggs & J.H.Leigh, *Rare or Threatened Australian Plants* (1995). It is known, though not common, in cultivation. The type (and the protologue) indicate a smooth-surfaced fruit. Collections from an extant population at Eradu have distinctly rugose fruits; plants at this location also have somewhat hairier branchlets and leaves than more easterly collections, possibly indicating an influence from *G. biternata*, which grows at the same site.

The upper leaf surface has a distinctive concavity immediately distal to the junction of the midvein and the primary lateral veins; the vein junction sometimes has a pimple-like pulvinus.

333. **Grevillea amplexans** F.Muell. ex Benth., *Fl. Austral.* 5: 488 (1870)

T: northern district, W.A., *date & coll. unknown*; holo: MEL; iso: K, PERTH.

Shrub 1–3 m high with arching branches. Branchlets glabrous and sometimes glaucous or occasionally subsericeous. Leaves 0.7–2.6 cm long, 1.0–3.5 cm wide, basally cordate-auriculate and stem-clasping, star-like or angularly ovate, 3–7-fid or marginally dentate with 5–11 teeth; teeth or lobes triangular, 2–10 mm long; lobes not divaricate; apices pungent; margins flat or slightly recurved; lower surface exposed, glabrous, sometimes glaucous. Unit conflorescence acropetal, subglobose to domed-subumbelloid; floral rachis glabrous. Flower colour: perianth white to cream (sometimes tinged pink), with yellow-green to pink limb; style white. Perianth glabrous outside. Pistil 2.5–5.8 mm long; stylar swelling ovoid; pollen-presenter conical, base markedly wider than apex of style. Follicle obloid-ellipsoidal, 9–12 mm long, smooth.

Occurs in W.A. from E of Geraldton S to Coomberdale, N of Moora. There are three subspecies.

McGillivray & Makinson (*Grevillea* 176, 1993) recognised a broad *G. amplexans* with a 'glaucous form' (including the type of *G. amplexans*) having glabrous branchlets and lower leaf surfaces, and a 'hairy-leaved form' with subsericeous branchlets and lower leaf surfaces. He also noted some unplaced 'intermediate' material, with subsericeous branchlets and glabrous lower leaf surfaces, as possibly representing a 'third distinct form'.

The 'hairy-leaved form' has since been named as *G. adpressa* Olde and Marriott (*Nuytsia* 9: 250, 1993), the authors reporting an interdigitating distribution of this taxon with *G. amplexans s. str.* 'in an irregular, geographical replacement pattern', with separate homogeneous populations of each and no sign of intergrades. They assigned McGillivray & Makinson's intermediate material to *G. amplexans*. Notwithstanding this distributional data, in view of the shared synapomorphic character-states and the relatively slight differences between the taxa on other characters, three subspecies within *G. amplexans* are recognised here, with *G. adpressa* Olde & Marriott reduced to subspecies and the 'intermediate' material being recognised (below) as a distinct taxon, subsp. *semivestita*. All three taxa are very closely related, and further research is warranted. Subspecies *semivestita* and *adpressa* are distinct from *G. uniformis*, which has the leaf base usually truncate to very broadly cuneate.

1 Branchlets glabrous, sometimes glaucous; lower leaf surface glabrous, sometimes glaucous ..... **333a.** subsp. **amplexans**

1: Branchlets hairy; lower leaf surface either hairy or glabrous

  2 Lower leaf surface glabrous; floral bracts usually persistent at least to anthesis, sometimes beyond ..... **333b.** subsp. **semivesitita**

  2: Lower leaf surface subsericeous; floral bracts caducous in early bud stage ..... **333c.** subsp. **adpressa**

333a. **Grevillea amplexans** F.Muell. ex Benth. subsp. **amplexans**

G. amplexans 'glaucous form', of D.J.McGillivray & R.O.Makinson, *Grevillea* 176 (1993).

Illustrations: P.M.Olde & N.R.Marriott, *Grevillea Book* 2: 24 (15A, B, & bottom right) (1995).

Branchlets glabrous, sometimes glaucous. Leaves 0.8–2.6 cm long; lower surface glabrous, sometimes glaucous. Floral bracts minute, 0.2–0.3 mm long, variably persistent, sometimes caducous in mid-bud stage or sometimes persistent to anthesis. Fig. 36H–J.

Occurs from E of Geraldton S to Watheroo and Coomberdale. Grows in heath or mallee-shrubland on sandplain, usually in deep yellow sand over laterite; has some propensity to colonise disturbed ground. Probably regenerates from seed only. Flowers July–Nov. Map 424.

W.A.: 4 km NE of Arrino, *J.S.Beard 7240* (PERTH); Nangade, Mingenew, Nov. 1907, *W.D.Campbell 86* (K); Eradu, *C.A.Gardner 2632* (PERTH); near Mingenew, *E.Pritzel 367* (A *n.v.*, B *n.v.*, E, G *n.v.*, GH *n.v.*, L *n.v.*, NSW, PERTH); Coomberdale, 10 Aug. 1949, *E.Salisbury* (K).

In addition to the key characters, subsp. *amplexans* has a tendency to have larger, less deeply lobed leaves than subsp. *adpressa*, the leaf margins more often flat, and the nectary prominent above the dorsal rim; there is however considerable overlap on these characters.

333b. Grevillea amplexans subsp. **semivestita** Makinson, *Fl. Australia* 17A: 505 (2000)

T: Marchagee Track, c. 45 km E of Brand Hwy, W.A., 30°07'S, 115°55'E, 1 Sept. 1984, *D.Foreman 487*; holo: MEL; iso: HO, NSW, PERTH.

Branchlets densely subsericeous. Leaves 0.8–1.7 cm long; lower surface glabrous, sometimes glaucous. Floral bracts minute, 0.2–0.3 mm long, ciliate, persistent to anthesis or beyond.

Occurs in south-western W.A., in the area from Watheroo Natl Park to Yarra Yarra Lakes (just W of Carnamah). Grows in shrubland or ?heath. Regeneration mode not known. Flowers c. Aug.–Oct. Map 425.

W.A.: c. 240 km N of Perth on Geraldton road, *E.M.Bennett 1364* (K, PERTH); 48 km from Eneabba on road to Carnamah, *R.W.Purdie 5166* (CANB); Watheroo Natl Park, *R.D.Royce 9687* (PERTH).

Distinguished from subsp. *amplexans*, which has glabrous branchlets, and from subsp. *adpressa*, which has the lower leaf surfaces hairy and the floral bracts caducous in early bud.

Subsp. *semivestita* is known from few collections, but these are consistent in form, including for the characters separating it from the other two taxa; I believe it to be a taxon and not a chance series of intermediates or hybrids.

333c. Grevillea amplexans subsp. **adpressa** (Olde & Marriott) Makinson, *Fl. Australia* 17A: 506 (2000)

G. adpressa Olde & Marriott, *Nuytsia* 9: 250 (1993). T: 5.6 km W of Arrino on Arrino West Rd, W.A., 16 Sept. 1991, *P.M.Olde 91/112*; holo: NSW.

Illustrations: P.M.Olde & N.R.Marriott, *op. cit.* 9: 251, fig. 4 I–L; P.M.Olde & N.R.Marriott, *Grevillea Book* 2: 16 (bottom right), 17 (9A, B) (1995).

Branchlets densely subsericeous. Leaves 0.7–1.2 cm long; lower surface ±densely subsericeous. Floral bracts minute, 0.2–0.3 mm long, ciliate, caducous in early bud stage.

Occurs in south-western W.A., between Mingenew and Watheroo. Grows in low heath in sandy or loamy soils. Probably regenerates from seed only. Flowers Aug.–Oct. Map 426.

W.A.: 20 km from Three Springs on road to Eneabba, *D.J.McGillivray 3307 & A.S.George* (CANB, K, NSW, PERTH, US); Arrino, 23 Sept. 1953, *N.H.Speck s.n.* (CANB, PERTH); 9 km N of Watheroo, 15 Aug. 1985, *D.Woolcock G26* (PERTH).

This subspecies tends to have smaller leaves than subsp. *amplexans*, and often has shortly recurved leaf margins and a relatively obscure nectary; these differences, however, are not fully consistent.

334. Grevillea uniformis (McGill.) Olde & Marriott, *Nuytsia* 9: 252 (1993)

G. acrobotrya subsp. *uniformis* McGill., *New Names Grevillea* 1 (1986), as subsp. *uniforma*. T: 8 km SW of Mt Lesueur, 8.8 km along track from Cockleshell Gully (Padbury) to main Jurien road, W.A., 26 Sept. 1976, *B.G.Briggs 6369*; holo: NSW; iso: K, PERTH.

Illustrations: P.M.Olde & N.R.Marriott, *op. cit.* 9: 251 fig. 4E–H; P.M.Olde & N.R.Marriott, *Grevillea Book* 3: 215 (centre right & 177B), 216 (177A) (1995).

Shrub to 1.5 m tall. Branchlets tomentose. Leaves 0.5–2 cm long, 8–30 mm wide, similar on all parts of plant, broadly ovate to fan-shaped in general outline, basally truncate to widely cuneate (rarely a few leaves slightly cordate-amplexicaul), with 7–11 regular marginal teeth or shallow triangular lobes to 5 mm long; apices of teeth or lobes pungent; margins shortly recurved; lower surface fully exposed, subsericeous. Unit conflorescence subglobose or loosely subumbelloid, with opening pattern uncertain; floral rachis glabrous. Flower colour: early buds brick-red, later white to cream below limb; limb brick-red becoming yellow just before anthesis; style white or cream. Perianth glabrous outside. Pistil 2.5–3.5 mm long; style scarcely swollen, narrowly ovoid; pollen-presenter conical, base broader than apex of style. Follicle obloid-ellipsoidal, 8–10 mm long, ±smooth.

Occurs in south-western W.A., E of Jurien Bay from Mt Lesueur to about Eneabba. Grows in sandy heath over lateritic gravel, creek banks in red sandy loam and sandstone slopes. Probably regenerates only from seed. Flowers July–Nov. Map 427.

W.A.: 1.5 km N of Mt Lesueur, *E.A.Griffin 1973* (PERTH); 2 km E of Western Titanium Leases, 8 km S of Eneabba, Sept. 1977, *R.Hnatiuk* (PERTH); Cockleshell Gully, *P.Olde 86/611* (NSW); Pen Rd, 2 km N of road to Greenhead, SW of Eneabba, *P.Olde 91/97* (NSW, PERTH); Jurien Bay, *R.D.Royce 7743* (PERTH).

Grevillea uniformis has similarities to both *G. acrobotrya* (which also has a scarcely swollen style but differs in having dimorphic foliage with the basal leaves cuneate-rhombate and the floral-branch leaves deeply tripartite), and to the group of taxa here classified under *G. amplexans* (which have the leaf base strongly cordate-amplexicaul and a conspicuous ovoid stylar swelling). *Grevillea stenogyne* is also somewhat similar but has the leaves deeply 3–5-partite with the primary lobes again trifid, and the leaf lower surface tomentose to subvillous.

This species is recognised as 'Poorly Known' in J.D.Briggs & J.H.Leigh, *Rare or Threatened Australian Plants* (1995).

335. Grevillea acrobotrya Meisn., *Hooker's J. Bot. Kew Gard. Misc.* 7: 74 (1855)

T: interior north of Swan River, A. [W.A.], 1850–51, [*J.*] *Drummond Coll. V. 185*; lecto: NY *n.v.*, *fide* D.J.McGillivray & R.O.Makinson, *Grevillea* 402 (1993); isolecto: [most as *Drummond VI: 185*] A, B [as *Diels 7582*], CGE, G, K, LD, MEL, P (all *n.v.* except K, MEL, P).

G. acrobotrya Meisn. subsp. *acrobotrya*, *sensu* D.J.McGillivray & R.O.Makinson, *Grevillea* 183–184 (1993).

Illustrations: D.J.McGillivray & R.O.Makinson, *Grevillea* 183 (1993), as *G. acrobotrya* subsp. *acrobotrya*; P.M.Olde & N.R.Marriott, *Nuytsia* 9: 251, fig. 4A–D (1993); P.M.Olde & N.R.Marriott, *Grevillea Book* 2: 12 (top centre & 6B), 13 (6C, D) (1995).

Shrub 0.6–2 m high, with emergent flowering branches. Branchlets (basal, vegetative) subsericeous to tomentose, or (flowering zone) glabrous or rarely tomentose. Leaves 0.6–4 cm long, 10–35 mm wide, dimorphic, on vegetative branches obovate to obtrullate, basally cuneate and apically 5–9-toothed or -fid, on flowering branches much smaller, deeply tripartite with spreading simple narrowly triangular lobes to c. 7 mm long, 1.3 mm wide; apices pungent; margins shortly refracted to recurved; lower surface exposed, sericeous. Unit conflorescence ?basipetal, subglobose to domed-subumbelloid; floral rachis glabrous or rarely tomentose. Flower colour: perianth white to cream with limb chocolate-brown; style white. Perianth glabrous outside. Pistil 3–4 mm long; style scarcely swollen, subcylindrical; pollen-presenter narrowly truncate-conical to fusiform, base slightly wider than apex of style. Follicle obloid-ellipsoidal, 8–11 mm long, smooth or obscurely colliculate. Fig. 38F–I.

Occurs in south-western W.A., between Eneabba and Badgingarra. Grows in open heath associations in sandy soils over laterite. Probably regenerates from seed only. Flowers much of the year, peaking June–Sept. Map 428.

W.A.: between Moore [R.] and Murchison [R.], 1853, *Js Drummond* [*s.n.*] (K); near Jurien Bay, *A.R.Fairall 2479* (CANB, KPBG); W of Coomallo Ck, E of Jurien, *E.A.Griffin 2706* (PERTH); 0.3 km S of Boothendarra Ck on Brand Hwy, *D.J.McGillivray 3285 & A.S.George* (NSW, PERTH); on hill crest c. 3.2 km SE of Mt Lesueur, 7 Oct. 1961, *J.H.Willis* (MEL).

Over most of the range, the flowering branchlets are glabrous and often slightly glaucous, and the floral rachises glabrous. In the southern part of the range (e.g. near Badgingarra) some specimens have densely tomentose flowering branchlets and floral rachises, similar to those of the related *G. stenogyne* (see under that species for differences). The populational basis of this variation needs assessment; recognition at subspecific level may be warranted.

McGillivray & Makinson (*Grevillea* 184 (1993)) treated the taxon here recognised as *G. uniformis* (McGill.) Olde & Marriott, as a subspecies of *G. acrobotrya*. McGillivray & Makinson (*Grevillea* 402 (1993)) and Olde & Marriott (*Nuytsia* 9: 252–254 (1993); *Grevillea Book* 2: 12 (1995)) regarded the taxon here recognised as *G. stenogyne* (Benth.) Makinson as a synonym of *G. acrobotrya*.

336. Grevillea metamorpha Makinson, *Fl. Australia* 17A: 506 (2000)

T: off Green Head–Coorow road, W.A., 16 Sept. 1995, *M.Hislop 88*; holo: PERTH.

Erect spindly shrub c. 1.5 m tall. Branchlets softly tomentose. Leaves 1.4–2.4 cm long, 8–15 mm wide, trimorphic; leaves of vegetative branches cuneate to ±obtrullate and 5-fid, primary lobes sometimes toothed, teeth or lobes subtriangular, 1–4 mm long, 1–3mm wide; leaves of flower-bearing branches below the inflorescent zone more deeply divided, 3-fid to 3-partite, primary lobes again 3-fid to 3-partite, ultimate lobes triangular, 2–6 mm long, 1–2 mm wide; leaves in flower-bearing zone as previous and/or more deeply 3–5-partite with narrowly triangular to sublinear lobes 4–10 mm long, c. 1 mm wide; apices pungent; margins almost flat on lower, broader leaves, very shortly recurved and somewhat thickened on upper leaves; lower surface exposed, subsericeous to softly tomentose. Unit conflorescence acropetal, domed-subumbelloid; floral rachis openly to sparsely tomentose. Flower colour: perianth white (limb dark in bud); style white. Perianth loosely silky outside. Pistil 3 mm long; style scarcely swollen, very narrowly ovoid; pollen-presenter narrowly truncate-conical to subcylindrical, base slightly wider than apex of style. Follicle ellipsoidal, 9–10 mm long, 5 mm wide, obscurely colliculate to almost smooth.

Occurs in south-western W.A. where known only from the type locality. Ecology uncertain. Flowers Sept –?. Map 429.

Known only from the Type collection.

The transition of leaf form from the vegetative to inflorescent branches, and the scarcely swollen style with narrow pollen-presenter, indicate a close relationship with *G. acrobotrya*, which differs in its glabrous perianth and pedicels and in having a more abrupt transition from basal toothed leaves to strictly 3-partite leaves in the floral zone. *Grevillea metamorpha* is one of only two species (with *G. erinacea*) in the *Triloba* group to consistently have hairs on the pedicels and outer surface of the perianth; *G. vestita* subsp. *vestita* occasionally has pedicel hairs.

337. Grevillea roycei McGill., *New Names Grevillea* 13 (1986)

T: 5 miles [c. 8 km] E of Goomalling, W.A., 18 Aug. 1962, *R.D.Royce 7502*; holo: PERTH.

Illustrations: D.J.McGillivray & R.O.Makinson, *Grevillea* 174, 175, fig. 44 (1993); P.M.Olde & N.R.Marriott, *Grevillea Book* 3: 150 (top centre & 118A, B) (1995).

Open shrub 1–2 m high, with emergent arching flowering branches. Branchlets glabrous, glaucous. Leaves 0.7–3 cm long, deeply and divaricately tripartite to biternate, often with terminal lobe simple and lateral lobes 2- or 3-lobed; ultimate lobes linear to subulate, 0.5–2 cm long, 0.7–1.8 mm wide; leaf rachis reflexed at lobe node; apices pungent; margins angularly revolute; lower surface subsericeous, enclosed except for midvein, or sometimes partly exposed at sinuses. Unit conflorescence acropetal, loosely subglobose; floral rachis glabrous. Flower colour: perianth green in bud becoming cream with a green to yellow limb; style white. Perianth glabrous outside. Pistil 3.3–4.5 mm long; stylar swelling narrowly ovoid to -ellipsoidal; pollen-presenter broadly conical, base much wider than apex of style. Follicle obloid-ellipsoidal, c. 10 mm long, ±smooth. Fig. 37E–G.

Occurs in south-western W.A., in the central wheatbelt from Amery and Goomalling to Cunderdin, and S to Lake Mears and Brookton. Grows in heath, shrubland, and shrubby woodland in sandy soils, sometimes over laterite. Regenerates from seed. Flowers Aug.–Oct. Map 430.

W.A.: 17 km NE of Brookton, between Jurakine Pool and Yenyening Lakes, *R.J.Hnatiuk 790137* (PERTH); near Waeel, *E.Pritzel 821* (A *n.v.*, AD, B *n.v.*, BR *n.v.*, E *n.v.*, G, M *n.v.*, MO *n.v.*, P *n.v.*, PERTH, S *n.v.*); Cunderdin Cemetery, 5 Oct. 1977, *J.H.Willis s.n.* (MEL, NSW).

Grevillea roycei has crowded leaves except on flowering branches, where they become sparse. It can be confused with *G. spinosissima* which has densely hairy branchlets; with *G. acrobotrya* which has dimorphic foliage (broad obovate to rhombate leaves on vegetative branches) and a narrower pollen-presenter; and with forms of *G. paniculata* and *G. levis*, which have the basal internode of the leaf > 10 mm long (⩽ 5 mm long in *G. roycei*).

This species is recognised as 'Poorly Known' in J.D.Briggs & J.H.Leigh, *Rare or Threatened Australian Plants* (1995).

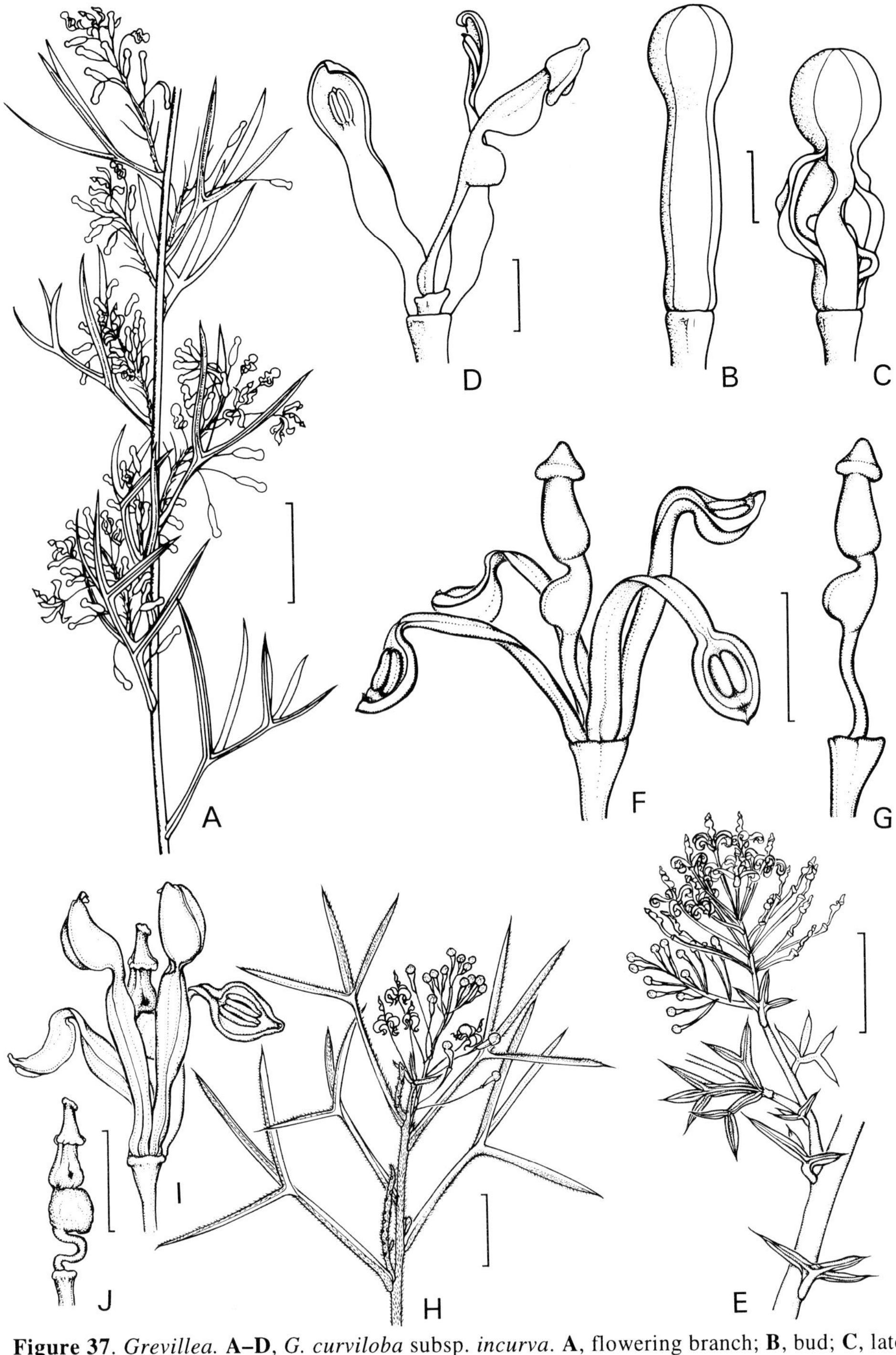

Figure 37. *Grevillea*. **A–D**, *G. curviloba* subsp. *incurva*. **A**, flowering branch; **B**, bud; **C**, late bud; **D**, pistil and partial perianth (**A–D**, Sept. 1927, H.Steedman, PERTH). **E–G**, *G. roycei*. **E**, flowering branch; **F**, flower; **G**, pistil (**E–G**, E.Pritzel *s.n.*, NSW125523, NSW). **H–J**, *G. paniculata*. **H**, flowering branch; **I**, flower; **J**, pistil (**H–J**, M.E.Phillips 833A, CANB). Scale bars: **A**, **E**, **H** = 1 cm; **B–D**, **F–G** = 1 mm; **I–J** = 2 mm. Drawn by: **A–D**, D.Mackay; **E–J**, C.Payne.

338. Grevillea stenogyne (Benth.) Makinson, *Fl. Australia* 17A: 506 (2000)

G. vestita var. *stenogyne* Benth., *Fl. Austral.* 5: 488 (1870). T: W. Australia, [*J.*] *Drummond* [*s.n.*], [per] F.Mueller 1870; holo: K, (sheet Neg. no. 2350); iso: MEL, ?P (*s. leg.* ex Mueller), PERTH.

Shrub, size unknown. Branchlets tomentose. Leaves 1.5–2 cm long, 13–20 mm wide, deeply tripartite with lateral lobes ascending; primary lobes again more shallowly 3-partite to 3-fid; ultimate lobes (2–) 3–7 (–10) mm long, 1–3 mm wide, triangular to narrowly so; apices pungent; margins shortly revolute; lower surface mostly or wholly exposed, tomentose to subvillous. Unit conflorescence acropetal, subglobose ?to shortly conical; floral rachis glabrous. Flower colour unknown, probably whitish. Perianth glabrous outside. Pistil 4 mm long; stylar swelling slight, narrowly ovoid to subcylindrical; pollen-presenter subcylindrical to narrowly truncate-conical, base scarcely wider than apex of style. Follicle not seen.

Occurs in south-western W.A.; locality not known. Map 431.

Known only from the Type collection.

This name of this taxon was treated as a synonym of *G. acrobotrya* by McGillivray & Makinson (*Grevillea* 402 (1993)) and by Olde & Marriott (*Grevillea Book* 2: 12 (1995)). *Grevillea stenogyne* is closely related to *G. acrobotrya* but does not share the reduction in leaf size and increased depth of leaf dissection from vegetative to flowering branches of that species; *G. acrobotrya* also has the flower-bearing branchlets usually ±glabrous (rarely densely hairy), whereas in *G. stenogyne* the branchlets are consistently tomentose throughout. In the context of the group the differences are likely to be stable and it is therefore here treated as a distinct taxon. A collection from near Eneabba (*D.J.McGillivray 3301*, NSW, PERTH) bears a strong resemblance to the type, except in having a more pronounced stylar swelling and a more conical pollen-presenter. This population should be investigated further for material matching the type exactly.

339. Grevillea curviloba McGill., *New Names Grevillea* 4 (1986)

Replaced synonym: *G. vestita* var. *angustata* Meisn., *in* J.G.C.Lehmann, *Pl. Preiss.* 1: 549 (1845). T: Swan River, [W.A.], 1839, [*J.*] *Drummond*; lecto: G-DC, *fide* D.J.McGillivray & R.O.Makinson, *Grevillea* 412 (1993); isolecto: [most as *Drummond I: 622*] CGE *n.v.*, G, K, MEL, P *n.v.*

Prostrate to erect shrub, 0.1–2.5 m high, lower-growing forms often with ascending to erect flowering branches. Branchlets shortly and irregularly pubescent to almost glabrous. Leaves 1–5 cm long, 10–15 mm wide across lobes, usually pinnatipartite with 3–5 ascending and slightly to strongly divaricate primary lobes, the basal ones sometimes again 2- or 3-partite, sometimes a few leaves simple and sublinear to narrowly oblanceolate and up to 3 cm long and 4 mm wide; ultimate lobes 4–20 mm long, 0.7–2.0 mm wide, linear to oblong or narrowly triangular, often upcurved; apices scarcely pungent; margins recurved or revolute; lower surface sometimes enclosed, glabrous or with a sparse indumentum of short ascending hairs. Unit conflorescence acropetal, dome-shaped to shortly subcylindrical; floral rachis loosely tomentose at least near the base, sometimes glabrous above. Flower colour: perianth white to cream; style white. Perianth glabrous outside. Pistil 3.5–6.5 mm long; stylar swelling ovoid; pollen-presenter conical, base broader than apex of style. Follicle obloid-ellipsoidal, 10–13 mm long, strongly rugose, thick-walled.

Occurs in south-western W.A., N of Perth between Bullsbrook and Badgingarra; there are two subspecies.

Distinct from but perhaps allied to *G. vestita*, which has a conspicuous tomentose to subvillous indumentum on the branchlets, a denser and longer indumentum on the leaves, and floral bracts usually persistent to anthesis (caducous in early bud stage in *G. curviloba*). The combination of leaf lobes often curved, leaf upper surface somewhat concave and with obscure venation, leaf lower surface exposed, and follicle rugose and thick-walled (pericarp c. 1 mm thick at suture), serve to distinguish this from most other species in the group, which have crustaceous pericarps (easily snapped) < 5 mm thick.

Leaf margins shortly recurved, exposing much or all of lower leaf surface; leaves ±flat or weakly secund, usually lacking secondary division; leaf lobes obovate-cuneate to oblong or broadly linear, > 1.5 mm wide **339a.** subsp. **curviloba**

Leaf margins tightly revolute to a prominent abaxial midrib, concealing lower surface of lamina on either side of midribs of leaf and veins (except sometimes at sinuses); leaves strongly secund, and usually with some secondary division; ultimate lobes linear to subulate, 0.7–1.2 mm wide **339b.** subsp. **incurva**

339a. Grevillea curviloba McGill. subsp. curviloba

G. diversifolia var. *rigida* Meisn., *in* A.L.P.P. de Candolle, *Prodr.* 14: 368 (1856). T: [*J.*]*Drummond Coll. [4] 1848 n. 286*! hb. Shuttl. [W.A.]; lecto: NY *n.v.*, *fide* D.J.McGillivray & R.O.Makinson, *Grevillea* 412 (1993); isolecto: [most as *J.Drummond IV: 286*] BM, CGE *n.v.*, G *n.v.*, K, MEL, P *n.v.*, TCD *n.v.*

Illustrations: P.M.Olde & N.R.Marriott, *Nuytsia* 9: 241, fig. 1A–G (1993); P.M.Olde & N.R.Marriott, *Grevillea Book* 2: 107 (86A), 108 (top left & 86B) (1995); A.Brown *et al.* (eds), *W. Australia's Threatened Flora* 142 (1998).

Leaves 1.5–5 cm long, mostly pinnatifid, cuneate below the basal lobes, weakly secund with upper surface somewhat concave, usually lacking secondary division; leaf lobes oblong or broadly linear, > 1.5 mm wide; margins shortly recurved; lower surface mostly or wholly exposed, glabrous or sparsely tomentose. Flower colour: perianth and style white to cream.

Occurs in south-western W.A., where restricted to a small area near Bullsbrook. Grows in open shrub associations in deep humic sand. Regenerates from seed. Flowers mainly Aug.–Oct. Map 432.

W.A.: Bullsbrook, *H.Demarz 3945* (PERTH); 2 km N of Bullsbrook, *P.Olde 86/283* (NSW); Warbrook Rd, S of Bullsbrook, *P.Olde 91/78* (NSW); 'Blackwood River', 1883, *Mrs McHard* (MEL).

This subspecies is recognised as 'Poorly Known' in J.D.Briggs & J.H.Leigh, *Rare or Threatened Australian Plants* (1995).

339b. Grevillea curviloba subsp. incurva Olde & Marriott, *Nuytsia* 9: 243 (1993)

T: Muchea, W.A., 26 Sept. 1992, *P.M.Olde 92/108*; holo: NSW.

Illustrations: P.M.Olde & N.R.Marriott, *op cit.* 9: 241, fig. 1H–J; P.M.Olde & N.R.Marriott, *Grevillea Book* 2: 108 (bottom right), 109 (87A, B) (1995); A.Brown *et al.* (eds), *W. Australia's Threatened Flora* 142 (1998).

Leaves 2–5 cm long, 3–5-partite (almost -sect), sublinear below the basal lobes, secund, often with lower lateral lobes 2- or 3-partite; leaf lobes ±linear, ⩽ 1.2 mm wide; margins tightly revolute to prominent abaxial midrib; lower surface 2-grooved (lamina sometimes exposed at sinuses). Flower colour: perianth and style white to cream. Fig. 37A–D.

Occurs in south-western W.A., near Muchea and with one collection from near Badgingarra. Grows in heath associations in sandy soils. Probably regenerates from seed only. Flowers mainly July–Sept. Map 433.

W.A.: Muchea Townsite Reserve, *G.Keighery & J.Alford 83* (PERTH); near Muchea, *D.J.McGillivray 3273 & A.S.George* (NSW, PERTH); c. 1.6 km N of Muchea, *K.Newbey 1674* (PERTH); Brand Hwy, Muchea, *P.Olde 86/632* (NSW); Badgingarra, Sept. 1960, *L.Steenbohm* (PERTH).

Close to *G. rara*, which has the branchlets glabrous to sparsely hairy, the leaf lobes wider (0.7–1.5 mm wide), the margins more consistently revolute to the midveins (no laminal lower surface exposed at sinuses), and pedicels 2–5 mm long (in subsp. *incurva* 7–10 mm long).

A selection of *G. curviloba* subsp. *incurva* is widely cultivated as an ornamental shrub and ground cover; it has long been misidentified and sold as '*G. biternata*'.

This subspecies is recognised as 'Poorly Known' in J.D.Briggs & J.H.Leigh, *Rare or Threatened Australian Plants* (1995).

340. Grevillea rara Olde & Marriott, *Nuytsia* 9: 244 (1993)

T: 14.5 km N of Collie on road to Tallanalla, W.A., 33°11'S, 116°08'E, 31 Oct. 1986, *P.M.Olde 86/1008*; holo: PERTH; iso: NSW.

Illustrations: P.M.Olde & N.R.Marriott, *op cit.* 9: 245, fig. 2A–F; P.M.Olde & N.R.Marriott, *Grevillea Book* 3: 130 (top right & 102A, B) (1995); A.Brown *et al.* (eds), *W. Australia's Threatened Flora* 121 (1998).

Shrub, prostrate to sprawling when young, ascending to 2 m tall when adult. Branchlets densely pubescent to tomentose. Leaves 2–2.5 cm long, secund, pinnatisect with usually 5 divaricate primary lobes, these often again 2- or 3-partite; ultimate lobes divaricate, linear, sometimes gently curved, 4–15 (–20) mm long, (0.3–) 0.5–0.8 mm wide; apices weakly pungent not or weakly pungent; margins tightly and smoothly revolute; lower surface completely enclosed, 2-grooved. Unit conflorescence ?basipetal, subglobose to dome-shaped; floral rachis tomentose. Flower colour: perianth and style white, sometimes tinged pink. Perianth glabrous outside. Pistil 3.5 mm long; stylar swelling ovoid to narrowly so; pollen-presenter conical, base slightly broader than apex of style. Follicle (only very immature seen) rugose.

Occurs in south-western W.A. where known only from a small area N of Collie, now partly submerged by the Harris River reservoir. Grows in Jarrah forest in lateritic loam, moist situations along creek lines. Probably regenerates from seed only. Flowers Aug.–Nov. Map 434.

Recent collections at PERTH have not been seen for this treatment.

Grevillea rara appears to be most closely related to *G. curviloba*; fruit characters (if confirmed as including a thick pericarp and rugose surface) would support this. Also narrowly distinct from *G. corrugata* which has somewhat longer and wider straight leaf lobes (usually 1.5–3.5 cm long, 0.7–1.2 mm wide), more loosely revolute leaf margins with a narrow exposure of the lower surface at least at the nodes and sinuses, and the basal internode with the lamina continuing as a narrow wing right to the leaf base (leaf sessile). In *G. rara* the basal internode has the lamina gradually disappearing on the adaxial surface, with a short petiole below.

This species is recognised as 'Poorly Known' in J.D.Briggs & J.H.Leigh, *Rare or Threatened Australian Plants* (1995).

341. Grevillea corrugata Olde & Marriott, *Nuytsia* 9: 247 (1993)

T: c. 10 km S of Bindoon, W.A., 4 Oct. 1992, *P.Olde 92/230*; holo: NSW; iso: CANB, PERTH.

Illustrations: P.M.Olde & N.R.Marriott, *op cit.* 9: 248, fig. 3A–E; P.M.Olde & N.R.Marriott, *Grevillea Book* 2: 100 (bottom right), 101 (80A–C) (1995).

Dense shrub 0.5–1.5 m tall. Branchlets open tomentose. Leaves 4–6 cm long, to 9 cm wide (flattened), usually deeply 3–5-partite or -sect; lateral lobes spreading and divaricate, with some or all lobes again 3-partite (rarely some tertiary division); ultimate lobes straight, linear, (0.8–) 1.5–3.5 cm long, 0.7–0.8 (–1.2) mm wide; basal internode with lamina continuing as a narrow wing right to leaf base; apices pungent; margins loosely revolute to midveins or almost so; lower surface 2-grooved, often a narrow area of lower surface exposed at least at nodes. Unit conflorescence acropetal, subglobose to obovoid; floral rachis loosely tomentose. Flower colour: perianth and style white to cream. Perianth glabrous outside. Pistil 3.5–4 mm long; stylar swelling ovoid; pollen-presenter conical to bluntly fusiform, base slightly wide than apex of style. Follicle obloid-ellipsoidal, 7–11 mm long, rugose with short rounded ridges.

Occurs near Bindoon in south-western W.A. Known only from the type locality, where growing in disturbed eucalypt woodland in gravelly loam soil. Probably regenerates from seed only. Flowers ?Aug.–Sept. Map 435.

No specimens cited in view of rarity.

Similar and probably closely related to *G. rara* (see under that species for differences), and perhaps to *G. paniculata* which has usually glabrous branchlets and floral rachises and lacks a laminal 'wing' on the basal section of the leaf.

This species is recognised as 'Poorly Known' in J.D.Briggs & J.H.Leigh, *Rare or Threatened Australian Plants* (1995).

342. Grevillea paniculata Meisn., *in* J.G.C.Lehmann, *Pl. Preiss.* 1: 550 (1845)

T: 'prope oppidum York', [W.A.] [protologue]; ...ad urbicula York, d. 10 Sept. 1839, *L.Preiss 617a*; lecto: NY *n.v.*, *fide* D.J.McGillivray & R.O.Makinson, *Grevillea* 432 (1993); isolecto [some as Preiss 617 or 617b]: B *n.v.*, G, G-DC, LD *n.v.*, LE *n.v.*, MEL, P *n.v.*; interior W.A., 10 Apr. 1840, *L.Preiss 705*; remaining syntypes: LD, LE, NY, all *n.v.*

G. paniculata populations g, j, l, m, n, and (in part) o, of McGillivray & R.O.Makinson, *Grevillea* 178–179 (1993).

Illustrations: D.J.McGillivray & R.O.Makinson, *Grevillea* 176, fig. 45 (1993); P.M.Olde & N.R.Marriott, *Grevillea Book* 3: 72 (top right & 52A–C) (1995).

Dense shrub 0.7–3 m tall. Branchlets glabrous or occasionally densely subsericeous. Leaves (1–) 2.5–6 cm long, divaricately divided with 3 (5) primary lobes, these simple or occasionally lateral lobes again 2- or 3-partite; ultimate lobes (1–) 2–4 cm long, 0.5–0.8 mm wide, straight, linear, with terminal lobe deflexed; apices subpungent; margins tightly and angularly to smoothly revolute; lower surface completely enclosed except for prominent midribs, the 2 grooves almost lateral. Unit conflorescence acropetal, subumbelloid to subglobose; floral rachis glabrous or occasionally subsericeous. Flower colour: perianth and style white to cream; limb of bud pale yellow. Perianth glabrous outside. Pistil 3–4 mm long; stylar swelling ovoid; pollen-presenter conical to truncate-conical, base rimmed, slightly wider than apex of style. Follicle obloid-ellipsoidal, (6–) 8–9 mm long, strongly rugose with low longitudinal ridges drawn into short irregular peaks, or sometimes strongly muricate. *Kerosene Bush*. Fig. 37H–J.

Occurs in south-western W.A., widespread from Mundaring area E to Hyden and Merredin, and N to Pithara and Bonnie Rock. Grows in sclerophyll shrubland or open heath in sandy or gravelly soils, often in low-lying sites or near runoff from granite outcrops. Regenerates from seed and (at least some populations) rhizomes. Flowers May–Oct. Map 436.

W.A.: 10 km NE of Westonia, Sanford Rock, *M.D.Crisp 6578* (CANB, NSW, PERTH); B.Smith's property, Manmanning, *D.J.McGillivray 3416 & A.S.George* (CANB, K, MEL, NSW, PERTH); c. 1.6 km N of Merredin on road to Nungarin, *M.D.Tindale 80 & E.M.Bennett* (K, NSW, PERTH, RSA); Goomalling district, *M.D.Tindale 113 & E.M.Bennett* (K, NSW, PERTH, RSA).

McGillivray & Makinson (*Grevillea* 176–180 (1993)) recognised a broad-concept *G. paniculata* with many informally delimited populations, including the taxa here recognised (following Olde & Marriott, *op. cit.* 1994, 1995) as *G. levis*, *G. biternata*, *G. xiphoidea* and *G. elongata*. Diagnostic features for the species can be fairly cryptic, and there are numerous unassigned specimens and populations; more are to be expected, and further research is desirable. *Grevillea biternata*, *G. paniculata*, and to a lesser extent *G. levis* all encompass significant internal variation.

The strongest diagnostic features of *G. paniculata* are: fruits highly rugose to muricate, floral bracts glabrous except for ciliate margins, and floral rachis usually glabrous. Additional diagnostic characters include: branchlets usually glabrous and slightly glaucous, rarely subsericeous; leaves usually with only primary (occasionally some secondary) division; lobes linear, straight, divaricate, and 0.5–0.8 mm wide; leaf lower surface enclosed except for the very prominent midribs; pedicels 8–11 mm long; pollen-presenter conical. Plants with glabrous, non-glaucous branchlets are easily confused with *G. levis* unless fruit is available.

The 'type form' ('form n' of McGillivray & Makinson, *loc. cit.*) occurs around Burracoppin, Kellerberrin, Yorkrakine, Cunderdin, Quairading, and Beverley, and from Wongan Hills to Ballidu; it has glabrous and glaucous branchlets, glabrous floral rachises, and leaves tripartite or sometimes biternate. The 'hairy form' (forms 'm' and 'o' of McGillivray & Makinson, *loc. cit.*), occurs around Goomalling and Manmanning, and between Pingelly, Quairading and Northam; it has subsericeous to tomentose branchlets and (usually) floral

Occurs in south-western W.A., possibly restricted to the Muntadgin–Tandegin area. Grows in open shrubland in granitic sandy-loam soils, often around granite outcrops. Regenerates from seed. Flowers mainly June–Sept. Map 439.

W.A.: Henderson Rd, 6.4 km E of rail crossing, Tandegin, *P.M.Olde 92/214* (NSW); Bissenden Rd, 5.9 km E of railway, Muntadgin, *P.M.Olde 92/218* (NSW); Mt Cramphorne, *R.D.Royce 7860* (PERTH).

Grevillea xiphoidea differs from *G. levis* in its loose subcylindrical conflorescences with rachises 20–30 mm long, the pedicels almost spreading from the rachis and 3–5 mm long, and its slightly wider leaf lobes. *Grevillea levis* has shorter, denser, more subglobose unit conflorescences (rachises c. 10 mm long, rarely to 20 mm), the pedicels ascending and 5–7 mm long, and leaf lobes 0.3–0.8 mm wide. Nearby populations of *G. paniculata* can be distinguished by their strongly rugose fruits, shorter denser unit conflorescences, and tightly revolute leaf margins enclosing the lamina on either side of the abaxial midribs. Some populations of *G. biternata* with long floral rachises can be distinguished by having coarser leaf lobes (usually 1–2.8 mm wide), subvillous floral rachises, and floral bracts densely tomentose on the outer surface, persistent almost to anthesis (glabrous with ciliate margins and falling early in *G. xiphoidea*). A collection (*Beard 5941*, CANB, KPBG) from Weowanie Rock, N of Yellowdine, well E of the main range, may belong to this species.

346. Grevillea elongata Olde & Marriott, *Grevillea Book* 1: 175 (1994)

T: Tutunup Rd, Ruabon, W.A., 10 Oct. 1991, *P.M.Olde 91/271*; holo: NSW; iso: PERTH.

G. paniculata population 'r', 'Busselton form', of D.J.McGillivray & R.O.Makinson, *Grevillea* 179 (1993).

Illustrations: P.M.Olde & N.R.Marriott, *Grevillea Book* 2: 142 (top right & 115A), 143 (115B, C) (1995); A.Brown *et al.* (eds), *W. Australia's Threatened Flora* 143 (1998).

Shrub 1.5–2 m tall. Branchlets glabrous or with a sparse appressed indumentum. Leaves 2.5–5 cm long, tripartite with lateral lobes often again tripartite, occasionally biternate; ultimate lobes linear, 0.5–3 cm long, c. 0.8 mm wide; apices pungent; margins tightly revolute to prominent midrib; lower surface 2-grooved. Unit conflorescence acropetal, narrowly conical to subcylindrical; floral rachis glabrous or sometimes sparsely subsericeous near base. Flower colour: perianth and style white. Perianth glabrous outside. Pistil 3.5–4.5 mm long; stylar swelling ovoid; pollen-presenter truncate-conical to sub-cylindrical, scarcely wider than apex of style. Follicle obloid-ellipsoidal, c. 8 mm long, rugulose. Fig. 38J.

Occurs in south-western W.A., where restricted to the Busselton–Ruabon area. Grows in heathy associations in poorly drained sandy soils, often beside creeks. Regenerates from seed. Flowers Oct.–Dec. Map 440.

W.A.: Ruabon Rd, 7 km along the old railway track, *Nunn 536* (PERTH); Oates Rd, Ruabon, *P.M.Olde 86/1021* (NSW, PERTH); Abba R., Busselton District, *R.D.Royce 3391* (PERTH); Ruabon, Busselton district, *R.D.Royce 4868* (PERTH).

Grevillea elongata has a distinctively long unit conflorescence 2–5.5 cm long, (equalled among related species only occasionally, notably in *G. xiphoidea* which has more loosely revolute leaf margins, smaller floral bracts ⩽ 1 mm long, and smooth fruits). *Grevillea elongata* also has larger floral bracts and shorter pedicels than any similar or closely related species (bracts 2.8–3.4 mm long, 3–4 mm wide, glabrous except for ciliate margins, persistent almost to anthesis; pedicels 2.2–3.2 mm long).

A collection from 12 km E of Harvey (*P.G.Wilson 6555*, PERTH) has some similarities to *G. elongata*; it has branchlets with scattered appressed hairs, coarser leaf lobes (0.8–1.1 mm wide), unit conflorescences shortly ?cylindrical with glabrous or sparsely hairy floral rachises 1.5–2 cm long, (bracts not seen). Flower colour (buds?) is recorded as 'red'. This specimen may have affinities to *G. elongata* or to the hairy form of *G. paniculata*; it is not known whether it represents a consistent population.

This species is recognised as 'Poorly Known' in J.D.Briggs & J.H.Leigh, *Rare or Threatened Australian Plants* (1995).

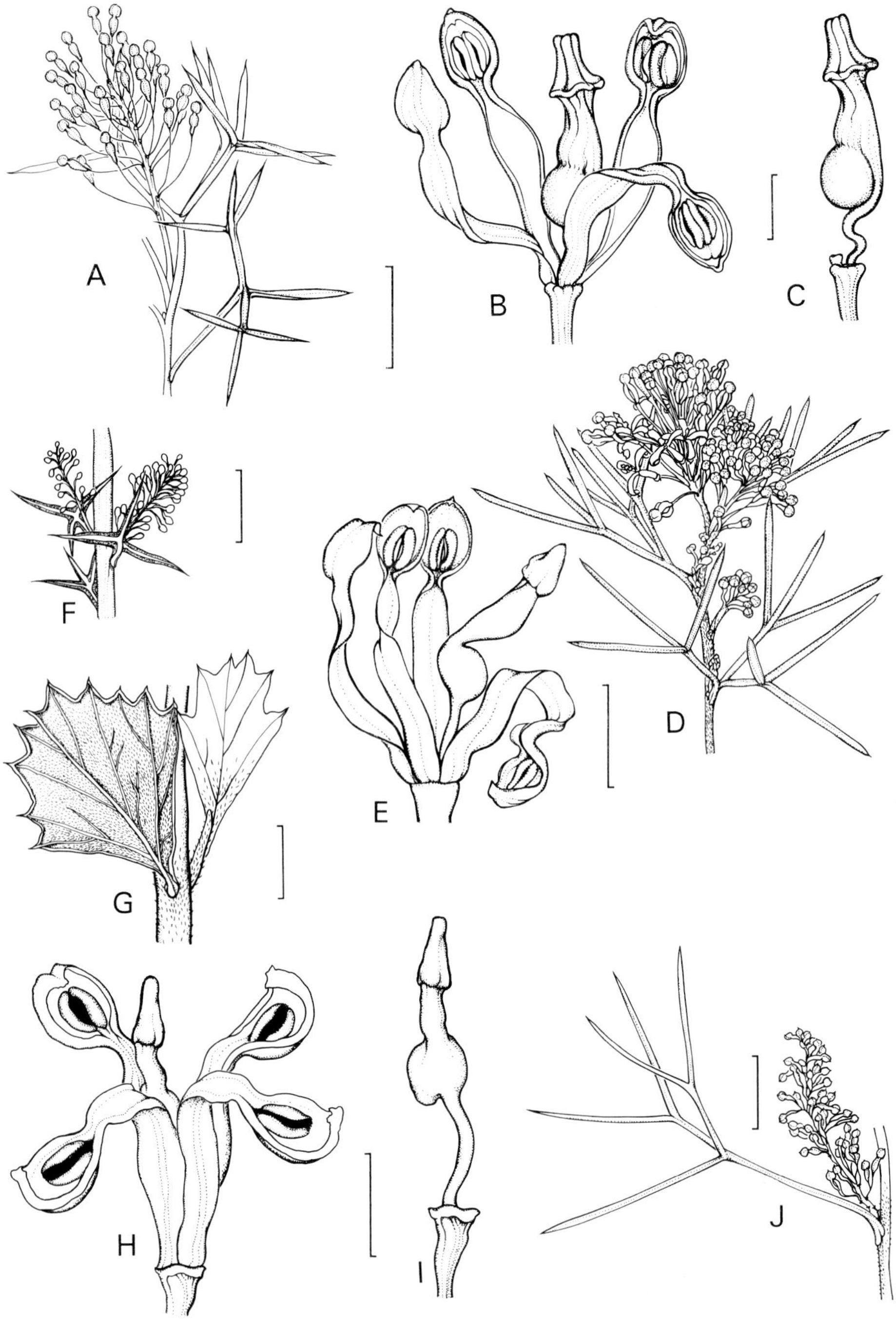

Figure 38. *Grevillea*. **A–C**, *G. levis*. **A**, flowering branch; **B**, flower; **C**, pistil (**A–C**, M.E.Phillips, 18 Sept. 1962, CBG017096, CANB). **D–E**, *G. anethifolia*. **D**, flowering branch; **E**, flower (**D–E**, J.W.Wrigley, 4 Oct. 1971, CBG041302, CANB). **F–I**, *G. acrobotrya*. **F**, flowering branch; **G**, leaves of non-flowering portion (**F–G**, N.H.Speck *s.n.*, 19 June 1952, CANB); **H**, flower; **I**, pistil (**H–I**, A.R.Fairall 2479, CANB). **J**, *G. elongata,* flowering branch (P.Olde 1021,CANB). Scale bars: **A**, **D**, **F–G**, **J** = 1 cm; **B–C**, **H–I** = 1 mm; **E** = 2 mm. Drawn by C.Payne.

347. Grevillea anethifolia R.Br., *Suppl. Prodr. Fl. Nov. Holl.* 21 (1830)

T: interior Port Jackson, N.S.W., 1817, *A.Cunningham & C.Fraser*, Oxley's 1st Exped.; lecto: BM, *fide* D.J.McGillivray & R.O.Makinson, *Grevillea* 404 (1993); south coast New Holland, W.A., 1823, *W.Baxter*; remaining syntype: BM.

[*G. biternata auct. non* Meisn.: J.M.Black, *Trans. Roy. Soc. S. Australia* 67: 36 (1943)]

Illustrations: G.M.Cunningham *et al.*, *Pl. W New South Wales* 212 (1982); D.J.McGillivray & R.O.Makinson, *Grevillea* 180 (1993); P.M.Olde & N.R.Marriott, *Grevillea Book* 2: 25 (top right & 16A), 26 (16B, C) (1995).

Shrub 0.5–2 m high, often rhizomatous. Branchlets subsericeous or occasionally tomentose. Leaves 1.5–5 cm long, with 3–7 divaricate primary lobes, these often again 2- or 3-partite; leaf rachis deflexed at each node; ultimate lobes linear to subterete, 0.5–4.5 cm long, 0.4–1.2 mm wide; apices pungent; margins tightly and usually angularly revolute, enclosing lower surface except for midveins; lower surface 2-grooved. Unit conflorescence acropetal, dome-shaped to loosely subglobose, shortly cylindrical or subconical; floral rachis tomentose at base, often glabrous above. Flower colour: perianth cream with limb yellow; style white. Perianth glabrous outside. Pistil 4–6 mm long; stylar swelling ovoid to ellipsoidal; pollen-presenter slightly oblique at base, truncate-conical to subcylindrical, base distinctly wider than apex of style. Follicle obloid-ellipsoidal to slightly obovoid, (5–) 7.5–9 mm long, rugose and colliculate. Fig. 38D–E.

Occurs in south-western W.A. (Southern Cross S to Lake Grace, and E to Zanthus and Cape Arid), in S.A. (Eyre Peninsula from Minnipa to Gawler Ra. and Kimba), and in N.S.W. (Hillston, Griffith, Rankins Springs and N nearly to Mount Hope). Grows in heath, shrubland or mallee shrubland, usually in sandy calcareous to acid soils; some tendency to colonise disturbed ground. Regenerates from seed and/or rhizomes. Flowers July–Dec. Map 441.

W.A.: c. 22 km ESE of Pingaring on road to Newdegate via L. Biddy, *L.Haegi 1078* (AD, PERTH); 3 km NNW of Uraryie Rock, 22 km SW of Zanthus, *K.Newbey 7156* (PERTH). S.A.: E slopes of Mt Nott, 9 km S of Thurlga Stn, Gawler Ra., *D.E.Symon 8057* (AD, CANB, NSW); Gawler Ra., 13 km SW of Yardea HS, *D.E.Symon 8330* (AD, CANB, NSW). N.S.W.: 17.5 km by road from Newell Hwy at Ardlethan along road to Griffith, *R.O.Makinson 1305 & D.Mallinson* (AD, CANB, K, NSW, PERTH).

Grevillea anethifolia may be distinguished from its close relatives by the following suite of characters: branchlets subsericeous to tomentose; ultimate leaf-lobes ascending, usually less than 3 cm long, usually rigid, pungent; perianth glabrous outside; pollen-presenter truncate-conical to subcylindrical with stigma 0.2–0.3 mm across; follicle surface rugose with discontinuous ridges and irregularly colliculate.

There is some variation, with small-fruited plants or populations (follicles 5–7.5 mm long) scattered throughout the range. Plants from the Mt Arid area in W.A. have conspicuous floral bracts 3–5 mm long and 2–4 mm wide.

348. Grevillea erinacea Meisn., *Hooker's J. Bot. Kew Gard. Misc.* 7: 74 (1855)

T: north of Swan River, W.A., 1850–51, *J.Drummond colln 4: 186*; holo: NY *n.v.*; iso: B *n.v.*, CGE *n.v.*, E, FI *n.v.*, G *n.v.*, K, LD *n.v.*, MEL, NSW, P, PERTH.

Illustrations: P.M.Olde & N.R.Marriott, *Grevillea Book* 2: 147 (bottom right), 148 (119A, B) (1995).

Spreading shrub 0.3–1.8 m high. Branchlets tomentose to subsericeous. Leaves secund, 1–3.5 cm long, deeply 3–5-partite (almost -sect); primary lobes divaricate, simple to tripartite; ultimate lobes linear, 0.5–2 cm long, 0.6–1.0 mm wide; apices usually pungent; margins very tightly revolute, enclosing lower surface on either side of midvein; lower surface 2-grooved. Unit conflorescence acropetal, subglobose to dome-shaped; floral rachis tomentose. Flower colour: perianth greenish white in late bud, becoming cream; style white. Perianth subsericeous outside. Pistil 2.5–4.5 mm long, usually glabrous, occasionally with biramous hairs on ovary; stylar swelling ovoid; pollen-presenter conical, base slightly broader than apex of style. Follicle obloid to obovoid, 7.5–10 mm long, almost smooth or with scattered tubercles. Fig. 40A–C.

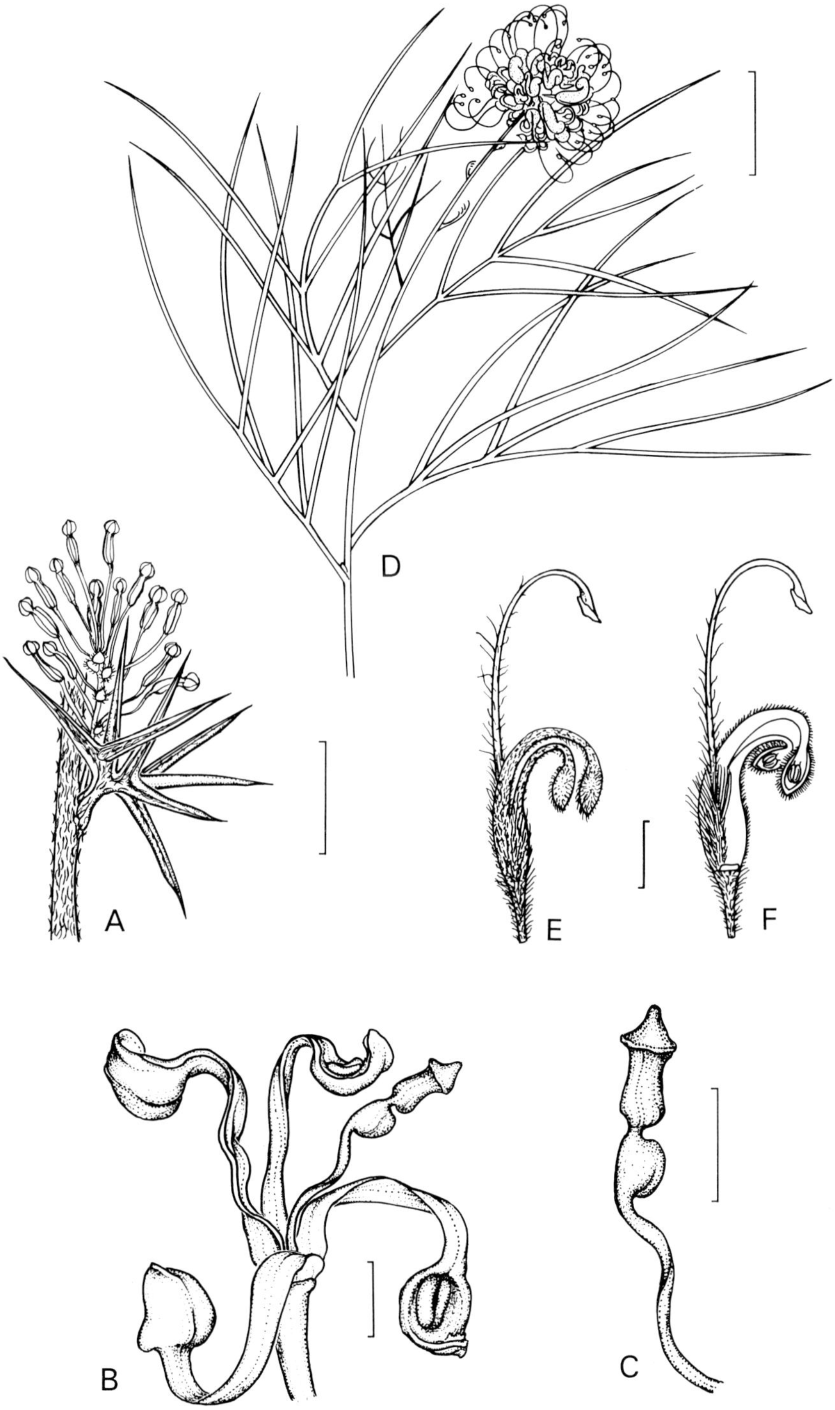

Figure 39. *Grevillea*. **A–C**, *G. spinosissima*. **A**, flowering branch; **B**, flower; **C**, pistil (**A–C**, J.Taylor 2179, CANB). **D–F**, *G. globosa*. **D**, flowering branch; **E**, flower; **F**, pistil and half perianth (**D–F**, D.J.McGillivray 3391 & A.S.George, NSW). Scale bars: **A**, **D** = 1 cm; **B–C** = 1 mm; **E–F** = 3 mm. Drawn by: **A–C**, C.Payne; **D–F**, D.Mackay.

Occurs in south-western W.A., SW of Geraldton in the area between Ellendale, Three Springs and Arrowsmith. Grows in heath or shrubland in sandy or lateritic gravel soils. Regenerates from seed. Flowers mainly July–Oct. Map 442.

W.A.: c. 14 km W of Walkaway, *J.S.Beard 6921* (NSW, PERTH); Burma Rd, c. 29 km SE of Walkaway, *R.Filson 8548* (MEL); c. 48 km W from Three Springs, *C.A.Gardner 9140* (PERTH); c. 16 km N of Strawberry Siding, *R.Hnatiuk 760365* (PERTH); 30.5 km from Eneabba on road to Three Springs, *D.J.McGillivray 3305* (NSW, PERTH).

Grevillea erinacea is distinctive for the biramous hairs on the pedicels and the outer surface of the perianth; the only related species sharing these features is *G. metamorpha*, which has triangular leaf lobes with the laminal lower surface exposed.

349. Grevillea spinosissima McGill., *New Names Grevillea* 14 (1986)

T: Manmanning, W.A., 17 June 1976, *D.J.McGillivray 3418 & A.S.George*; holo: NSW; iso: PERTH.

Illustrations: P.M.Olde & N.R.Marriott, *Grevillea Book* 3: 181 (top right & 147A, B) (1995).

Shrub 0.3–1.5 m high with arching columnar branches. Branchlets tomentose to pubescent. Leaves 0.7–2 cm long; tripartite with lateral lobes simple to tripartite; leaf rachis sharply deflexed at node; ultimate lobes subulate to sublinear, divaricate, 0.4–1.5 cm long, 0.7–1.4 mm wide; apices pungent; margins revolute, enclosing lower surface except for midveins; lower surface 2-grooved. Unit conflorescence acropetal, loosely subglobose; floral rachis loosely tomentose to almost glabrous. Flower colour: perianth creamy white with limb green to mauve; style white. Perianth glabrous outside. Pistil 3.4–4.1 mm long; stylar swelling flask-shaped; pollen-presenter conical, base much broader than apex of style. Follicle obovoid or obloid-ellipsoidal, 7–8 mm long, glaucous, rugulose. Fig. 39A–C.

Occurs in south-western W.A., in the Wongan Hills area, and patchy occurrences S to Quairading and York. Grows in heath, shrubland, or open shrubby woodland in well-drained sandy to gravelly lateritic loams; colonises disturbed ground. Regenerates from seed, some suggestion (Olde & Marriott, *op. cit.* 182) of rhizomes. Flowers mainly June–Sept. Map 443.

W.A.: near York, *W.E.Blackall 3296* (PERTH); E sources of Swan R., 1889, *Miss Eaton* (MEL); 1 km NW of Mawson Siding, 19 km W of Quairading, *B.G.Muir 393* (PERTH); c. 11 km E of Wongan Hills, *K.Newbey 1976* (PERTH).

This species is recognised as 'Poorly Known' in J.D.Briggs & J.H.Leigh, *Rare or Threatened Australian Plants* (1995).

Rudis Group

Shrubs. Leaves divided or toothed, dorsiventral with surfaces dissimilar or terete-unifacial with surfaces similar; margins flat or recurved, rarely obscure. Conflorescence terminal or axillary, erect or decurved to pendulous, usually simple, cylindrical or shortly so, acropetal or basipetal. Flowers obscurely acroscopic. Torus straight. Perianth actinomorphic, hairy or glabrous outside, usually glabrous inside (papillose to hairy in *G. althoferorum*); tepals separating to base and individually lax or recoiled. Pistil 3.8–11 mm long; ovary hairy, with biramous and usually some glandular hairs, sessile to shortly stipitate; style folded above ovary, sparsely glandular-pubescent on filiform portion, gradually dilated in apical half and papillose, not or scarcely exserted from late bud; pollen-presenter conical to conico-cylindrical with cup-shaped apex. Follicle glabrous or with erect simple glandular hairs soon falling (lacking indumental markings), viscid; pericarp thin, crustaceous. Seed ellipsoidal, with a short revolute waxy margin along one side.

A group of five species, endemic to south-western W.A. Insect pollinated. The nectary is apparently absent. Closest affinities are probably to the *Triloba* group, and probably also the *Integrifolia* group; the *Rudis* group may be seen as forming a bridge between the former (postulated here to be a highly derived group) and the latter (a more basal lineage).

1	Leaf lobes terete; most conflorescences branched	**352. G. stenostachya**
1:	Leaf lobes not terete; most conflorescences simple	
2	Ovary sessile, villous; conflorescence acropetal	
3	Most leaves trifid to pinnatifid (rarely bipinnatifid), usually obovate-cuneate with apical teeth or lobes; perianth glabrous and smooth inside; conflorescences on emergent canes above foliage	**350. G. rudis**
3:	Most leaves pinnatisect with 3–7 distant, primary lobes again once or twice divided; perianth papillose to hairy inside; conflorescences not emergent	**351. G. althoferorum**
2:	Ovary stipitate, glandular-pubescent; conflorescence basipetal	
4	Pedicels ⩽ 2 mm long; floral bracts falling early; ovary stipe 0.6–0.9 mm long	**354. G. pulchella**
4:	Pedicel > 2 mm long; floral bracts persistent; ovary stipe 0.8–1.6 mm long	**353. G. tenuiflora**

350. **Grevillea rudis** Meisn., *Hooker's J. Bot. Kew Gard. Misc.* 7: 73 (1855)

T: interior north of Swan River, W.A., 1850–51, *J.Drummond coll 6, 180*; holo: NY *n.v.*; iso: BM, CGE *n.v.*, G-DC, K, LD *n.v.*, MEL, NSW, P, PERTH.

Illustrations: D.J.McGillivray & R.O.Makinson, *Grevillea* 391, fig. 105 & col. pl. (1993); P.M.Olde & N.R.Marriott, *Grevillea Book* 3: 152 (bottom right), 153 (fig. 121A, B) (1995).

Erect to spreading shrub, 0.2–0.9 m high. Leaves mostly obovate-cuneate in outline, trifid to pinnatifid or rarely pinnatipartite, 1.5–3 (–5) cm long, 0.5–2 cm wide; basal and subfloral leaves sometimes entire, narrowly obovate; primary lobes or teeth 3–5 (–7), broadly triangular or oblong, sometimes divided; ultimate lobes or teeth subtriangular to oblong-ovate, (2–) 5–10 mm long, pungent; margins thickened, flat or slightly incurved; lower surface scabrous, loosely villous, becoming ±glabrous. Conflorescence on emergent cane, terminal, erect, usually simple, 3–12 (–50) cm long, acropetal, loosely cylindrical to subconical. Flower colour: perianth yellowish green in bud, becoming cream at anthesis; style creamy white, becoming pinkish below pollen-presenter with age. Perianth ±villous outside with simple glandular and biramous hairs, glabrous inside; tepals independently recoiled after anthesis. Pistil 5–6.5 mm long; ovary sessile, glandular-villous; style folded to deflexed above ovary, sparsely glandular-pubescent above ovary, becoming tuberculose on swollen upper fifth; pollen-presenter ±erect, conical. Follicles ovoid to slightly obovoid, 11–12 mm long, glandular-pubescent, slightly viscid.

Occurs in south-western W.A., between Eneabba and Jurien Bay, inland to Watheroo, growing in well-drained lateritic areas. Regenerates from seed. Flowers sporadically throughout the year, mainly in spring. Map 444.

W.A.: Eneabba and Three Springs road (c. 35 km W of Three Springs), *A.M.Ashby 3243* (AD, C *n.v.*, CANB, FI *n.v.*, L *n.v.*, MEL); 2 km from Badgingarra towards Moora, 25 Sept. 1968. *E.M.Canning CBG038725* (CANB, NSW); Greenhead road, *J.D'Alonzo 230* (CANB, PERTH); c. 12 km W of Brand Hwy, along Coomallo–Jurien road, *R.W.Purdie 5135* (CANB); 23 airmiles SE of Eneabba (Toolbardie Rd), *B.H.Smith 1153* (CANB, LE, MEL, NBG, NSW).

Grevillea rudis is very similar to *G. althoferorum*, which differs in its much more deeply divided leaves, its fewer-flowered conflorescences scarcely exceeding the foliage, and its papillose inner perianth surface.

This species is recognised as 'Rare' in J.D.Briggs & J.H.Leigh, *Rare or Threatened Australian Plants* (1995).

351. **Grevillea althoferorum** Olde & Marriott, *Nuytsia* 9: 295 (1993), as *G. althoferi*

T: near Eneabba, W.A., 15 Sept. 1991, *P.M.Olde 91/102*; holo: NSW; iso: PERTH. Orthographic correction of name made in *Nuytsia* 9(3): 438 (1994) by K.F.Kenneally.

Illustrations: P.M.Olde & N.R.Marriott, *Grevillea Book* 2: 23 (centre right & 14A, B) (1995), as *G. althoferi*; A.Brown *et al.* (eds), *W. Australia's Threatened Flora* 142 (1998).

10: Conflorescences either erect and then subglobose to ovoid or very few-flowered, or conflorescences decurved to deflexed and then ovoid to shortly cylindrical or secund; leaves variously shaped (often cuneate to rhomboid in general outline, rarely elliptic but then never falcate), and usually with 1–many shallow triangular lobes or teeth (rarely entire); flowers red, orange, greenish, cream or yellow

11 Leaves elliptic or narrowly so, entire or with 1–3 short spreading triangular teeth well below the apex; style with minute hairs or papillae in apical 2–5 mm, otherwise glabrous; dorsal edge of torus neither elongated along nor adnate to the ovary stipe — **Agrifolia Group** (*G. brevis*) (p. 131)

11: Leaves cuneate-rhomboid or rarely oblong in general outline, with (1–) 2–9 (–15) obtuse broad teeth or broadly triangular lobes, mostly in the apical half of the leaf; style ±hairy throughout, rarely with a few hairs near the apex only, or rarely completely glabrous; dorsal edge of torus ±elongated, adnate to stipe of ovary

12 Unit conflorescences usually decurved, elongate, subsecund or conico-cylindrical; floral rachises 15–120 mm long; ±robust shrubs or small treees 1–6 m tall — **Wickhamii Group** (p. 133)

12: Unit conflorescences erect, subglobose to ovoid; floral rachises 5–20 mm long; low multistemmed shrublet to 0.6 m tall — **Adenotricha Group** (*G. longicuspis*) (p. 142)

2: Ovary with hairs

13 Unit conflorescences erect, cylindrical; leaves entire and linear, or divided with linear lobes, all leaf segments < 2 mm wide — **Pterosperma Group** (p. 352)

13: Unit conflorescences either decurved to deflexed (and then ovoid to shortly cylindrical or subsecund), or if erect or straight on prostrate branches, then secund or ovoid (never cylindrical); leaves or leaf segments either not linear and usually > 10 mm wide, or if linear then only with secund conflorescences

14 Pistils ⩽ 25 mm long; flowers acroscopic; unit conflorescences decurved

15 Pistils 5–10 mm long, scarcely projecting beyond curve of perianth after release of style-end; style densely hairy — **Wickhamii Group** (*G. wickhamii* subsp. *aprica*) (p. 136)

15: Pistils 15–20 mm long, projecting well beyond curve of perianth after release of style-end; style glabrous or sparsely hairy — **Agrifolia Group** (*G. agrifolia* subsp. *agrifolia*) (p. 128)

14: Pistils 35–55 mm long; flowers twisting in bud stage to basiscopic or transverse orientation on the rachis; unit conflorescences erect, or straight on prostrate branchlets

16 All leaves entire, ovate or narrowly to broadly elliptic, 15–85 mm wide — **Goodii Group** (p. 111)

16: Most or all leaves pinnatisectly divided with numerous linear to narrowly elliptic lobes — **Heliosperma Group** (*G. dryandri*) (p. 118)

1: Inner surface of perianth below the anthers glabrous

17 Ovary with few to many hairs

18 Flowers ±sessile (pedicels 0–0.5 mm long), basiscopic (concave curve of style facing base of conflorescence); pistils 5–7 mm long; style villous or sparsely villous on the dorsal side over most of its length; nectary of 4 separate erect lobes spaced around the torus **Rubicunda Group** (p. 108)

18: Flowers pedicellate (pedicels 2–20 mm long), acroscopic (curve of style facing apex of conflorescence) or sometimes transversely oriented, or basiscopic; pistils 9–60 mm long; style glabrous except sometimes for minute simple erect hairs; nectary integral, arcuate or reniform

19 Unit conflorescences strongly secund; flowers orange or yellow (and then pistils > 12 mm long) or green to brownish (and then pistils < 12 mm long) **Pteridifolia Group** (p. 39)

19: Unit conflorescences regular, cylindrical or conico-cylindrical; flowers orange to deep yellow, or white to cream or pale yellow

20 Pedicels 8–20 mm long; unit conflorescences conico-cylindrical, acropetal; flowers orange to deep yellow **Pteridifolia Group** (*G. juncifolia*) (p. 102)

20: Pedicels 2–4 mm long; unit conflorescences cylindrical, basipetal or subsynchronous; flowers white to cream **Pterosperma Group** (*G. pterosperma*) (p. 353)

17: Ovary completely glabrous

21 Pistils 34–47 mm long; flowers red or deep pink; leaves with numerous non-pungent elliptic to narrowly elliptic or obovate or sublinear lobes, the laminal tissue of lobes not decurrent to leaf rachis **Heliosperma Group** (*G. heliosperma*) (p. 117)

21: Pistils 5–26 mm long; flowers white to cream or pale yellow or greenish white, or sometimes red, pink or orange; leaves entire or if divided then the lobes linear or subterete or strap-like or triangular, pungent or non-pungent, the laminal tissue of lobes decurrent to leaf rachis

22 Leaves entire, or divided and then with linear or subterete or strap-like lobes > 25 mm long, not pungent; conflorescences usually erect, simple or paniculately branched; unit conflorescences regular-cylindrical **Hilliana Group** (p. 361)

22: Leaves divided, with 2–50 shallow to deep short (≤ 15 mm long) triangular to linear often pungent teeth or lobes; conflorescences usually decurved, simple or basally few-branched; unit conflorescences secund, or regular and then conical or shortly cylindrical

23 Leaves with 2–10 shallow broad lobes or teeth; upper and lower leaf surfaces similar **Agrifolia Group** (*G. glabrescens*) (p. 131)

23: Leaves with 6–30 (–50) pungent teeth or lobes; upper and lower leaf surfaces clearly dissimilar **Heliosperma Group** (*G. pungens*) (p. 121)

South Australia

1 Ovary with few to many hairs	
2 Inner surface of perianth glabrous	
3 Unit conflorescences dense, cylindrical; flower opening basipetal or subsynchronous; flowers oriented transversely to rachis (flowers white to cream; pedicels 2–4 mm long)	**Pterosperma Group** (*G. pterosperma*) (p. 353)
3: Unit conflorescences dense to loose; flower opening acropetal, either secund with the flowers acroscopic, or conico-cylindrical (but then loose, the flowers orange and basiscopic, and the pedicels 8–20 mm long)	**Pteridifolia Group** (p. 39)
2: Inner surface of perianth with few to many hairs	
4 Pistils 5–10 mm long, scarcely projecting beyond curve of perianth after release of style-end; styles ±densely hairy to the tip, with short stiff hairs; leaves cuneate-rhomboid, broadly and shallowly toothed, the upper and lower surfaces similar	**† Wickhamii Group** (*G. wickhamii* subsp. *aprica*) (p. 136)
4: Pistils 12–30 mm long, projecting well beyond the curve of the perianth after release of style-end; styles loosely hairy, often less so (or glabrous) in apical half; leaves entire or deeply divided	
5 Conflorescences simple or paniculately branched; unit conflorescences dense, many-flowered, cylindrical; floral rachises 50–80 mm long; flower opening basipetal or subsynchronous; flowers white to cream	**Pterosperma Group** (*G. pterosperma*) (p. 353)
5: Conflorescences simple; unit conflorescences loose 3–16-flowered clusters, or 1- or 2-flowered; floral rachises ⩽ 10 mm long; flower opening acropetal or pattern obscure; flowers red, pink or white, or bicolorous	**Floribunda Group** (p. 268)
1: Ovary glabrous	
6 Pistils 19–29 mm long, reflexed at ovary; stipe of ovary 7–14 mm long	**Huegelii Group** (p. 144)
6: Pistils 4–15 mm long, not reflexed at ovary; stipe of ovary < 5 mm long	
7 Perianth regular, the apical knob-like limb of the bud remaining erect; pistils 4–6 mm long; style with a flask-shaped swelling over the upper half; leaves divaricately divided with fine linear lobes	**Triloba Group** (*G. anethifolia*) (p. 438)
7: Perianth zygomorphic, the apical limb of the bud becoming deflexed or decurved; pistils 5–15 mm long; style slender except right at the apex; leaves various, but not with divaricate linear lobes	
8 Styles ±densely hairy to the tip; leaves cuneate-rhomboid, with broad shallow marginal teeth or lobes	**† Wickhamii Group** (*G. wickhamii* subsp. *aprica*) (p. 136)
8: Styles glabrous, or with minute hairs near apex only; leaves entire, or deeply divided and then with the lobes linear or subterete	
9 Conflorescences often paniculately branched, sometimes simple; unit conflorescences regular-cylindrical; floral rachises 50–140 mm long; flowers white to cream; leaves entire or divided, with all leaves or leaf segments linear or subterete or strap-like (margins parallel)	**Hilliana Group** (p. 361)

9: Conflorescences simple or rarely few- and basally branched; unit conflorescences secund, or regular and then umbelloid, or 1–4-flowered; floral rachises ⩽ 50 mm long; flowers red, pink or white; all leaves entire, linear to elliptic or narrowly obovate

10 Styles projecting well beyond the curve of the perianth after release of style-end; unit conflorescences erect, umbelloid; all 4 tepals separating to level of ovary and spreading widely and independently **Linearifolia Group** (p. 196)

10: Styles scarcely projecting beyond the curve of the perianth after release of style-end; unit conflorescences *either* secund and decurved, many-flowered, with the tepals remaining loosely connate and held ventral to the style, *or* conflorescences 1–4-flowered, erect, with the tepals shortly recoiled in 2 opposing pairs held dorsal and ventral to the style **Aspera Group** (p. 264)

†This key for South Australia provides for identification of the Wickhamii group. *Grevillea wickhamii* subsp. *aprica* has not been confirmed as occurring in the State, but does occur very close to the S.A. borders in N.T. and W.A.

New South Wales

1 Ovary with hairs	
2 Leaves *either* entire and linear, 60–300 mm long, 1–2 mm wide, *or* 2–10-partite with similarly long and narrow lobes; leaf margin revolute, enclosing all or nearly all of lower surface (except for midvein)	
3 Conflorescences paniculately several-branched or occasionally simple; unit conflorescence dense, neatly cylindrical; ultimate floral rachises 50–100 mm long; flowers white to cream; pistils 10–22 mm long	**Pterosperma Group** (p. 352)
3: Conflorescences simple or basally few-branched; unit conflorescences loose clusters or shortly and untidily subcylindrical; ultimate floral rachises 4–12 mm long; flowers red to pink or orange; pistils 26–37 mm long	**Longistyla Group** (*G. johnsonii*) (p. 342)
2: Leaves entire, or toothed or divided; variously shaped, if entire leaves or ultimate lobes linear then < 60 mm long and/or > 3 mm wide and/or the lower surface exposed on either side of the midvein	
4 Stipe of ovary 10–13 mm long; ovary with scattered minute simple erect hairs only	**Shiressii Group** (p. 124)
4: Stipe of ovary < 5 mm long, or ovary sessile; ovary usually densely clad, with non-glandular hairs only, or rarely glabrous in upper half but then with a tuft of ascending hairs near base on the ventral side	
5 Inner surface of perianth glabrous	
6 Pistils 3–8 mm long; pollen-presenter ±erect and strongly conical; conflorescences ovoid or loosely conico-cylindrical; all leaves deeply divided, usually with secondary and sometimes tertiary division	**Pteridifolia Group** (*Ramosissima* subgroup) (p. 91)
6: Pistils 9–32 mm long; pollen-presenter oblique on the style, discoid or convex or weakly conical; conflorescences strongly secund, or rarely cylindrical to conico-cylindrical and then the pistils ⩾ 18 mm long; leaves variously divided, or entire	**Pteridifolia Group** (p. 39)
5: Inner surface of perianth with few to many hairs	
7 Style apex with a hairy, slender antrorse to retrorse dorsal appendage 1–4 mm long	**Buxifolia Group** (p. 327)
7: Style apex lacking appendage	
8 Unit conflorescences erect, regular-umbelloid, styles curved inwards or outwards from rachis axis; tepals remaining coherent (except along dorsal suture) after release of style-end, and everting widely to form a round platform held ventrally to style, with most of the hairy inner surface displayed; pistils 9–13 mm long	**Buxifolia Group** (p. 327)
8: Unit conflorescences erect to decurved or pendulous, 1- to few-flowered or an irregular cluster or shortly and broadly secund, or rarely subumbelloid but then usually decurved; tepals *either* separating along dorsal suture and then held loosely together ventral to the style, not everting widely to form a platform, *or* sometimes partially everting (but then conflorescences usually decurved to pendulous); pistils 9–32 mm long	**Floribunda Group** (p. 268)

1: Ovary glabrous

9 Stipe of ovary short (< 1.5 mm long), and swollen, almost as wide as ovary, and with a tuft of hairs on the ventral side — **Floribunda Group** (*Rosmarinifolia* subgroup) (p. 299)

9: Stipe of ovary 1–12 mm long, slender, glabrous

10 Unit conflorescences long-secund; floral rachises 120–160 mm long; styles yellow-orange; all leaves divided, deeply lobed; tree 8–40 m tall — **Robusta Group** (p. 148)

10: Unit conflorescences variously shaped (if floral rachises > 20 mm long then unit conflorescences regular-cylindrical, not secund); styles white to cream, or pink or red; leaves entire or variously divided; prostrate to erect shrubs, or trees

11 Stipe of ovary 4–13 mm long (including portion adnate to inner wall of the strongly oblique torus), *and* pistils 10–32 mm long and glabrous

12 Leaves divided; apices of lobes pungent — **Huegelii Group** (*G. huegelii*) (p. 144)

12: Leaves entire; apex pungent or not

13 Leaves linear, rigid, pungent, 10–40 mm long, 1–2 mm wide; flowers red or deep pink — **Acuaria Group** (*G. kennedyana*) (p. 196)

13: Leaves elliptic to lanceolate, 60–210 mm long, 9–30 mm wide; flowers with green to cream perianth and maroon style — **Shiressii Group** (p. 124)

11: Stipe of ovary ⩽ 5 mm long, free, on a transverse to slightly oblique torus; pistils 4–36 mm long, glabrous or with minute hairs (these sometimes inconspicuous and confined to the apical 1–3 mm or the back of the pollen-presenter

14 Conflorescences paniculately several- to many-branched, rarely simple; unit conflorescences long-cylindrical — **Hilliana Group** (p. 361)

14: Conflorescences simple, or basally 1–3-branched; unit conflorescences regular-umbelloid or shortly and broadly secund or a loose short cluster

15 Leaves deeply and divaricately divided; pistils 4–6 mm long, glabrous; style with an ovoid swelling; perianth actinomorphic — **Triloba Group** (*G. anethifolia*) (p. 438)

15: Leaves entire; pistils 6–36 mm long, with minute hairs at least near the apex (sometimes on back of pollen-presenter only); style slender except right at the apex; perianth zygomorphic — **Linearifolia Group** (*Speciosa* subgroup)(p. 205)

Victoria

1 Ovary glabrous

2 Pistils strongly refracted at ovary; leaves divided; torus strongly oblique; pistils entirely glabrous; (semi-arid habitats) **Huegelii Group** (p. 144)

2: Pistils not strongly refracted at ovary; leaves entire; torus transverse on the pedicel or only weakly oblique; pistils with tiny hairs or granular papillae near apex (sometimes on rear of pollen-presenter only) (mostly mesic habitats) **Linearifolia Group** (p. 196)

1: Ovary with hairs

3 Inner surface of perianth glabrous

4 Unit conflorescences usually secund (toothbrush-like, the flowers crowded to one side) or occasionally regular-ovoid (and then the pistils < 6 mm long); flowers acroscopic (ovary and concave arch of style pointing towards apex of conflorescence); fruits hairy with red to purplish markings, lacking viscid glandular hairs; leaves entire and variously shaped, or toothed to deeply divided with the lobes variously shaped **Pteridifolia Group** (p. 39)

4: Unit conflorescences regular, cylindrical; flowers transversely oriented (axis of flowers at right-angles to rachis); fruits hairy, lacking red to purplish markings, usually with viscid glandular hairs; leaves entire and long-linear, or with a few long-linear lobes **Pterosperma Group** (p. 352)

3: Inner surface of perianth with few to many hairs

5 Unit conflorescences regular, dense, cylindrical; flowers transversely oriented (axis of flowers at right-angles to rachis); fruits hairy, usually with viscid glandular hairs; leaves entire and long-linear, or with a few long-linear lobes; seed-body with a narrow flat papery wing projecting all around **Pterosperma Group** (p. 352)

5: Unit conflorescences regular to irregular or few-flowered, usually a loose (rarely dense) short cluster, rarely shortly sub-cylindrical; flowers usually acroscopic (ovary and concave arch of style facing apex of conflorescence), never transversely oriented; fruits hairy but never with viscid glandular hairs, or occasionally glabrous; leaves always entire, linear or variously shaped; seed-body with a very short waxy margin and tip **Floribunda Group** (p. 268)

MAPS

Number in brackets refers to the page on which the taxon is described.

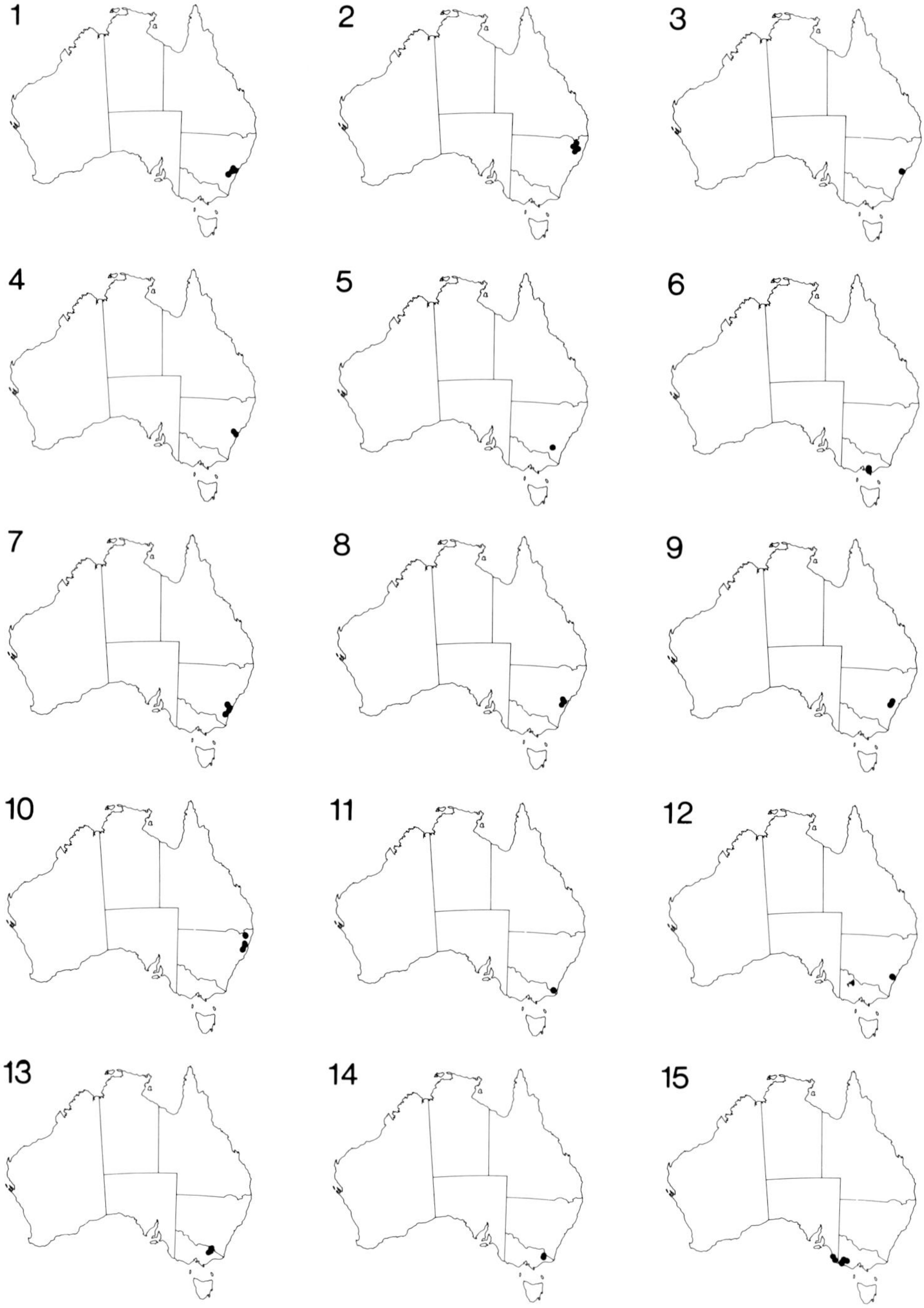

1. Grevillea aspleniifolia (50)
2. Grevillea beadleana (51)
3. Grevillea caleyi (53)
4. Grevillea longifolia (53)
5. Grevillea wilkinsonii (54)
6. Grevillea barklyana (54)
7. Grevillea macleayana (55)
8. Grevillea laurifolia (55)
9. Grevillea acanthifolia subsp. acanthifolia (57)
10. Grevillea acanthifolia subsp. stenomera (57)
11. Grevillea acanthifolia subsp. paludosa (57)
12. Grevillea rivularis (58)
13. Grevillea willisii (58)
14. Grevillea pachylostyla (59)
15. Grevillea aquifolium (60)

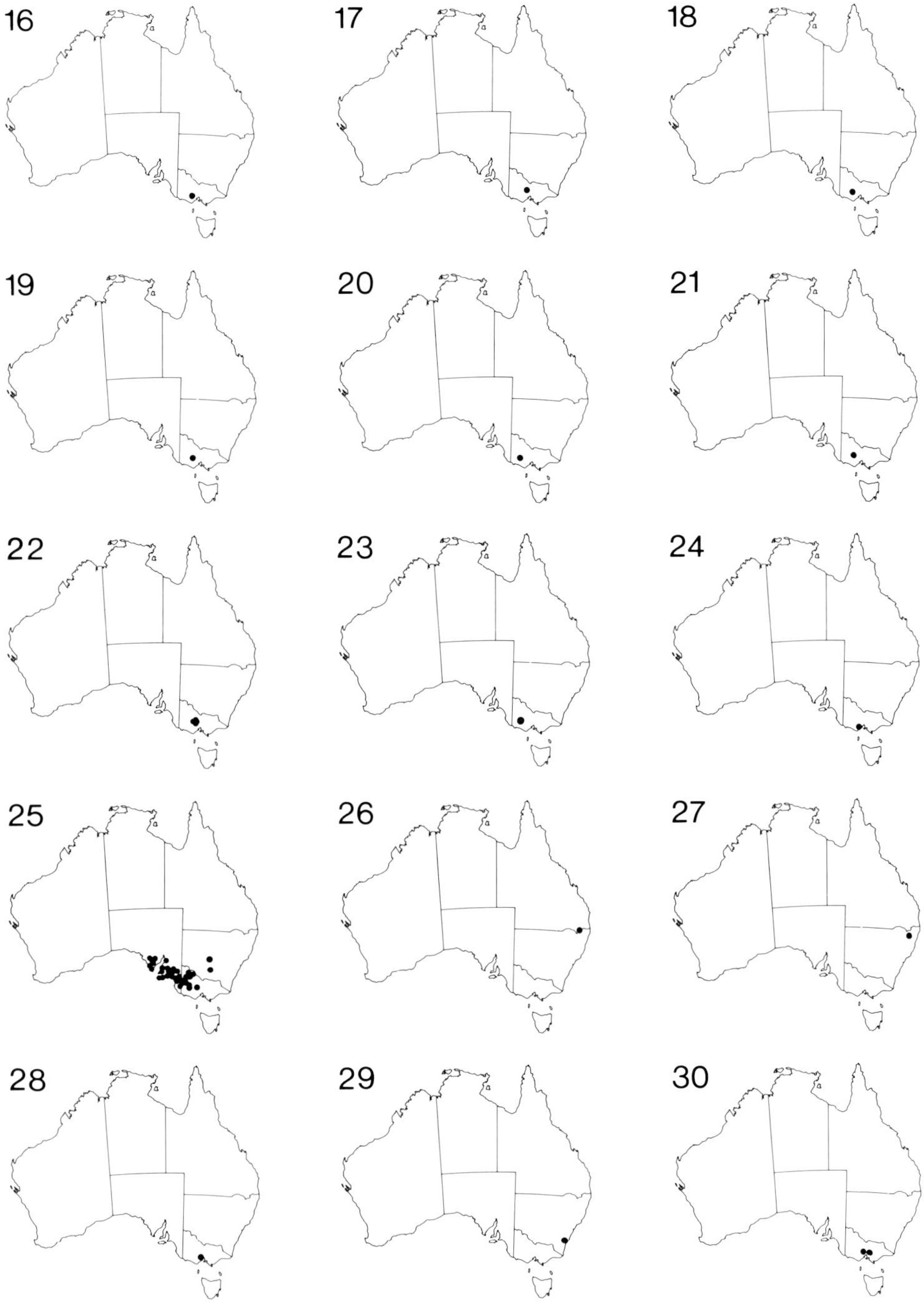

16. Grevillea bedggoodiana (60)
17. Grevillea obtecta (61)
18. Grevillea montis-cole subsp. montis-cole (62)
19. Grevillea montis-cole subsp. brevistyla (62)
20. Grevillea microstegia (63)
21. Grevillea floripendula (63)
22. Grevillea dryophylla (64)
23. Grevillea williamsonii (64)
24. Grevillea infecunda (65)
25. Grevillea ilicifolia (66)
26. Grevillea scortechinii subsp. scortechinii (68)
27. Grevillea scortechinii subsp. sarmentosa (68)
28. Grevillea steiglitziana (69)
29. Grevillea renwickiana (69)
30. Grevillea repens (70)

31

32

33

34

35

36

37

38

39

40

41

42

43

44

45

31. Grevillea cagiana (70)
32. Grevillea baxteri (71)
33. Grevillea tetragonoloba (72)
34. Grevillea rigida subsp. rigida (73)
35. Grevillea rigida subsp. distans (74)
36. Grevillea fastigiata (74)
37. Grevillea wittweri (74)
38. Grevillea coccinea subsp. coccinea (76)
39. Grevillea coccinea subsp. lanata (76)
40. Grevillea concinna subsp. concinna (77)
41. Grevil!ea concinna subsp. lemanniana (77)
42. Grevillea beardiana (77)
43. Grevillea armigera (78)
44. Grevillea hookeriana subsp. hookeriana (81)
45. Grevillea hookeriana subsp. apiciloba (82)

46

47

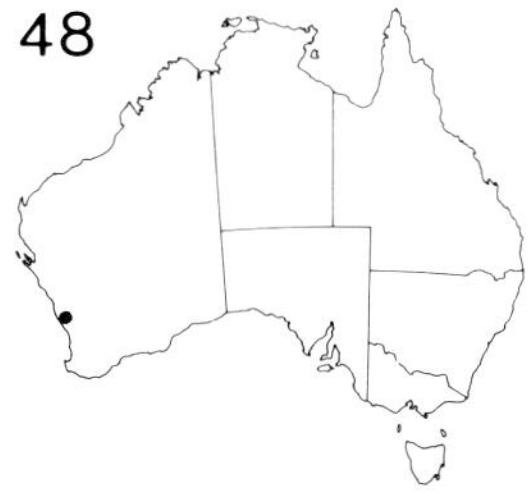
48

49

50

51

52

53

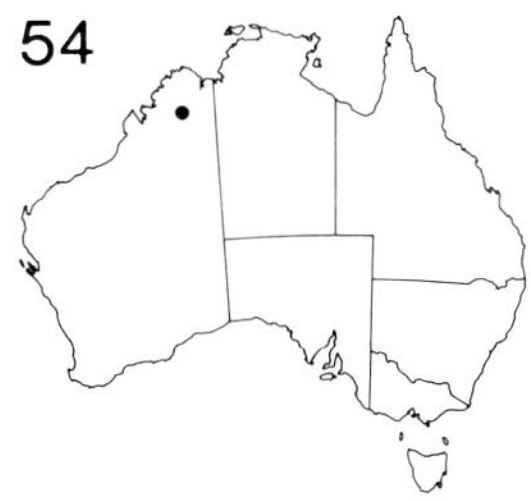
54

55

56

57

58

59

60

46. Grevillea hookeriana subsp. digitata (82)

47. Grevillea crowleyae (83)

48. Grevillea calliantha (83)

49. Grevillea treueriana (84)

50. Grevillea aneura (84)

51. Grevillea tenuiloba (85)

52. Grevillea nana subsp. nana (86)

53. Grevillea nana subsp. abbreviata (86)

54. Grevillea maherae (88)

55. Grevillea cravenii (88)

56. Grevillea dryandroides subsp. dryandroides (89)

57. Grevillea dryandroides subsp. hirsuta (90)

58. Grevillea thyrsoides subsp. thyrsoides (90)

59. Grevillea thyrsoides subsp. pustulata (91)

60. Grevillea ramosissima subsp. ramosissima (92)

91

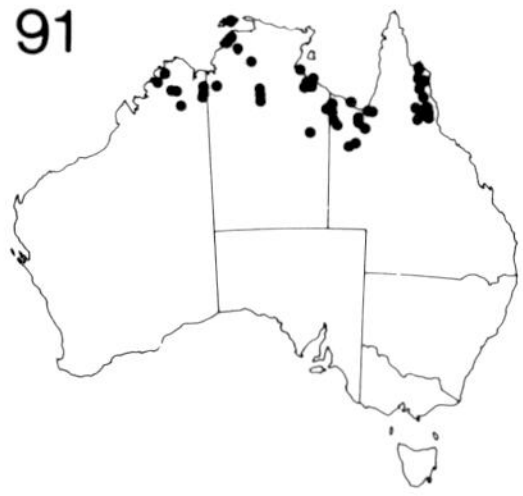

92

93

94

95

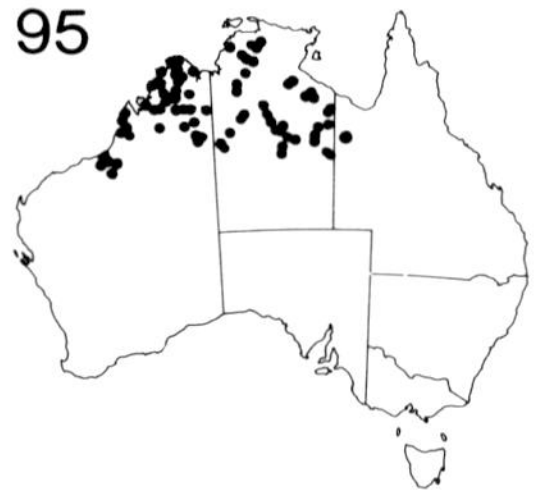

96

97

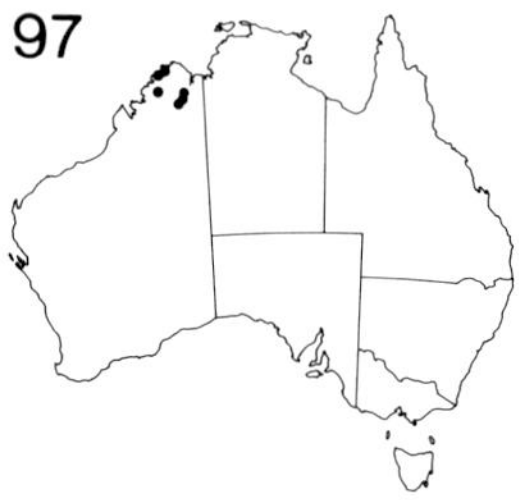

98

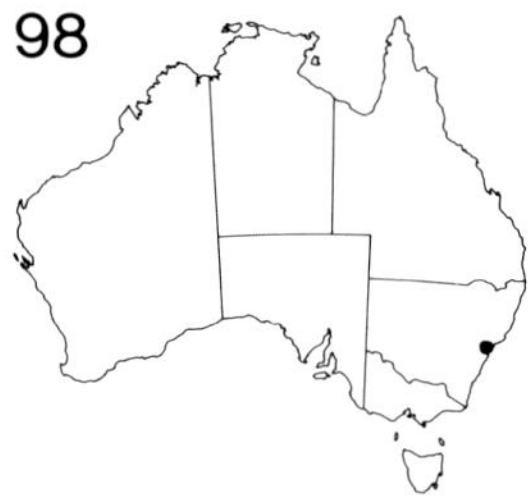

99

100

101

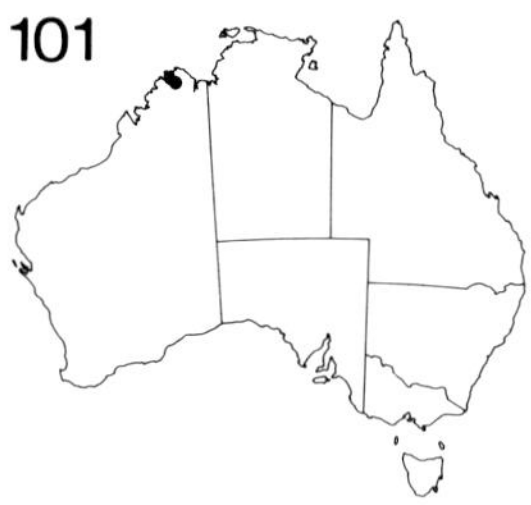

102

103

104

105

91. Grevillea dryandri subsp. dryandri (119)

92. Grevillea dryandri subsp. dasycarpa (119)

93. Grevillea pungens (121)

94. Grevillea versicolor (121)

95. Grevillea refracta subsp. refracta (122)

96. Grevillea refracta subsp. glandulifera (123)

97. Grevillea latifolia (123)

98. Grevillea shiressii (124)

99. Grevillea singuliflora (126)

100. Grevillea agrifolia subsp. agrifolia (128)

101. Grevillea agrifolia subsp. microcarpa (129)

102. Grevillea prasina (129)

103. Grevillea angulata (130)

104. Grevillea glabrescens (131)

105. Grevillea brevis (131)

106

107
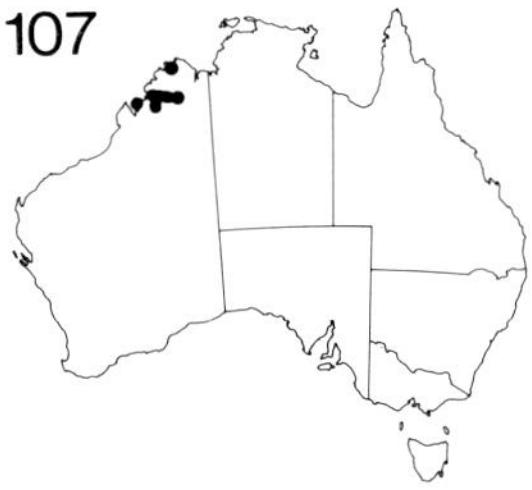
108
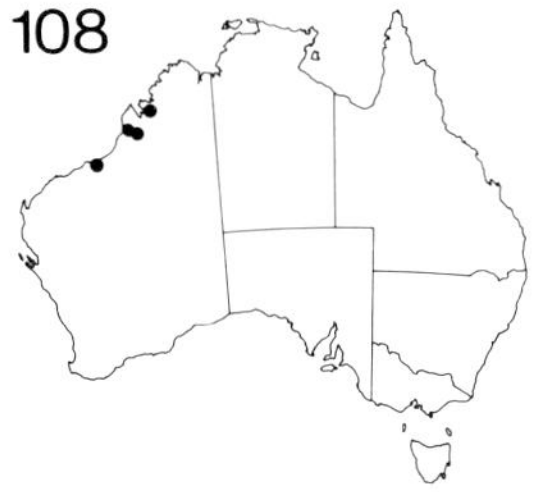
109
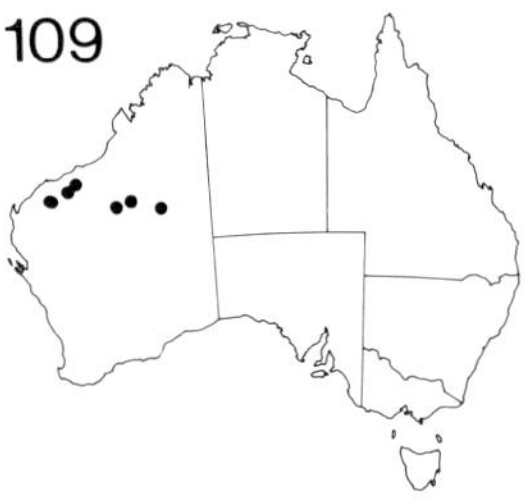
110
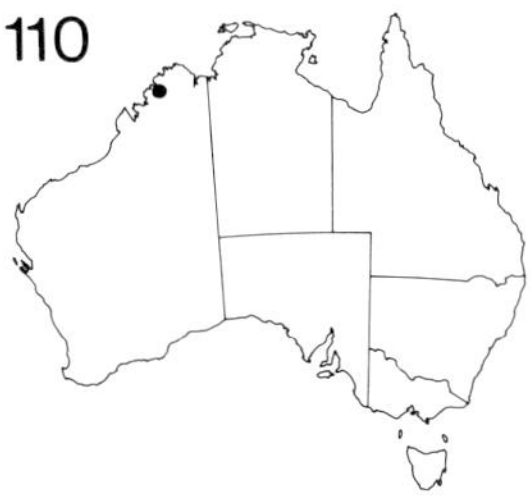
111
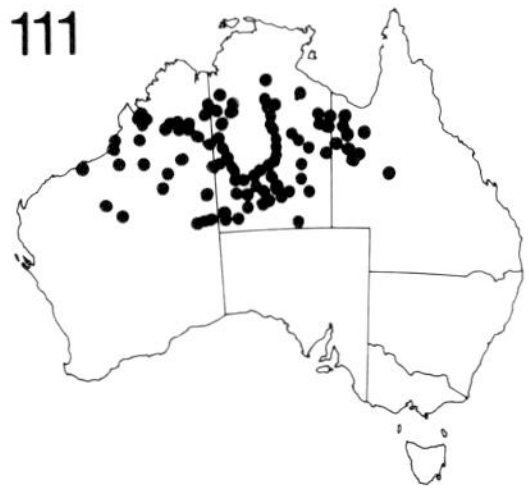
112
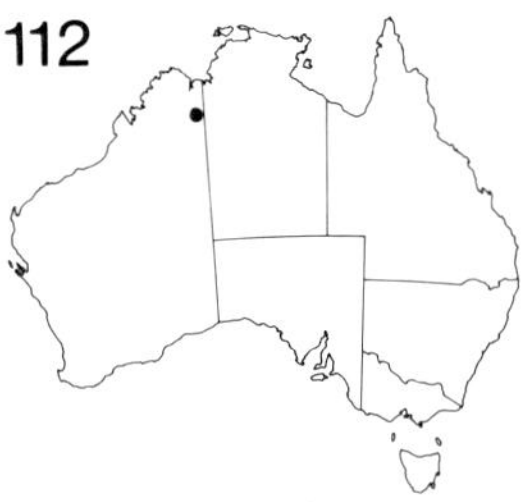
113

114

115

116

117

118

119

120

106. Grevillea aurea (132)

107. Grevillea wickhamii subsp. wickhamii (134)

108. Grevillea wickhamii subsp. macrodonta (135)

109. Grevillea wickhamii subsp. hispidula (135)

110. Grevillea wickhamii subsp. pallida (136)

111. Grevillea wickhamii subsp. aprica (136)

112. Grevillea wickhamii subsp. cratista (137)

113. Grevillea byrnesii (138)

114. Grevillea velutinella (138)

115. Grevillea miniata (139)

116. Grevillea adenotricha (140)

117. Grevillea longicuspis (142)

118. Grevillea microstyla (142)

119. Grevillea cunninghamii (143)

120. Grevillea benthamiana (143)

121

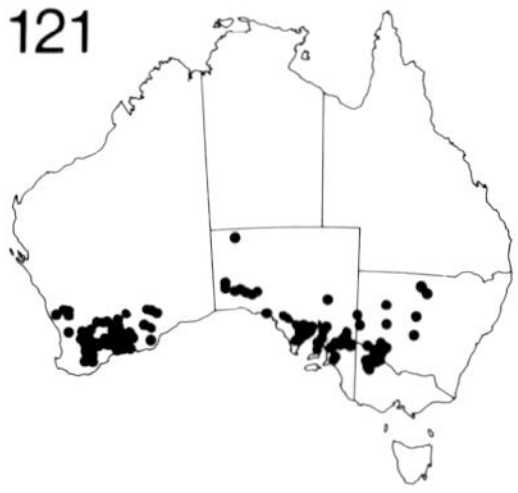

122

123

124

125

126

127

128

129

130

131

132

133

134

135

121. Grevillea huegelii (144)

122. Grevillea sarissa subsp. sarissa (146)

123. Grevillea sarissa subsp. umbellifera (147)

124. Grevillea sarissa subsp. succincta (147)

125. Grevillea sarissa subsp. anfractifolia (147)

126. Grevillea sarissa subsp. rectitepala (148)

127. Grevillea sarissa subsp. bicolor (148)

128. Grevillea robusta (149)

129. Grevillea thelemanniana (152)

130. Grevillea hirtella (153)

131. Grevillea fililoba (153)

132. Grevillea humifusa (154)

133. Grevillea delta (154)

134. Grevillea obtusifolia (155)

135. Grevillea exposita (155)

136

137

138

139

140

141

142

143

144

145

146

147

148

149

150

136. Grevillea evanescens (156)
137. Grevillea pinaster (156)
138. Grevillea preissii subsp. preissii (160)
139. Grevillea preissii subsp. glabrilimba (160)
140. Grevillea ripicola (161)
141. Grevillea acropogon (161)
142. Grevillea mccutcheonii (162)
143. Grevillea stenomera (162)
144. Grevillea variifolia subsp. varifolia (163)
145. Grevillea variifolia subsp. bundera (163)
146. Grevillea olivacea (164)
147. Grevillea oncogyne (166)
148. Grevillea tripartita subsp. tripartita (167)
149. Grevillea tripartita subsp. macrostylis (167)
150. Grevillea newbeyi (168)

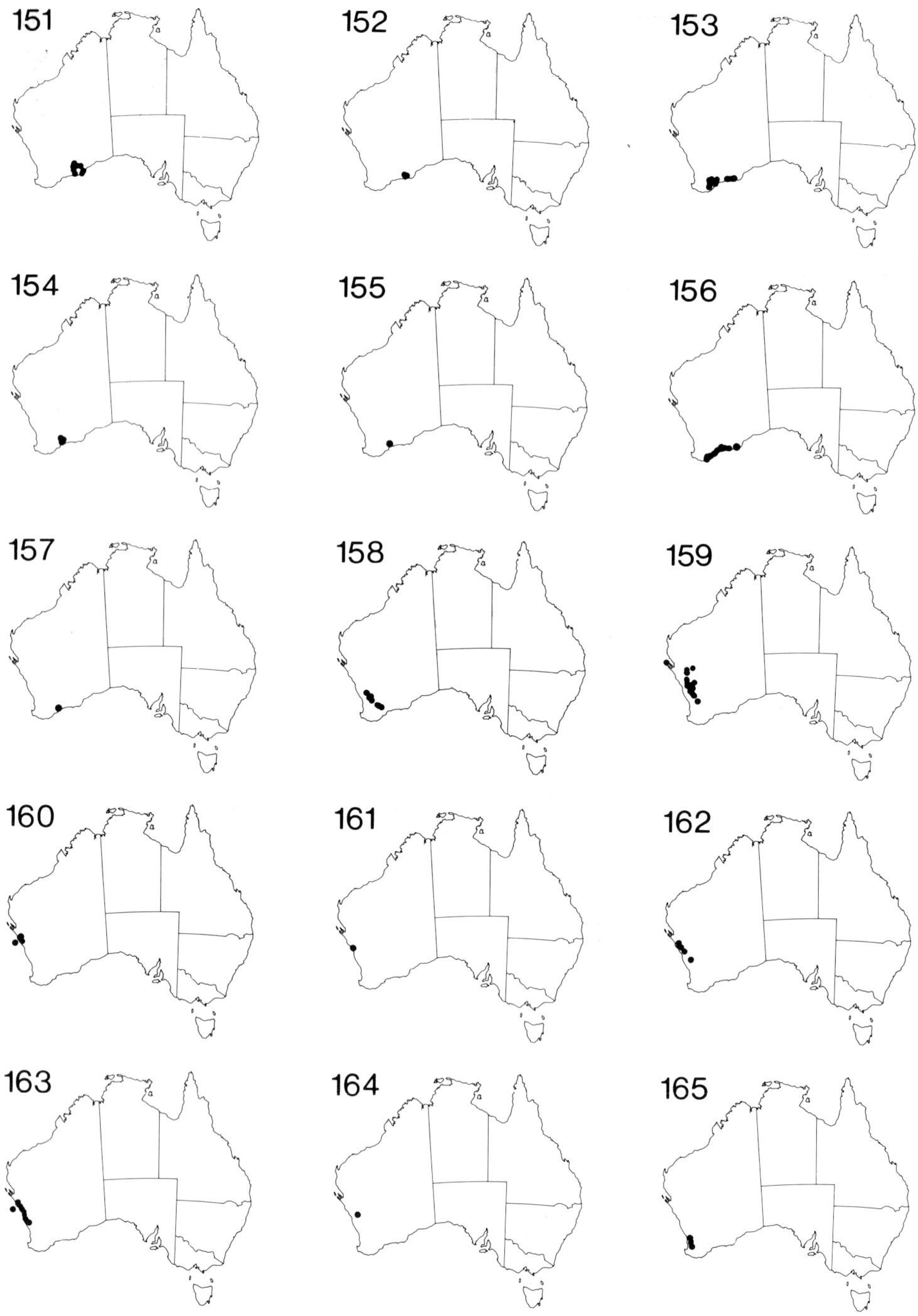

151. Grevillea plurijuga subsp. plurijuga (169)

152. Grevillea plurijuga subsp. superba (171)

153. Grevillea pectinata (171)

154. Grevillea patentiloba subsp. patentiloba (172)

155. Grevillea patentiloba subsp. platypoda (173)

156. Grevillea nudiflora (173)

157. Grevillea infundibularis (174)

158. Grevillea hakeoides subsp. hakeoides (177)

159. Grevillea hakeoides subsp. stenophylla (177)

160. Grevillea commutata subsp. commutata (178)

161. Grevillea commutata subsp. pinnatisecta (179)

162. Grevillea brachystachya (179)

163. Grevillea argyrophylla (181)

164. Grevillea murex (181)

165. Grevillea diversifolia subsp. diversifolia (182)

166

167
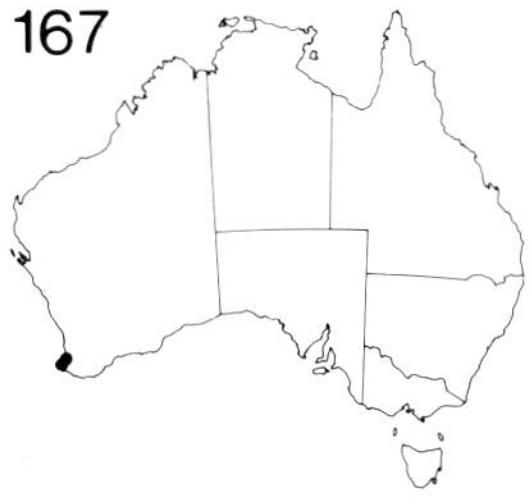
168
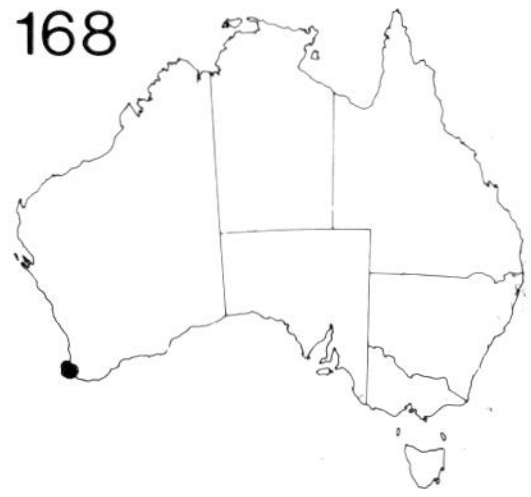
169

170
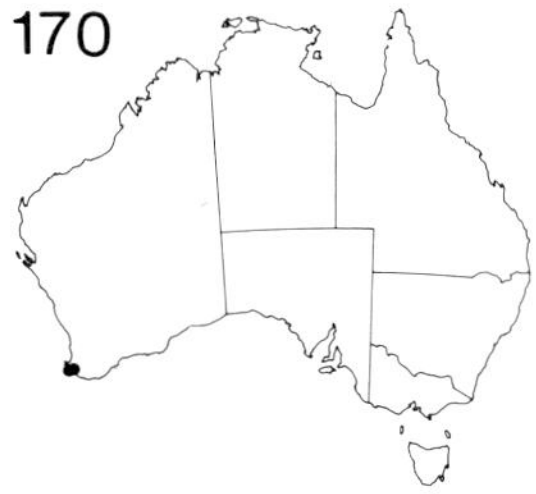
171

172

173

174

175

176
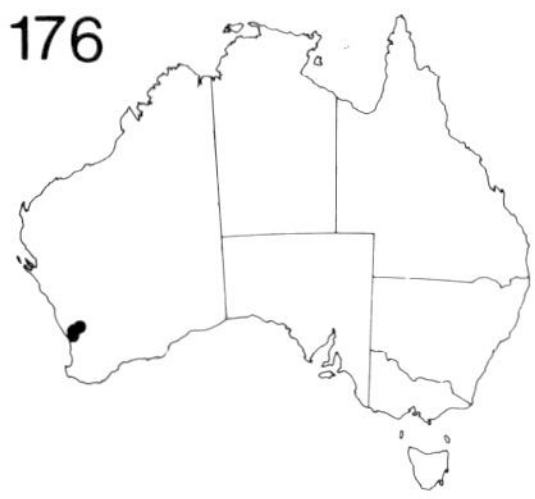
177
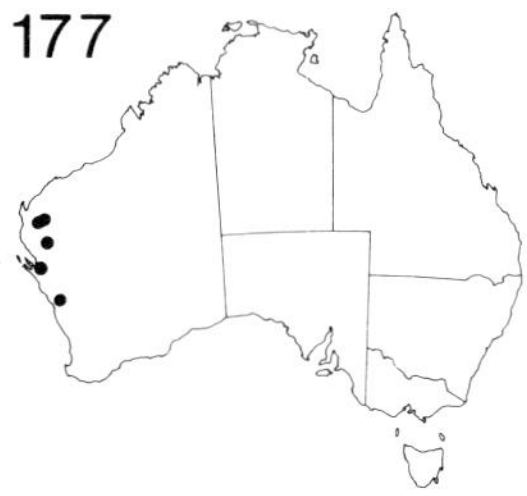
178

179

180

166. Grevillea diversifolia subsp. subtersericata (182)
167. Grevillea manglesioides subsp. manglesioides (183)
168. Grevillea manglesioides subsp. metaxa (184)
169. Grevillea manglesioides subsp. ferricola (184)
170. Grevillea papillosa (185)
171. Grevillea lullfitzii (187)
172. Grevillea subterlineata (187)
173. Grevillea scapigera (187)
174. Grevillea bracteosa (188)
175. Grevillea acacioides (189)
176. Grevillea endlicheriana (190)
177. Grevillea gordoniana (190)
178. Grevillea acuaria (191)
179. Grevillea punctata (192)
180. Grevillea sulcata (193)

211

212

213

214

215

216

217

218

219

220

221

222

223

224

225

211. Grevillea leiophylla (230)

212. Grevillea reptans (232)

213. Grevillea parviflora subsp. parviflora (233)

214. Grevillea parviflora subsp. supplicans (235)

215. Grevillea patulifolia (235)

216. Grevillea neurophylla subsp. neurophylla (237)

217. Grevillea neurophylla subsp. fluviatilis (238)

218. Grevillea alpivaga (238)

219. Grevillea micrantha (240)

220. Grevillea gariwerdensis (240)

221. Grevillea wiradjuri (241)

222. Grevillea imberbis (243)

223. Grevillea halmaturina subsp. halmaturina (244)

224. Grevillea halmaturina subsp. laevis (244)

225. Grevillea australis (245)

226

227

228

229

230

231

232

233

234

235

236

237

238

239

240

226. Grevillea quinquinervis (246)
227. Grevillea costata (247)
228. Grevillea inconspicua (247)
229. Grevillea christinae (248)
230. Grevillea victoriae subsp. victoriae (250)
231. Grevillea victoriae subsp. nivalis (250)
232. Grevillea brevifolia subsp. brevifolia (252)
233. Grevillea brevifolia subsp. polychroma (252)
234. Grevillea parvula (253)
235. Grevillea epicroca (254)
236. Grevillea monslacana (255)
237. Grevillea miqueliana subsp. miqueliana (256)
238. Grevillea miqueliana subsp. moroka (256)
239. Grevillea irrasa subsp. irrasa (257)
240. Grevillea irrasa subsp. didymochiton (258)

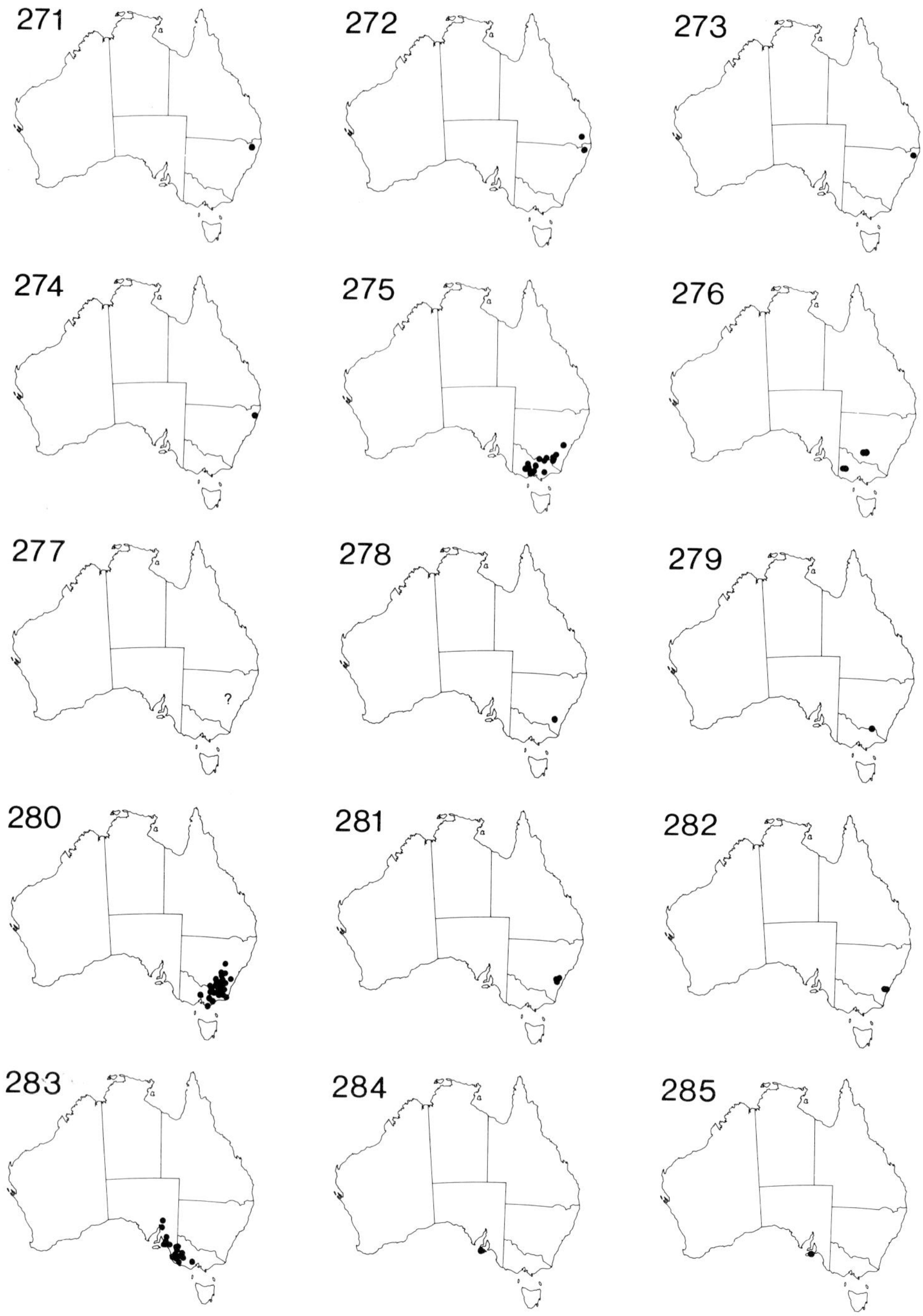

271. Grevillea rhizomatosa (296)
272. Grevillea quadricauda (297)
273. Grevillea banyabba (298)
274. Grevillea masonii (298)
275. Grevillea rosmarinifolia subsp. rosmarinifolia (300)
276. Grevillea rosmarinifolia subsp. glabella (301)
277. Grevillea divaricata (302)
278. Grevillea iaspicula (302)
279. Grevillea jephcottii (303)
280. Grevillea lanigera (303)
281. Grevillea baueri subsp. baueri (306)
282. Grevillea baueri subsp. asperula (307)
283. Grevillea lavandulacea subsp. lavandulacea (308)
284. Grevillea lavandulacea subsp. rogersii (309)
285. Grevillea muricata (310)

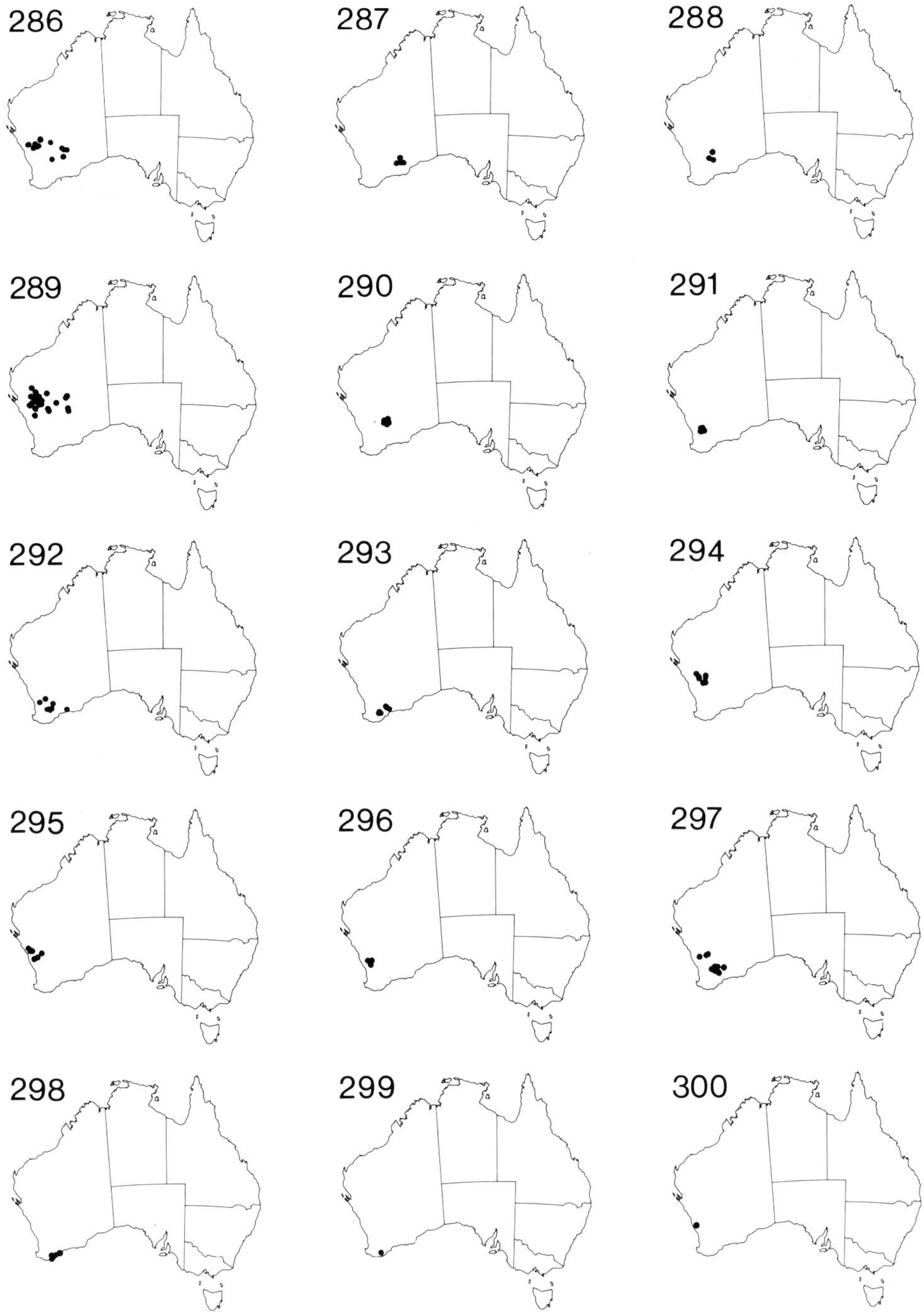

286. Grevillea extorris (310)
287. Grevillea phillipsiana (311)
288. Grevillea tetrapleura (311)
289. Grevillea deflexa (313)
290. Grevillea haplantha subsp. haplantha (314)
291. Grevillea haplantha subsp. recedens (314)
292. Grevillea disjuncta (314)
293. Grevillea dolichopoda (315)
294. Grevillea pityophylla (315)
295. Grevillea granulosa (316)
296. Grevillea rosieri (317)
297. Grevillea yorkrakinensis (317)
298. Grevillea fasciculata (318)
299. Grevillea crassifolia (319)
300. Grevillea saccata (321)

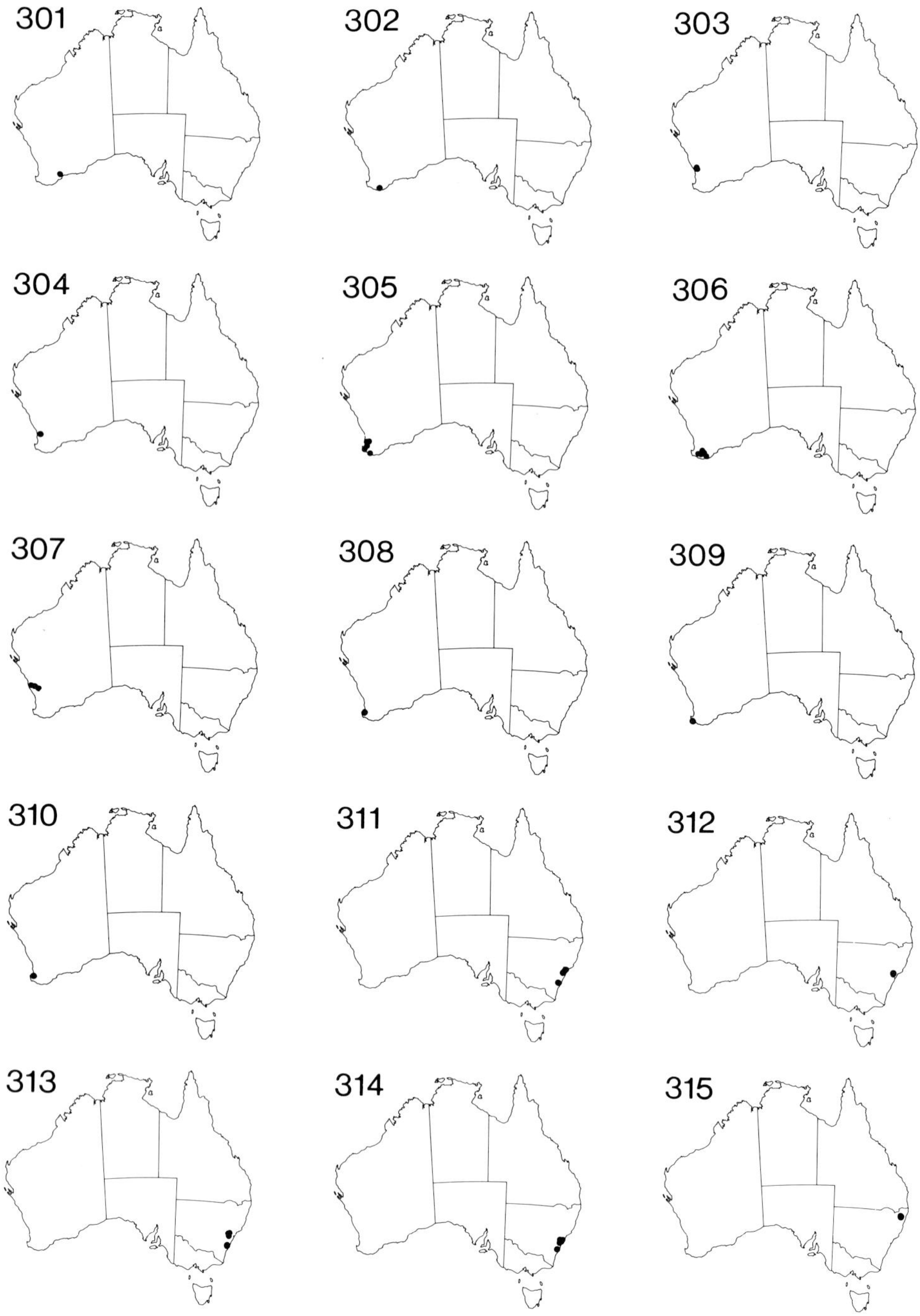

301. Grevillea fistulosa (321)
302. Grevillea fuscolutea (322)
303. Grevillea drummondii (322)
304. Grevillea pimeleoides (323)
305. Grevillea centristigma (324)
306. Grevillea depauperata (324)
307. Grevillea pinifolia (325)
308. Grevillea brachystylis subsp. brachystylis (326)
309. Grevillea brachystylis subsp. australis (326)
310. Grevillea bronwenae (327)
311. Grevillea buxifolia subsp. buxifolia (330)
312. Grevillea buxifolia subsp. ecorniculata (331)
313. Grevillea phylicoides (331)
314. Grevillea sphacelata (332)
315. Grevillea acerata (333)

316

317
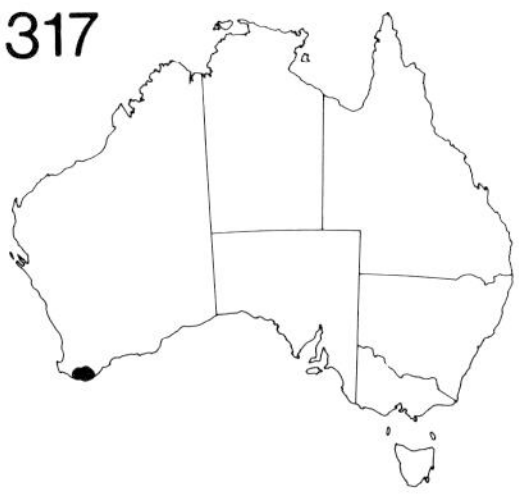
318

319

320

321

322

323

324

325

326

327

328

329

330

316. Grevillea umbellulata (333)
317. Grevillea occidentalis (334)
318. Grevillea pilulifera (335)
319. Grevillea uncinulata (336)
320. Grevillea florida (337)
321. Grevillea candolleana (338)
322. Grevillea scabra (338)
323. Grevillea longistyla (341)
324. Grevillea johnsonii (342)
325. Grevillea erectiloba (342)
326. Grevillea wilsonii (343)
327. Grevillea georgeana (343)
328. Grevillea dissecta (344)
329. Grevillea pilosa subsp. pilosa (345)
330. Grevillea pilosa subsp. redacta (345)

331

332

333

334

335

336

337

338

339

340

341
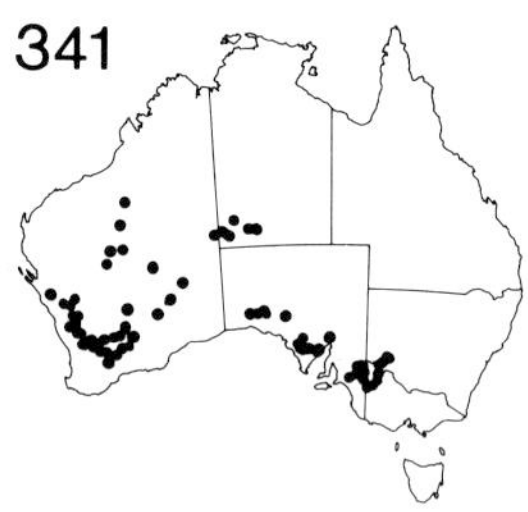
342

343

344

345

331. Grevillea asteriscosa (346)

332. Grevillea insignis subsp. insignis (347)

333. Grevillea insignis subsp. elliotii (347)

334. Grevillea fulgens (348)

335. Grevillea involucrata (348)

336. Grevillea marriottii (349)

337. Grevillea lissopleura (350)

338. Grevillea scabrida (350)

339. Grevillea cirsiifolia (351)

340. Grevillea pythara (352)

341. Grevillea pterosperma (353)

342. Grevillea albiflora (354)

343. Grevillea eriobotrya (354)

344. Grevillea globosa (355)

345. Grevillea petrophiloides subsp. petrophiloides (357)

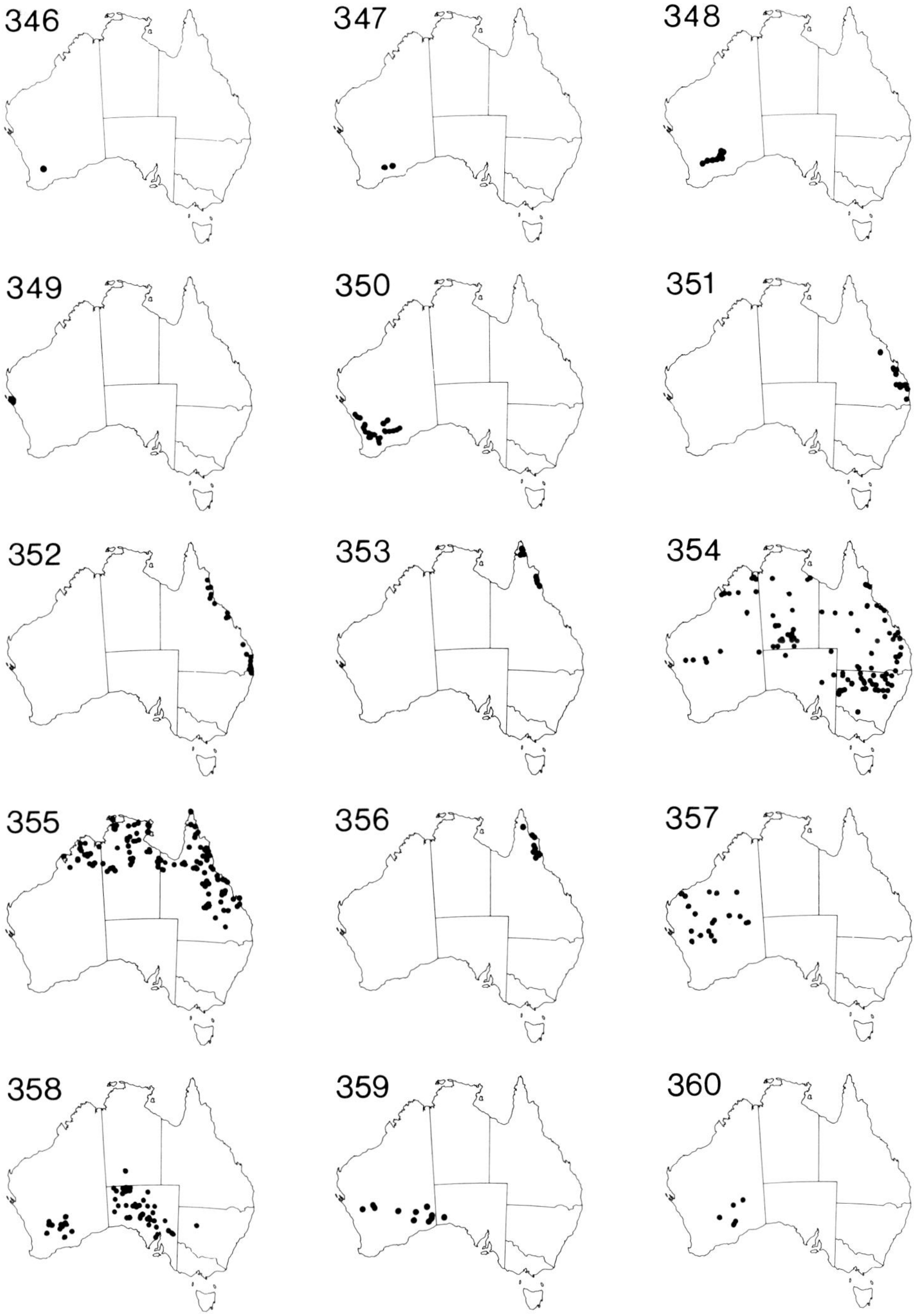

346. Grevillea petrophiloides subsp. magnifica (357)
347. Grevillea petrophiloides subsp. remota (358)
348. Grevillea oligomera (358)
349. Grevillea rogersoniana (360)
350. Grevillea paradoxa (360)
351. Grevillea helmsiae (364)
352. Grevillea hilliana (365)
353. Grevillea baileyana (365)
354. Grevillea striata (366)
355. Grevillea parallela (367)
356. Grevillea coriacea (367)
357. Grevillea berryana (368)
358. Grevillea nematophylla subsp. nematophylla (370)
359. Grevillea nematophylla subsp. supraplana (371)
360. Grevillea nematophylla subsp. planicosta (371)

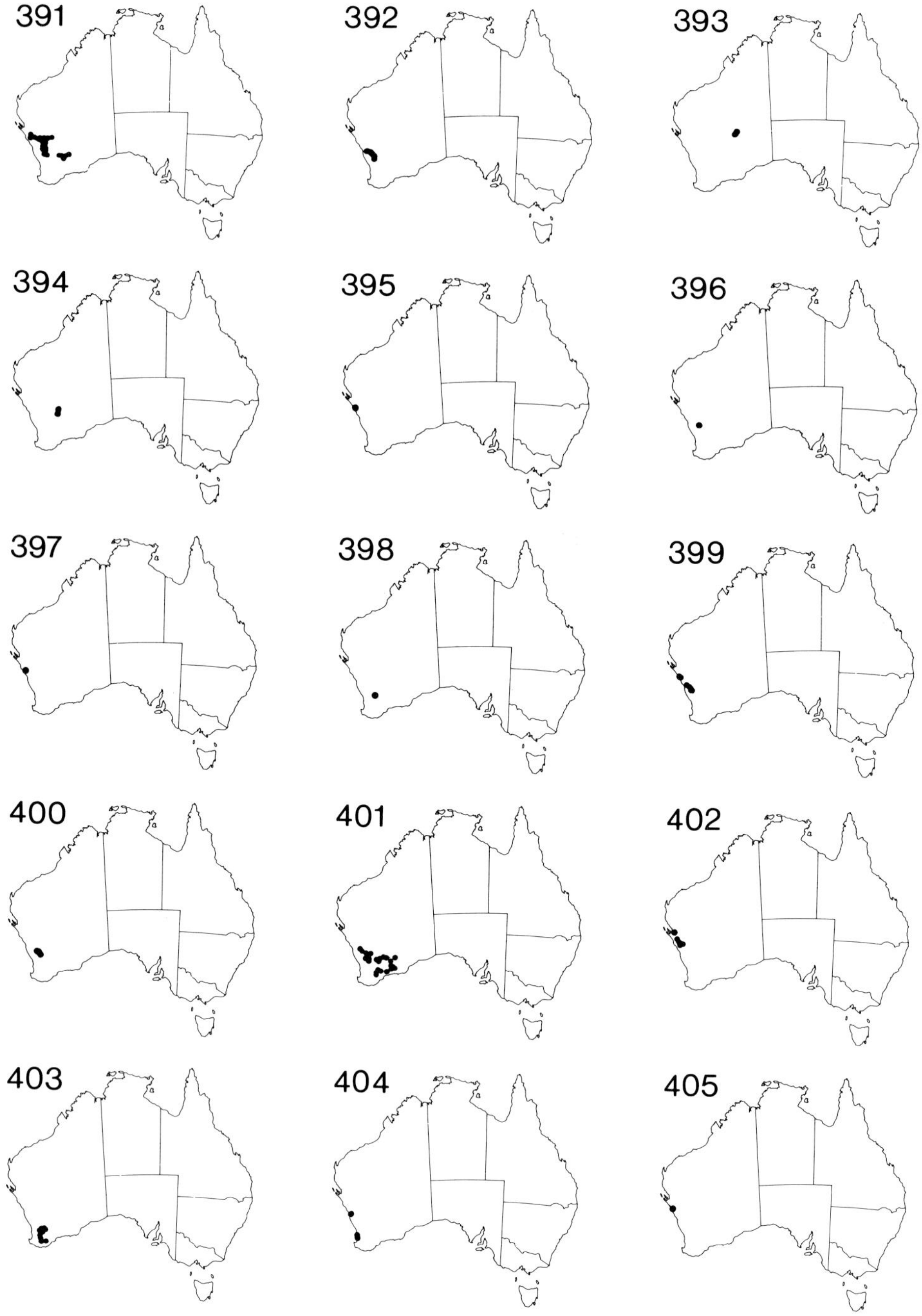

391. Grevillea obliquistigma subsp. obliquistigma (397)
392. Grevillea obliquistigma subsp. funicularis (398)
393. Grevillea obliquistigma subsp. cullenii (398)
394. Grevillea zygoloba (399)
395. Grevillea leucoclada (399)
396. Grevillea subtiliflora (400)
397. Grevillea intricata (400)
398. Grevillea minutiflora (401)
399. Grevillea leptopoda (401)
400. Grevillea kenneallyi (402)
401. Grevillea teretifolia (402)
402. Grevillea dielsiana (404)
403. Grevillea leptobotrys (404)
404. Grevillea crithmifolia (405)
405. Grevillea trachytheca (406)

406

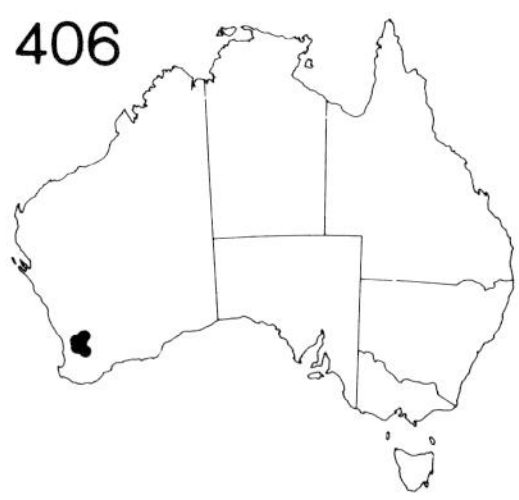

407

408

409

410

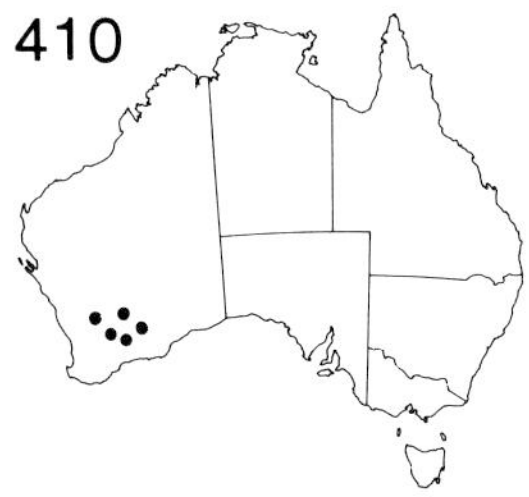

411

412

413

414

415

416

417

418

419

420

406. Grevillea integrifolia (407)

407. Grevillea biformis subsp. biformis (408)

408. Grevillea biformis subsp. cymbiformis (409)

409. Grevillea ceratocarpa (409)

410. Grevillea eremophila (410)

411. Grevillea incrassata (410)

412. Grevillea incurva (411)

413. Grevillea shuttleworthiana subsp. shuttleworthiana (412)

414. Grevillea shuttleworthiana subsp. canarina (413)

415. Grevillea shuttleworthiana subsp. obovata (413)

416. Grevillea psilantha (414)

417. Grevillea triloba (419)

418. Grevillea vestita subsp. vestita (420)

419. Grevillea vestita subsp. isopogoides (421)

420. Grevillea manglesii subsp. manglesii (423)

421

422

423

424

425

426

427

428

429

430

431

432

433

434

435

421. Grevillea manglesii subsp. ornithopoda (423)

422. Grevillea manglesii subsp. dissectifolia (424)

423. Grevillea phanerophlebia (424)

424. Grevillea amplexans subsp. amplexans (425)

425. Grevillea amplexans subsp. semivestita (426)

426. Grevillea amplexans subsp. adpressa (426)

427. Grevillea uniformis (426)

428. Grevillea acrobotrya (427)

429. Grevillea metamorpha (428)

430. Grevillea roycei (428)

431. Grevillea stenogyne (430)

432. Grevillea curviloba subsp. curviloba (431)

433. Grevillea curviloba subsp. incurva (431)

434. Grevillea rara (432)

435. Grevillea corrugata (432)

436

437

438

439

440

441
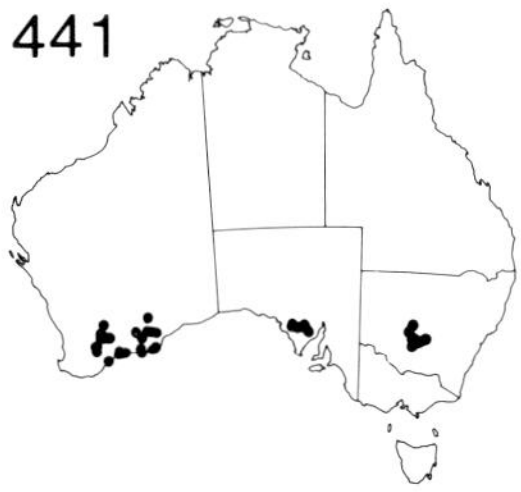
442

443

444

445

446

447

448

449

450

436. Grevillea paniculata (433)
437. Grevillea biternata (434)
438. Grevillea levis (435)
439. Grevillea xiphoidea (435)
440. Grevillea elongata (436)
441. Grevillea anethifolia (438)
442. Grevillea erinacea (438)
443. Grevillea spinosissima (440)
444. Grevillea rudis (441)
445. Grevillea althoferorum (441)
446. Grevillea stenostachya (442)
447. Grevillea tenuiflora (443)
448. Grevillea pulchella subsp. pulchella (444)
449. Grevillea pulchella subsp. ascendens (444)
450. Grevillea eryngioides (446)

451. Grevillea prostrata (446)

452. Grevillea quercifolia (448)

APPENDIX

New taxa, combinations and lectotypifications

New taxa, combinations and lectotypifications occurring in this volume of the *Flora of Australia* are formally published here. Taxa are arranged in the order they appear in the text. For economy the entries are brief; the treatment in the main text is more comprehensive. Accepted names are in **bold**, basionyms and synonyms in *italic*. The date of publication of this Volume will be given in Volume 11.

Subfam. GREVILLEOIDEAE

GREVILLEA

R.O.Makinson[1] (unless otherwise acknowledged)

Grevillea laurifolia Sieber ex Spreng., *Syst. Veg.* 4(2), *Cur. Post.* 46 (1827)

G. laurifolia Sieber ex Schult. & Schult.f., *Mant.* 3: 279 (1827), *nom. superfl.* T: 'Nov. Holl.' *s.d.*, [protologue]; lecto: Blue Mounts N. Holl. [N.S.W.], *F.W.Sieber, Fl. Novae Holl. No. 26*; lecto (here chosen): K; isolecto: A *n.v.*, B *n.v.*, G, HAL *n.v.*, K, LD *n.v.*, P.

Anadenia ilicifolia R.Br., *Trans. Linn. Soc London* 10: 167 (1810)

Grevillea ilicifolia var. *attenuata* R.Br., *Suppl. Prodr. Fl. Nov. Holl.* 21 (1830), as α *attenuata*, *nom. illeg.* (type var.). T: coast of southern Australia, Bay X [Port Lincoln, S.A.], 1802, *R.Brown Iter. Austral. 3314*; lecto (here chosen): BM; isolecto: E, G-DC, K, MEL, NSW.

Grevillea hookeriana subsp. **apiciloba** (F.Muell.) Makinson, *comb. nov.*

G. apiciloba F.Muell., *Fragm.* 10: 45 (1876); *G. apiciloba* subsp. *apiciloba* (F.Muell.) Olde & Marriott, *Grevillea Book* 1: 185 (1994); 2: 30 (1995). T: Ularing and Mt Jackson, W.A., 10–15 Oct. 1875, *J.Young*; holo: MEL.

Grevillea hookeriana subsp. **digitata** (F.Muell.) Makinson, *comb. nov.*

G. apiciloba var. *digitata* F.Muell., *Fragm.* 10: 46 (1876); *G. apiciloba* subsp. *digitata* (F.Muell.) Olde & Marriott, *Grevillea Book* 1: 185 (1994). T: near Mt Churchman, W.A., [1875?], [*J.*]*Young*; holo: MEL.

Grevillea maherae Makinson & M.D.Barrett, *sp. nov.*

Species aliquet formis *G. aquifolio* similis sed pistillo longiore (29–30 mm longo), foliis isobilateralibus, et pagina exteriore perianthii trichomatibus conspicuis purpureis glandulosis inter pilos 2-ramosos, differt.

T: c. 10 km west of new Mount Elizabeth Homestead, W.A., 13 Mar. 1998, *M.D.Barrett 340*; holo: PERTH05060729 (sheet annotated 'sheet 1 of 3'). [Other sheets of this collection number are from separate plants and are excluded from Type status.]

The epithet is given in recognition of the discovery of this species by Robyn Maher of Derby, W.A., who was assisted by her son Joshua Maher.

---

[1] Australian National Herbarium, Centre for Plant Biodiversity Research, CSIRO, GPO Box 1600, Canberra, Australian Capital Territory 2601

Grevillea cravenii Makinson, *sp. nov.*

Species *G. maherae* affinis et similis sed pistillo longiore (33–40 mm longo), conflorescentio non vel vix secundo, et fructibus axeis valde ferrugineis intra indumentum, differt.

T: c. 10 km E of Purulba massif, Prince Regent River Reserve, W.A., 2 Feb. 1999, *M.D.Barrett 706*; holo: PERTH; iso: CANB, DNA, K, KPBG, MEL.

The epithet is in recognition of Lyndley Alan (Lyn) Craven (1945–), of the Australian National Herbarium, whose initial collections brought this species to notice.

Grevillea dunlopii Makinson, *sp. nov.*

Affinis et similis *G. rubicundae*, sed pilis glandulosis in foliis, lobis foliis brevioribus et latioribus (2–5 cm longis, 2–4 mm latis), et indumento stylis sparsiore, differt.

T: Mt Gilruth area, 13°03'S, 133°01'E, N.T., 5 June 1978, *C.R.Dunlop 4901*; holo: DNA; iso: BRI, CANB, K, NSW.

Named in honour of Clyde R. Dunlop (1946–), curator of the Northern Territory Herbarium in Darwin, whose collections have brought to light many new species from the Kakadu Escarpment.

Grevillea decora subsp. **telfordii** Makinson, *subsp. nov.*

Affinis et similis *G. decorae* subsp. *decorae*, sed foliis obtusis in basim, indumento diluto (non ferrugineo) in ramulis et rhachidibus floralis, rhachidibus floralis brevioribus (1–2.5 cm longis), floribus parvioribus, pistilis 35–40 mm longis, et fructu parviore, c. 8 mm longo, differt.

T: c. 14 km from Laura on Cairns road, 15°40'S, 144°30'E, Qld, 11 Apr. 1975, *L.A.Craven 3235*; holo: CANB; iso: BRI, DNA, L *n.v.*, NSW, PERTH *n.v.*, RSA *n.v.*

The epithet is given in acknowledgment of the wide-ranging collecting and botanical activities of Ian Robert Hall Telford (1941–) now retired from the Australian National Botanic Gardens and Australian National Herbarium.

Grevillea agrifolia subsp. **microcarpa** (Olde & Marriott) Makinson, *comb. et stat. nov.*

G. microcarpa Olde & Marriott, *Telopea* 5: 415 (1993). T: Northern Kimberley: King Edward Crossing on road to Mitchell Plateau, W.A., July 1991, *P. & G.Keane s.n.*; holo: NSW; iso: CANB, DNA, PERTH.

Grevillea wickhamii subsp. **macrodonta** Makinson, *subsp. nov.*

Affinis et similis *G. wickhamii* subsp. *wickhamii*, sed nectario prominentiore et valde dentato latrorso (vix prominenti et plerumque integro ad undulatum in subsp. *wickhamii*), et inflorescentiis plerumque longioribus 30–60 mm longis (10–40 mm in subsp. *wickhamii*), differt; a subspeciebus ceteris (subsp. *hispidula*, subsp. *aprica*, subsp. *pallida*, subsp. *cratista*) rhachidi florali et pagina exteriore perianthii glabra, differt.

T: 65 km along Gibb River Rd from turnoff just E of Derby, c. 5 km W of Kimberley Downs Stn western boundary, W.A., 28 Apr. 1992, *R.O.Makinson 1088 et al.*; holo: CANB; iso: AD, BRI, DNA, K, NSW, PERTH.

The epithet, from the Greek *macros* (large or great) and *odont-* (relating to teeth), refers to the large lateral teeth of the prominent nectary.

Grevillea wickhamii subsp. **hispidula** Makinson, *subsp. nov.*

Similis *G. wickhamii* subsp. *wickhamii*, sed indumento aperto pilorum minutorum erectorum simplicium in rhachidi florali, in pedicellis, et in pagina exteriore perianthii, differt.

T: Mt Windell Rd corridor, 10.6 km ENE of Mt Windell, 20.4 km ESE of Karijini Natl Park H.Q., W.A., 29 July 1991, *S. van Leeuwen 862*; holo: PERTH; iso: CANB, Pilbara Regional Herbarium (Karratha, W.A.).

The epithet is a diminutive form of the Latin *hispidus* (covered with coarse erect rigid hairs or bristles), in reference to the minutely hispid indumentum of the flowers.

Grevillea wickhamii subsp. **pallida** Makinson, *subsp. nov.*

Similis *G. wickhamii* subsp. *wickhamii*, sed pilis appressis biramosis in rhachidi florali, in pedicillis et in pagina exteriore perianthii, et periantho luteolo differt; a subsp. *aprica* et subsp. *cratista* indumento styli minore denso et pilis biramosis destitutis, et praebitoribus pollinis minoribus oblongo-ellipticis et 1 mm diam. (circularibus ad obovatos et 1–2 mm diam. in subsp. *apricae* et subsp. *cratistae*), differt.

T: Prince Regent Nature Reserve, unnamed northern tributary of Prince Regent R., c. 800 m upstream of tidal limit, W Kimberley, W.A., 3 June 1998, *R.O.Makinson 1714*; holo: CANB; iso: AD, BRI, DNA, K, MEL, NSW, PERTH.

The epithet, from the Latin *pallidus* (pale), refers to the pale yellowish perianth, in contrast to the stronger colours of the other subspecies.

Grevillea wickhamii subsp. **cratista** Makinson, *subsp. nov.*

Affinis et similis *G. wickhamii* subsp. *apricae*, sed habitu robustiore, usque ad 8 m altum, caule principali uno, et pedicellis 6–9 mm longis (3–5 mm longis in subsp. *aprica*), et rhachidibus floralibus longioribus 30–100 mm longis (15–50 mm longis in subsp. *aprica*), differt.

T: 19.2 km SSW of Bungle Bungle outcamp, near Bellburn Ck, 8.2 km S of Tickalara Track en route to Piccaninny Creek, W.A., 7 July 1984, *S.J.Forbes 2557*; holo: PERTH; iso: CANB, MEL, NSW.

The epithet, from the Greek *kratistos* (strongest), refers to the robust nature of the subspecies.

Grevillea microstyla M.D.Barrett & Makinson, *sp. nov.*

Affinis et similis *G. longicuspi*, sed pilis biramosis in pedicillis, pilis glandulosis absentibus in partibus omnibus praeter stylum, pistillo breviore (5.5–7 mm longo), et conflorescentiis subglobosis densioribus, differt.

T: 4.6 km by road from Bachsten Ck (South Arm) campsite above falls, via side road to Wren Gorge ..., West Kimberley, W.A., 1 June 1998, *R.O.Makinson 1687 & H.Nicholson*; holo: CANB; iso: DNA, MEL, NSW, PERTH.

The epithet, from the Greek *micro-* (small or little) and *Stylos* (style or pillar), refers to the short style relative to some related species.

Grevillea sarissa var. *brevifolia* S.Moore, *J. Linn. Soc. Bot.* 34: 223 (1899)

T: '... The Elder Expedition specimens ... may be distinguished as var. brevifolia ...' [protologue]; lecto (here chosen): R.Helms, Elder Exploring Expedition, Victoria Desert Camp 53, 17 Sept. [18]91; lecto: K, sheet Kew Neg. No. 2365, piece second from right at top of sheet; isolecto: BM, K [as for lecto, piece third from right at top of sheet].

The name Grevillea *sarissa* var. *brevifolia* S.Moore is here regarded as a synonym of *G. sarissa* subsp. *sarissa*, sensu McGillivray (*New Names Grevillea* (1986), and in McGillivray & Makinson, *Grevillea* (1993)). McGillivray did not have the type material of the varietal name to hand after developing his infraspecific classification, and left the name unassigned to subspecies and unlectotypified. Material at K and BM has been seen for the present work. The K sheet also bears two pieces assignable to *G. sarissa* subsp. *anfractifolia* McGill.

Grevillea robusta A.Cunn. ex R.Br., *Suppl. Prodr. Fl. Nov. Holl.* 24 (1830)

Stylurus robusta (A.Cunn. ex R.Br.) O.Deg., *Fl. Hawaiiensis*: Family no. 98 (1932). T: Qld, 'Ora orient. Moreton Bay, 1827. *D.Cunningham.*' [protologue]; lecto: 1827, *A.Cunningham*; lecto (here chosen): K; isolecto: BM (1827, *A.Cunningham*), ?G-DC.

D.J.McGillivray (in McGillivray & Makinson, *Grevillea* 438 (1993)) neglected to lectotypify this taxon. While the syntypes at K and BM lack locality data, their status and suitability as lectotypes seem clear.

Grevillea acropogon Makinson, *sp. nov.*

Affinis *G. ripicolae*, sed pistillis brevioribus 20–22 mm longis, lobis ultimis foliorum brevioribus et angustioribus (10–15 mm longis, 0.8–1.1 mm latis), foliis pilosis in pagina inferiore, et segmentis perianthii barba sparsa ad apices, differt.

T: c. 6 km NNE of Lake Unicup, W.A., 7 July 1996, *E.Middleton ARA5755*; holo: PERTH.

The epithet, from the Greek *acros* (summit) and *pogon* (bearded), refers to the hairs near the apices of the tepals.

Grevillea tripartita subsp. **macrostylis** (F.Muell.) Makinson, *comb. et stat. nov.*

G. macrostylis F.Muell., *Fragm.* 1: 137 (1859). T: East Mt Barren, W.A., *s.d.*, *coll. unknown*; lecto: MEL, *fide* D.J.McGillivray & R.O.Makinson, *Grevillea* 445 (1993); isolecto: ?K; remaining syntypes: East Mt Barren, W.A., *s.d.*, *coll. unknown*; syn: ?K, MEL.

Grevillea plurijuga subsp. **superba** (Olde & Marriott) Makinson, *comb. et stat. nov.*

G. superba Olde & Marriott, *Nuytsia* 9: 298 (1993). T: Norwood Rd, E of Scaddan, W.A., 13 Oct. 1991, *P.M.Olde 91/332*; holo: NSW; iso: NSW, PERTH.

Grevillea commutata subsp. **pinnatisecta** (F.Muell.) Makinson, *stat. nov.*

G. commutata var. *pinnatisecta* F.Muell., *Fragm.* 6: 208 (1868); *G. pinnatisecta* (F.Muell.) Benth., *Fl. Austral.* 5: 473 (1870). T: between Moore and Murchison Rivers, W.A., *J.Drummond (6th coll.) 184*; lecto: K, *fide* D.J.McGillivray & R.O.Makinson, *Grevillea* 419 (1993); isolecto: BM, CGE *n.v.*, G *n.v.*, K, MEL, NY *n.v.*, PERTH.

This taxon was recognised as *G. hakeoides* subsp. *commutata*, 'divided-leaved form' by D.J.McGillivray & R.O.Makinson, *Grevillea* 258 (1993), and as *G. commutata* 'divided-leaved form' by P.M.Olde & N.Marriott, *The Grevillea Book* 2: 95 (1995). The diagnostic leaf form, and the geographical integrity of populations answering the circumscription, make subspecies recognition appropriate. Occasional specimens morphologically intermediate between the subspecies are known.

McGillivray & Makinson (*loc. cit.*) treated the base name as *G. pinnatisecta* F.Muell. ex Benth. Bentham (1870) attributes authorship at species rank to Mueller, perhaps without Mueller's agreement since he then refers to Mueller's (1868) view that the taxon is 'a variety only of *G. commutata*'. Both Bentham and subsequent writers have treated Mueller's discussion of a pinnatisect-leaved variety of *G. commutata* as purely descriptive, but it is better read as constituting a varietal protologue: 'Hujus speciei varietatem sistere videtur planta Drummondi 184... a planta typica me nunc descripta foliis pinnatisectis diversa. Hisce autem foliis interspersa sunt alia... Fructum *varietatis pinnatisectae* invenio vix semiuncialem...' [ital. added]. This does not affect lectotypification, but does dictate that the varietal combination be viewed as the basionym, and a different author citation be adopted for the combination at species rank, i.e. *Grevillea pinnatisecta* (F.Muell.) Benth.

Grevillea loboana Domin, *Věstn. Král. Ceské Společn. Nauk, Tř. Mat.-Přír* 1921–1922(2): 11 (1923)

T: Slab Hut Creek to Cranbrook, W.A., 1910, *A.A.Dorrien-Smith*; lecto (here chosen): K; isolecto PR.

Grevillea manglesioides subsp. **metaxa** Makinson, *subsp. nov.*

Affinis et similis *G. manglesioidi* subsp. *manglesioidi* sed pagina inferiore folii dense subsericea vel sericea, differt; et similis *G. manglesioidi* subsp. *ferricolae* sed foliis latioribus cuneatis (5–25 mm latis) dentibus vel lobis apicalibus 2–9, differt; et similis *G. diversifoliae* subsp. *subtersericata* sed inflorescentiis regularibus et foliis integris, differt.

T: Rapids Crossing, Margaret River, W.A., 6 Dec. 1974, *R.Pullen 9884*; holo: PERTH; iso: BRI, CANB, NSW.

The epithet, from the Greek *metaxa* (raw silk), refers to the subsericeous to sericeous indumentum of the leaf lower surface.

Grevillea manglesioides subsp. **ferricola** Keighery, *subsp. nov.*

T: Beenyup, Scott Plains, W.A., 5 Oct. 1997, *G.J.Keighery 15158*; holo: PERTH; iso: CANB, MEL, NSW.

Similis *G. manglesioidi* subsp. *manglesioidi* sed pilis in pagina inferiore folii appressis, differt; similis etiam *G. manglesioidi* subsp. *metataxae* sed foliis latioribus (5–25 mm latis), dentibus 9 in apicem et rhachidibus floralibus dense tomentosis, differt.

The epithet is derived from the Latin *ferreus* (iron) and *-cola* (dweller), in reference to the taxon's occurrence on ironstone substrates.

Grevillea manglesioides var. *sericea* Meisn., *in* J.G.C.Lehmann, *Pl. Preiss.* 2: 255 (1848), as β *sericea*

T: SW Australia, [W.A.], *s.d.*, *[J.] Drummond [2nd coll.] 318*; lecto (here chosen): K; isolecto: A *n.v.*, BM, CGE *n.v.*, G, K, LD *n.v.*, LE *n.v.*, MEL, P *n.v.*

Grevillea oleoides Sieber ex Schult. & Schult.f., *Mant.* 3: 277 (1827)

G. speciosa subsp. *oleoides* (Sieber ex Schult. & Schult.f.) McGill., *New Names Grevillea* 14 (1986). T: 'Sieber Herb. nov. Holl. Nro. 35' [protologue], Fl. Novae Holl. [N.S.W.], *s.d.*, *F.W.Sieber 35*; lecto (here chosen): NSW; isolecto: A *n.v.*, B *n.v.*, BM, G, G-DC, K *p.p.?*, LE *n.v.*, NY *n.v.*, P.

McGillivray (*New Names Grevillea* (1986)) and in McGillivray & Makinson (*Grevillea* (1993)) chose not to designate lectotypes for several type collections made by Sieber, apparently on the basis that no sheets had been seen with the exact form of words cited in the protologues. There is no doubt as to the status of the collection and components (at least those seen), and lectotypification is accordingly made here.

Grevillea juniperina subsp. **trinervis** (R.Br.) Makinson, *comb. et stat. nov.*

G. trinervis R.Br., *Suppl. Prodr. Fl. Nov. Holl.* 18 (1830). T: near Port Jackson, N.S.W., [?]*Fraser*; lecto: BM, *fide* D.J.McGillivray & R.O.Makinson, *Grevillea* 424 (1993): isolecto: BM; remaining syntypes: 'Country west from Bathurst in the vale at the source of Bell's River Nov. 1822 C. Fraser spcn from J. Smith'; syn: K; 'Bed of the Fish river near old fish R. bridge and in the vale at the source of the Bell's river N.W. from Batht [Bathurst] very rare Fraser'; syn: K.

Grevillea juniperina subsp. **allojohnsonii** Makinson, *subsp. nov.*

Similis *G. juniperinae* subsp. *trinervi*, sed plerumque prostrata, et foliis latioribus et longioribus, pari interiore venarum lateralium costam convenienti ad angulo acuto, differt.

T: c. 30 km (direct) NE of Guyra, Pheasant Mtn (near Backwater), N.S.W., 17 Nov. 1993, *R.O.Makinson 1449, I.R.Telford & J.Nightingale*; holo: CANB; iso: K, NSW.

The epithet is given in honour of Dr Lawrence Alexander Sidney Johnson (1925–1997), former Director (1972–1985) of the Royal Botanic Gardens, Sydney, in grateful acknowledgment of his encouragement of this author's work on *Grevillea*. Numerous discussions with him helped to clarify delimitation of this and the other subspecies in the complex.

The combination *G. johnsonii* McGill. already exists. There are also recorded usages, both horticultural nomina nuda, of the combinations *G. lawrenceana* P.N.Don (*Hortus Cantabrig.* edn 13: 65 (1845)), and *G. lawrenciana* Hort. ex Regel (*Cat. Pl. Hort. Asakov* 65 (1860)). These cases, while not invalidating use in a new epithet of Johnson's first given name or unmodified surname at subspecies rank, make it undesirable to do so.

In the interests of maintaining availability of the epithet in the event of future re-reranking to species level, the epithet given here is based on the surname plus Greek 'allo-' ('other'), with a respectful dig at Johnson's erection of the genus *Allocasuarina* (Casuarinaceae), viewed by some as controversial.

Grevillea juniperina subsp. **amphitricha** Makinson, *subsp. nov.*

Affinis *G. juniperinae* subsp. *trinervi* foliis convexis insuper, venis longitudinalibus vix prominentibus, foliis juvenalibus dense pilosis subter, pagina exteriore perianthii pilis appressis biramosis et pilis minutis erectis simplicibus, differt.

T: Southern Tablelands, Windellama Cemetery, N.S.W., 30 Nov. 1954, *C.W.E.Moore 3004*; holo: CANB.

The epithet, derived from the Greek prefix *amphi-* (both, double) and *trichion* (small hair), alludes to the presence of two types of trichome on the outer surface of the perianth.

Grevillea juniperina subsp. **fortis** Makinson, *subsp. nov.*

Frutex robustus 1–3 m altus, foliis 1 mm latis excedentibus, angulariter deltoideis ad trigonis in sectione transversali, venis 1–3 prominentalibus, subsericeis subter.

T: Pine Island, Lanyon district, A.C.T., 28 Oct. 1961, *R.D.Hoogland 8422*; holo: CANB; iso: A *n.v.*, AD, B *n.v.*, BH *n.v.*, BM, BO *n.v.*, BRI, CANB, E *n.v.*, FI *n.v.*, G *n.v.*, K, L *n.v.*, MEL, NE *n.v.*, NSW, P *n.v.*, UC *n.v.*, US *n.v.*

The epithet, from the Latin *fortis* (strong, powerful, vigorous), refers to the robust habit and vigorous growth of the individual plants.

Grevillea juniperina subsp. **villosa** Makinson, *subsp. nov.*

Affinis *G. juniperinae* subsp. *forti*, sed indumento villoso ad tomentosum in gemmis floralibus, et pilis ascendentibus in pagina inferiore folii, differt.

T: Southern Tablelands, banks of Corang River, at crossing on Braidwood to Nerriga road, c. 35 km (direct) NNE of Braidwood, N.S.W., 31 Aug. 1994, *R.O.Makinson 1533 et al.*; holo: CANB; iso: NSW.

The epithet is from the Latin *villosus*, alluding to the shaggy indumentum on the outer surface of the perianth, best seen in the bud stage and particularly on the limb of the bud. It should be noted that what may appear to be single ascending hairs are in fact single arms of biramous hairs.

Grevillea juniperina subsp. **sulphurea** (A.Cunn.) Makinson, *stat. nov.*

G. sulphurea A.Cunn., *in* B.Field, *Geogr. Mem. New South Wales* 329 (1825); *G. juniperina* var. *sulphurea* (A.Cunn.) Benth., *Fl. Austral.* 5: 469 (1870); *G. juniperina* f. *sulphurea* (A.Cunn.) I.K.Ferguson, *Bot. Mag.* 182: t. 761 (1978). T: Cox R., N.S.W., 1822, *A.Cunningham 46/1822*; holo: K; iso: A *n.v.*, K.

Grevillea acifolia Sieber ex Spreng., *Syst. Veg.* 4(2), *Cur. Post.* 46 (1827)

T: *'Nov. Holl.' Sieber* [protologue], *Sieber Fl. Nov. Holl. No. 28*; lecto (here chosen): MEL; isolecto: A *n.v.*, B *n.v.*, C *n.v.*, G *n.v.*, G-DC *n.v.*, K, LE *n.v.*, NY *n.v.*, P *n.v.*

D.J.McGillivray (*New Names Grevillea* (1986), and in McGillivray & Makinson, *Grevillea* (1993)) chose not to designate lectotypes for several type collections made by Sieber, apparently on the basis that no sheets had been seen with the exact form of words cited in the protologues. There is no doubt as to the status of the collection and components (at least those seen), and lectotypification is accordingly made here.

Grevillea acicularis Schult. & Schult.f., *Mant.* 3: 278 (1827)

T: *Sieber Fl. Nov. Holl. No. 28*; lecto (here chosen): MEL; isolecto: A *n.v.*, B *n.v.*, C *n.v.*, G *n.v.*, G-DC *n.v.*, K, LE *n.v.*, NY *n.v.*, P *n.v.*

D.J.McGillivray (*New Names Grevillea* (1986), and in McGillivray & Makinson, *Grevillea* (1993)) chose not to designate lectotypes for several type collections made by Sieber, apparently on the basis that no sheets had been seen with the exact form of words cited in the protologues. There is no doubt as to the status of the collection and components (at least those seen), and lectotypification is accordingly made here.

Grevillea diffusa subsp. **constablei** Makinson, *subsp. nov.*

Affinis *G. diffusa* subsp *diffusa*, sed frutice altiore, ramulis angularibus sericeis vel glabris, foliis longioribus (4.5–7 (–10) cm longis), et pedunculis longioribus sericeis, differt.

T: c. 1.5 miles [2.4 km] from waterfall on E side of line, N.S.W., 2 July 1950, *T.M.Whaite 722*; holo: NSW.

Named for Ernest F. Constable (fl. 1950–1970), former plant and seed collector for the National Herbarium of New South Wales and Royal Botanic Gardens, Sydney.

Grevillea evansiana MacKee, *Proc. Linn. Soc. New South Wales* 78: 49 (1953)

G. diffusa subsp. *evansiana* (MacKee) McGill., *New Names Grevillea* 4 (1986). T: Currant Mtn Gap, 5 ml [8 km] E of Olinda, N.S.W., 2 Sept. 1951, *L.A.S.Johnson & H.S.McKee NSW22651*; lecto (here chosen): NSW; isolecto: BM, K; ?isolecto: BRI, MEL, SYD, US *n.v.*

Grevillea linearis var. *incarnata* Sims, *Bot. Mag.* 53: t. 2661 (1826)

T: 'cult. by C.Loddiges, Mar. 1824, propagated from collection made in Australia by J.Banks' [protologue]. lecto (here chosen): *op. cit.*, t. 2661.

No specimen has been found answering to protologue, plate, and provenance given, making lectotypification on the plate desirable. Assignation of this name to species is not entirely certain, as the plate depicts shorter leaves than is usual in *G. linearifolia*, and pink flowers (as shown) are rare in that species; in these respects the plate could equally represent *G. humilis* subsp. *humilis*. Nevertheless, assignation of the name to *G. linearifolia* is provisionally made, given Sims' statement that the cultivated material was propagated from a Banks (1770) collection (Banks having collected very close to the known historic range of *G. linearifolia* and not at all near that of *G. humilis*). Numerous early collections of *G. linearifolia* from the Sydney area have shorter leaves than most extant populations, although most of these collections are from the Parramatta R. and Port Jackson, just N of Banks' collecting area. The possibility cannot be entirely excluded that the plate represents a stock of a later, non-Banks provenance, either of *G. linearifolia* or of *G. humilis* subsp. *humilis*. A hybrid of *G. linearifolia* with G. *sericea* is also possible, although there are no positive indications in the protologue of this, but such hybrids do occur and sometimes answer the features of the plate. Assignation of *G. linearis* var. *incarnata* to *G. linearifolia* is therefore made tentatively.

Grevillea humilis Makinson, *sp. nov.*

Affinis et similis *G. linearifoliae* sed frutice rhizomatoso et humiliore, plerumque pedicellis pistilis et fructibus parvulis, differt; similis *G. viridiflavae* sed stylo roseo vel albo, non citrino, differt.

T: 5 km N of Bucketts ['Buckets'] Way along Scotts trail, Wallaroo State Forest, N.S.W., 10 Aug. 1985, *P.G.Abell 57*; holo: NSW.

The epithet, from the Latin (humilis), refers to the usually low-growing habit of the species.

Grevillea humilis subsp. **lucens** Makinson, *subsp. nov.*

Similis *G. humili* subsp. *humili*, sed pistillo longiore (9–16 mm longo), et pilis ramulinorum et paginis inferioribus foliorum lucentibus, differt; similis *G. leiophyllae* sed paginis inferioribus foliorum dense pilosis, differt.

T: Glass House Mountains State Forest, foot of main peak of Mt Tunbubudla, south side, Qld, 1 July 1997, *R.O.Makinson 1618*; holo: CANB; iso: BRI, K, NSW.

The epithet, from the Latin *lucens* (glistening), refers to the highly reflective hairs of the leaves and stems.

Grevillea humilis subsp. **maritima** Makinson, *subsp. nov.*

Affinis *G. humili*, sed foliis saepe latioribus (usque ad 10 mm lata) et obovatis et saepe 3-fasciculatis, pedicellis longioribus (5–9 mm longis), differt.

T: Yuragir Natl Park, c. 8 km (direct) S of Yamba, Shelley Beach walking track c. 1.5 km S of Mara Ck picnic area, N.S.W., 29 June 1997, *R.O.Makinson 1614*; holo: CANB; iso: BRI, MEL, NE, NSW.

The epithet, from the Latin *maritimus* (growing by the sea), refers to the habitat.

Grevillea viridiflava Makinson, *sp. nov.*

Affinis et similis *G. humili*, sed perianthio albo et stylo viridiflavo, differt; similis *G. linearifoliae* sed rhizomate habitu inferiore, pistillo et fructu plerumque parviore, differt.

T: 12.3 km N of Torrington on road to Silent Grove, N.S.W., 21 Oct. 1988, *R.O.Makinson 576 & S.Krauss*; holo: CANB; iso: AD, BRI, K, MEL, NE, NSW.

The epithet is from the Latin *viridiflavus* (greenish yellow), in reference to the colour of the style.

Grevillea virgata Makinson, *sp. nov.*

Affinis et similis *G. humili*, sed habitu altiore erecto usque ad 2 m altum, foliis remotioribus. foliis multis patentibus et 3-fasciculatis, et ramis rubellis, differt.

T: 5.3 km S of Bulahdelah on Pacific Hwy, N.S.W., 4 July 1997, *R.O.Makinson 1627*; holo: CANB; iso: BRI, K, NSW.

The epithet, from the Latin *virgatus* (twiggy or long and slender), refers to the gracile habit.

Grevillea leiophylla F.Muell. ex Benth., *Fl. Austral.* 5: 471 (1870)

T: 'Glasshouse ranges, Moreton Bay, F.Mueller, and probably from the same neighbourhood, Leichhardt' [protologue]; lecto (here chosen): Glass houses Moreton Bay [Glass House Mtns, Qld], *s.d.*, *F.Mueller*; lecto: K; remaining syntypes: Moreton Bay [Qld], *s.d.*, *Hill & Mueller*; syn: K; round Tibburrocan [Qld], *s.d.*, *Leichhardt* [per] *Mueller*; syn: MEL 65617, ?65618, 65619, 65620.

Grevillea reptans Makinson, *sp. nov.*

Affinis *G. leptophyllae s. str.* sed saepe habitu scandenti vel prostrato, ramis saepe longis et repentibus, foliis longioribus (6–17 cm longis), perianthio tomentoso ad villosum externo, et pistillis brevibus (7–9 mm longis), differt.

T: Burrum Heads road, 18 miles [c. 29 km] NNE of Howard, Qld, Sept. 1966, *C.H.Gittins 1215*; holo: BRI; iso: K.

The epithet, from the Latin *reptans* (creeping — without the sometimes associated notion of rooting along the stems), alludes to the often scandent to prostrate habit of the long, slender wandering branches.

Grevillea parviflora subsp. **supplicans** Makinson, *subsp. nov.*

Affinis et similis *G. parviflorae* subsp. *parviflorae* sed frutice compactiore, floribus stipite ovarii brevioribus (c. 0.5 mm longis), ramulis secundis foliis erectis, differt.

T: Berrilee, c. 32 km (direct) NNW of Sydney GPO, 1.9 km by road from Berowra Waters Ferry (W bank terminus) towards Glenorie, N.S.W., 24 Sept. 1993, *R.O.Makinson 1294*; holo: NSW; iso: AD, BRI, CANB, G, HO, K, MEL, MO, NE, PERTH.

The epithet, from the Latin *supplicans* (beseeching, entreating, worshipping), refers to the leaves held skywards as though in supplication.

Grevillea neurophylla subsp. **fluviatilis** Makinson, *subsp. nov.*

Similis *G. neurophyllae* subsp. *neurophyllae* sed foliis congestioribus rectioribus longioribus (4–8 cm longis), et fructibus leviter minoribus (6.5–8 mm longis), differt.

T: upper Genoa R., Vic., *A.C.Beauglehole 35004 & K.C.Rogers*; holo: MEL.

The epithet, from the Latin *fluviatilis* (pertaining to rivers), refers to the riparian, rheophytic habitat of this taxon.

Grevillea gariwerdensis Makinson, *sp. nov.*

Affinis et similis *G. micranthae*, sed barba sparsa in pagina interiore perianthii ovarium oppositum, plerumque pistillis longioribus (3–7.5 mm longis), et venis folii vix granularibus, differt.

T: 2 miles [3.2 km] NE of Halls Gap on Pomonal Rd, Vic., 14 Nov. 1966, *A.C.Beauglehole 22204*; holo: MEL; iso: NSW.

The epithet is given as a noun in apposition, being the Aboriginal name for the Grampians Range, now applied to the Gariwerd National Park within which the species mostly occurs.

Grevillea wiradjuri Makinson, *sp. nov.*

Similis *G. neurophyllae*, sed foliis longioribus et angustioribus, plerumque 3–7.5 cm longis et 0.7–1.3 mm latis, pedicellis brevioribus (c. 3.5 mm longis), pistillis parum longioribus (7–9 mm longis), et margine folii angulariter revoluto ad 90°, differt.

T: 19 km W of Temora [towards] Griffith, N.S.W., 13 Nov. 1975, *M.D.Crisp 1529*; holo: CANB; iso: A *n.v.*, AD, NSW.

The epithet is given as a noun in apposition, relating to the occurrence of the species in the country historically occupied by the Wiradjuri aboriginal people.

Grevillea imberbis Makinson, *sp. nov.*

Affinis *G. linearifoliae* et *G. patulifolia*, sed foliis brevioribus ellipticis ad obovata 1–4 cm longis, 2–6.5 mm latis, et barba densa destituta in pagina interiora perianthii, differt.

T: Kanangra Walls, c. 47 km S of Mount Victoria, N.S.W., 6 Nov. 1973, *R.Coveny 5310*; holo: NSW.

The epithet, from the Latin *imberbis* (lacking a beard), refers to the absence of a dense beard on the inner surface of the perianth.

Grevillea halmaturina subsp. **laevis** Makinson, *subsp. nov.*

Affinis et similis *G. halmaturinae* subsp. *halmaturinae*, sed pagina superiora folii adulti sola costa visibili, cristae destitutae, margine folii rotundato-revoluto, et pedicellis non glabrescentibus basi, differt.

T: 0.5 mile [0.8 km] from Edilillie towards Cummins, S.A., 23 Sept. 1965, *M.E.Phillips & Carroll CBG020662*; holo: CANB; iso: AD.

The epithet, from the Latin *laevis* (smooth), refers to the rounded leaf margins and the absence of mutiple longitudinal ridges on the upper surface of the adult leaf.

Grevillea victoriae subsp. **nivalis** Stajsic & Molyneux, *sp. nov.*

T: near Tooma Reservoir, N.S.W., 29 Apr. 1973, *R.J.Chinnock 298*; holo: CANB; iso: AD.

Similis *G. victoriae* subsp. *victoriae*, sed foliis plerumque proportione latioribus, rhachidibus floralibus plerumque brevioribus, pagina superiore foliorum nitida, et margine folii plusminusve plano, differt.

The epithet is from the Latin *nivalis* (snowy), in reference to the occurrence of this taxon high in the Snowy Mtns of N.S.W.

Grevillea brevifolia subsp. **polychroma** Molyneux & Stajsic, *subsp. nov.*

T: Betts Creek Track, 12 km E of Tulloch Ard Rd, 9.5 km E of Mt Murrindal, Vic., 9 Feb. 1980, *S.J.Forbes 264*; holo: MEL.

Similis *G. brevifoliam* ssp. *brevifoliam*, sed foliis majibus (plerumque 30–50 mm longis, 10–20 mm latis), colore floris variabili, et bracteis floralibus spathulatis et majibus, differt.

The epithet, from the Greek *poly-* (many) and *chroma* (colour), refers to the propensity of this taxon for variable flower colour.

Grevillea parvula Molyneux & Stajsic, *nom. et stat. nov.*

G. victoriae var. *leptoneura* Benth., *Fl. Austral.* 5: 468 (1870), as var. *?leptoneura*. T: sources of the Genoa river, Vic., *F.Mueller* recd 1870; lecto: K, *fide* D.J.McGillivray & R.O.Makinson, *Grevillea* 447 (1993); remaining syntypes: K, MEL (2 sheets).

The epithet is from the diminutive of Latin *parvus* (small), in reference to the leaves, flowers and follicles of many populations which are unusually small compared with closely related species.

Grevillea epicroca Stajsic & Molyneux, *sp. nov.*

T: Merricumbene Fire Trail, 18.4 km S of Batemans Bay road, c. 19 km SW of Braidwood, N.S.W., 30 Mar. 1976, *M.D.Crisp 2008*; holo: CANB; iso: NSW.

Affinis *G. victoriae*, sed pagina inferiore foliorum pilis sparsioribus, rhacidi florali breviore (5–20 mm longo), et apice gemmae floralis quadrato, differt; similis *G. monslacanae*, sed apice gemmae floralis quadrato et praebitoribus pollinis ventraliter concavis, differt.

The epithet, from the Latin *epicrocum* (a transparent woman's garment), refers to the thin, open cladding of hairs on the lower surface of the leaf.

Grevillea monslacana Molyneux & Stajsic, *sp. nov.*

T: Rubicon State Forest, NE junction of Ruoaks Rd and Boundary Trail West, Blue Ra., Vic., 10 Dec. 1995, *N.H.Sinnot 3136*; holo: MEL.

Similis *G. victoriae*, sed indumento aperto in pagina inferiore folii, differt; et similis *G. epicrocam* sed apice albastri subgloboso et praebitoribus pollinis convexis ventraliter, differt.

The epithet, from the Latin *mons* (mountain) and *lacus* (lake), refers to the locality Lake Mountain around which the plant is found.

Grevillea miqueliana subsp. **moroka** Molyneux & Stajsic, *subsp. nov.*

T: The Sentinels, 4 mls [6.4 km] SW of Mt Wellington, Vic., 2 Jan. 1964, *T.B.Muir 3044*; holo: MEL.

Similis *G. miquelianam* subsp. *miquelianum*, sed foliis coriaceis brevioribus et angustioribus (ad 50 mm longis et 25 mm latis), et pedunculis crassioribus (c. 1 mm crassis), differt.

The epithet is taken as a noun in apposition from the parish name Moroka, which occupies part of the range of the taxon, and is derived from an Aboriginal word for sky.

Grevillea irrasa Makinson, *sp. nov.*

T: Nullica State Forest, Yowaka R. catchment, Mine Rd, 500 m uphill along side track from pyrophyllite mine entrance, N.S.W., 23 Nov. 1997, *R.O.Makinson 1649, W.Molyneux, S.Forrester & B.J.Harrison*; holo: CANB; iso: AD, BRI, HO, K, MEL, NE, NSW.

Affinis *G. miquelianae*, sed foliis anguste obovatis (5–8 mm latis) et pistillo breviore (14–17 mm longo), differt.

The epithet, from the Latin *irrasus* (unpolished, unshaven), refers to both the rough upper leaf surfaces and the subvillous branchlets, lower leaf surfaces, and perianths.

Grevillea irrasa subsp. **didymochiton** Makinson, *subsp. nov.*

T: 22.1 km W from Nerrigundah on road to Belowra, N.S.W., 21 Nov. 1997, *R.O.Makinson 1646 & B.J.Harrison*; holo: CANB; iso: AD, BRI, HO, K, MEL, NE, NSW.

Affinis *G. irrasae* subsp. *irrasae*, sed foliis latioribus (8–22 mm latis), et pagina inferiore foliorum pilis appressis et strato sparsiore emergente pilorum patentium, differt.

The epithet is from the Greek *didymos-* (double) and *chiton* (tunic or garment worn next to the skin), in reference to the two-layered indumentum on the lower surfaces of the leaves.

Grevillea pauciflora R.Br., *Trans. Linn. Soc. London* 10: 171 (1810)

T: Flinders' Land [S.A.], 1802, *R.Brown*; lecto (here chosen): *R.Brown Iter Austral 3340*, specimen at top right of sheet; lecto: BM; isolecto: BM (same sheet, top left), BM (*R.Brown 3340*, mixed with Baxter, 1824), E (2 sheets), FI *n.v.*, G *n.v.*, G-DC, K (2 sheets), LE *n.v.*, MEL.

Grevillea obtusiflora R.Br., *Suppl. Prodr. Fl. Nov. Holl.* 19 (1830)

T: '... mont. Port Jackson, 1822, D.Cunningham' [protologue]; lecto (here chosen): Brushy hills north of Bathurst, N.S.W., [Dec.] 1822, *[A.]Cunningham 197*: lecto: BM; isolecto: K; ?remaining syntype: on the hills north of the settlement of Bathurst, N.S.W., 1822, *[A.]Cunningham 157*; syn: BM.

D.J.McGillivray & Makinson (*Grevillea* 431 (1993)) indicated as holotype the sheet here designated as lectotype; the presence of at least one other collection by Cunningham (157) at BM, collected in the same month and with an identical locality, makes lectotypification desirable.

Grevillea ferruginea Sieber ex Spreng., *Syst. Veg.* 4(2), *Cur. Post.* 46 (1827)

G. ferruginea Sieber ex Schult. & Schult.f., *Mant.* 3: 280 (1827), *nom. superfl.* T: Australia, N.S.W., 1823, *F.W.Sieber 27*; lecto (here chosen): BM; isolecto: B *n.v.*, G, G-DC, K, LE *n.v.*, NSW, NY *n.v.*, P, PR, TCD *n.v.*

D.J.McGillivray (*New Names Grevillea* (1986), and in McGillivray & Makinson, *Grevillea* (1993)) chose not to designate lectotypes for several type collections made by Sieber, apparently on the basis that no sheets had been seen with the exact form of words cited in the protologues. There is no doubt as to the status of the collection and components (at least those seen), and lectotypification is accordingly made here.

Grevillea rosmarinifolia subsp. **glabella** (R.Br). Makinson, *comb. et stat. nov.*

G. glabella R.Br., *Suppl. Prodr. Fl. Nov. Holl.* 20 (1830). T: N.S.W., '... mon. prope Port Jackson 1817. D. Cunningh.' [protologue]; lecto: Oxley's 1st Expedition [R.Brown script; leg. prob. A.Cunningham]; lecto: BM, *fide* D.J.McGillivray, *Telopea* 1: 28 (1975); isolecto: K, NY *n.v.*, both as '31 May 1817, *A.Cunningham 28*'.

Grevillea lavandulacea subsp. **rogersii** (Maiden) Makinson, *comb. et stat. nov.*

G. rogersii Maiden, *Trans. Proc. & Rep. Roy. Soc. S. Australia* 32: 257 (1908), as *G. Rogersi.* T: Cape Borda, Kangaroo Is., S.A., Sept. 1907, *R.S.Rogers*; lecto: NSW, *fide* D.J.McGillivray & R.O.Makinson, *Grevillea* 425 (1993).

Grevillea phylicoides R.Br., *Trans. Linn. Soc. London* 10: 174 (1810)

G. buxifolia subsp. *phylicoides* (R.Br.) McGill., *New Names Grevillea* 3 (1986), *p.p.* T: '... prope Port Jackson; in montibus saxosis', N.S.W. [protologue]; lecto (here chosen): BM (*R.Brown Iter Austral.* 3323, '22 Grevillea phylicoides prodr. 379 Hills near the banks of the Grose'; [piece at centre of sheet, marked 'd' in pencil]); isolecto: BM [sheet as for lecto, pieces marked 'c' at lower left of sheet, and 'e' at upper right; excluding pieces marked 'a', 'b' which are *G. buxifolia* (Sm.) R.Br. subsp. *buxifolia*], ?E *n.v.*, K (*Brown [Iter Austral.] 3323*, *s. loc.*, whole sheet), ?LE *n.v.*, NSW.

Grevillea phylicoides has been regarded by McGillivray (*New Names Grevillea*, 3 (1986)), McGillivray & Makinson (*Grevillea*, 312 (1993)), and Olde & Marriott (*Telopea* 5: 707 (1994); *Grevillea Book* 2: 75 (1995)), and implicitly by Makinson (*Fl. New South Wales* 2: 48 (1991)) as comprising two races or forms ('race b' and 'race d' sensu McGillivray & Makinson). In the present treatment, agreeing with Hart & Henwood (*Telopea* 7: 65–76 (1996)), 'race b' is regarded as assignable to *G. buxifolia* subsp. *buxifolia*, and McGillivray & Makinson's 'race d' is regarded as corresponding exactly to *G. phylicoides* as here lectotypified and circumscribed. Two sheets at BM, long regarded as types of *G. phylicoides* R.Br., are mixed, bearing specimens assignable to each of these 'races'. Brown's protologue is somewhat ambiguous for diagnosis between these taxa on leaf characters, but refers to the pollen-presenter as '*stigmatibus ovalibus appendice duplo longioribus*', ('stigma oval, twice as long as the appendage'). On this last character, the 'race d' material on the type sheets at BM and K conforms well with the protologue, the stylar appendages being c. 0.7–1.0 mm long, as compared with a face-length for the narrowly elliptic pollen-presenters of 1.4–1.9 mm. The pieces assignable to 'race b' on the same sheets have appendages c. 1.5–2 mm long, with pollen-presenters 1.5–1.7 mm long.

A second sheet at BM, labelled in Brown's hand 'Grevillia [sic] phylicoides prodr. Hills near Port Jackson', is also mixed, with pieces marked 'A' in pencil assignable to *G. phylicoides* and possibly belonging to the lectotype collection. A single piece marked 'B' has leaf indumentum similar to *G. phylicoides* but the flowers conforming to *G. buxifolia* subsp. *buxifolia*; it is here regarded as an intermediate between these taxa.

The locality given on the lectotype sheet, 'Hills near the banks of the Grose [R.]', if strictly interpreted, is consistent with the known distribution of 'race d', whereas 'race b' occurs somewhat to the north (but could have been encountered by Brown in the Windsor area *en route* to the Grose). While an element of uncertainty remains as to Brown's intent, and as to whether he recognised the taxa as separate, lectotypification is made here on the above grounds and allows a fixing of the name, used *s. str.*, to the Blue Mountains taxon ('race d') to which it has been long and consistently applied.

Grevillea florida (McGill.) Makinson, *comb. et stat. nov.*

G. uncinulata subsp. *florida* McGill., *New Names Grevillea* 15 (1986). T: 74 mile peg on Perth to New Norcia road [c. 118 km N of Perth], W.A., 11 Sept. 1963, *F.G.Smith 1732*; holo: PERTH.

Grevillea petrophiloides subsp. **remota** (Olde & Marriott) Makinson, *comb. nov.*

G. magnifica subsp. *remota* Olde & Marriott, *Grevillea Book* 1: 179 (1994). T: Cave Hill Nature Reserve, W.A., 30 Sept. 1988, *P.M.Olde 88/18 & N.R.Marriott*; holo: PERTH; iso: NSW.

Grevillea nematophylla subsp. **supraplana** Makinson, *subsp. nov.*

Arcte affinis *G. nematophyllae* subsp. *nematophyllae*, sed foliis interdum divisis, et lobis vel foliis simplicibus pro parte maxima subteretibus costa abaxiali distentissima, pagina adaxiali lata et plana vel vadosissima concava c. latitudo folii aequans, et labio prominenti sulcum vadosum superpendenti secus quoque marginem, differt.

T: 30 miles [50 km] N from Mt Churchman, W.A., 10 Dec. 1891, *R.Helms* ['14' on some sheets]; holo: K, *Helms 14*, excluding detached inflorescence at right of sheet; iso: AD, MEL, NSW, PERTH.

The epithet is from Latin *supra* (upper or above), and *planus* (flat), referring to the ±flat to slightly concave adaxial surface of the otherwise subterete leaf, in contradistinction to the adaxial ridge of subsp. *nematophylla*.

Grevillea nematophylla subsp. **planicosta** Makinson, *subsp. nov.*

Arcte affinis *G. nematophyllae* subsp. *nematophyllae*, sed foliis interdum divisis et lobis vel foliis simplicibus linearibus, costa abaxiali lateraliter complanata et verticaliter extensa, anguste elliptica ad oblongam in sectione transversali, differt.

T: Comet Vale, W.A., 14 Dec. 1961, *C.A.Gardner 13862*; holo: K; iso: PERTH.

The epithet, from the Latin *planus* (flat) and *costa* (rib or midrib, as a noun in apposition), refers to the laterally flattened abaxial midrib of the leaf.

Grevillea pyramidalis subsp. **leucadendron** (A.Cunn. ex R.Br.) Makinson, *stat. nov.*

G. leucadendron A.Cunn. ex R.Br., *Suppl. Prodr. Fl. Nov. Holl.* 25 (1830); *G. pyramidalis* var. *leucadendron* (A.Cunn. ex R.Br.) C.A.Gardner ex E.M.Watson, *J. Roy. Soc. W. Australia* 30: 89 (1946). T: Cambridge Gulf, [W.A.], Sept. 1819, *A.Cunningham 412* (2nd Voyage of *Mermaid*); lecto: BM, *fide* D.J.McGillivray & R.O.Makinson, *Grevillea* 436 (1993); isolecto: BM, K.

Grevillea cheilocarpa Makinson, *sp. nov.*

Affinis et similis *G. polybotryae*, sed foliis dense obtectis pilis robustis hyalinis, et pagina exteriore perianthii et ovario et stipite laxe villoso, differt.

T: Dragon Rock Nature Reserve, 36128, 75 km E of Kulin on roadside, W.A., 32°38'30"S, 119°01'30"E, 7 Sept. 1984, *J.M.Brown 089*; holo: PERTH.

The epithet, from the Greek *cheilos* (lip) and *carpos* (fruit), refers to the pronounced lip-like flange along the ventral margin of each follicle valve.

Grevillea synapheae subsp. **latiloba** (Meisn.) Makinson, *stat. nov.*

G. synapheae var. *latiloba* Meisn., *in* J.G.C.Lehmann, *Pl. Preiss.* 2: 259 (1848). T: Swan River, W.A., *[J.] Drummond [2nd coll.] 313*; neo: NY *n.v.*, *fide* D.J.McGillivray & R.O.Makinson, *Grevillea* 443 (1993); isoneo: A *n.v.*, E, G, K, LD *n.v.*, LE *n.v.*, MEL, NSW.

Similar to *G. synapheae* subsp. *synapheae* but lacking the minute ascending hairs on the leaf undersurface characteristic of that taxon, and with broader primary leaf lobes 5–15 mm wide. Also similar to *G. synapheae* subsp. *pachyphylla*, but with thinner-textured leaves, broader primary lobes, floral bracts falling earlier (in early bud stage), and unit conflorescences shorter (rachises 1–2 cm long).

Grevillea synapheae subsp. **minyulo** Makinson, *subsp. nov.*

Similis *G. synapheae* subsp. *synapheae*, sed conflorescentiis longioribus emergentibus, rhachidibus floralibus 3–6 cm longis, foliis pinnatisectis lobis angustis curvis, et praebitoribus pollinis conicis annulo basali, differt.

T: Walyering Road, 1 km S of Minyulo Brook Reserve, W.A., 20 Sept. 1990, *M.Pieroni 90/4*; holo: PERTH; iso: NSW [plus 2 at PERTH yet to be allocated].

The epithet is taken from Minyulo, W.A., a locality near the site of one of the first collections of the taxon.

Grevillea amplexans subsp. **semivestita** Makinson, *subsp. nov.*

Affinis et similis *G. amplexantem* subsp. *amplexantem* et *adpressam*, sed ramulis subsericeis, foliis glabris, bracteis floralis persistentibus, differt.

T: Marchagee Track, c. 45 km E of Brand Highway, W.A., 30°07'S, 115°55'E, 1 Sept. 1984, *D.Foreman 487*; holo: MEL; iso: HO, NSW, PERTH.

The epithet, from the Latin *semi-* (half) and *vestitus* (clothed, clad), refers to the presence of indumentum on the branchlets but not on the leaf lower surface.

Grevillea amplexans subsp. **adpressa** (Olde & Marriott) Makinson, *comb. et stat. nov.*

G. adpressa Olde & Marriott, *Nuytsia* 9: 250 (1993). T: 5.6 km W of Arrino on Arrino West Rd, W.A., 16 Sept. 1991, *P.M.Olde 91/112*; holo: NSW.

Grevillea metamorpha Makinson, *sp. nov.*

Affinis *G. acrobotryae*, sed pedicellis et pagina exteriore perianthii pilosa, et transito a foliis dentatis ad foliis penitus divisis plus gradatim versus inflorescentias, differt.

T: off Green Head–Coorow road, W.A., 16 Sept. 1995, *M.Hislop 88*; holo: PERTH.

The epithet, from the Greek *meta-* (changing) and *morphos* (form), refers to the transition in leaf form from vegetative to floral branches.

Grevillea stenogyne (Benth.) Makinson, *comb. et stat. nov.*

G. vestita var. *stenogyne* Benth., *Fl. Austral.* 5: 488 (1870). T: W. Australia, [*J.*] *Drummond* [*s.n.*], [per] F.Mueller 1870; holo: K, (sheet Neg. no. 2350); iso: MEL, ?P (*s. leg.* ex Mueller), PERTH.

This name of this taxon was treated as a synonym of *G. acrobotrya* Meisn. by McGillivray & Makinson (1993) and by Olde & Marriott (1995). *Grevillea stenogyne* is closely related to *G. acrobotrya* but does not share the reduction in leaf size and increased depth of leaf dissection from vegetative to flowering branches of that species; *G. acrobotrya* also has the flower-bearing branchlets more or less glabrous, whereas in *G. stenogyne* the branchlets are tomentose throughout. In the context of the group the differences are likely to be stable.

SUPPLEMENTARY GLOSSARY

acroscopic: *of a flower*, with the ventral suture of the perianth and the concave curve of the style facing towards the apex of the floral rachis; **adaxially acroscopic**, with the flower declined so that the ventral sutures of the perianth and ovary/fruit are facing the floral rachis; **abaxially acroscopic**, with the flower reclinate so that the ventral sutures of the perianth and ovary/fruit are facing away from the floral rachis.

asperities: the peaks on an asperous leaf surface.

asperous: rough with small relatively sharp peaks.

basiscopic: *of a flower*, with the ventral suture of the perianth and the concave curve of the style facing towards the base of the floral rachis.

biramous hair: a multicellular hair with the terminal cell elongate, medifixed, and more or less transverse; the 'arms' formed by this terminal cell may be even or uneven in length, and vary in posture.

diplural: *of a leaf*, subterete with a lateral groove along each side (the leaf cross-section like a fat figure-8). The lower (abaxial) bulge is the lower midvein, and the upper (adaxial) bulge is the lamina and upper part of the midvein; tissue apparently homologous with the abaxial surface of the lamina is located in the grooves.

limb: *of a flower bud*, swollen knoblike area at apex of flower bud containing anthers and style end and consisting of the connate apical segments of the tepals.

limb segment: that part of the tepals that houses the anthers on the inner surface.

papilloid: *of a hair*, very short and papilla-like.

peripterous: with a continuous wing around the margin of the seed body.

ramet: a daughter-plantlet arising from subterranean roots or stems (rhizomes) at some distance from the parent plant.

refracted: *of leaf margins*, sharply or angularly bent through a single angle inwards over the lower (abaxial) surface.

torus: the rim around the apex of the pedicel, on which the perianth is situated, together with the hypothetical plane that the rim surrounds. The toral cavity is the cupped space surrounding the base of the style.

unifacial: *of a leaf*, with the adaxial lamina reduced to a narrow vestigial ridge along either side of the upper (dorsal) edge, and the bulk of the vertically-oriented leaf blade derived from a ventrally extended and often flattened midrib.

Abbreviations and Contractions

Literature

Author abbreviations follow R.K.Brummitt & C.E.Powell, *Authors of Plant Names* (Royal Botanic Gardens, Kew, 1992).

Journal titles are abbreviated in accordance with G.H.M.Lawrence *et al.*, *Botanico-Periodicum-Huntianum* (Hunt Botanical Library, Pittsburgh, 1968) and G.D.R.Bridson & E.R.Smith, *Botanico-Periodicum-Huntianum/Supplementum* (Hunt Institute for Botanical Documentation, Pittsburgh, 1991).

Other literature is abbreviated in accordance with F.A.Stafleu & R.S.Cowan, *Taxonomic Literature*, 2nd edn (Bohn, Scheltema & Holkema, Utrecht, 1976–1987), except that upper case initial letters are used for proper names and significant words. The *Flora of Australia* is abbreviated to *Fl. Australia.*

Herbaria

Abbreviations of herbaria are in accordance with P.K.Holmgren, N.H.Holmgren & L.C.Barnett, *Index Herbariorum* Part I, 8th edn (New York Botanical Garden, 1990). Those most commonly cited in the *Flora* are:

AD	State Herbarium of South Australia, Adelaide
BM	The Natural History Museum, London
BRI	Queensland Herbarium, Brisbane
CANB	Australian National Herbarium, Canberra
CBG	Australian National Botanic Gardens Herbarium, Canberra
DNA	Northern Territory Herbarium, Darwin
HO	Tasmanian Herbarium, Hobart
K	Royal Botanic Gardens, Kew
MEL	National Herbarium of Victoria, Melbourne
NSW	National Herbarium of New South Wales, Sydney
PERTH	Western Australian Herbarium, Perth
QRS	Australian National Herbarium, Atherton

States, Territories

Abbreviations of Australian States and Territories as used in statements of distribution and citation of collections are:

A.C.T.	Australian Capital Territory
N.S.W.	New South Wales
N.T.	Northern Territory
Qld	Queensland
S.A.	South Australia
Tas.	Tasmania
Vic.	Victoria
W.A.	Western Australia

General abbreviations

add.	addendum
agg.	aggregate species
alt.	altitude
app.	appendix
auct.	*auctoris/auctorum* (of an author or authors)
auct. mult.	*auctorum multorum* (of many authors)
auct. non	*auctorum non* (of authors [but] not....), used for misapplied names
c.	*circa* (about)
cf.	*confer* (compare)
Ck	Creek
cm	centimetre
col.	colour
coll.	collector
colln	collection
comb.	*combinatio* (combination)
cons.	*conservandus*
cult.	cultivated
cv.	cultivar
d.b.h.	diameter at breast height
Dept	Department
descr.	*descriptio*
diam.	diameter
E	east
ed./eds	editor/editors
edn	edition
e.g.	*exempli gratia* (for example)
et al.	*et alii/et aliorum*; and others/and of others
f.	*forma* (form)
fam.	*familia* (family)
fig./figs	figure/figures (in other works)
Fig./Figs	Figure/Figures (in this volume of the *Flora*)
gen.	*genus* (genus)
gen. nov.	*genus novus* (new genus)
Gt	Great
holo	holotype
hort.	*hortus* (garden) or *hortensis* (of a garden)
HS	Homestead
Hwy	Highway
i.e.	*id est* (that is)
ined.	*ineditus* (unpublished)
in litt.	*in litteris* (in correspondence)
in obs.	*in observatio* (in observation)
Is.	Island/s
iso	isotype
isolecto	isolectotype
km	kilometre
kt	kilotonne
L.	Lake
L.A.	Logging Area
lat.	latitude
lecto	lectotype
loc. cit.	*loco citato* (in bibliographic citations: in the same work and page as just cited)

Publication date of previous volumes

Volume 1	22 August 1981 (1st edn)
Volume 1	2 March 1999 (2nd edn)
Volume 3	24 April 1989
Volume 4	12 November 1984
Volume 8	9 December 1982
Volume 12	4 May 1998
Volume 16	30 November 1995
Volume 17B	26 May 1999
Volume 18	8 June 1990
Volume 19	27 June 1988
Volume 22	17 May 1984
Volume 25	25 December 1985
Volume 28	28 June 1996
Volume 29	27 July 1982
Volume 35	6 August 1992
Volume 45	15 May 1987
Volume 46	2 May 1986
Volume 48	27 October 1998
Volume 49	3 May 1994
Volume 50	29 July 1993
Volume 54	4 September 1992
Volume 55	21 December 1994

For the publication date of Volume 17A, see Volume 11.

INDEX

Accepted names are in roman, synonyms and doubtful names in *italic*.

Principal page references are in **bold**, figures and plates in *italic*.